EARTH SCIENCE

EARTH SCIENCE

EIGHTH EDITION

Edward J. Tarbuck
Frederick K. Lutgens

Illinois Central College

Illustrated by
Dennis Tasa

PRENTICE HALL Upper Saddle River, New Jersey 07458

Library of Congress Cataloging-in-Publication Data

Tarbuck, Edward J.
   Earth Science/Edward J. Tarbuck, Frederick K. Lutgens; illustrated by Dennis Tasa.—8th ed.
      p. cm.
   Includes index.
   ISBN 0–13–570839–7
   1. Earth sciences. I. Lutgens, Frederick K. II. Title.
QE26.2.T38 1997   96–23755
550—dc20   CIP

Acquisitions Editor : *Dan Kaveney*
Editorial Director: *Tim Bozik*
Editor in Chief: *Paul Corey*
A.V. P. of Production and Manufacturing: *David W. Riccardi*
Executive Managing Editor: *Kathleen Schiaparelli*
Assistant Managing Editor: *Shari Toron*
Production Editor: *Karen M. Malley*
Creative Director: *Paula Maylahn*
Art Director: *Joseph Sengotta*
Interior and Cover Design: *Amy Rosen*
Manufacturing Manager: *Trudy Pisciotti*
Page Formatter: *Eric Hulsizer*
Art Manager: *Gus Vibal*
Photo Research: *Clare Maxwell*
Photo Editor: *Debbie Hewitson*
Cover Photo: *Tibet, Tsangpo Gorge TS 2629s Tsangpo R. (Brahmaputra) flows below
   Mt. Gvala Pheri in this deep Himalayan cyn., Gordon Wiltsie/AlpenImage Ltd.*
Illustrations by: *Dennis Tasa, Tasa Graphic Arts*

© 1997 by Prentice-Hall, Inc.
Simon & Schuster / A Viacom Company
Upper Saddle River, New Jersey 07458

Printed in the United States of America.

10   9   8   7   6   5

ISBN 0-13-570839-7

Prentice-Hall International (UK) Limited, *London*
Prentice-Hall of Australia Pty. Limited, *Sydney*
Prentice-Hall Canada, Inc., *Toronto*
Prentice-Hall Hispanoamericana, S.A., *Mexico*
Prentice-Hall of India Private Limited, *New Delhi*
Prentice-Hall of Japan, Inc., *Tokyo*
Simon & Schuster Asia Pte. Limited, *Singapore*
Editora Prentice-Hall do Brasil, Ltda., *Rio de Janeiro*

BRIEF CONTENTS

CONTENTS

4 RUNNING WATER AND GROUNDWATER   87

5 GLACIERS, DESERTS, AND WIND 121

6 EARTHQUAKES AND EARTH'S INTERIOR 153

7 PLATE TECTONICS 179

8 IGNEOUS ACTIVITY 211

9 MOUNTAIN BUILDING   237

10 GEOLOGIC TIME   261

11 EARTH HISTORY: A BRIEF SUMMARY   285

Part Two
The Oceans

Part Three
The Atmosphere

15   MOISTURE   387

16   AIR PRESSURE AND WIND   419

17 WEATHER PATTERNS AND SEVERE STORMS 437

18 CLIMATE 463

Part Four
Astronomy

19 EARTH'S PLACE IN THE UNIVERSE 489

PREFACE

Earth is a very small part of a vast universe, but it is our home. It provides the resources that support our modern society and the ingredients necessary to maintain life. Therefore, a knowledge and understanding of our planet is critical to our social well being and indeed, vital to our survival.

In recent years, media reports have made us increasingly aware of our place in the universe and the forces at work in our physical environment. We are also beginning to learn that human interactions with natural systems can upset delicate balances. News stories inform us of new discoveries in the solar system and beyond. Daily reports remind us of the destruction caused by hurricanes, earthquakes, floods, and landslides. We have been made aware of ozone depletion, potential global warming, and growing environmental concerns about the oceans. To comprehend, prepare for, and solve these and other concerns requires an awareness of how science is done and the scientific principles that influence our planet, its rocks, mountains, atmosphere, and oceans.

Earth Science is a college-level text designed for an introductory course that often has the same name as this text. It consists of four units that emphasize broad and up-to-date coverage of basic topics and principles in geology, oceanography, meteorology, and astronomy. The book is intended to be a meaningful, nontechnical survey for undergraduate students with little background in science. Usually these students are taking an Earth science class to meet a portion of their college's or university's general requirements.

In addition to being informative and up-to-date, a major goal of *Earth Science* is to meet the need of beginning students for a readable and user-friendly text, a book that is a highly usable "tool" for learning basic Earth science principles and concepts.

The language of this book is straightforward and *written to be understood*. Clear, readable discussions with a minimum of technical language are the rule. When new terms are introduced, they are placed in **boldface** and defined. The frequent headings and subheadings help students follow discussions and identify the important ideas presented in each chapter. A list of key terms with page references is found at the end of each chapter, and a glossary is included at the conclusion of the text for easy reference to important terms. Review questions conclude each chapter to help the student prepare for exams and quizzes.

Illustrations and Photographs

The Earth sciences are highly visual. Therefore, photographs and artwork are a very important part of an introductory book. *Earth Science*, eighth edition, contains hundreds of new high-quality photographs that were carefully selected to aid understanding, add realism, and heighten the interest of the reader.

The illustrations in each new edition of *Earth Science* keep getting better and better. In the eighth edition *every* piece of existing art was redesigned. The new art illustrates ideas and concepts more clearly and realistically than ever before. The new art program was carried out by Dennis Tasa, a gifted artist and respected Earth science illustrator.

Earth as a System

An important occurrence in modern science has been the realization that Earth is a giant multidimensional system. Our planet consists of many separate but interacting parts. A change in any one part can produce changes in any or all of the other parts—often in ways that are neither obvious nor immediately apparent. Although it is not possible to study the entire system at once, it is possible to develop an awareness and appreciation for the concept and for many of the system's important interrelationships. With this as a goal, a number of new headings and boxes have been added to the eighth edition of *Earth Science*.

Environmental Issues

Because knowledge about our planet and how it works is necessary to our survival and well being, the treatment of environmental and resource topics has always been an important part of *Earth Science*. These issues serve to illustrate the relevance and application of Earth science knowledge. With each new edition this focus has been given ever-greater emphasis. This is certainly the case with the eighth edition. The text integrates a great deal of information about the relationship between people and the physical environment and explores applications of the Earth sciences to understanding and solving problems that arise from these interactions. In addition to the Introduction, 17 of the book's 22 chapters now contain discussions relating to environmental and resource topics.

Organizational Changes

- The new edition has separate chapters on minerals (Chapter 1) and rocks (Chapter 2). What had been the longest chapter in the book, has been divided into two smaller more manageable chapters.

- A new chapter entitled, "Earth History: A Brief Summary" (Chapter 11) now compliments the chapter on "Geologic Time" (Chapter 10)

- A new chapter on "Climate" (Chapter 18) completes the atmosphere unit. A survey of world climates and an examination of the impact of people on global climate are treated.

- Although *Earth Science*, eighth edition contains more chapters than its predecessors, the length of the text remains about the same.

- Improved readability was achieved by reducing sentence and paragraph length, omitting unnecessary details, examining chapter organization and flow, and writing in a more personal style.

- Special interest boxes allow us to explore applications, interesting examples, and related scientific principles without significantly disrupting the flow of basic text discussions. This popular feature has been retained. The total number of boxes remains the same as the previous edition, 62. However, 27 of the boxes in the eighth edition are new.

Maintaining a Focus on Basic Principles and Instructor Flexibility

Although many new topical issues are treated in *Earth Science*, eighth edition, it should be emphasized that the main focus of this new edition remains the same as its predecessors—to foster student understanding of basic Earth science principles. Whereas student use of the text is a primary concern, the book's adaptability to the needs and desires of the instructor is equally important. Realizing the broad diversity of Earth science courses in both content and approach, we have continued to use a relatively nonintegrated format to allow maximum flexibility for the instructor. Each of the four major units stands alone; hence, they can be taught in any order. A unit can be omitted entirely without appreciable loss of continuity, and portions of some chapters may be interchanged or excluded at the instructor's discretion.

Supplements

In addition to the text itself, we have worked with a number of talented people to produce an excellent supplements package. This package includes the traditional supplements that students and professors have come to expect from authors and publishers, as well as some new kinds of supplements that involve electronic media. We always strive to produce the highest quality materials, and hope the supplements described here will make the study of Earth science more fulfilling for both students and instructors.

- *Study Guide*: Written in conjunction with the authors by an experienced college educator, Ken Pinzke. Professor Pinzke's publications have helped thousands of students master the Earth sciences. The study guide complements the text by providing students with additional tools for learning Earth science more effectively. Each chapter of the *Earth Science* study guide contains a chapter overview, learning objectives, a comprehensive chapter review, key terms, a vocabulary review, and a practice test with corresponding answers.

- *Applications and Investigations in Earth Science*: Written by Ed Tarbuck, Fred Lutgens, and Ken Pinzke, this full color laboratory manual contains 22 exercises that provide students with hands-on experiences in geology, oceanography, meteorology, astronomy, and Earth science skills.

- *Instructor's Resource Manual with Tests*: In this manual Ken Pinzke gives new instructors the benefit of his of years of teaching experience and experienced instructors a ready source of new ideas to complement their teaching style. The manual contains a variety of lecture outlines, teaching tips, advice on how to integrate visual supplements, and many suggested test questions.

- *Prentice Hall Custom Test*: Based on the powerful testing technology developed by Engineering Software Associates, Inc., this supplement, available for both Mac and Windows, allows instructors to tailor exams to their own needs. With the On-line Testing option, exams can be administered on-line so data can be automatically transferred for evaluation. A comprehensive desk reference guide is included, along with on-line assistance.

- *Transparencies*: More than 100 full color acetates of illustrations from the text.

- *Slides*: One hundred fifty full-color slides of figures from the text and from other sources.

- *The New York Times* **Themes of the Times— Changing Earth**: This unique newspaper-format supplement features recent articles on dynamic Earth science applications from the pages of the *New York Times*. This free supplement, available in quantity from your local Prentice Hall representative, encourages students to make connections between the classroom and the world around them.

- *Life on the Internet: Geosciences*: Written by Andrew Stull of California State University–Fullerton and Duane Griffin of the University of Wisconsin–Madison, this guide to the internet helps Earth science students use the internet to enrich their studies. It includes general information on internet and WWW basics, as well as specific material about how to use the internet for the Geosciences. *Life on the Internet* is free to qualified adopters of *Earth Science*. Please contact your local Prentice Hall representative for details.

- *World Wide Web Resources*: Designed to give the students an interactive way to further their studies, the *Earth Science* eighth edition WWW site reaches out into cyberspace and brings the myriad Earth science resources available there closer to you and your students. Check out our interactive study resources, interesting links, and items for further thought at: http://www.prenhall.com/tarbuck.

- *Prentice Hall Geodisc*: This laser disk allows instructors to bring a wealth of images into the classroom. It contains more than 50 minutes of motion video and more than 900 still images and animations. The disk is accompanied by a bar code manual to access images easily.

- *Director Academic and Authorware Academic*: Distributed exclusively by Prentice Hall, these educational adaptations of the leading multimedia authoring tools provide instructors with high-power multimedia at a minimal cost. Both products contain templates designed specifically for academic uses. For further information contact the Prentice Hall Multimedia Group at 1-800-887-9998.

Acknowledgments

Writing a college textbook requires the talents and cooperation of many individuals. Once again we benefited from the skills of our Developmental Editor Fred Schroyer. He helped us make the eighth edition a more readable, user-friendly text. We are also grateful to Professor Ken Pinzke at Belleville Area College. In addition to his many helpful suggestions regarding the manuscript, Ken prepared the instructor's guide and revised the student study guide and laboratory manual that are available to accompany *Earth Science*. Working with Dennis Tasa, who is responsible for all of the outstanding illustrations, is always special for us. We not only value his outstanding artistic talents and imagination, but his friendship as well.

Special thanks goes to those colleagues who prepared in-depth reviews. Their critical comments and thoughtful input helped guide our work and clearly strengthened the text. We wish to thank: James R. Albanese, State University of New York at Oneonta; Charles Dunlap, University of California–Santa Cruz; James R. Ebert, State University of New York at Oneonta; J. Mark Erickson, St. Lawrence University; Don Greene, Baylor University; Gary Houlette, Oklahoma City Community College; Michael E. Lewis, University of North Carolina at Greensboro; Brendan McNulty, University of California–Santa Cruz; James Shea, University of Wisconsin–Parkside; John Reed, Hampshire College; Justin Revenaugh, University of California–Santa Cruz; Robert A. Schiffman, Bakersfield College; Parvinder Sethi, North East Illinois University.

We would also like to give our sincere thanks to the *Earth Science* eighth edition advisory board. Each of these individuals was kind enough to review the entire text and supplements package and provided us with valuable input, some of which we were able to put to work in this edition and some of which we plan to incorporate in future editions of the book. Here we wish to thank: Latsey Best, Palm Beach Community College; Matthew Werhner, Hillsborough Community College; Patricia Lee, University of Hawaii; Claire Correale, Burlington County College; Doug Sherman, College of Lake County; Adrienne Laracque, University of Manitoba.

In addition we would like to acknowledge the aid of our students. Their comments and criticisms sometimes challenge us to search for a deeper understanding and always help us maintain our focus on readability.

We also want to acknowledge the team of professionals at Prentice Hall. Thanks to Editor in Chief Paul Corey. We sincerely appreciate his continuing strong support for excellence and innovation. Thanks also to our editor Dan Kaveney. His strong communication skills and energetic style contributed greatly to the project. The production team, led by Karen Malley, has done an outstanding job. They are true professionals with whom we are very fortunate to be associated.

INTRODUCTION

The spectacular eruption of a volcano, the magnificent scenery of a rocky coast, and the destruction created by a hurricane are all subjects for the Earth scientist. The study of Earth science deals with many fascinating and practical questions about our environment (Figure I.1). What forces produce mountains? Why is weather so variable? Is the climate really changing? How old is Earth and how is it related to the other planets in the solar system? What causes ocean tides? What was the Ice Age like? Will there be another? Can a successful well be dug at this site?

The subject of this text is **Earth science.** To understand Earth is not an easy task because our planet is not a static and unchanging mass. Rather, it is a dynamic body with many interacting parts and a long and complex history.

Figure I.1
This eroded sandstone monument in Arizona's Vermillion Cliffs Wilderness was once a sand dune. Earth science studies the processes that created the ancient dune, turned it into solid rock, and then formed the present natural sandstone sculpture. (Photo by Jack W. Dykinga)

A View of Earth

A view of Earth from space gives us a unique perspective of our planet (Figure I.2). At first, it may strike us that Earth is a fragile-appearing sphere surrounded by the blackness of space. In fact, it is just a speck of matter in a vast universe. As we look more closely at our planet from space, it becomes apparent that Earth is much more than rock and soil. Indeed, the most conspicuous features are not the continents but the swirling clouds suspended above the surface and the vast global ocean. These features emphasize the importance of water to our planet.

From such a vantage point we can appreciate why Earth's physical environment is traditionally divided into three major parts: the solid earth; the water portion of our planet, the hydrosphere; and Earth's gaseous envelope, the atmosphere. It should be emphasized that our environment is highly integrated and is not dominated by rock, water, or air alone. Rather, it is characterized by continuous interactions as air comes in contact with rock, rock with water, and water with air. Moreover, the biosphere, the totality of life-forms on our planet, extends into each of the three physical realms and is an equally integral part of Earth.

The interactions among the spheres of Earth's environment are uncountable. Figure I.3 provides us with one easy-to-visualize example. The shoreline is an obvious meeting place for rock, water, and air. In this scene, ocean waves that were created by the drag of air moving across the water are breaking against the rocky shore. The force of the water can be powerful and the erosional work that is accomplished can be great.

Hydrosphere

Earth is sometimes called the *blue* planet. Water more than anything else makes Earth unique. The **hydrosphere** is a dynamic mass of liquid that is continually on the move, from the oceans to the atmosphere, to the land, and back again. The global ocean is certainly the most prominent feature of the hydrosphere, blanketing nearly 71 percent of Earth's surface and accounting for about 97 percent of Earth's water. However, the hydrosphere also includes the freshwater found in streams, lakes, and glaciers, as well as that found underground.

Although these latter sources constitute just a tiny fraction of the total, they are much more important than their meager percentage indicates. In addition to providing the

Figure I.2

Africa and Arabia are prominent in this image of Earth taken from Apollo 17. The dark blue of the oceans and the swirling cloud patterns remind us of the importance of the oceans and atmosphere. Antarctica, a continent covered by glacial ice, is visible at the South Pole. (Courtesy of NASA/Science Source/Photo Researchers, Inc.)

Figure I.3
The shoreline is one obvious meeting place for the hydrosphere, atmosphere, and solid Earth. In this scene, along California's Big Sur coastline, ocean waves that were created by the force of moving air break against the rocky shore. The force of the water can be powerful and the erosional work that is accomplished can be great. (Photo by David Muench)

freshwater that is so vital to life on the continents, streams, glaciers, and groundwater are responsible for sculpturing and creating many of our planet's varied landforms.

Atmosphere

Earth is surrounded by a life-giving gaseous envelope called the **atmosphere.** This thin blanket of air is an integral part of the planet. It not only provides the air that we breathe but also acts to protect us from the sun's intense heat and dangerous radiation. The energy exchanges that continually occur between the atmosphere and the surface and between the atmosphere and space produce the effects we call weather. If, like the moon, Earth had no atmosphere, our planet would not only be lifeless but also

many of the processes and interactions that make the surface such a dynamic place could not operate. Without weathering and erosion, the face of our planet might more closely resemble the lunar surface, which has not changed appreciably in nearly three billion years.

Biosphere

The **biosphere** includes all life on Earth and consists of the parts of the solid Earth, hydrosphere, and atmosphere in which living organisms can be found. Plants and animals depend on the physical environment for the basics of life. However, it is important to note that organisms do more than just respond to their physical environment. Indeed, through countless interactions, life-forms help maintain and alter their physical environment. Without life, the makeup and nature of the lithosphere, hydrosphere, and atmosphere would be very different (Figure I.4).

Lithosphere

Lying beneath the atmosphere and the ocean is the solid Earth. It is divided into three principal units: the dense **core;** the less dense **mantle;** and the **crust,** which is the light and very thin outer skin of Earth (Figure I.5). The term **lithosphere** refers to a rigid outer layer of Earth, which includes the crust and part of the upper mantle. The crust is not a layer of uniform thickness; rather, it is characterized by many irregularities. It is thinnest beneath the oceans and thickest where continents exist. Although the crust may seem insignificant when compared with the other units of the solid Earth, it was created by the same general processes that were responsible for Earth's present structure. By examining the most prominent surface features and their global extent, we can obtain clues to the dynamic processes that shape our planet.

The two principal divisions of Earth's surface are the continents and the ocean basins. It is important to realize that the present shoreline is not the boundary between these distinct regions. Rather, along most coasts a gently sloping platform of continental material, called the *continental shelf,* extends seaward from the shore. A glance at Figure I.6 shows that there can be considerable variation in the extent of the continental shelf from one region to another. For example, the shelf is broad along the East and Gulf coasts of the United States, but relatively narrow along the Pacific margin of the continent. The extent of the continental shelf also varies greatly from one time to another. For instance, during the most recent ice age, when more of the world's water was stored on land in the form of glacial ice, the level of the sea was about 150 meters (500 feet) lower than it is today. Consequently, during this period, more of the surface was dry land. The boundary between the continents and the deep-ocean basins is perhaps best placed about halfway down the *continental slopes,* which are steep dropoffs that lead from the edge of the continental shelves to the deep-ocean basins. Using this as the di-

Figure I.4

This schematic diagram illustrates the dynamic nature of Earth. The planet's four "spheres" constantly and vigorously interact with each other to produce a highly complex system.

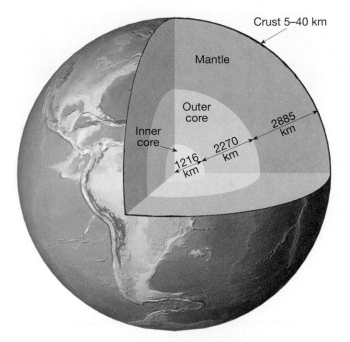

Crust 5–40 km

Mantle

Outer core

Inner core

1216 km

2270 km

2885 km

Figure I.5
View of Earth's layered structure. The inner core, outer core, and mantle are drawn to scale but the thickness of the crust is exaggerated by about five times.

viding line, we find that about 60 percent of Earth's surface is represented by the ocean basins, whereas the remaining 40 percent exists as continental masses.

The most obvious difference between the continents and the ocean basins is their relative levels. The average elevation of the continents above sea level is about 840 meters (2750 feet), whereas the average depth of the oceans is about 3800 meters (12,500 feet). Thus, the continents stand on the average 4640 meters (about 4.6 kilometers) above the level of the ocean floor.

Within these two diverse provinces, great variations in elevation exist. The most prominent features of the continents are elevated and deformed mountain belts (see Figure I.6). The most striking feature on the ocean floor is the *mid-ocean ridge system.* In Figure I.6 the Mid-Atlantic Ridge and the East Pacific Rise are parts of this system. This broad elevated zone forms a continuous belt that winds for nearly 65,000 kilometers (40,000 miles) around the globe in a manner similar to the seam of a baseball. Rather than consisting of highly deformed rock, such as most of the mountains found on the continents, the oceanic ridge system consists of layer upon layer of once molten rock that has been fractured and uplifted.

The ocean floor also contains extremely deep grooves that are occasionally more than 11,000 meters (36,000 feet) deep. Although these deep-ocean *trenches* are relatively narrow and represent only a small portion of the ocean

floor, they are nevertheless very important features that mark zones of significant geological activity. Some trenches are located adjacent to young mountains which flank the continents. For example, in Figure I.6 the Peru-Chile trench off the west coast of South America parallels the Andes Mountains. Other trenches are found adjacent to island chains called volcanic island arcs.

To summarize this brief view of Earth, we have seen that the physical environment consists of the hydrosphere, atmosphere, and the solid Earth. These three parts are characterized by countless interactions. In addition, the biosphere extends into each of the spheres of the physical environment, and must also be considered an integral part of our planet (see Box I.1). Finally, we have seen that the ocean basins are not featureless plains, but, like the continents, are characterized by a broad diversity of features.

The Earth Sciences

Earth science is the name for all the sciences that collectively seek to understand Earth and its neighbors in space. It includes geology, oceanography, meteorology, and astronomy.

The focus of Part One of this text, *The Solid Earth,* is the science of **geology,** a word that literally means "study of Earth." Geology is traditionally divided into two broad areas—physical and historical.

Physical geology examines the materials composing Earth and seeks to understand the many processes that operate beneath and upon its surface. Earth is a dynamic, ever-changing planet. Internal forces create earthquakes, build mountains, and produce volcanic structures. At the surface, external processes break rock apart and sculpture a broad array of landforms. The erosional effects of water, wind, and ice result in a great diversity of landscapes. Since rocks and minerals form in response to Earth's internal and external processes, their interpretation is basic to an understanding of our planet.

In contrast to physical geology, the aim of *historical geology* is to understand the origin of Earth and the development of the planet through its 4.6-billion-year history. It strives to establish an orderly chronological arrangement of the multitude of physical and biological changes that have occurred in the geologic past (Figure I.7).

Part Two of the text, *The Oceans,* is devoted to **oceanography.** Oceanography is actually not a separate and distinct science. Rather, it involves the application of all sciences in a comprehensive and interrelated study of the oceans in all their aspects and relationships. Oceanography integrates chemistry, physics, geology, and biology. It includes the study of the composition and movements of seawater, as well as coastal processes, seafloor topography, and marine life.

Figure I.6
Major physical features of the continents and ocean basins. The diversity of features on the ocean floor is as varied as on the continents.

Part Three, *The Atmosphere,* examines the mixture of gases that is held to the planet by gravity and thins rapidly with altitude. Acted on by the combined effects of Earth's motions and energy from the sun, the formless and invisible atmosphere reacts by producing an infinite variety of weather, which, in turn, creates the basic pattern of global climates. **Meteorology** is the study of the atmosphere and the processes that produce weather and climate. Like oceanography, meteorology involves the application of other sciences in an integrated study of the thin layer of air that surrounds Earth.

Part Four, *Astronomy,* demonstrates that the study of Earth is not confined to investigations of its four interacting "spheres." The Earth sciences also attempt to relate our planet to the larger universe. Because Earth is related to all of the other objects in space, the science of **astronomy**—the study of the universe—is very useful in probing the origins of our own environment. Since we

Box I.1

Earth as a System

A *system* is a group of interrelated, interacting, or interdependent parts that form a complex whole. Most of us hear and use the term frequently. We may service our car's cooling *system*, make use of the city's transportation *system*, and be a participant in the political *system*. A news report may inform us of an approaching weather *system*.

We know that Earth is just a small part of a large system known as the *solar system*. As we study Earth, it also becomes clear that our planet can be viewed as a system with many separate but interacting parts or subsystems. The hydrosphere, atmosphere, biosphere, and solid Earth and all of their components can be studied separately. However, the parts are not isolated. Each is related in some way to the others to produce a complex and continuously interacting whole that we call the *Earth system*.

The parts of the Earth system are linked so that a change in one part can produce changes in any or all of the other parts. For example, when a volcano erupts, lava from Earth's interior may flow out at the surface and block a nearby valley. This new obstruction influences the region's drainage system by creating a lake or causing streams to change course. The large quantities of volcanic ash and gases that can be emitted during an eruption may be blown high into the atmosphere and influence the amount of solar energy that can reach the surface. The result could be a drop in air temperatures over the entire hemisphere. Where the surface is covered by lava flows or a thick layer of volcanic ash, existing soils are buried. This causes the soil-forming processes to begin anew to transform the new surface material into soil (Figure I.A). The soil that eventually forms will reflect the interaction among many parts of the Earth system. Of course, there would also be significant changes in the biosphere. Some organisms and their habitats would be eliminated by the lava and ash, while new settings for life, such as the lake, would be created. The potential climate change could also impact sensitive life-forms.

The Earth system is powered by energy from two sources. The sun drives external processes that occur in the atmosphere, hydrosphere, and at Earth's surface. Weather and climate, ocean circulation, and erosional processes are driven by energy from the sun. Earth's interior is the second source of energy. Heat remaining from when our planet formed and heat that is continuously generated by radioactive decay powers the internal processes that produce volcanoes, earthquakes, and mountains.

Humans are *part of* the Earth system, a system in which the living and nonliving components are entwined and interconnected. Therefore, our actions produce changes in all of the other parts. When we burn gasoline and coal, build breakwaters along the shoreline, dispose of our wastes, and clear the land, we cause other parts of the system to respond, often in unforeseen ways. Throughout this book you will learn about many of Earth's subsystems: the hydrologic system, the tectonic (mountain-building) system, and the climate system, to name a few. Remember that these components *and we humans* are all part of the complex interacting whole we call the Earth system.

Figure I.A
When Mount St. Helens erupted in May 1980, the area shown here was buried by a volcanic mudflow. Now, plants are reestablished and new soil is forming. (Photo by Jack Dykinga)

are so closely acquainted with the planet on which we live, it is easy to forget that Earth is just a tiny object in a vast universe. Indeed, Earth is subject to the same physical laws that govern the great many other objects that populate the great expanses of space. Thus, to understand explanations of our planet's origin, it is useful to learn something about the other members of our solar system. Moreover, it is helpful to view the solar system as a part of the great assemblage of stars that comprise our galaxy, which, in turn, is but one of many galaxies (see Box I.2).

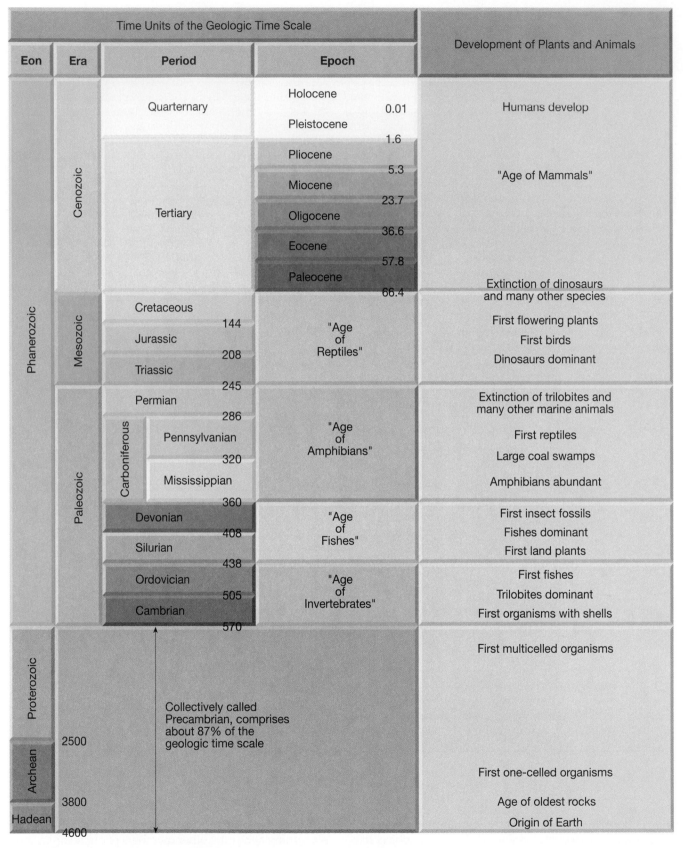

Figure I.7

The geological time scale. Numbers on the time scale represent time in millions of years before the present. These dates were added long after the time scale had been established using relative dating techniques. The Precambrian accounts for more than 85 percent of geologic time. (Data from Geological Society of America)

Box I.2

Earth's Place in the Cosmos*

For centuries, those who have gazed at the night sky have wondered about the nature of the universe, Earth's place within it, and whether or not we are "alone." Today, many exciting discoveries in astronomy are beginning to provide some answers concerning the origin of the universe, the formation and evolution of stars, and how Earth and its materials came into existence.

The realization that the universe is immense and orderly began in the early 1900s when Edwin Hubble and other dedicated scientists demonstrated that the Milky Way Galaxy is one of hundreds of billions of galaxies, each of which contains billions of stars. Today we understand that Earth, its materials, and indeed all living things are the consequence of a sequence of events, governed by natural laws, that occurred within the universe during the past 20 billion years.

Astronomical evidence supports the theory, called the Big Bang, that the universe began perhaps as long as 20 billion years ago when a dense, hot, supermassive concentration of material exploded with cataclysmic force (Figure I.B). With-

*This box was prepared by Kenneth Pinzke, Belleville Area College.

in about one second, the temperature of the expanding universe cooled to approximately 10 billion degrees and fundamental atomic particles called protons and neutrons began to appear. After a few minutes, atoms of the least complex elements—hydrogen and helium—had formed and the initial conversion of energy to matter in the young universe was over.

During the first billion years or so, matter (essentially hydrogen and to a lesser extent helium) in the expanding universe became clumpy and fragmented into enormous clouds and groups of clouds that eventually collapsed to become galaxies and clusters of galaxies. Inside these collapsing clouds, smaller concentrations of matter formed into stars. One of the billions of galaxies to form was the one we call the Milky Way.

The Milky Way is a collection of several hundred billion stars, the oldest of which is about 10 billion years. It is one of a cluster of approximately 28 galaxies, called the Local Group, that exist in our region of the universe. Initially, the oldest stars in the Milky Way formed from nearly pure hydrogen. Later, succeeding generations of younger stars, including our sun, would have heavier, more complex atoms available for their formation.

After forming, all stars go through a series of changes before finally dying out. During the life of most stars, energy is produced as hydrogen nuclei (protons) fuse with other hydrogen nuclei to form helium. During this process, called nuclear fusion, matter is converted to

energy. Stars begin to die when their nuclear fuel becomes exhausted. Massive stars often have indescribably explosive deaths. During these violent events, called supernovas, nuclear fusion advances beyond helium and produces heavier, more complex atoms such as oxygen, carbon, and iron. In turn, these products of stellar death may become the materials that compose future generations of stars. It was from the debris scattered during the death of a preexisting star, or stars, that our sun, as well as the rest of the solar system, formed.

Our star, the sun, is at the very least a second-generation star. It, along with the planets and other members of the solar system, began forming nearly 5 billion years ago, some 15 billion years after the Big Bang, from a large, interstellar cloud, called a nebula, which consisted of dust particles and gases enriched in heavy elements from supernova explosions. Gravitational energy caused the nebula to contract and in so doing, it began to rotate and flatten. Inside, smaller concentrations of matter began condensing to form the planets. At the center of the nebula there was sufficient pressure and heat to initiate hydrogen nuclear fusion, and our sun was born.

It has been said that all life on Earth is related to the stars. This is true because the very atoms in our bodies, as well as the atoms composing every other thing on Earth, owe their origin to a supernova event that occurred billions of years ago, trillions of kilometers away.

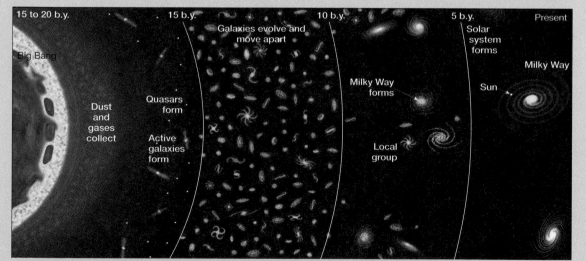

Figure I.B
About twenty billion years ago, an incomprehensibly large explosion sent all the matter of the universe flying outward at incredible speeds. After a few billion years, the material cooled and condensed into the first stars and galaxies. Because the universe is expanding, the evolving galaxies continue to move away from one another. About five billion years ago, our solar system began forming in one of these galaxies, the Milky Way.

Earth Science and the Environment

Environment refers to everything that surrounds and influences an organism. Some of these things are biological and social, but others are nonliving. The factors in this latter category are collectively called our *physical environment.* The physical environment encompasses water, air, soil, and rock, as well as conditions such as temperature, humidity, and sunlight. The phenomena and processes studied by the Earth sciences are basic to an understanding of the physical environment. In this sense, most of Earth science may be characterized as environmental science.

However, when the term *environmental* is applied to Earth science today, it is usually reserved for those aspects that focus on the relationships between people and the physical environment. Application of the Earth sciences is necessary to understand and solve problems that arise from these interactions.

Resources

Resources are an important environmental concern. Resources range from water and soil to metallic and nonmetallic minerals and energy. These materials are the very basis of modern civilization. The mineral and energy resources that are extracted from the crust are the raw materials from which the products used by society are made (Figure I.8).

Few people who live in highly industrialized nations realize the quantity of resources needed to maintain their present standard of living. For example, the annual per capita consumption of metallic and nonmetallic mineral resources for the United States is nearly 10,000 kilograms (11 tons). This is each person's prorated share of the materials required by industry to provide the vast array of products modern society demands. Figures for other highly industrialized countries are comparable.

Resources are commonly divided into two broad categories. Some are classified as **renewable,** which means that they can be replenished over relatively short time spans. Common examples are plants and animals for food, natural fibers for clothing, and forest products for lumber and paper. Energy from flowing water, wind, and the sun are also considered renewable.

By contrast, many other basic resources are classified as **nonrenewable.** Important metals such as iron, aluminum, and copper fall into this category, as do our most important fuels: oil, natural gas, and coal. Although these and other resources continue to form, the processes that create them are so slow that significant deposits take millions of years to accumulate. In essence, Earth contains fixed quantities of these substances. When the present supplies are mined or pumped from the ground, there will be no more. Al-

Figure I.8

As world population grows, demand for mineral and energy resources climbs. Earth scientists must deal with the search for additional supplies of traditional and alternative resources as well as the environmental impact of their extraction and use. These offshore oil rigs are in California's Santa Barbara channel. (Photo by Glenn Short/Liaison International)

though some nonrenewable resources, such as aluminum, can be used over and over again, others, such as oil, cannot be recycled.

Population

A glance at Figure I.9 shows that the population of our planet is growing rapidly. Although it took until the beginning of the nineteenth century for the number to reach one billion, just 130 years were needed for the population to double to two billion. Between 1930 and 1975 the figure doubled again to four billion, and by the year 2000 nearly seven billion people will inhabit the planet. Clearly, as population grows, the demand for resources expands as well. However, the rate of mineral and energy resource usage has climbed more rapidly than the overall growth of population. This fact results from an increasing standard of living. This is certainly true in the United States, where about 6 percent of the world's population uses approximately 30 percent of the annual production of mineral and energy resources.

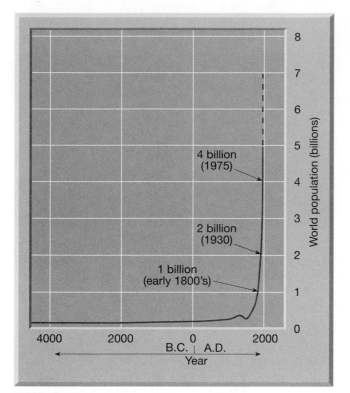

Figure I.9

Growth of world population. It took until 1800 for the number to reach one billion. By the year 2000, nearly seven billion people will inhabit the planet. The demand for basic resources is growing faster than the rate of population increase. (Data from the Population Reference Bureau)

How long will the remaining supplies of basic resources last? How long can we sustain the rising standard of living in today's industrialized countries and still provide for the growing needs of developing regions? How much environmental deterioration are we willing to accept in pursuit of basic resources? Can alternatives be found? If we are to cope with an increasing per capita demand and a growing world population, it is important that we have some understanding of our present and potential resources.

Environmental Problems

In addition to the quest for adequate mineral and energy resources, the Earth sciences must also deal with a broad array of other environmental problems. Some are local, some are regional, and still others are global in extent. Developed and developing nations alike face serious difficulties. Urban air pollution, acid rain, ozone depletion, and global warming are just a few that pose significant threats (Figure I.10). Other problems involve the loss of fertile soils to erosion, the disposal of toxic wastes, and the contamination and depletion of water resources. The list continues to grow.

In addition to human-induced and human-accentuated problems, people must also cope with the many natural hazards posed by the physical environment (Figure I.11). Earthquakes, landslides, floods, and hurricanes are just four of the many risks. Others such as drought, although not as spectacular, are nevertheless equally important environmental concerns. In many cases, the threat of natural hazards is aggravated by increases in population as more people crowd into places where an impending danger exists or attempt to cultivate marginal lands that should not be farmed.

It is clear that as world population continues its rapid growth, pressures on the environment will increase as well. Therefore, an understanding of Earth is not only essential for the location and recovery of basic resources, but also for dealing with the human impact on the environment and minimizing the effects of natural hazards. Knowledge about our planet and how it works is necessary to our survival and well-being. Earth is the only suitable habitat we have, and its resources are limited.

Figure I.10

Air pollution in downtown Los Angeles. Air quality problems affect many cities. Fuel combustion by motor vehicles and power plants provide a high proportion of the pollutants. Meteorological factors determine whether pollutants remain "trapped" in the city or are dispersed. (Photo by Ted Spiegel/Black Star)

A.

B.

Figure I.11

There are many Earth processes and phenomena that are hazardous to people, including volcanoes, earthquakes, landslides, floods, hurricanes, and tornadoes. **A.** Hurricane Andrew devastated South Florida in August 1992. (Photo by John Lopinot/*Palm Beach Post/Sygma*) **B.** This elevated highway collapsed during a severe quake in Kobe, Japan, January 1995. (Photo by Oshiaki Nagashima/*Black Star*)

The Nature of Scientific Inquiry

As members of a modern society, we are constantly reminded of the benefits derived from science. But what exactly is the nature of scientific inquiry?

All science is based on the assumption that the *natural world behaves in a consistent and predictable manner.* This implies that the physical laws which govern the smallest atomic particles also operate in the largest, most distant galaxies. Evidence for the existence of these underlying patterns is abundant in the physical and biological worlds. For example, the same biochemical processes and the same genetic codes that are found in bacterial cells are also found in human cells. The overall goal of science is to discover the underlying patterns in the natural world and then to use this knowledge to predict what will or will not happen, given certain facts or circumstances.

Collecting Facts

The development of new scientific knowledge involves some basic, logical processes that are universally accepted. To determine what is occurring in the natural world, scientists collect *facts* through observation and measurement (Figure I.12). These data are the raw material from which scientific theories and laws are formulated.

Hypothesis

Once facts have been gathered and principles have been formulated to describe a natural phenomenon, investigators try to explain how or why things happen in the manner observed. They can do this by constructing a preliminary, untested explanation, which we call a scientific **hypothesis.** Often, several different hypotheses are advanced to explain the same facts and observations.

For example, there are currently five major hypotheses that have been proposed to explain the origin of Earth's moon. Until recently, the most widely held hypothesis argued that the moon and Earth formed simultaneously from the same cloud of nebular dust and gases. A newer hypothesis suggests that a Mars-sized body impacted Earth. The collision ejected a huge quantity of material into Earth's orbit that eventually accumulated into the moon. Both hypotheses will be submitted to rigorous testing that will undoubtedly result in the modification or rejection of one, or perhaps both.

The history of science is littered with discarded hypotheses. One of the best known is the idea that Earth was at the center of the universe, a proposal that was supported by the apparent daily motion of the sun, moon, and stars around Earth.

Figure I.12

This scientist is gathering data on soil processes. (Photo by Richard T. Nowitz/Photo Researchers, Inc.)

Theory

When a hypothesis has survived extensive scrutiny, and when competing hypotheses have been eliminated, a hypothesis may be elevated to the status of a scientific **theory.** In everyday language, we may say "that's only a theory." But a scientific theory is a well-tested and widely accepted view that scientists agree best explains certain observable facts.

Scientific theories, like scientific hypotheses, are accepted only provisionally. It is always possible that a theory which has withstood previous testing may eventually be disproven. As theories survive more testing, they are regarded with higher levels of confidence. Theories that have withstood extensive testing, such as the theory of plate tectonics or the theory of evolution, are held with a very high degree of confidence.

Scientific Law

Some concepts in science are formulated into scientific laws. A scientific **law** is a generalization about the behavior of nature from which there has been no known deviation after numerous observations or experiments. Scientific laws are generally narrower in scope than theories and can usually be expressed mathematically. Examples include Newton's laws of motion and the laws of thermodynamics.

Scientific laws describe what happens in nature, but they do not explain how or why things happen this way. Also, unlike hypotheses, which are inventions of the mind to explain scientific facts, scientific laws are based on observations and measurements and thus are rarely overthrown or seriously modified. Nevertheless, even scientific laws are not necessarily "perpetual truths." As the mathematician Jacob Bronowski so aptly stated, "Science is a great many things, but in the end they all return to this: Science is the acceptance of what works and the rejection of what does not."

The processes just described, in which scientists gather facts through observations and formulate scientific hypotheses, theories, and laws, is called the *scientific method*. Contrary to popular belief, the scientific method is not a standard recipe that scientists apply in a routine manner to unravel the secrets of our natural world. Neither is this process a haphazard one.

Some scientific knowledge is gained through the following steps: (1) collection of scientific facts (data) through observation and measurement (Figure I.13); (2) development of a working hypothesis to explain these facts; (3) construction of experiments to validate the hypothesis; and (4) acceptance, modification, or rejection of the hypothesis on the basis of extensive testing.

Other scientific discoveries represent purely theoretical ideas, which then stood up to extensive examination. Still other scientific advancements have been made when a totally unexpected happening occurred

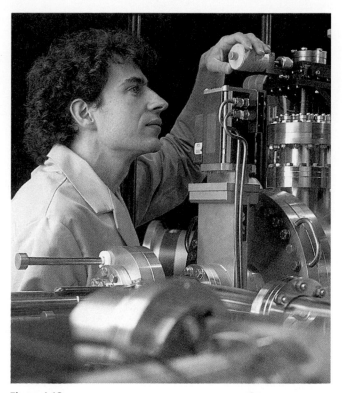

Figure I.13
Many Earth scientists work in the laboratory as well as in the field. (Photo by Jean Miele/The Stock Market)

during an experiment. These so-called serendipitous discoveries are more than pure luck, for as Louis Pasteur said, "In the field of observation, chance favors only the prepared mind."

Because scientific knowledge is acquired through several avenues, it might be best to describe the nature of scientific inquiry as the *methods* of science, rather than *the* scientific method.

Studying Earth Science

In this book, you will discover the results of centuries of scientific work. You will see the end product of millions of observations, thousands of hypotheses, hundreds of theories, and dozens of laws. We have distilled all of this to give you a "briefing" on Earth science.

But realize that our knowledge of Earth is changing daily, as thousands of scientists worldwide make satellite observations, analyze drill cores from the sea floor, measure earthquakes, develop computer models to predict climate, examine the genetic codes of organisms, and discover new facts about our planet's long history. This new knowledge often updates hypotheses and theories. Expect to see many new discoveries and changes in scientific thinking in your lifetime.

Review Questions

1. List and briefly define the four "spheres" that constitute our environment.

2. Name the three principal divisions of Earth. Which one is thinnest?

3. The oceans cover nearly _____ percent of Earth's surface and contain about _____ percent of the planet's total water supply.

4. Is the present shoreline the boundary between the continents and ocean basins? Briefly explain.

5. Name the specific Earth science described by each of the following statements.
   (a) The science that deals with the dynamics of the oceans.
   (b) This word literally means "the study of Earth."
   (c) An understanding of the atmosphere is the primary focus of this science.
   (d) This science helps us understand Earth's place in the universe.

6. Contrast renewable and nonrenewable resources. Give one or more examples of each.

7. World population is estimated to reach nearly _____ billion by the year 2000. How does this compare to the world population near the beginning of the nineteenth century?

8. How is a scientific hypothesis different from a scientific theory?

9. A _____ is a generalization about the behaviour of nature from which there has been no known deviation.

10. Box I.A suggests that Earth can be considered a system. What is a system? Provide a brief example of how Earth acts as a system.

Key Terms

astronomy (p. 6)
atmosphere (p. 3)
biosphere (p. 3)
core (p. 4)
crust (p. 4)
geology (p. 5)
hydrosphere (p. 2)
hypothesis (p. 12)
law (p. 13)
lithosphere (p. 4)
mantle (p. 4)
meteorology (p. 6)
nonrenewable resource (p. 10)
oceanography (p. 5)
renewable resource (p. 10)
theory (p. 13)

PART 1
The Solid Earth

Liberty Bell Mountain, Washington. (Photo by Larry
Ulrich/Tony Stone Images)

CHAPTER 1
Minerals: Building Blocks of Rocks

Emerald variety of beryl on calcite. (Photo by Jeffrey Scovil)

The study of geology appropriately begins with the study of rocks and minerals, because a knowledge of the materials that make up Earth is basic to our understanding of much of the rest of geology. In this chapter we shall examine atoms and elements, the building blocks of minerals, and then investigate the major mineral groups. Our inquiry shall attempt to answer some important questions: How are minerals identified? How are rocks and minerals similar and how are they different? What are the major rock-forming minerals? What defines a mineral resource?

Earth's crust is the source of a wide variety of minerals, many of which are useful and essential to people. In fact, practically every manufactured product contains materials obtained from minerals. Most people are familiar with the common uses of basic metals, including aluminum in beverage cans, copper in electrical wiring, gold in jewelry, and silicon in computer chips. But fewer are aware that pencil "lead" does not contain lead metal but is really made of the soft black mineral called *graphite*. Baby powder ("talcum powder") is ground-up rock made of the mineral *talc*, with perfume added. Drill bits impregnated with pieces of *diamond* (a mineral) are used to cut through rock layers in search of oil. The common mineral *quartz* is the main ingredient in ordinary glass. And on and on. As the material demands of modern society grow, the need to locate additional supplies of useful minerals also grows, and becomes more challenging as the easily mined sources become depleted.

The economic uses of rocks and minerals are important to us. Also important to Earth scientists is that every process they study depends in some way on the properties of these basic Earth materials. Events such as volcanic eruptions, mountain building, weathering and erosion, and even earthquakes involve rocks and minerals (Figure 1.1). Consequently, a basic knowledge of Earth materials is essential to the understanding of all Earth science phenomena.

Figure 1.1
These mountain peaks are part of the Alaska Range in Denali National Park. (Photo by Carr Clifton)

Minerals Versus Rocks

Many people think of rocks as nondescript, hard, often dirty objects. Minerals are considered by many to be part of the human diet ("vitamins and minerals"), or possibly rare ores or precious gems. However, these common perceptions are not entirely correct.

Figure 1.2 shows the relation between rocks and minerals. A **rock** can be defined simply as an aggregate of minerals. Minerals are the *building blocks* that make up rocks. In a rock, the minerals occur together as a *mixture*, so each mineral retains its distinctive properties, such as color, as you can see in Figure 1.2.

Although most rocks are composed of more than one mineral, certain minerals are commonly found by themselves in large, relatively pure quantities. In these instances, they are still considered to be rocks. A common example is the mineral *calcite*, which frequently is the dominant constituent in widespread rock layers where it is given the name *limestone*.

Minerals, the building blocks of rocks, have a very precise definition. A **mineral** is a naturally occurring inorganic solid that possesses a definite chemical structure, which gives it a unique set of physical properties. Thus, for any Earth material to be considered a mineral, it must exhibit the following characteristics:

1. It must be naturally occurring.
2. It must be inorganic (was never alive).
3. It must be a solid.
4. It must possess a definite chemical structure.

When the term mineral is used by geologists, only those substances that fulfill these precise conditions are considered minerals. Consequently, synthetic diamonds, although chemically the same as natural diamonds, are human-made, and therefore are not considered minerals. Further, oil and natural gas, which are not solids, and which once were "living" (they are the remains of ancient organisms), are also not classified as minerals. An exception is material formed by the activity of animals, such as clam shells, which are considered minerals.

Composition and Structure of Minerals

Each of Earth's nearly 4000 minerals is a unique substance that is defined by its chemical composition and internal structure. What elements make up a particular

Figure 1.2
Rocks are aggregates of one or more minerals.

Granite
(Rock)

Quartz
(Mineral)

Hornblende
(Mineral)

Feldspar
(Mineral)

mineral? In what pattern are its elements bonded together? And what important properties does this give each mineral?

At this point, let us briefly review the basic building blocks of minerals, the chemical **elements**. At present, over 100 elements are known, although a dozen and a half have been produced only in the laboratory. Some minerals, such as gold and sulfur, are made entirely from one element. But most elements are not stable, and thus most minerals are a combination of two or more elements, joined to form a chemically stable compound.

To better understand how elements combine to form minerals, we must first consider the atom. The **atom** is the smallest particle of matter that has all the characteristics of an element. It is this extremely small particle that does the combining.

How Atoms Are Constructed

A simplified model of an atom is shown in Figure 1.3. Each atom has a central region, called the **nucleus**, which contains very dense **protons** (particles with positive electrical charges) and equally dense **neutrons** (particles with neutral electrical charges). Orbiting the nucleus are **electrons**, which have negative electrical charges. For convenience, we often diagram atoms to show the electrons orbiting the nucleus, like the orderly orbiting of the planets around the sun. However, electrons move so rapidly that they create a sphere-shaped negative zone around the nucleus. Hence, a more realistic picture of the positions of electrons can be obtained by envisioning a cloud of negatively charged electrons surrounding the nucleus.

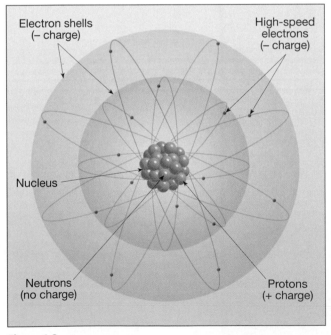

Figure 1.3

Simplified model of an atom. An atom consists of a central nucleus composed of protons and neutrons that is encircled by electrons.

It is also known that individual electrons are located at given distances from the nucleus in regions called **energy levels**, or **shells**. Two such shells are shown in Figure 1.3. As we shall see, an important fact about these shells is that each can have only a specific number of electrons.

The number of protons in an atom's nucleus determines its **atomic number** and name of the element. For example, all atoms with six protons are carbon atoms, all those with eight protons are oxygen atoms, and so forth. "Free" atoms (those not combined with other atoms) have the same number of electrons as protons, so the atomic number also equals the number of electrons surrounding the nucleus. Therefore, carbon has six electrons to match its six protons, and oxygen has eight electrons to match its eight protons. Neutrons have no charge, so the positive charge of the protons is exactly balanced by the negative charge of the electrons. Consequently, uncombined atoms are neutral electrically and have no overall electrical charge. Thus, the *element* is a large collection of electrically neutral atoms, all having the same atomic number.

How Atoms Bond Together

Elements combine with each other to form a wide variety of more complex mineral compounds. A **compound** is composed of two or more elements bonded together in definite proportions. When the elements separate, the bonds are broken and the compound no longer exists; only its component elements do. Through experimentation, scientists have learned that the forces bonding atoms together are electrical. Further, it is known that chemical bonding results in a change in the arrangement of electrons in the shells of the bonded atoms.

When an atom combines chemically, it either gains, loses, or shares electrons with another atom. An atom that gains electrons becomes negatively charged because it has more electrons than protons. Atoms that lose electrons become positively charged. Atoms that have an electrical charge because of a gain or loss of electrons are called **ions**. Simply stated, oppositely charged ions attract one another to produce a neutral chemical compound.

An example of chemical bonding involves sodium (Na) atoms and chlorine (Cl) atoms to produce sodium chloride molecules (NaCl). Sodium chloride is common table salt (Figure 1.4). When the sodium atom loses one electron it becomes a positive ion, and when the chlorine atom gains one electron it becomes a negative ion. These opposite charges act as an "electrical glue" that bonds the atoms together.

The properties of a chemical compound are dramatically different from the properties of the elements composing it. For example, chlorine is a green, poisonous gas that is so toxic it was used as a weapon during World War I. Sodium is a soft, silvery metal that reacts vigorously with water and, if held in your hand,

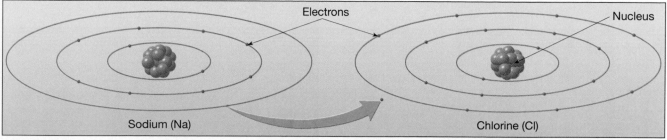

Figure 1.4

Chemical bonding of sodium and chlorine atoms to produce sodium chloride. Through the transfer of one electron in the outer shell of a sodium atom to the outer shell of a chlorine atom, the sodium becomes a positive ion and chlorine a negative ion.

could burn it severely. Together, however, these atoms produce the compound sodium chloride (table salt), the familiar clear, crystalline solid that is essential for human life.

This example also illustrates an important difference between a rock and a mineral. In a mineral, the constituent elements lose their properties. In a *rock*, on the other hand, which is a *mixture* of minerals, each mineral retains its own distinctive properties.

Isotopes and Radioactivity.   As noted, the atomic number is the number of protons in an atom's nucleus. The **mass number** of an atom is higher, because it is the *sum of the neutrons and protons* in the nucleus. Atoms of the same element always have the same number of protons, but they commonly have varying numbers of neutrons. This means that an element can have more than one mass number. These variants of the same element are called **isotopes** of that element.

For example, carbon has two well-known isotopes, one having a mass number of 12 (carbon-12), the other a mass number of 14 (carbon-14). All atoms of the same element must have the same number of protons, and carbon always has six (atomic number = 6). Hence, carbon-12 must have six protons *plus six neutrons* to give it a mass number of 12, whereas carbon-14 must have six protons *plus eight neutrons* to give it a mass number of 14.

The *average* atomic mass of any random sample of carbon is much closer to 12 than 14, because carbon-12 is the more common isotope. This average is called *atomic weight.**

In chemical behavior, all isotopes of the same element are nearly identical. To distinguish among them is like trying to differentiate identical twins, with one slightly heavier. Because isotopes of an element react the same chemically, different isotopes can become parts of the same mineral. For example, when calcite is formed from calcium, carbon, and oxygen, some of its carbon atoms are carbon-12 and some are carbon-14.

Although the vast majority are stable, many elements do have isotopes that are unstable. "Unstable" means that the isotopes decay naturally through a process called **radioactivity**. It is so called because a decaying atom actively radiates energy and particles. Radioactivity occurs when the forces that bind the nucleus are not strong enough to keep it together.

The rate at which the unstable nuclei decay is steady and measurable. This makes such isotopes useful "clocks" for dating the events of Earth history. A discussion of radioactivity and its applications in dating past geologic events is found in Chapter 10.

Properties of Minerals

Minerals occur in a fascinating palette of colors, shapes, and lusters. They vary in hardness and may even have a distinctive taste or scent. We will now look at these properties, which are useful to help identify minerals.

Crystal Form

Crystal form is the external expression of a mineral's internal orderly arrangement of atoms. Generally, when a mineral forms without space restrictions, it will develop individual crystals with well-formed crystal faces. Figure 1.5A illustrates this for pyrite, an iron-and-sulfur mineral with cubic crystals. Figure 1.5B shows the distinctive hexagonal crystals of quartz that grow when space and time permit. However, most of the time, crystal growth is severely constrained. It is stunted because of competition for space, resulting in an intergrown mass of small, jammed crystals, none of which exhibits its crystal form. This is what happened to the minerals in the granite in Figure 1.2. Thus, most inorganic solid objects are composed of crystals, but they are not clearly visible to the unaided eye.

Luster

Luster is the appearance or quality of light reflected from the surface of a mineral. Minerals that have the appearance of metals, regardless of color, are said to have a *metallic luster*, like the pyrite crystals in Figure 1.5A. Minerals

---

*The term *weight* is a misnomer that has resulted from long use. The correct term is *atomic mass*.

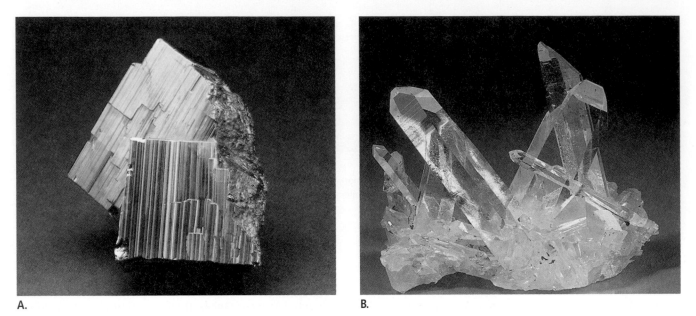

A.

B.

Figure 1.5
Crystal form is the external expression of a mineral's orderly internal structure. **A.** Pyrite, commonly known as "fool's gold," often forms cubic crystals that may contain parallel lines called striations. (Photo by GeoScience/PH) **B.** Quartz sample that exhibits well-developed hexagonal (six-sided) crystals with pyramidal-shaped ends. (Photo by Breck P. Kent)

with a *nonmetallic luster* are described by various adjectives. These include vitreous (glassy, like the quartz crystals in Figure 1.5B), pearly, silky, resinous, and earthy (dull).

Color

Although **color** is an obvious feature of a mineral, it is often an unreliable diagnostic property. Slight impurities in the common mineral quartz, for example, give it a variety of colors, including pink, purple (amethyst), milky white, and even black.

Streak

Streak is the color of a mineral in its powdered form, which is a much more reliable indication of color. Streak is obtained by rubbing the mineral across a piece of hard, unglazed porcelain, termed a *streak plate*. Whereas the color of a mineral may vary from sample to sample, the streak usually does not, and is therefore the more consistent property. Streak can also help to dis-

tinguish minerals with metallic lusters from those having nonmetallic lusters. Metallic minerals generally have a dense, dark streak, whereas minerals with nonmetallic lusters do not.

Hardness

One of the most useful diagnostic properties is **hardness**, a measure of the resistance of a mineral to abrasion or scratching. This property is determined by rubbing the mineral to be identified against another mineral of known hardness. One will scratch the other (unless they have the same hardness).

Geologists use a standard hardness scale, called the **Mohs scale**. It consists of ten minerals arranged in order from 10 (hardest) to 1 (softest), as shown in Table 1.1. Any mineral of unknown hardness can be rubbed against these to determine its hardness. In the field, other handy objects work, too. For example, your fingernail has a hardness of 2.5, a copper penny 3, and a

Table 1.1 Mohs scale of mineral hardness.

Relative Scale		Mineral	Hardness of Some Common Objects
Hardest	10	Diamond	
	9	Corundum	
	8	Topaz	
	7	Quartz	
	6	Potassium Feldspar	
	5	Apatite	5.5 Glass, Pocketknife
	4	Fluorite	
	3	Calcite	3.5 Copper Penny
	2	Gypsum	2.5 Fingernail
Softest	1	Talc	

piece of glass 5.5. The mineral gypsum, which has a hardness of 2, can be easily scratched by your fingernail. On the other hand, the similar-looking mineral calcite, which has a hardness of 3, cannot be scratched by your fingernail. Calcite cannot scratch glass, because its hardness is less than 5.5. Quartz, the hardest of the common minerals at 7, will scratch a glass plate. Diamonds, hardest of all, scratch anything. You can see that by rubbing an unknown mineral against objects of known hardness, you can quickly narrow its hardness range.

Cleavage

In the crystal structure of a mineral, some bonds are weaker than others. These bonds are where a mineral will break when it is stressed. **Cleavage** is the tendency of a mineral to cleave, or break, along planes of weak bonding. Not all minerals have definite planes of weak bonding, but those that possess cleavage can be identified by the distinctive smooth surfaces that are produced when the mineral is broken.

The simplest type of cleavage is exhibited by minerals called micas (Figure 1.6). Because the micas have weak bonds in one direction, they cleave to form thin, flat sheets. Some minerals have several cleavage planes which produce smooth surfaces when broken, while others exhibit poor cleavage, and still others have no cleavage at all. When minerals break evenly in more than one direction, cleavage is described by the *number of planes* exhibited and the *angles at which they meet* (Figure 1.7).

Do not confuse *cleavage* with *crystal form*! When a mineral exhibits cleavage, it will break into pieces *that have the same shape as the original sample*. By contrast, the quartz crystals shown in Figure 1.5B do not have cleavage. If broken, they fracture into pieces that do not resemble each other or the original crystals.

Figure 1.6

Sheet-type cleavage common to the micas. (Photo by Chip Clark)

Fracture

Minerals that do not exhibit cleavage when broken, such as quartz, are said to **fracture**. Even fracturing has variety: minerals that break into smooth curved surfaces like those seen in broken glass have a *conchoidal fracture* (Figure 1.8). Others break into splinters or fibers, like asbestos, but most minerals fracture irregularly.

Specific Gravity

Specific gravity compares the weight of a mineral to the weight of an equal volume of water. For example, if a cubic centimeter of a mineral weighs three times as much as a cubic centimeter of water, its specific gravity is 3. With a little practice, you can estimate the specific gravity of minerals by hefting them in your hand. For example, if a mineral feels as heavy as the common rocks you have handled, its specific gravity will probably be somewhere between 2.5 and 3. Some metallic

Figure 1.7

Smooth surfaces produced when a mineral with cleavage is broken. The sample on the left (flourite) exhibits four planes of cleavage (eight sides), whereas the other two samples exhibit three planes of cleavage (six sides). Also notice that the mineral in the center (halite) has cleavage planes that meet at 90-degree angles, whereas the mineral on the right (calcite) has cleavage planes that meet at 75-degree angles. (Photo by GeoScience/PH)

Figure 1.8
Conchoidal fracture. The smooth curved surfaces result when minerals break in a glasslike manner. (Photo by E. J. Tarbuck)

minerals have a specific gravity noticeably greater than that of common rock-forming minerals. Galena, which is an ore of lead, has a specific gravity of roughly 7.5 (Figure 1.9). The specific gravity of pure 24-karat gold is a very hefty 20 (see Box 1.1).

Other Properties of Minerals

In addition, some minerals can be recognized by other distinctive properties. For example, halite is ordinary salt, so it is quickly identified with your tongue. Thin sheets of mica will bend and elastically snap back. Gold is malleable, which means it can be easily hammered or shaped. Talc and graphite both have distinctive feels. Talc feels soapy. Graphite feels greasy (it is a principal ingredient in dry lubricants).

A few minerals, such as magnetite, have a high iron content and can be picked up with a magnet. Some varieties of magnetite (lodestone) are natural magnets and will pick up small iron-based objects such as pins and paper

clips. Some minerals exhibit special optical properties. For example, when a transparent piece of calcite is placed over printed material, the letters appear doubled. This optical property is known as *double refraction*. In addition, the streak of many sulfur-bearing minerals smells like rotten eggs.

One very simple chemical test involves placing a drop of dilute hydrochloric acid from a dropper bottle on a freshly broken mineral surface. Certain minerals, called carbonates, will effervesce (fizz) with hydrochloric acid. This test is useful in identifying the mineral calcite, which is a common carbonate mineral.

In summary, a number of special physical and chemical properties are useful in identifying particular minerals. These include taste, smell, elasticity, malleability, feel, magnetism, double refraction, and chemical reaction to hydrochloric acid. Remember that every one of these properties depends on the composition (elements) of a mineral and its structure (how the atoms are arranged).

Mineral Groups

Nearly 4000 minerals have been named and about 40 to 50 new ones are identified each year. Fortunately, for students who are beginning to study minerals, no more than a few dozen are abundant! Collectively, these few make up most of the rocks of Earth's crust and as such, are classified as the *rock-forming minerals*. It is also interesting to note that *only eight elements* compose the bulk of these minerals and represent over 98 percent (by weight) of the continental crust (Table 1.2).

The two most abundant elements are silicon and oxygen, comprising nearly three-fourths of Earth's continental crust. Silicon and oxygen combine to form the framework of the most common mineral group, the **silicates**. Perhaps the next most common mineral group is the carbonates (carbon plus oxygen plus other elements), of which calcite is the most prominent member. Other common rock-forming minerals include gypsum and halite.

Table 1.2 Relative abundance of the most common elements in Earth's crust.

Element	Approximate Percentage by Weight
Oxygen (O)	46.6
Silicon (Si)	27.7
Aluminum (Al)	8.1
Iron (Fe)	5.0
Calcium (Ca)	3.6
Sodium (Na)	2.8
Potassium (K)	2.6
Magnesium (Mg)	2.1
All others	1.7
Total	100

SOURCE: Data from Brian Mason.

Figure 1.9
Galena is lead sulfide. Like other metallic ores, it has a relatively high specific gravity and therefore feels heavy when hefted. (Photo by Geo-Science/PH)

Box 1.1

Carrot, Karat, or Carat

Like many other words in the English language, the words *carrot, karat,* and *carat* all have the same sound but are different in meaning (Figure 1.A). Such words are called *homonyms.* We all know that a carrot is a crunchy, orange vegetable that is supposed to be good for our eyesight. But what about the words *karat* and *carat*?

The term *karat* is used to indicate the purity of gold. Pure or fine gold is 24 karats. Gold that is less than 24 karats is actually an alloy (mixture) of gold and another metal, usually copper or silver. For example, 14-karat gold contains 14 parts of gold (by weight) mixed with 10 parts of other metals. Obviously, gold with a higher karat number is more expensive. Because 24-karat gold is much softer and more malleable than a gold-silver alloy, the alloy forms are more durable.

The third of our homonyms, *carat,* is a unit of weight used for precious gems such as diamonds, emeralds, and rubies. Throughout history, the size of a carat has varied somewhat. However, early in the twentieth century, a carat weight was established at 200 milligrams (or 0.2 gram). To put this in everyday terms, a 1-ounce diamond would be roughly 142 carats. The terms *karat* and *carat* are both derived from the Greek word *keration,* meaning "carob bean." The ancient Greeks used carob beans as a standard of weight.

Figure 1.A
The purity of gold, such as this 82-ounce gold nugget, is measured in units called karats. Pure gold is 24 karats. By comparison, the weight of gemstones is measured in carat weight. Each carat equals 0.2 gram. This diamond ring is 29.3 carats. (Courtesy of Smithsonian Institution)

We will first discuss the most common mineral group, the silicates, and then consider other prominent mineral groups.

Rock-Forming Silicates

Each of the silicate minerals contains oxygen and silicon atoms. Except for a few "pure" silicate minerals such as quartz, every other silicate mineral also contains one or more additional elements that are needed to produce electrical neutrality. These elements give rise to the great variety of silicate minerals and their varied properties.

All silicates have the same fundamental building block, the **silicon-oxygen tetrahedron**. This structure consists of four oxygen atoms surrounding a much smaller silicon atom as shown in Figure 1.10. Thus, a typical hand-size silicate mineral specimen contains millions of these silicon-oxygen tetrahedra, joined together in a variety of ways.

In some minerals, the tetrahedra are joined into chains, sheets, or three-dimensional networks by sharing oxygen atoms (Figure 1.11). These larger silicate structures are then connected to one another by other elements. The primary elements that join silicate structures are iron (Fe), magnesium (Mg), potassium (K), sodium (Na), and calcium (Ca).

Major groups of silicate minerals and common examples are given in Figure 1.11. The feldspars are by far the most plentiful group, comprising over 50 percent of Earth's crust. Quartz, the second most abundant mineral in the continental crust, is the only common mineral made completely of silicon and oxygen.

Notice in Figure 1.11 that each mineral *group* has a particular silicate *structure*. A relationship exists between this internal structure of a mineral and the *cleavage* it exhibits. Because the silicon-oxygen bonds are strong, silicate minerals tend to cleave between

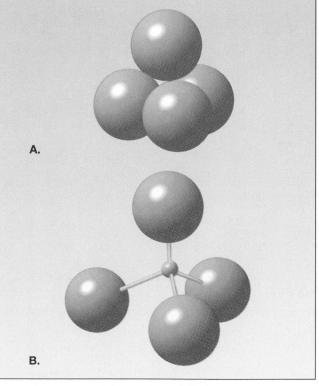

Figure 1.10
Two representations of the silicon-oxygen tetrahedron. **A.** The four large spheres represent oxygen atoms and the blue sphere represents a silicon atom. The spheres are drawn in proportion to the radii of the atoms. **B.** A model of the tetrahedron using rods to depict the bonds that connect the atoms.

Mineral		Idealized Formula	Cleavage	Silicate Structure
Olivine		$(Mg, Fe)_2SiO_4$	None	Single tetrahedron
Pyroxene group (Augite)		$(Mg,Fe)SiO_3$	Two planes at right angles	Single chains
Amphibole group (Hornblende)		$Ca_2(Fe,Mg)_5Si_8O_{22}(OH)_2$	Two planes at 60° and 120°	Double chains
Micas	Biotite	$K(Mg,Fe)_3AlSi_3O_{10}(OH)_2$	One plane	Sheets
	Muscovite	$KAl_2(AlSi_3O_{10})(OH)_2$		
Feld-spars	Orthoclase	$KAlSi_3O_8$	Two planes at 90°	Three-dimensional networks
	Plagioclase	$(Ca,Na)AlSi_3O_8$		
Quartz		SiO_2	None	(Expanded view)

Figure 1.11

Common silicate minerals. Note that the complexity of the silicate structure increases down the chart.

the silicon-oxygen structures rather than across them. For example, the micas have a sheet structure and thus tend to cleave into flat plates (see muscovite in Figure 1.12). Quartz, which has equally strong silicon-oxygen bonds in all directions, has no cleavage, but fractures instead.

How do silicate minerals form? Most crystallize from molten rock as it cools. This cooling can occur at or near Earth's surface (low temperature and pressure) or at great depths (high temperature and pressure). The *environment* during crystallization and the *chemical composition of the molten rock* mainly determine which minerals are pro-

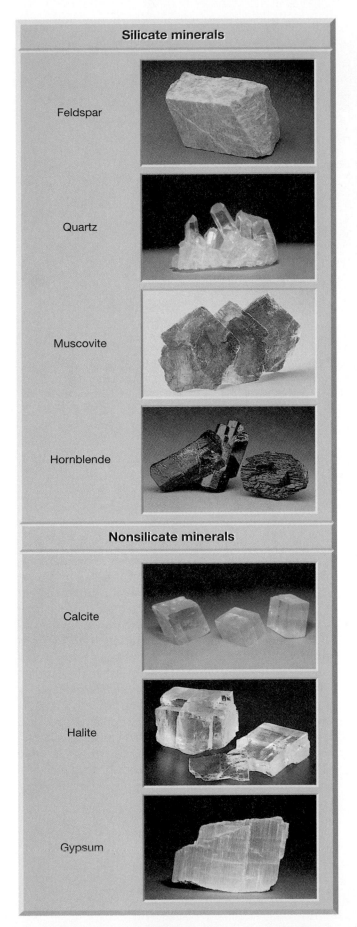

Silicate minerals

Feldspar

Quartz

Muscovite

Hornblende

Nonsilicate minerals

Calcite

Halite

Gypsum

duced. For example, the silicate mineral, olivine, crystallizes at high temperatures (around 1200°C), while quartz crystallizes at much lower temperatures (around 700°C).

In addition, some silicate minerals form at Earth's surface from the weathered products of older silicate minerals. Still other silicate minerals are formed under the extreme pressures associated with mountain building. Each silicate mineral, therefore, has a structure and a chemical composition that *indicate the conditions under which it formed*. Thus, by carefully examining the mineral makeup of rocks, geologists can often determine the circumstances under which the rocks formed.

Important Nonsilicate Minerals

Nonsilicate minerals make up only about one-fourth of the continental crust, so they can be considered scarce compared to the silicates. However, many are important economically. Table 1.3 lists examples of nonsilicate minerals of economic value: oxides, sulfides, sulfates, halides, and "native" elements.

A major rock-forming group worldwide is the *carbonates*, which includes the mineral calcite ($CaCO_3$). This mineral is the major constituent in two well-known rocks: limestone and marble. Limestone is used commercially for road paving, building stone, and as the main ingredient in portland cement. Marble is used decoratively.

Two other nonsilicate minerals frequently found in sedimentary rocks are *halite* and *gypsum*. Both minerals are commonly found in thick layers, which are the last vestiges of ancient seas that long ago evaporated, leaving behind their salts (Figure 1.13). Like limestone, halite and gypsum both are important nonmetallic resources. Halite is the mineral name for common table salt, sodium chloride. Gypsum is the mineral from which plaster and other similar building materials are composed.

In addition, a number of other minerals have economic value (Table 1.3). Ores of metals include hematite (iron), sphalerite (zinc), and galena (lead). Native (free-occurring, not in compounds) elements include gold, silver, copper, and carbon (diamonds and graphite). Other important minerals are fluorite, corundum, and sylvite.

Mineral Resources

Mineral resources are Earth's storehouse of useful minerals that can be recovered for use. Resources include already identified deposits from which minerals can be extracted profitably, called **reserves**, as well as known deposits that are not yet recoverable under present

Figure 1.12

Some common rock-forming minerals. Silicate minerals are common constituents of igneous rocks, while the nonsilicates along with quartz and clay are prominent minerals in many sedimentary rocks.

Table 1.3 Common nonsilicate mineral groups.

Mineral Group	Name	Chemical Formula	Economic Use
Oxides	Hematite	Fe_2O_3	Ore of iron, pigment
	Magnetite	Fe_3O_4	Ore of iron
	Corundum	Al_2O_3	Gemstone, abrasive
	Ice	H_2O	Solid form of water
Sulfides	Galena	PbS	Ore of lead
	Sphalerite	ZnS	Ore of zinc
	Pyrite	FeS_2	Sulfuric acid production
	Chalcopyrite	$CuFeS_2$	Ore of copper
	Cinnabar	HgS	Ore of mercury
Sulfates	Gypsum	$CaSO_4 \cdot 2H_2O$	Plaster
	Anhydrite	$CaSO_4$	Plaster
	Barite	$BaSO_4$	Drilling mud
Native elements	Gold	Au	Trade, jewelry
	Copper	Cu	Electrical conductor
	Diamond	C	Gemstone, abrasive
	Sulfur	S	Sulfa drugs, chemicals
	Graphite	C	Pencil lead, dry lubricant
	Silver	Ag	Jewelry, photography
	Platinum	Pt	Catlyst
Halides	Halite	NaCl	Common salt
	Fluorite	CaF_2	Used in steel making
	Sylvite	KCl	Fertilizer
Carbonates	Calcite	$CaCO_3$	Portland cement, lime
	Dolomite	$CaMg(CO_3)_2$	Portland cement, lime

economic conditions or technology. Deposits inferred to exist, but not yet discovered, are also considered as mineral resources.

The term **ore** denotes useful metallic minerals that can be mined at a profit. In common usage, the term *ore* is also applied to some nonmetallic minerals, such as fluorite and sulfur. However, materials used for such purposes as building stone, road paving, abrasives, ceramics, and fertilizers are not usually called ores; rather, they are classified as *industrial rocks and minerals*.

Recall that more than 98 percent of Earth's crust is composed of only eight elements. Except for oxygen and silicon, all other elements make up a relatively small fraction (about one-fourth) of common crustal rocks (see Table 1.2, p. 24). Indeed, the natural concentrations of many elements are exceedingly small. A deposit containing only

Figure 1.13

Thick beds of halite (salt) are being drilled at an underground mine near Grand Saline, Texas. (Courtesy of Morton International, Inc., Chicago, IL 60606)

Box 1.2

Gemstones

Precious stones have been prized by people since antiquity. This tradition survives today. Gems are not just prized keepsakes of the rich but are possessed by people of modest means as well. Misinformation abounds about the nature of gems and the minerals of which they are composed. Part of the misinformation stems from the ancient practice of grouping precious stones by color rather than mineral make-up. For example, the more common red spinels were often passed off to royalty as rubies, which are more valuable gems. Even today, when modern techniques of mineral identification are commonplace, yellow quartz is frequently sold as topaz.

Naming Gemstones

Compounding the confusion is the fact that many gems have names that are different from their mineral names. For example, diamond is composed of the mineral of the same name, whereas sapphire is a form of corundum, an aluminum oxide-rich mineral. Although pure aluminum oxide is colorless, a minute amount of a foreign element can produce a vividly colored gemstone. Hence, depending on the impurity, sapphires of nearly every color are known to exist. Pure aluminum oxide with trace amounts of titanium and iron produce the most prized blue sapphires. When the mineral corundum contains sufficient quantities of chromium, it exhibits a brilliant red color, and the gem is called *ruby*. Large, gem-quality rubies are much rarer than diamonds and thus can command a very high price.

To summarize, when a gem-quality sample of corundum exhibits a red hue, it is called *ruby*, but if it exhibits any other color, the gem is called *sapphire*. If the specimen is not suitable as a gem, it simply goes by the mineral name *corundum*. Although common corundum is not a gemstone, it does have value as an abrasive material. Whereas two gems, rubies and sapphires, are composed of the mineral corundum, quartz is the parent mineral of more than a dozen gems. Table 1.A lists some well-known gemstones and their mineral names.

Table 1.A Important gemstones.

Gem	Mineral Source	Prized Hues
Precious		
Diamond	Diamond	Colorless, yellows
Emerald	Beryl	Greens
Opal	Opal	Brilliant hues
Ruby	Corundum	Reds
Sapphire	Corundum	Blues
Semiprecious		
Alexandrite	Chrysoberyl	Variable
Amethyst	Quartz	Purples
Cat's-eye	Chrysoberyl	Yellows
Chalcedony	Quartz (agate)	Banded
Citrine	Quartz	Yellows
Garnet	Garnet	Reds, greens
Jade	Jadeite or nephrite	Greens
Moonstone	Feldspar	Transparent blues
Peridot	Olivine	Olive greens
Smoky quartz	Quartz	Browns
Spinel	Spinel	Reds
Topaz	Topaz	Purples, reds
Tourmaline	Tourmaline	Reds, blue-greens
Turquoise	Turquoise	Blues
Zircon	Zircon	Reds

What Constitutes a Gemstone?

In essence, certain mineral specimens when cut and polished posses beauty of such quality that they can command a price that makes the process of producing the gem profitable. Gemstones can be divided into two categories: precious and semiprecious. A *precious* gem has beauty, durability, size, and rarity, whereas a *semiprecious* gem generally has only one or two of these qualities. The gems that have traditionally enjoyed the highest esteem are diamonds, rubies, sapphires, emeralds, and some varieties of opal (Table 1.A). All other gemstones are classified as semiprecious. It should be noted, however, that large, high-quality specimens of the so-called semiprecious stones can often command a very high price.

Obviously, beauty is the most important quality that a gem can possess. Today, we prefer translucent stones with evenly tinted colors. The most favored hues appear to be red, blue, green, purple, rose, and yellow. The most prized stones are pigeon-blood rubies, blue sapphires, grass-green emeralds, and canary-yellow diamonds. Colorless gems are generally less than desirable except in the case of diamonds that display "flashes of color" known as *brilliance*. Note that gemstones in the "rough" are dull and would be passed over by most laypersons as "just another rock." Gemstones must be cut and polished by experienced craftsmen before their true beauty can be displayed (Figure 1.B).

(continues)

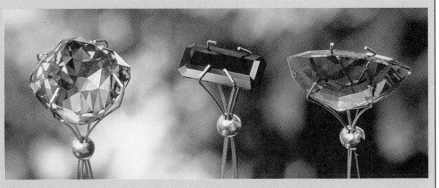

Figure 1.B
Australian sapphires depicting variations in cuts and colors. (Photo by Fred Ward, Black Star)

(Box 1.2, continued)

The durability of a gem depends on its hardness; that is, its resistance to abrasion by objects normally encountered in everyday living. For good durability, gems should be as hard or harder than quartz as defined by the Mohs scale of hardness, which consists of ten minerals arranged in order from 1 (softest) to 10 (hardest). One notable exception is opal, which is comparatively soft (hardness 5 to 6.5) and brittle. Opal's esteem comes from its "fire," which is a display of a variety of brilliant colors including greens, blues, and reds.

It seems to be human nature to treasure that which is rare. In the case of gemstones, large, high-quality specimens are much rarer than smaller stones. Thus, large rubies, diamonds, and emeralds, which are rare in addition to being beautiful and durable, command the very highest prices.

the average crustal percentage of a valuable element like gold is worthless if the cost of extracting it greatly exceeds the value of the material recovered.

To be considered of value, an element must be concentrated above the level of its average crustal abundance. For example, copper makes up about 0.0135 percent of the crust. However, for a material to be considered as copper ore, it must contain a concentration that is about 100 times this amount. Aluminum, on the other hand, represents 8.13 percent of the crust and must be concentrated to only about four times its average crustal percentage before it can be extracted profitably.

It is important to realize that economic changes may make a deposit profitable to extract, or lose its profitability. If demand for a metal increases and prices rise, the status of a previously unprofitable deposit changes, and it becomes an ore. The status of unprofitable deposits may also change if a technological advance allows the useful element to be extracted at a lower cost than before.

Conversely, changing economic factors can turn a once-profitable ore deposit into an unprofitable deposit that can no longer be called an ore. This situation is illustrated by a copper mine at Bingham Canyon, Utah, the largest open-pit mine on Earth (Figure 1.14). Mining was halted there in 1985 because outmoded equipment had driven the cost of extracting the copper beyond the current selling price. The owners responded by replacing an antiquated 1000-car railroad with conveyor belts and pipelines for transporting the ore and waste. These devices achieved a cost reduction of nearly 30 percent and returned this mining operation to profitability.

Figure 1.14

Aerial view of Bingham Canyon copper mine near Salt Lake City, Utah. This huge open-pit mine is about 4 kilometers across and 900 meters deep. Although the amount of copper in the rock is less than 1 percent, the huge volumes of material removed and processed each day (about 200,000 tons) yield significant quantities of metal. (Photo by Michael Collier)

Review Questions

1. Define the term *rock*.

2. List the three main particles of an atom and explain how they differ from one another.

3. If the number of electrons in an atom is 35 and its mass number is 80, calculate the following:
   (a) The number of protons.
   (b) The atomic number.
   (c) The number of neutrons.

4. What occurs in an atom to produce an ion?

5. What is an isotope?

6. Although all minerals have an orderly internal arrangement of atoms (crystalline structure), most mineral samples do not visibly demonstrate their crystal form. Why?

7. Why might it be difficult to identify a mineral by its color?

8. If you found a glassy-appearing mineral while rock hunting and had hopes that it was a diamond, what simple test might help you make a determination?

9. Table 1.3 (p. 28) lists a use for corundum as an abrasive. Explain why it makes a good abrasive in terms of Mohs hardness scale.

10. Gold has a specific gravity of almost 20. If a 25-liter pail of water weighs about 25 kilograms, how much would a 25-liter pail of gold weigh?

11. What are the two most common elements in Earth's crust?

12. What is the term used to describe the basic building block of all silicate minerals?

13. What are the two most common silicate minerals?

14. List three nonsilicate minerals that are commonly found in rocks.

15. Contrast a mineral *resource* and a mineral *reserve*.

16. What might cause a mineral deposit that had not been considered an ore to become reclassified as an ore?

Key Terms

atom (p. 20)
atomic number (p. 20)
cleavage (p. 23)
color (p. 22)
compound (p. 20)
crystal form (p. 21)
electron (p. 20)
energy levels (shells) (p. 20)

element (p. 20)
fracture (p. 23)
hardness (p. 22)
ions (p. 20)
isotope (p. 21)
luster (p. 21)
mass number (p.21)
mineral (p. 19)
mineral resource (p. 27)
Mohs hardness scale (p. 22)
neutron (p. 20)

nucleus (p. 20)
ore (p. 28)
proton (p. 20)
radioactivity (p. 21)
reserve (p. 27)
rock (p. 19)
silicate (p. 24)
silicon-oxygen tetrahedron (p. 25)
specific gravity (p. 23)
streak (p. 22)

CHAPTER 2
Rocks: Materials of the Lithosphere

Sedimentary rocks compose the Mittens, Monument Valley, Arizona.
(Photo by Paul Chesley/Tony Stone Worldwide)

Why study rocks? You have already learned that rocks and minerals have great economic value. Furthermore, all Earth processes in some way depend on the properties of these basic materials. Events such as volcanic eruptions, mountain building, weathering, erosion, and even earthquakes involve rocks and minerals. Consequently, a basic knowledge of Earth materials is essential to understanding Earth phenomena.

Every rock contains clues about the environments in which it formed (Figure 2.1). For example, some rocks are composed entirely of small shell fragments. This tells Earth scientists that the particles composing the rock originated in a shallow marine environment. Other rocks contain clues that indicate they formed from a volcanic eruption, or deep in Earth during mountain building. Thus, rocks contain a wealth of information about events that have occurred over Earth's long history.

We divide rocks into three groups, based on their mode of origin. The groups are igneous, sedimentary, and metamorphic. Before examining each group, we will view the *rock cycle*, which depicts the interrelationships among these rock groups.

Earth as a System: The Rock Cycle

Earth is a system. This means that our planet consists of many interacting parts that form a complex whole. Nowhere is this idea better illustrated than when we examine the rock cycle (Figure 2.2). The **rock cycle** allows us to view many of the interrelationships among different parts of the Earth system. It helps us understand the origin of igneous, sedimentary, and metamorphic rocks and to see that each type is linked to the others by the processes that act upon and within the planet. Learn the

Figure 2.1
Rocks contain information about the processes that produce them. This brightly colored sandstone exposed in northern Arizona was once a sand dune. (Photo by Carr Clifton/Minden Pictures)

rock cycle well; you will be examining its interrelationships in greater detail throughout this chapter and the other chapters in the unit on the solid Earth.

The Basic Cycle

Let us begin at the top of Figure 2.2. **Magma** is molten material that forms inside Earth. Eventually magma cools and solidifies. This process, called **crystallization**, may occur either beneath the surface or, following a volcanic eruption, at the surface. In either situation, the resulting rocks are called **igneous rocks**.

If igneous rocks are exposed at the surface, they will undergo **weathering**, in which the day-in and day-out influences of the atmosphere slowly disintegrate and decompose rocks. The materials that result are often moved downslope by gravity before being picked up and transported by any of a number of erosional agents—running water, glaciers, wind, or waves. Eventually these particles and dissolved substances, called **sediment**, are deposited. Although most sediment ultimately comes to rest in the ocean, other sites of deposition include river floodplains, desert basins, swamps, and dunes.

Next the sediments undergo **lithification**, a term meaning "conversion into rock." Sediment is usually lithified into **sedimentary rock** when compacted by the weight of overlying layers or when cemented as percolating groundwater fills the pores with mineral matter.

If the resulting sedimentary rock is buried deep within Earth and involved in the dynamics of mountain building, or intruded by a mass of magma, it will be subjected to great pressures and/or intense heat. The sedimentary rock will react to the changing environment and turn into the third rock type, **metamorphic rock**. When metamorphic rock is subjected to additional pressure changes or to still higher temperatures, it will melt, creating magma, which will eventually crystallize into igneous rock.

Processes driven by heat from Earth's interior are responsible for creating igneous and metamorphic rocks. Weathering and erosion, external processes powered by energy from the sun, produce the sediment from which sedimentary rocks form.

Alternative Paths

The paths shown in the basic cycle are not the only ones that are possible. To the contrary, other paths are just as likely to be followed as those described in the preceding section. These alternatives are indicated by the blue arrows in Figure 2.2.

Igneous rocks, rather than being exposed to weathering and erosion at Earth's surface, may remain deeply buried. Eventually these masses may be subjected to the strong compressional forces and high temperatures associated with mountain building. When this occurs, they are transformed directly into metamorphic rocks.

Metamorphic and sedimentary rocks, as well as sediment, do not always remain buried. Rather, overlying layers may be stripped away, exposing the once-buried rock. When this happens, the material is attacked by weathering processes and turned into new raw materials for sedimentary rocks.

Although rocks may seem to be unchanging masses, the rock cycle shows that they are not. The changes, however, take time—great amounts of time.

Igneous Rocks: "Formed by Fire"

In our discussion of the rock cycle, we pointed out that igneous rocks form as magma cools and crystallizes. This molten rock, which originates at depths as great as 200 kilometers within Earth, consists primarily of the elements found in silicate minerals (silicon and oxygen, plus aluminum, iron, calcium, sodium, potassium, magnesium, and others). Magma also contains some gases, particularly water vapor, which are confined within the magma by the surrounding rocks. Because the magma body is less dense than the surrounding rocks, it works its way toward the surface over time spans of thousands to millions of years. On occasion, the magma breaks through the surface, producing a volcanic eruption.

The spectacular explosions that sometimes accompany an eruption are produced by the gases that escape as the confining pressure lessens near the surface. Sometimes blockage of the volcanic vent, coupled with surface-water seepage into the magma chamber, can produce catastrophic explosions.

Along with ejected rock fragments, a volcanic eruption often generates extensive lava flows. **Lava** is similar to magma, except that most of the gaseous component has escaped (Figure 2.3). The rocks that result when lava solidifies are classified as **volcanic** or **extrusive**, because they are *extruded* onto the surface. The magma not able to reach the surface eventually crystallizes at depth. Igneous rocks produced in this manner are termed **plutonic**, or **intrusive**, because they *intrude* existing rocks. These deep rocks would never be observed were it not for erosion stripping away the overlying rock.

Magma Crystallizes to Form Igneous Rocks

Magma is basically a very hot, thick fluid, but it also contains solids and gases. The solids are mineral crystals. The liquid portion of a magma body is composed of ions that move about freely. However, as magma cools, the random movements of the ions slow, and the ions begin to arrange themselves into orderly patterns. This process is called *crystallization*. Usually all of the molten material does not solidify at the same time. Rather, as it cools, numerous small crystals develop. In

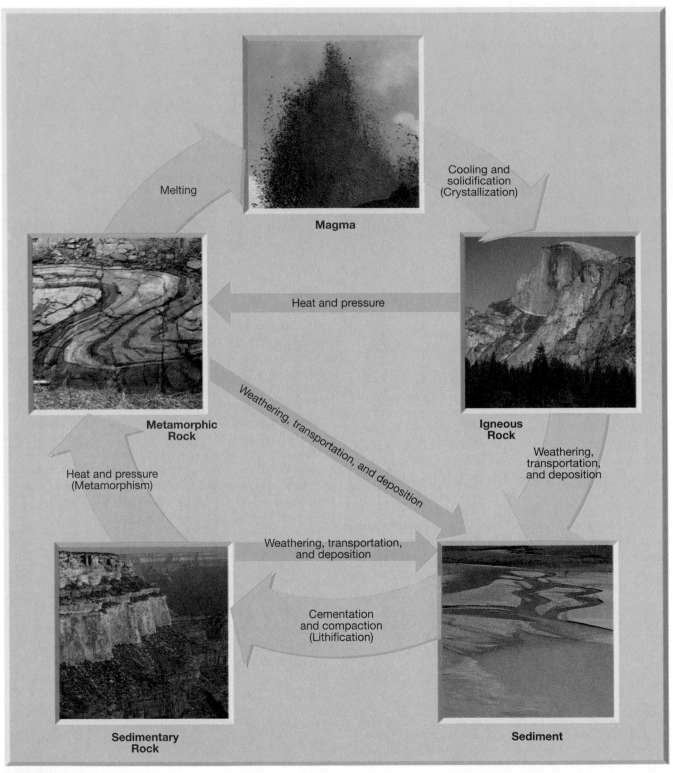

Figure 2.2
The rock cycle is one way of viewing many of the interrelationships among different parts of the Earth system. It shows us that Earth's materials (boxes) are all linked to each other by processes (arrows). Every rock contains clues to the processes that formed it.

a systematic fashion, ions are added to these centers of crystal growth. When the crystals grow large enough for their edges to meet, their growth ceases for lack of space, and crystallization continues elsewhere. Eventually, all of the liquid is transformed into a solid mass of interlocking crystals.

Figure 2.3

Lava of basaltic composition flowing from a lava tube, Hawaii Volcanoes National Park. (Photo by J. D. Griggs, U.S. Geological Survey)

The rate of cooling strongly influences crystal size. If a magma cools very slowly, relatively few centers of crystal growth develop. Slow cooling also allows ions to migrate over relatively great distances. Consequently, *slow cooling results in the formation of large crystals*. On the other hand, if cooling occurs quite rapidly, the ions quickly lose their motion and quickly combine. This results in a large number of tiny crystals that all compete for the available ions. Therefore, the outcome of rapid cooling is the formation of a solid mass of *very small intergrown crystals*.

Thus, if a geologist discovers rock containing large, finger-sized crystals, it means the rock cooled very slowly. But if the crystals can be seen only with a microscope, the geologist knows that the magma cooled very quickly.

If the molten material is quenched almost instantly, there is not sufficient time for the ions to arrange themselves into a crystalline network at all. Therefore, solids produced in this manner consist of randomly distributed ions. Such rocks are called *glass* and are quite similar to ordinary manufactured glass. "Instant" quenching occurs during violent volcanic eruptions that produce tiny shards of glass called volcanic ash.

Classifying Igneous Rocks

In addition to the rate of cooling, the composition of a magma and the amount of gases influence crystallization. Because magmas differ in each of these aspects, the physical appearance and mineral composition of igneous rocks vary widely. Nevertheless, it is possible to classify igneous rocks based on their *texture* and *mineral constituents*. We will now look at both features.

Igneous Textures. **Texture** describes the overall appearance of an igneous rock, based on the *size* and *arrangement* of its interlocking crystals. Texture is a very important characteristic, because it reveals a great deal about the environment in which the rock formed. You learned that rapid cooling produces small crystals, whereas very slow cooling produces much larger crystals. As you might expect, the rate of cooling is quite slow in magma chambers lying deep within the crust, whereas a thin layer of lava extruded upon Earth's surface may chill "rock solid" in a matter of hours. Small molten blobs ejected into the air during a violent eruption can solidify almost instantly.

Igneous rocks that form rapidly at the surface or as small masses within the upper crust have a very **fine-grained texture**, with the individual crystals too small to be seen with the unaided eye (Figure 2.4A). Common in many fine-grained igneous rocks are voids left by gas bubbles that escape as the magma solidifies. These openings are called *vesicles* and are limited to the fast-cooling upper portion of lava flows (Figure 2.5).

When large masses of magma solidify far below the surface, they form igneous rocks that exhibit a **coarse-grained texture**. These coarse-grained rocks have the appearance of a mass of intergrown crystals, which are roughly equal in size and large enough that the individual minerals can be identified with the unaided eye. Granite is a classic example (Figure 2.4B).

A large mass of magma located at depth may require tens of thousands of years to solidify. Since all materials within a magma do not crystallize at the same rate or at the same time during cooling, it is possible for some to become quite large before others even start to form. If magma that already contains some large crystals suddenly erupts at the surface, this environmental change would make the molten portion of the lava cool quickly. The resulting rock, which has large crystals embedded in a matrix of smaller crystals, is said to have a **porphyritic texture** (Figure 2.6).

During some volcanic eruptions, molten rock is ejected into the atmosphere, where it is quenched very quickly. Rapid cooling of this type may generate rock with a **glassy texture**. As was indicated, glass results when the ions do not have sufficient time to unite into an orderly crystalline structure. *Obsidian*, a common type of natural glass, is similar in appearance to a dark chunk of manufactured glass (Figure 2.7A). Another volcanic rock that exhibits a glassy texture is *pumice*.

A.   **B.**

Figure 2.4
A. Rhyolite exhibits a fine-grained texture. **B.** Granite is a common igneous rock that has a coarse-grained texture. (Photos by E. J. Tarbuck)

Usually found with obsidian, pumice forms then large amounts of gas escape through lava to generate a gray, frothy mass (Figure 2.7B). In some samples, the vesicles are quite noticeable, while in others, the pumice resembles fine shards of intertwined glass. Because of the large volume of air-filled voids, many samples of pumice will float in water.

Mineral Composition.   The mineral makeup of an igneous rock depends on the chemical composition of the magma from which it crystallizes. Such a large variety of igneous rocks exists that you might assume an equally large variety of magmas must also exist. However, various eruptive stages of the same volcano often extrude lavas exhibiting somewhat different mineral compositions, particularly if long periods

separate the eruptions. Such evidence led geologists to examine the possibility that a single magma might produce rocks of varying mineral content.

N. L. Bowen studied this idea in the first quarter of the twentieth century. Bowen discovered that, as magma cools in the laboratory, certain minerals crystallize first, at very high temperatures (top of Figure 2.8). At successively lower temperatures, other minerals crystallize.

Bowen also demonstrated that if a mineral remains in the molten solution after crystallization, it will react with the remaining liquid to produce the next mineral in the sequence shown in Figure 2.8. For this reason, this arrangement of minerals became known as *Bowen's reaction series.*

On the upper left branch of this reaction series, olivine, the first mineral to form, will react with the remaining melt to become pyroxene. This reaction will

Figure 2.5
Scoria is a volcanic rock that exhibits vesicles, which form as gas bubbles escape near the top of a lava flow. (Photo by GeoScience/PH)

Figure 2.6
Andesite porphyry. Notice the two distinctively different sizes of crystals. (Photo by E. J. Tarbuck)

A.

B.

Figure 2.7

Igneous rocks that exhibit a glassy texture. **A.** Obsidian, a glassy volcanic rock. **B.** Pumice, a glassy rock containing numerous tiny voids. (Photos by GeoScience/PH)

continue until the last mineral in the series, biotite, is formed. The right branch of the reaction series is a continuum in which the earliest-formed calcium-rich feldspar crystals react with the sodium ions contained in the melt to become progressively more sodium-rich. Ordinarily, these reactions are not complete, so that various amounts of several of these minerals may exist at any given time.

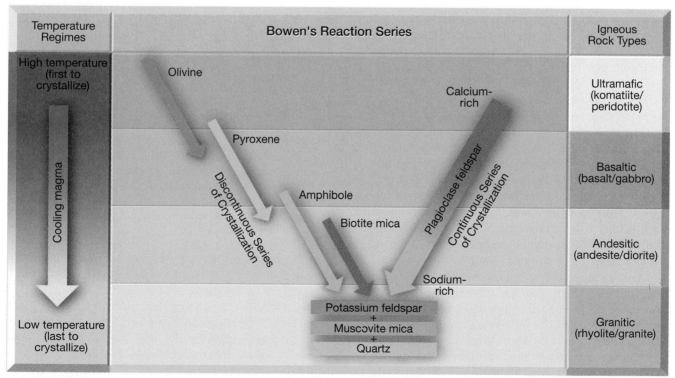

Figure 2.8

Bowen's reaction series shows the sequence in which minerals crystallize from a magma. Compare this figure to the mineral composition of the rock groups in Table 1.3. Note that each rock group consists of minerals that crystallize at the same time.

During the last stage of crystallization, after most of the magma is solidified, the minerals muscovite and potassium feldspar form. Finally, the remaining melt (if any) will have a high silica content, which eventually crystallizes as quartz.

A consequence of this crystallization scheme is that minerals that form at about the same temperature are found together in the same igneous rock. For example, notice that the minerals potassium feldspar, muscovite mica, and quartz are found in the same region of the diagram, and they are the major constituents of the igneous rock granite.

How does Bowen's reaction series account for the great diversity of igneous rocks? During crystallization, a separation of the solid and liquid components of a magma can occur. This happens if the earlier formed minerals are denser (heavier) than the liquid portion and settle to the bottom as shown in Figure 2.9A. This *crystal settling* is thought to occur frequently with the dark, dense silicates, such as olivine and pyroxene. When the remaining melt solidifies it will form a rock with a chemical composition much different from the parent magma (Figure 2.9B). Consequently, a wide variety of igneous rocks results.

Naming Igneous Rocks

Igneous rocks are classified by their texture and mineral composition. Various igneous textures result from different cooling histories, while the mineral compositions are a consequence of the chemical makeup of the parent magma and the environment of crystallization. A general classification scheme based on texture and mineral composition is provided in Table 2.1.

The rocks on the right side of Table 2.1 consist of the *first* minerals to crystallize. They contain more iron (Fe) and magnesium (Mg) and less silica (SiO_2) than later-formed minerals. Their iron content makes them darker in color and slightly more dense than the other rocks. Basalt is a common rock of this composition, so geologists use the term *basaltic* to indicate any igneous rock having this composition. Basalt is the most common extrusive igneous rock. Many volcanic islands, such as the Hawaiian Islands and Iceland, are composed mainly of basalt. Further, the upper layers of the oceanic crust consist primarily of basalt.

The rocks on the left side of Table 2.1 contain the *last* minerals to crystallize and consist mainly of feldspar and quartz. These rocks have a *granitic* composition. *Granite* is perhaps the best-known igneous rock (see Figure 2.4B). This is partly because of its natural beauty, which is enhanced when polished, and partly because of its abundance. Slabs of polished granite are commonly used for tombstones, monuments, and as building stones. Granite is often produced by the processes that generate mountains.

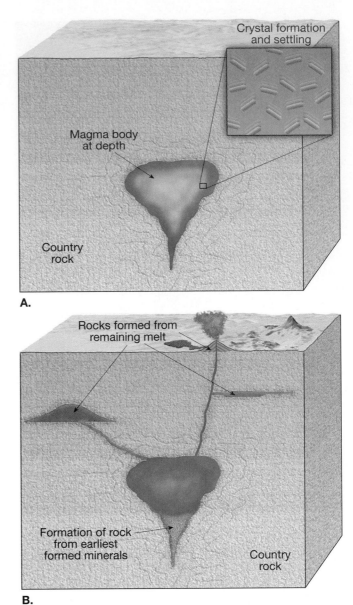

Figure 2.9

Separation of minerals by crystal settling. **A.** Illustration of how the earliest-formed minerals can be separated from a magma by settling. **B.** The remaining melt could migrate to a number of different locations and, upon further crystallization, generate rocks having compositions much different from that of the parent magma.

Because granite is a by-product of mountain building and is very resistant to weathering and erosion, it frequently forms the core of eroded mountains: Pikes Peak in the Rockies; Mount Rushmore in the Black Hills; the White Mountains of New Hampshire; Stone Mountain, Georgia; and Yosemite National Park in the Sierra Nevada are all areas where large quantities of granite are exposed at the surface (see Figure 2.10).

Please note that two rocks may have the same mineral constituents, but have *different textures and hence different names*. For example, the coarse-grained intrusive

Table 2.1 Classification of igneous rocks.

		Granitic (felsic)	Andesitic (intermediate)	Basaltic (mafic)	Ultramafic*
Dominant Minerals		Quartz Potassium feldspar	Amphibole Intermediate plagioclase feldspar	Pyroxene Calcium-rich plagioclase feldspar	Olivine Pyroxene
Color		Light-colored Less than 15% dark minerals	Medium-colored 15-40% dark minerals	Dark gray to black More than 40% dark minerals	Dark-green to black Nearly 100% dark minerals
T e x t u r e	**Coarse-grained**	Granite	Diorite	Gabbro	Peridotite
	Fine-grained	Rhyolite	Andesite	Basalt	Komatiite
	Porphyritic	"Porphry" follows any of the above names whenever there are appreciable phenocrysts			
	Glassy	Obsidian (compact glass) Pumace (frothy glass)			

* Ultramific rocks are dark, dense rocks, composed almost entirely of minerals containing iron and magnesium. Although relatively rare on Earth's surface, these rocks are believed to be major constituents of the upper mantle.

rock granite has a fine-grained volcanic equivalent called *rhyolite* (see Figure 2.4A). Although these rocks contain the same minerals, they have different textures and do not look at all alike. Further, in contrast to granite, rhyolite is rather uncommon.

Note that gradations exist (Table 2.1). For example, an abundant extrusive igneous rock called *andesite* has a mineral composition between that of granitic and basaltic rocks. Andesite is a fine-grained volcanic rock whose name derives from the Andes Mountains, where numerous volcanoes are composed of this rock type.

Sedimentary Rocks: Compacted and Cemented Sediment

Recall the rock cycle, which shows the origin of sedimentary rocks. Weathering begins the process. Next, gravity and erosional agents (running water, wind, waves, and glacial ice) remove the products of weathering and carry them to a new location where they are deposited. Usually the particles are broken down further during this transport phase. Following deposition, this *sediment* may become lithified, or "turned to rock." Commonly, *compaction* and *cementation* transform the sediment into solid sedimentary rock.

The word *sedimentary* indicates the nature of these rocks, for it is derived from the Latin *sedimentum*, which means "settling," a reference to a solid material settling out of a fluid. Most sediment is deposited in this fashion. Weathered debris is constantly being swept from

bedrock and carried away by water, ice, or wind. Eventually the material is deposited in lakes, river valleys, seas, and countless other places. The particles in a desert sand dune, the mud on the floor of a swamp, the gravels in a stream bed, and even household dust are examples of sediment produced by this never-ending process.

The weathering of bedrock and the transport and deposition of the weathering products are continuous. As piles of sediment accumulate, the materials near the bottom are compacted by the weight of the overlying layers. Over long periods, these sediments are cemented together by mineral matter deposited from water in the spaces between particles. This forms solid sedimentary rock.

Geologists estimate that sedimentary rocks account for only about 5 percent (by volume) of Earth's outer 16 kilometers (10 miles). However, the importance of this group of rocks is far greater than this percentage implies. If you sampled the rocks exposed at Earth's surface, you would find that the great majority are sedimentary (Figure 2.11). Indeed, about 75 percent of all rock outcrops on the continents are sedimentary. Therefore, we can think of sedimentary rocks as comprising a relatively thin and somewhat discontinuous layer in the uppermost portion of the crust. This makes sense because sediment accumulates at the surface.

It is from sedimentary rocks that geologists reconstruct many details of Earth's history. Because sediments are deposited in a variety of different settings at the surface, the rock layers that they eventually form hold many clues to past surface environments. They may also exhibit characteristics that allow geologists to decipher information

Figure 2.10
El Capitan is a massive granitic structure in Yosemite National Park, California. (Photo by Lewis Kemper/DRK Photo)

about the method and distance of sediment transport. Furthermore, it is sedimentary rocks that contains fossils, which are vital evidence in the study of the geologic past.

Finally, many sedimentary rocks are very important economically. Coal, for example, is a sedimentary rock that presently is the source for over half the electric power generated in the United States. Other major energy resources (petroleum and natural gas) occur in pores within sedimentary rocks. Other sedimentary rocks are major sources of iron, aluminum, manganese, fertilizer, and sand and gravel.

Classifying Sedimentary Rocks

Materials accumulating as sediment have two principal sources. First, sediments may originate as solid particles from weathered rocks, such as the igneous rocks earlier described. These particles are called *detritus*, and the sedimentary rocks that they form are called **detrital sedimentary rocks.**

The second major source of sediment is soluble material produced largely by chemical weathering. When these dissolved substances are precipitated back as solids, they are called chemical sediment, and they form **chemical sedimentary rocks**. We will now look at detrital and chemical sedimentary rocks.

Detrital Sedimentary Rocks.   Though a wide variety of minerals and rock fragments may be found in detrital rocks, clay minerals and quartz dominate. Clay minerals are the most abundant product of the chemical weathering of silicate minerals, especially the feldspars. Quartz, on the other hand, is abundant because it is extremely durable and very resistant to chemical weathering. Thus, when igneous rocks such as granite are weathered, individual quartz grains are set free.

Geologists use particle size to distinguish among detrital sedimentary rocks. Table 2.2 presents the four size categories for particles making up detrital rocks. When gravel-sized particles predominate, the rock is called *con-*

Figure 2.11
Sedimentary rocks exposed in Canyonlands National Park, Utah. About 75 percent of all rock outcrops on the continents are sedimentary rocks. (Photo by Carr Clifton/Minden Pictures)

glomerate if the sediment is rounded (Figure 2.12A) and *breccia* if the pieces are angular (Figure 2.12B). Angular fragments indicate that the particles were not transported very far from their source prior to deposition, and so have not had corners and rough edges abraded. *Sandstone* is the name given rocks when sand-sized grains prevail (Figure 2.12C). *Shale*, the most common sedimentary rock, is made of very fine-grained sediment (Figure 2.12D). *Silt-stone*, another rather fine-grained rock, is sometimes difficult to differentiate from rocks such as shale that are composed of even smaller clay-sized sediment.

Particle size is not only a convenient method of dividing detrital rocks; the sizes of the component grains also provide useful information about the environment in which the sediment was deposited. Currents of water or air sort the particles by size. The stronger the current, the larger the particle size carried. Gravels, for example, are moved by swiftly flowing rivers, rockslides, and glaciers. Less energy is required to transport sand; thus it is common in windblown dunes, river deposits, and beaches.

Because silts and clays settle very slowly, accumulations of these materials are generally associated with the quiet waters of a lake, lagoon, swamp, or marine environment.

Table 2.2 Particle size classification for detrital rocks.

Size Range (millimeters)	Particle Name	Common Sediment Name	Detrital Rock
>256 64–256 4–64 2–4	Boulder Cobble Pebble Granule	Gravel	Conglomerate or breccia
1/16–2	Sand	Sand	Sandstone
1/256–1/16 <1/256	Silt Clay	Mud	Shale

| 10 | 20 | 30 | 40 | 50 | 60 |

Scale in millimeters

A.

B.

C.

D.

Figure 2.12
Common detrital sedimentary rocks. **A.** Conglomerate (rounded particles). **B.** Breccia (angular particles). **C.** Sandstone **D.** Shale with plant fossil. (Photos by E. J. Tarbuck)

Although detrital sedimentary rocks are classified by particle size, in certain cases the mineral composition is also part of naming a rock. For example, most sandstones are predominantly quartz-rich, and are often referred to as quartz sandstone. In addition, rocks consisting of detrital sediments are rarely composed of grains of just one size. Consequently, a rock containing quantities of both sand and silt can be correctly classified as sandy siltstone or silty sandstone, depending on which particle size dominates.

Chemical Sedimentary Rocks. In contrast to detrital rocks, which form from the solid products of weathering, chemical sediments are derived from material that is carried in solution to lakes and seas. This material does not remain dissolved in the water indefinitely. When conditions are right, it precipitates to form chemical sediments. This precipitation may occur directly as the result of physical processes, or indirectly through life processes of water-dwelling organisms. Sediment formed in this second way has a *biochemical* origin.

An example of a deposit resulting from physical processes is the salt left behind as a body of salt water evaporates. In contrast, many water-dwelling animals and plants extract dissolved mineral matter to form

shells and other hard parts. After the organisms die, their skeletons may accumulate on the floor of a lake or ocean.

Figure 2.13
This rock, called coquina, consists of shell fragments; therefore, it has a biochemical origin.

Figure 2.14
The White Chalk Cliffs of Dover. (Photo by John Heseltine/Science Photo Library/Photo Researchers, Inc.)

Limestone is the most abundant chemical sedimentary rock. It is composed chiefly of the mineral calcite ($CaCO_3$). Ninety percent of limestone is biochemical sediment. The rest precipitates directly from water.

One easily identified biochemical limestone is *coquina*, a coarse rock composed of loosely cemented shells and shell fragments (Figure 2.13). Another less obvious but familiar example is *chalk*, a soft, porous rock made up almost entirely of the hard parts of microscopic organisms that are no larger than the head of a pin (Figure 2.14).

Inorganic limestones form when chemical changes or high water temperatures increase the concentration of calcium carbonate to the point that it precipitates. *Travertine*, the type of limestone that decorates caverns, is one example. Groundwater is the source of travertine that is deposited in caves. As water drops reach the air in a cavern, some of the carbon dioxide dissolved in the water escapes, causing calcium carbonate to precipitate.

Dissolved silica (SiO_2) precipitates to form varieties of microcrystalline quartz. Minerals composed of microcrystalline quartz include chert (light color), flint (dark), jasper (red), and agate (banded). These chemical sedimentary rocks may have either an inorganic or biochemical origin, but the mode of origin is usually difficult to determine.

Very often, evaporation causes minerals to precipitate from water. Such minerals include halite, the chief component of *rock salt*, and gypsum, the main ingredient of *rock gypsum*. Both materials have significant commercial importance. Halite is familiar to everyone as the common salt used in cooking and seasoning foods. Of course, it has many other uses and has been considered important enough that people have sought, traded, and fought over it for much of human history. Gypsum is the basic ingredient of plaster of Paris. This material is used most extensively in the construction industry for "drywall" and plaster.

In the geologic past, many areas that are now dry land were covered by shallow arms of the sea that had only narrow connections to the open ocean. Under these conditions, water continually moved into the bay to replace water lost by evaporation. Eventually the waters of the bay became saturated and salt deposition began. Today, these arms of the sea are gone, and the remaining deposits are called **evaporites**.

On a smaller scale, evaporite deposits may be seen in such places as Death Valley, California. Here, following rains or periods of snowmelt in the mountains, streams flow from surrounding mountains into an enclosed basin. As the water evaporates, *salt flats* form from dissolved materials left behind as a white crust on the ground (Figure 2.15).

Figure 2.15
Bonneville Salt Flats, Utah. (Photo by Scott T. Smith)

Coal is quite different from other chemical sedimentary rocks. Unlike other rocks in this category, which are calcite- or silica-rich, coal is made mostly of organic matter. Close examination of a piece of coal under a microscope or magnifying glass often reveals plant structures such as leaves, bark, and wood that have been chemically altered but are still identifiable. This supports the conclusion that coal is the end product of the burial of large amounts of plant material over extended periods.

Coal is commonly called a *fossil fuel*. This is certainly appropriate, since each time we burn coal we are using energy from the sun that was stored by plants many millions of years ago. We are indeed burning a "fossil."

The initial stage in coal formation is the accumulation of large quantities of plant remains. However, special conditions are required for such accumulations, because dead plants normally decompose when exposed to the atmosphere. An ideal environment that allows for the buildup of plant material is a swamp. Because stagnant swamp water is oxygen-deficient, complete decay (oxidation) of the plant material is not possible. At various times during Earth history, such environments have been common. Coal

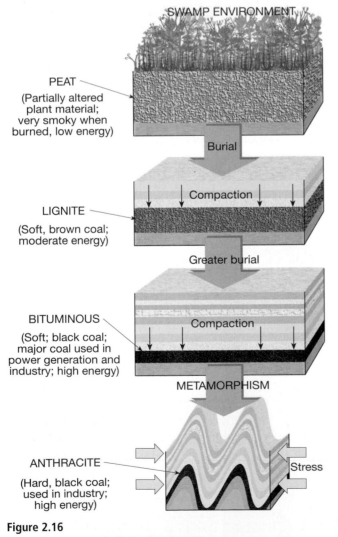

SWAMP ENVIRONMENT

PEAT
(Partially altered plant material; very smoky when burned, low energy)

Burial

Compaction

LIGNITE
(Soft, brown coal; moderate energy)

Greater burial

Compaction

BITUMINOUS
(Soft; black coal; major coal used in power generation and industry; high energy)

METAMORPHISM

ANTHRACITE
(Hard, black coal; used in industry; high energy)

Stress

Figure 2.16
Successive stages in the formation of coal.

undergoes successive stages of formation. With each successive stage, higher temperatures and pressures drive off impurities and volatiles, as shown in Figure 2.16.

Lignite and bituminous coals are sedimentary rocks, but anthracite is a metamorphic rock. Anthracite forms when sedimentary layers are subjected to the folding and deformation associated with mountain building.

In summary, we divide sedimentary rocks into two major groups: detrital and chemical. The main criterion for classifying detrital rocks is particle size, whereas chemical rocks are distinguished by their mineral composition. The categories presented here are more rigid than is the actual state of nature. Many detrital sedimentary rocks are a mixture of more than one particle size. Furthermore, many sedimentary rocks classified as chemical also contain at least small quantities of detrital sediment, and practically all detrital rocks are cemented with material that was originally dissolved in water.

Lithification of Sediment

Lithification refers to the processes by which sediments are transformed into solid sedimentary rocks. One of the most common processes is *compaction*. As sediments accumulate through time, the weight of overlying material compresses the deeper sediments. As the grains are pressed closer and closer, pore space is greatly reduced. For example, when clays are buried beneath several thousand meters of material, the volume of the clay may be reduced as much as 40 percent. Compaction is most significant in fine-grained sedimentary rocks such as shale, because sand and other coarse sediments compress little.

Cementation is another important means by which sediments are converted to sedimentary rock. The cementing materials are carried in solution by water percolating through the pore spaces between particles. Through time, the cement precipitates onto the sediment grains, fills the open spaces, and joins the particles. Calcite, silica, and iron oxide are the most common cements. Identification of the cementing material is simple. Calcite cement will effervesce (fizz) with dilute hydrochloric acid. Silica is the hardest cement and thus produces the hardest sedimentary rocks. When a sedimentary rock has an orange or red color, this usually means iron oxide is present.

Features of Sedimentary Rocks

Sedimentary rocks are particularly important evidence of Earth's long history. These rocks form at Earth's surface, and as layer upon layer of sediment accumulates, each records the nature of the environment at the time the sediment was deposited. These layers, called **strata**, or **beds**, are the *single most characteristic feature of sedimentary rocks* (see Figure 2.11).

The thickness of beds range from microscopically thin to tens of meters thick. Separating the strata are *bedding planes*, flat surfaces along which rocks tend to separate or break. Generally, each bedding plane marks the end of one episode of sedimentation and the beginning of another.

A.

B.

Figure 2.17

A. Ripple marks preserved in sedimentary rocks may indicate a beach or stream channel environment. (Photo by Stephen Trimble) **B.** Mud cracks form when wet mud or clay dries and shrinks, perhaps signifying a tidal flat or desert basin. (Photo by Gary Yeowell/Tony Stone Images)

As geologists examine sedimentary rocks, they deduce a lot. A conglomerate, for example, may indicate a high-energy environment, such as a rushing stream, where only the coarse materials can settle out. If the rock is arkose (rich in feldspar), it may signify a dry climate, where little chemical alteration of feldspar is possible. Black shale and coal are associated with a low-energy, organic-rich environment such as a swamp or lagoon. Other features found in some sedimentary rocks also give clues to past environments (Figure 2.17).

Fossils, the traces or remains of prehistoric life, are perhaps the most important inclusions found in some sedimentary rock (Figure 2.18). Knowing the nature of the life-forms that existed at a particular time may help to answer many questions about the environment. Was it land or ocean? A lake or swamp? Was the climate hot or cold, rainy or dry? Was the ocean water shallow or deep, turbid or clear? Furthermore, fossils are important time indicators and play a key role in matching up rocks from different places that are the same age. Fossils are important tools used in interpreting the geologic past and will be examined in some detail in Chapter 10.

Metamorphic Rocks: Changed in Form

Metamorphism literally means to "change form." Thus, metamorphism changes existing rocks in size, shape, texture, and even the minerals they contain. Metamorphic rocks can form from igneous, sedimentary, or even from other metamorphic rocks. The agents of change include heat, pressure, and chemically active fluids.

Metamorphism varies broadly in degree. In some instances, rocks are only slightly changed. For example, under *low-grade metamorphism*, the common sedimentary rock shale becomes the metamorphic rock called *slate*. Although slate is more compact than shale, hand samples of these rocks are sometimes difficult to distinguish. In other cases, the transformation is so complete that the identity of the original rock cannot be determined. In such *high-grade metamorphism*, features like bedding planes, fossils, and vesicles that may have existed in the parent rock are completely obliterated. Further, when rocks at depth are subjected to directional pressure, they flow and bend into intricate folds.

Figure 2.18

Fossil of a trilobite, an ancient marine organism, preserved in shale. Fossils are important tools used to determine past environmental conditions. They are also valuable time indicators. (Photo by Geo-Science/PH)

In the most extreme metamorphic environments, the temperatures approach those at which rocks melt. However, during metamorphism the deformed material must remain solid, for once melting occurs, we enter the realm of igneous activity.

Metamorphism takes place when rock is subjected to conditions unlike those in which it originally formed. As a result, the rock becomes unstable and gradually changes until a state of equilibrium with the new environment is reached. The changes occur at the elevated temperatures and pressures of the region extending from a few kilometers below Earth's surface to the crust-mantle boundary. Since the formation of metamorphic rocks is completely hidden from view (which is not the case for many sedimentary and some igneous rocks), metamorphism is one of the most difficult processes for geologists to study.

Metamorphism most often occurs in one of two settings. First, during mountain building, great quantities of rock are subjected to the intense stresses and high temperatures associated with large-scale deformation. The end result may be extensive areas of metamorphic rocks that are said to have undergone **regional metamorphism**. The greatest volume of metamorphic rock is produced in this fashion. Second, when rock is in contact with, or near, a mass of magma, **contact metamorphism** takes place. In this circumstance, the changes are caused primarily by the high temperatures of the molten material, which in effect "bake" the surrounding rock.

Extensive areas of metamorphic rocks are exposed on every continent. Metamorphic rocks are an important component of many mountain belts, where they make up a large portion of a mountain's crystalline core. Even the stable continental interiors, which are generally covered by sedimentary rocks, are underlain by metamorphic basement rocks (Figure 2.19). In all of these settings, the metamorphic rocks are usually highly deformed and intruded by igneous masses. Indeed, significant parts of Earth's continental crust are composed of metamorphic and associated igneous rocks.

Three Metamorphic Agents

The agents of metamorphism include *heat, pressure,* and *chemically active fluids.* During metamorphism, rocks are often subjected to all three agents simultaneously. However, the degree of metamorphism and the contribution of each agent vary greatly from one environment to another. In low-grade metamorphism, rocks are subjected to temperatures and pressures only slightly greater than those associated with the lithification of sediments. High-grade metamorphism, on the other hand, involves extreme conditions closer to those at which rocks melt.

In addition, the minerals making up the parent rock determine, to a large extent, the degree to which various metamorphic agents will cause change. For example, when an intruding igneous mass enters a rock unit, hot, ion-rich fluids circulate through the host rock. If the host rock is quartz sandstone, very little alteration may take place. On the other hand, if the host rock is limestone, the impact of these fluids can be dramatic and the effects of metamorphism may extend for several kilometers from the igneous mass.

Heat is perhaps the most important agent of metamorphism. It provides the energy to drive chemical reactions that recrystallize minerals. Rocks formed near Earth's surface may be subjected to intense heat when they are intruded by molten material rising from below. Because temperature increases with depth, rocks that originate in a surface environment may also be subjected to extreme temperatures if they are subsequently

Figure 2.19

In the stable interior of a continent, metamorphic rocks are generally covered by layers of sedimentary rocks. The Precambrian Vishnu Schist shown here exposed in the inner gorge of the Grand Canyon is an example of metamorphic basement rocks. (Photo by E. J. Tarbuck)

buried deep within Earth. When buried to a depth of only a few kilometers, certain minerals, such as clay, become unstable and recrystallize into minerals that are stable in this more intense environment. Other minerals, particularly those in crystalline igneous rocks, are stable at relatively high temperatures and pressures and therefore require burial to 20 kilometers or more before metamorphism will occur.

Pressure, like temperature, also increases with depth. Buried rocks are subjected to the force exerted by the load above. This confining pressure is analogous to air pressure, where the force is applied equally in all directions. In addition to the pressure exerted by the load of material above, rocks are also subjected to *stress* during the process of mountain building (Figure 2.20). In this situation the applied force is directional, and the material is squeezed as if it had been placed in a vise. Rock located at great depth is quite warm and behaves plastically during deformation. This accounts for its ability to flow and bend into intricate folds (Figure 2.21).

Chemically active fluids also influence the metamorphic process. Most common is the water contained in the pore spaces of virtually every rock. Water that surrounds the crystals acts as a catalyst by aiding the migration of ions. In some instances, the minerals recrystallize into

more stable crystal shapes. In other cases, ion exchange among minerals forces the formation of completely different minerals.

Metamorphism Changes Texture

The degree of metamorphism is reflected in the rock's texture and mineralogy. When rocks are subjected to low-grade metamorphism, they become more compact and thus more dense. A common example is the metamorphic rock slate, which forms when shale is subjected to temperatures and pressures only slightly greater than those associated with the compaction that lithifies sediment. In this case, directed pressure causes the microscopic clay minerals in shale to align into the more compact arrangement found in slate. Under more extreme conditions, pressure causes certain minerals to recrystallize. In general, recrystallization encourages the growth of larger crystals. Consequently, many metamorphic rocks consist of visible crystals, much like coarse-grained igneous rocks. The crystals of some minerals will recrystallize with a preferred orientation, essentially perpendicular to the direction of the compressional force. The resulting mineral alignment usually gives the rock a layered or banded appearance termed **foliated texture** (Figure 2.22). Simply, foliation results whenever the minerals of a rock are brought into parallel alignment.

Not all metamorphic rocks have a foliated texture. Such rocks are said to exhibit a **nonfoliated texture**. Metamorphic rocks composed of only one mineral that forms equidimensional crystals are, as a rule, not visibly foliated. For example, limestone, if pure, is only a single mineral, calcite. When a fine-grained limestone is metamorphosed, the small calcite crystals combine to

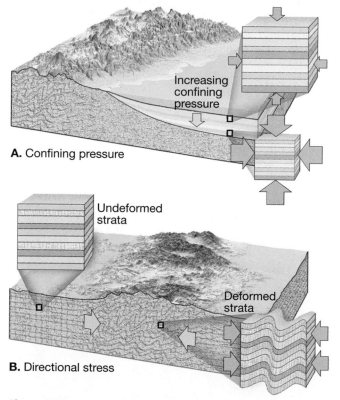

A. Confining pressure

B. Directional stress

Figure 2.20

Pressure (stress) as a metamorphic agent. **A.** In a depositional environment, as confining pressure increases, rocks deform by decreasing in volume. **B.** During mountain building, directional stress shortens and deforms rock strata.

Figure 2.21

Deformed metamorphic rocks exposed in a road cut in the Eastern Highland of Connecticut. Imagine the tremendous force required to fold rock in this manner. (Photo by Phil Dombrowski)

Before metamorphism | After metamorphism

Stress | Stress | Stress | Stress

Figure 2.22
Under directed pressure, planar minerals, such as the micas, become reoriented or recrystallized so that their surfaces are aligned at right angles to the stress. The resulting planar orientation of mineral grains gives the rock a foliated texture. If the coarse-grained igneous rock (granite) on the left underwent intense metamorphism, it could end up closely resembling the metamorphic rock the right (gneiss). (Photos by E. J. Tarbuck)

form larger interlocking crystals. The resulting rock resembles a coarse-grained igneous rock. This nonfoliated metamorphic equivalent of limestone is called *marble*.

In some environments, new materials are actually introduced during the metamorphic process. For example, host rock adjacent to a large intruding magma body may be altered by *hydrothermal* (hot-water) *solutions* released during the latter stages of crystallization. Hydrothermal solutions are rich in ions, including those of valuable metals. Many metallic ore deposits are formed by the precipitation of minerals from hydrothermal solutions.

Classifying Metamorphic Rocks

To review, metamorphic processes cause many changes in existing rocks, including increased density, growth of larger crystals, foliation (reorientation of the mineral grains into a layered or banded appearance), and the transformation of low-temperature minerals into high-temperature minerals. Further, the introduction of ions generates new minerals, some of which are economically important.

Here is a brief look at common rocks produced by metamorphic processes.

Foliated Rocks. *Slate* is a very fine-grained foliated rock composed of minute mica flakes (Figure 2.23). The most noteworthy characteristic of slate is its excellent rock cleavage, meaning that it splits easily into flat slabs.

Figure 2.23
Slate, a common metamorphic rock produced by the low-grade metamorphism of shale. (Photo by E. J. Tarbuck)

This property has made slate a most useful rock for roof and floor tile, chalkboards, and billiard tables. Slate is most often generated by the low-grade metamorphism of shale, although less frequently it forms from the metamorphism of volcanic ash. Slate can be almost any color, depending on its mineral constituents. Black slate contains organic material; red slate gets its color from iron oxide; and green slate is usually composed of chlorite, a micalike mineral.

Schists are strongly foliated rocks, formed by regional metamorphism. They are "platy" and can be readily split into thin flakes or slabs. Like slate, the parent material from which many schists originate is shale, but in the case of schist, the metamorphism is more intense.

The term *schist* describes the *texture* of a rock regardless of composition. For example, schists composed primarily of muscovite and biotite are called *mica schists* (Figure 2.24).

Gneiss (pronounced "nice") is the term applied to banded metamorphic rocks that contain mostly elongated and granular, as opposed to platy, minerals (see Figure 2.22, right). The most common minerals in gneisses are quartz and feldspar, with lesser amounts of muscovite, biotite, and hornblende. Gneisses exhibit strong segregation of light and dark silicates, giving them a characteristic banded texture. While in a plastic state, these banded gneisses can be deformed into intricate folds.

Nonfoliated Rocks. *Marble* is a coarse, crystalline rock whose parent rock is limestone (Figure 2.25). Marble is composed of large interlocking calcite crystals, which form from the recrystallization of smaller grains in the parent rock.

Figure 2.24

Mica shist, a common metamorphic rock composed of shiny mica flakes. (Photo by E. J. Tarbuck)

Because of its color and relative softness (hardness of only 3 on the Mohs scale) marble is a popular building stone. White marble is particularly prized as a stone from which to carve monuments and statues, such as the famous statue of David by Michelangelo. Often the limestone from which marble forms contains impurities that color the marble. Thus, marble can be pink, gray, green, or even black.

Quartzite is a very hard metamorphic rock most often formed from quartz sandstone. Under moderate-to-high-grade metamorphism, the quartz grains in sandstone fuse. Quartzite is typically white, but iron oxide may produce reddish or pinkish stains and dark minerals may impart a gray color.

Figure 2.25

Marble, a crystalline rock formed by the metamorphism of limestone. (Photo by E. J. Tarbuck)

Resources from Rocks and Minerals

The outer layer of Earth, which we call the crust, is only as thick when compared to the remainder of the Earth as a peach skin is to a peach, yet it is of supreme importance to us. We depend on it for fossil fuels and as a source of such diverse minerals as the talc for baby powder, salt to flavor food, and gold for world trade. In fact, on occasion, the availability or absence of certain Earth materials has altered the course of history. As the material requirements of modern society grow, the need to locate additional supplies of useful minerals also grows, and becomes more challenging as well (see Box 2.1).

Metallic Mineral Resources

Some of the most important accumulations of metals, such as gold, silver, copper, mercury, lead, platinum, and nickel, are produced by igneous and metamorphic processes (Table 2.3). These mineral resources, like most others, result from processes that concentrate desirable materials to the extent that extraction is economically feasible.

The igneous processes that generate some metal deposits are quite straightforward. For example, as a large magma body cools, the heavy minerals that crystallize early tend to settle to the lower portion of the magma chamber. This type of magmatic segregation is particularly active in large basaltic magmas where chromite (ore of chromium), magnetite, and platinum are occasionally generated. Layers of chromite, interbedded with other heavy minerals, are mined from such deposits in the Bushveld Complex in South Africa, which contains over 70 percent of the world's known reserves of platinum.

Among the best-known and most important ore deposits are those generated from **hydrothermal** (hot-water) **solutions**. Included in this group are the gold deposits of the Homestake mine in South Dakota; the lead, zinc, and silver ores near Coeur d'Alene, Idaho; the silver deposits of the Comstock Lode in Nevada; and the copper ores of the Keweenaw Peninsula in Michigan.

Table 2.3 Ore minerals of important metals.

Mineral	Uses
Aluminum	Bauxite
Chromium	Chromite
Copper	Chalcopyrite
	Bornite
	Chalcocite
Gold	Native gold
Iron	Hematite
	Magnetite
	Limonite
Lead	Galena
Magnesium	Magnesite
	Dolomite
Manganese	Pyrolusite
Mercury	Cinnabar
Molybdenum	Molybdenite
Nickel	Pentlandite
Platinum	Native platinum
Silver	Native silver
	Argentite
Tin	Cassiterite
Titanium	Ilmenite
	Rutile
Tungsten	Wolframite
	Scheelite
Uranium	Uraninite (pitchblende)
Zinc	Sphalerite

The majority of hydrothermal deposits are thought to originate from hot, metal-rich fluids that are associated with cooling magma bodies. During solidification, liquids plus various metallic ions accumulate near the top to the magma chamber. Because these hot fluids are very mobile, they can migrate great distances through the surrounding rock before they are eventually deposited. Some of this fluid moves along fractures or bedding planes, where it cools and precipitates the metallic ions to produce **vein deposits** (Figure 2.26). Many of the most productive deposits of gold, silver, and mercury occur as hydrothermal vein deposits.

Figure 2.26
Light-colored vein deposits emplaced along a series of fractures in dark-colored igneous rock. (Photo by James E. Patterson)

Box 2.1

Asbestos: What Are the Risks?

You have probably heard that health risk is associated with asbestos. In 1986, the Environmental Protection Agency (EPA) instituted the Asbestos Hazard Emergency Response Act, a mandate that requires inspection of the nation's public and private schools for asbestos. This action brought asbestos to the attention of the public and instilled fear in parents that their children could contract asbestos-related cancers because of high levels of airborne fibers in schools.

Since that time, billions of dollars have been spent on testing and asbestos removal. In addition, there have been significant uncalculated costs in not being able to use facilities during the rehabilitation process. Further, the EPA proposed a ban on all asbestos products that was intended to stop the use of these materials by 1996. Such actions fueled support for misconceptions such as the "one-fiber theory," which maintains that a single fiber of inhaled asbestos can cause a malignancy. Such is not the case.

A number of researchers familiar with the "asbestos problem" disagree with the actions taken by the EPA. Although they admit that some forms of asbestos are a serious health risk in an occupational setting, they argue that most of the asbestos used in the United States is relatively harmless.

What Is the Nature of Asbestos?

Asbestos is not a single mineral. "Asbestos" is a general term applied to a group of silicate minerals that readily separate into thin, strong fibers (Figure 2.A). Because these fibers are flexible, heat resistant, chemically inert, and nonconducting, they have many uses. Asbestos has been widely used to strengthen concrete, make fireproof fabrics, and insulate around boilers and pipes. It is also employed as a component in floor tiles, and it constitutes the major ingredient in brake linings. In addition, wall coatings rich in asbestos fibers were used extensively during the 1950s and 1960s as substitutes for plaster.

Exposure and Risk

It is estimated that over 30 million metric tons of asbestos have been used in the United States since the turn of the century. Most of the world's production of asbestos comes from three minerals. The most important, *chrysotile*, or "white asbestos," is a fibrous form of the mineral serpentine and is the only asbestos mineral mined in North America. *Crocidolite*, "blue asbestos," and *amosite*, "brown asbestos," are currently being mined in South Africa and make up about 5 percent of the world production.

The dangers of prolonged exposure to air that is laden with asbestos dust in an unregulated work place are well established. When very thin, rodlike fibers are inhaled into the lungs, they are neither broken down nor easily expelled but can remain for life. Three lung diseases may result: (1) asbestosis, a scarring of the tissue that decreases the lung's ability to absorb oxygen; (2) mesothelioma, a rare cancer that develops in the lining of the lungs; and (3) lung cancer.

Evidence

Evidence that incriminates one form of asbestos comes from health studies conducted at asbestos mines in South Africa and western Australia. Miners and millers who worked extensively with crocidolite showed an extremely high incidence of mesothelioma. In some individuals, exposure of less than one year resulted in the disease. As a result, the Australian mining operation was closed.

Despite this bleak picture, Malcolm Ross of the U.S. Geological Survey and others who have studied the issue conclude that the most widely used form of asbestos, chrysotile, is relatively harmless, unless it is breathed in high concentrations for long periods of time. They cite studies of miners and millers of white asbestos in Canada and northern Italy where mortality rates from mesothelioma and lung cancer differ very little from rates for the general public.

One study that supports the nontoxicity of chrysotile was conducted on miner's wives in the area of Thetford Mines, Quebec. For many years when there were no dust controls on the mining and milling operations, these women were exposed to extremely high levels of airborne asbestos. Nevertheless, the subjects of the study exhibited below-normal levels of the diseases thought to be associated with asbestos exposure.

The various types of asbestos fibers differ in their chemical composition, shape, and durability. The thin, rodlike crocidolite fibers, which can easily penetrate the lining of the lungs, are certainly the most pathogenic. Chrysotile, which accounts for over 90 percent of the world's production, consists of

Figure 2.A
Chrysotile asbestos. This sample is a fibrous form of the mineral serpentine. (Photo by E. J. Tarbuck)

(continues)

(Box 2.1, continued)

fibers that are curly, occur in bundles, and can be intercepted in air passageways. Further, if inhaled, chrysotile fibers are expelled more rapidly from human lungs than crocidolite fibers. These differences are thought to explain the fact that the mortality rates for chrysotile workers differ very little from the rates for the general population.

What Risk?

Does asbestos present a risk to the health of the nation's students? Available data indicate that the levels of airborne asbestos in schools are approximately 0.01 of the permissible exposure levels for the U.S. work place. In addi-

tion, the indoor concentrations of the most biologically active fibers are comparable to outdoor levels. With few exceptions, the type of asbestos fibers found in schools is chrysotile, which in low concentrations has been shown to be relatively harmless. A comparison of

the risk from asbestos exposure in schools to other risks in society is shown in Table 2.A. Playing high school football poses over 100 times the risk as exposure to airborne asbestos. These data clearly demonstrate that the asbestos panic was and is unwarranted.

Table 2.A Estimates of risk from asbestos exposure in schools in comparison with other risks.

Cause	Annual Death Rate (per million)
Asbestos exposure in schools	0.005–0.093
Whooping cough vaccination (1970–80)	1–6
Aircraft accidents (1979)	6
High school football (1970–80)	10
Drowning (ages 5–14)	27
Motor vehicle accident, Pedestrian (ages 5–14)	32
Home accidents (ages 1–14)	60
Long-term smoking	1200

SOURCE: Data from Weill and Hughes.

Another important type of accumulation generated by hydrothermal activity is called a **disseminated deposit**. Rather than being concentrated in narrow veins and dikes, these ores are distributed as minute masses throughout the entire rock mass. Much of the world's copper is extracted from disseminated deposits, including the huge Bingham Canyon copper mine in Utah. Because these accumulations contain only 0.4 to 0.8 percent copper, between 125 and 250 metric tons of ore must be mined for every ton of metal recovered. The environmental impact of these large excavations, including the problems of waste disposal, is significant.

Nonmetallic Mineral Resources

Mineral resources that are not used as fuels or processed for the metals they contain are referred to as *nonmetallic mineral resources*. These materials are extracted and processed either to make use of the nonmetallic elements they contain or for the physical and chemical properties they possess (Table 2.4). Nonmetallic mineral resources are commonly divided into two broad groups—*building materials* and *industrial minerals*. Since some substances have many different uses, they are found in both categories. Limestone, perhaps the most versatile and widely used rock of all, is the best example (Figure 2.27). As a building material, it is used not only as crushed rock and building stone, but in the making of cement as well. Moreover, as an industrial mineral, it is an ingredient in the manufacture of steel and is used in agriculture to neutralize acidic soils.

Besides aggregate (sand, gravel, and crushed rock) and cut stone, the other important building materials include gypsum for plaster and wallboard, clay for tile and bricks, and cement, which is made from limestone

and shale. Cement and aggregate go into the making of concrete, a material that is essential to practically all construction.

There is a wide variety of nonmetallic resources that are classified as industrial minerals. People often do not realize the importance of industrial minerals because

Table 2.4 Uses of nonmetallic minerals.

Mineral	Uses
Apatite	Phosphorus fertilizers
Asbestos (chrysotile)	Incombustible fibers
Calcite	Aggregate; steelmaking; soil conditioning; chemicals; cement; building stone
Clay minerals (kaolinite)	Ceramics; china
Corundum	Gemstones; abrasives
Diamond	Gemstones; abrasives
Fluorite	Steelmaking; aluminum refining; glass; chemicals
Garnet	Abrasives; gemstones
Graphite	Pencil lead; lubricant; refractories
Gypsum	Plaster of Paris
Halite	Table salt; chemicals; ice control
Muscovite	Insulator in electrical applications
Quartz	Primary ingredient in glass
Sulfur	Chemicals; fertilizer manufacture
Sylvite	Potassium fertilizers
Talc	Powder used in paints, cosmetics, etc.

Figure 2.27
Limestone processing in southern Virginia. Limestone has many uses and is both a building material and an industrial mineral. (Photo by Holt Confer/DRK Photo)

they see only the products that resulted from their use and not the minerals themselves. That is, many non-metallics are used up in the process of creating other products. Examples include fluorite and limestone, which are part of the steel making process; corundum and garnet, which are used as abrasives to make a piece of machinery; and sylvite, which is used in the production of the fertilizers used to grow a food crop.

Review Questions

1. Explain the statement "One rock is the raw material for another" using the rock cycle.
2. If a lava flow at Earth's surface had a basaltic composition, what rock type would the flow likely be (see Table 2.1, p. 41)? What igneous rock would form from the same magma if it did not reach the surface but instead crystallized at great depth?
3. What does a porphyritic texture indicate about the history of an igneous rock?
4. How are granite and rhyolite different? The same? (See Table 2.1, p. 41.)
5. Relate the classification of igneous rocks to Bowen's reaction series.
6. What minerals are most common in detrital sedimentary rocks? Why are these minerals so abundant?
7. What is the primary basis for distinguishing among various detrital sedimentary rocks?
8. Distinguish between the two categories of chemical sedimentary rocks.
9. What are evaporite deposits? Name a rock that is an evaporite.
10. Compaction is an important lithification process with which sediment size?
11. What is probably the single most characteristic feature of sedimentary rocks?
12. What is metamorphism? What are the *agents of change*?
13. Distinguish between regional and contact metamorphism.
14. What feature would easily distinguish schist and gneiss from quartzite and marble?
15. In what ways do metamorphic rocks differ from the igneous and sedimentary rocks from which they formed?
16. List two general types of hydrothermal deposits.
17. Nonmetallic resources are commonly divided into two broad groups. List the two groups and some examples of materials that belong to each.

Key Terms

chemical sedimentary rock (p. 42)
coarse-grained texture (p. 37)
contact metamorphism (p. 48)
crystallization (p. 35)
detrital sedimentary rock (p. 42)
disseminated deposit (p. 54)
evaporite (p. 45)
extrusive (volcanic) (p. 35)

fine-grained texture (p. 37)
foliated texture (p. 49)
glassy texture (p. 37)
hydrothermal solution (p. 52)
igneous rock (p. 35)
intrusive (plutonic) (p. 35)
lava (p. 35)
lithification (p. 35)
magma (p. 35)
metamorphic rock (p. 35)

nonfoliated texture (p. 49)
porphyritic texture (p. 37)
regional metamorphism (p. 48)
rock cycle (p. 34)
sediment (p. 35)
sedimentary rock (p. 35)
strata (beds) (p. 46)
texture (p. 37)
vein deposit (p. 52)
weathering (p. 35)

CHAPTER 3
Weathering, Soil, and Mass Wasting

■ Box 3.1 Acid Precipitation: A Human Impact on the Earth System
■ Box 3.2 Laterites and the Clearing of the Rain Forest
■ Box 3.3 The Gros Ventre Slide: A Classic Case Study

Differerential weathering exhibited by Delicate Arch, Arches National Park, Utah.
(Photo by David Muench)

Earth's surface is constantly changing. Rock is disintegrated and decomposed, moved to lower elevations by gravity, and carried away by water, wind, or ice. In this manner Earth's physical landscape is sculptured. This chapter focuses on the first two steps of this never-ending process—weathering and mass wasting—probing into how and why rock disintegrates and decomposes and what mechanisms act to move it downslope. Soil, an important product of the weathering process and a vital resource, is also examined.

Earth's External Processes

Weathering, mass wasting, and erosion are called *external processes* because they occur at or near Earth's surface and are powered by energy from the sun. External processes are a basic part of the rock cycle because they are responsible for transforming solid rock into sediment.

To the casual observer, the face of Earth may appear to be without change, unaffected by time. In fact, less than 200 years ago, most people believed that mountains, lakes, and deserts were permanent features of an Earth that was thought to be no more than a few thousand years old. Today, however, we know that mountains eventually succumb to weathering and erosion, lakes fill with sediment and vegetation or are drained by streams, and deserts come and go as relatively minor climatic changes occur.

Earth is a dynamic body. Some parts of Earth's surface are gradually elevated by mountain building and volcanic activity. Meanwhile opposing processes are continually removing materials from higher elevations and transporting them to lower elevations (Figure 3.1). The latter processes include:

1. **Weathering**—disintegration and decomposition of rock at or near Earth's surface.
2. **Mass wasting**—transfer of rock material downslope under the influence of gravity.
3. **Erosion**—incorporation and transportation of material by a mobile agent, usually water, wind, or ice.

We will first turn our attention to the process of weathering and the products generated by this activity. However, weathering cannot be easily separated from the other two processes because, as weathering breaks rocks apart, it facilitates the movement of rock debris by mass wasting and erosion. On the other hand, the transport of material by mass wasting and erosion furthers the disintegration and decomposition of rock.

Weathering

All materials are susceptible to weathering. Consider, for example, the synthetic rock we call concrete, which closely resembles the sedimentary rock called conglomerate. A newly poured concrete sidewalk has a smooth, fresh look. However, not many years later, the same sidewalk will appear chipped, cracked, and rough, with pebbles exposed at the surface. If a tree is nearby, its roots may grow under the concrete, heaving and buckling it. The same natural processes that eventually break apart a concrete sidewalk also act to disintegrate natural rocks, regardless of their type or strength.

Why does rock weather? Simply, weathering is the response of Earth materials to a changing environment. For instance, after millions of years of uplift and erosion, the rocks overlying a large body of intrusive igneous rock may be removed. This exposes the rock to a whole new environment at the surface. The mass of crystalline rock, which formed deep below ground where temperatures and pressures are much greater than at the surface, is now subjected to very different and comparatively hostile surface conditions. In response, this rock mass will gradually change until it is once again in equilibrium, or balance, with its new environment. Such transformation of rock is what we call *weathering*.

In the following sections we will discuss two kinds of weathering—mechanical and chemical. Mechanical weathering is the physical breaking up of rocks into smaller pieces. Chemical weathering actually alters a rock's chemistry, changing it into different substances. Although we will consider these two processes separately, keep in mind that they usually work simultaneously in nature.

Mechanical Weathering

When a rock undergoes **mechanical weathering** it is broken into smaller and smaller pieces, each retaining the characteristics of the original material. The end result is many small pieces from a single large one. Figure 3.2 shows that breaking a rock into smaller pieces increases the surface area available for chemical attack. An analogous situation occurs when a sugar cube is added to water. A cube of sugar will dissolve much more slowly than will an equal volume of loose granules because of the vast difference in surface area. Hence, by breaking rocks into smaller pieces, mechanical weathering increases the amount of surface area available for chemical weathering.

In nature, four important physical processes break rocks into smaller fragments: frost wedging, expansion resulting from unloading, thermal expansion, and biological activity.

Frost Wedging

Alternate freezing and thawing of water is one of the most important processes of mechanical weathering. Water has the unique property of expanding about 9 percent when it freezes. This increase in volume occurs

Figure 3.1
Slopes here, as elsewhere, are places where materials are continually moving from higher to lower elevations. Weathering begins the process by attacking the solid rock exposed at the surface. Next, gravity moves the weathered debris downslope. This step, termed mass wasting, may range from a slow and gradual creep to a thundering landslide. Eventually, the material that was once high up the slope reaches the stream at the bottom. The moving water then transports the debris away. (Photo by Carr Clifton)

because, as ice forms, the water molecules arrange themselves into a very open crystalline structure. As a result, when water freezes, it expands and exerts a tremendous outward force. This can be verified by completely filling a container with water and freezing it. The formation of ice will rupture the container.

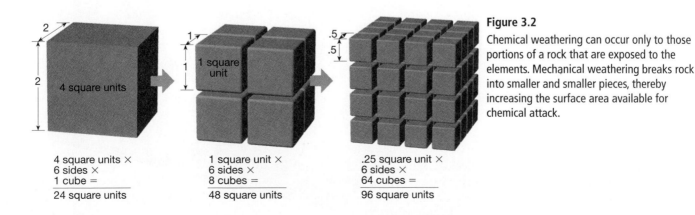

Figure 3.2
Chemical weathering can occur only to those portions of a rock that are exposed to the elements. Mechanical weathering breaks rock into smaller and smaller pieces, thereby increasing the surface area available for chemical attack.

4 square units ×
6 sides ×
1 cube =
_____
24 square units

1 square unit ×
6 sides ×
8 cubes =
_____
48 square units

.25 square unit ×
6 sides ×
64 cubes =
_____
96 square units

In nature, water works its way into every crack or void in rock and, upon freezing, expands and enlarges the opening. After many freeze-thaw cycles, the rock is broken into pieces. This process is appropriately called **frost wedging** (Figure 3.3). Frost wedging is most pronounced in mountainous regions in the middle latitudes where a daily freeze-thaw cycle often exists. Here, sections of rock are wedged loose and may tumble into large piles called **talus slopes** that often form at the base of steep rock outcrops (Figure 3.3).

Unloading

When large masses of igneous rock, particularly those composed of granite, are exposed by erosion, slabs begin to break loose like the layers of an onion. The process is called **sheeting** and is thought to occur, at least in part, because of the great reduction in pressure when the overlying rock is eroded away. Accompanying this unloading, the outer layers expand more than the rock below, and thus separate from the rock body. Continued weathering eventually causes the slabs to separate and spall off, creating **exfoliation domes.** Excellent examples of exfoliation domes include Stone Mountain, Georgia, and Half Dome (Figure 3.4) and Liberty Cap in Yosemite National Park.

Deep underground mining provides us with another example of how rocks behave once the confining pressure is removed. Large rock slabs sometimes explode off the walls of newly cut mine tunnels because of the abrupt-ly reduced pressure. Evidence of this type, plus the fact that fracturing occurs parallel to the floor of a quarry when large blocks of rock are removed, strongly supports the process of unloading as the cause of sheeting.

Although many fractures are created by expansion, others are produced by contraction during the crystallization of magma, and still others by tectonic forces during mountain building. Fractures produced by these activities generally form a definite pattern and are called *joints.* Joints are important rock structures that allow water to penetrate to depth and start the process of weathering long before the rock is exposed at the surface.

Thermal Expansion

The daily cycle of temperature change is thought to weaken rocks, particularly in deserts where daily variations may exceed 30°C (54°F). Heating a rock causes it to expand, and cooling causes it to contract. Repeated swelling and shrinking of minerals that have different expansion and contraction rates should exert some stress on the rock's outer shell. However, laboratory experiments have not substantiated this. In one test, unweathered rocks were heated much hotter than is normally experienced on Earth's surface, and then cooled. This procedure was repeated many times to simulate hundreds of years of weathering, but the rocks showed little apparent change. Nevertheless, in desert areas pebbles do exhibit unmistakable evidence of shattering from what appear to be temperature changes.

Figure 3.3

Frost wedging. As water freezes it expands to 109 percent of its original volume, exerting a force great enough to break rock. When frost wedging occurs in a setting such as this, the broken rock fragments fall to the base of the cliff and create a cone-shaped accumulation known as talus.

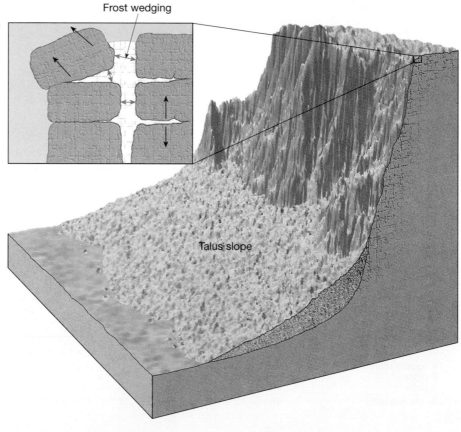

Frost wedging

Talus slope

Figure 3.4
Summit of Half Dome, an exfoliation dome in Yosemite National Park. (Photo by Breck Kent)

Biological Activity

Weathering is also accomplished by the activities of organisms, including plants, burrowing animals, and humans. Plant roots in search of minerals and water grow into fractures, and as the roots grow they wedge the rock apart (Figure 3.5). Burrowing animals further break down the rock by moving fresh material to the surface, where physical and chemical processes can more effectively attack it. Decaying organisms also produce acids, which contribute to chemical weathering. Where rock has been blasted in search of minerals or for road construction, the impact is quite noticeable, but on a worldwide scale human activity probably ranks behind burrowing animals in Earth-moving accomplishments.

Although usually considered separately from mechanical weathering, the activities of the erosional agents—wind, water, and glaciers—are nonetheless important. For as these mobile agents move rock debris, they relentlessly disintegrate the Earth materials they carry.

Chemical Weathering

Chemical weathering involves the complex processes that alter the internal structures of minerals by removing and/or adding elements. During this transformation, the original rock decomposes into substances that are stable in the surface environment. Consequently, the

Figure 3.5
Root wedging widens fractures in rock and aids the process of mechanical weathering. (Photo by Tom Bean/DRK Photo)

products of chemical weathering will remain essentially unchanged as long as they remain in an environment similar to the one in which they formed.

Water and Carbonic Acid

Water is by far the most important agent of chemical weathering. Although pure water is nonreactive, a small amount of dissolved material is generally all that is needed to activate it. Oxygen dissolved in water will *oxidize* some materials. For example, when an iron nail is found in moist soil, it will have a coating of rust (iron oxide), and if the time of exposure has been long, the nail will be so weak that it can be broken as easily as a toothpick. When rocks containing iron-rich minerals oxidize, a yellow to reddish-brown rust will appear on the surface.

Carbon dioxide (CO_2) dissolved in water (H_2O) forms carbonic acid (H_2CO_3), the same weak acid produced when soft drinks are carbonated. Rain dissolves some carbon dioxide as it falls through the atmosphere, and additional amounts released by decaying organic matter are acquired as the water percolates through the soil. Carbonic acid ionizes to form the very reactive hydrogen ion (H^+) and the bicarbonate ion (HCO_3^-).

How Granite Weathers

To illustrate how rock chemically weathers when attacked by carbonic acid, we will consider the weathering of granite, the most abundant continental rock. Recall that granite consists mainly of quartz and potassium feldspar. The weathering of the potassium feldspar component of granite takes place as follows:

$$2KAlSi_3O_8 + 2(H^+ + HCO_3^-) + H_2O \longrightarrow$$

potassium feldspar   carbonic acid   water

$$Al_2Si_2O_5(OH)_4 + 2KHCO_3 + 4SiO_2$$

clay mineral   potassium bicarbonate   silica

In this reaction, the hydrogen ions (H^+) attack and replace potassium ions (K^+) in the feldspar structure, thereby disrupting the crystalline network. Once removed, the potassium is available as a nutrient for plants. Or it becomes potassium bicarbonate ($KHCO_3$), a soluble salt that may be incorporated into other minerals or carried to the ocean in dissolved form by streams.

The most abundant products of the chemical breakdown of feldspar are residual clay minerals. Because clay minerals are the end product of weathering, they are very stable under surface conditions. Consequently, clay minerals make up a high percentage of the inorganic material in soils. Moreover, the most abundant sedimentary rock, shale, also contains a high proportion of clay minerals.

In addition to the formation of clay minerals during this reaction, some silica is removed from the feldspar structure and is carried away by groundwater (water beneath Earth's surface). This dissolved silica will eventually precipitate to produce nodules of chert or flint, fill in the port spaces between sediment grains, or be carried to the ocean, where microscopic animals will remove it to build hard silica shells.

To summarize, the weathering of potassium feldspar generates a residual clay mineral, a soluble salt (potassium bicarbonate), and some silica which enters into solution.

Quartz, the other main component of granite, is very resistant to chemical weathering; hence, it remains substantially unaltered when attacked by weakly acidic solutions. As a result, when granite weathers, the feldspar crystals dull and slowly turn to clay, releasing the once-interlocked quartz grains, which still retain their fresh, glassy appearance. Although some quartz remains in the soil, much is transported to the sea or to other sites of deposition, where it becomes the main constituent of such features as sandy beaches and sand dunes. In time it may become lithified to form the sedimentary rock *sandstone*.

Table 3.1 Products of weathering.

Mineral	Residual Products	Material in Solution
Quartz	Quartz grains	Silica
Feldspars	Clay minerals	Silica K^+, Na^+, Ca^{2+}
Amphibole (hornblende)	Clay minerals Limonite Hematite	Silica Ca^{2+}, Mg^{2+}
Olivine	Limonite Hematite	Silica Mg^{2+}

Weathering of Silicate Minerals

Table 3.1 lists the weathered products of some of the most common silicate minerals. Remember that silicate minerals make up most of Earth's crust and that these minerals are composed essentially of only eight elements. When chemically weathered, silicate minerals yield sodium, calcium, potassium, and magnesium ions. These form soluble products that may be removed by groundwater. The element iron combines with oxygen, producing relatively insoluble iron oxides, which give soil a reddish-brown or yellowish color. Under most conditions the three remaining elements, aluminum, silicon, and oxygen, join with water to produce residual clay minerals. However, even the highly insoluble clay minerals are very slowly removed by subsurface water.

Spheroidal Weathering

In addition to altering the internal structure of minerals, chemical weathering causes physical changes as well. For instance, when angular rock masses are attacked by water that enters along joints, the rocks tend to take on a spherical shape. Gradually the corners and edges of the angular blocks become more rounded. The corners are attacked most readily because of their greater surface area, as compared to the edges and faces. This process, called **spheroidal weathering**, gives the weathered rock a more rounded or spherical shape (Figure 3.6).

Sometimes during the formation of spheroidal boulders, successive shells separate from the rock's main body. Eventually the outer shells spall off, allowing the chemical weathering activity to penetrate deeper into the boulder. This spherical scaling results because, as the minerals in the rock weather to clay, they increase in size through the addition of water to their structure. This increased bulk exerts an outward force that causes concentric layers of rock to break loose and fall off. Hence, chemical weathering does produce forces great enough to cause mechanical weathering.

This type of spheroidal weathering, in which shells spall off, should not be confused with the phenomenon of sheeting discussed earlier. In sheeting, the fracturing occurs as a result of unloading and the rock layers which separate from the main body are largely unaltered at the time of separation.

Rates of Weathering

Several factors influence the type and rate of rock weathering. We have already seen how mechanical weathering affects the rate of weathering. By breaking rock into smaller pieces, the amount of surface area exposed to chemical weathering is increased. The presence or absence of joints (cracks) can be significant because they influence the ability of water to penetrate the rock (Figure 3.6). Other important factors include the mineral makeup of rocks and climate.

Mineral Makeup

The variations in weathering rates, attributable to the mineral constituents, can be demonstrated by comparing old headstones made from different rock types. Headstones of granite, which is composed of silicate minerals, are relatively resistant to chemical weathering. We can see this by examining the inscriptions on the headstones shown in Figure 3.7. This is not true of the marble headstone, which shows signs of extensive chemical alteration over a relatively short period. Marble is composed of calcite (calcium carbonate), which readily dissolves even in a weakly acidic solution.

The silicates, the most abundant mineral group, weather in essentially the same order as their order of crystallization. The minerals that crystallize first form under much higher temperatures than those that crystallize last. Consequently, the early formed minerals are not as stable at Earth's surface, where the temperature and pressure are drastically different from the environment in which they formed. By examining Bowen's reaction series (see Figure 2.8, p. 39), we see that olivine crystallizes first and is therefore the least resistant to chemical weathering, whereas quartz, which crystallizes last, is the most resistant.

Climate

Climatic factors, particularly temperature and moisture, are crucial to the rate of rock weathering. These climatic elements largely determine the weathering rate and strongly influence the kind and amount of vegetation present. Regions with lush vegetation generally have a thick mantle of soil rich in decayed organic matter from which chemically active fluids such as carbonic and humic acids are derived.

The optimum environment for chemical weathering is a combination of warm temperatures and abundant moisture. In polar regions chemical weathering

Figure 3.6
Spheroidal weathering is evident in this exposure of granite in California's Joshua Tree National Monument. Because the rocks are attacked more vigorously on the corners and edges, they take on a spherical shape. The lines visible in the rock are called *joints*. Joints are important rock structures that allow water to penetrate and start the weathering process long before the rock is exposed. (Photo by E. J. Tarbuck)

is ineffective because frigid temperatures keep the available moisture locked up as ice, whereas in arid regions there is insufficient moisture to foster rapid chemical weathering.

A classic example of how climate affects the rate of weathering was provided when Cleopatra's Needle, a granite obelisk, was moved from Egypt to New York City. After withstanding approximately 3500 years of exposure in the dry climate of Egypt, the hieroglyphics were almost completely removed from substantial portions of the monument in less than 75 years in the moist New York climate (Figure 3.8). Acid precipitation has also contributed to the weathering of the obelisk (see Box 3.1).

The sum of these factors determines the type and rate of rock weathering for a given place. However, there is generally enough variation, even within a relatively small area, for the rocks to exhibit some differential weathering. **Differential weathering** simply relates to the fact that rocks exposed at Earth's surface usually do not weather at the same rate. Because of variations in

such factors as mineral makeup, degree of jointing, and exposure to the elements, significant differences occur. Consequently, differential weathering and subsequent erosion create many unusual and often spectacular rock formations and landforms. Included are features such as the natural bridges found in Arches National Park (see chapter-opening photo) and the sculptured rock pinnacles found in Bryce Canyon National Park (Figure 3.9).

Soil

Soil covers most land surfaces. Along with air and water, it is one of our most indispensable resources (Figure 3.10). Also like air and water, soil is taken for granted by many of us. The following quote helps put this vital layer in perspective.

> Science, in recent years, has focused more and more on the Earth as a planet, one that for all we know is unique—where a thin blanket of air, a

Figure 3.7
An examination of headstones reveals the rate of chemical weathering on diverse rock types. The granite headstone (left) was erected three years after the marble headstone (right). The inscription date of 1885 on the marble monument is nearly illegible. (Photos by E. J. Tarbuck)

A.

B.

Figure 3.8
Chemical weathering of Cleopatra's Needle, a granite obelisk. **A.** Before it was removed from Egypt. (Courtesy of the Metropolitan Museum of Art) **B.** After a span of 75 years in New York City's Central Park. After surviving intact for about 35 centuries in Egypt, significant portions have been almost completely defaced in less than a century. (Photo by Catherine Ursillo/Photo Researchers, Inc.)

thinner film of water, and the thinnest veneer of soil combine to support a web of life of wondrous diversity in continuous change.[*]

Soil has accurately been called "the bridge between life and the inanimate world." All life—the entire biosphere—owes its existence to a dozen or so elements that must ultimately come from Earth's crust. Once weathering and other processes create soil, plants carry out the intermediary role of assimilating the necessary elements and making them available to animals including humans.

An Interface in the Earth System

When Earth is viewed as a system, soil is referred to as an *interface*—a common boundary where different parts of a system interact. This is an appropriate designation because soil forms where the solid Earth, the atmosphere, the hy-

drosphere, and the biosphere meet. Soil is a material that develops in response to complex environmental interactions among different parts of the Earth system. Over time, soil gradually evolves to a state of equilibrium or balance with the environment. Soil is dynamic and sensitive to almost every aspect of its surroundings. Thus, when environmental changes occur, such as climate, vegetative cover, and animal (including human) activity, the soil responds. Any such change produces a gradual alteration of soil characteristics until a new balance is reached. Although thinly distributed over the land surface, soil functions as a fundamental interface; providing an excellent example of the integration among many parts of the Earth system.

What Is Soil?

With few exceptions, Earth's land surface is covered by **regolith,** the layer of rock and mineral fragments produced by weathering. Some would call this material soil, but soil is more than an accumulation of weathered debris. **Soil** is a combination of mineral and organic matter,

---

[*]Jack Eddy. "A Fragile Seam of Dark Blue Light," in *Proceedings of the Global Change Research Forum.* U.S. Geological Survey Circular 1086, 1993, p. 15.

Box 3.1

Acid Precipitation: A Human Impact on the Earth System

Humans are part of the complex interacting whole we call the Earth system (see Box I.1, p. 7). As such, our actions cause changes to all the other parts of the system. For example, by going about our normal routine, we humans modify the composition of the atmosphere. These atmospheric modifications, in turn, cause unintended and unwanted changes to occur in the hydrosphere, biosphere, and solid Earth. Acid precipitation is one small but significant example.

Decomposed stone monuments and structures are common sights in many cities (Figure 3.A). Although we expect rock to gradually decompose, many of these monuments have succumbed prematurely. An important cause for this accelerated chemical weathering is acid precipitation.

Rain is naturally somewhat acidic. When carbon dioxide from the atmosphere dissolves in water, it becomes weak carbonic acid. However, the term *acid precipitation* refers to precipitation that is much more acidic than natural, unpolluted rain and snow.

As a consequence of burning large quantities of fossil fuels like coal and petroleum products, nearly 40 million tons of sulfur and nitrogen oxides are released into the atmosphere each year in the United States. The major sources of these emissions include power-generating plants, industrial processes, such as ore smelting and petroleum refining, and vehicles of all kinds. Through a series of complex chemical reactions, some of these pollutants are converted into acids that then fall to Earth's surface as rain or snow. Another portion is deposited in dry form and subsequently converted into acid after coming in contact with precipitation, dew, or fog.

Widespread acid rain has been known in northern Europe and eastern North America for some time. Studies have also shown that acid rain occurs in many other regions, including western North America, Japan, China, Russia, and South America. In addition to local pollution sources, a portion of the acidity found in the northeastern United States and eastern Canada originates hundreds of kilometers away in industrialized regions to the south and southwest. This situation occurs because many pollutants remain in the atmosphere for periods as long as five days, during which time they may be transported great distances.

The damaging environmental effects of acid rain are thought to be considerable in some areas and imminent in others. The best-known effect is an increased acidity in thousands of lakes in Scandinavia and eastern North America. Accompanying this have been substantial increases in dissolved aluminum that is leached from the soil by the acidic water and that, in turn, is toxic to fish. Consequently, some lakes are virtually devoid of fish, and others are approaching this condition. Ecosystems are characterized by many interactions at many levels of organization, which means that evaluating the effects of acid precipitation on these complex systems is difficult and expensive, and far from complete.

In addition to the thousands of lakes that can no longer support fish, research indicates that acid precipitation may also reduce agricultural crop yields and impair the productivity of forests. Acid rain not only harms the foliage, but also damages roots and leaches nutrient minerals from the soil. Finally, acid precipitation is known to promote the corrosion of metals and contributes to the destruction of stone structures.

Figure 3.A
Acid rain accelerates the chemical weathering of stone monuments and structures, including this building facade in Leipzig, Germany. (Photo by Doug Plummer/Photo Researchers, Inc.)

water, and air—that portion of the regolith that supports the growth of plants. Although the proportions of the major components in soil vary, the same four components always are present to some extent (Figure 3.11). About one-half of the total volume of a good quality surface soil is a mixture of disintegrated and decomposed rock (mineral matter) and **humus,** the decayed remains of animal and plant life (organic matter). The remaining half consists of pore spaces among the solid particles where air and water circulate.

Although the mineral portion of the soil is usually much greater then the organic portion, humus is an essential component. In addition to being an important source of plant nutrients, humus enhances the soil's

Figure 3.9
Differential weathering is illustrated by these sculpted rock pinnacles in Bryce National Park, Utah. (Photo by Adam Jones/Photo Researchers, Inc.)

ability to retain water. Because plants require air and water to live and grow, the portion of the soil consisting of pore spaces that allow for the circulation of these fluids is as vital as the solid soil constituents.

Soil water is far from "pure" water; instead, it is a complex solution containing many soluble nutrients. Soil water not only provides the necessary moisture for the chemical reactions that sustain life; it also supplies

Figure 3.10
Soil is an essential resource that we often take for granted. Soil is not a living entity, but it contains a great deal of life. Moreover, this complex medium supports nearly all plant life, which, in turn, supports animal life. (Photo by James E. Patterson)

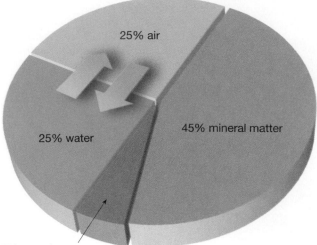

Figure 3.11

Composition (by volume) of a soil in good condition for plant growth. Although the percentages vary, each soil is composed of mineral and organic matter, water, and air.

plants with nutrients in a form they can use. The pore spaces not filled with water contain air. This air is the source of necessary oxygen and carbon dioxide for most microorganisms and plants that live in the soil.

Soil Texture and Structure

Most soils are far from uniform, and contain particles of different sizes. **Soil texture** refers to the proportions of different particle sizes. Texture is a very basic soil property because it strongly influences the soil's ability to retain and transmit water and air, both of which are essential to plant growth. Sandy soils may drain too rapidly and dry out quickly. At the opposite extreme, the pore spaces of clay-rich soils may be so small that they inhibit drainage, and long-lasting puddles result. Moreover, when the clay and silt content is very high, plant roots may have difficulty penetrating the soil.

Because soils rarely consist of particles of only one size, *textural categories* have been established, based upon the varying proportions of clay, silt, and sand. The standard system of classes used by the U.S. Department of Agriculture is shown in Figure 3.12. For example, point A on this triangular diagram (left center) represents a soil composed of 10 percent silt, 40 percent clay, and 50 percent sand. Such a soil is called a *sandy clay.* The soils called *loam,* which occupy the central portion of the diagram, are those in which no single particle size predominates over the other two. Loam soils are best suited to support plant life because they generally have better moisture characteristics and nutrient storage ability than do soils composed predominantly of clay or coarse sand.

Figure 3.12

The texture of any soil can be represented by a point on this soil texture diagram. Soil texture is one of the most significant factors used to estimate agricultural potential and engineering characteristics. (After U.S. Department of Agriculture)

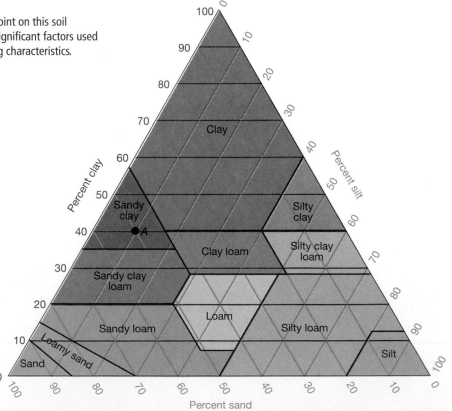

Soil particles are seldom completely independent of one another. Rather, they usually form clumps called *peds* that give soils a particular structure. Four basic soil structures are recognized: platy, prismatic, blocky, and spheroidal. Soil structure is important because it influences the ease of a soil's cultivation as well as the susceptibility of a soil to erosion. In addition, soil structure affects the porosity and permeability of soil (that is, the ease with which water can penetrate). This, in turn, influences the movement of nutrients to plant roots. Prismatic and blocky peds usually allow for moderate water infiltration, whereas platy and spheroidal structures are characterized by slower infiltration rates.

Controls of Soil Formation

Soil is the product of the complex interplay of several factors. The most important of these are parent material, time, climate, plants, animals, and slope. Although all of these factors are interdependent, their roles will be examined separately.

Parent Material

The source of the weathered mineral matter from which soils develop is called the **parent material** and is a major factor influencing a newly forming soil. Gradually it undergoes physical and chemical changes as the processes of soil formation progress. Parent material may be the underlying bedrock or it can be a layer of unconsolidated deposits, as in a stream valley. When the parent material is bedrock, the soils are termed *residual soils.* By contrast, those developed on unconsolidated sediment are called *transported soils* (Figure 3.13). Note that transported soils form *in place* on parent materials that have been carried from elsewhere and deposited by gravity, water, wind, or ice.

The nature of the parent material influences soils in two ways. First, the type of parent material affects the rate of weathering, and thus the rate of soil formation. (Consider the weathering rates of granite versus limestone.) Also, because unconsolidated deposits are already partly weathered and provide more surface area for chemical weathering, soil development on such material usually progresses more rapidly. Second, the chemical makeup of the parent material affects the soil's fertility. This influences the character of the natural vegetation the soil can support.

At one time the parent material was believed to be the primary factor causing differences among soils. Today soil scientists realize that other factors, especially climate, are more important. In fact, it has been found that similar soils are often produced from different parent materials and that dissimilar soils have developed from the same parent material. Such discoveries reinforce the importance of the other soil-forming factors.

Time

Time is an important component of every geological process, and soil formation is no exception. The nature of soil is strongly influenced by the length of time processes have been operating. If weathering has been going on for a comparatively short time, the parent material determines to a large extent the characteristics of the soil. As weathering processes continue, the influence of parent material on soil is overshadowed by the other soil-forming factors, especially climate. The amount of time required for various soils to evolve cannot be specified because the soil-forming processes act at varying rates under different circumstances. However, as a rule the longer a soil has been forming, the thicker it becomes and the less it resembles the parent material.

Climate

Climate is the most influential control of soil formation. Just as temperature and precipitation are the climatic elements that influence people the most, so too are they the elements that exert the strongest impact on soil formation. Variations in temperature and precipitation determine whether chemical or mechanical weathering predominates. They also greatly influence the rate and depth of weathering. For instance, a hot, wet climate may produce a thick layer of chemically weathered soil in the same amount of time that a cold, dry climate produces a thin mantle of mechanically weathered debris. Also, the amount of precipitation influences the degree to which various materials are removed (leached) from the soil, thereby affecting soil fertility. Finally, climatic conditions create an important control of the type of plant and animal life present.

Plants and Animals

Plants and animals play a vital role in soil formation. The types and abundance of organisms present have a strong influence on the physical and chemical properties of a soil. In fact, for well-developed soils in many regions, the significance of natural vegetation is frequently implied in the description used by soil scientists. Such phrases as *prairie soil, forest soil,* and *tundra soil* are common.

The chief function of plants and animals is to furnish organic matter to the soil. Certain bog soils are composed almost entirely of organic matter, whereas desert soils may contain as little as a small fraction of 1 percent. Although the quantity of organic matter varies substantially among soils, it is the rare soil that completely lacks it.

The primary source of organic matter is plants, although animals and the uncountable microorganisms also contribute. When organic matter decomposes, important nutrients are supplied to plants, as well as to animals and microorganisms living in the soil. Consequently, soil fertility depends in part on the amount of organic matter

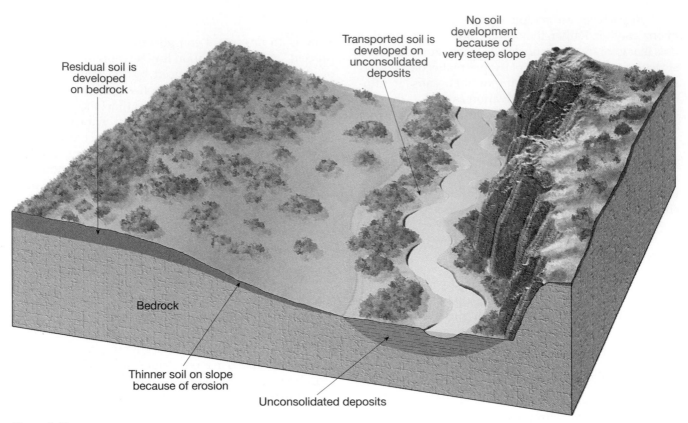

Residual soil is developed on bedrock

Transported soil is developed on unconsolidated deposits

No soil development because of very steep slope

Bedrock

Thinner soil on slope because of erosion

Unconsolidated deposits

Figure 3.13
The parent material for residual soils is the underlying bedrock, whereas transported soils form on unconsolidated deposits. Also note that as slopes become steeper, soil becomes thinner.

present. Furthermore, the decay of plant and animal remains causes the formation of various organic acids. These complex acids hasten the weathering process. Organic matter also has a high water-holding ability and thus aids water retention in a soil.

Microorganisms, including fungi, bacteria, and single-celled protozoa, play an active role in the decay of plant and animal remains. The end product is *humus*, a material that no longer resembles the plants and animals from which it is formed. In addition, certain microorganisms aid soil fertility because they have the ability to *fix* (alter) atmospheric nitrogen gas into soil nitrogen compounds.

Earthworms and other burrowing animals act to mix the mineral and organic portions of a soil. Earthworms, for example, feed on organic matter and thoroughly mix soils in which they live, often moving and enriching many tons per acre each year. Burrows and holes also aid the passage of water and air through the soil.

Slope

The slope of the land can vary greatly over short distances. Such variations, in turn, can lead to widely varied local soil types. Many of the differences exist because slope has a significant impact on the amount of erosion and the water content of soil.

On steep slopes soils are often poorly developed. In such situations little water can soak in, and as a result, soil moisture may be insufficient for vigorous plant growth. Further, because of accelerated erosion on steep slopes, the soils are thin, or nonexistent (Figure 3.13).

On the other hand, poorly drained and waterlogged soils in bottomlands have a much different character. Such soils are usually thick and dark. The dark color results from generous organic matter that accumulates because saturated conditions retard the decay of vegetation. The optimum terrain for soil development is a flat-to-undulating upland surface. Here we find good drainage, minimum erosion, and sufficient infiltration of water into the soil.

Slope orientation, the direction the slope is facing, also is significant. In the mid-latitudes of the Northern Hemisphere, a south-facing slope receives a great deal more sunlight than a north-facing slope. In fact, a steep north-facing slope may receive no direct sunlight at all. The difference in the amount of solar radiation received causes substantial differences in soil temperature and moisture, which in turn may influence the nature of the vegetation and the character of the soil.

Although we have dealt separately with each of the soil-forming factors, remember that *all work together* to form soil. No single factor is responsible for a soil being as it is. Rather, it is the combined influence of parent material, time, climate, plants, animals, and slope that determines a soil's character.

The Soil Profile

It is important to realize that soil-forming processes *operate from the surface downward.* Thus, variations in composition, texture, structure, and color gradually evolve at varying depths. These vertical differences, which usually become more pronounced as time passes, divide the soil into zones or layers that soil scientists call **horizons.** If you were to dig a trench in soil, you would see that its walls are layered. Such a vertical section through all of the soil horizons constitutes the **soil profile** (Figure 3.14).

Figure 3.15 presents an idealized view of a well-developed soil profile in which five horizons are identified. From the surface downward, they are designated as *O, A, E, B,* and *C,* respectively. These five horizons are common to soils in temperate regions. The characteristics and extent of development of horizons vary in different environments. Thus, different localities exhibit soil profiles that can contrast greatly with one another.

The *O* horizon consists largely of organic material, unlike the layers beneath it that consist mainly of mineral matter. The upper portion of the *O* horizon is primarily plant litter such as loose leaves and other organic debris that are still recognizable. By contrast, the lower portion of the *O* horizon is made up of partly decomposed organic matter (humus) in which plant structures can no longer be identified. In addition to plants, the *O* horizon is teeming with microscopic life including bacteria, fungi, algae, and insects. All of these organisms contribute oxygen, carbon dioxide, and organic acids to the developing soil.

Underlying the organic-rich *O* horizon is the *A* horizon. This zone is largely mineral matter, yet biological activity is high and humus is generally present—up to 30 percent in some instances. Together the *O* and *A* horizons make up what is commonly called *topsoil.* Below the *A* horizon, the *E* horizon is a light-colored layer that contains little organic material. As water percolates downward through this zone, finer particles are carried away. This washing out of the fine soil components is termed **eluviation.** Water percolating downward also dissolves soluble inorganic soil components and carries them to deeper zones. This depletion of soluble materials from the upper soil is termed **leaching.**

Immediately below the *E* horizon is the *B* horizon, or *subsoil.* Much of the material removed from the *E* horizon by eluviation is deposited in the *B* horizon, which is often referred to as the *zone of accumulation.* The accumulation of the fine clay particles enhances water retention in the subsoil. However, in extreme cases, clay accumulation can form a very compact and impermeable layer called

Figure 3.14

A soil profile is a vertical cross-section from the surface down to the parent material. Well-developed soils show distinct layers called horizons. (Photo by E. J. Tarbuck)

hardpan. The *O, A, E,* and *B* horizons together constitute the **solum,** or "true soil." It is in the solum that the soil-forming processes are active and that living roots and other plant and animal life are largely confined.

Below the solum and above the unaltered parent material is the *C* horizon, a layer characterized by partially altered parent material. Whereas the *O, A, E,* and *B* horizons bear little resemblance to the parent material, it is easily identifiable in the *C* horizon. Although this material is undergoing changes that will eventually transform it into soil, it has not yet crossed the threshold that separates regolith from soil.

Soil characteristics and development can vary greatly in different environments. The boundaries between soil horizons may be very distinct, as in Figures 3.14 and 3.15, or the horizons may blend gradually from one to another. Consequently, a well-developed soil profile indicates that environmental conditions have been relatively stable over an extended time span and that the soil is *mature.* By contrast, some soils lack horizons altogether.

Such soils are called *immature,* because soil building has been going on for only a short time. Immature soils are also characteristic of steep slopes where erosion continually strips away the soil, preventing full development.

O horizon
Loose and partly
decayed
organic matter

A horizon
Mineral matter
mixed with
some humus

E horizon
Light colored
mineral particles.
Zone of eluviation
and leaching

B horizon
Accumulation of
clay transported
from above

C horizon
Partially altered
parent material

Unweathered
parent material

Figure 3.15
Idealized soil profile from a humid climate in the middle latitudes.

Soil Types

There are hundreds of soil types and subtypes world-wide, far too many to consider here. Instead, we shall look at just three very generic types—pedalfers, pedo-cals, and laterites. As you read, notice that the characteristics of each soil primarily reflect the prevailing climatic conditions. A summary of the characteristics is provided in Table 3.2. (For more information on different soil groups and their distribution, see Appendix F.)

Pedalfer

The term **pedalfer** is derived from the Greek **ped**on, meaning "soil," and the chemical symbols **Al** (aluminum) and **Fe** (Iron). Pedalfers are characterized by an accumulation of iron oxides and aluminum-rich clays in the B horizon. In mid-latitude areas where the annual rainfall exceeds 63 centimeters (25 inches), most of the soluble materials, such as calcium carbonate, are leached from the soil and carried away by underground water. The less soluble iron oxides and clays are carried from the E horizon and deposited in the B horizon, giving it a brown to red-brown color. These soils are best developed under forest vegetation where large quantities of decomposing organic matter provide the acid conditions necessary for leaching. In the United States pedalfers are found east of a line from northwestern Minnesota to south-central Texas.

Pedocal

The term **pedocal** is derived from the Greek **ped**on, meaning "soil," and the first three letters of **cal**cite (calcium carbonate). As the name implies, pedocals are characterized by an accumulation of calcium carbonate. This soil type is found in the drier western United States in association with grassland and brush vegetation. Because chemical weathering is less intense in drier areas, pedocals generally contain a smaller percentage of clay materials than pedalfers.

In the arid and semiarid western states, a calcite-enriched layer called *caliche* may be present in the soils. In these areas little of the rain that falls penetrates to great depths. Rather, it is held by the soil particles near the surface until it evaporates. As a result, the soluble materials, chiefly calcium carbonate, are removed from the uppermost layer and redeposited below, forming the caliche layer.

Laterite

In hot, wet tropical climates, soils called **laterites** may develop. Chemical weathering is intense under such climate conditions, so these soils are usually deeper than soils developing over a similar period in the mid-latitudes. Not only does leaching remove the soluble materials such as calcite, but the great quantities of percolating water also remove much of the silica, with the result that oxides of iron and aluminum become concentrated in the soil. The iron gives the soil a distinctive red color.

Since bacterial activity is great in the tropics, laterites contain practically no humus. This fact, coupled with the highly leached and bricklike nature of these soils, makes laterites poor for growing crops. Their infertility has been demonstrated repeatedly in tropical countries where cultivation has been attempted (see Box 3.2).

In sharp contrast, cold or dry climates generally have very thin and poorly developed soils. The reasons for this are fairly obvious. Chemical weathering progresses very slowly in such climates, and the scant plant life yields very little organic matter.

Table 3.2 Summary of soil types.

	Temperate humid (>63 cm rainfall)	Temperate dry (<63 cm rainfall)	Tropical (heavy rainfall)	Extreme arctic or desert
Climate	Temperate humid (>63 cm rainfall)	Temperate dry (<63 cm rainfall)	Tropical (heavy rainfall)	Extreme arctic or desert
Vegetation	Forest	Grass and brush	Grass and trees	Almost none, so no humus develops
Typical Area	Eastern U.S.	Western U.S.		
Soil Type	Pedalfer	Pedocal	Laterite	
Topsoil	Sandy, light colored; acid	Commonly enriched in calcite; whitish color	*[Zones not developed]* Enriched in iron (and aluminum); brick red color	No real soil forms because there is no organic material. Chemical weathering is very slow
Subsoil	Enriched in aluminum, iron and clay; brown color	Enriched in calcite; whitish color	All other elements removed by leaching	
Remarks	Extreme development in conifer forests, because abundant humus makes groundwater very acidic. Produces light gray soil because of removal of iron	*Caliche* is name applied to the accumulation of calcite	Apparently bacteria destroy humus, so no acid is available to remove iron	

Soil Erosion

Soils are just a tiny fraction of all Earth materials, yet they are a vital resource. Because soils are necessary for the growth of rooted plants, they are the very foundation of the human life-support system. Just as human ingenuity can increase the agricultural productivity of soils through fertilization and irrigation, soils can be damaged or destroyed by careless activities. Despite their basic role in providing food, fiber, and other basic materials, soils are among our most abused resources.

Perhaps this neglect and indifference has occurred because a substantial amount of soil seems to remain even where soil erosion is serious. Nevertheless, although the loss of fertile topsoil may not be obvious to the untrained eye, it is a growing problem as human activities expand and disturb more and more of Earth's surface.

How Soil Is Eroded

Soil erosion is a natural process; it is part of the constant recycling of Earth materials that we call the *rock cycle*. Once soil forms, erosional forces, especially water and wind, move soil components from one place to another. Every time it rains, raindrops strike the land with surprising force (Figure 3.16). Each drop acts like a tiny bomb, blasting movable soil particles out of their positions in the soil mass. Then, water flowing across the surface carries away the dislodged soil particles. Because the soil is moved by thin sheets of water, this process is termed *sheet erosion*.

After flowing as a thin, unconfined sheet for a relatively short distance, threads of current typically develop and tiny channels called *rills* begin to form. Still deeper cuts in the soil, known as *gullies*, are created as rills enlarge (Figure 3.17). When normal farm cultivation cannot eliminate the channels, we know the rills have grown large enough to be called gullies. Although most dislodged soil particles move only a short distance during each rainfall, substantial quantities eventually leave the fields and make their way downslope to a stream. Once in the stream channel, these soil particles, which can now be called *sediment*, are transported downstream and eventually deposited.

Rates of Erosion

We know that soil erosion is the ultimate fate of practically all soils. In the past, erosion occurred at slower rates than it does today because more of the land surface was covered and protected by trees, shrubs, grasses, and other plants. However, human activities such as farming, logging, and construction, which remove or disrupt the natural vegetation, have greatly accelerated the rate of soil erosion. Without the stabilizing effect of plants, the soil is more easily swept away by the wind or carried downslope by sheet wash.

Natural rates of soil erosion vary greatly from one place to another and depend on soil characteristics as well as such factors as climate, slope, and type of vegetation. Over a broad area, erosion caused by surface

Box 3.2

Laterites and the Clearing of the Rain Forest

Laterites are thick red soils that form in the wet tropics and subtropics and are the end product of extreme chemical weathering. Because lush tropical rain forests have lateritic soils, it might seem as though they would have great potential for agriculture. However, this is not the case. In fact, just the opposite is true; laterites are among the poorest soils for farming. Why is this the case?

Because laterites develop under conditions of high temperatures and heavy rainfall, they are severely leached. Leaching leads to low fertility because plant nutrients are removed by the large volume of downward-percolating water. Therefore, even though the vegetation may be dense and luxuriant, the soil itself contains few available nutrients. Most of the nutrients that support the rain forest are locked up in the trees themselves. The forest is maintained by the decay of plant material. That is, as vegetation dies and decomposes, the roots of the rain-forest trees quickly absorb the nutrients

Figure 3.C
This ancient temple at Angor Wat, Cambodia, was built of bricks made of laterite. (Photo by R. Ian Lloyd/The Stock Market)

before they are leached from the soil. The nutrients are continuously recycled as trees die and decompose.

Therefore, when forests are cleared to provide land for farming or to harvest the timber, most of the nutrients are removed as well (Figure 3.B). What remains is a soil that contains little to nourish the crop growth.

The clearing of rain forests not only removes the supply of plant nutrients, but also leads to accelerated erosion. When vegetation is present, its roots an-

chor the soil while its leaves and branches provide a canopy that protects the ground from the full force of the frequent heavy rains. The removal of the vegetation also exposes the ground to strong direct sunlight. When baked by the sun, laterites can harden to a bricklike consistency and become practically impenetrable to water and crop roots. In only a few years, lateritic soils in a freshly cleared area may no longer be cultivable.

The term *laterite* is derived from the Latin word *latere*, meaning "brick," and was first applied to the use of this material for brick making in India and Cambodia. Laborers simply excavated the soil, shaped it, and allowed it to harden in the sun. Ancient but still well-preserved structures built of laterite remain standing today in the wet tropics (Figure 3.C). Such structures have withstood centuries of weathering because all of the original soluble materials were already removed from the soil by chemical weathering. Laterites are therefore virtually insoluble and thus very stable.

In summary, we have seen that laterites are highly leached soils that are the products of extreme chemical weathering in the warm, wet tropics. Although they may be associated with lush tropical rain forests, these soils are unproductive when vegetation is removed. Moreover, when cleared of plants, laterites are subject to accelerated erosion and can be baked to brick-like hardness by the sun.

Figure 3.B
Clearing the Amazon rain forest in Surinam. The thick lateritic soil is highly leached. (Photo by Wesley Bocxe/Photo Researchers)

runoff may be estimated by determining the sediment loads of the streams that drain the region. When studies of this kind were made on a global scale they indicated that, prior to the appearance of humans, sediment transport by rivers to the oceans amounted to just over 9 billion metric tons per year. By contrast, the amount of

water except during periods of prolonged drought. When dry conditions prevail, strong winds can remove large quantities of soil from unprotected fields. Such was the case in the 1930s in the portions of the Great Plains that came to be called the Dust Bowl (see Box 5.4, p. 145).

In many regions the rate of soil erosion is significantly greater than the rate of soil formation. This means that a renewable resource has become nonrenewable in these places. At present, it is estimated that topsoil is eroding faster than it forms on more than one-third of the world's croplands. The result is lower productivity, poorer crop quality, reduced agricultural income, and an ominous future.

Sedimentation and Chemical Pollution

Another problem related to excessive soil erosion involves the deposition of sediment. Each year in the United States hundreds of millions of tons of eroded soil are deposited in lakes, reservoirs, and streams. The detrimental impact of this process can be significant. For example, as more and more sediment is deposited in a reservoir, the capacity of the reservoir is diminished, limiting its usefulness for flood control, water supply, and/or hydroelectric power generation. In addition, sedimentation in streams and other waterways can restrict navigation and lead to costly dredging operations.

In some cases soil particles are contaminated with pesticides used in farming. When these chemicals are introduced into a lake or reservoir, the quality of the water supply is threatened and aquatic organisms may be endangered. In addition to pesticides, nutrients found naturally in soils as well as those added by agricultural fertilizers make their way into streams and lakes, where they stimulate the growth of plants. Over a period of time, excessive nutrients accelerate the process by which plant growth leads to the depletion of oxygen and an early death of the lake.

The availability of good soils is critical if the world's rapidly growing population is to be fed. On every continent unnecessary soil loss is occurring because appropriate conservation measures are not being used. Although it is a recognized fact that soil erosion can never be completely eliminated, soil conservation programs can substantially reduce the loss of this basic resource. Windbreaks (rows of trees), and terracing, and plowing along the contours of hills, are some of the effective measures, as are special tillage practices and crop rotation.

Weathering Creates Ore Deposits

Weathering creates many important mineral deposits. It does so by concentrating minor amounts of metals that are scattered through unweathered rock into economically valuable concentrations. Such a transformation is

Figure 3.16
When it is raining, millions of water drops are falling at velocities approaching 10 meters per second (35 kilometers per hour). When water drops strike an exposed surface, soil particles may splash as high as 1 meter into the air and land more than a meter away from the point of raindrop impact. Soil dislodged by splash erosion is more easily moved by sheet erosion. (Photo courtesy of U.S. Department of Agriculture)

material currently transported to the sea by rivers is about 24 billion metric tons per year, or more than two and one-half times the earlier rate.

It is more difficult to measure the loss of soil due to wind erosion. However, the removal of soil by wind is generally much less significant than erosion by flowing

Figure 3.17
Gully erosion in poorly protected soil. (Photo by John C. Coulter/Visuals Unlimited)

often termed **secondary enrichment** and takes place in one of two ways. In one situation, chemical weathering coupled with downward-percolating water removes undesired materials from decomposing rock, leaving the desired elements enriched in the upper zones of the soil. The second way is basically the reverse of the first. That is, the desirable elements that are found in low concentrations near the surface are removed and carried to lower zones, where they are redeposited and become more concentrated.

Bauxite

The formation of *bauxite,* the principal ore of aluminum, is one important example of an ore created as a result of enrichment by weathering processes (Figure 3.18). Although aluminum is the third most abundant element in Earth's crust, economically valuable concentrations of this important metal are not common, because most aluminum is tied up in silicate minerals, from which it is extremely difficult to extract.

Bauxite forms in rainy tropical climates in association with laterites. (In fact, bauxite is sometimes called aluminum laterite.) When aluminum-rich source rocks are subjected to the intense and prolonged chemical weathering of the tropics, most of the common elements, including calcium, sodium, and silicon, are removed by leaching. Because aluminum is extremely insoluble, it becomes concentrated in the soil (as bauxite, a hydrated aluminum oxide). Thus, the formation of bauxite depends on climatic conditions in which chemical weathering and leaching are pronounced, plus of course the presence of aluminum-rich source rock. Important deposits of nickel and cobalt are also found in laterite soils that develop from igneous rocks rich in other silicate minerals.

Other Deposits

Many copper and silver deposits result when weathering processes concentrate metals that are dispersed through a low-grade primary ore. Usually such enrichment occurs in deposits containing pyrite (FeS_2), the most common and widespread sulfide mineral. Pyrite is important because when it chemically weathers, sulfuric acid forms, which enables percolating waters to dissolve the ore metals. Once dissolved, the metals gradually migrate downward through the primary ore body until they are precipitated. Deposition takes place because of changes that occur in the chemistry of the solution when it reaches the groundwater zone (the zone beneath the surface where all pore spaces are filled with water). In this manner, the small percentage of dispersed metal can be removed from a large volume of rock and redeposited as a higher-grade ore in a smaller volume of rock.

Mass Wasting

Earth's surface is never perfectly flat but instead consists of slopes. Some are steep and precipitous; others are moderate or gentle. Some are long and gradual; others are short and abrupt. Some slopes are mantled with soil and covered by vegetation; others consist of barren rock and rubble. Their form and variety are great. Taken together, slopes are the most common elements in our physical landscape. Although most slopes appear to be stable and unchanging, they are not static features because the force of gravity causes material to move downslope. At one extreme, the movement may be gradual and practically imperceptible. At the other extreme, it may consist of a thundering landslide or rockfall.

Rapid forms of mass wasting are significant geologic hazards (Figure 3.19). Each year landslides in the United States are responsible for 25-50 deaths and produce more than $1.5 billion in losses. Worldwide, landslides cause an average of 600 deaths each year.

Most mass wasting, whether spectacular or subtle, is the result of circumstances that are completely independent of human activities. Very few landslides occur where they cannot be anticipated. In places where mass wasting is a recognized threat, steps can often be taken to control downslope movements or limit the damages that such movements can cause. If the potential for mass wasting goes unrecognized or is ignored, the results can be costly and dangerous. We should also note that, although most downslope movements occur whether people are present or not, many occurrences are aggravated or even triggered by human actions.

Figure 3.18
Bauxite is the ore of aluminum and forms as a result of weathering processes under tropical conditions. Its color varies from red or brown to nearly white. (Photo by E. J. Tarbuck)

Figure 3.19
This home in Pacific Palisades, California, was destroyed by a landslide triggered by the January 1994 Northridge earthquake. In some cases, damages from earthquake-induced mass wasting are greater than damages caused directly by an earthquake's ground vibrations. (Photo by Chromo Sohm/The Stock Market)

Mass Wasting and Landform Development

Landslides are spectacular examples of a basic geologic process called *mass wasting*. Mass wasting refers to the downslope movement of rock, regolith, and soil under the direct influence of gravity. It is distinct from the erosional processes that are examined in subsequent chapters because mass wasting does not require a transporting medium.

In the evolution of most landforms, mass wasting is the step that follows weathering. By itself, weathering does not produce significant landforms. Rather, landforms develop as the products of weathering are removed from the places where they originate. Once weathering weakens and breaks rock apart, mass wasting transfers the debris downslope, where a stream, acting as a conveyor belt, usually carries it away (see Figure 3.1). Although there may be many intermediate stops along the way, the sediment is eventually transported to its ultimate destination, the sea.

The combined effects of mass wasting and running water produce stream valleys, which are the most common and conspicuous of Earth's landforms. If streams alone were responsible for creating the valleys in which they flow, the valleys would be very narrow features. However, the fact that most river valleys are much wider than they are deep is a strong indication of the significance of mass wasting processes in supplying material to streams. The walls of a canyon extend far from the river because of the transfer of weathered debris downslope to the river and its trib-utaries by mass wasting processes. In this manner, streams and mass wasting combine to modify and sculpture the surface. Of course, glaciers, groundwater, waves, and wind are also important agents in shaping landforms and developing landscapes.

Controls and Triggers of Mass Wasting

Gravity is the controlling force of mass wasting, but several factors play an important role in overcoming inertia and triggering downslope movements. Among these factors are saturation of material with water, oversteepening of slopes, removal of anchoring vegetation, and ground vibrations from earthquakes.

The Role of Water

When the pores in sediment become filled with water, the cohesion among particles is destroyed, allowing them to slide past one another with relative ease. For example, when sand is slightly moist, it sticks together quite well. However, if enough water is added to fill the openings between the grains, the sand will ooze out in all directions. Thus, saturation reduces the internal resistance of materials, which are then easily set in motion by the force of gravity. When clay is wetted, it becomes very slick—another example of the "lubricating" effect of water. Water also adds considerable weight to a mass of material. The added weight in itself may be enough to cause the material to slide or flow downslope.

Oversteepened Slopes

Oversteepening of slopes is another cause of many mass movements. Unconsolidated, granular sand-sized or coarser particles assume a stable slope called the **angle of repose,** the steepest angle at which material remains stable. Depending on the size and shape of the particles, the angle varies from 25 to 40 degrees. The larger, more angular particles maintain the steepest slopes. If the angle is increased, the rock debris will adjust by moving downslope.

Oversteepening is not just important because it triggers movements of unconsolidated granular materials. Oversteepening also produces unstable slopes and mass movements in cohesive soils, regolith, and bedrock. The response will not be immediate, as with loose, granular material, but sooner or later, one or more mass wasting processes will eliminate the oversteepening and restore stability to the slope.

There are many situations in nature where this takes place. A stream undercutting a valley wall and waves pounding against the base of a cliff are but two familiar examples. Furthermore, through their activities, people often create oversteepened and unstable slopes that become prime sites for mass wasting.

Vegetation

Plants protect against erosion and contribute to the stability of slopes because their root systems bind soil and regolith together. Where plants are lacking, mass wasting is enhanced, especially if slopes are steep and water is plentiful. When anchoring vegetation is removed by forest fires or by people (for timber, farming, or development), surface materials frequently move downslope.

An unusual but well-known example occurred several decades ago on steep slopes near Menton, France. Farmers replaced olive trees, which have deep roots, with a more profitable, but shallow-rooted crop, carnations. When the less stable slope failed, the landslide took 11 lives.

Earthquakes as Triggers

Conditions favoring mass wasting may exist in an area for a long time without movement occurring. An additional factor is sometimes necessary to trigger the movement. Among the more important and dramatic triggers are earthquakes. An earthquake and its aftershocks can dislodge enormous volumes of rock and unconsolidated material. The mass wasting events shown in Figures 3.19 and 3.22 were both triggered by earthquakes. In

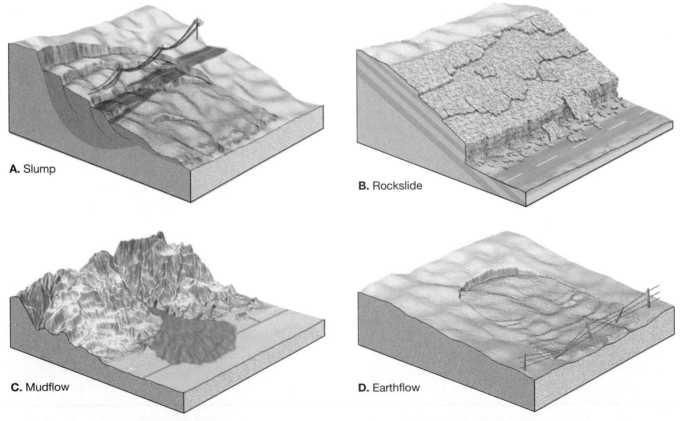

A. Slump

B. Rockslide

C. Mudflow

D. Earthflow

Figure 3.20
The four processes illustrated here are all considered to be relatively rapid forms of mass wasting. Because material in slumps **A.** and rockslides **B.** move along well-defined surfaces, they are said to move by sliding. By contrast, when material moves downslope as a viscous fluid, the movement is described as a flow. Mudflow **C.** and earthflow **D.** advance downslope in this manner.

many areas that are jolted by earthquakes, it is not ground vibrations directly, but landslides and ground subsidence triggered by the vibrations that cause the greatest damage.

Falls, Slides, and Flows

Geologists include several processes under the name of mass wasting. Four are illustrated in Figure 3.20. Generally, each process is defined by the type of material involved, the kind of motion, and the velocity of the movement.

If soil and regolith dominate, terms such as *debris, mud,* or *earth* are used. On the other hand, when a mass of bedrock breaks loose and moves downslope, the term *rock* may be part of the description. Generally, the kind of motion is described as either a fall, a slide, or a flow.

When the movement involves the free-fall of detached individual pieces of any size, it is termed a **fall.** Falls are common on slopes that are too steep for loose material to remain on the surface (Figure 3.21). Many falls result when freeze and thaw cycles or the action of plant roots loosen rock to the point that gravity takes over. Rockfall is the primary way in which talus slopes are built and maintained (see Figure 3.3). Sometimes falls may trigger other forms of downslope movement.

Many mass wasting processes are **slides,** which occur whenever material remains fairly coherent and moves along a well-defined surface that is approximately parallel to the slope. Sometimes the surface is a joint, a fault, or a bedding plane. Sometimes the word *slide* is used as a synonym for the word *landslide.* It should be pointed out that, although many people, including geologists, use the term, the world *landslide* has no specific definition in geology. Rather, it should be considered as a popular nontechnical term to describe all relatively rapid forms of mass wasting, including those in which sliding does not occur.

A third type of movement common to mass wasting is termed **flow.** Flow occurs when material moves downslope as a viscous fluid. Most flows are saturated with water and typically move as lobes or tongues.

When mass wasting events make the news, a large quantity of material has in all likelihood moved rapidly downslope and has had a disastrous effect upon people and property. Indeed, during events called *rock avalanches,* rock and debris can hurtle downslope at speeds exceeding 200 kilometers (125 miles) per hour. Many researchers believe that rock avalanches, such as the one that produced the scene in Figure 3.22, must literally "float on air" as they move downslope. That is, high velocities result when air becomes trapped and compressed beneath the falling mass of debris, allowing it to move as a buoyant, flexible sheet across the surface.

Most mass movements, however, do not move with the speed of a rock avalanche. In fact, a great deal of mass wasting is imperceptibly slow. One process that we will examine later, termed *creep,* results in particle movements that are usually measured in millimeters or centimeters per year. Thus, as you can see, rates of movement can be spectacularly sudden or exceptionally gradual. Even the velocity of a single process at a particular site can vary considerably from one time to another.

Slump

Slump refers to the downward slipping of a mass of rock or unconsolidated material moving as a unit along a *curved* surface (see Figure 3.20A). Usually the slumped material does not travel spectacularly fast nor very far. This is a common form of mass wasting, especially in thick accumulations of cohesive materials such as clay. As the movement occurs, a crescent-shaped scarp (cliff) is created at the head and the block's upper surface is sometimes tilted backward.

Slump commonly occurs because a slope has been oversteepened. The material on the upper portion of a slope is held in place by the material at the bottom of the slope. As this anchoring material at the base is removed, the material above is made unstable and reacts to the pull of gravity. A common example is a valley wall that becomes oversteepened by a meandering river. Another is a coastal area where a cliffed seashore is undercut by wave activity at its base. Slumping may also occur when a slope is overloaded, causing internal stress on the material below. This type of slump often occurs where weak, clay-rich material underlies layers of stronger, more resistant rock such as sandstone. The seepage of water through the upper layers reduces the strength of the clay below, and slope failure results.

Figure 3.21
Rockfall blocking a highway near Bozeman, Montana. (Photo by Phil Farnes/Photo Researcher)

Figure 3.22
Debris deposited atop Sherman Glacier in Alaska by a rock avalanche. The event was triggered by a tremendous earthquake in March 1964. (Photo by Austin Post, U.S. Geological Survey)

Rockslide

Rockslides occur when blocks of bedrock break loose and slide down a slope (see Figure 3.20B). If the material involved is largely unconsolidated, the term **debris slide** is used instead. Such events are among the fastest and most destructive mass movements. Usually rockslides take place in a geologic setting where the rock strata are inclined, or where joints and fractures exist parallel to the slope. When such a rock unit is undercut at the base of the slope, it loses support and the rock eventually gives way. The Gros Ventre slide described in Box 3.3 is a classic example. Sometimes the rockslide is triggered when rain or melting snow lubricates the underlying surface to the point that friction is no longer sufficient to hold the rock unit in place. As a result, rockslides tend to be more common during the spring, when heavy rains and melting snow are most prevalent.

Earthquakes may trigger rockslides and other mass movements. The 1811 earthquake at New Madrid, Missouri, for example, caused slides in an area of more than 13,000 square kilometers (5000 square miles) along the Mississippi River valley. A more recent example occurred on August 17, 1959, when a severe earthquake west of Yellowstone National Park triggered a massive slide in the canyon of the Madison River in southwestern Mon-

tana. In a matter of moments an estimated 27 million cubic meters of rock, soil, and trees slid into the canyon. The debris dammed the river and buried a campground and highway. More than 20 unsuspecting campers perished.

Mudflow

Mudflow is a relatively rapid type of mass wasting that involves a flowage of debris containing a large amount of water (Figure 3.20C). Mudflows are most characteristic of semiarid mountainous regions and are also common on the slopes of some volcanoes. Because of their fluid properties, mudflows follow canyons and stream channels.

Mudflows in Semiarid Regions

Although rains in semiarid regions are infrequent, they are typically heavy when they occur. When a cloudburst or rapidly melting mountain snows create a sudden flood, large quantities of soil and regolith are washed into nearby stream channels because there is usually little or no vegetation to anchor the surface material. The end product is a flowing tongue of well-mixed mud, soil, rock, and water. Its consistency may range from that of wet concrete to a soupy mixture not much thicker than muddy water. The rate of flow therefore depends not

Box 3.3

The Gros Ventre Slide: A Classic Case Study

The Gros Ventre River flows west from the northernmost part of the Wind River Range in northwestern Wyoming, through Grand Teton National Park, and eventually empties into the Snake River. On June 23, 1925, a classic rockslide took place in its valley, just east of the small town of Kelly. In the span of just a few minutes a great mass of sandstone, shale, and soil crashed down the south side of the valley, carrying with it a dense pine forest. The volume of debris, estimated at 38 million cubic meters (50 million cubic yards), created a 70-meter-high dam on the Gros Ventre River (Figure 3.D). Because the river was completely blocked, a lake was created. It filled so quickly that a house

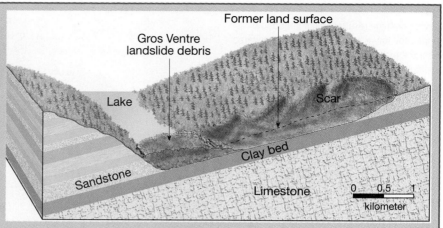

Figure 3.E

Cross-sectional view of the Gros Ventre rockslide. The slide occurred when the tilted and undercut sandstone bed could no longer maintain its position atop the saturated bed of clay. (After W. C. Alden, "Landslide and Flood at Gros Ventre, Wyoming," Transactions (AIME) 76 (1928): 348)

that had been 18 meters (60 feet) above the river was floated off its foundation 18 hours after the slide. In 1927, the lake

Figure 3.D

The scar, 2.4 kilometers (1.5 miles) long and 0.8 kilometer (0.5 mile) wide, left on the side of Sheep Mountain by the Gros Ventre rockslide. (Photo by Stephen Trimble)

overflowed the dam, partially draining the lake and resulting in a devastating flood downstream.

Why did the Gros Ventre rockslide take place? Figure 3.E is a diagrammatic cross-sectional view of the geology of the valley. Notice the following points: (1) the sedimentary strata in this area dip (tilt) 15–21 degrees; (2) underlying the bed of sandstone is a relatively thin layer of clay; and (3) at the bottom of the valley the river had cut through much of the sandstone layer. During the spring of 1925, water from heavy rains and melting snow seeped through the sandstone, saturating the clay below. Because much of the sandstone layer had been cut through by the Gros Ventre River, the layer had virtually no support at the bottom of the slope. Eventually the sandstone could no longer hold its position on the wetted clay, and gravity pulled the mass down the side of the valley. The circumstances at this location were such that the event was inevitable.

only on the slope but on the water content as well. When dense, mudflows are capable of carrying or pushing large boulders, trees, and even houses with relative ease.

Mudflows pose a serious hazard to development in dry mountainous areas such as southern California. Here construction of homes on canyon hillsides and the removal of native vegetation by brush fires and other means have increased the frequency of these destructive events (Figure 3.23).

Lahars

Mudflows are also common on the slopes of some volcanoes, in which case they are termed **lahars.** The word originated in Indonesia, a volcanic region that has experienced many of these often destructive events. Lahars result when highly unstable layers of ash and debris become saturated with water and flow down steep volcanic slopes, generally following existing stream channels.

Figure 3.23

Severe damage resulted when a mudflow buried the lower portion of this house located near the mouth of a canyon in southern California. (Photo by James E. Patterson)

Some are initiated when heavy rainfalls erode volcanic deposits. Others are triggered when large volumes of ice and snow are suddenly melted by heat flowing to the surface from within the volcano or by the hot gases and near-molten debris emitted during a violent eruption.

In November 1985, lahars were produced when Nevado del Ruiz, a 5300-meter (17,400-foot) volcano in the Andes Mountains of Colombia erupted. The eruption melted much of the snow and ice that capped the uppermost 600 meters of the peak, producing torrents of hot viscous mud, ash, and debris. The lahars moved outward from the volcano, following the valleys of three rain-swollen rivers that radiate from the peak. The flow that moved down the valley of the Lagunilla River was the most destructive, devastating the town of Armero, 48 kilometers from the mountain. Most of the more than 25,000 deaths caused by the event occurred in this once-thriving agricultural community. Death and property damage also occurred in 13 other villages within the 180-square-kilometer disaster area. Although a great deal of volcanic debris was explosively ejected from Nevado del Ruiz, it

was the lahars triggered by this eruption that made this such a devastating natural disaster. In fact, it was the worst volcanic disaster since 28,000 people died following the 1902 eruption of Mount Pelée on the Caribbean island of Martinique. This event is described in Chapter 8.

Earthflow

Unlike mudflows, which are usually confined to channels in semiarid regions, **earthflows** most often form on hillsides in humid areas during times of heavy precipitation or snowmelt (see Figure 3.20D). When water saturates the soil and regolith on a hillside, the material may break away, leaving a scar on the slope and forming a tongue- or teardrop-shaped mass that flows downslope (Figure 3.24). The materials most commonly involved are rich in clay and silt and contain only small proportions of sand and coarser particles. Earthflows range in size from bodies a few meters long, a few meters wide, and less than 1 meter deep to masses more than 1 kilometer long, several hundred meters wide, and more than 10 meters deep.

Figure 3.24

This small, tongue-shaped earthflow occurred on a newly formed slope along a recently constructed highway. It formed in clay-rich material following a period of heavy rain. Notice the small slump at the head of the earthflow. (Photo by E. J. Tarbuck)

Because earthflows are quite viscous, they generally move much slower than the more fluid mudflows. They move slowly and persistently for periods ranging from days to years. Depending on slope steepness and the material's consistency, velocities range from less than 1 millimeter per day up to several meters per day. Movement is typically faster during wet periods. In addition to occurring as isolated hillside phenomena, earthflows commonly take place in association with large slumps. In this situation, they may be seen as tonguelike flows at the base of the slump.

A special type of earthflow, known as *liquefaction*, sometimes occurs in association with earthquakes. Porous, clay- to sand-sized sediments that are saturated with water are most vulnerable. When shaken suddenly, the grains lose cohesion and the ground flows. Liquefaction can cause buildings to sink or tip on their sides and underground storage tanks and sewer lines to float upward. To say the least, damage can be substantial. You will learn more about this in Chapter 6.

Slow Movements

Movements such as rockslides, rock avalanches, and lahars are certainly the most spectacular and catastrophic forms of mass wasting. These dangerous events deserve intensive study to enable more effective prediction, timely warnings and better controls to save lives. However, their spectacular nature gives us a false impression of their importance as a mass wasting process. Indeed, these sudden movements transport less material than the slow, subtle action of creep. Whereas rapid types of mass wasting are characteristic of mountains and steep hillsides, creep can take place on gentle slopes and is thus much more widespread.

Creep

Creep is a type of mass wasting that involves the gradual downhill movement of soil and regolith. A primary cause is the alternate expansion and contraction of surface material caused by freezing and thawing or wetting and drying. As shown in Figure 3.25, freezing or wetting lifts particles at right angles to the slope, and thawing or

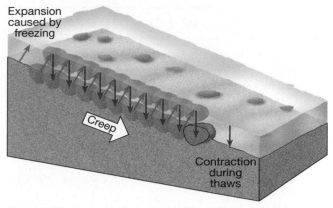

Figure 3.25
The repeated expansion and contraction of the surface material causes a net downslope migration of rock particles—a process called creep.

drying allows the particles to fall back to a slightly lower level. Each cycle therefore moves the material a short distance downhill. Creep may also be initiated if the ground becomes saturated with water. Following a heavy rain or snowmelt, a waterlogged soil may lose its internal cohesion, allowing gravity to pull the material downslope. Because creep is imperceptibly slow, the process cannot be observed in action. What can be observed, however, are the effects of creep. Creep causes fences and utility poles to tilt and retaining walls to be displaced.

Solifluction

Solifluction is a common form of mass wasting in regions underlain by **permafrost**. Permafrost refers to the permanently frozen ground that occurs in Earth's harsh tundra and ice cap climates. Solifluction is a form of creep in which unconsolidated, water-saturated material gradually moves downslope. It occurs in a zone above the permafrost called the *active layer*, which thaws in summer and refreezes in winter. During summer, water is unable to percolate into the impervious permafrost layer below. As a result, the active layer becomes saturated and slowly flows. The process can occur on slopes as gentle as 2–3 degrees. Where there is a well-developed mat of vegetation, a solifluction sheet may move in a series of well-defined lobes or overriding folds (Figure 3.26).

Figure 3.26
Solifluction lobes northeast of Fairbanks, Alaska. (Photo by James E. Patterson)

Review Questions

1. Describe the role of external processes in the rock cycle.
2. If two identical rocks were weathered, one mechanically and the other chemically, how would the products of weathering for the two rocks differ?
3. How does mechanical weathering add to the effectiveness of chemical weathering?
4. Describe the formation of an exfoliation dome. Give an example of such a feature.
5. Granite and basalt are exposed at the surface in a hot, wet region.
   (a) Which type of weathering will predominate?
   (b) Which of the rocks will weather most rapidly? Why?
6. Heat speeds up a chemical reaction. Why then does chemical weathering proceed slowly in a hot desert?
7. How is carbonic acid (H_2CO_3) formed in nature? What results when this acid reacts with potassium feldspar?
8. List some possible environmental effects of acid precipitation (see Box 3.1).
9. Relate soil to the Earth system.
10. Using the soil texture diagram (Figure 3.12), name the soil that consists of 60 percent sand, 30 percent silt, and 10 percent clay.
11. What factors might cause different soils to develop from the same parent material, or similar soils to form from different parent materials?
12. Which of the controls of soil formation is most important? Explain.
13. How can slope affect the development of soil? What is meant by the term *slope orientation*?
14. List the characteristics associated with each of the horizons in a well-developed soil profile. Which of the horizons constitute the solum? Under what circumstances do soils lack horizons?
15. Distinguish between pedalfers and pedocals.
16. Soils formed in the humid tropics and the Arctic both contain little organic matter. Do both lack humus for the same reasons?
17. What soil type is associated with tropical rainforests? Since this soil supports the growth of lush natural vegetation, is it also excellent for growing crops? Briefly explain. (See Box 3.2.)
18. Before soil particles are carried by water moving in tiny channels called rills, they are moved by an unconfined movement of water across the surface termed _____.
19. List three detrimental effects of soil erosion other than the loss of topsoil from croplands.
20. Name the primary ore of aluminum and describe its formation.
21. What role does mass wasting play in sculpting Earth's landscape?
22. What is the controlling force of mass wasting? What other factors are important?
23. Distinguish among fall, slide, and flow.
24. Why can rock avalanches move at such great speeds?
25. What factors led to the massive rockslide at Gros Ventre, Wyoming (Figure 3.E)?
26. What type of mass wasting event killed thousands of people in the vicinity of the Colombian volcano Nevado del Ruiz in 1985?
27. Compare and contrast mudflow and earthflow.
28. Since creep is an imperceptibly slow process, what evidence might indicate to you that this phenomenon is affecting a slope? Describe the mechanism that creates this slow movement (Figure 3.25).
29. Why is solifluction only a summertime phenomenon?

Key Terms

angle of repose (p. 78)
chemical weathering (p. 61)
creep (p. 83)
debris slide (p. 80)
differential weathering (p. 64)
earthflow (p. 82)
eluviation (p. 71)
erosion (p. 58)
exfoliation dome (p. 60)
fall (p. 79)
flow (p. 79)

frost wedging (p. 60)
horizon (p. 71)
humus (p. 66)
lahar (p. 81)
laterite (p. 72)
leaching (p. 71)
mass wasting (p. 58)
mechanical weathering (p. 58)
mudflow (p. 80)
parent material (p. 69)
pedalfer (p. 72)
pedocal (p. 72)
permafrost (p. 83)
regolith (p. 65)

rockslide (p. 80)
secondary enrichment (p. 76)
sheeting (p. 60)
slide (p. 79)
slump (p. 79)
soil (p. 65)
soil profile (p. 71)
soil texture (p. 68)
solifluction (p. 83)
solum (p. 71)
spheroidal weathering (p. 63)
talus slope (p. 60)
weathering (p. 58)

CHAPTER 4

Running Water and Groundwater

Warren Creek Falls, Columbia River Gorge, Oregon. (Photo by Craig Tuttle/The Stock Market)

All the rivers run into the sea; yet the sea is not full; unto the place from whence the rivers come, thither they return again.

(Ecclesiastes 1:7)

As the perceptive writer of Ecclesiastes indicated, water is continually on the move, from the ocean to the land and back again in an endless cycle. This chapter deals with that part of the water cycle that returns water to the sea. Some travels quickly via a rushing stream, and some moves more slowly below the surface. We shall examine the factors that influence the distribution and movement of water, as well as look at how water sculptures the landscape. To a great extent, the Grand Canyon, Niagara Falls, Old Faithful, and Mammoth Cave all owe their existence to the action of water on its way to the sea.

The amount of water on Earth is immense, an estimated 1.36 billion cubic kilometers (326 million cubic miles). Of this total, the vast bulk—97.2 percent—is part of the world ocean. Ice sheets and glaciers account for another 2.15 percent, leaving only 0.65 percent to be divided among lakes, streams, subsurface water, and the atmosphere (Figure 4.1). Although the percentages of Earth's total water found in each of the latter sources is but a small fraction of the total inventory, the absolute quantities are great.

Earth as a System: The Hydrologic Cycle

The water found in each of the reservoirs depicted in Figure 4.1 does not remain in these places indefinitely. Water can readily change from one state of matter (solid, liquid, or gas) to another at the temperatures and pressures that occur at Earth's surface. Therefore, water is constantly moving between the hydrosphere, the atmosphere, the solid Earth, and the biosphere. This unending circulation of Earth's water supply is called the **hydrologic cycle**. The cycle shows us many critical interrelationships among different parts of the Earth system.

The hydrologic cycle is a gigantic worldwide system powered by energy from the sun in which the atmosphere provides the vital link between the oceans and continents (Figure 4.2). Water evaporates into the atmosphere from the ocean and to a much lesser extent from the continents. Winds transport this moisture-laden air, often great distances, until conditions cause the moisture to condense into clouds, and precipitation to fall. The precipitation that falls into the ocean has completed its cycle and is ready to begin another. The water that falls on the continents, however, must make its way back to the ocean.

What happens to precipitation once it has fallen on land? A portion of the water soaks into the ground (called **infiltration**), slowly moving downward, then laterally, finally seeping into lakes, streams, or directly into the ocean. When the rate of rainfall exceeds Earth's ability to absorb it, the surplus water flows over the surface into lakes and streams, a process called **runoff**. Much of the water which infiltrates or runs off eventually returns to the atmosphere because of evaporation from the soil, lakes, and streams. Also, some of the water that infiltrates the ground surface is absorbed by plants, which then release it into the atmosphere. This process is called **transpiration**. Each year a field of crops may transpire the equivalent of a water layer 60 centimeters (2 feet) deep over the entire field. The same area of trees may pump twice this amount into the atmosphere.

When precipitation falls in very cold areas—at high elevations or high latitudes—the water may not immediately soak in, run off, or evaporate. Instead, it may become part of a snowfield or a glacier. In this way, glaciers store large quantities of water on land. If present-day glaciers were to melt and release all their water, sea level would rise by several tens of meters. This would submerge many heavily populated coastal areas. As we shall see in Chapter 5, over the past two million years, huge ice sheets have formed and melted on several occasions, each time changing the balance of the hydrologic cycle.

Figure 4.2 also shows Earth's overall *water balance*, or the volume of water that passes through each part of the cycle annually. The amount of water vapor in the air at any one time is just a tiny fraction of Earth's total water supply. But the *absolute* quantities that are cycled through the atmosphere over a 1-year period are immense—some 380,000 cubic kilometers—enough to cover Earth's entire surface to a depth of about 1 meter (39 inches).

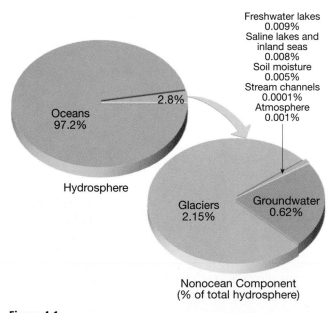

Figure 4.1
Distribution of Earth's water.

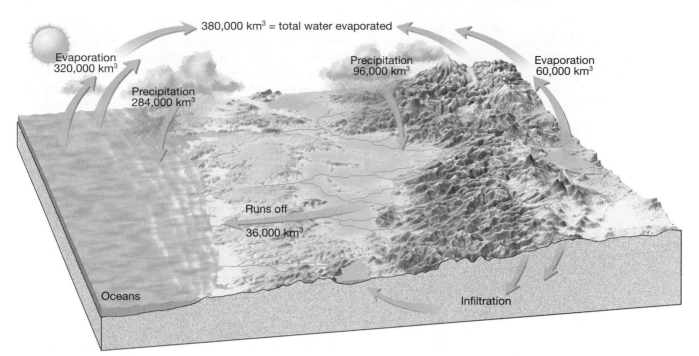

380,000 km³ = total water evaporated

Evaporation
320,000 km³

Precipitation
284,000 km³

Precipitation
96,000 km³

Evaporation
60,000 km³

Runs off

36,000 km³

Oceans

Infiltration

Figure 4.2
Earth's water balance. About 320,000 cubic kilometers of water are evaporated each year from the oceans, while evaporation from the land (including lakes and streams) contributes 60,000 cubic kilometers of water. Of this total of 380,000 cubic kilometers of water, about 284,000 cubic kilometers fall back to the ocean, and the remaining 96,000 cubic kilometers fall on Earth's land surface. Since 60,000 cubic kilometers of water evaporate from the land, 36,000 cubic kilometers of water remain to erode the land during the journey back to the oceans.

It is important to know that the water cycle is *balanced*. Because the total amount of water vapor in the atmosphere remains about the same, the average annual precipitation worldwide must be equal to the quantity of water evaporated. However, for all of the continents taken together, precipitation exceeds evaporation. Conversely, over the oceans, evaporation exceeds precipitation. Since the level of the world ocean is not dropping, the system must be in balance. In Figure 4.2, the 36,000 cubic kilometers of water that annually runs off from the land to the ocean causes enormous erosion. In fact, this immense volume of moving water is *the single most important agent sculpturing Earth's land surface*.

To summarize, the hydrologic cycle is the continuous movement of water from the oceans to the atmosphere, from the atmosphere to the land, and from the land back to the sea. The land-back-to-the-sea step is the primary action that wears down Earth's land surface. In this chapter, we will first observe the work of water running over the surface, including floods, erosion, and the formation of valleys. Then we will look underground at the slow labors of groundwater as it forms springs and caverns, and provides drinking water on its long migration to the sea.

Running Water

Running water is of great importance to people. We depend on rivers for energy, transportation, and irrigation. Their fertile floodplains have been favored sites

for agriculture and industry since the dawn of civilization. As the dominant agent of erosion, running water has shaped much of our physical environment.

Although people have always depended on running water, its source eluded them for centuries. It was not until the sixteenth century that they first realized streams were supplied by runoff and underground water, which ultimately had their sources as rain and snow.

Streamflow

Water makes its way to the sea under the influence of gravity. The time required for the journey depends on the velocity of the stream. Velocity is the distance that water travels in a unit of time. Water in some sluggish streams travels at less than 0.8 kilometer (0.5 mile) per hour, whereas water in a few rapid streams reaches speeds as high as 32 kilometers (20 miles) per hour. Velocities are measured at gauging stations. Along straight stretches, the highest velocities are near the center of the channel just below the surface, where friction is lowest. But when a stream curves, its zone of maximum speed shifts toward its outer bank (Figure 4.3).

The ability of a stream to erode and transport materials depends on its velocity. Even slight variations in velocity can lead to significant changes in how much sediment can be transported by the water. Several factors determine the velocity of a stream, including (1) gradient; (2) shape, size, and roughness of the channel; and (3) discharge.

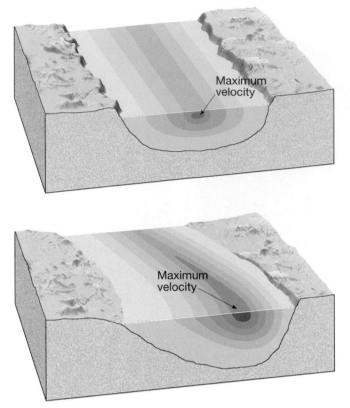

Figure 4.3
Along straight stretches, stream velocity is highest at the center of the channel. When a stream curves, its zone of maximum speed shifts toward the outer bank.

Gradient is the slope of a stream channel expressed as the vertical drop of a stream over a specified distance. For example, portions of the lower Mississippi River have very low gradients of 10 centimeters per kilometer or less. By contrast, some mountain stream channels decrease in elevation at a rate of more than 40 meters

per kilometer, or a gradient 400 times steeper than the lower Mississippi (Figure 4.4). Gradient varies not only between streams, but over a stream's length. The steeper the gradient, the more energy available for streamflow. If two streams were identical in every respect except gradient, the stream with the higher gradient would obviously have the greater velocity.

A stream's channel is a conduit that guides the flow of water, but the water encounters friction as it flows. The shape, size, and roughness of the channel affect the amount of friction. Larger channels have more efficient flow because a smaller proportion of water is in contact with the channel. A smooth channel promotes a more uniform flow, whereas an irregular channel filled with boulders creates enough turbulence to significantly slow the stream.

The **discharge** of a stream is the volume of water flowing past a certain point in a given unit of time. This is usually measured in cubic meters per second or cubic feet per second. Table 4.1 lists the world's largest rivers in terms of discharge. The largest river in North America, the Mississippi, discharges an average of 17,300 cubic meters (611,000 cubic feet) per second. Although this is a huge quantity of water, it is nevertheless dwarfed by the mighty Amazon in South America, the world's largest river. Draining a rainy region that is nearly three-fourths the size of the conterminous United States, the Amazon discharges 12 times more water than the Mississippi.

The discharge of an individual stream is far from constant. This is true because of such variables as rainfall and snowmelt. If discharge changes, the factors noted earlier adjust. Indeed, measurements show that when the amount of water in a stream increases, the width, depth, and velocity all increase. To handle the additional water, the stream will increase the size of its channel by widening and deepening it. The larger the channel, the less friction, and the more swiftly the water will flow (see Box 4.1).

Figure 4.4
Rapids can form where a portion of a stream channel has a steep gradient. (Photo by Bruce Gaylord/Visuals Unlimited)

Table 4.1 World's largest rivers ranked by discharge.

Rank	River	Country	Drainage Area		Average Discharge	
			Square kilometers	Square miles	Cubic meters per second	Cubic feet per second
1	Amazon	Brazil	5,778,000	2,231,000	212,400	7,500,000
2	Congo	Zaire	4,014,500	1,550,000	39,650	1,400,000
3	Yangtze	China	1,942,500	750,000	21,800	770,000
4	Brahmaputra	Bangladesh	935,000	361,000	19,800	700,000
5	Ganges	India	1,059,300	409,000	18,700	660,000
6	Yenisei	Russia	2,590,000	1,000,000	17,400	614,000
7	Mississippi	United States	3,222,000	1,244,000	17,300	611,000
8	Orinoco	Venezuela	880,600	340,000	17,000	600,000
9	Lena	Russia	2,424,000	936,000	15,500	547,000
10	Parana	Argentina	2,305,000	890,000	14,900	526,000

Box 4.1

The Effect of Urbanization on Discharge

When rains occur, stream discharge increases. If the rains are sufficiently heavy, the ability of the channel to contain the discharge is exceeded, and water spills over the banks as a flood. Floods are natural events that should be expected. However, when cities are built, the magnitude and frequency of flooding increases. The top portion of Figure 4.A is a hypothetical hydrograph that shows the time relationship between a rainstorm and the occurrence of flooding. Notice that the water level in the stream does not rise at the onset of precipitation because time is needed for water to move from the place where it fell to the stream. This time difference is called the *lag time*.

When an area changes from being predominantly rural to largely urban, streamflow is affected. The effect of urbanization on streamflow is illustrated by the bottom hydrograph in Figure 4.A. Notice that after urbanization the peak discharge during a flood is greater, and that the lag time between precipitation and flood peak is shorter than before urbanization. The explanation for this effect is relatively simple. The construction of streets, parking lots, and buildings covers over the ground that once soaked up water. Thus, less water infiltrates the ground, and the rate and amount of runoff increase. Further, since much less water soaks into the ground, the low-water (dry-season) flow in urban streams, which is maintained by the seepage of groundwater into the channel, is greatly reduced. As one might expect, the magnitude of these effects is a function of the percentage of land that is covered by impermeable surfaces.

Urbanization is just one example of human interference with streams. There are many other ways that land use inadvertently influences the flow of streams and the work they carry out. Moreover, there are also many ways by which people intentionally attempt to manipulate and control streams. Some of these are discussed at appropriate points in this chapter.

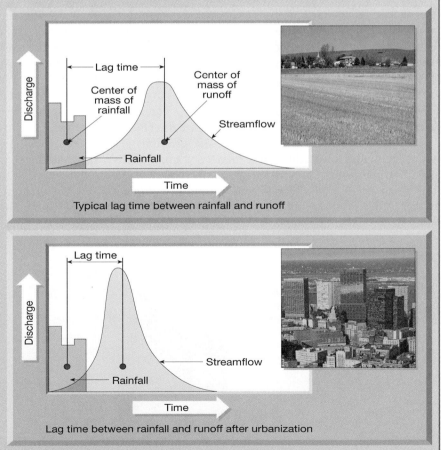

Typical lag time between rainfall and runoff

Lag time between rainfall and runoff after urbanization

Figure 4.A

When an area changes from rural to urban, the lag time between rainfall and flood peak is shortened. The flood peak is also higher following urbanization. (After L. B. Leopold, U.S. Geological Survey)

Changes from Upstream to Downstream

One useful way of studying a stream is to examine its *profile*. A profile is simply a cross-sectional view of a stream from its source area (called the *head* or *headwaters*) to its *mouth*, the point downstream where the river ends by emptying into another water body. By examining Figure 4.5, you can see that the most obvious feature of a typical profile is a constantly decreasing gradient from the head to the mouth. Although many local irregularities may exist, the overall profile is a smooth curve.

The profile shows that the gradient decreases downstream. To see how other factors change in a downstream direction, observations and measurements must be made. When data are collected from several gauging stations along a river, they show that discharge increases toward the mouth. This should come as no surprise because, as we move downstream, more and more tributaries contribute water to the main channel (Figure 4.5). Furthermore, in most humid regions, rainwater that soaks into the ground continually migrates into streams. Thus, as you move downstream, the stream's width, depth, and velocity change in response to the increased volume of water carried by the stream.

The observed increase in velocity that occurs downstream contradicts our impressions about wild, rushing mountain streams and wide, placid rivers. The mental picture that we may have of "old man river just rollin' along" is just not so. Although a mountain stream may have the appearance of a raging torrent, its average velocity is often less than that of the river near its mouth. The difference is primarily attributable to the greater efficiency of the larger channel in a downstream direction.

In the headwaters region where the gradient is steep, the water must flow in a relatively small and often boulder-strewn channel. The small channel and rough bed create great friction and inhibit movement by scattering water in all directions, with almost as much backward motion as forward motion. However, downstream, the material on the bed of the stream becomes much smaller, offering less resistance to flow, and the width and depth of the channel increase to accommodate the greater discharge. These factors, especially the wider and deeper channel, permit the water to flow more freely and hence more rapidly.

In summary, you have seen an inverse relationship between gradient and discharge. Where gradient is steep, discharge is small, and where discharge is great, gradient is small. Stated another way, a stream can maintain a higher velocity near its mouth even though it has a lower gradient than upstream because of the greater discharge, larger channel, and smoother bed.

Base Level

Streams cannot endlessly erode their channels deeper and deeper. There is a lower limit to how deep a stream can erode, and that limit is the stream's base level. **Base level** is the lowest point to which a stream can erode its channel. Two general types of base level exist. Sea level is considered the *ultimate base level*, because it is the lowest level to which stream erosion could lower the land. *Temporary*, or *local, base levels* include lakes, resistant layers of rock, and main streams which act as base level for their tributaries. For example, when a stream enters a lake, its velocity quickly approaches zero and its ability to erode decreases. Thus, the lake prevents the stream from eroding below its level at any point upstream from the lake. However, because the outlet of the lake can cut downward and drain the lake, the lake is only a temporary hindrance to the stream's ability to downcut its channel. In a similar manner, the layer of resistant rock at the lip of the waterfall in Figure 4.6 acts as a temporary base level. Until the ledge of hard rock is eliminated, it will limit the amount of downcutting upstream.

Any change in base level will cause a corresponding readjustment of stream activities. When a dam is built along a stream, the reservoir that forms behind it

Figure 4.5

A longitudinal profile is a cross-section along the length of a stream. Note the concave-upward curve of the profile, with a steeper gradient upstream and a gentler gradient downstream.

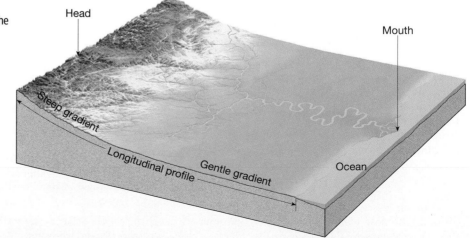

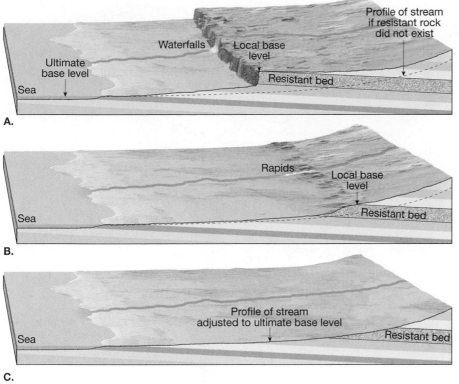

raises the base level of the stream (Figure 4.7). Upstream from the dam the stream gradient is reduced, lowering its velocity and, hence, its sediment-transporting ability. The stream, now having too little energy to transport all of its load, will deposit sediment. This builds up its channel. This process continues until the stream again has a gradient sufficient to carry its load.

If, on the other hand, the base level were lowered, either by an uplift of the land or by a drop in the base level, the stream would readjust. The stream, now above base level, would have excess energy and would downcut its channel to establish a balance with its new base level. Erosion would progress most rapidly near the mouth, then work upstream until the stream profile was adjusted along its full length.

Work of Streams

The work of streams includes erosion, transportation, and deposition. These activities go on simultaneously in all stream channels, even though they are presented individually here.

Erosion

Erosion is the removal of rock and soil. Much of the material carried by streams reaches them through underground water, overland flow, and mass wasting. Streams also create some of their load by eroding their own channels. If a channel is composed of bedrock, most of the erosion is accomplished by the abrasive action of water armed with sediment, a process analogous to sandblasting. Pebbles caught in swirling eddies act like cutting tools and bore circular "potholes" into the channel floor. In channels composed of loose material, considerable lifting and removal can be accomplished by the impact of water alone.

Transportation

Streams transport their load of sediment in three ways: (1) in solution (**dissolved load**); (2) in suspension (**suspended load**); and (3) scooting or rolling along the bottom (**bed load**).

Most of the *dissolved load* is brought to the stream by groundwater and to a lesser degree it is acquired by dissolving rock along the stream's course. The quantity of material carried in solution is highly variable and depends on climate and the geologic setting. Usually the dissolved load is expressed as parts of dissolved material per million parts of water (parts per million, or ppm). Although some rivers may have a dissolved load of 1000 ppm or more, the average figure for the world's rivers is estimated at 115 to 120 ppm. Almost 4 billion metric tons of dissolved mineral matter are supplied to the oceans each year by streams.

Most streams (but not all) carry the largest part of their load in *suspension*. Indeed, the visible cloud of sediment suspended in the water is the most obvious portion of a stream's load. Usually only sand, silt, and clay

Figure 4.7
When a dam is built and a reservoir forms, the stream's base level is raised. This reduces the stream's velocity and leads to deposition and a reduction of the gradient upstream from the reservoir.

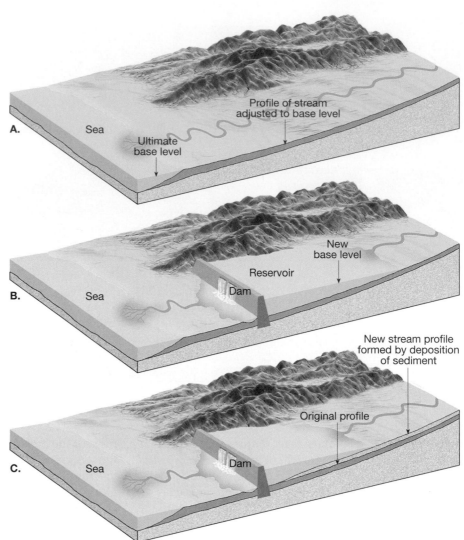

can be carried this way, but during a flood larger particles are transported as well. Also, during a flood, the total quantity of material carried in suspension increases dramatically, as can be verified by anyone whose home has been a site for the deposition of this material.

A portion of a stream's load of solid material consists of sediment that is too large to be carried in suspension. These coarser particles move along the bottom of the stream and constitute the *bed load*. Unlike the suspended and dissolved loads, which are constantly in motion, the bed load is in motion only intermittently, when the force of the water is sufficient to move the larger particles. In terms of the erosional work accomplished by a downcutting stream, the grinding action of the bed load is of great importance.

Each year, the Mississippi River transports about 750 million tons of material to the Gulf of Mexico. Of this total, it is estimated that approximately 500 million tons are carried in suspension, 200 million tons in solution, and the remaining 50 million tons as bed load; however, such proportions vary widely from stream to stream.

Streams vary in their ability to carry a load. Their ability is determined by two criteria. First, the **competence** of a stream measures the maximum size of particles it is capable of transporting. The stream's velocity determines its competence. If the velocity of a stream doubles, its competence increases four times; if the velocity triples, its competence increases nine times; and so forth. This explains how large boulders that seem immovable can be transported during a flood, which greatly increases a stream's velocity.

Second, the **capacity** of a stream is the maximum load it can carry. The capacity of a stream is directly related to its discharge. The greater the volume of water flowing in a stream, the greater is its capacity for hauling sediment.

It should now be clear why the greatest erosion and transportation of sediment occur during a flood. The increase in discharge results in a greater capacity and the increase in velocity results in greater competence. With rising velocity the water becomes more turbulent, and larger and larger particles are set in motion. In just a few days, or perhaps a few hours, a stream in flood stage can erode and transport more sediment than it does during months of normal flow (Figure 4.8).

Deposition

Whenever a stream slows down, the situation reverses. As its velocity decreases, its competence is reduced and sediment begins to drop out, largest particles first. Each particle size has a *critical settling velocity*. As streamflow drops below the critical settling velocity of a certain particle size, sediment in that category begins to settle out. Thus, stream transport provides a mechanism by which solid particles of various sizes are separated. This process, called **sorting**, explains why particles of similar size are deposited together.

The well-sorted material typically deposited by a stream is called **alluvium**, the general term for any stream-deposited sediment. Many different depositional features are composed of alluvium. Some occur within stream channels, some occur on the valley floor adjacent to the channel, and some exist at the mouth of the stream.

Deltas.   When a stream enters the relatively still waters of an ocean or lake, its velocity drops abruptly, and the resulting deposits form a **delta** (Figure 4.9). As the delta

Figure 4.8

The suspended load is clearly visible because it gives this flooding river a brown "muddy" appearance. During floods, both capacity and competency increase. Therefore, the greatest erosion and sediment transport occur during these high-water periods. This muddy torrent washed out a section of Interstate 5 near Coalinga, California, in March 1995. (Photo by AP/Wide World)

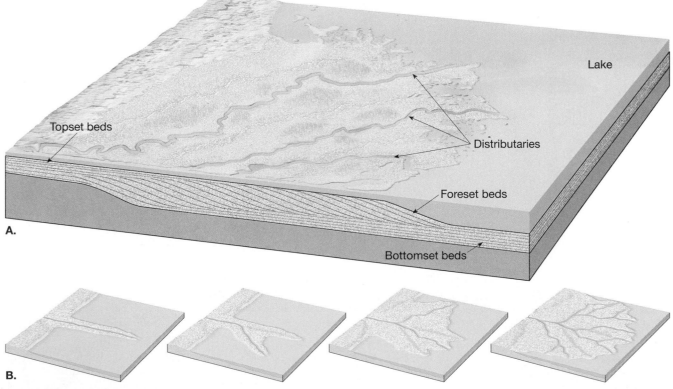

Figure 4.9

A. Structure of a simple delta that forms in the relatively quiet waters of a lake. The finer silts and clays will settle out some distance from the mouth into nearly horizontal layers called *bottomset beds*. Prior to the accumulation of bottomset beds, *foreset beds* begin to form. These beds are composed of coarse sediment, which is dropped almost immediately upon entering a lake or ocean, forming sloping layers. The foreset beds are usually covered by thin, horizontal *topset beds* deposited during floodstage. **B.** Growth of a simple delta. As a stream extends its channel, the gradient is reduced. Frequently, during flood stage the river is diverted to a higher-gradient route, forming a new distributary. Old abandoned distributaries gradually fill with sediment and vegetation. (After Ward's Natural Science Establishment, Inc., Rochester, N.Y.)

grows outward, effectively lengthening the river, the gradient of the river continually lessens. This causes the stream to seek a shorter route to base level. Frequently the main channel divides into several smaller ones called **distributaries**. These shifting channels act in an opposite way from tributaries, *distributing* water instead of contributing it. Rather than carrying water into the main channel, distributaries carry water away from the main channel. After numerous shifts of the channel, a delta may grow into a rough triangular shape like the Greek letter delta (Δ), for which it is named. Note, however, that many deltas do not exhibit the idealized shape. Differences in the configurations of shorelines and variations in the nature and strength of wave activity result in many shapes.

Many large rivers have deltas extending over thousands of square kilometers. The delta of the Mississippi River is one example. It resulted from the accumulation of huge quantities of sediment derived from the vast region drained by the river and its tributaries. Today, New Orleans rests where there was ocean less than 5000 years ago. Figure 4.10 shows that portion of the Mississippi delta that has been built over the past 5000–6000 years. As shown, the delta is actu-

ally a series of seven coalescing subdeltas. Each formed when the river left its existing channel in favor of a shorter, more direct path to the Gulf of Mexico. The individual subdeltas interfinger and partially cover one another to produce a very complex structure. The present subdelta, called a *bird-foot* delta because of the configuration of its distributaries, has been built by the Mississippi in the last 500 years.

Natural Levees.   Some rivers occupy valleys with broad, flat floors and build **natural levees** that parallel their channels on both banks (Figure 4.11). Natural levees are built by successive floods over many years. When a stream overflows its banks, its velocity immediately diminishes, leaving coarse sediment deposited in strips bordering the channel. As the water spreads out over the valley, a lesser amount of fine sediment is deposited over the valley floor. This uneven distribution of material produces the very gentle slope of the natural levee.

The natural levees of the lower Mississippi rise 6 meters (20 feet) above the valley floor. The area behind the levee is characteristically poorly drained for the obvious reason that water cannot flow up the levee and into the river. Marshes called **backswamps** result. A trib-

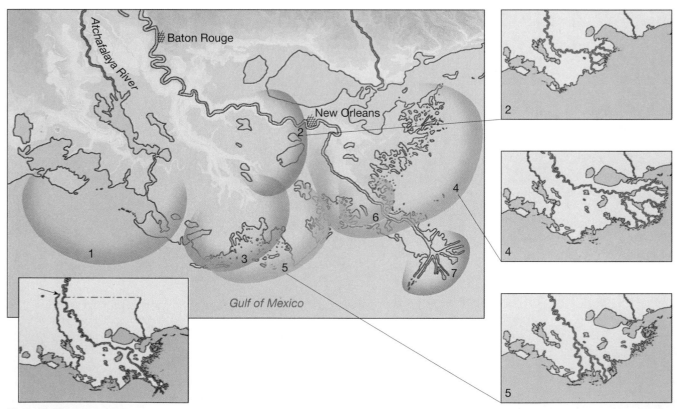

Figure 4.10

During the past 5000–6000 years, the Mississippi River has built a series of seven coalescing subdeltas. The numbers indicate the order in which the subdeltas were deposited. The present bird-foot delta (number 7) represents the activity of the past 500 years. Without ongoing human efforts, the present course of the Mississippi will shift and follow the path of the Atchafalaya River (see arrows in inset). (After C. R. Kolb and J. R. Van Lopik, *Depositional Environments of the Mississippi River Deltaic Plain*, p. 22. Copyright © 1966 by the Houston Geological Society.)

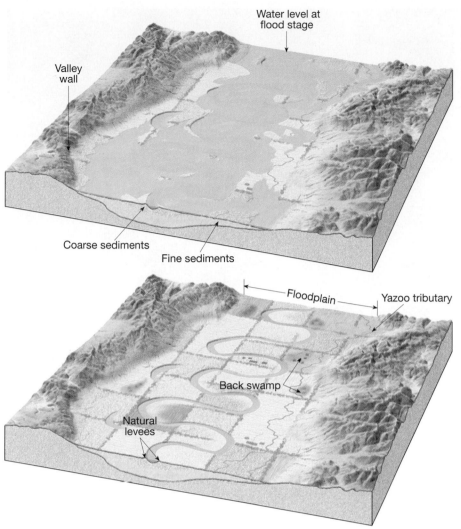

Figure 4.11
Natural levees are gently sloping structures that are created by repeated floods. Because the ground next to the stream channel is higher than the adjacent floodplain, backswamps and yazoo tributaries may develop

utary stream that cannot enter a river because levees block the way often has to flow parallel to the river until it can breach the levee. Such streams are called **yazoo tributaries** after the Yazoo River, which parallels the Mississippi for over 300 kilometers.

Artificial Levees.   Sometimes *artificial levees* are built along rivers to control flooding. Artificial levees are usually easy to distinguish from natural levees because their slopes are much steeper. When a river is confined by levees during periods of high water, it deposits material in its channel as the discharge diminishes. This is sediment that otherwise would have been dropped on the floodplain. Thus, each time there is a high flow, deposits are left on the river bed and the bottom of the channel is built up. With the buildup of the bed, less water is required to overflow the original levee. As a result, the height of the levee may have to be raised periodically to protect the floodplain. Moreover, many artificial levees are not built to withstand periods of

extreme flooding (Figure 4.12). For example, levee failures were numerous in the Midwest during the summer of 1993, when the upper Mississippi and many of its tributaries experienced record floods (see Box 4.2).

Stream Valleys

Stream valleys can be divided into two general types. Narrow V-shaped valleys and wide valleys with flat floors exist as the ideal forms, with many gradations between.

Most stream valleys are much broader at the top than is the width of their channel at the bottom. This would not be the case if the only agent responsible for eroding valleys were the streams flowing through them. The sides of most valleys are shaped by a combination of weathering, overland flow, and mass wasting. In some arid regions, where downcutting is rapid and weathering is slow, and in places where rock is particularly resistant, narrow valleys may not be V-shaped but rather may have nearly vertical walls.

Figure 4.12
Water rushes through a break in an artificial levee in Monroe County, Illinois. During the record-breaking 1993 Midwest floods, many artificial levees could not withstand the force of the floodwaters. Sections of many weakened structures were overtopped or simply collapsed. (Photo by James A. Finley / Wide World Photos)

Narrow Valleys

The Yellowstone River provides an excellent example of a narrow valley (Figure 4.13). A narrow V-shaped valley indicates that the primary work of the stream has been downcutting toward base level. The most prominent features of a narrow valley are *rapids* and *waterfalls*. Both occur where the stream profile drops rapidly, a situation usually caused by variations in the erodibility of the bedrock into which a stream channel is cutting. Resistant beds create rapids by acting as a temporary base level upstream while allowing downcutting to continue downstream. Once erosion has eliminated the resistant rock, the stream profile smooths out again. Waterfalls are places where the stream profile makes a vertical drop.

Wide Valleys

Once a stream has cut its channel closer to base level, downward erosion becomes less dominant. At this point more of the stream's energy is directed from side to side. The result is a widening of the valley as the river cuts away first at one bank and then at the other (Figure 4.14).

Floodplains. The side-to-side cutting of a stream eventually produces a flat valley floor, or **floodplain**. It is appropriately named because the river is confined to its channel, except during flood stage, when it overflows its banks and inundates the floodplain.

When a river erodes laterally, creating a floodplain as just described, it is called an *erosional floodplain*. Floodplains can be depositional in nature as well. *Depositional floodplains* are produced by a major fluctuation in conditions, such as a change in base level. The floodplain in California's Yosemite Valley is one such feature, and was produced when a glacier gouged the former

stream valley deeper by about 300 meters (1000 feet). After the glacial ice melted, the stream readjusted to its former base level by refilling the valley with alluvium.

Meanders. Streams that flow on floodplains move in sweeping bends called **meanders** (Figure 4.14). Meanders continually change position by eroding sideways and slightly downstream. The sideways movement occurs because the maximum velocity of the stream shifts toward the outside of the bend, causing erosion of the outer bank (Figure 4.15). At the same time, the reduced current at the inside of the meander results in the deposition of coarse sediment, especially sand, called a *point bar*. Thus, by eroding its outer bank and depositing material along its inner bank, a stream moves sideways without changing its channel size.

Because of the slope of the channel, erosion is more effective on the downstream side of a meander. Therefore, in addition to migrating laterally, the bends also gradually migrate down the valley. Sometimes the downstream migration of a meander is slowed when it reaches a more resistant portion of the floodplain. This allows the next meander upstream to overtake it. Gradually the neck of land between the meanders is narrowed. When they get close enough, the river erodes through the narrow neck of land to the next loop (Figure 4.16). The new, shorter channel segment is called a **cutoff** and, because of its shape, the abandoned bend is called an **oxbow lake**.

Artificial Cutoffs. One method of flood control involves straightening a channel by creating *artificial cutoffs*. The idea is that by shortening the stream, the gradient, and hence the velocity, are increased. By increasing velocity, the larger discharge associated with flooding can be dispersed more rapidly. Since the early 1930s, the Army Corps of Engineers

Box 4.2

The Midwest Floods of 1993

The devastation caused by the floods in the upper Mississippi River valley during the summer of 1993 was unprecedented in the history of the United States. "The Great Flood of 1993," as it was aptly named, rose to 7 meters (23 feet) above flood stage, advancing into areas never before flooded in historical time. Towns and croplands were damaged extensively. Afterward, some towns were relocated rather than being rebuilt.

Although casualties were relatively light when compared to similar events in other parts of the world, property damage approached $15 billion (Figure 4.B). In all, nearly 14 million acres were inundated by the flood, displacing at least 50,000 people and causing 50 deaths.

Exceptional rainfall produced the wettest spring and early summer of the century for the upper Mississippi River basin. An examination of Figure 4.C shows that between April 1 and July 31, the region received up to twice the normal rainfall. The magnitude of rainfall over such a vast area of the Upper Midwest resulted in flooding of extraordinary and catastrophic proportions on the Mississippi and many of its tributaries, affecting large portions of Illinois, Iowa, Kansas, Minnesota, Missouri, Nebraska, South Dakota, and Wisconsin.

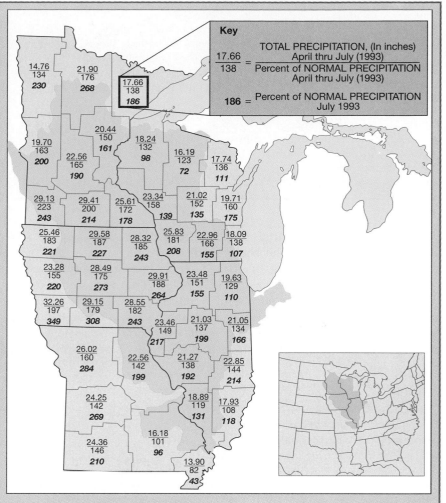

Figure 4.C

Precipitation data for the upper Mississippi River basin between April 1 and July 31, 1993. (Data from Illinois State Water Survey)

Figure 4.B

The flooding Mississippi River near Columbia, Illinois in August 1993. (Photo by James A. Finley/Wide World Photos)

Soils throughout the Midwest were already saturated from ample rainfall in summer and fall 1992, and soils remained moist as winter began. This pattern continued into spring and summer 1993.

A stationary weather pattern over the United States was responsible for the Midwest's persistent, drenching rains. Compared to the long-term average, rainfall in the upper Mississippi River basin during April and May was 40 percent higher than average, and June rainfall was double the average. As the deluge continued through July, much of the basin received rainfall between two and three times the norm.

Some of the individual station reports were nothing short of astounding. In northwestern Missouri, Skidmore reported 25.35 inches of rain in July, and Worth County reported 30.3 inches

(continues)

(Box 4.2, continued)

through July 25. Normal July rainfall for this area is about 4 inches, while the average annual rainfall is 35 inches.

At the height of the flood, the flow of the Mississippi River at St. Louis was more than six times the normal discharge. In several instances the river was so high that water from its tributaries had nowhere to go. Dammed by the Mississippi's flood waters, the tributaries overflowed their banks and flooded their valleys. In some places, streamflow actually reversed, and rivers temporarily flowed upstream.

The "Great Flood of 1993" had four characteristics that made it an unprecedented event:

1. Rivers remained above flood stage for months rather than for days or weeks.
2. Flooding, which is a common springtime phenomenon, lasted through the summer.
3. Multiple flood crests occurred at most locations.
4. Flood crests set new highs for the historical flood record.

The severity of the 1993 Midwest floods raised many issues concerning how human modifications of floodplains and river channels may adversely affect flood events. The issues are not new. However, because of the magnitude of this flood, new approaches to floodplain management and land use may result.

---

has created many artificial cutoffs on the Mississippi for the purpose of increasing the channel efficiency and reducing the threat of flooding. In all, the river has been shortened more than 240 kilometers (150 miles). The program has been somewhat successful in reducing the height of the river in flood. However, since the river's tendency to meander still exists, preventing the river from returning to its previous condition has been difficult.

Drainage Basins and Patterns

Every stream, no matter how large or small, has a **drainage basin**. This is the land area that contributes water to the stream. The drainage basin of one stream is separated from the drainage basin of another by an imaginary line called a **divide** (Figure 4.17). Divides range in scale from a ridge separating two small gullies on a hill-

Figure 4.13
V-shaped valley of the Yellowstone River in Wyoming. The rapids and waterfalls indicate that the river is vigorously downcutting. (Photo by John M. Roberts/The Stock Market)

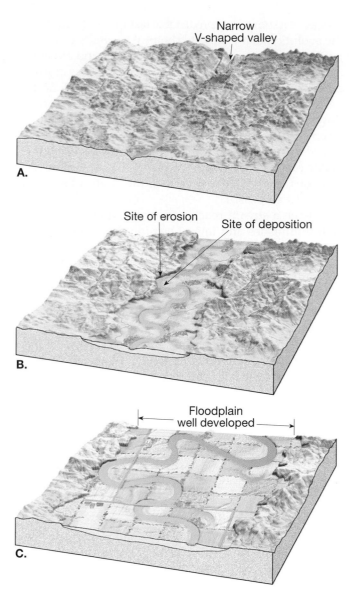

Figure 4.14

Stream eroding its floodplain.

side to a *continental divide* that splits continents into enormous drainage basins. For example, the continental divide that runs somewhat north–south through the Rocky Mountains separates the drainage which flows west to the Pacific Ocean from that which flows to the Gulf of Mexico. Although divides separate the drainage of two streams, if both streams are tributaries of the same river, they are both a part of the river's drainage system.

Drainage systems are networks of streams that together form distinctive patterns. The nature of a drainage pattern can vary greatly from one type of terrain to another, primarily in response to the kinds of rock on which the streams developed or the structural pattern of faults and folds.

Certainly the most commonly encountered drainage pattern is the **dendritic** pattern (Figure 4.18A). This pattern of irregularly branching tributary streams resembles the branching pattern of a deciduous tree. In fact, the word *dendritic* means "treelike." The dendritic pattern forms where the underlying material is relatively uniform. Because the surface material is essentially uniform in its resistance to erosion, it does not control the pattern of streamflow. Rather, the pattern is determined chiefly by the direction of slope of the land.

When streams diverge from a central area like spokes from the hub of a wheel, the pattern is said to be **radial** (Figure 4.18B). This pattern typically develops on isolated volcanic cones and domal uplifts.

Figure 4.18C illustrates a **rectangular** pattern, in which many right-angle bends can be seen. This pattern develops when the bedrock is crisscrossed by a series of joints and/or faults. Because these structures are eroded more easily than unbroken rock, their geometric pattern guides the directions of valleys.

Figure 4.18D illustrates a **trellis** drainage pattern, a rectangular pattern in which tributary streams are nearly parallel to one another and have the appearance of a garden trellis. This pattern forms in areas underlain by alternating bands of resistant and less resistant rock.

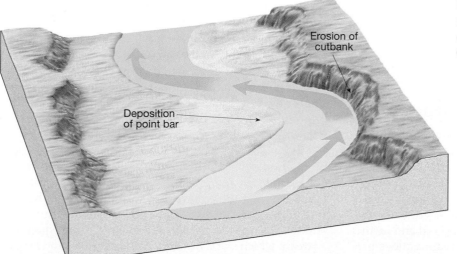

Figure 4.15

Lateral movement of meanders. By eroding its outer bank and depositing material on the inside of the bend, a stream is able to shift its channel.

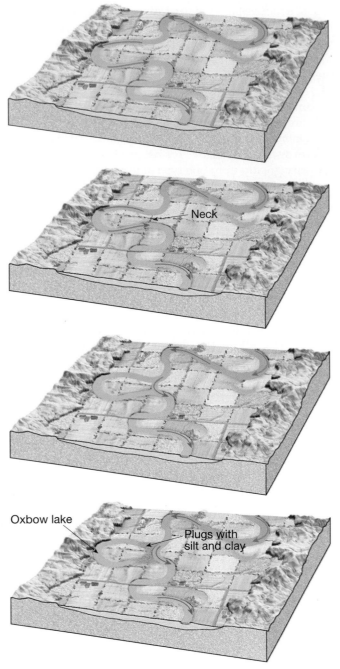

Figure 4.16
Formation of a cutoff and oxbow lake.

Stages of Valley Development

Two centuries ago, many people believed Earth to be only a few thousand years old, and that rainwater flowed down valleys that had been there since the beginning of time. An early geologist named James Hutton believed otherwise and proposed that streams were responsible for cutting the valleys in which they flowed. Later geologic work substantiated Hutton's proposal and further revealed that the development of stream valleys pro-

gresses in a somewhat predictable fashion. To learn about the evolution of a valley, it is helpful to divide its development into three stages: youth, maturity, and old age.

As long as the stream is downcutting, it is considered youthful. Rapids, an occasional waterfall, and a narrow V-shaped valley are all visible signs of vigorous downcutting. Other characteristics of youth include a steep gradient, little or no floodplain, and a relatively straight course without meanders (Figure 4.19A). The valley of the Yellowstone River pictured in Figure 4.13 provides an excellent example of the youthful stage of valley development.

When a stream reaches maturity, downward erosion diminishes and lateral erosion dominates. Thus, the mature stream begins to create a floodplain and meander upon it (Figures 4.19B and C). In contrast to the gradient of a youthful stream, the gradient of a mature stream is much lower and the profile is much smoother because all rapids and waterfalls have been eliminated.

A stream enters old age after it has cut its floodplain several times wider than its *meander belt*, which is the width of the meander (Figure 4.19D). When this stage is reached, the stream is rarely near the valley walls; hence it ceases to significantly enlarge the floodplain. Thus, the primary work of a river in an old-age valley is the reworking of unconsolidated floodplain deposits. Because this task is easier than cutting bedrock, a stream in an old-age valley shifts more rapidly than a stream in a mature valley. For example, some meanders of the lower Mississippi move about 20 meters (over 60 feet) a year, and its large floodplain is dotted with oxbow lakes and old cutoffs. Natural levees are also common features of old-age valleys and, when present, are accompanied by backswamps and yazoo tributaries.

Thus far we have assumed that the base level of a stream remains constant as a river progresses from youth to old age. On many occasions, however, the land is uplifted or the base level is lowered. The effect of uplifting on a youthful stream is to increase its gradient and accelerate its rate of downcutting. However, uplifting of a mature stream would cause it to abandon lateral erosion and revert to downcutting. Rivers of this type are said to be **rejuvenated** ("made young again"). The meanders stay in the same place but become deeper, and are called **entrenched meanders** (Figure 4.20). Mature streams may eventually readjust to uplift by cutting a new floodplain at a level below the old one. The remnants of the old higher floodplain often remain as flat surfaces called *terraces* (Figure 4.21).

Two additional points concerning valley development should be made. First, the time required for a stream to reach any given stage depends on several factors, including the erosive ability of the stream, the nature of the material through which the stream must cut, and the stream's height above base level. Consequently, a stream which starts out very near base level and has to

Figure 4.17

A *drainage basin* is the land area drained by a stream and its tributaries. The drainage basin of the Mississippi River, North America's largest river, covers about 3 million square kilometers. *Divides* are the boundaries that separate drainage basins from each other. Drainage basins and divides exist for all streams.

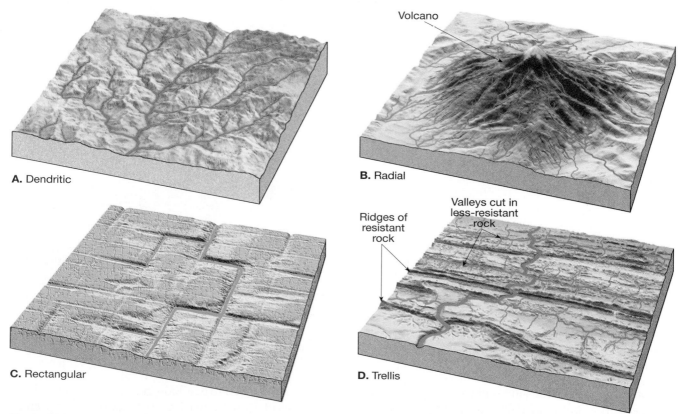

Figure 4.18

Drainage patterns. **A.** Dendritic. **B.** Radial. **C.** Rectangular. **D.** Trellis.

Figure 4.19
Stages of valley development. **A.** Youth. The youthful stage is characterized by downcutting and a V-shaped valley. **B.** and **C.** Maturity. Once a stream has sufficiently lowered its gradient, it begins to erode laterally, producing a wide valley. **D.** Old age. After the valley has been cut several times wider than the width of the meander belt, it has entered old age.

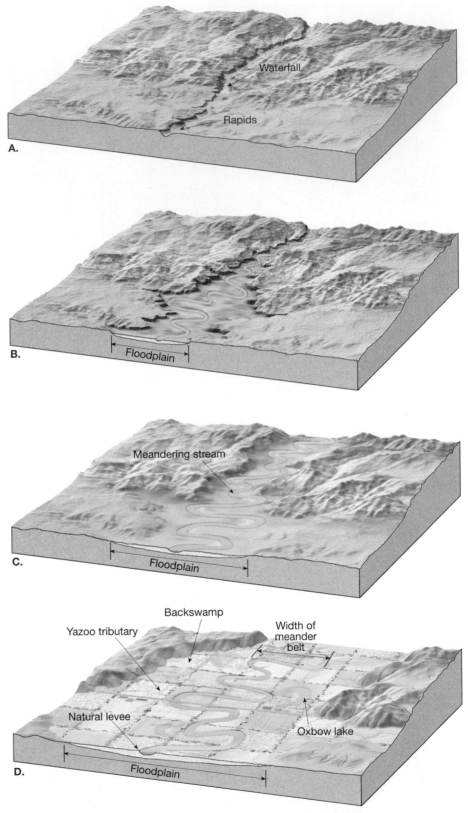

cut only through unconsolidated sediments may reach maturity in a matter of a few hundred years. On the other hand, the Colorado River, where it is actively cutting the Grand Canyon, has retained its youthful nature for an estimated 5 to 6 million years. Second, individual portions of a stream reach each stage at different times. Often the lower reaches of a stream attain old age while the headwaters are still youthful in character.

Figure 4.20
A close-up view of entrenched meanders in Canyonlands National Park, Utah. The meandering river became entrenched because of the uplift of the Colorado Plateau. (Photo by Michael Collier)

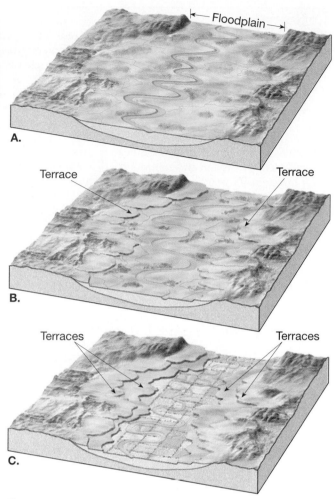

Figure 4.21
A. A stream near base level has developed a floodplain. **B.** Because of a lowering of base level or regional uplift, the river downcuts through the former floodplain and creates a new floodplain. The remnants of the former floodplain now exist as *terraces*. **C.** A second set of terraces form in response to another change in base level.

Water Beneath the Surface

Groundwater, water that has soaked into the ground, is one of our most important and widely available resources. Yet people's perceptions of groundwater are often unclear and incorrect (see Box 4.3). The reason is that groundwater is hidden from view except in caves and mines, and the impressions people gain from these subsurface openings are often misleading. Observations on the land surface give an impression that Earth is "solid." This view is not changed very much when we enter a cave and see water flowing in a channel that appears to have been cut into solid rock. Because of such observations many people believe that groundwater occurs only in underground "rivers." But actual rivers underground are extremely rare.

In reality, most of the subsurface environment is not "solid" at all but contains a huge volume of openings that exist as spaces between grains of soil and sediment and occur as narrow joints and fractures in bedrock. It is in these tiny openings that groundwater collects and moves.

The Importance of Groundwater

The importance of groundwater can be demonstrated by comparing its volume with the quantity of water in other parts of the hydrosphere. Of all the world's water, only about six-tenths of 1 percent occurs underground. Nevertheless, the amount of water stored in the rocks and sediments beneath Earth's surface is vast. When the oceans are excluded and only sources of freshwater are considered, the significance of groundwater becomes more apparent. Table 4.2 contains estimates of the distribution of freshwater in the hydrosphere. Clearly, the largest volume occurs as glacial ice. Second in

rank is groundwater, with slightly more than 14 percent of the total. However, when ice is excluded and just liquid water is considered, more than 94 percent is groundwater. Without question, *groundwater represents the largest reservoir of freshwater that is readily available to humans.* Its value in terms of economics and human well-being is incalculable.

In many parts of the world, wells and springs provide the water needed not only for great numbers of people, but also for crops, livestock, and industry. In the United States, groundwater is the source of about 40 percent of the water used for all purposes except hydroelectric power generation and power plant cooling. Groundwater provides drinking water for more than 50 percent of the population, as well as 40 percent of the water for irrigation and 26 percent of industry's needs. In some areas, however, overuse of this basic resource has caused serious problems, including streamflow depletion,

Box 4.3

Dowsing for Water*

"Water dowsing" refers in general to the practice of using a forked stick, rod, pendulum, or similar device to locate underground water, minerals, or other hidden or lost substances, and has been a subject of discussion and controversy for hundreds of years.

Although tools and methods vary widely, most dowsers (also called diviners or water witches) probably still use the traditional forked stick, which may come from a variety of trees, including the willow, peach, and witch hazel (Figure 4.D). Other dowsers may use keys,

---

*Much of this information is based on material prepared by the U.S. Geological Survey.

Figure 4.D
Most dowsers still use the traditional forked stick.

wire coat hangers, pliers, wire rods, pendulums, or various kinds of elaborate boxes and electrical instruments.

In the classic method of using a forked stick, one fork is held in each hand with the palms upward and the bottom of the "Y" pointed skyward. The dowser then walks back and forth over the area to be tested. When a source of water is detected, the butt end of the stick is supposed to rotate or be attracted downward.

According to dowsers, the attraction of the water may be so great that the bark peels off as the rod twists in the hands. Some dowsers are said to have suffered blistered or bloody hands from the twisting.

The exact origin of the divining rod in Europe is not known. The first detailed description of it is in Johannes Agricola's *De Re Metallica* (1556), a description of German mines and mining methods. The device was introduced into England during the reign of Elizabeth I (1558–1603) to locate mineral deposits, and soon afterward it was adopted as a water finder throughout Europe.

Water dowsing seems to be a mainly European cultural phenomenon, completely unknown to New World Indians and Eskimos. It was carried to America by some of the earliest settlers from England and Germany. Although the published record was very slight at first, water dowsing or witching began to be mentioned after 1675 in connection with witches and witchcraft. Two articles condemning it appeared in the 1821 and 1826 issues of the *American Journal of Science* and were among the first in a long line of treatises on water witching.

Despite almost unanimous condemnation by geologists and technicians, the practice of water dowsing has spread throughout America. Thousands of dowsers are currently active in the United States.

Case histories and demonstrations of dowsers may seem convincing, but when dowsing is exposed to scientific examination, it presents a very different picture. A study in Australia compared geologist's successes at locating groundwater with those of dowsers. The dowsers caused twice as many dry holes to be dug as the geologists. At Iowa State University, dowsers were given the task of finding water along a prescribed path on campus. They were not able to "discover" the water mains directly beneath them.

What does it mean to say that a dowser is successful? The dowser may find water, but how much? And of what quality? At what rate can it be withdrawn? For how long and with what impact on other wells and on nearby streams?

The natural explanation of "successful" water dowsing is that in many areas water would be hard to miss. The dowser commonly implies that the spot indicated by the rod is the *only* one where water could be found, but this is not necessarily true. In a region of adequate rainfall and favorable geology, it is difficult *not* to drill and find water!

Some water exists under the surface almost everywhere. This explains why many dowsers appear to be successful. To locate groundwater accurately, however, as to depth, quantity, and quality, a number of techniques must be used. Hydrologic, geologic, and geophysical knowledge is needed to determine the depths and extent of the different water-bearing strata and the quantity and quality of water found in each. The area must be thoroughly tested and studied to determine these facts.

---

land subsidence, and increased pumping costs. In addition, groundwater contamination due to human activities is a real and growing threat in many places.

Groundwater's Geological Roles

Geologically, groundwater is important as an erosional agent. The dissolving action of groundwater slowly removes soluble rock, allowing surface depressions known as sinkholes to form as well as creating subterranean caverns (Figure 4.22). Groundwater is also an equalizer of streamflow. Much of the water that flows in rivers is not direct runoff from rain and snowmelt. Rather, a large percentage of precipitation soaks in and

then moves slowly underground to stream channels. Groundwater is thus a form of storage that sustains streams during periods when rain does not fall. When we see water flowing in a river during a dry period, it is water from rain that fell at some earlier time and was stored underground.

Distribution and Movement of Groundwater

When rain falls, the water may run off immediately, evaporate, be taken up by plants and transpired back into the air, or soak into the ground. This last path is the

Table 4.2 Fresh water of the hydrosphere.

Parts of the Hydrosphere	Volume of Freshwater (km³)	Share of Total Volume of Freshwater (percent)	Rate of Water Exchange
Ice sheets and glaciers	24,000,000	84.945	8000 years
Groundwater	4,000,000	14.158	280 years
Lakes and reservoirs	155,000	0.549	7 years
Soil moisture	83,000	0.294	1 year
Water vapor in the atmosphere	14,000	0.049	9.9 days
River water	1,200	0.004	11.3 days
Total	28,253,200	100.000	

SOURCE: U.S. Geological Survey Water Supply Paper 2220, 1987

primary source of practically all subsurface water. The amount of water that takes each of these paths, however, varies considerably both in time and space. For any location, the steepness of the slope, the nature of the surface material, the intensity of the rainfall, and the type and amount of vegetation are all influential factors. Heavy rains that fall on steep slopes underlain by impervious materials will obviously result in a high percentage of the water running off. On the other hand, a gentle, steady rain that falls on more gradual slopes composed of materials more easily penetrated by water would result in a much larger percentage of the water soaking into the ground.

Distribution

Some of the water that soaks in does not travel far, because it is held by molecular attraction as a surface film on soil particles. A portion of this moisture evaporates back into the atmosphere. Much of the remainder is used by plants between rains. But water that is not held near the surface penetrates downward until it reaches a zone where all of the open spaces in sediment and rock are completely filled with water. This is called the **zone of saturation**. Water within it is called **groundwater**. The upper limit of this zone is known as the **water table**. The area above the water table where the

Figure 4.22

The dissolving action of groundwater creates caves. Three Fingers Cave, Lincoln County, New Mexico. (Photo by Harris Photographic/Tom Stock and Associates)

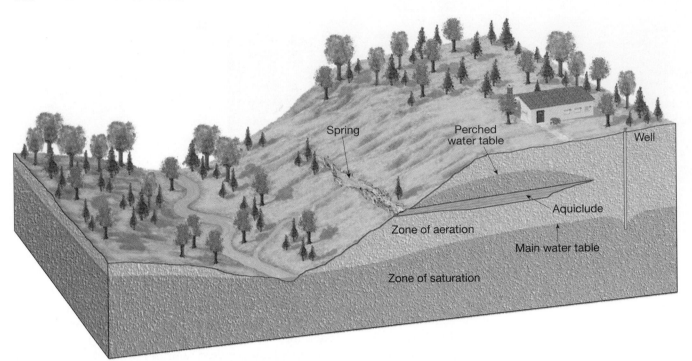

Figure 4.23
This diagram illustrates the relative positions of many features associated with subsurface water.

soil, sediment, and rock are not saturated is called the **zone of aeration** (Figure 4.23). The open spaces here are filled mainly with air.

The water table is rarely level as we might expect a table to be. Instead, its shape is usually a subdued replica of the surface, reaching its highest elevations beneath hills and decreasing in height toward valleys (Figure 4.23). When you see a wetland (swamp), it tells you that the water table is right at the surface. Lakes and streams generally occupy areas where the land surface is below the water table.

Although many factors contribute to the irregular surface of the water table, the most important cause is that groundwater moves very slowly. For this reason, the water tends to "pile up" beneath hills. If rainfall were to cease completely, these water "hills" would slowly subside and gradually approach the level of the valleys. However, new supplies of rainwater are usually added often enough to prevent this. Nevertheless, in times of extended drought, the water table may drop enough to fall below otherwise productive wells.

Movement

Depending on the nature of the subsurface material, the flow of groundwater and the amount of water that can be stored vary greatly. The quantity of groundwater that can be stored depends on the **porosity** of the material; that is, the percentage of the total volume of rock or sediment that consists of pore spaces. Although these openings often consist of spaces between particles of sediment, such features as vesicles (voids left by gases escaping from lava), joints and faults, and cavities formed by the solution of soluble rocks such as limestone are also common.

Rock or sediment may be very porous and still not allow water to move through it. The **permeability** of a material indicates its ability to transmit water through interconnected pore spaces. In rock or sediment, the smaller the pore spaces, the slower the groundwater moves. If the spaces between particles are too small, the films of water clinging to the grains will combine. As a result, the force of molecular attraction that binds the water to the particles holds the water firmly in place. Clay exemplifies this circumstance. Although the ability of clay to store water is often high, its pore spaces are so small that water is unable to move through it. Thus, we say that clay is *impermeable*.

Impermeable layers such as clay that hinder or prevent water movement are termed **aquicludes** (literally, "water closed"). On the other hand, larger particles, such as sand or gravel, have larger pore spaces. Therefore, the water can move with relative ease. Such permeable rock strata or sediments that transmit groundwater freely are called **aquifers** ("water carriers"). Aquifers are important because they are the water-bearing layers that well drillers seek. A good aquifer can provide water for numerous wells, homes, cities, and even irrigation and industries.

Springs

Springs have aroused the curiosity and wonder of people for thousands of years. The fact that springs were (and to some people still are) rather mysterious phenomena is not difficult to understand, for here is water flowing freely from the ground in all kinds of weather in seemingly inexhaustible supply but with no obvious source. Today we know that the source of springs is water from the zone of saturation and that the ultimate source of this water is precipitation.

Whenever the water table intersects the ground surface, a natural flow of groundwater results, which we call a **spring**. There are many circumstances that create springs. Springs such as those in Figure 4.24 form when an aquiclude blocks the downward movement of groundwater and forces it to move laterally. When the permeable bed (aquifer) outcrops in a valley, a spring or series of springs results.

Another situation that can produce a spring is illustrated in Figure 4.23. Here an aquiclude is situated above the main water table. As water percolates downward, a portion accumulates above the aquiclude to create a localized zone of saturation and a *perched water table*. Springs, however, are not confined to places where a perched water table creates a flow at the surface. Indeed, there are a wide variety of spring types because subsurface conditions vary greatly from place to place.

Hot Springs

By definition, the water in **hot springs** is 6–9°C (10–15°F) warmer than the mean annual air temperature for the localities where they occur. In the United States alone, there are well over 1000 such springs.

Figure 4.24
Minnie Miller Springs, Thousand Springs Preserve, along the Snake River near Hagerman, Idaho. (Photo by William H. Mullins/Photo Researchers, Inc.)

Mineral explorations over the world have shown that temperatures in deep mines and oil wells usually rise with an increase in depth below the surface. Temperatures in such situations increase an average of about 2°C per 100 meters (1°F per 100 feet). Therefore, when groundwater circulates at great depths, it becomes heated, and if it rises to the surface, the water may emerge as a hot spring. The water of some hot springs in the United States, particularly in the East, is heated in this manner. However, the great majority (over 95 percent) of the hot springs (and geysers) in the United States are found in the West. The reason for such a distribution is that the source of heat for most hot springs is cooling igneous rock, and it is in the West that igneous activity has been most recent (see Box 4.4).

Geysers

Geysers are intermittent hot springs or fountains in which columns of water are ejected with great force at various intervals, often rising 30–60 meters (100–200 feet; Figure 4.25). After the jet of water ceases, a column of steam rushes out, usually with a thundering roar. Perhaps the most famous geyser in the world is Old Faithful in Yellowstone National Park, which erupts about once each hour. Geysers are also found in other parts of the world, including New Zealand and Iceland, where the term *geyser*, meaning "spouter" or "gusher," was coined.

Geysers occur where extensive underground chambers exist within hot igneous rocks. As relatively cool groundwater enters the chambers, it is heated by the surrounding rock. At the bottom of the chamber, the water is under great pressure because of the weight of the overlying water. This great pressure prevents the water from boiling at the normal surface temperature of 100°C (212°F). For example, at the bottom of a 300-meter (1000-foot) chamber, water must attain a temperature of nearly 230°C (450°F) before it will boil. The heating causes the water to expand, with the result that some is forced out at the surface. This loss of water reduces the pressure on the remaining water in the chamber. The reduced pressure lowers the boiling point and a portion of the water deep within the chamber quickly turns to steam and causes the geyser to erupt (Figure 4.26). Following the eruption, cool groundwater again seeps into the chamber and the cycle begins anew.

Wells

The most common method for removing groundwater is the **well**, a hole bored into the zone of saturation (see Figure 4.23). Wells serve as small reservoirs into which groundwater moves and from which it can be pumped

Box 4.4

Geothermal Energy: An Alternative

Approximately 85 percent of the energy consumed in the United States comes from coal, petroleum, and natural gas. Hydroelectric and nuclear power account for most of the rest (Figure 4.E). The remaining 2 percent is provided by *alternative energy sources* that include solar power, wind energy, tidal power, and geothermal energy. Geothermal energy is the subject of this box. The other alternatives are treated elsewhere in the text.

Geothermal energy is produced by tapping naturally occurring steam and hot water located beneath the surface in regions where subsurface temperatures are high as a result of recent volcanic activity. Electrical power generated from geothermal sources is a relatively clean and nonpolluting alternative to other power plants that use steam. Geothermal plants do not produce the air pollution associated with burning fossil fuels; nor do they produce the radioactive wastes associated with nuclear power plants.

By the mid-1990s, nearly 200 geothermal facilities were operating in 17 countries. In addition to the United

Figure 4.F

The Geysers, a field of steaming vents 115 kilometers north of San Francisco, California. The natural steam beneath these hills was first tapped to power electrical-generating plants in 1960. (Photo courtesy of Pacific Gas and Electric)

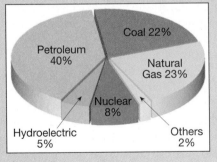

Figure 4.E

Consumption of energy in the United States, 1996. Alternative energy sources account for only about 2 percent of the total.

States, countries that lead in the use of geothermal energy to produce electricity are the Philippines, Indonesia, Mexico, Italy, and New Zealand.

The first commercial geothermal power plant in the United States was built in 1960 at The Geysers, north of San Francisco (Figure 4.F). By the late 1980s, electricity from The Geysers met most of the needs for San Francisco and Oakland. However, this peak soon passed. Although still significant, electricity generated at The Geysers has gradually been declining. In addition to The Geysers, geothermal development is occurring elsewhere in the western United States, including Nevada, Utah, and the Imperial Valley in southern California.

Geothermal energy is not used exclusively for generating electricity. In Iceland's capital, Reykjavik, steam and hot water are pumped into buildings throughout the city for space heating and are used to warm greenhouses where

fruits and vegetables are grown all year. In the United States, localities in several western states use hot water from geothermal sources for space heating.

It should be emphasized that geothermal power is not an inexhaustible source of energy. When hot fluids are pumped from volcanically heated reservoirs, water cannot be replaced and then heated sufficiently to recharge the reservoir. Experience has shown that the output of steam and hot water from individual wells usually does not last for more than 10 to 15 years. Therefore, more wells must be drilled to maintain power production. Eventually, of course, the field is depleted.

As with other alternate methods of power production, geothermal sources are not expected to provide a high percentage of the world's growing energy needs. Nevertheless, in regions where its potential can be developed, its use will no doubt continue to grow.

to the surface. The use of wells dates back many centuries and continues to be an important method of obtaining water. By far the single greatest use of this water in the United States is irrigation for agriculture. More than 65 percent of the groundwater used each year is for this purpose. Industrial uses rank a distant second, followed by the amount used in city water systems and rural homes.

The level of the water table may fluctuate considerably during the course of a year, dropping during dry seasons and rising following periods of rain. Therefore, to ensure a continuous supply of water, a well must penetrate below the water table. Whenever water is withdrawn from a well, the water table around the well is lowered. The extent of this effect, which is termed **draw-**

Figure 4.25

Old Faithful, one of the world's most famous geysers, emits as much as 45,000 liters (almost 12,000 gallons) of hot water and steam about once each hour. (Photo by Dallas and John Heaton/Westlight)

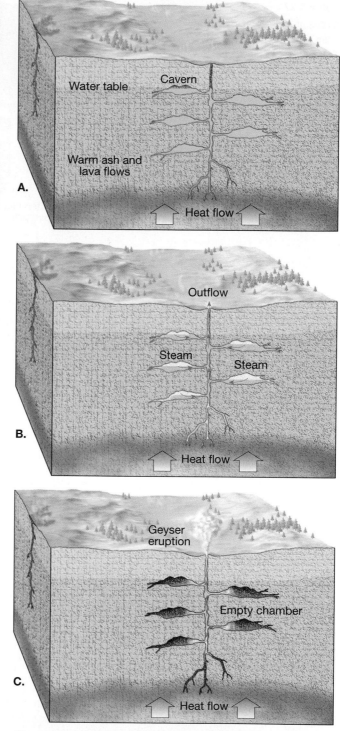

down, decreases with increasing distance from the well. The result is a depression in the water table, roughly conical in shape, known as a **cone of depression** (Figure 4.27). For most small domestic wells, the cone of depression is negligible. However, when wells are used for irrigation or for industrial purposes, the withdrawal of water can be great enough to create a very wide and steep cone of depression that may substantially lower the water table in an area and cause nearby shallow wells to become dry. Figure 4.27 illustrates this situation.

Artesian Wells

In most wells, water does not rise on its own. If water is first encountered at 30 meters depth, it remains at that level, fluctuating perhaps a meter or two because of rainfall variations. However, in some wells, water rises, sometimes overflowing at the surface.

Figure 4.26

Idealized diagrams of a geyser. A geyser can form if the heat is not distributed by convection. **A.** In this figure, the water near the bottom is heated to near its boiling point. The boiling point is higher there than at the surface because the weight of the water above increases the pressure. **B.** The water higher in the geyser system is also heated; therefore, it expands and flows out at the top, reducing the pressure on the water at the bottom. **C.** At the reduced pressure on the bottom, boiling occurs. Some of the bottom water flashes into steam, and the expanding steam causes an eruption.

Figure 4.27
A cone of depression in the water table often forms around a pumping well. If heavy pumping lowers the water table, some wells may be left dry.

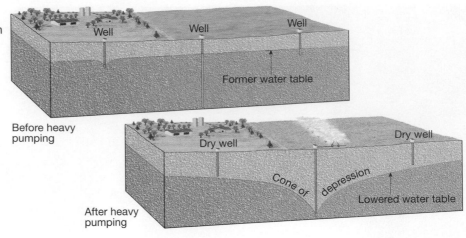

The term **artesian** is applied to any situation in which groundwater rises in a well above the level where it was initially encountered. For such a situation to occur, two conditions must exist (Figure 4.28): (1) water must be confined to an aquifer that is inclined so that one end is exposed at the surface, where it can receive water; and (2) impermeable layers (aquicludes), both above and below the aquifer, must be present to prevent the water from escaping. When such a layer is tapped, the pressure created by the weight of the water above will force the water to rise. If there were no friction the water in the well would rise to the level of the water at the top of the aquifer. However, friction reduces the height of this pressure surface. The greater the dis-

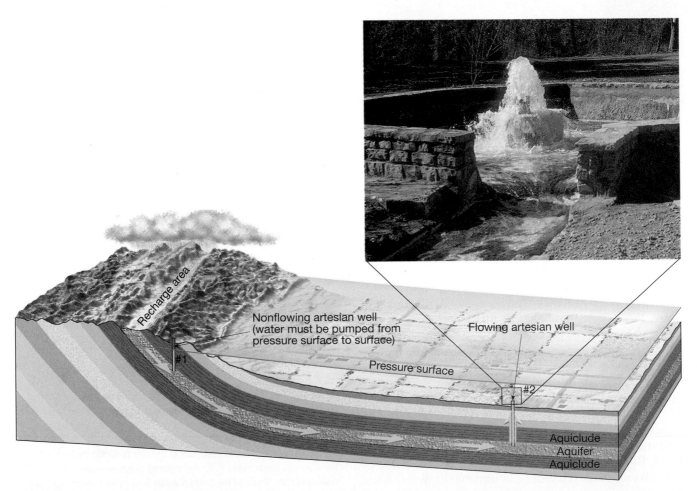

Figure 4.28
Artesian systems occur when an inclined aquifer is surrounded by impermeable beds. (Photo by James Patterson)

tance from the recharge area (area where water enters the inclined aquifer), the greater the friction and the less the rise of water.

In Figure 4.28, Well 1 is a *nonflowing artesian well*, because at this location the pressure surface is below ground level. When the pressure surface is above the ground and a well is drilled into the aquifer, a *flowing artesian well* is created (Well 2, Figure 4.28).

Artesian systems act as conduits, transmitting water from remote areas of recharge great distances to the points of discharge. In this manner, water that fell in central Wisconsin years ago is now taken from the ground and used by communities many kilometers away in Illinois. In South Dakota, such a system brings water from the Black Hills in the west, eastward across the state.

On a different scale, city water systems may be considered as examples of artificial artesian systems. The water tower, into which water is pumped, may be considered the area of recharge, the pipes the confined aquifer, and the faucets in homes the flowing artesian wells.

Environmental Problems of Groundwater

As with many of our valuable natural resources, groundwater is being exploited at an increasing rate. In some areas, overuse threatens the groundwater supply. In other places, groundwater withdrawal has caused the ground and everything resting upon it to sink. Still other localities are concerned with the possible contamination of their groundwater supply.

Treating Groundwater as a Nonrenewable Resource

For many, groundwater appears to be an endlessly renewable resource, for it is continually replenished by rainfall and melting snow. But in some regions, groundwater has been and continues to be treated as a *nonrenewable* resource. Where this occurs, the amount of water available to recharge the aquifer is significantly less than the amount being withdrawn. The High Plains, a relatively dry region that extends from South Dakota to western Texas, provides one example. Here an extensive agricultural economy is largely dependent on irrigation (Figure 4.29). As a result, there are an estimated 168,000 wells being used to irrigate more than 65,000 square kilometers (16 million acres) of land. In the southern part of this region, which includes the Texas panhandle, the natural recharge of the aquifer is very slow and the problem of declining groundwater levels is acute. In fact, in years of average or below average precipitation, recharge is negligible because all or nearly all of the meager rainfall is returned to the atmosphere by evaporation and transpiration.

Figure 4.29
In some agricultural regions water is pumped from the ground faster than it is replenished. In such instances, groundwater is being treated as a nonrenewable resource. (Photo by Steve Welsh/Liaison International)

Therefore, where intense irrigation has been practiced for an extended period, depletion of groundwater can be severe. Declines in groundwater levels at rates as great as 1 meter per year have led to an overall drop in the water table of between 15 and 60 meters (50 and 200 feet) in some areas. Under these circumstances, it can be said that the groundwater is literally being "mined." Even if pumping were to cease immediately, it could take hundreds or thousands of years for the groundwater to be fully replenished.

Land Subsidence Caused by Groundwater Withdrawal

As you shall see later in this chapter, surface subsidence can result from natural processes related to groundwater. However, the ground may also sink when water is pumped from wells faster than natural recharge processes can replace it. This effect is particularly pronounced in areas underlain by thick layers of loose sediments. As water is withdrawn, the weight of the overburden packs the sediment grains more tightly together and the ground subsides.

Many areas may be used to illustrate such land subsidence. A classic example in the United States occurred in the San Joaquin Valley of California (Figure 4.30). This important agricultural region relies heavily on irrigation. Land subsidence due to groundwater withdrawal began in the valley in the mid-1920s and

Figure 4.30
The shaded area on the map shows California's San Joaquin Valley. The marks on the utility pole in the photo indicate the level of the surrounding land in preceding years. Between 1925 and 1975 this part of the San Joaquin Valley subsided almost 9 meters because of the withdrawal of groundwater and the resulting compaction of sediments. (Photo courtesy of U.S. Geological Survey)

locally exceeded 8 meters (28 feet) by 1970. Then, because of the importation of surface water and a decrease in groundwater pumping, water levels in the aquifer recovered and subsidence ceased. However, during a drought in 1976–1977, heavy groundwater pumping led to renewed subsidence. This time, water levels dropped at a much faster rate than during the previous period because of the reduced storage capacity caused by earlier compaction of material in the aquifer. In all, more than 13,400 square kilometers (5200 square miles) of irrigable land, one-half the entire valley, were affected by subsidence. Damage to structures, including highways, bridges, water lines, and wells, was extensive. Because subsidence changed the gradients of some streams, flooding also became a costly problem. Many other examples of land subsidence due to groundwater pumping occur in the United States and elsewhere in the world.

Groundwater Contamination

The pollution of groundwater is a serious matter, particularly in areas where aquifers provide a large part of the water supply. One common source of groundwater pollution is sewage which emanates from an ever-increasing number of septic tanks. Other sources are inadequate or broken sewer systems, and farm wastes.

If sewage water, which is contaminated with bacteria, enters the groundwater system, it may become purified through natural processes. The harmful bacteria may be mechanically filtered by the sediment through which the water percolates, destroyed by chemical oxidation, and/or assimilated by other organisms. For purification to occur, however, the aquifer must be of the correct composition. For example, extremely permeable aquifers (such as highly fractured crystalline rock, coarse gravel, or cavernous limestone) have such large openings that contaminated groundwater may travel long distances without being cleansed. In this case, the water flows too rapidly and is not in contact with the surrounding material long enough for purification to occur. This is the problem at Well 1 in Figure 4.31A.

On the other hand, when the aquifer is composed of sand or permeable sandstone, the water can sometimes be purified after traveling only a few tens of meters through it. The openings between sand grains are large enough to permit water movement, yet the movement of the water is slow enough to allow ample time for its purification (Well 2, Figure 4.31B).

Other sources and types of contamination also threaten groundwater supplies (Figure 4.32). These include widely used substances such as highway salt, fertilizers that are spread across the land surface, and pesticides. In addition, a wide array of chemicals and industrial materials may leak from pipelines, storage tanks, landfills, and holding ponds. Some of these pollutants are classified as *hazardous*, meaning that they are either flammable, corrosive, explosive, or toxic. As rainwater percolates through the soil, it can carry pollutants to the water table. Here they mix with the groundwater and contaminate the supply.

Because groundwater movement is usually slow, polluted water may go undetected for a long time. In fact, most contamination is discovered only after drinking water has been affected and people become ill. By this time, the volume of polluted water may be very large, and even if the source of contamination is removed immediately, the problem is not solved. Although the sources of groundwater contamination are numerous, the solutions are relatively few.

Once the source of the problem has been identified and eliminated, the most common practice is simply to abandon the water supply and allow the pollutants to be flushed away gradually. This is the least costly and

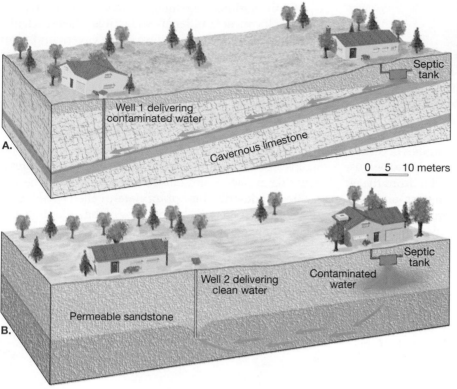

Figure 4.31
A. Although the contaminated water has traveled more than 100 meters before reaching Well 1, the water moves too rapidly through the cavernous limestone to be purified. **B.** As the discharge from the septic tank percolates through the permeable sandstone, it is purified in a relatively short distance.

easiest solution, but the aquifer must remain unused for many years. To accelerate this process, polluted water is sometimes pumped out and treated. Following removal of the tainted water, the aquifer is allowed to recharge naturally or, in some cases, the treated water or other freshwater is pumped back in. This process is costly, time consuming, and it may be risky because there is no way to be certain that all of the contamination has been removed. Clearly, the most effective solution to groundwater contamination is prevention.

Figure 4.32
Sometimes, agricultural chemicals **A.** and materials leached from landfills **B.** find their way into the groundwater. These are two of the potential sources of groundwater contamination. (Photo A by Roy Morsch/The Stock Market; Photo B by F. Rossotto/The Stock Market)

The Geologic Work of Groundwater

Groundwater dissolves rock. This fact is the key to understanding how caverns and sinkholes form. Because soluble rocks, especially limestone, underlie millions of square kilometers of Earth's surface, it is here that groundwater carries on its important role as an erosional agent. Limestone is nearly insoluble in pure water, but is quite easily dissolved by water containing small quantities of carbonic acid. Most natural water contains this weak acid because rainwater readily dissolves carbon dioxide from the air and from decaying plants. Therefore, when groundwater comes in contact with limestone, the carbonic acid reacts with calcite in the rocks to form calcium bicarbonate, a soluble material that is then carried away in solution.

Caverns

Among the most spectacular results of groundwater's erosional handiwork is the creation of limestone **caverns**. In the United States alone about 17,000 caves have been discovered. Although most are relatively small, some have spectacular dimensions. Carlsbad Caverns in southeastern New Mexico and Mammoth Cave in Kentucky are famous examples. One chamber in Carlsbad Caverns has an area equivalent to 14 football fields and enough height to accommodate the U.S. Capitol Building. At Mammoth Cave, the total length of interconnected caverns extends for more than 540 kilometers (340 miles).

Most caverns are created at or below the water table in the zone of saturation. Here acidic groundwater follows lines of weakness in the rock, such as joints and bedding planes. As time passes, the dissolving process slowly creates cavities and gradually enlarges them into caverns. Material that is dissolved by the groundwater is eventually discharged into streams and carried to the ocean.

Certainly the features that arouse the greatest curiosity for most cavern visitors are the stone formations that give some caverns a wonderland appearance (Figure 4.33). These are not erosional features, like the caverns in which they reside, but depositional features. They are created by the seemingly endless dripping of water over great spans of time. The calcite that is left behind produces the limestone we call travertine. These cave deposits, however, are also commonly called *dripstone*, an obvious reference to their mode of origin.

Although the formation of caverns takes place in the zone of saturation, the deposition of dripstone is not possible until the caverns are above the water table, in the zone of aeration. This commonly occurs as nearby streams cut their valleys deeper, lowering the water table as the elevation of the rivers drops. As soon as the chamber is filled with air, the conditions are right for the decoration phase of cavern building to begin.

Of the various dripstone features found in caverns, perhaps the most familiar are **stalactites**. These iciclelike pendants hang from the ceiling of the cavern and form where water seeps through cracks above. When water reaches air in the cave, some of the dissolved carbon dioxide escapes from the drop and calcite begins to precipitate. Deposition occurs as a ring around the edge of the water drop. As drop after drop follows, each leaves an infinitesimal trace of calcite behind, and a hollow limestone tube is created. Water then moves through the tube, remains suspended momentarily at the end, contributes a tiny ring of calcite, and falls to the cavern floor. The stalactite just described is appropriately called a *soda straw* (Figure 4.34). Often the hollow tube of the soda straw becomes plugged or its supply of water increases. In either

Figure 4.33

Speleothems are of many types, including stalactites, stalagmites, and columns. Temple of the Sun Carlsbad Caverns National Park. (Photo by Russ Finley)

Figure 4.34

A "live" solitary soda straw stalactite. (Photo by Clifford Stroud, National Park Service)

case, the water is forced to flow and deposit along the outside of the tube. As deposition continues, the stalactite takes on the more common conical shape.

Formations that develop on the floor of a cavern and reach upward toward the ceiling are called **stalagmites**. The water supplying the calcite for stalagmite growth falls from the ceiling and splatters over the surface. As a result, stalagmites do not have a central tube and are usually more massive in appearance and more rounded on their upper ends than stalactites.

Karst Topography

Many areas of the world have landscapes that, to a large extent, have been shaped by the dissolving power of groundwater. Such areas are said to exhibit **karst topography**, named for the *Krs* region in the border area between Slovenia (formerly a part of Yugoslavia) and Italy where such topography is strikingly developed. In the United States, karst landscapes occur in many areas that are underlain by limestone, including portions of Kentucky, Tennessee, Alabama, southern Indiana, and central and northern Florida. Generally, arid and semiarid areas do not develop karst topography because there is insufficient groundwater. When solution features exist

in such regions, they are likely to be remnants of a time when rainfall was more abundant.

Karst areas typically have irregular terrain punctuated with many depressions called **sinkholes** or, simply, **sinks**. In the limestone areas of Florida, Kentucky, and southern Indiana, there are literally tens of thousands of these depressions varying in depth from just a meter or two to a maximum of more than 50 meters (Figure 4.35).

Sinkholes commonly form in one of two ways. Some develop gradually over many years without any physical disturbance to the rock. In these situations, the limestone immediately below the soil is dissolved by downward-seeping rainwater that is freshly charged with carbon dioxide. These depressions are usually not deep and are characterized by relatively gentle slopes. By contrast, sinkholes can also form suddenly and without warning when the roof of a cavern collapses under its own weight. Typically, the depressions created in this manner are steep-sided and deep. When they form in populous areas, they may represent a serious geologic hazard.

In addition to a surface pockmarked by sinkholes, karst regions characteristically show a striking lack of surface drainage (streams). Following a rainfall, runoff is quickly funneled below ground through sinks. It then flows through caverns until it finally reaches the water table. Where streams do exist at the surface, their paths are usually short. The names of such streams often give a clue to their fate. In the Mammoth Cave area of Kentucky, for example, there is Sinking Creek, Little Sinking Creek, and Sinking Branch. Some sinkholes become plugged with clay and debris, creating small lakes or ponds.

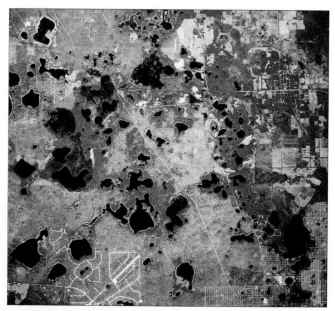

Figure 4.35

This high-altitude infared image shows an area of karst topography in central Florida. The numerous lakes occupy sinkholes. (Courtesy of USDA-ASCS)

Review Questions

1. Describe the movement of water through the hydrologic cycle. Once precipitation has fallen on land, what paths are available to it?

2. A stream starts out 2000 meters above sea level and travels 250 kilometers to the ocean. What is its average gradient in meters per kilometer?

3. Suppose that the stream mentioned in Question 2 developed extensive meanders so that its course was lengthened to 500 kilometers. Calculate its new gradient. How does meandering affect gradient?

4. When the discharge of a stream increases, what happens to the stream's velocity?

5. Define *base level*. Name the main river in your area. For what streams does it act as base level? What is the base level for the Mississippi River? The Missouri River?

6. In what three ways does a stream transport its load?

7. If you collect a jar of water from a stream, what part of its load will settle to the bottom of the jar? What portion will remain in the water?

8. Differentiate between competency and capacity.

9. Why must the height of many artificial levees be increased periodically?

10. What is an artificial cutoff? What is its purpose?

11. What is a divide?

12. Each of the following statements refers to a particular drainage pattern. Identify the pattern.
    (a) Streams diverge from a central high area such as a volcano.
    (b) Streams form a branching, "treelike" pattern
    (c) A pattern that develops when bedrock is crisscrossed by joints and faults

13. Rivers are often used as political boundaries. Do rivers in mature and old-age valleys make good political boundaries? Explain.

14. What percentage of freshwater is groundwater (see Table 4.2)? If glacial ice is excluded and only liquid freshwater is considered, about what percentage is groundwater?

15. Geologically, groundwater is important as an erosional agent. Name another significant geological role of groundwater.

16. Define groundwater and relate it to the water table.

17. How do porosity and permeability differ?

18. What is the source of heat for most hot springs and geysers? How is this reflected in the distribution of these features?

19. What is meant by the term *artesian*? Under what circumstances do artesian wells form?

20. What problem is associated with the pumping of groundwater for irrigation in the southern part of the High Plains?

21. Briefly explain what happened in the San Joaquin Valley of California as the result of excessive groundwater withdrawal.

22. Which would be most effective in purifying polluted groundwater: an aquifer composed mainly of coarse gravel, or sand, or cavernous limestone?

23. Is geothermal power considered an inexhaustible energy source? Explain. (see Box 4.4.)

24. List two conditions required for the development of karst topography.

25. Differentiate between stalacites and stalagmites. How do these features form?

Key Terms

alluvium (p. 95)
aquiclude (p. 108)
aquifer (p. 108)
artesian well (p. 112)
backswamp (p. 96)
base level (p. 92)
bed load (p. 93)
capacity (p. 94)
cavern (p. 116)
competence (p. 94)
cone of depression (p. 111)
cutoff (p. 98)
delta (p. 95)
dendritic pattern (p. 101)
discharge (p. 90)

dissolved load (p. 93)
distributary (p. 96)
divide (p. 100)
drainage basin (p. 100)
drawdown (p. 110)
entrenched meander (p. 102)
floodplain (p. 98)
geyser (p. 109)
gradient (p. 90)
groundwater (p. 107)
hot spring (p. 109)
hydrologic cycle (p. 88)
infiltration (p. 88)
karst topography (p. 117)
meander (p. 98)
natural levee (p. 96)
oxbow lake (p. 98)
permeability (p. 108)

porosity (p. 108)
radial pattern (p. 101)
retangular pattern (p. 101)
rejuvenation (p. 102)
runoff (p. 88)
sinkhole (sink) (p. 117)
sorting (p. 95)
spring (p. 109)
stalactite (p. 116)
stalagmite (p. 117)
suspended load (p. 93)
transpiration (p. 88)
trellis pattern (p. 101)
water table (p. 107)
well (p. 109)
yazoo tributary (p. 96)
zone of aeration (p. 108)
zone of saturation (p. 107)

CHAPTER 5
Glaciers, Deserts, and Wind

Sand dunes, Death Valley National Monument, California. (Photo by Carr Clifton)

Today, glaciers cover nearly 10 percent of Earth's land surface; however, in the recent geologic past ice sheets were three times more extensive, covering vast areas with ice thousands of meters thick. Many regions still bear the mark of these glaciers. The first part of this chapter examines glaciers and the erosional and depositional features they create. The second part is devoted to dry lands and the geologic work of wind. Because desert and near-desert conditions prevail over an area as large as that affected by the massive glaciers of the Ice Age, the nature of such landscapes is indeed worth investigating.

Many present-day landscapes were modified by the widespread glaciers of the most recent Ice Age and still strongly reflect the handiwork of ice. The basic character of such diverse places as the Alps, Cape Cod, and Yosemite Valley was fashioned by now-vanished masses of glacial ice. Moreover, Long Island, the Great Lakes, and the fiords of Norway and Alaska all owe their existence to glaciers. Glaciers, of course, are not just a phenomenon of the geologic past. As we shall see, they are still sculpting and depositing debris in many regions today.

Glaciers: A Part of the Hydrologic Cycle

Earlier we learned that Earth's water is in constant motion. Time and time again the same water is evaporated from the oceans into the atmosphere, precipitated upon the land, and carried by rivers and underground flow back to the sea. However, when precipitation falls at high elevations or high latitudes, the water may not immediately make its way toward the sea. Instead, it may become part of a glacier. Although the ice will eventually melt allowing the water to continue its path to the sea, water can be stored as glacial ice for many tens, hundreds, or even thousands of years.

A **glacier** is a thick ice mass that forms over hundreds or thousands of years. It originates on land from the accumulation, compaction, and recrystallization of snow. A glacier appears to be motionless, but it is not—glaciers move very slowly. Like running water, groundwater, wind, and waves, glaciers are dynamic erosional agents that accumulate, transport, and deposit sediment. Although glaciers are found in many parts of the world today, most are located in remote areas, either near Earth's poles or in high mountains.

Literally thousands of relatively small glaciers exist in lofty mountain areas, where they usually follow valleys originally occupied by streams. Unlike the rivers that previously flowed in these valleys, the glaciers advance slowly, perhaps only a few centimeters per day. Because of their setting, these moving ice masses are termed **valley glaciers** or **alpine glaciers** (Figure 5.1). Each glacier is a stream of ice, bounded by precipitous

rock walls, that flows downvalley from a snow accumulation center near its head. Like rivers, valley glaciers can be long or short, wide or narrow, single or with branching tributaries. Generally the widths of alpine glaciers are small compared to their lengths. In length, some extend for just a fraction of a kilometer, whereas others go on for many tens of kilometers. The west branch of the Hubbard Glacier, for example, runs through 112 kilometers of mountainous terrain in Alaska and the Yukon Territory.

In contrast to valley glaciers, **ice sheets** exist on a much larger scale. These enormous masses flow out in all directions from one or more centers and completely obscure all but the highest areas of underlying terrain. Although many ice sheets have existed in the past, just two achieve this status at present (Figure 5.2). In the Northern Hemisphere, Greenland is covered by an imposing ice sheet averaging nearly 1500 meters thick. It occupies 1.7 million square kilometers, or about 80 percent of this large island. In the Southern Hemisphere, the huge Antarctic Ice Sheet attains a maximum thickness of nearly 4300 meters and covers an area of more than 13.9 million square kilometers. Because of the proportions of these huge features, they often are called *continental ice sheets*. Indeed, the combined areas of present-day continental ice sheets represent almost 10 percent of Earth's land area.

In addition to valley glaciers and ice sheets, other types of glaciers are also identified. Covering some uplands and plateaus are masses of glacial ice called **ice caps**. They resemble ice sheets but are much smaller than the continental-scale features. Ice caps occur in many places, including Iceland and several of the large islands in the Arctic Ocean. Another type, known as **piedmont glaciers**, occupy broad lowlands at the bases of steep mountains and form when one or more valley glaciers emerge from the confining walls of mountain valleys. Here the advancing ice spreads out to form a broad sheet. The size of individual piedmont glaciers varies greatly. Among the largest is the broad Malaspina Glacier along the coast of southern Alaska. It covers more than 5000 square kilometers of the flat coastal plain at the foot of the lofty St. Elias range.

How Glaciers Move

The movement of glacial ice is generally referred to as *flow*. The fact that glacial movement is described in this way seems paradoxical—how can a solid flow? Glacial ice flows in two ways. One mechanism involves plastic movement within the ice. Ice behaves as a brittle solid until the pressure upon it is equivalent to the weight of about 50 meters (165 feet) of ice. Once that load is surpassed, ice behaves as a plastic mater-

Figure 5.1
Gornergrat Glacier in the Swiss Alps is an example of a valley glacier. (Photo by RAGA/The Stock Market)

ial and flow begins. A second and often equally important mechanism of glacial movement consists of the whole ice mass slipping along the ground. The lowest portions of most glaciers are thought to move by this sliding process.

The uppermost 50 meters of a glacier are appropriately referred to as the *zone of fracture*. Since there is not enough overlying ice to cause plastic flow, this upper part of the glacier consists of brittle ice. Consequently, the ice in this zone is carried along piggyback style by the ice below. When the glacier moves over irregular terrain, the zone of fracture is subjected to tension, with cracks called **crevasses** resulting (Figure 5.3). These gaping cracks, which often make travel across glaciers dangerous, may extend to depths of 50 meters (165 feet). Beyond this depth, plastic flow seals them off.

Unlike streamflow, glacial movement is not obvious (see Box 5.1). If we could watch a valley glacier move, we would see that, like the water in a river, all of the ice does not move downstream at the same rate. Flow is greatest in the center of the glacier because of the drag created by the walls and floor of the valley.

How rapidly does glacial ice move? Average rates vary considerably from one glacier to another. Some move so slowly that trees and other vegetation may become well established in the debris that accumulates on the glacier's surface. Others advance up to several meters per day. The movement of some glaciers is characterized by periods of extremely rapid advance followed by periods during which movement is practically nonexistent.

Snow is the raw material from which glacial ice originates. Therefore, glaciers form in areas where more snow falls in winter than can melt during the summer. Glaciers

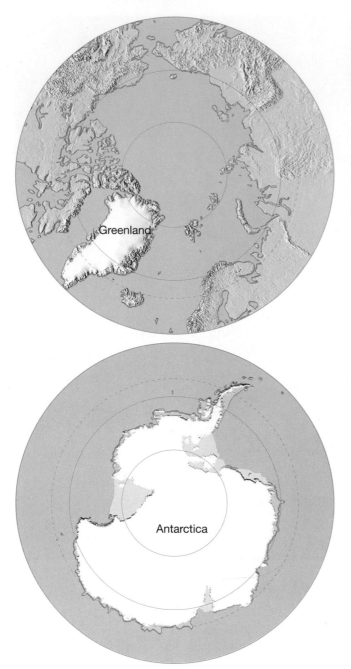

Figure 5.2
The only present-day continental ice sheets are those covering Greenland and Antarctica. Their combined areas represent almost 10 percent of Earth's land area. Greenland's ice sheet occupies 1.7 million square kilometers, or about 80 percent of the island. The area of the Antarctic Ice Sheet is almost 14 million square kilometers.

Figure 5.3
Crevasses are clearly visible in this aerial view of Mendenhall Glacier near Juneau, Alaska. Crevasses form in the brittle ice of the zone of fracture. (Photo by Wolfgang Kaehler)

are constantly gaining and losing ice. Snow accumulation and ice formation occur in the **zone of accumulation** (Figure 5.4). Here the addition of snow thickens the glacier and promotes movement. Beyond this area of ice formation is the **zone of wastage**. Here there is a net loss to the glacier as all of the snow from the previous winter melts, as does some of the glacial ice (Figure 5.4).

In addition to melting, glaciers also waste as large pieces of ice break off the front of a glacier in a process called *calving*. Where glaciers reach the sea, calving creates icebergs (Figure 5.5). Because icebergs are just slightly less dense than seawater, they float very low in the water, with more than 80 percent of their mass submerged. The margins of the Greenland Ice Sheet produce thousands of icebergs each year. Many drift southward and find their way into the North Atlantic, where they are a hazard to navigation.

Whether the margin of a glacier is advancing, retreating, or remaining stationary depends on the budget of the glacier. That is, it depends on the balance or lack of balance between accumulation on the one hand and wastage on the other. If ice accumulation exceeds wastage, the glacial front advances until the two factors balance. At this point, the terminus of the glacier becomes stationary. At a later time, when wastage exceeds accumulation, the ice front will retreat until a balance is again reached.

Box 5.1

Do Glaciers Move? An Application of the Scientific Method

Today we know that glaciers can do extraordinary erosional work and that in the past glacial ice has affected vast areas that are now ice-free. We understand much about how glaciers form and move, as well as how they erode and deposit. This and other knowledge about glaciers has been acquired gradually over the past 200 years, yet scientists still have much more to learn about these moving masses of ice.

The study of glaciers provides an early application of the scientific method. High in the Alps of Switzerland and France, small glaciers exist in the upper portions of some valleys. In the late eighteenth and early nineteenth centuries, people who farmed and herded animals in these valleys suggested that glaciers in the upper reaches of the valleys had previously been much larger and had occupied downvalley areas. They based their belief on the fact that the valley floors were littered with angular boulders and other rock debris that seemed identical to the materials that they could see in and near the glaciers at the heads of the valleys.

Although the explanation of these observations seemed logical, others did not accept the notion that masses of ice hundreds of meters thick were capable of movement. The disagreement was settled after a simple experiment was designed and carried out to test the hypothesis that glacial ice can move.

Markers were placed in a straight line completely across an alpine glacier. The position of the line was marked on the valley walls so that if the ice moved,

the change in position could be detected. After a year or two the results were clear: the markers on the glacier had advanced down the valley, proving that glacial ice indeed moves. In addition, the experiment demonstrated that ice within a glacier does not move at a uniform rate, because the markers in the center advanced farther than did those along the margins. Although most glaciers move too slowly for direct visual detection, the experiment succeeded in demonstrating that movement nevertheless occurs. In the years that followed, this experiment was repeated many times with greater accuracy using

more modern surveying techniques. Each time, the basic relationships established by earlier attempts were verified.

The experiment illustrated in Figure 5.A was carried out at Switzerland's Rhone Glacier later in the nineteenth century. It not only traced the movement of markers within the ice, but also mapped the position of the glacier's terminus. Notice that even though the ice within the glacier was advancing, the ice front was retreating. As often occurs in science, experiments and observations designed to test one hypothesis yield new information that requires further analysis and explanation.

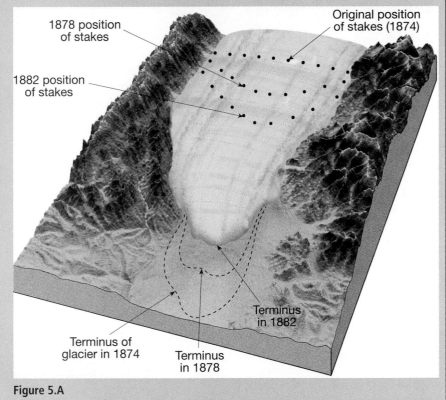

Figure 5.A

Ice movement and changes in the terminus at Rhone Glacier, Switzerland. In this classic study of a valley glacier, the movement of stakes clearly showed that ice along the sides of the glacier moves slowest. Also notice that even though the ice front was retreating, the ice within the glacier was advancing.

Whether the margin of a glacier is advancing, retreating, or stationary, the ice within the glacier continues to flow forward. In the case of a receding glacier, the ice simply does not flow forward rapidly enough to offset wastage. This point is illustrated in Figure 5.A. While the line of stakes within the Rhone Glacier continued to move downstream, the terminus of the glacier slowly retreated upstream.

Glacial Erosion

Glaciers erode tremendous volumes of rock. For anyone who has observed the terminus of an alpine glacier, the evidence of its erosive force is plain. You can witness firsthand the melting ice unlocking rock material of all sizes from huge boulders to tiny grit. All signs lead to the conclusion that the ice has scraped, scoured,

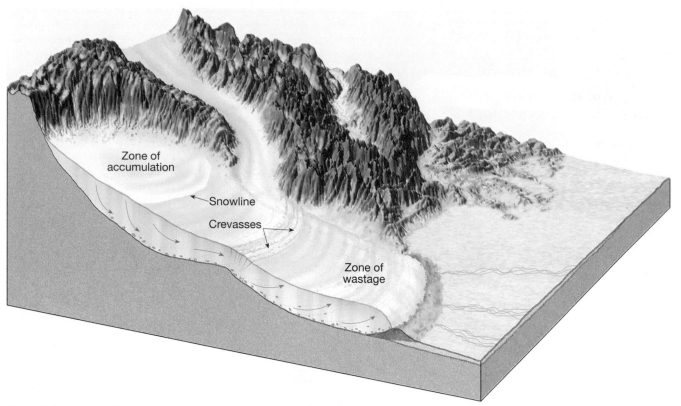

Figure 5.4
The snowline separates the zone of accumulation and the zone of wastage. Above the snowline, more snow falls each winter than melts each summer. Below the snowline, the snow from the previous winter completely melts, as does some of the underlying ice. Whether the margin of a glacier advances, retreats, or remains stationary depends on the balance or lack of balance between accumulation and wastage. When a glacier moves across irregular terrain, *crevasses* form in the brittle portion.

and torn rock debris from the floor and walls of the valley and carried it downslope.

Indeed, as a transporter of sediment, ice has no equal, because once the debris is acquired by the ice, it cannot settle out as does the load carried by a stream or by the wind. Consequently, glaciers can carry huge blocks that no other erosional agent could possibly budge. Although glaciers are of limited importance as

Figure 5.5
Icebergs are created when large pieces break off the front of a glacier after it reaches a water body. These formed along the coast of the Antarctic Peninsula. (Photo by Barbara Gerlach/DRK Photo)

erosional agents today, many landscapes that were modified by the widespread glaciers of the recent Ice Age still reflect to a high degree the work of ice.

How Glaciers Erode

Glaciers erode land primarily in two ways. First, as a glacier flows over a fractured bedrock surface, it loosens and lifts blocks of rock, incorporates them into the ice, and carries them off. This process, known as **plucking**, occurs when meltwater penetrates the cracks and joints along the rock floor of the glacier and refreezes. As the water expands, it exerts tremendous leverage that pries the rock loose. In this manner, sediment of all sizes becomes part of the glacier's load.

The second major erosional process is *abrasion*. As the ice with its load of rock fragments moves along, it acts as a giant rasp or file and grinds the surface below as well as the rocks within the ice. The pulverized rock produced by the glacial "grist mill" is appropriately called **rock flour**. So much rock flour may be produced that meltwater streams leaving a glacier often have the grayish appearance of skimmed milk—visible evidence of the grinding power of the ice. When the embedded material includes large fragments, long scratches and grooves called **glacial striations** may be gouged into the bedrock passed over by the glacier (Figure 5.6). These linear scratches on the bedrock surface provide clues to the direction of glacial movement. By mapping the striations over large areas, glacial flow patterns can often be reconstructed. On the other hand, not all abrasive action produces striations. When the sediment consists primarily of fine silt-sized particles, the rock surfaces over which the glacier moves may become highly polished.

Although the erosional accomplishments of ice sheets can be tremendous, landforms carved by these huge ice masses usually do not inspire the same awe as do the erosional features created by valley glaciers. In regions where the erosional effects of ice sheets are significant, glacially scoured surfaces and subdued terrain are the rule. By contrast, in mountainous areas, erosion by valley glaciers produces many truly spectacular features. Much of the rugged mountain scenery so celebrated for its majestic beauty is the product of erosion by valley glaciers.

Features of Valley Glaciers

Take a moment to study Figure 5.7 which shows a mountain setting before, during, and after glaciation. You will refer to this often in the following discussion.

Prior to glaciation, alpine valleys are characteristically V-shaped because streams are well above base level and are therefore downcutting (Figure 5.7A). However, in mountainous regions that have been glaciated, the valleys are no longer narrow. As a glacier moves down a valley once occupied by a stream, the ice modifies it in three ways: The glacier widens,

Figure 5.6
Glacial abrasion created the scratches and grooves in this bedrock. Glacier Bay National Park, Alaska. (Photo by Carr Clifton)

deepens, and straightens the valley, so that what was once a narrow V-shaped valley is transformed into a U-shaped **glacial trough** (Figures 5.7C and 5.8).

Since the amount of glacial erosion depends in part on the thickness of the ice, the main or trunk glacier cuts its valley deeper than can the smaller tributary glaciers. Thus, after the ice has receded, the valleys of tributary glaciers are left standing above the main trough and are termed **hanging valleys**. Rivers flowing through hanging valleys may produce spectacular waterfalls, such as those in Yosemite National Park, California (Figure 5.9).

At the head of a glacial valley is a characteristic and often imposing feature associated with an alpine glacier—a **cirque**. As Figure 5.10 illustrates, these hollowed-out, bowl-shaped depressions have precipitous walls on three sides but are open on the downvalley side. The cirque represents the focal point of the glacier's source, that is, the area of snow accumulation and ice formation. Cirques begin as irregularities in the mountainside that are subsequently enlarged by frost wedging and plucking along the sides and bottom of the glacier. The glacier, in turn, acts as a conveyor belt that carries the debris away. After the glacier has melted away, the cirque basin is often occupied by a small lake.

Figure 5.7

Erosional landforms created by alpine glaciers. The unglaciated landscape in part **A** is modified by valley glaciers in part **B**. After the ice recedes, in part **C**, the terrain looks very different than before glaciation.

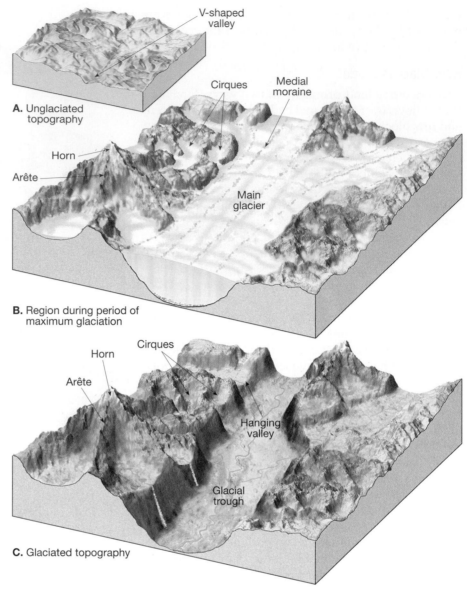

A. Unglaciated topography

B. Region during period of maximum glaciation

C. Glaciated topography

The Alps, Northern Rockies, and many other mountain landscapes carved by valley glaciers reveal more than glacial troughs and cirques. In addition, sinuous, sharp-edged ridges called **arêtes** and sharp, pyramid-like peaks called **horns** project above the surroundings (Figure 5.7C). Both features can originate from the same basic process—the enlargement of cirques produced by plucking and frost action. A group of cirques around a single high mountain create the spires of rock called horns. As the cirques enlarge and converge, an isolated horn is produced. The most famous example is the Matterhorn in the Swiss Alps (Figure 5.11).

Arêtes can form in a similar manner except that the cirques are not clustered around a point but rather exist on opposite sides of a divide. As the cirques grow, the divide separating them is reduced to a very narrow, knifelike partition. An arête may also be created in another way. When two glaciers occupy parallel valleys, an arête can form when the divide separating the moving tongues of ice is progressively narrowed as the glaciers scour and widen their valleys.

Fiords are deep, often spectacular, steep-sided inlets of the sea that exist in many high-latitude areas of the world where mountains are adjacent to the ocean (Figure 5.12). Norway, British Columbia, Greenland, New Zealand, Chile, and Alaska all have coastlines characterized by fiords. They are glacial troughs that became submerged as the ice left the valley and sea level rose following the Ice Age. The depth of some fiords exceeds 1000 meters (3300 feet).

However, the great depths of these flooded troughs are only partly explained by the post-Ice Age rise in sea level. Unlike the situation governing the downward erosional work of rivers, sea level does not act as base level for glaciers. As a consequence, glaciers are capable of eroding their beds far below the surface of the sea. For

Figure 5.8
Prior to glaciation, a mountain valley is typically narrow and V-shaped. During glaciation, an alpine glacier widens, deepens, and straightens the valley, creating the U-shaped glacial trough seen here. The string of lakes is called pater noster lakes. This valley is in Glacier National Park, Montana. (Photo by John Montagne)

example, a valley glacier 300 meters (1000 feet) thick can carve its valley floor more than 250 meters (800 feet) below sea level before downward erosion ceases and the ice begins to float.

Glacial Deposits

Glaciers pick up and transport a huge load of debris as they slowly advance across the land. Ultimately these materials are deposited when the ice melts. In regions where glacial sediment is deposited, it can play a truly significant role in forming the physical landscape. For example, in many areas once covered by the ice sheets of the recent Ice Age, the bedrock is rarely exposed because glacial deposits that are tens or even hundreds of meters thick completely mantle the terrain. The general effect of these deposits is to reduce the local relief and thus level the topography. Indeed, much of the familiar country scenery today—rocky pastures in New England, wheat fields in the Dakotas, rolling farmland in the Midwest—results directly from glacial deposition.

Types of Glacial Drift

Long before the theory of an extensive Ice Age was proposed, much of the soil and rock debris covering portions of Europe was recognized as coming from elsewhere. At the time, these foreign materials were believed to have been "drifted" into their present positions by floating ice during an ancient flood. As a consequence, the term *drift* was applied to this sediment. Although rooted in a concept that was not correct, this term was

so well established by the time the true glacial origin of the debris became widely recognized that it remained in the glacial vocabulary. Today, **drift** is an all-embracing term for sediments of glacial origin, no matter how, where, or in what form they were deposited.

Glacial drift is divided into two distinct types: (1) materials deposited directly by the glacier, which are known as *till*, and (2) sediments laid down by glacial meltwater, called *stratified drift*. Here is the difference: **till** is deposited as glacial ice melts and drops its load of rock fragments. Unlike moving water and wind, ice cannot sort the sediment it carries; therefore, deposits of till are characteristically unsorted mixtures of many particle sizes (Figure 5.13). **Stratified drift** is sorted according to the size and weight of the fragments. Since ice is not capable of such sorting activity, these sediments are not deposited directly by the glacier. Rather, they reflect the sorting action of glacial meltwater. Some deposits of stratified drift are made by streams issuing directly from the glacier. Other stratified deposits involve sediment that was originally laid down as till and later picked up, transported, and redeposited by meltwater beyond the margin of the ice. Accumulations of stratified drift often consist largely of sand and gravel, because the meltwater is not capable of moving larger material and because the finer rock flour remains suspended and is commonly carried far from the glacier. An indication that stratified drift consists primarily of sand and gravel can be seen in many areas where these deposits are actively mined as aggregate for road work and other construction projects.

Figure 5.9
Bridalveil Falls in Yosemite National Park cascades from a hanging valley into the glacial trough below. (Photo by E. J. Tarbuck)

When boulders are found in the till or lying free on the surface, they are called **glacial erratics** if they are different from the bedrock below. Of course, this means that they must have been derived from a source outside the area where they are found (Figure 5.14). Although the locality of origin for most erratics is unknown, the origin of some can be determined. Therefore, by studying glacial erratics as well as the mineral composition of the till, geologists can sometimes trace the path of a lobe of ice. In portions of New England as well as other areas, erratics may be seen dotting pastures and farm fields. In some places, these rocks were cleared from fields and piled to make fences and walls.

Moraines, Outwash Plains, and Kettles

Perhaps the most widespread features created by glacial deposition are *moraines* which are simply layers or ridges of till. Several types of moraines are identified; some are common only to mountain valleys, and others are associated with areas affected by either ice sheets or valley glaciers. Lateral and medial moraines fall in the first category, whereas end moraines and ground moraines are in the second.

The sides of a valley glacier accumulate large quantities of debris from the valley walls. When the glacier wastes away, these materials are left as ridges, called

Figure 5.10
Aerial view of bowl-shaped depressions called cirques in the Uinta Range, Utah. (Photo by John S. Shelton)

lateral moraines, along the sides of the valley (Figure 5.15). **Medial moraines** are formed when two valley glaciers coalesce to form a single ice stream. The till that was once carried along the edges of each glacier joins to form a single dark stripe of debris within the newly enlarged glacier. The creation of these dark stripes within the ice stream is one obvious proof that glacial ice moves, because the medial moraine could not form if the ice did not flow downvalley (Figure 5.15).

End moraines, as the name implies, form at the terminus of a glacier. Here, while the ice front is stationary, the glacier continues to carry in and deposit large quantities of rock debris, creating a ridge of till tens to hundreds of meters high. The end moraine marking the farthest advance of the glacier is called the *terminal moraine*, and those moraines that formed as the ice front

Figure 5.11
Horns are sharp, pyramid-like peaks that are fashioned by alpine glaciers. This example is the famous Matterhorn in the Swiss Alps. (Photo by E. J. Tarbuck)

Figure 5.12
Like other fiords, this one near Ketchikan in southeastern Alaska is a drowned glacial trough. (Photo by F. Stuart Westmorland/Photo Researchers, Inc.)

Figure 5.13

Glacial till is an unsorted mixture of many different sediment sizes. (Photo by E. J. Tarbuck)

Figure 5.15

Lateral moraines form from the accumulation of debris along the sides of a valley glacier. Medial moraines form when the lateral moraines of merging valley glaciers join. Medial moraines could not form if the ice did not advance downvalley. Therefore, these dark stripes are proof that glacial ice moves. (Photo by Austin Post, U.S. Geological Survey)

periodically became stationary during retreat are termed *recessional moraines*. As the glacier recedes, a layer of till is laid down, forming a gently undulating surface of **ground moraine**. Ground moraine has a leveling effect, filling in low spots and clogging old stream channels, often leading to a disruption of drainage.

End moraines deposited by the most recent major stage of Ice Age glaciation are prominent features in many parts of the Midwest and Northeast. In Wisconsin, the wooded, hilly terrain of the Kettle Moraine near Milwaukee is a particularly picturesque example. A well-known example in the Northeast is Long Island. This linear strip of glacial sediment that extends north-

eastward from New York City is part of an end moraine complex that stretches from eastern Pennsylvania to Cape Cod, Massachusetts (Figure 5.16).

At the same time that an end moraine is forming, meltwater emerges from the ice in rapidly moving streams. Often they are choked with suspended material and carry a substantial bed load. As the water leaves the glacier, it rapidly loses velocity and much of its bed

Figure 5.14

Land cleared of glacial erratics which were then piled atop one another to build this stone wall near West Bend, Wisconsin. (Photo by Tom Bean)

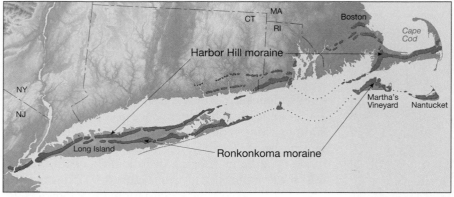

Figure 5.16
End moraines make up substantial parts of Long Island, Cape Cod, Martha's Vineyard, and Nantucket. Although portions are submerged, the Ronkonkoma moraine (a terminal moraine) extends through central Long Island, Martha's Vineyard, and Nantucket. It was deposited about 20,000 years ago. The recessional Harbor Hill moraine, which formed about 14,000 years ago, extends along the north shore of Long Island, through southern Rhode Island and Cape Cod.

load is dropped. In this way a broad, ramplike surface of stratified drift is built adjacent to the downstream edge of most end moraines. When the feature is formed in association with an ice sheet, it is termed an **outwash plain**, and when it is confined to a mountain valley, it is usually referred to as a **valley train**. Figure 5.17 shows an outwash plain and other common depositional features.

Often end moraines, outwash plains, and valley trains are pockmarked with basins or depressions known as **kettles** (Figures 5.17 and 5.18). Kettles form when blocks of stagnant ice become buried in drift and melt, leaving pits in the glacial sediment. Most kettles do not exceed 2 kilometers in diameter and the typical depth of most kettles is less than 10 meters (33 feet).

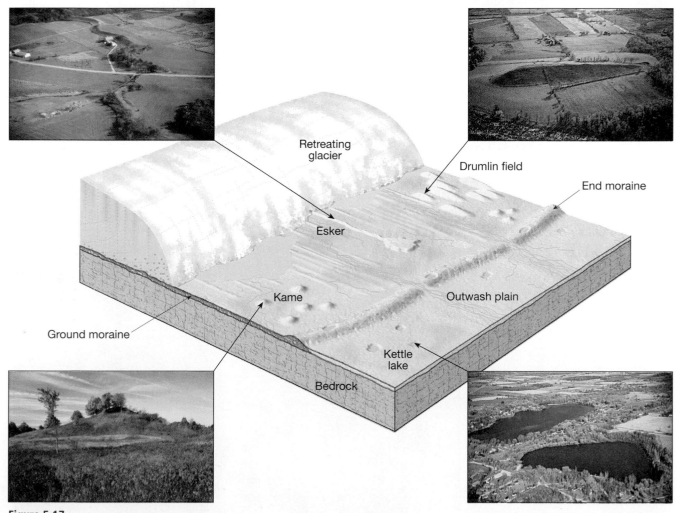

Figure 5.17
This hypothetical area illustrates many common depositional landforms.

Figure 5.18
These ponds occupy depressions called kettles. Kettles form when blocks of ice that were buried in drift melt and leave pits. Chippewa Moraine, Wisconsin. (Photo by Tom Bean)

Water often fills the depression and forms a pond or lake. One well-known example is Walden Pond near Concord, Massachusetts. It is here that Henry David Thoreau lived alone for two years in the 1840s and about which he wrote *Walden*, his classic of American literature.

Drumlins, Eskers, and Kames

Drumlins are streamlined asymmetrical hills composed of till (Figures 5.17 and 5.19). They range in height from 15 to 60 meters (50 to 200 feet) and average 0.4–0.8 kilometer (0.25–0.50 mile) in length. The steep side of the hill faces the direction from which the ice advanced, while the gentler slope points in the direction the ice moved.

Drumlins are not found singly, but rather occur in clusters, called *drumlin fields*. One such cluster, east of Rochester, New York, is estimated to contain about 10,000 drumlins. Their streamlined shape indicates that they were molded in the zone of flow within an active glacier. It is thought that drumlins originate when glaciers advance over previously deposited drift and reshape the material.

In some areas that were once occupied by glaciers, sinuous ridges composed largely of sand and gravel may be found. These ridges, called **eskers**, are deposits made by streams flowing in tunnels beneath the ice, near the terminus of a glacier (Figure 5.17). They may be several meters high and extend for many kilometers. In some areas they are mined for sand and gravel, and for this reason, eskers are disappearing in some localities.

Kames are steep-sided hills that, like eskers, are composed of sand and gravel (Figure 5.17). Kames originate when glacial meltwater washes sediment into openings and depressions in the stagnant wasting terminus of a glacier. When the ice eventually melts away, the stratified drift is left behind as mounds or hills.

Glaciers of the Past

At various points in the preceding pages, we mentioned the Ice Age, a time when ice sheets and alpine glaciers were far more extensive than they are today. There was a time when the most popular explanation for drift was that the material had been drifted in by means of icebergs or perhaps simply swept across the landscape by a catastrophic flood. However, during the nineteenth century, field investigations by many scientists provided convincing proof that an extensive Ice Age was responsible for these deposits, and for many other features.

Figure 5.19
Drumlins, such as this one in upstate New York, are depositional features associated with continental ice sheets. (Courtesy of Ward's Natural Science Establishment, Inc., Rochester, N.Y.)

By the beginning of the twentieth century, geologists had largely determined the extent of Ice Age glaciation. Further, they discovered that many glaciated regions had not one layer of drift, but several. Close examination of these older deposits showed well-developed zones of chemical weathering and soil formation as well as the remains of plants that require warm temperatures. The evidence was clear: there had not been just one glacial advance but several, each separated by extended periods when climates were as warm or warmer than at present. The Ice Age had not simply been a time when the ice advanced over the land, lingered for a while, and then receded. Rather, the period was a very complex event characterized by a number of advances and withdrawals of glacial ice.

The glacial record on land is punctuated by many erosional gaps. This makes it difficult to clearly reconstruct the episodes of the Ice Age. But sediment on the ocean floor provides an uninterrupted record of climate cycles for this period.* Studies of cores drilled from these seafloor sediments show that glacial/interglacial cycles have occurred about every 100,000 years. About 20 such cycles of cooling and warming were identified for the span we call the Ice Age.

During the glacial age, ice left its imprint on almost 30 percent of Earth's land area, including about 10 million square kilometers of North America, 5 million square kilometers of Europe, and 4 million square kilometers of Siberia (Figure 5.20). The amount of glacial ice in the Northern Hemisphere was roughly twice that of the Southern Hemisphere. The primary reason is that the Southern Hemisphere has little land in the middle latitudes and therefore the southern polar ice could not spread far beyond the margins of Antarctica. By contrast, North America and Eurasia provided great expanses of land for the spread of ice sheets.

Today we know that the Ice Age began between two and three million years ago. This means that most of the major glacial episodes occurred during a division of the geologic time scale called the **Pleistocene epoch**. Although the Pleistocene is commonly used as a synonym for the Ice Age, this epoch does not encompass it all. The Antarctic ice sheet, for example, formed at least 14 million years ago, and, in fact, might be much older.

---

*For more on this topic, see "Seafloor Sediments and Climatic Change" in Chapter 12, p. 329.

Figure 5.20

Maximum extent of glaciation in the Northern Hemisphere during the Ice Age.

Glaciers have not been ever-present features throughout Earth's long history. In fact, for most of geologic time, glaciers have been absent. Evidence does indicate that, in addition to the Pleistocene epoch, there were at least three earlier periods of glacial activity: 2 billion, 600 million, and 250 million years ago. However, the most recent period of glaciation is of greatest interest, because the features of many present-day landscapes are a reflection of the work of Pleistocene glaciers.

Some Indirect Effects of Ice Age Glaciers

In addition to the massive erosional and depositional work carried on by Pleistocene glaciers, the ice sheets had other, sometimes profound, effects on the landscape. For example, as the ice advanced and retreated, animals and plants were forced to migrate. This led to stresses that some organisms could not tolerate. Furthermore, many present-day stream courses bear little resemblance to their preglacial routes. The Missouri River once flowed northward toward Hudson Bay in Canada. The Mississippi River followed a path through central Illinois, and the head of the Ohio River reached only as far as Indiana. Some rivers that today carry only a trickle of water but occupy broad channels are a testament to the fact that they once carried torrents of glacial meltwater.

In areas that were centers of ice accumulation, such as Scandinavia and northern Canada, the land has been slowly rising for the past several thousand years. The land had downwarped under the tremendous weight of three-kilometer-thick masses of ice. Following the removal of this immense load, the crust has been adjusting by gradually rebounding upward ever since.

A far-reaching effect of the Ice Age was the worldwide change in sea level that accompanied each advance and retreat of the ice sheets (see Box 5.2). The snow that nourishes glaciers ultimately comes from moisture evaporated from the oceans. Therefore, when the ice sheets increased in size, sea level fell and the shoreline shifted seaward (see Figure 5.B). Estimates suggest that sea level was as much as 100 meters (330 feet) lower than today. Consequently, the Atlantic Coast

---

Box 5.2

What If the Ice Melted?

How much water is stored as glacial ice? Estimates by the U.S. Geological Survey indicate that only slightly more than 2 percent of the world's water is accounted for by glaciers. But this small figure may be misleading when the actual amounts of water are considered. The total volume of all valley glaciers is about 210,000 cubic kilometers, comparable to the combined volume of the world's largest saline and freshwater lakes. Furthermore, 80 percent of the world's ice and nearly two-thirds of Earth's freshwater are represented by Antarctica's ice sheet, which covers an area almost one and one-half times that of the United States. If this ice melted, sea level would rise an estimated 60 to 70 meters, and the ocean would inundate many densely populated coastal areas (Figure 5.B). The hydrologic importance of the continent and its ice can be illustrated in another way. If Antarctica's ice sheet were melted at a uniform rate, it could feed (1) the Mississippi

River for more than 50,000 years, (2) all the rivers in the United States for about 17,000 years, (3) The Amazon River for approximately, 5000 years, or (4) all the rivers of the world for about 750 years.

As the foregoing discussion illustrates, the quantity of ice on Earth today is truly immense. However, present glaciers occupy only about one-third the area they did in the very recent geologic past.

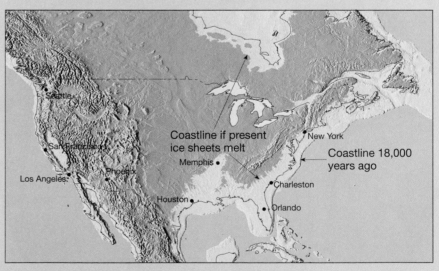

Figure 5.B

This map of a portion of North America shows the present-day coastline compared to the coastline that existed during the last ice-age maximum (18,000 years ago) and the coastline that would exist if present ice sheets in Greenland and Antarctica melted. (After R. H. Dott, Jr. and R. L. Battan, *Evolution of the Earth*, New York: McGraw-Hill, 1971. Reprinted by permission of the publisher.)

of the United States was located more than 100 kilometers (60 miles) to the east of New York City. Moreover, France and Britain were joined where the English Channel is today. Alaska and Siberia were connected across the Bering Strait, and Southeast Asia was tied by dry land to the islands of Indonesia.

The formation and growth of ice sheets was an obvious response to significant changes in climate. But the existence of the glaciers themselves triggered climatic changes in the regions beyond their margins. In arid and semiarid areas on all continents, temperatures were lowered, which meant evaporation rates were lowered, too. At the same time, precipitation was moderate. This cooler, wetter climate resulted in the formation of many lakes called **pluvial lakes** (from the Latin term *pluvia* meaning "rain"). In North America, pluvial lakes were concentrated in the vast Basin and Range region of Nevada and Utah (Figure 5.21). Although most are now gone, there are a few remnants, the largest being Utah's Great Salt Lake.

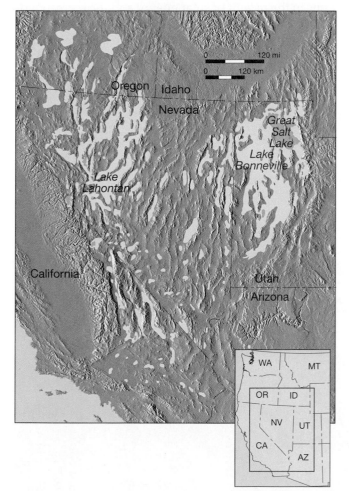

Figure 5.21

Pluvial lakes of the Western United States. (After R. F. Flint, *Glacial and Quaternary Geology*, New York: John Wiley & Sons)

Causes of Glaciation

A great deal is known about glaciers and glaciation. Much has been learned about glacier formation and movement, the extent of glaciers past and present, and the features created by glaciers, both erosional and depositional. However, scientists have not yet developed a completely satisfactory explanation for the causes of glacial ages.

Any theory that attempts to explain the causes of glacial ages must successfully answer two basic questions. (1) *What causes the onset of glacial conditions?* For continental ice sheets to have formed, average temperatures must have been somewhat lower than at present and perhaps substantially lower than throughout much of geologic time. Thus, a successful theory would have to account for the cooling that finally leads to glacial conditions. (2) *What caused the alternation of glacial and interglacial stages that have been documented for the Pleistocene epoch?* The first question deals with long-term trends in temperature on a scale of millions of years but this second question relates to much shorter-term changes.

Although the literature of science contains a vast array of hypotheses relating to the possible causes of glacial periods, we will discuss only a few major ideas to summarize current thought.

Plate Tectonics

Probably the most attractive proposal for explaining the fact that extensive glaciations have occurred only a few times in the geologic past comes from the theory of plate tectonics.[*] Not only does this theory provide geologists with explanations about many previously misunderstood processes and features, but it also provides a possible explanation for some hitherto unexplainable climatic changes, including the onset of glacial conditions. Because glaciers can form only on land, we know that landmasses must exist somewhere in the higher latitudes before an ice age can commence. Many believe that ice ages have occurred only when Earth's shifting crustal plates have carried the continents from tropical latitudes to more poleward positions.

Glacial features in present-day Africa, Australia, South America, and India indicate that these regions, which are now tropical or subtropical, experienced an Ice Age near the end of the Paleozoic era, about 250 million years ago. However, there is no evidence that ice sheets existed during this same period in what are today the higher latitudes of North America and Eurasia. For many years this puzzled scientists. Was the climate in these relatively tropical latitudes once like it is today in Greenland and Antarctica? Why did glaciers

_____
[*] A complete discussion of plate tectonics is presented in Chapter 7.

not form in North America and Eurasia? Until the plate tectonics theory was formulated, there had been no reasonable explanation.

Today, scientists realize that the areas containing these ancient glacial features were joined together as a single supercontinent located at latitudes far to the south of their present positions. Later, this landmass broke apart, and its pieces, each moving on a different plate, drifted toward their present locations (Figure 5.22). It is now understood that during the geologic past, plate movements accounted for many dramatic climatic changes as landmasses shifted in relation to one another and moved to different latitudinal positions. Changes in oceanic circulation also must have occurred, altering the transport of heat and moisture and consequently the climate as well. Because the rate of plate movement is very slow—a few centimeters per year—appreciable changes in the positions of the continents occur only over great spans of geologic time. Thus, climatic changes brought about by shifting plates are extremely gradual and happen on a scale of millions of years.

Variations in Earth's Orbit

Because climatic changes brought about by moving plates are extremely gradual, the plate tectonics theory cannot be used to explain the alternation between glacial and interglacial climates that occurred during the Pleis-tocene epoch. Therefore, we must look to some other triggering mechanism that may cause climatic change on a scale of thousands rather than millions of years. Today many scientists strongly suspect that the climatic oscillations that characterized the Pleistocene may be linked to variations in Earth's orbit. This hypothesis was first developed and strongly advocated by the Yugoslavian scientist Milutin Milankovitch and is based on the premise that variations in incoming solar radiation are a principal factor in controlling Earth's climate.

Milankovitch formulated a comprehensive mathematical model based on the following elements (Figure 5.23):

1. Variations in the shape *(eccentricity)* of Earth's orbit about the sun;

2. Changes in *obliquity*; that is, changes in the angle that the axis makes with the plane of Earth's orbit; and

3. The wobbling of Earth's axis, called *preccession*.

Using these factors, Milankovitch calculated variations in the receipt of solar energy and the corresponding surface temperature of Earth back into time in an attempt to correlate these changes with the climatic fluctuations of the Pleistocene. In explaining climatic changes that result from these three variables, note that

Figure 5.22

A. The supercontinent Pangaea showing the area covered by glacial ice 300 million years ago. **B.** The continents as they are today. The white areas indicate where evidence of the old ice sheets exists.

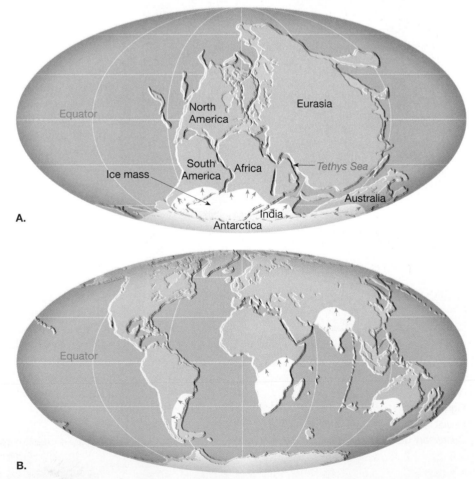

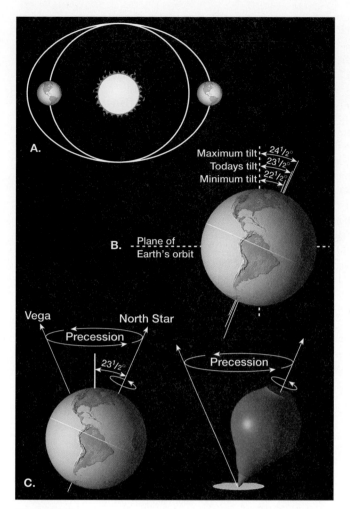

Figure 5.23

Orbital variations. **A.** The shape of Earth's orbit changes during a cycle that spans about 100,000 years. It gradually changes from nearly circular to one that is more elliptical and then back again. This diagram greatly exaggerates the amount of change. **B.** Today the axis of rotation is tilted about 23.5° to the plane of Earth's orbit. During a cycle of 41,000 years, this angle varies from 21.5° to 24.5°. **C.** Precession. Earth's axis wobbles like that of a spinning top. Consequently, the axis points to different spots in the sky during a cycle of about 26,000 years.

they cause little or no variation in the total solar energy reaching the ground. Instead, their impact is felt because they change the degree of contrast between the seasons. Somewhat milder winters in the middle to high latitudes means greater snowfall totals, while cooler summers would bring a reduction in snowmelt.

Among the studies that have added credibility to this astronomical hypothesis is one in which deep-sea sediments containing certain climatically sensitive microorganisms were analyzed to establish a chronology of temperature changes going back nearly one-half million years.[*] This time scale of climatic change was then compared to astronomical calculations of eccentricity, obliquity, and precession to determine if a correlation did indeed exist. Although the study was very involved and mathematically complex, the conclusions were straightforward. The authors found that major variations in climate over the past several hundred thousand years were closely associated with changes in the geometry of Earth's orbit; that is, cycles of climatic change were shown to correspond closely with the periods of obliquity, precession, and orbital eccentricity. More specifically, the authors stated: "It is concluded that changes in the earth's orbital geometry are the fundamental cause of the succession of Quaternary ice ages."[†]

Let us briefly summarize the ideas that were just described. The theory of plate tectonics provides us with an explanation for the widely spaced and nonperiodic onset of glacial conditions at various times in the geologic past, while the astronomical model proposed by Milankovitch and supported by the work of J. D. Hays and his colleagues furnishes an explanation for the alternating glacial and interglacial episodes of the Pleistocene.

In conclusion, we emphasize that the ideas just discussed do not represent the only possible explanations for glacial ages. Although interesting and attractive, these proposals are certainly not without critics; nor are they the only possibilities currently under study. Other factors may be, and probably are, involved.

Deserts

Climate has a strong influence on the nature and intensity of Earth's external processes. This was clearly demonstrated in the preceding section on glaciers and the Ice Age. Another excellent example of the strong link between climate and geology is seen when we examine the development of arid landscapes.

The dry regions of the world encompass about 42 million square kilometers, nearly 30 percent of Earth's land surface. No other climatic group covers so large a land area.[‡] The word *desert* literally means "deserted" or "unoccupied." For many dry regions this is a very appropriate description. Yet where water is available in deserts, many people thrive. Nevertheless, the world's dry regions are among the least familiar land areas on Earth outside of the polar realm (see Box 5.3).

Desert landscapes frequently appear stark. Their profiles are not softened by a carpet of soil and abundant plant life. Instead, barren rocky outcrops with steep, angular slopes are common. At some places the rocks are tinted orange and red. At others they are gray and brown and streaked with black. For many visitors desert scenery exhibits a striking beauty; to others, the

---

[*] J. D. Hays, John Imbrie, and N. J. Shackelton, "Variations in the Earth's Orbit: Pacemaker of the Ice Ages," *Science* 194 (1976): 1121–32

[†] J. D. Hays et al., p. 1131. The term *Quaternary* refers to the period on the geologic time scale that encompasses the last 1.6 million years.

[‡] An examination of dry climates is found in Chapter 18.

Box 5.3

Common Misconceptions About Deserts

Deserts are hot, lifeless, sand-covered landscapes shaped largely by the force of wind. The preceding statement summarizes the image of arid regions that many people hold, especially those living in more humid places. Is it an accurate view? The answer is no. Although there are clearly elements of reality in such an impression, it is a generalization that contains a number of misconceptions (Figure 5.C).

One common fallacy about deserts is that they are lifeless or almost lifeless. Although reduced in amount and different in character, plant and animal life are usually present. Desert plants all have one characteristic in common: they have developed adaptations that make them highly tolerant of drought. Many have waxy leaves, stems, or branches or a thickened cuticle (outermost protective layer) to reduce water loss. Others have very small leaves or no leaves at all. Also, the roots of some species often extend to great depths in order to tap the moisture found there, whereas others produce a shallow but widespread root

system that enables them to absorb great amounts of moisture quickly from the infrequent desert downpours. Thus, although widely dispersed and providing little ground cover, plants of many kinds flourish in the desert.

A second widely held belief about the world's dry lands is that they are always hot. This fact seems to be reinforced by commonly quoted temperature statistics. The highest accepted temperature record for the United States as well as the entire Western Hemisphere is 57°C (134°F). This long-standing record was set at Death Valley, California, on July 10, 1913. The nearly 59°C (137°F) reading in Azizia, Libya, in North Africa's Sahara Desert on September 13, 1922, is the world record. Despite these remarkably high figures, cold temperatures are also experienced in desert regions. For example, the average daily minimum in January at Phoenix, Arizona, is 17°C (35°F), just barely above freezing. At Ulan Bator in Mongolia's Gobi Desert, the average *high* temperature in January is only –19°C (–2°F)! Dry climates are found from the tropics to the high middle latitudes. Consequently, although tropical deserts lack a cold season, deserts in the middle latitudes do experience seasonal temperature changes.

The last two commonly held misconceptions about the world's deserts are more geologic than climatic. One mistaken assumption is that they consist of mile after mile of drifting sand. It is true that sand accumulations do exist in some areas and may be striking features, but they represent only a small percentage of the total desert area. For example, in the Sahara, the world's largest desert, accumulations of sand cover only one-tenth of its area. The sandiest of all deserts is the Arabian, one-third of which consists of sand. The final mistaken assumption is the seemingly logical idea that wind is the most important agent of erosion in deserts. Although wind is relatively more significant in dry areas than anywhere else, most erosional landforms in deserts are created by running water. When the rains come, they usually take the form of thunderstorms. Because the heavy rain associated with these storms cannot all soak in, rapid runoff results. Without a thick vegetative cover to protect the ground, erosion is great.

Figure 5.C
Snow blankets the rocky ground of the Sonoran Desert in southern Arizona. (Photo by Jack Dykinga)

terrain seems bleak. No matter which feeling is elicited, it is clear that deserts are very different from the more humid places where most people live.

As you shall see, arid regions are not dominated by a single geologic process. Rather, the effects of tectonic forces, running water, and wind are all apparent. Because these processes combine in different ways from place to place, the appearance of desert landscapes varies a great deal as well (Figure 5.24).

Geologic Processes in Arid Climates

The angular rock exposures, the sheer canyon walls, and the pebble- or sand-covered surface of the desert contrast sharply with the rounded hills and curving slopes of more humid places. To a visitor from a humid region, a desert landscape may seem to have been shaped by forces different from those operating in wetter areas. However, although the contrasts may be striking, they are not a reflection of different processes but merely the differing effects of the same processes operating under contrasting climatic conditions.

Weathering

In humid regions, relatively well-developed soils support an almost continuous cover of vegetation. Here the slopes and rock edges are rounded. Such a landscape reflects the strong influence of chemical weathering in a humid climate. By contrast, much of the weathered debris in deserts consists of unaltered rock and mineral fragments—the result of mechanical weathering processes. In dry lands rock weathering of any type is greatly reduced because of the lack of moisture and the scarcity of

Figure 5.24
Prickly pear and saguaro cactus beneath rock pinnacles in the arid Superstition Mountains near Phoenix, Arizona. The appearance of desert landscapes varies a great deal from place to place. (Photo by Carr Clifton)

organic acids from decaying plants. Chemical weathering, however, is not completely lacking in deserts. Over long spans of time clays and thin soils do form and many iron-bearing silicate minerals oxidize, producing the rust-colored stain found tinting some desert landscapes.

The Role of Water

Permanent streams are normal in humid regions but practically all desert streams are dry most of the time (Figure 5.25A). Desert streams are said to be **ephemeral**, which means that they carry water only in response to specific episodes of rainfall. A typical ephemeral stream may flow only a few days or perhaps just a few hours during the year. In some years the channel may carry no water at all.

This fact is obvious even to the casual observer who, while traveling in a dry region, notices the number of bridges with no streams beneath them or the number of dips in the road where dry channels cross. However, when the rare heavy showers do come, so much rain falls, in such a short time that all of it can not soak in. Because the vegetative cover is sparse, runoff is largely unhindered and consequently rapid, often creating flash floods along valley floors (Figure 5.25B). Such floods, however, are quite unlike floods in humid regions. A flood on a river like the Mississippi may take several days to reach its crest and then subside. But desert floods arrive suddenly and subside quickly. Because much of the surface material is not anchored by vegetation, the amount of erosional work that occurs during a single short-lived rain event is impressive. In the dry western United States a number of different names are used for ephemeral streams. Two of the most common are *wash* and *arroyo*. In other parts of the world, a dry desert stream may be called a *wadi* (Arabia and North Africa), a *donga* (South America), or a *nullah* (India).

Unlike the drainage in humid regions, stream courses in arid regions are seldom well integrated. That is, desert streams lack an extensive system of tributaries. In fact, a basic characteristic of deserts is that most of the streams that originate in them are small and die out before reaching the sea. Because the water table is usually far below the surface, few desert streams can draw upon it. Without a steady supply of water, the combination of evaporation and infiltration soon depletes the stream.

The few permanent streams that do cross arid regions, such as the Colorado and Nile rivers, originate *outside* the desert, often in well-watered mountains. Here the water supply must be great to compensate for the losses occurring as the stream crosses the desert. For example, after the Nile leaves the lakes and mountains of central Africa that are its source, it traverses almost 3000 kilometers of the Sahara *without a single tributary*.

It should be emphasized that running water, although an infrequent occurrence, nevertheless does most of the erosional work in deserts. This is contrary to a common belief that wind is the most important erosional agent sculpturing desert landscapes. Although wind erosion is indeed more significant in dry areas than elsewhere, most desert landforms are nevertheless carved by running water. As you will see shortly, the main role of wind is in the transportation and deposition of sediment, which creates and shapes the ridges and mounds we call dunes.

A.   B.

Figure 5.25
A. Most of the time, desert stream channels are dry. **B.** An ephemeral stream shortly after a heavy shower. Although such floods are short-lived, large amounts of erosion occur. (Photos by E. J. Tarbuck)

Basin and Range: The Evolution of a Desert Landscape

Because arid regions typically lack permanent streams, they are characterized as having **interior drainage**. This means that they have a discontinuous pattern of intermittent streams that do not flow out of the desert to the ocean. In the United States, the dry Basin and Range region provides an excellent example. The region includes southern Oregon, all of Nevada, western Utah, southeastern California, southern Arizona, and southern New Mexico. The name Basin and Range is an apt description for this almost 800,000-square-kilometer region, since it is characterized by more than 200 relatively small

mountain ranges which rise 900–1500 meters above the basins that separate them.

In this region, as in others like it around the world, most erosion occurs without reference to the ocean (ultimate base level), because the interior drainage never reaches the sea. Even where permanent streams flow to the ocean, few tributaries exist, and thus only a narrow strip of land adjacent to the stream has sea level as its ultimate level of land reduction.

The block models in Figure 5.26 depict how the landscape has evolved in the Basin and Range region. During and following uplift of the mountains, running water begins carving the elevated mass and depositing large quantities of debris in the basin. In this early stage relief is greatest, and as erosion lowers the mountains and sediment fills the basins, elevation differences diminish.

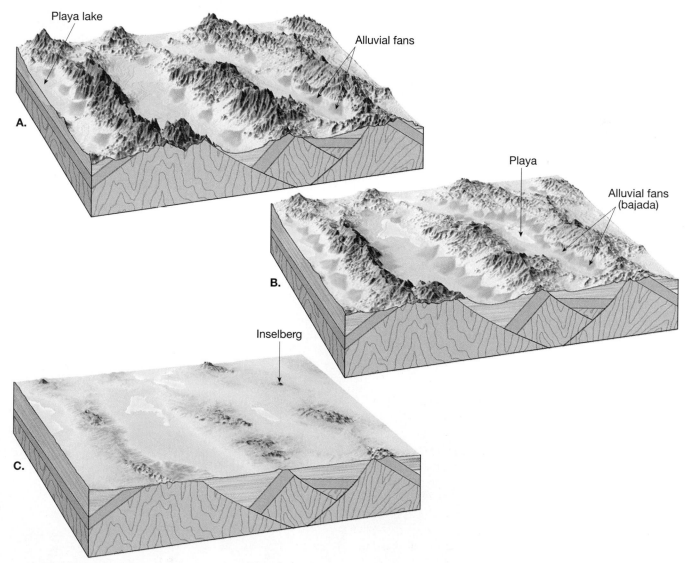

Figure 5.26
Stages of landscape evolution in a mountainous desert. As erosion of the mountains and deposition in the basins continue, relief diminishes.
A. Early stage. **B.** Middle stage. **C.** Late stage.

When the occasional torrents of water produced by sporadic rains move down the mountain canyons, they are heavily loaded with sediment. Emerging from the confines of the canyon, the runoff spreads over the gentler slopes at the base of the mountains and quickly loses velocity. Consequently, most of its load is dumped within a short distance. The result is a cone of debris known as an **alluvial fan** at the mouth of a canyon (Figure 5.27). Over the years, a fan enlarges, eventually coalescing with fans from adjacent canyons to produce an apron of sediment (*bajada*) along the mountain front.

On the rare occasions of abundant rainfall, streams may flow across the alluvial fans to the center of the basin, converting the basin floor into a shallow **playa lake**. Playa lakes last only a few days or weeks, before evaporation and infiltration remove the water. The dry, flat lake bed that remains is termed a *playa*.

Playas occasionally become encrusted with salts left behind by evaporation. These precipitated salts may be uncommon. A case in point is the sodium borate (better known as borax) mined from ancient playa lake deposits in Death Valley, California.

With the ongoing erosion of the mountain mass and the accompanying sedimentation, the local relief continues to diminish. Eventually nearly the entire mountain mass is gone. Thus, by the late stages of erosion, the mountain areas are reduced to a few large bedrock knobs (called *inselbergs*) projecting above the sediment-filled basin.

Each of the stages of landscape evolution in an arid climate depicted in Figure 5.26 can be observed in the Basin and Range region. Recently uplifted mountains in an early stage of erosion are found in southern Oregon and northern Nevada. Death Valley, California, and southern Nevada fit into the more advanced middle stage, while the late stage, with its inselbergs and extensive pediments, can be seen in southern Arizona.

Wind Erosion

In humid regions, moisture binds particles together and vegetation anchors the soil so that wind erosion is negligible. For wind to be an effective erosional force, dryness and scant vegetation are essential. When such circumstances exist, wind may pick up, transport, and deposit great quantities of fine sediment. During the 1930s, parts of the Great Plains experienced vast dust storms. The plowing under of the natural vegetative cover for farming, followed by severe drought, exposed the land to wind erosion and led to the area being labeled the Dust Bowl (see Box 5.4).

Moving air, like moving water, is turbulent and able to pick up loose debris and transport it to other locations. Just as in a stream, the velocity of wind increases with height above the surface. Also like a stream, wind transports fine particles in suspension while heavier ones are carried as bed load. However, the transport of sediment by wind differs from that by running water

Figure 5.27

Alluvial fans develop where the gradient of a stream changes abruptly from steep to flat. Such a situation exists in Death Valley, California, where streams emerge from the mountains into a flat basin. As a result, Death Valley has many large alluvial fans. (Photo by Michael Collier)

Box 5.4

Dust Bowl: Soil Erosion in the Great Plains

During a span of dry years in the 1930s, large dust storms plagued the Great Plains. Because of the size and severity of these storms, the region came to be called the "Dust Bowl," and the time period, the "dirty thirties." The heart of the Dust Bowl consisted of nearly 100 million acres in the panhandles of Texas and Oklahoma, as well as adjacent parts of Colorado, New Mexico, and Kansas (Figure 5.D). To a lesser extent, dust storms were also a problem over much of the Great Plains, from North Dakota to west central Texas.

At times dust storms were so severe that they were called "black blizzards" and "black rollers" because visibility was reduced to only a few feet. Examples of storms that lasted for hours and stripped huge volumes of topsoil from the land are numerous. In the spring of 1934, a wind storm that lasted for a day and a half created a dust cloud that extended for 2000 kilometers (1200 miles). As the sediment moved east, "muddy rains" were experienced in New York, and "black snows" in Vermont. Less than a year later, another storm carried dust more than 3 kilometers (2 miles) into the atmosphere and transported it 3000 kilometers from its source in Colorado to create twilight conditions in the middle of the day in parts of New England and New York.

What caused the Dust Bowl? Clearly, the fact that portions of the Great Plains experience some of North America's strongest winds is important. However, it was the expansion of agriculture that set the stage for the disastrous period of soil erosion. Mechanization allowed the rapid transformation of the grass covered prairies of this semiarid region into farms. Between the 1870s and 1930, the area of cultivation in the region expanded nearly tenfold, from about 10 million acres to more than 100 million acres.

As long as precipitation was adequate, the soil remained in place. However, when a prolonged drought struck in the 1930s, the unprotected fields were vulnerable to the wind. The results were severe soil loss, crop failures, and economic hardship.

Beginning in 1939, a return to rainier conditions brought relief. Moreover, farming practices that were designed to reduce soil loss by wind had also been instituted. Although dust storms are less numerous and not as severe as in the "dirty thirties," soil erosion by strong winds still occurs periodically whenever the combination of drought and unprotected soil exists.

Figure 5.D
Dust blackens the sky on May 21, 1937 near Elkhart, Kansas. It was because of storms like this that portions of the Great Plains were called the "Dust Bowl" in the 1930s. (Photo reproduced from the collection of the Library of Congress)

in two significant ways. First, wind has a low density compared to water; thus, it is not capable of picking up and transporting coarse materials. Second, because wind is not confined to channels, it can spread over large areas, as well as high into the atmosphere.

One way that wind erodes is by **deflation**, the lifting and removal of loose material. Because the competence (ability to transport different-size particles) of moving air is low, it can suspend only fine sediment such as clay and silt. Larger grains of sand are rolled or skipped along the surface (a process called *saltation*) and comprise the bed load. Particles larger than sand are usually not transported by wind. Deflation sometimes is difficult to notice because the entire surface is being lowered at the same time, but it can be significant. In portions of the 1930s Dust Bowl, the land was lowered by a meter or more in only a few years (Figure 5.28).

The most noticeable result of deflation in some places is shallow depressions called **blowouts**. In the Great Plains region, from Texas north to Montana, thousands of blowouts can be seen. They range from small dimples less than 1 meter deep and 3 meters wide to depressions that are over 45 meters deep and several kilometers across.

In portions of many deserts, the surface is characterized by a layer of coarse pebbles and cobbles that are too large to be moved by the wind. This stony veneer, called **desert pavement**, is created as deflation lowers the surface by removing sand and silt until eventually only a continuous cover of coarse particles remains (Figure 5.29). Once desert pavement becomes established, a process which may take hundreds of years, the surface is effectively protected from further deflation if left undisturbed. However, since the layer is only one or two stones thick, the passage of vehicles or animals can dislodge the pavement and expose the fine-grained material below. If this happens, the surface is once again subject to deflation.

Figure 5.28
This photo was taken north of Granville, North Dakota in July, 1936, during a prolonged drought. Strong winds removed the soil that was not anchored by vegetation. The mounds are 1.2 meters (4 feet) high and show the level of the land prior to deflation. (Photo courtesy of the State Historical Society of North Dakota)

Like glaciers and streams, wind erodes in part by *abrasion*. In dry regions as well as along some beaches, windblown sand will cut and polish exposed rock surfaces. However, abrasion is often given too much credit. Such features as balanced rocks that stand high atop narrow pedestals and intricate detailing on tall pinnacles are not the results of abrasion. Sand seldom travels more than a meter above the surface, so the wind's sandblasting effect is obviously limited in vertical extent. However, in areas prone to such activity, telephone poles have actually been cut through near their bases. For this reason, collars are often fitted on the poles to protect them from being "sawed" down.

Wind Deposits

Although wind is relatively unimportant as a producer of erosional landforms, wind deposits are significant features in some regions. Accumulations of windblown sediment are particularly conspicuous landscape elements in the world's dry lands and along many sandy coasts. Wind deposits are of two distinctive types: (1) extensive blankets of silt that once were carried in suspension, and (2) mounds and ridges of sand from the wind's bed load.

Loess

In some parts of the world the surface topography is mantled with deposits of windblown silt. Over thousands of years dust storms deposited this material, called **loess**. When loess is breached by streams or road cuts, it tends to maintain vertical cliffs and lacks any visible layers (Figure 5.30).

The distribution of loess indicates two primary sources for this sediment: deserts and glacial deposits of stratified drift. The thickest and most extensive loess deposits occur in western and northern China, where accumulations of 30 meters are not uncommon and thicknesses of more than 100 meters have been measured. It is this fine, buff-colored sediment that gives the Yellow River (Hwang Ho) and the adjacent Yellow Sea their names. The sources of China's 800,000 square kilometers of loess are the extensive desert basins of central Asia.

In the United States, deposits of loess are significant in many areas, including South Dakota, Nebraska, Iowa, Missouri, and Illinois as well as portions of the Columbia Plateau in the Pacific Northwest. Unlike the deposits in China, the loess in the United States, as well as in Europe, is an indirect product of glaciation, for its source was deposits of stratified drift. During the retreat of the glacial ice, many river valleys were choked

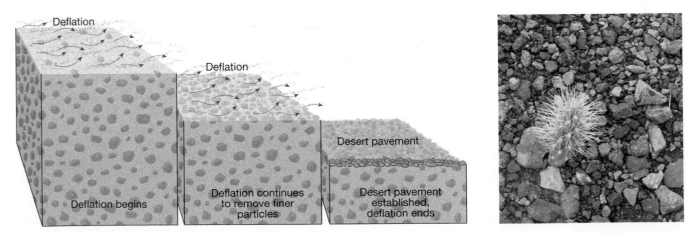

Figure 5.29
Formation of desert pavement. Coarse particles gradually become concentrated into a tightly packed layer as deflation lowers the surface by removing sand and silt. If left undisturbed, desert pavement will protect the surface from further deflation. (Photo by Scott T. Smith)

The diagram labels: Deflation; Deflation; Deflation begins; Deflation continues to remove finer particles; Desert pavement; Desert pavement established, deflation ends

with glacial sediment. Strong winds sweeping across the barren floodplains picked up the finer sediment and dropped it as a blanket on areas adjacent to the valleys.

Sand Dunes

Like running water, wind releases its load of sediment when its velocity falls and the energy available for transport diminishes. Thus, sand begins to accumulate wherever an obstruction across the path of the wind slows the movement of the air. Unlike deposits of loess, which form blanketlike layers over broad areas, winds commonly deposit sand in mounds or ridges called *dunes* (Figure 5.31).

As moving air encounters an object, such as a clump of vegetation or a rock, the wind sweeps around and over it, leaving a shadow of more slowly moving air behind the obstacle as well as a smaller zone of quieter air just in front of the obstacle. Some of the sand grains moving with the wind come to rest in these "wind shadows." As the accumulation of sand continues, it forms an increasingly efficient wind barrier to trap even more sand. If there is a sufficient supply of sand and the wind blows steadily long enough, the mound of sand grows into a dune.

Figure 5.30
A vertical loess bluff near the Mississippi River in southern Illinois. (Photo by James E. Patterson)

Figure 5.31
Sand sliding down the steep slip face of a dune in White Sands National Monument, New Mexico. (Photo by Michael Collier)

The profile of many dunes is asymmetrical, with the leeward slope being steep and the windward slope more gently inclined (Figure 5.32). Sand is rolled up the gentle slope on the windward side by the force of the wind. Just beyond the crest of the dune, the wind velocity is reduced and the sand accumulates. As more sand collects, the slope steepens and eventually some of it slides or slumps under the pull of gravity. In this way, the leeward slope of the dune, called the **slip face**, maintains an angle of about 34 degrees. Continued sand accumulation, coupled with periodic slides down the slip face, results in the slow migration of the dune in the direction of air movement.

As sand is deposited on the slip face, it forms layers inclined in the direction the wind is blowing. These sloping layers are called **cross beds** (Figure 5.32). When the dunes are eventually buried under layers of sediment and become part of the sedimentary rock record, their asymmetrical shape is destroyed, but the cross beds remain as a testimony to their origin. Nowhere is cross-bedding more prominent than in the sandstone walls of Zion Canyon in Utah (Figure 5.32 photo).

Types of Sand Dunes

Dunes are not just random heaps of wind-blown sediment. Rather, they are accumulations that usually assume surprisingly consistent patterns (Figure 5.33) A broad assortment of dune forms exists; so to simplify and provide some order, several major types are recognized. Of course, there are gradations among different forms as well as irregularly shaped dunes that do not fit easily into any category. Several factors influence the form and size that dunes ultimately assume. These include wind direction and velocity, availability of sand, and the amount of vegetation present.

Barchan Dunes. Solitary sand dunes shaped like crescents and with their tips pointing downwind are called **barchan dunes** (Figure 5.33A). These dunes form where supplies of sand are limited and the surface is relatively flat, hard, and lacking vegetation. They migrate slowly with the wind at a rate of up to 15 meters per year. Their size is usually modest, with the largest barchans reaching heights of about 30 meters while the maximum spread between their horns approaches 300 meters. When the wind direction is nearly constant, the crescent form of these dunes is nearly symmetrical. However, when the wind direction is not perfectly fixed, one tip becomes larger than the other.

Transverse Dunes. In regions where the prevailing winds are steady, sand is plentiful, and vegetation is sparse or absent, the dunes form a series of long ridges that are separated by troughs and oriented at right angles to the prevailing wind. Because of this orientation, they

Figure 5.32

Dunes commonly have an asymmetrical shape. The steeper leeward side is called the slip face. Sand grains deposited on the slip face create the cross-bedding of the dunes. A complex pattern develops in response to changes in prevailing winds. The sandstone walls in Zion Canyon, Utah, exhibit excellent cross-bedding.

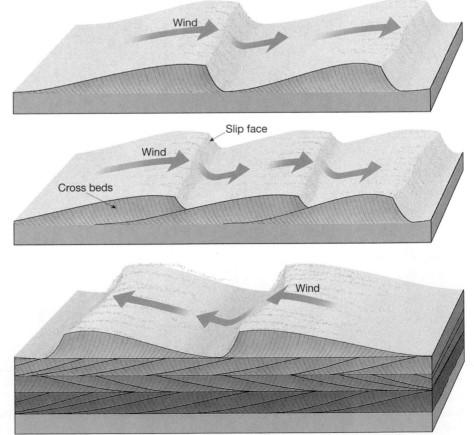

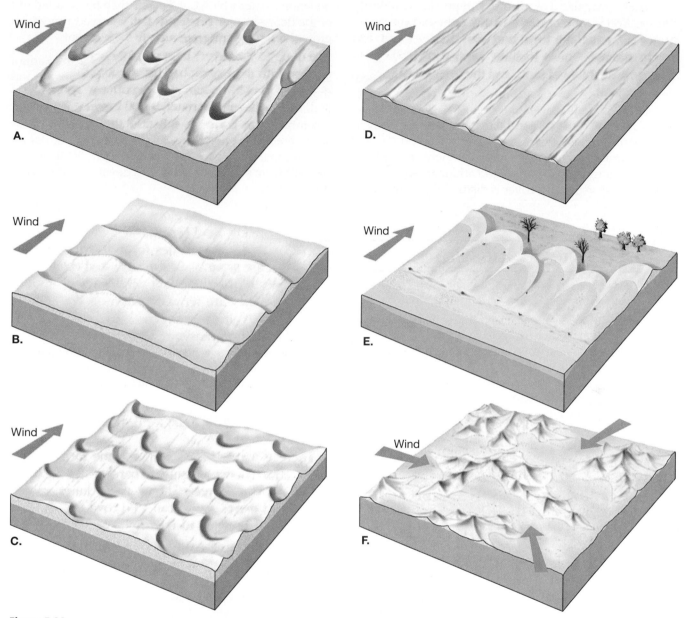

Figure 5.33

Sand dune types. **A.** Barchan dunes. **B.** Transverse dunes. **C.** Barchanoid dunes. **D.** Longitudinal dunes. **E.** Parabolic dunes. **F.** Star dunes.

are termed **transverse dunes** (Figure 5.33B). Typically, many coastal dunes are of this type. In addition, transverse dunes are common in many arid regions where the extensive surface of wavy sand is sometimes called a *sand sea*. In some parts of the Sahara and Arabian deserts, transverse dunes reach heights of 200 meters, are 1 to 3 kilometers across, and can extend for distances of 100 kilometers or more.

There is a relatively common dune form that is intermediate between isolated barchans and extensive waves of transverse dunes. Such dunes, called **barchanoid dunes**, form scalloped rows of sand oriented at right angles to the wind (Figure 5.33C). The

rows resemble a series of barchans that have been positioned side by side. Visitors exploring the gypsum dunes at White Sands National Monument, New Mexico, will recognize this form.

Longitudinal Dunes.   **Longitudinal dunes** are long ridges of sand that form more or less parallel to the prevailing wind and where sand supplies are limited (Figure 5.33D). Apparently the prevailing wind direction must vary somewhat, but still remain in the same quadrant of the compass. Although the smaller types are only three or four meters high and several tens of meters long, in some large deserts longitudinal dunes can reach great size. For example, in portions of North Africa, Arabia, and central Australia, these dunes may approach a height of 100 meters and extend for distances of more than 100 kilometers (62 miles).

Parabolic Dunes.   Unlike the other dunes that have been described thus far, **parabolic dunes** form where vegetation partially covers the sand. The shape of these dunes resembles the shape of barchans except that their tips point into the wind rather than downwind (Figure 5.33E). Parabolic dunes often form along coasts where there are strong onshore winds and abundant sand. If the sand's sparse vegetative cover is disturbed at some spot,

deflation creates a blowout. Sand is then transported out of the depression and deposited as a curved rim which grows higher as deflation enlarges the blowout.

Star Dunes.   Confined largely to parts of the Sahara and Arabian deserts, **star dunes** are isolated hills of sand that exhibit a complex form (Figure 5.33F). Their name is derived from the fact that the bases of these dunes resemble multipointed stars. Usually three or four sharp-crested ridges diverge from a central high point that in some cases may approach a height of 90 meters. As their form suggests, star dunes develop where wind directions are variable.

Review Questions

1. What is a glacier? What percentage of Earth's land area do glaciers cover?

2. Each of the following statements refers to a particular type of glacier. Name the type of glacier.
   (a) The term *continental* is often used to describe this type of glacier.
   (b) This type of glacier is also called an *alpine glacier*.
   (c) This is a glacier formed when one or more valley glaciers spreads out at the base of a steep mountain front.
   (d) Greenland is the only example in the Northern Hemisphere.

3. Describe how glaciers fit into the hydrologic cycle. What role do they play in the rock cycle?

4. Describe the two components of glacial flow. At what rates do glaciers move? In a valley glacier does all of the ice move at the same rate? Explain.

5. Why do crevasses form in the upper portion of a glacier but not below a depth of about 50 meters?

6. Under what circumstances will the front of a glacier advance? Retreat? Remain stationary?

7. Describe two basic processes of glacial erosion.

8. How does a glaciated mountain valley differ from a mountain valley that was not glaciated?

9. List and describe the erosional features you might expect to see in an area where alpine glaciers exist or have recently existed.

10. What is glacial drift? What is the difference between till and stratified drift? What general effect do glacial deposits have on the landscape?

11. List the four basic moraine types. What do all moraines have in common? Distinguish between terminal and recessional moraines.

12. List and briefly describe four depositional features other than moraines.

13. Examine the drumlin in Figure 5.19. From what direction (right or left) did the ice sheet advance in this area?

14. How does a kettle form?

15. About what percentage of Earth's land surface was covered at some time by Pleistocene glaciers? How does this compare to the area presently covered by ice sheets and glaciers? (Check your answer with Question 1).

16. List three indirect effects of Ice Age glaciers.

17. How might plate tectonics help us understand the cause of ice ages? Can plate tectonics explain the alternation between glacial and interglacial climates during the Pleistocene?

18. List four common misconceptions about deserts (see Box 5.3).

19. How extensive are the desert and steppe regions of Earth?

20. Describe the characteristics associated with each of the stages in the evolution of a mountainous desert.

21. Why is wind erosion relatively more important in arid regions than in humid areas?

22. Briefly describe the conditions that led to the Dust Bowl of the 1930s (see Box 5.4).

23. Although sand dunes are the best-known wind deposits, accumulations of loess are very significant in some parts of the world. What is loess? Where are such deposits found? What are the origins of this sediment?

24. How do sand dunes migrate?

25. Identify each of the dunes described in the following statements.
    (a) Long ridges of sand oriented parallel to the prevailing wind.
    (b) Solitary, crescent-shaped dunes oriented with their tips pointing downwind.
    (c) Ridges of sand oriented at right angles to the prevailing wind.
    (d) Dunes whose tips point into the wind.
    (e) Scalloped rows of sand oriented at right angles to the wind.
    (f) Isolated dunes consisting of three or four sharp-crested ridges diverging from a central high point.

Key Terms

alluvial fan (p. 144)
alpine glacier (p. 122)
arête (p. 128)
barchan dune (p. 148)
barchanoid dune (p. 149)
blowout (p. 145)
cirque (p. 127)
crevasse (p. 123)
cross beds (p. 148)
deflation (p. 145)
desert pavement (p. 145)
drift (p. 129)
drumlin (p. 134)
end moraine (p. 131)
ephemeral (p. 142)

esker (p. 134)
fiord (p. 128)
glacial erratic (p. 130)
glacial striations (p. 127)
glacial trough (p. 127)
glacier (p. 122)
ground moraine (p. 132)
hanging valley (p. 127)
horn (p. 128)
ice cap (p. 122)
ice sheet (p. 122)
interior drainage (p. 143)
kame (p. 134)
kettle (p. 133)
lateral moraine (p. 131)
loess (p. 146)
longitudinal dune (p. 149)
medial moraine (p. 131)

outwash plain (p. 133)
parabolic dune (p. 150)
piedmont glacier (p. 122)
playa lake (p. 144)
Pleistocene epoch (p. 135)
plucking (p. 127)
pluvial lake (p. 137)
rock flour (p. 127)
slip face (p. 148)
star dune (p. 150)
stratified drift (p. 129)
till (p. 129)
transverse dune (p. 149)
valley glacier (p. 122)
valley train (p. 133)
zone of accumulation (p. 124)
zone of wastage (p. 124)

CHAPTER 6

Earthquakes and Earth's Interior

In January 1995. a strong earthquake toppled this elevated expressway near Kobe, Japan. (Photo by Noboru Hashimoto/Sygma)

What is an earthquake? How does a seismograph record the location of a "quake"? Can earthquakes ever be predicted? If we could view the Earth's interior, what would it look like? In this chapter we shall answer these and other questions. The study of earthquakes is important, not only because of the devastating effect that some earthquakes have on us, but also because they furnish clues about the structure of Earth's interior.

On October 17, 1989, at 5:04 P.M. Pacific Daylight Time, millions of television viewers around the world were settling in to watch the third game of the World Series. Instead, they saw their television sets go black as tremors hit San Francisco's Candlestick Park. Although the earthquake was centered in a remote section of the Santa Cruz Mountains, 100 kilometers to the south, major damage occurred in the Marina District of San Francisco.

The most tragic result of the violent shaking was the collapse of some double-decked sections of Interstate 880, also known as the Nimitz Freeway. The ground motions caused the upper deck to sway, shattering the concrete support columns along a mile-long section of the freeway. The upper deck then collapsed onto the lower roadway, flattening cars as if they were aluminum beverage cans. This earthquake, named the Loma Prieta quake for its point of origin, claimed 67 lives.

In mid-January 1994, less that five years after the Loma Prieta earthquake devastated portions of the San Francisco Bay area, a major earthquake struck the Northridge area of Los Angeles. Although it was not the fabled "Big One," the moderate 6.6 to 6.9 magnitude earthquake left 51 dead, over 5000 injured, and tens of thousands of households without water and electricity. In total, damage in excess of $15 billion was attributed to an apparently unknown fault that ruptured at a depth of 14 kilometers (9 miles) beneath Northridge (Figure 6.1).

The Northridge earthquake began at 4:31 A.M. and lasted roughly 40 seconds. During this brief period, the quake terrorized the entire Los Angeles area. In the three-story Northridge Meadows apartment complex, 16 people died when sections of the upper floors collapsed onto the first-floor units. Nearly 300 schools were seriously damaged and a dozen major roadways buckled. Among these were two of California's major arteries—the Golden State Freeway (Interstate 5), where an

Figure 6.1
This parking deck in Northridge, California, collapsed during an earthquake in January 1994. (Photo by Spencer Grant/Liaison International)

overpass collapsed completely and blocked the roadway, and the Santa Monica Freeway. Fortunately, these roadways had practically no traffic at this early morning hour.

In nearby Granada Hills, broken gas lines were set ablaze while the streets were flooded from broken water mains. Seventy homes burned in the Sylmar area. A 64-car freight train derailed, including some cars carrying hazardous cargo. But it is remarkable that the destruction was not greater. Unquestionably, the upgrading of structures to meet the requirements of building codes developed for this earthquake-prone area helped minimize what could have been a much greater human tragedy.

Across the Pacific from California lies Japan, no stranger to earthquakes, and among the most "quake-proofed" countries in the world. Yet at 5:46 A.M. on January 24, 1995, much of the "quake-proofing" proved futile as more than 5000 people perished in a 7.2-magnitude tremor centered near Kobe, the country's sixth-largest city (see chapter-opening photo).

Over 30,000 earthquakes that are strong enough to be felt occur worldwide annually. Fortunately, most are minor tremors and do very little damage. Generally, only about 75 significant earthquakes take place each year, and many of these occur in remote regions. However, occasionally a large earthquake occurs near a large population center. Under these conditions, an earthquake is among the most destructive natural forces on Earth (see Box 6.1).

The shaking of the ground, coupled with the liquefaction of some soils, wreaks havoc on buildings and other structures. In addition, when a quake occurs in a populated area, power and gas lines are often ruptured, causing numerous fires. In the famous 1906 San Francisco earthquake, much of the damage was caused by fires (Figure 6.2). They quickly became uncontrollable when broken water mains left firefighters with only trickles of water.

What Is an Earthquake?

An **earthquake** is the vibration of Earth produced by the rapid release of energy. Most often earthquakes are caused by slippage along a fault in Earth's crust. The energy released radiates in all directions from its source, the **focus**, in the form of waves. These waves are analogous to those produced when a stone is dropped into

Figure 6.2
San Francisco in flames after the 1906 earthquake. (Reproduced from the collection of the Library of Congress)

Box 6.1

Damaging Earthquakes East of the Rockies

When you think "earthquake," you probably think of California and Japan. However, recent estimates by seismologists indicate that a damaging earthquake east of the Rocky Mountains is roughly two-thirds as likely as an earthquake of comparable damage in California during the next 30 years. Like all earthquake risk assessments, this prediction considers the geographic distribution and average rate of earthquake occurrences in these regions.

At least six major earthquakes have occurred in the central and eastern United States since colonial times. Three of these had estimated Richter magnitudes of 7.5, 7.3, and 7.8, and they were centered near the Mississippi River Valley in southeastern Missouri. Occurring on December 16, 1811, January 23, 1812, and February 7, 1812, these earthquakes, plus numerous smaller tremors, destroyed the town of New Madrid, Missouri, triggered massive landslides, and caused damage over a six-state area. The course of the Mississippi River was altered, and Tennessee's Reelfoot Lake was enlarged. The distances over which these earthquakes were felt are truly remarkable. Chimneys were reported downed in Cincinnati, Ohio, and Richmond, Virginia, while Boston residents, located 1770 kilometers (1100 miles) to the northeast, felt the tremor.

Although the destruction caused by the New Madrid earthquake was slight compared to that caused by the Loma Prieta and Northridge earthquakes, remember that in the early 1800s the Midwest was sparsely populated. Memphis, Tennessee, which is located near the epicenter, had not yet been established. St. Louis was but a small frontier town of a few thousand inhabitants.

Damaging earthquakes that occurred in Aurora, Illinois (1909), and Valentine, Texas, (1931), also remind us that many areas in the central United States are vulnerable.

The greatest historical earthquake in the eastern states occurred August 31, 1886, in Charleston, South Carolina. The event, which spanned one minute, caused 60 deaths, numerous injuries, and great economic loss within a radius of 200 kilometers (120 miles) of Charleston. Within eight minutes, effects were felt as far away as Chicago and St. Louis, where strong vibrations shook the upper floors of buildings, causing people to rush outdoors. In Charleston alone, over 100 buildings were destroyed and 90 percent of the remaining structures were damaged. It was difficult to find a chimney still standing (Figure 6.A).

Numerous other strong earthquakes have been recorded in the central and eastern United States. New England and adjacent areas have experienced sizable shocks since colonial times. The first reported earthquake in the Northeast took place in Plymouth, Massachusetts, in 1683, and was followed in 1755 by the destructive Cambridge, Massachusetts, earthquake. Moreover, since records have been kept, New York State alone has experienced over 300 earthquakes large enough to be felt.

Earthquakes in the central and eastern United States occur far less frequently than in California. Yet history indicates that the East is vulnerable. Further, these shocks east of the Rockies have generally produced structural damage over a larger area than their counterparts of similar magnitude in California. The reason is that the underlying bedrock in the central and eastern United States is older and more rigid. As a result, seismic waves are able to travel greater distances with less attenuation than in the western United States. It is estimated that for earthquakes of similar magnitude, the region of maximum ground motion in the East may be up to ten times larger than in the West. Consequently, the higher rate of earthquake occurrence in the western United States is balanced somewhat by the fact that central and eastern U.S. quakes can damage larger areas.

Despite the history of the New Madrid earthquake, Memphis, Tennessee, the largest population center in the area, does not have adequate earthquake provisions in its building code. Further, because Memphis is located on unconsolidated floodplain deposits, buildings are more susceptible to damage than similar structures built on bedrock. It has been estimated that if an earthquake the size of 1811–1812 New Madrid event were to strike in the next decade, it would result in casualties in the thousands and damages in tens of billions of dollars.

Figure 6.A

Damage to Charleston, South Carolina, caused by the August 31, 1886, earthquake. Damage ranged from toppled chimneys and broken plaster to total collapse. (Photo courtesy of U.S. Geological Survey)

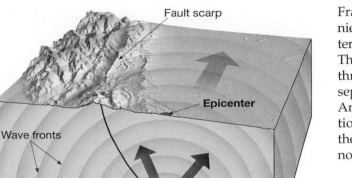

Figure 6.3
The focus of all earthquakes is located at depth. The surface location directly above it is called the epicenter.

a calm pond (Figure 6.3). Just as the impact of the stone sets water waves in motion, an earthquake generates seismic waves that radiate throughout the Earth. Even though the energy dissipates rapidly with increasing distance from the focus, sensitive instruments located throughout the world record the event.

Earthquakes and Faults

The tremendous energy released by atomic explosions or by volcanic eruptions can produce an earthquake, but these events are relatively weak and infrequent. What mechanism produces a destructive earthquake? Ample evidence exists that Earth is not a static planet. We know that Earth's crust has been uplifted at times, because we have found numerous ancient wave-cut benches many meters above the level of the highest tides. Other regions exhibit evidence of extensive subsidence. In addition to these vertical displacements, offsets in fence lines, roads, and other structures indicate that horizontal movement is common (Figure 6.4). These movements are usually associated with large fractures in Earth's crust called **faults**.

Most of the motion along faults can be satisfactorily explained by the plate tectonics theory. This theory states that large slabs of Earth's crust are in continual slow motion. These mobile plates interact with neighboring plates, straining and deforming the rocks at their edges. In fact, it is along faults associated with plate boundaries that most earthquakes occur. Furthermore, earthquakes are repetitive: as soon as one is over, the continuous motion of the plates resumes, adding strain to the rocks until they fail again.

Elastic Rebound

The actual mechanism of earthquake generation eluded geologists until H. F. Reid of Johns Hopkins University conducted a study following the great 1906 San Francisco earthquake. The earthquake was accompanied by horizontal surface displacements of several meters along the northern portion of the San Andreas fault. This 1300-kilometer (780-mile) fracture runs north-south through southern California. It is a large fault zone that separates two great sections of Earth's crust, the North American plate and the Pacific plate. Field investigations determined that, during this single earthquake, the Pacific plate lurched as much as 4.7 meters (15 feet) northward past the adjacent North American plate.

The mechanism for earthquake formation which Reid deduced form this information is illustrated in Figure 6.5. In part A of the figure, you see an existing fault, or break in the rock. In part B, tectonic forces ever so slowly deform the crustal rocks on both sides of the fault, as demonstrated by the bent features. Under these conditions, rocks are bending and storing elastic energy, much like a wooden stick does if bent. Eventually, the frictional resistance holding the rocks together is overcome. As slippage occurs at the weakest point (the focus), displacement will exert stress farther along the fault, where additional slippage will occur until most of the built-up strain is released (Figure 6.5C). This slippage allows the deformed rock to "snap back." The vibrations we know as an earthquake occur as the rock elastically returns to its original shape. The "springing back" of the rock was termed **elastic rebound** by Reid, because the rock behaves elastically, much like a stretched rubber band does when it is released.

In summary, most earthquakes are produced by the rapid release of elastic energy stored in rock that has been subjected to great stress. Once the strength of the rock is exceeded, it suddenly ruptures, causing the vibrations of an earthquake. Earthquakes also occur along existing fault surfaces whenever the frictional forces on the fault surfaces are overcome.

Figure 6.4
Slippage along a fault produced an offset in this orange grove east of Calexico, California. (Photo by John S. Shelton)

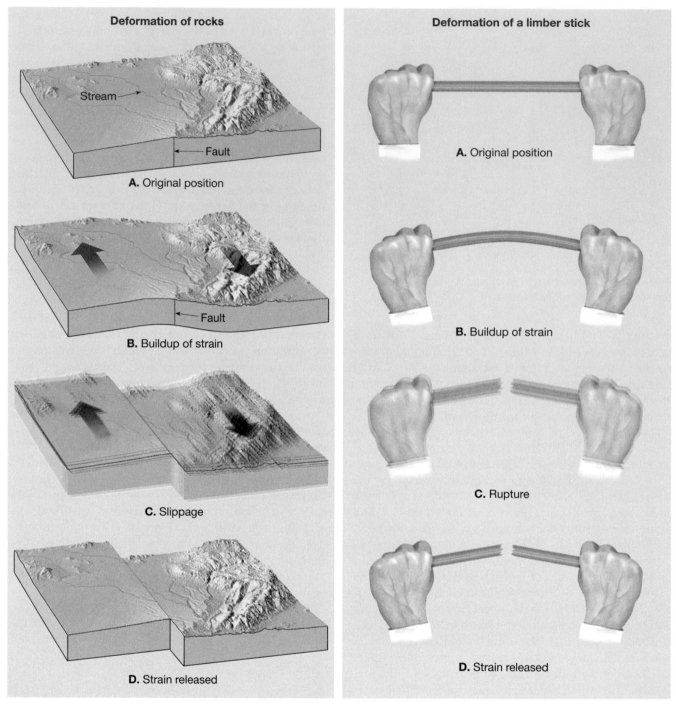

Figure 6.5

Elastic rebound. As rock is deformed it bends, storing elastic energy. Once the rock is strained beyond its breaking point it ruptures, releasing the stored-up energy in the form of earthquake waves.

The San Andreas is undoubtedly the most studied fault system in the world. Over the years, investigations have shown that displacement occurs along discrete segments that are 100 to 200 kilometers long. Further, each fault segment behaves somewhat differently from the others. Some portions of the San Andreas exhibit a slow, gradual displacement known as *fault creep*, which occurs relatively smoothly, and therefore with little noticeable seismic activity. Other segments regularly slip, producing small earthquakes.

Still other segments remain locked and store elastic energy for hundreds of years before rupturing in great earthquakes. The latter process is described as *stick-slip* motion, because the fault exhibits alternating periods of locked behavior followed by sudden slippage. It is estimated that great earthquakes should occur about every 50 to 200 years along those sections of the San Andreas fault that exhibit stick-slip motion. This knowledge is useful when assigning a potential earthquake risk to a given segment of the fault zone.

Not all movement along faults is horizontal. Vertical displacement, in which one side is lifted higher in relation to the other, is also common. Figure 6.6 shows a *fault scarp* (cliff) produced by such vertical movement. Further, many earthquakes occur at such great depths that no displacement is evident at the surface.

Foreshocks and Aftershocks

The intense vibrations of the 1906 San Francisco earthquake lasted about 40 seconds. Although most of the displacement along the fault occurred in this rather short period, additional movements along this and other nearby faults occurred for several days following the main quake. The adjustments that follow a major earthquake often generate smaller earthquakes called **aftershocks**. Although these aftershocks are usually much weaker than the main earthquake, they can sometimes destroy already badly weakened structures. This occurred, for example, during a 1988 earthquake in Armenia. A large aftershock of magnitude 5.8 collapsed many structures that had been weakened by the main tremor. That disaster killed about 25,000 people.

In addition, small earthquakes called **foreshocks** often precede a major earthquake by days or, in some cases, by as much as several years. Monitoring of these foreshocks has been used as a means of predicting forthcoming major earthquakes, with mixed success. We will consider the topic of earthquake prediction in a later section of this chapter.

Tectonic Forces and Earthquakes

It is important to understand that the tectonic forces creating the strain that was eventually released during the 1906 San Francisco earthquake are still active. Currently, laser beams are used to measure the relative motion between the opposite sides of this fault. These measurements reveal a displacement of 2 to 5 centimeters (1 to 2 inches) per year. Although this seems slow, it produces substantial movement over millions of years. To illustrate, in 30 million years, this rate of displacement would slide the western portion of California northward so that Los Angeles, on the Pacific plate, would be adjacent to San Francisco on the North American plate! More important in the short term, a displacement of just 2 centimeters per year produces 2 meters of offset every 100 years. Consequently, the 4 meters of displacement produced during the 1906 San Francisco earthquake should occur at least every 200 years along this segment of the fault zone. This fact lies behind California's concern for making buildings earthquake-resistant, in anticipation of the inevitable "big one."

Earthquake Waves

The study of earthquake waves, **seismology**, dates back to attempts by the Chinese almost 2000 years ago to determine the direction to the source of each earthquake. Modern **seismographs** are instruments that record earthquake waves. Their principle is simple. A weight is freely suspended from a support that is attached to bedrock (Figure 6.7). When waves from an earthquake reach the instrument, the inertia of the weight keeps it stationary, while Earth and the support vibrate. The movement of Earth in relation to the stationary weight is recorded on a rotating drum. (*Inertia* is the tendency of a stationary object to hold still, or a moving object to stay in motion.)

Modern seismographs amplify and record ground motion, producing a trace as shown in Figure 6.8. These records, called **seismograms**, reveal that seismic waves are elastic energy. This energy radiates outward in all directions from the focus, as you saw in Figure 6.3. The transmission of this energy can be compared to the shaking of gelatin in a bowl that is jarred. Seismograms reveal

Figure 6.6

Fault scarp produced from vertical movement during the 1964 Alaskan earthquake. (Courtesy of U.S. Geological Survey)

Figure 6.7

Principle of the seismograph. The inertia of the suspended mass tends to keep it motionless, while the recording drum, which is anchored to bedrock, vibrates in response to seismic waves. Thus, the stationary mass provides a reference point from which to measure the amount of displacement occurring as the seismic wave passes through the ground below.

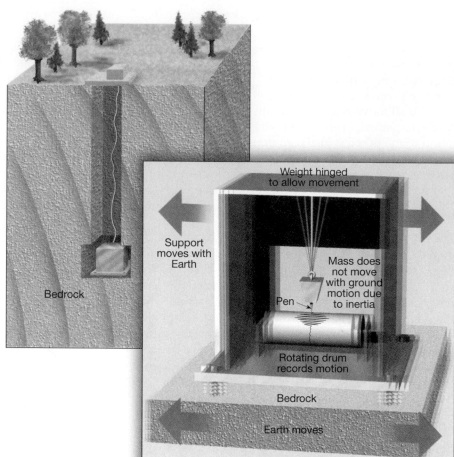

that two main types of seismic waves are generated by the slippage of a rock mass. Some travel along Earth's outer layer, and are called **surface waves**. Others travel through Earth's interior and are called **body waves**. Body waves are further divided into **primary waves (P waves)** and **secondary waves (S waves)**.

Figure 6.8

Typical seismic record. Note the time interval (about 5 minutes) between the arrival of the first P waves and the arrival of the first S waves.

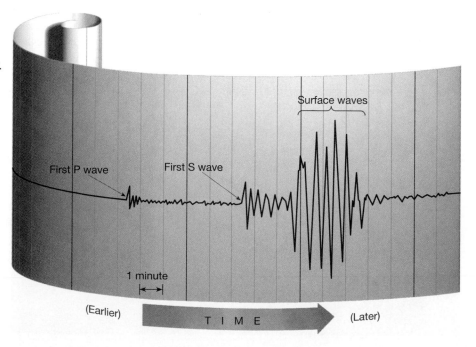

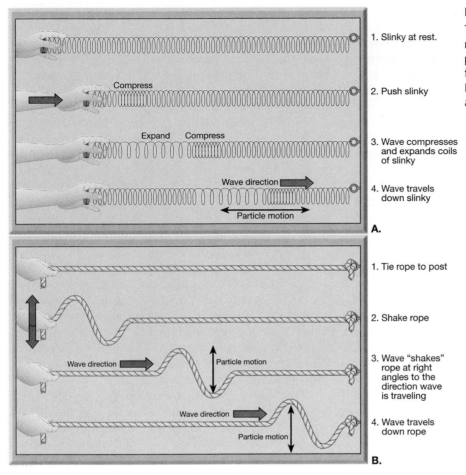

1. Slinky at rest.

2. Push slinky

3. Wave compresses and expands coils of slinky

4. Wave travels down slinky

A.

1. Tie rope to post

2. Shake rope

3. Wave "shakes" rope at right angles to the direction wave is traveling

4. Wave travels down rope

B.

Figure 6.9

Types of seismic waves and their characteristic motion. **A.** P waves compress, causing the particles in the material to vibrate back and forth in the same direction as the waves move. **B.** S waves cause particles to oscillate at right angles to the direction of wave motion.

Body waves are divided into P and S waves by their mode of travel through intervening materials. P waves are "push-pull" waves—they push (compress) and pull (expand) rocks in the direction the wave is traveling (Figure 6.9A). Imagine holding someone by the shoulders and shaking them. This push-pull movement is how P waves move through the Earth. This wave motion is analogous to that generated by human vocal cords as they move air to create sound. Solids, liquids, and gases resist a change in volume when compressed and will elastically spring back once the force is removed. Therefore, P waves, which are compressional waves, can travel through all these materials.

S waves, on the other hand, "shake" the particles at right angles to their direction of travel. This can be illustrated by fastening one end of a rope and shaking the other end, as shown in Figure 6.9B. Unlike P waves, which temporarily change the *volume* of the intervening material by alternately compressing and expanding it, S waves temporarily change the *shape* of the material that transmits them. Because fluids (gases and liquids) do not respond elastically to changes in shape, they will not transmit S waves.

The motion of surface waves is somewhat more complex. As surface waves travel along the ground, they cause the ground and anything resting upon it to move, much like ocean swells toss a ship. In addition to their up-and-down motion, surface waves have a side-to-side motion similar to an S wave oriented in a horizontal plane. This latter motion is particularly damaging to the foundations of structures.

By observing a "typical" seismic record, as shown in Figure 6.8, you can see a major difference among these seismic waves: P waves arrive at the recording station first; then S waves; and then surface waves. This is a consequence of their speeds. To illustrate, the velocity of P waves through granite within the crust is about 6 kilometers per second. S waves under the same conditions travel at 3.5 kilometers per second. Differences in density and elastic properties of the rock greatly influence the velocities of these waves. Generally, in any solid material, P waves travel about 1.7 times faster than S waves, and surface waves can be expected to travel at 90 percent of the velocity of the S waves.

As you shall see, seismic waves allow us to determine the location and magnitude of earthquakes. In addition, seismic waves provide us with a tool for probing Earth's interior.

Finding Earthquake Epicenters

Recall that the *focus* is the place within Earth where earthquake waves originate. The **epicenter** is the location on the surface directly above the focus (see Figure 6.3).

The difference in velocities of P and S waves provides a method for locating the epicenter. The principle used is analogous to a race between two autos, one faster than the other. The P wave always wins the race, arriving ahead of the S wave. But, the greater the length of the race, the greater will be the difference in the arrival times at the finish line (the seismic station). Therefore, the greater the interval measured on a seismogram between the arrival of the first P wave and the first S wave, the greater the distance to the earthquake source.

A system for locating earthquake epicenters was developed by using seismograms from earthquakes whose epicenters could be easily pinpointed from physical evidence. From these seismograms, travel-time graphs were constructed (Figure 6.10). The first travel-time graphs were greatly improved when seismograms became available from nuclear explosions, because the precise location and time of detonation were known.

Using the sample seismogram in Figure 6.8 and the travel-time curves in Figure 6.10, we can determine the distance separating the recording station from the

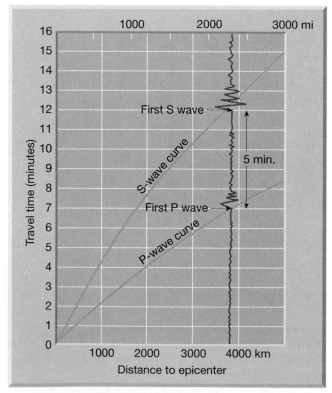

Figure 6.10

A travel-time graph is used to determine the distance to the epicenter. The difference in arrival times of the first P wave and the first S wave in the example is 5 minutes. Thus, the epicenter is roughly 3800 kilometers (2350 miles) away.

earthquake in two steps: (1) determine the time interval between the arrival of the first P wave and the first S wave, and (2) find on the travel-time graph the equivalent time spread between the P and S wave curves. From this information, we can determine that this earthquake occurred 3800 kilometers (2350 miles) from the recording instrument.

Now we know the *distance*, but what *direction*? The epicenter could be in any direction from the seismic station. As shown in Figure 6.11, the precise location can be found when the distance is known from three or more different seismic stations, On a globe, we draw a circle around each seismic station, Each circle represents the epicenter distance for each station. The point where the three circles intersect is the epicenter of the quake. This method is called triangulation.

About 95 percent of the energy released by earthquakes originates in a few relatively narrow zones (Figure 6.12). The greatest energy is released along a path around the outer edge of the Pacific Ocean known as the *circum-Pacific belt*. Included in this zone are regions of great seismic activity, such as Japan, the Philippines, Chile, and numerous volcanic island chains, as exemplified by Alaska's Aleutian Islands.

Figure 6.12 reveals another continuous belt which extends for thousands of kilometers through the world's oceans. This zone coincides with the oceanic ridge system, an area of frequent but low-intensity seismic activity. By comparing this figure with Figure 7.8 (pp. 186–87), you can see a close correlation between the location of earthquake epicenters and plate boundaries.

Earthquake Intensity and Magnitude

Until a century ago, earthquake size and strength were described subjectively, making accurate classification of earthquake intensity difficult. Then, in 1902, Giuseppe Mercalli developed a fairly reliable intensity scale based on damage to various types of structures. The U.S. Coast and Geodetic Survey uses a modification of this scale today (Table 6.1).

The **Mercalli intensity scale** assesses the damage from a quake at a specific location. Please note that earthquake intensity depends not only on the strength of the earthquake, but on other factors, such as distance from the epicenter, the nature of surface materials, and building design. A modest 6.9-Richter-magnitude earthquake in Armenia in 1988 was very destructive, mainly because of inferior construction. A 1985 Mexico City quake was deadly because of the soft sediment upon which the city rests. Thus, the destruction wrought by earthquakes is very meaningful to people living there, but it is not a true measure of the earthquake's actual strength. Further, many earthquakes occur beneath the sea or at great depths in the crust and are not felt.

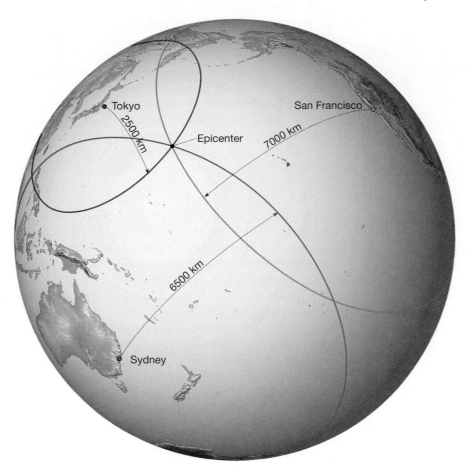

In 1935, Charles Richter of the California Institute of Technology introduced the concept of earthquake **magnitude**. Today, a refined **Richter scale** is used worldwide

to describe earthquake magnitude. Richter magnitude is determined by measuring the amplitude of the largest wave recorded on the seismogram (see Figure 6.8). For

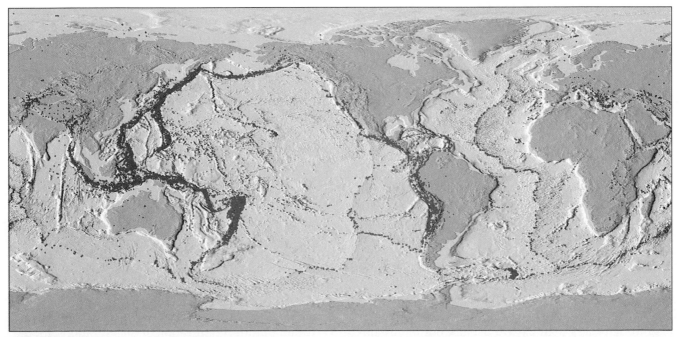

Figure 6.12
Distribution of the 14,229 earthquakes with magnitudes equal to or greater than 5 for the period of 1980–1990. (Data from National Geophysical Data Center/NOAA)

Table 6.1 Modified Mercalli intensity scale.

I   Not felt except by a very few under especially favorable circumstances.

II   Felt only by a few persons at rest, especially on upper floors of buildings.

III   Felt quite noticeably indoors, especially on upper floors of buildings, but many people do not recognize it as an earthquake.

IV   During the day felt indoors by many, outdoors by few. Sensation like heavy truck striking building.

V   Felt by nearly everyone, many awakened. Disturbances of trees, poles, and other tall objects sometimes noticed.

VI   Felt by all; many frightened and run outdoors. Some heavy furniture moved; few instances of fallen plaster or damaged chimneys. Damage slight.

VII   Everybody runs outdoors. Damage negligible in buildings of good design and construction; slight to moderate in well-built ordinary structures; considerable in poorly built or badly designed structures.

VIII   Damage slight in specially designed structures; considerable in ordinary substantial buildings with partial collapse; great in poorly built structures. (Fall of chimneys, factory stacks, columns, monuments, walls.)

IX   Damage considerable in specially designed structures. Buildings shifted off foundations. Ground cracked conspicuously.

X   Some well-built wooden structures destroyed. Most masonry and frame structures destroyed. Ground badly cracked.

XI   Few, if any (masonry) structures remain standing. Bridges destroyed. Broad fissues in ground.

XII   Damage total. Waves seen on ground surfaces. Objects thrown upward into air.

seismic stations worldwide to obtain the same magnitude for a given earthquake, adjustments are made for the weakening of seismic waves with distance as they move away form the focus and for the sensitivity of the recording instrument.

The largest earthquakes ever recorded had Richter magnitudes near 8.6. These great shocks released energy roughly equivalent to the detonation of one billion tons of TNT. Conversely, earthquakes with a Richter magnitude of less than 2.0 are usually not felt by humans. Table 6.2 shows how earthquake magnitudes and their effects are related.

Earthquakes vary enormously in strength, and great earthquakes produce traces having wave amplitudes that are thousands of times larger than those generated by weak tremors. To accommodate this wide variation, Richter could not use a linear scale, but instead used a *logarithmic scale* to express magnitude. On this scale, an increase of one magnitude means a *tenfold* increase in wave amplitude. Thus, the amplitude of the largest surface wave for a 5.3-magnitude earthquake is 10 times greater than the wave amplitude produce by an earthquake having a magnitude of 4.3.

More important, each unit of Richter magnitude equates to roughly a *30-fold energy increase*. Thus, an earthquake with a magnitude of 6.5 releases 30 times more energy than one with a magnitude of 5.5, and roughly 900 times (30×30) more energy than a 4.5-magnitude quake. A great earthquake with a magnitude of 8.5 releases millions of times more energy than the smallest earthquake felt by humans.

Knowing this should dispel the notion that a moderate earthquake acts as a "pressure relief valve," decreasing the chances for a major quake in the same region. A moderate quake certainly relieves some strain, but thousands of moderate tremors would be needed to release the equivalent energy of one "great" earthquake.

Some of the world's major earthquakes and their corresponding Richter magnitudes are listed in Table 6.3.

Destruction from Earthquakes

The most violent earthquake to jar North America this century—the Good Friday Alaskan Earthquake—occurred in 1964. Felt throughout that state, the earthquake had a Richter magnitude of 8.3–8.4 and reportedly lasted 3 to 4 minutes. This event left 131 persons dead, thousands homeless, and the economy of the state badly disrupted because it occurred near major towns and seaports (Figure 6.13). Had the schools and business districts been open on this holiday, the toll surely would have been higher. Within 24 hours of the initial shock, 28 aftershocks were recorded, 10 of which exceeded a Richter magnitude of 6.

Destruction from Seismic Vibrations

The 1964 Alaskan earthquake provided geologists with new insights into the role of ground shaking as a destructive force. Recall that the structural damage from earthquake waves depends on several factors, including (1) the amplitude, (2) duration of the vibrations, (3) the nature of the material upon which the structure rests, and (4) the design of the structure.

Table 6.2 Earthquake magnitudes and expected world incidence.

Richter Magnitudes	Effects Near Epicenter	Estimated Number per Year
<2.0	Generally not felt, but recorded.	600,000
2.0–2.9	Potentially perceptible.	300,000
3.0–3.9	Felt by some.	49,000
4.0–4.9	Felt by most.	6200
5.0–5.9	Damaging shocks.	800
6.0–6.9	Destructive in populous regions.	266
7.0–7.9	Major earthquakes. Inflict serious damage.	18
≥8.0	Great earthquakes. Destroy communities near epicenter.	1.4

SOURCE: *Earthquake Information Bulletin* and others.

Table 6.3 Some notable earthquakes.

Year	Location	Deaths (est.)	Magnitude	Comments
1290	Chihli (Hopei), China	100,000		
1556	Shensi, China	830,000		Possibly the greatest natural disaster.
1737	Calcutta, India	300,000		
1755	Lisbon, Portugal	70,000		Tsunami damage extensive.
*1811–1812	New Madrid, Missouri	Few		Three major earthquakes.
*1886	Charleston, South Carolina	60		Greatest historical earthquake in the eastern United States.
*1906	San Francisco, California	1500	8.1–8.2	Fires caused extensive damage.
1908	Messina, Italy	120,000		
1920	Kansu, China	180,000		
1923	Tokyo, Japan	143,000	7.9	Fire caused extensive destruction.
1960	Southern Chile	5700	8.5–8.6	Possibly the largest-magnitude earthquake ever recorded.
*1964	Alaska	131	8.3–8.4	
1970	Peru	66,000	7.8	Great rockslide.
*1971	San Fernando, California	65	6.5	Damage exceeded $1 billion.
1975	Liaoning Province, China	Few	7.5	First major earthquake to be predicted.
1976	Tangshan, China	240,000	7.6	Not predicted.
1985	Mexico City	9500	8.1	Major damage occurred 400 km from epicenter.
1988	Armenia	25,000	6.9	Poor construction practices contributed to destruction.
*1989	San Francisco Bay area	62	7.1	Damages exceeded $6 billion.
1990	Northwestern Iran	50,000	7.3	Landslides and poor construction practices caused great damage.
*1994	Northridge, California	61	6.7	Damages in excess of $15 billion.
1995	Kobe, Japan	5472	6.9	Damage estimated to exceed $100 billion.

*U.S. earthquakes.
SOURCE: U.S. National Oceanic and Atmospheric Administration.

Figure 6.13

Region most affected by the Good Friday earthquake of 1964. Note the epicenter (red dot). (After U.S. Geological Survey)

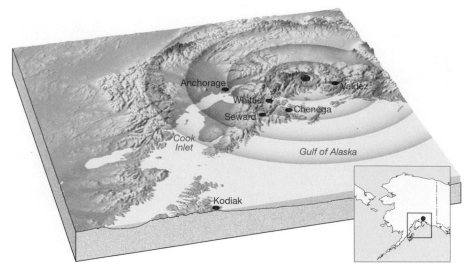

All multistory structures in Anchorage were damaged by the vibrations, but the more flexible wood-frame residential buildings fared best. Figure 6.14 offers a striking example of how construction variations affect earthquake damage; you can see that the steel-frame building on the left withstood the vibrations, whereas the relatively rigid concrete structure was badly damaged.

Most large structures in Anchorage were damaged even though they were built to conform to the earthquake provisions of the Uniform Building Code of California. Perhaps some of that destruction can be attributed to the unusually long duration of this earthquake (estimated at 3 to 4 minutes). (The San Francisco earthquake of 1906 was felt for about 40 seconds; the strong vibrations of the 1989 Loma Prieta earthquake lasted less than 15 seconds.)

The buildings in Anchorage are situated on soft, unconsolidated sediments, which amplify vibrations more than solid bedrock. Thus, damage in Anchorage was more severe than in Whittier, which rests on a firm foundation of granite. However, Whittier was damaged by a seismic sea wave (described in the next section).

The 1985 Mexican earthquake gave seismologists and engineers a vivid reminder of what had been learned following the 1964 Alaskan earthquake. In central Mexico City, nearly 400 kilometers (250 miles) from the epicenter, the vibrations intensified to five times that experienced in outlying districts, due to the city's construction on soft sediments, remnants of an ancient lake bed (Figure 6.15).

Where unconsolidated materials are saturated with water, earthquakes can generate a phenomenon known as **liquefaction**. Under these conditions, what had been a

Figure 6.14

Damage to the five-story JCPenney Co. building, Anchorage, Alaska. Very little structural damage was incurred by the adjacent building. (Courtesy of NOAA)

Figure 6.15
During the 1985 Mexican earthquake, multistory buildings swayed back and forth as much as one meter. Many, including the hotel shown here, collapsed or were seriously damaged. (Photo by James L. Beck)

stable soil turns into a fluid that is no longer capable of supporting buildings or other structures (Figure 6.16). As a result, underground objects such as storage tanks and sewer lines may float toward the surface of their newly liquefied environment. Buildings and other structures may settle and collapse. During the Loma Prieta earthquake, in San Francisco's Marina District, foundations failed and geysers of sand and water shot from the ground, indicating that liquefaction had occurred (Figure 6.17).

Tsunami

Most deaths associated with the 1964 Alaskan quake were caused by **seismic sea waves**, or **tsunami**. These destructive waves often are called "tidal waves" by the media. However, this name is wrong, for these waves are generated by earthquakes, not by the tidal effect of the moon or sun. The name *tsunami* is Japanese for "harbor wave," for Japanese harbors have suffered from many of them.

Most tsunami result from vertical displacement of the ocean floor during an earthquake (Figure 6.18). Once formed, a tsunami resembles the ripples created when a pebble is dropped into a pond. In contrast to ripples, tsunami advance across the ocean at speeds between 500 and 950 kilometers (300 and 600 miles) per hour. Despite this, a tsunami in the open ocean can pass undetected because its height is usually less than one meter and the distance between wave crests is great, ranging from 100 to 700 kilometers. However, upon entering shallower coastal water, these destructive waves are slowed and the water begins to pile up to heights that occasionally exceed 30 meters (100 feet), as shown in Figure 6.18. As the crest of a tsunami approaches shore, it appears as a rapid rise in sea level with a turbulent and chaotic surface. Tsunami can be very destructive (Figure 6.19).

Usually the first warning of an approaching tsunami is a rather rapid withdrawal of water from beaches. Coastal residents have learned to heed this warning and move to

Figure 6.16
Effects of liquefaction. This tilted building rests on unconsolidated sediment that imitated quicksand during the 1985 Mexican earthquake. (Photo by James L. Beck)

Figure 6.17
These "mud volcanoes" were produced by the Loma Prieta earthquake of 1989. They formed when geysers of sand and water shot from the ground, an indication that liquefaction occurred. (Photo by Richard Hilton, courtesy of Dennis Fox)

higher ground, for about 5 to 30 minutes later the retreat of water is followed by a surge capable of extending hundreds of meters inland. In a successive fashion, each surge is followed by a rapid oceanward retreat of the water. These waves are separated by intervals of between 10 and 60 minutes. They are able to traverse thousands of kilometers of the ocean before their energy is dissipated (see Box 6.2).

The tsunami generated in the 1964 Alaskan earthquake heavily damaged communities along the Gulf of Alaska and killed 107. By contrast, only 9 persons died in Anchorage as a direct result of the vibrations.

Tsunami damage following the Alaskan earthquake extended along much of the west coast of North America. Despite a one-hour warning, 12 persons perished in Crescent City, California, from the fifth wave. The first wave crested about 4 meters (13 feet) above low tide and was followed by three progressively smaller waves. Believing that the tsunami had ceased, people returned to the shore, only to be met by the fifth and most devastating wave. Superimposed upon high tide, it crested about 6 meters higher than the level of low tide.

Although most tsunami are generated by earthquakes, a volcanic eruption in the ocean can generate this destructive phenomenon as well. For example, the 1883 volcanic explosion of Krakatoa, an island in Indonesia, generated a tsunami that drowned some 36,000 coastal residents of Java and Sumatra.

Landslides and Ground Subsidence

In the 1964 Alaskan earthquake, the greatest damage to structures was from landslides and ground subsidence triggered by the vibrations. At Valdez and Seward, the violent shaking caused river-delta materials to experience liquefaction; the subsequent slumping carried both

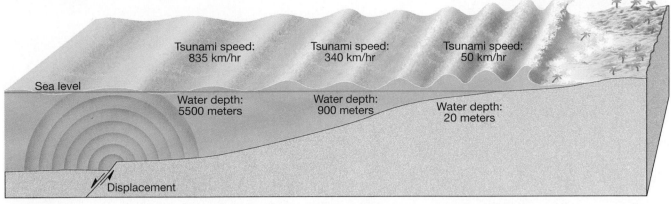

Figure 6.18
Schematic drawing of a tsunami generated by displacement of the ocean floor. The speed of a wave moving across the ocean correlates with ocean depth. As shown, waves moving in deep water advance at speeds in excess of 800 kilometers per hour. Speed gradually slows to 50 kilometers per hour at depths of 20 meters. Decreasing depth slows the movement of the wave column. As waves slow in shallow water, they grow in height until they topple and rush onto shore with tremendous force. The size and spacing of these swells are not to scale.

waterfronts away. Because the disaster could happen again, the entire town of Valdez was relocated about 7 kilometers away on more stable ground. In Valdez, 31 persons on a dock died when it slid into the sea.

Most of the damage in Anchorage was attributed to landslides. Many homes were destroyed in Turnagain Heights when a layer of clay lost its strength and over 200 acres of land slid toward the ocean (Figure 6.20). A

Box 6.2

Tsunami Warning System

Tsunamis traverse large stretches of ocean before their energy is fully dissipated. The tsunami generated by a 1960 Chilean earthquake, in addition to destroying villages along an 800-kilometer stretch of coastal South America, traveled 17,000 kilometers across the Pacific to Japan. Here, about 22 hours after the quake, considerable damage occurred in southern coastal villages. For several days afterward, tidal gauges in Hilo, Hawaii, detected these diminishing waves as they reverberated like echoes about the Pacific.

In 1946, a large tsunami struck the Hawaiian Islands without warning. A wave more than 15 meters (50 feet) high left several coastal villages in shambles. This destruction motivated the U.S. Coast and Geodetic Survey to establish a tsunami warning system for coastal areas of the Pacific. From seismic observatories throughout the region, large earthquakes are reported to the Tsunami Warning Center in Honolulu. Scientists at the Center use tidal gauges to determine whether a tsunami has been formed. Within an

hour a warning is issued. Although tsunamis travel very rapidly, there is sufficient time to evacuate all but the region nearest the epicenter (Figure 6.B). For example, a tsunami generated near the Aleutian islands would take 5 hours to reach Hawaii, and one generated near the coast of Chile would travel 15 hours before reaching Hawaii.

Fortunately, most earthquakes do not generate tsunamis. On the average, only about 1.5 destructive tsunamis are generated worldwide each year. Of these, only about one every ten years is catastrophic.

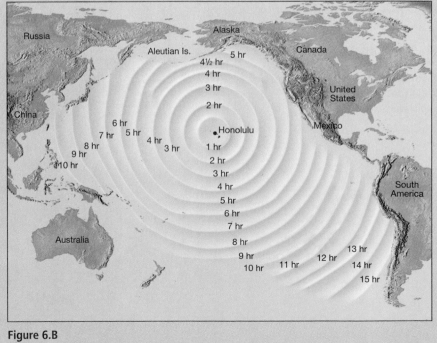

Figure 6.B
Tsunami travel times to Honolulu, Hawaii, from locations throughout the Pacific. (From NOAA)

Figure 6.19

A man stands before a wall of water about to engulf him at Hilo, Hawaii, on April 1, 1946. This tsunami, which originated in the Aleutian Islands near Alaska, was still powerful enough when it hit Hawaii to rise 9 to 16 meters (30 to 55 feet). The S.S. *Brigham Victory*, from which this photograph was taken, managed to survive the onslaught, but 159 people in Hawaii, including the man seen here, were killed. (Photo courtesy of Water Resources Center Archives, University of California, Berkeley.)

portion of this landslide has been left in its natural condition as a reminder of this destructive event. The site was named "Earthquake Park." Downtown Anchorage was also disrupted as sections of the main business district dropped by as much as 3 meters (10 feet).

Fire

The 1906 San Francisco earthquake reminds us of the formidable threat of fire. The central city contained mostly large, older wooden structures and brick build-

Figure 6.20

Photo of a small portion of the Turnagain Heights slide. (Photo courtesy of U.S. Geological Survey)

ings. The greatest destruction was caused by fires that started when gas and electrical lines were severed. The fires raged uncontrolled for three days and devastated over 500 city blocks (Figure 6.2). The problem was compounded by the initial ground shaking, which broke the city's water lines into hundreds of unconnected pieces.

The fire was finally contained when buildings were dynamited along a wide boulevard to create a *fire break*, the same strategy used in fighting a forest fire. Although only a few deaths were attributed to the fires, such is not always the case. A 1923 earthquake in Japan (their worst quake prior to the 1995 Kobe tremor) triggered an estimated 250 fires, which devastated the city of Yokohama and destroyed more than half the homes in Tokyo. Over 100,000 deaths were attributed to the fires, which were driven by unusually high winds.

Can Earthquakes Be Predicted?

The vibrations that shook Northridge, California, in 1994 inflicted 51 deaths and about $15 billion in damage (Figure 6.21). This was from a brief earthquake (about 40 seconds) of moderate rating (6.6 to 6.9 on the Richter scale). Seismologists warn that earthquakes of comparable or greater strength will occur along the San Andreas fault, which cuts a 1300-kilometer path through the state (Figure 6.22). Following the 1995 earthquake in Kobe, Japan, a U.S. Geologist Survey physicist cautioned: "Kobe is almost a dress rehearsal for an earthquake on the Hayward fault (a fault parallel to the San Andreas, near San Francisco)." The obvious question is, can earthquakes be predicted?

Substantial research to predict earthquakes is under way in Japan, the United States, China, and Russia—countries where earthquake risk is high. This research is striving to identify phenomena that precede major earthquakes. In California, for example, a pattern has been observed to precede some earthquakes: uplift or subsidence of the land, changes in the movements along a fault zone, and a period of seismic quiescence, often followed by renewed activity. Japanese scientists are studying peculiar animal behavior that may precede a quake. Others are examining changes in groundwater levels and radio waves.

Figure 6.21
Damage to Interstate 5 during the January 17, 1994 Northridge earthquake. (Photo by John Barr/Gamma Liaison)

Short-Range Predictions

Although no consistent method of short-range prediction has yet been devised, one notable success was the foretelling in 1975 of a 7.3-magnitude earthquake in the Lainoning Province of China. Here, for the first and only time, seismologists forecast a large earthquake that was about to destroy a major city. By evacuating some three million residents from their unreinforced masonry structures, tens of thousands of lives were spared. But tragically, one year later, a devastating earthquake in China was *not* predicted, and more than 200,000 people were killed.

The Chinese have also issued false alarms. In a province near Hong Kong, people left their dwellings for over a month, but no earthquake followed. Can you imagine the debate that would precede an order to evacuate a large city in the United States, such as Los Angeles or San Francisco? The cost of evacuating millions of people, arranging for living accomodations, and providing for their lost work time and wages, would have to be weighed against the earthquake's probability.

Long-Range Predictions

Long-range forecasts are based on the premise that earthquakes are repetitive or cyclical, like the weather. In other words, as soon as one earthquake is over, the continuing motions of Earth's plates begin to build strain in the rocks again, until they fail again. This has led seismologists to study the history of earthquakes for patterns, so their occurrences might be predicted.

One U.S. Geological Survey study gives the probability of a rupture occurring along various segments of the San Andreas fault for the 30 years between 1988 and 2018 (Figure 6.22). From this study, the Santa Cruz Mountains area was given a 30 percent probability of producing a 6.5 magnitude earthquake during this time period. In fact, it produced the Loma Prieta quake in 1989, of 7.1 magnitude.

The region given the highest probability (90 percent) of generating a quake is the Parkfield section. This area has been called the "Old Faithful" of earthquake zones because activity here has been very regular since record-keeping began in 1857. The U.S. Geological Survey has

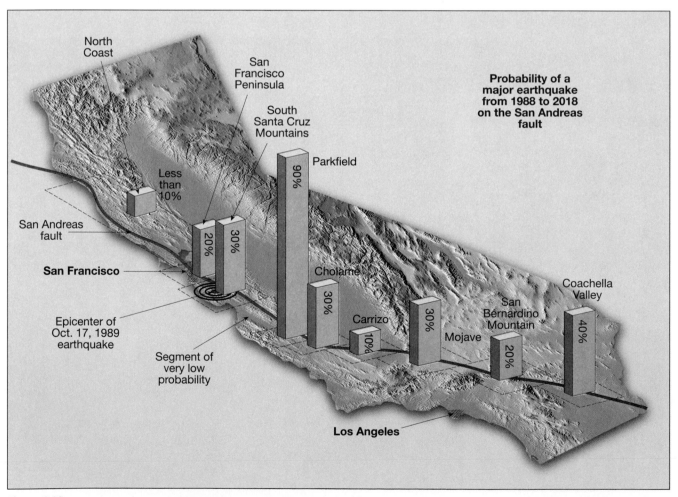

Figure 6.22
Probability of a major earthquake from 1988 to 2018 on the San Andreas fault.

14. List four factors that affect the amount of destruction caused by seismic vibrations.

15. In addition to the destruction created directly by seismic vibrations, list three other types of destruction associated with earthquakes.

16. Distinguish between the Mercalli scale and the Richter scale.

17. What is a tsunami? How is one generated?

18. Cite some reasons why an earthquake with a moderate magnitude might cause more extensive damage than a quake with a high magnitude.

19. What evidence do we have that Earth's outer core is molten?

20. Contrast the physical makeup of the asthenosphere and the lithosphere.

21. Why are meteorites considered important clues to the composition of Earth's interior?

22. Describe the composition (mineral makeup) of the following:

    (a) continental crust **(c)** mantle

    (b) oceanic crust   **(d)** core

Key Terms

aftershock (p. 159)
asthenosphere (p. 174)
body wave (p. 160)
crust (p. 173)
earthquake (p. 155)
elastic rebound (p. 157)
epicenter (p. 162)
fault (p. 157)

focus (p. 155)
foreshock (p. 159)
inner core (p. 173)
liquefaction (p. 166)
lithosphere (p. 174)
magnitude (p. 163)
mantle (p. 173)
Mercalli intensity scale (p. 162)
Mohorovičić discontinuity (Moho) (p. 173)

outer core (p. 173)
primary (P) wave (p. 160)
Richter scale (p. 163)
secondary (S) wave (p. 160)
seismic sea wave (tsunami) (p. 167)
seismogram (p. 159)
seismograph (p. 159)
seismology (p. 159)
shadow zone (p. 173)
surface wave (p. 160)

CHAPTER 7
Plate Tectonics

Composite satellite image of a part of North Africa and the Arabian Peninsula.
(©Worldsat International, Inc./Jim Knighton, 1995. Mississauga, Ontario, Canada. All Rights Reserved.)

Will California eventually "slide" into the ocean, as some predict? Have continents really "drifted" apart over the centuries? Answers to these questions and many others that have intrigued geologists for decades are now being provided by an exciting theory on large-scale movements taking place within Earth. This theory, called plate tectonics, represents the real frontier of the Earth sciences, and its implications are so far-reaching that it can be considered the framework from which most other geological processes should be viewed.

Early in this century, most geologists thought that the geographic positions of the ocean basins and continents were fixed. During the last few decades, however, vast amounts of new data have dramatically changed our understanding of the nature and workings of our planet. Earth scientists now realize that the continents gradually migrate across the globe. Where landmasses split apart, new ocean basins are created between the diverging blocks. Meanwhile, older portions of the seafloor are carried back into the mantle in regions where trenches occur in the deep ocean floor. Because of these movements, segments of continental material eventually collide and form Earth's great mountain ranges (Figure 7.1). In short, a revolutionary new model of Earth's tectonic* processes has emerged.

This profound reversal of scientific understanding has been appropriately described as a scientific revolution. Like other scientific revolutions, considerable time elapsed between the idea's inception and its general acceptance. The revolution began early in the twentieth century as a relatively straightforward proposal that the continents drift about the face of Earth. After many years of heated debate, the idea of drifting continents

*Tectonics refers to the deformation of Earth's crust and results in the formation of structural features such as mountains.

Figure 7.1
Lone Pine Peak in California's Sierra Nevada. (Photo by Carr Clifton/Minden Pictures)

was rejected by the vast majority of Earth scientists as improbable. However, during the 1950s and 1960s, new evidence rekindled interest in this proposal. By 1968, these new developments led to the unfolding of a far more encompassing theory than continental drift—a theory known as *plate tectonics*.

Continental Drift: An Idea Before Its Time

The idea that continents, particularly South America and Africa, fit together like pieces of a jigsaw puzzle originated with improved world maps. However, little significance was given this idea until 1915, when Alfred Wegener, a German climatologist and geophysicist, published *The Origin of Continents and Oceans*. In this book, Wegener set forth his radical hypothesis of **continental drift**.[*]

He suggested that a supercontinent he called **Pangaea** (meaning "all land") once existed (Figure 7.2). He further hypothesized that, about 200 million years ago,

[*] Wegener's ideas were actually preceded by those of an American geologist, F. B. Taylor, who in 1910 published a paper on continental drift. Taylor's paper provided little supporting evidence for continental drift, which may have been the reason that it had a relatively small impact on the scientific community.

this supercontinent began breaking into smaller continents, which then "drifted" to their present positions (see Figure 7.25).

Wegener and others collected substantial evidence to support these claims. The fit of South America and Africa, fossils, rock structures, and ancient climates all seemed to support the idea that these now-separate landmasses were once joined. Let us examine their evidence.

Evidence: The Continental Jigsaw Puzzle

Like a few others before him, Wegener first suspected that the continents might have been joined when he noticed the remarkable similarity between the coastlines on opposite sides of the South Atlantic. However, his use of present-day shorelines to make a fit of the continents was challenged immediately by other Earth scientists. These opponents correctly argued that shorelines are continually modified by erosional processes, and even if continental displacement had taken place, a good fit today would be unlikely. Wegener appeared to be aware of this problem, and, in fact, his original jigsaw fit of the continents was only very crude.

A much better approximation of the true outer boundary of the continents is the continental shelf. Today, the seaward edge of the continental shelf lies submerged, several hundred meters below sea level. In the early 1960s, scientists produced a map that attempted to fit the edges of the continental shelves at a

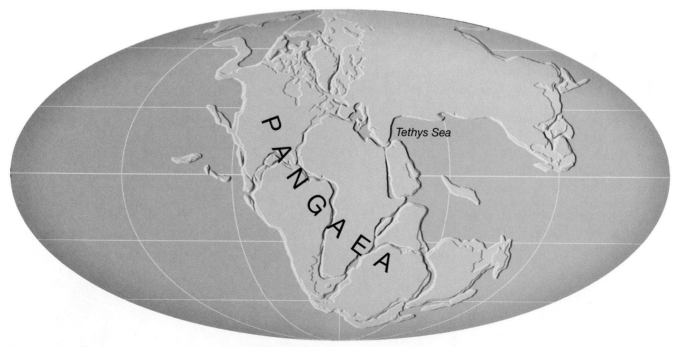

Figure 7.2

Reconstruction of Pangaea as it is thought to have appeared 200 million years ago. (After R. S. Deitz and J.C. Holden. *Journal of Geophysical Research* 75: 4943. Copyright by American Geophysical Union)

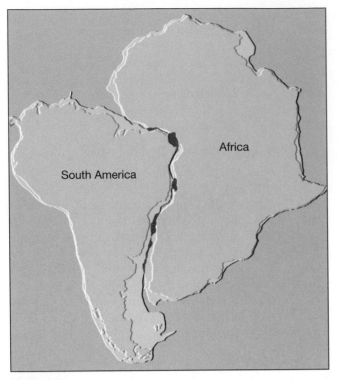

Figure 7.3

This shows the best fit of South America and Africa along the continental slope at a depth of 500 fathoms (about 900 meters). The areas where continental blocks overlap appear in brown. (After A. G. Smith. "Continental Drift." In *Understanding the Earth*, edited by I. G. Gass. Courtesy of Artemis Press.)

depth of 900 meters. The remarkable fit that was obtained is shown in Figure 7.3. Although the continents overlap in a few places, these are regions where streams have deposited large quantities of sediment, thus enlarging the continental shelves. The overall fit was even better than the supporters of continental drift suspected it would be.

Evidence: Fossils Match Across the Seas

Although Wegener was intrigued by the remarkable jigsaw fit on opposite sides of the Atlantic, he at first thought the idea of a mobile Earth improbable. Not until he came across an article citing fossil evidence for the existence of a land bridge connecting South America and Africa did he begin to take his own idea seriously. Through a search of the literature, Wegener learned that most paleontologists were in agreement that some type of land connection was needed to explain the existence of identical fossils on the widely separated landmasses.

To add credibility to his argument for the existence of the supercontinent of Pangaea, Wegener cited documented cases of several fossil organisms that had been found on different landmasses but which could not have crossed the vast oceans presently separating the continents. The classic example is *Mesosaurus*, a presumably aquatic, snaggle-toothed reptile whose fossil remains are limited to eastern South America and southern Africa (Figure 7.4). If *Mesosaurus* had been able to swim well enough to cross the vast South Atlantic Ocean, its remains should be more widely distributed. Since this is not the case, Wegener argued that South America and Africa must have been joined—somehow.

Wegener also cited the distribution of the fossil fern *Glossopteris* as evidence for the existence of Pangaea. This plant, identified by its large seeds that could not be blown very far, was known to be widely dispersed among Africa, Australia, India, and South America during the late Paleozoic era. Later, fossil remains of *Glossopteris* were discovered in Antarctica as well. Wegener knew that these seed ferns and associated flora grew only in a subpolar climate; therefore, he concluded that these landmasses must have been joined, since they presently include climatic regions that are too diverse to support such flora. For Wegener, fossils proved without question that a supercontinent had existed.

Figure 7.4

Fossils of *Mesosaurus* have been found on both sides of the South Atlantic and nowhere else in the world. Fossil remains of this and other organisms on the continents of Africa and South America appear to link these landmasses during the late Paleozoic and early Mesozoic eras.

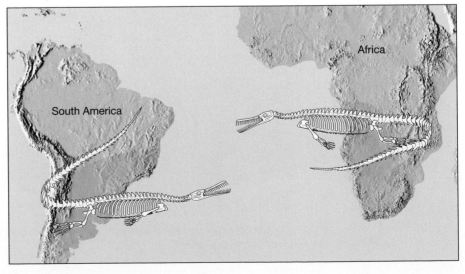

In his book, Wegener also cited the distribution of present-day organisms as evidence to support the concept of drifting continents. For example, modern organisms with similar ancestries clearly had to evolve in isolation during the last few tens of millions of years. Most obvious of these are the Australian marsupials, which have a direct fossil link to the marsupial opossums found in the Americas.

How did scientists explain the discovery of identical fossil organisms separated by thousands of kilometers of open ocean? The idea of land bridges was the most widely accepted solution to the problem of migration (Figure 7.5). We know, for example, that during the most recent glacial period, the lowering of sea level allowed animals to cross the narrow Bering Strait between Asia and North America. Was it possible, then, that one or more land bridges once connected Africa and South America? We are now quite certain that land bridges of this magnitude did not exist, for their remnants should still lie below sea level. But they are nowhere to be found.

Evidence: Rock Types and Structures Match

Anyone who has worked a picture puzzle knows that, in addition to the pieces fitting together, the picture must be continuous as well. The "picture" that must match in the "Continental Drift Puzzle" is one of rock types and mountain belts on the continents. If the continents were once together, the rocks found in a particular region on one continent should closely match in age and type those in corresponding positions on the matching continent.

Such evidence exists in the form of several mountain belts that terminate at one coastline, only to reappear on a landmass across the ocean. For instance, the mountain belt that includes the Appalachians trends northeastward through the eastern United States and disappears off the coast of Newfoundland (Figure 7.6A). Mountains of comparable age and structure are found in the British Isles and Scandinavia. When these landmasses are reassembled as in Figure 7.6B, the mountain chains form a nearly continuous belt. Numerous other rock structures exist that appear to have formed at the same time and were subsequently split apart.

Wegener was very satisfied that the similarities in rock structure on both sides of the Atlantic linked these landmasses. In his own words, "It is just as if we were to refit the torn pieces of a newspaper by matching their edges and then check whether the lines of print run smoothly across. If they do, there is nothing left but to conclude that the pieces were in fact joined in this way."[*]

Evidence: Ancient Climates

Because Alfred Wegener was a climatologist by training, he was keenly interested in obtaining paleoclimatic (ancient climatic) data in support of continental drift. His efforts were rewarded when he found evidence for dramatic climatic changes. For instance, glacial deposits indicate that, near the end of the Paleozoic era (between 220 and 300 million years ago), ice sheets covered extensive areas of the Southern Hemisphere. Layers of

[*] Alfred Wegener, *The Origin of Continents and Oceans*. Translated from the 4th revised German edition of 1929 by J. Birman (London: Methuen, 1966).

Figure 7.5

These sketches by John Holden illustrate various explanations for the occurrence of similar species on landmasses that are presently separated by vast oceans. (Reprinted with permission of John Holden)

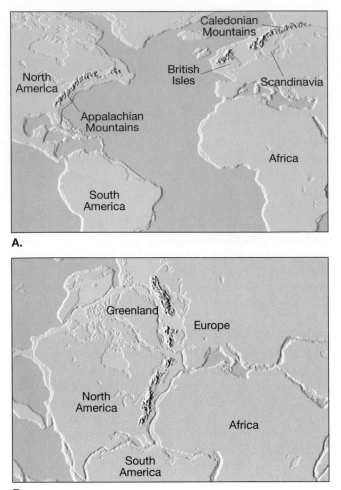

A.

B.

Figure 7.6
Matching mountain range across the North Atlantic. **A.** The Appalachian Mountains trend along the eastern flank of North America and disappear off the coast of Newfoundland. Mountains of comparable age and structure are found in the British Isles and Scandanavia. **B.** When these landmasses are placed in their predrift locations, these ancient mountain chains form a nearly continuous belt. These folded mountain belts formed roughly 300 million years ago as the landmasses collided during the formation of the supercontinent of Pangaea.

glacial till were found in southern Africa and South America, as well as in India and Australia. Below these beds of glacial debris lay striated and grooved bedrock. In some locations, the striations and grooves indicated that the ice had moved from what is now the sea onto land. Much of the land area containing evidence of this late Paleozoic glaciation presently lies within 30 degrees of the equator in a subtropical or tropical climate.

Could Earth have gone through a period sufficiently cold to have generated extensive continental glaciers in what is presently a tropical region? Wegener rejected this explanation because, during the late Pa-

leozoic, large swamps existed in the Northern Hemisphere. The lush vegetation of these swamps eventually became the major coal fields of the eastern United States, Europe, and Siberia.

Fossils from these coal fields indicate that the tree ferns which produced the coal deposits had large fronds. This indicates a tropical setting. Furthermore, unlike trees in colder climates, the tree trunks lacked growth rings. Growth rings do not form in tropical plants because there are minimal seasonal fluctuations in temperature.

Wegener believed that a better explanation for the paleoclimatic regimes he observed is provided by fitting together the landmasses as a supercontinent, with South Africa centered over the South Pole (Figure 7.7). This would account for the conditions necessary to generate extensive expanses of glacial ice over much of the Southern Hemisphere. At the same time, this geography would place the northern landmasses nearer the tropics and account for their vast coal deposits.

Wegener was so convinced that his explanation was correct that he wrote, "This evidence is so compelling that by comparison all other criteria must take a back seat."

How does a glacier develop in hot, arid Australia? How do land animals migrate across wide expanses of open water? As compelling as this evidence may have been, 50 years passed before most of the scientific community would accept it and the logical conclusions to which it led.

The Great Debate

Wegener's proposal did not attract much open criticism until 1924 when his book was translated into English. From this time on, until his death in 1930, his drift hypothesis encountered a great deal of hostile criticism. To quote the respected American geologist T. C. Chamberlin, "Wegener's hypothesis...takes considerable liberty with our globe, and is less bound by restrictions or tied down by awkward, ugly facts than most of its rival theories. Its appeal seems to lie in the fact that it plays a game in which there are few restrictive rules and no sharply drawn code of conduct."

One of the main objections to Wegener's hypothesis stemmed from his inability to provide a mechanism that was capable of moving the continents across the globe. Wegener proposed two possible energy sources. One of these, the tidal influence of the moon, was presumed by Wegener to be strong enough to give the continents a westward motion. However, the prominent physicist Harold Jeffreys quickly countered with the argument that tidal friction of the magnitude needed to displace the continents would bring Earth's rotation to a halt in a matter of a few years. Further, Wegener proposed that the larger and sturdier continents broke

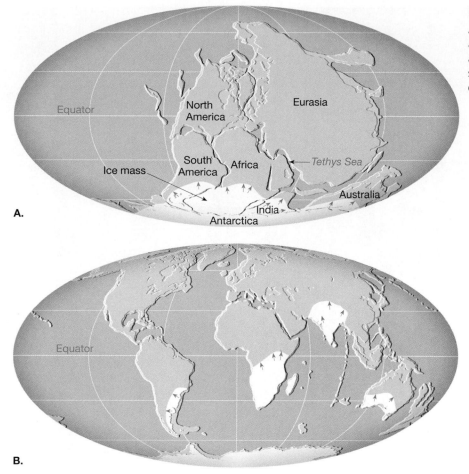

Figure 7.7
A. The supercontinent Pangaea showing the area covered by glacial ice 300 million years ago. **B.** The continents as they are today. The shading outlines areas where evidence of the old ice sheets exists.

through the oceanic crust, much like ice breakers cut through ice. However, no evidence existed to suggest that the ocean floor was weak enough to permit passage of the continents without themselves being appreciably deformed in the process.

Although most of Wegener's contemporaries opposed his views, even to the point of open ridicule, a few considered his ideas plausible. For these few geologists who continued the search for additional evidence, the exciting concept of continents in motion held their interest. Others viewed continental drift as a solution to previously unexplainable observations.

Plate Tectonics: A Modern Version of an Old Idea

During the years that followed Wegener's proposal, major strides in technology permitted mapping of the ocean floor. Moreover, extensive data on seismic activity and Earth's magnetic field became available. By 1968, these developments led to the unfolding of a far more

encompassing theory than continental drift, known as **plate tectonics**. The implications of plate tectonics are so far-reaching that this theory is today the framework within which to view most geologic processes.

The theory of plate tectonics holds that Earth's outer shell consists of about twenty rigid slabs called **plates**. They are in continuous slow motion relative to each other (Figure 7.8). The largest is the Pacific plate, which is located mostly beneath the ocean. An exception is a small sliver of North America that includes southwestern California and Mexico's Baja Peninsula.

Notice in Figure 7.8 that all of the other large plates include both continental and oceanic crust—a major departure from Wegener's continental drift hypothesis, which proposed that the continents moved through the ocean floor, not with it. Many smaller plates, on the other hand, consist exclusively of oceanic material; an example is the Nazca plate, located off the west coast of South America.

Recall that Earth's rigid outer shell is called the *lithosphere* and consists of both crustal rocks and a portion of the upper mantle. Also recall that the lithosphere

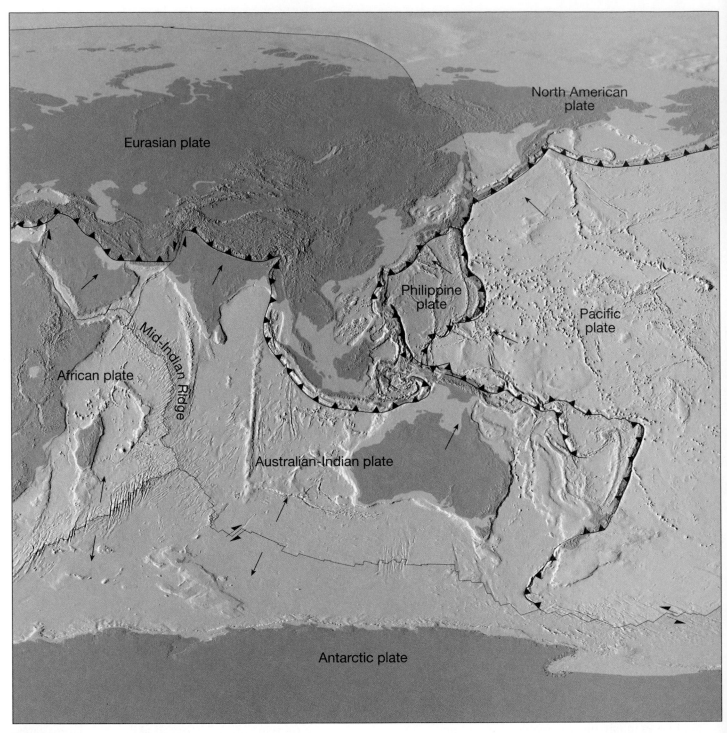

Figure 7.8
A mosaic of rigid plates constitutes Earth's outer shell. (After W. B. Hamilton, U.S. Geological Survey)

varies in thickness. In general, lithospheric plates are thinnest in the ocean basins, where their thicknesses vary from as little as 10 kilometers at the ocean ridges to as much as 100 kilometers in the deep-ocean basins. By contrast, continental lithosphere is generally 100 to 150 kilometers thick and may extend to 250 kilometers in some regions. Beneath the lithosphere is the hotter

and weaker zone known as the *asthenosphere*. The weak nature of the rock within the asthenosphere allows for motion in Earth's rigid outer shell.

A main assumption of plate tectonics theory is that lithospheric plates are rigid. Therefore, the distance between two places on the same plate does not change. For example, as the plates move, the distance between

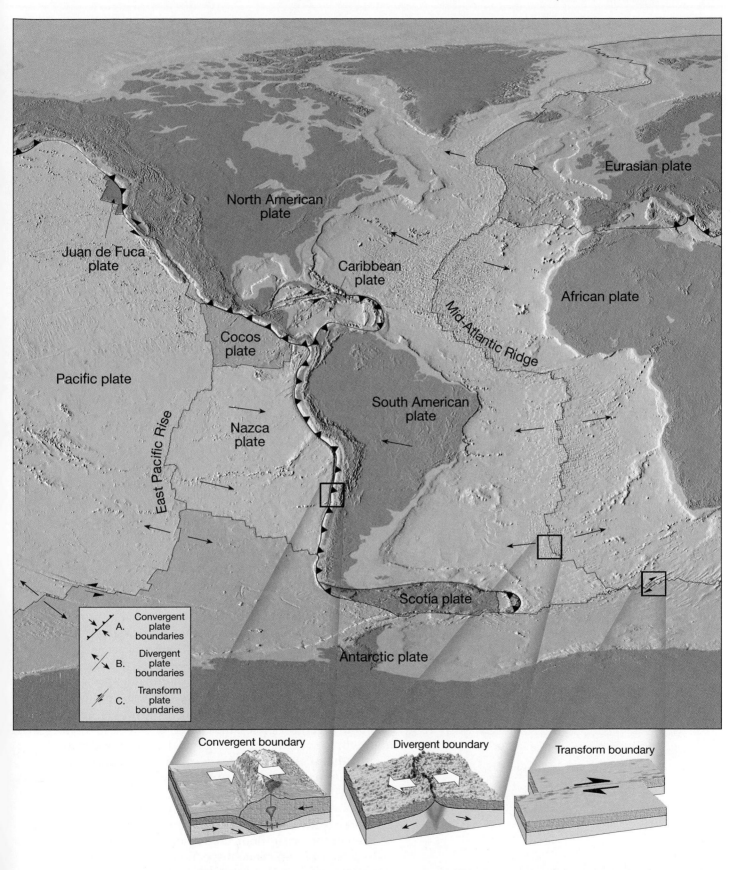

Convergent boundary

Divergent boundary

Transform boundary

New York and Denver, located on the same plate, remains unchanged. However, the distance between New York and London, which are located on different plates, is continually changing, by centimeters per year. Because each plate moves as a distinct unit, all major interactions between plates occur along *plate boundaries*.

Not surprisingly, most of Earth's seismic activity, volcanism, and mountain building occur along these dynamic margins, which we examine next.

Plate Boundaries

Plate boundaries are clearly shown in Figure 7.8. But all of them are concealed under the ocean or beneath rocks and soil on land. How were the boundaries located?

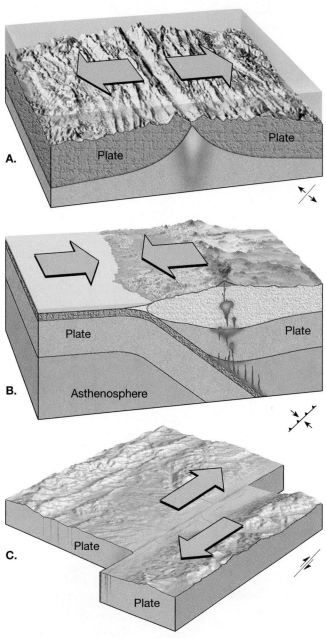

Figure 7.9
The three types of plate boundaries. **A.** Divergent boundary. **B.** Convergent boundary. **C.** Transform boundary.

For some time now, earthquake and volcanic activity have been known to be concentrated along narrow zones around the globe. Best known is the so-called *Ring of Fire* that encircles the Pacific. Thus, the first approximations of plate margins relied on the distribution of earthquake and volcanic activity. Later work revealed three distinct types of plate boundaries, differentiated by the movement each exhibits (Figure 7.9). These are:

1. **Divergent boundaries**—where plates move apart, resulting in upwelling of material from the mantle to create new sea floor.
2. **Convergent boundaries**—where plates move together, causing one of the slabs of lithosphere to be consumed into the mantle as it descends beneath an overriding plate.
3. **Transform boundaries**—where plates grind past each other without creating or destroying lithosphere.

Each plate is bounded by a combination of these zones, as you can see in Figure 7.8. For example, the Nazca plate has a divergent boundary on the west, a convergent boundary on the east, and numerous small transform faults that offset segments of divergent boundaries on the north and south.

Divergent Boundaries

Most divergent boundaries, where plate spreading occurs, are situated along the crests of oceanic ridges (Figure 7.10). Here, as the plates move away from the ridge axis, the fractures created are immediately filled with molten rock that oozes up from the hot asthenosphere. This material cools slowly to produce new slivers of sea floor. In a continuous manner, successive plate spreading and upwelling of magma add new oceanic crust (lithosphere) between the diverging plates.

This mechanism is called **seafloor spreading**. It has produced the floor of the Atlantic Ocean during the past 165 million years. The typical rate of spreading at these ridges ranges between 2 and 10 centimeters per year, and averages about 6 centimeters (2 inches) per year. Because new rock is added equally to the trailing edges of both diverging plates, the overall rate of ocean floor growth is twice the spreading rate. Despite these slow rates, the Atlantic Ocean basin could have opened and closed *more than ten times* during the nearly 5-billion-year history of our planet!

Our knowledge of oceanic ridge systems, where seafloor spreading occurs, comes from depth soundings taken of the ocean floor, core samples obtained from deep-sea drilling, visual inspection using submersible seacraft, and even first-hand inspection of slices of ocean floor which have been shoved up onto dry land. Be-

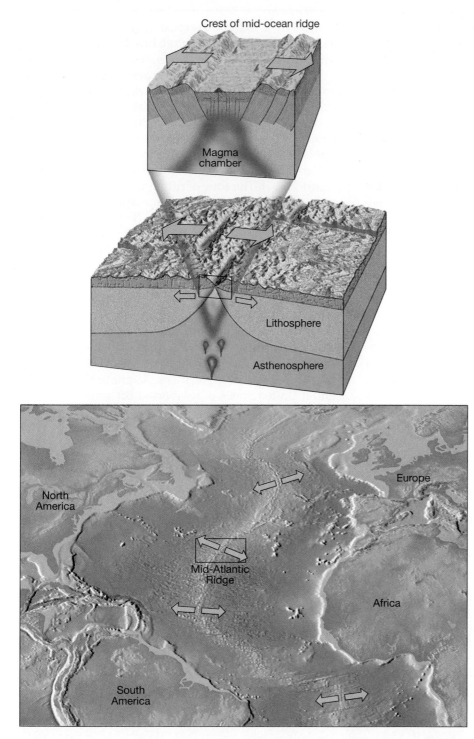

Crest of mid-ocean ridge

Magma chamber

Lithosphere

Asthenosphere

North America

Europe

Mid-Atlantic Ridge

Africa

South America

Figure 7.10
Most divergent plate boundaries are situated along the crests of oceanic ridges.

cause of its accessibility, the Mid-Atlantic Ridge has been studied more thoroughly than other ridge systems. The Mid-Atlantic Ridge is a gigantic submerged mountain range standing 2500–3000 meters (8200–10,000 feet) above the adjacent deep-ocean basins. It extends southward from the Arctic Ocean to beyond the southern tip of Africa. In a few places, the Mid-Atlantic Ridge has actually grown above sea level to form islands, the largest of which is Iceland. Throughout most of its length, however, this divergent boundary lies 2500 meters below sea level.

The buoyant nature of the upwelling magma is the primary reason for the elevated position of the Mid-Atlantic Ridge. As the newly formed lithosphere travels away from the spreading center, it gradually cools and contracts. This thermal contraction accounts in part for the greater ocean depths that exist away from the ridge. Almost 100 million years must pass before cooling and contraction cease completely. By this time, rock that was once a part of the majestic ocean mountain system becomes part of the deep-ocean basin.

Not all spreading centers have existed as long as the Mid-Atlantic Ridge and not all are found in the middle of large oceans. The Red Sea is believed to be the site of a recently formed divergent boundary. Here, the Arabian Peninsula separated from Africa and began to move toward the northeast (Figure 7.11). Consequently, the Red Sea is providing oceanographers with a view of how the Atlantic Ocean may have looked in its infancy. Another narrow, linear sea produced by seafloor spreading in the recent geologic past is the Gulf of California.

Most spreading centers are somewhere on the sea floor, but a few exist on continents. When a spreading center develops within a continent, the landmass may split into smaller segments, just as Wegener had proposed for the breakup of Pangaea. The fragmentation of a continent is thought to be associated with the upward movement of hot rock from below. The effect of

this activity is to force the crust upward directly above the hot rising plume. This stretches the crust and causes numerous tensional cracks, as shown in Figure 7.12A.

As the plates move from the area of upwelling, the broken slabs are displaced downward, creating downfaulted valleys called **rifts** or **rift valleys** (Figure 7.12B). As the spreading continues, the rift valley will lengthen and deepen, eventually extending out into the ocean. At this point the valley will become a narrow linear sea with an outlet to the ocean, similar to the Red Sea today (Figure 7.12C). The zone of rifting will remain the site of igneous activity, continually generating new sea floor in an ever-expanding ocean basin (Figure 7.12D).

The East African rift valleys represent the initial stage in the breakup of a continent as just described (Figure 7.11). The extensive volcanic activity believed to accompany continental rifting is exemplified by large volcanic mountains such as Kilimanjaro and Mount Kenya. If the rift valleys in Africa remain active, East Africa will eventually part from the mainland in much the same way the Arabian Peninsula did just a few million years ago. However, not all rift valleys develop into full-fledged spreading centers. Running through the central United States is an aborted rift zone extending

Figure 7.11

East African rift valleys and associated features.

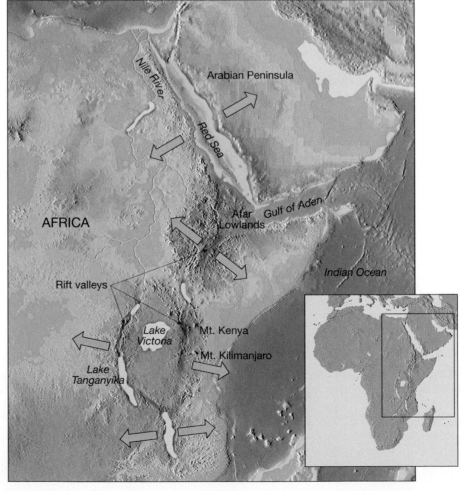

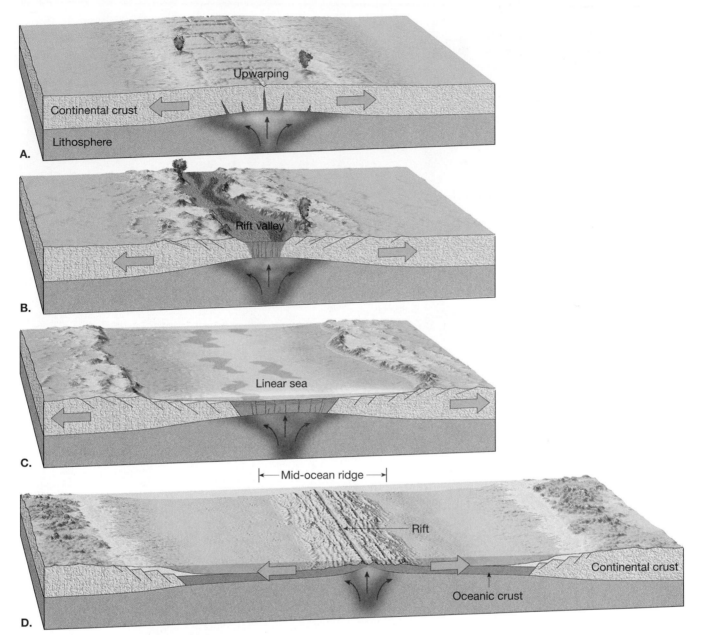

Figure 7.12
A. Rising magma forces the crust upward, causing numerous cracks in the rigid lithosphere. **B.** As the crust is pulled apart, large slabs of rock sink, generating a rift zone. **C.** Further spreading generates a narrow sea. **D.** Eventually, an expansive ocean basin and ridge system are created.

from Lake Superior to Kansas. This once-active rift valley is filled with rock that was extruded onto the crust more than one billion years ago. Why one rift valley continues to develop while others are abandoned is not yet known.

Convergent Boundaries

At spreading centers, new lithosphere is continually being generated. However, because the total surface area of Earth remains essentially constant, lithosphere must also be consumed. The zone of plate convergence is the site where lithosphere is reabsorbed, or subducted, into the mantle. When two plates collide, the lead-

ing edge of one is bent downward, allowing it to descend beneath the other. The typical angle of descent ranges from 35 to nearly 90 degrees from the surface.

Although all convergent zones are basically similar, the nature of plate collisions is influenced greatly by the type of crustal material involved. Convergence can occur between one oceanic and one continental plate, between two oceanic plates, or between two continental plates. All these situations are shown in Figure 7.13.

Oceanic-Continental Convergence. Whenever the leading edge of a plate capped with continental crust converges with oceanic crust, the less dense continental

Figure 7.13
Three types of convergent plate boundaries.
A. Oceanic-continental. **B.** Oceanic-oceanic.
C. Continental-continental.

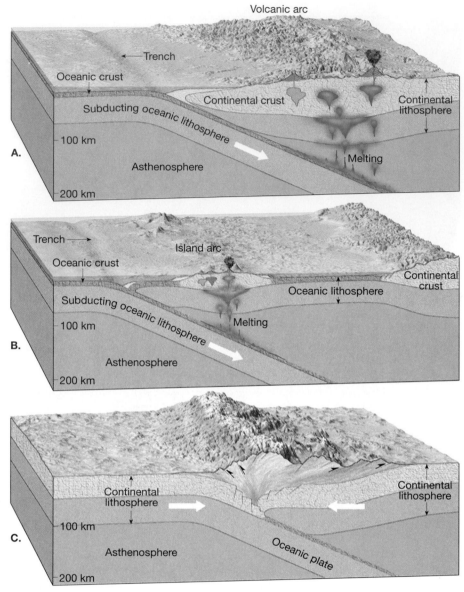

material remains "floating," while the more dense oceanic slab sinks into the asthenosphere. The region where an oceanic plate descends into the asthenosphere because of convergence is called a **subduction zone**. As the oceanic plate slides beneath the overriding plate, the oceanic plate bends, thereby producing a **deep-ocean trench** adjacent to the zone of subduction (Figure 7.13A). Trenches formed in this manner may be thousands of kilometers long and 8 to 11 kilometers deep.

When the descending oceanic plate reaches a depth of about 100 kilometers, partial melting of the water-rich oceanic crust and some of the overlying mantle takes place. The newly formed magma created in this manner is less dense than the surrounding mantle rocks and, when sufficient quantities have gathered, the

molten rock will slowly rise (Figure 7.13B). Most of the rising magma will intrude into the overlying continental crust where it will cool and crystallize at depth. However, some of the magma may migrate to the surface where it can give rise to numerous and occasionally explosive volcanic eruptions. The volcanic Andes Mountains are believed to have been produced by such activity when the Nazca plate melted as it plunged beneath the continent of South America (see Figure 7.8). The frequent earthquakes that occur within the Andes testify to the activity beneath our view.

Mountains such as the Andes that are believed to be produced in part by volcanic activity associated with the subduction of oceanic lithosphere are called **volcanic arcs** (Figure 7.13A). Two of these volcanic arcs are in the western United States. One, the Cascade Range of Washington and Oregon, is composed of several well-known volcanic mountains, including Mounts Rainier, Shasta, and St. Helens. As continuing eruptions of Mount St. Helens testify, the Cascade Range is still active. The magma here arises from the melting of a small remaining segment of the Juan de Fuca plate (see Figure 7.8). The second volcanic arc is the Sierra Nevada, in which Yosemite National Park is located. The Sierra Nevada system is the older of the two and has been inactive for several million years, as evidenced by the absence of volcanic cones. Here erosion has stripped away most of the obvious traces of volcanic activity and left exposed the large, crystallized magma chambers that once fed lofty volcanoes.

Oceanic-Oceanic Convergence.   When two oceanic slabs converge, one descends beneath the other, initiating volcanic activity in a manner similar to that which occurs at an oceanic-continental convergent boundary. However, in this case, the volcanoes form on the ocean floor rather than on the continents (Figure 7.13B). If this volcanic activity is sustained, it will eventually build crust upward until it emerges from the ocean depths as dry land. In the early stages, this newly formed land consists of a chain of small volcanic islands called an **island arc**. The Aleutian, Mariana, and Tonga islands exemplify such features. Island arcs such as these are generally located a few hundred kilometers from an ocean trench where active subduction of the lithosphere is occurring. Adjacent to the island arcs just mentioned are the Aleutian trench, Mariana trench, and the Tonga trench (Figure 7.14).

Over an extended period, numerous episodes of volcanic activity build large piles of lava on the ocean floor. This gradually increases the size and elevation of the developing arc. This growth, in turn, increases the amount of eroded sediments added to the sea floor. Some of these sediments reach the trench. Here, they are deformed and metamorphosed by the compressional forces exerted by the two converging plates. The result of these diverse activities is the development of a mature island arc composed of a complex system of volcanic rocks, folded and metamorphosed sedimentary rocks, and intrusive igneous rocks. Examples of mature island arc systems are the Alaskan Peninsula, the Philippines, and Japan (Figure 7.14).

Continental-Continental Convergence.   When two plates carrying continental crust converge, neither plate will subduct beneath the other because of the low density, and thus the buoyant nature, of continental rocks. The result is a collision between the two continental blocks (Figure 7.13C). Such a collision occurred when the once-separated continent of India "rammed" into Asia and produced the Himalayas,

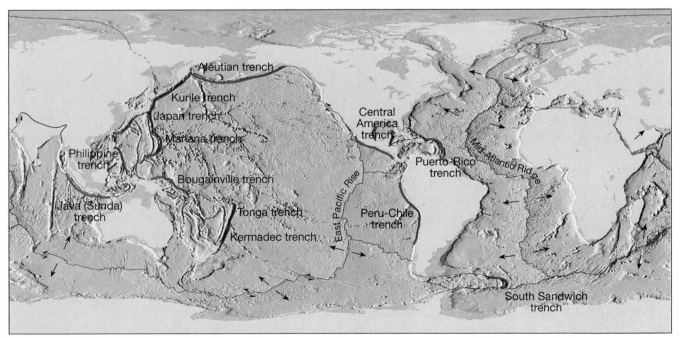

Figure 7.14

Distribution of the world's oceanic trenches, ridge system, and transform faults. Where transform faults offset ridge segments, they permit the ridge to change direction (curve) as can be seen in the Atlantic Ocean.

Figure 7.15
The ongoing collision of India and Asia, starting about 45 million years ago, produced the majestic Himalayas.

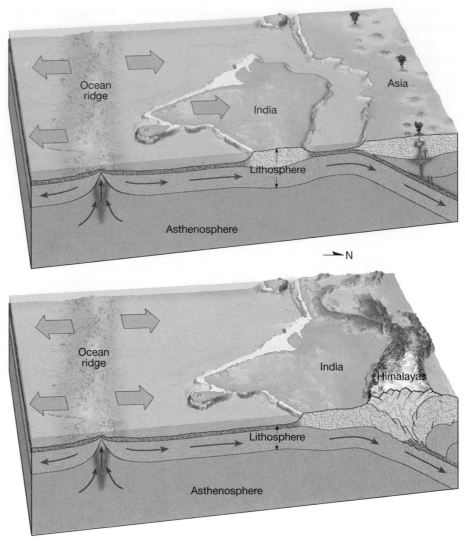

perhaps the most spectacular mountain range on Earth (Figure 7.15). During this collision, the continental crust buckled, fractured, and was generally shortened. In addition to the Himalayas, several other complex mountain systems, including the Alps, Appalachians, and Urals, are thought to have formed in this manner.

Prior to a continental collision, the landmasses involved are separated by an ocean basin (Figure 7.15, top). As the continental blocks converge, the intervening sea floor is subducted beneath one of the plates. The partial melting of the descending oceanic slab and mantle rocks generates a volcanic arc. Erosion of the newly formed volcanic arc adds large quantities of sediment to the already sediment-laden continental margin. Eventually, as the intervening sea floor is consumed, these continental masses collide. This squeezes, folds, and generally deforms the sediments as if they were placed

in a gigantic vise. The result is the formation of a new mountain range, composed of deformed sedimentary rocks and fragments of the volcanic arc.

Transform Boundaries

The third type of plate boundary is the transform fault, where plates grind past one another without the production of new crust, as occurs along oceanic ridges, or without the destruction of crust, as occurs at subduction zones. Transform faults roughly parallel the direction of plate movement. They were first identified where they join offset segments of the oceanic ridge system (see Figure 7.14).

Transform faults provide the means by which the oceanic crust created at the ridge crests can be transported to its site of destruction: the deep-ocean trenches. Figure 7.16 illustrates this activity. Notice that the

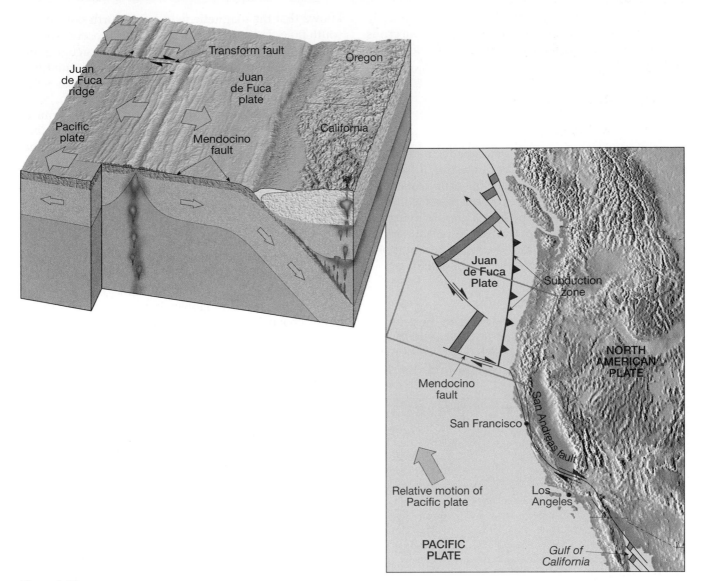

Figure 7.16
The role of transform faults. The Mendocino transform fault permits sea floor generated at the Juan de Fuca ridge to move southeastward past the Pacific plate and beneath the North American plate. Thus, this transform fault connects a divergent boundary to a subduction zone. Futhermore, the San Andreas fault, also a transform fault, connects two spreading centers—the Juan de Fuca ridge and a divergent zone located in the Gulf of California.

Juan de Fuca plate moves in a southeasterly direction, eventually being subducted under the west coast of the United States. The southern end of this relatively small plate is bounded by the Mendocino transform fault. This transform boundary connects the active spreading center to the subduction zone. Therefore, the fault facilitates the movement of the crustal material created at the ridge crest to its destination beneath the North American continent.

These special faults are called transform faults because the relative motion of the plates can be changed, or transformed, along them. As we saw in the preceding example, divergence occurring at a spreading cen-

ter can be transformed into convergence at a subduction zone. Because transform faults connect convergent and divergent boundaries in various combinations, other changes in relative plate motion are possible along transform faults.

Most transform faults are located in oceanic crust. However, a few, including California's famous San Andreas fault, are situated within continents (Figure 7.16). Along the San Andreas fault, the Pacific plate is moving toward the northwest, past the North American plate. If this movement continues for millions of years, that part of California west of the fault zone, including the Baja Peninsula, will become an island off the west coast of the United States and Canada. It could eventually reach Alaska. However, a more immediate concern is the earthquakes triggered by movements along this fault system.

Testing the Plate Tectonics Model

With the birth of the plate tectonics model, researchers from all of the Earth sciences began testing it (see Box 7.1). Some of the evidence supporting continental drift and seafloor spreading has already been presented. Some of the evidence that was instrumental in solidifying the support for this new concept follows. Note that some of the evidence was not new; rather, it was a new interpretation of old data that swayed the tide of opinion.

Evidence: Paleomagnetism

Probably the most persuasive evidence to the geologic community for the acceptance of the plate tectonics theory comes from the study of Earth's magnetic field. Anyone who has used a compass to find direction knows that the magnetic field has a north pole and a south pole. These magnetic poles align closely, but not exactly, with the geographic poles. (The geographic poles are simply the top and bottom of the spinning sphere we live on, the points through which passes the imaginary axis of rotation.)

In many respects the magnetic field is very much like that produced by a simple bar magnet. Invisible lines of force pass through Earth and extend from one pole to the other. A compass needle, itself a small magnet free to move about, becomes aligned with these lines of force and thus points toward the magnetic poles.

The technique used to study ancient magnetic fields relies on the fact that certain rocks contain minerals which serve as fossil compasses. These iron-rich minerals, such as magnetite, are abundant in lava flows of basaltic composition. When heated above a certain temperature called the *Curie point*, these magnetic minerals lose their magnetism. However, when these iron-rich grains cool below their Curie point (about 580°C) they become magnetized in the direction parallel to the existing magnetic field. Once the minerals solidify, the magnetism they possess will remain "frozen" in this po-

---

Box 7.1

A New Test for Plate Tectonics

Until the late 1980s the evidence supporting the theory of plate tectonics was acquired from the study of geologic phenomena such as volcanoes, earthquakes, and sea floor sediments. Recently, however, it became possible to test the theory directly. Specifically, scientists are now able to confirm the fact that the plates shift in relation to one another in the way that the plate tectonics theory predicts.

The new evidence comes from two different techniques that allow distances between widely separated points on Earth's surface to be measured with unprecedented accuracy. Called *Satellite Laser Ranging* (SLR) and *Very Long Baseline Interferometry* (VLBI), these methods can detect the motion of any one site with respect to another at a level of better than 1 centimeter per year. Thus, for the first time, scientists can directly measure the relative motions of Earth's plates. Further, because these techniques are quite different, scientists use them to cross-check one against the other by comparing measurements made for the same sites.

The Satellite Laser Ranging system employs ground-based stations that bounce laser pulses off satellites whose orbital positions are well established. Precise timing of the round-trip travel of these pulses allow scientists to calculate the precise locations of the ground-based stations. By monitoring these stations over time, researchers can establish the relative motions of the sites.

The Very Long Baseline Interferometry system uses large radio telescopes to record signals from very distant quasars (Figure 7.A). Since quasars (quasi-stellar objects) lie billions of light years from Earth, they act as stationary reference points. The millisecond differences in the arrival time of the same signal at different Earth-bound observational sites provide a means of establishing the distance between receivers.

Confirming data from these two techniques leaves little double that real plate motion has been detected. Calculations show that Hawaii is moving in a northwesterly direction and approaching Japan at a rate of 8.3 centimeters per year. Moreover, a site located in Maryland is retreating from one in England at a rate of about 1.7 centimeters per year. This rate is roughly equal to the 2.2 centimeters per year of seafloor spreading that was established from paleomagnetic evidence.

Figure 7.A
Radio telescopes like these located at Socorro, New Mexico, are used to accura tely determine the distance between two distant sites. Data collected by repeated measurements have detected relative plate motions of 1 to 12 centimeters per year between various sites worldwide. (Photo by Geoff Chester/ Smithsonian Institution)

sition. In this regard, they behave much like a compass needle inasmuch as they "point" toward the existing magnetic poles. Then, if the rock is moved, or if the magnetic pole changes position, the rock magnetism will, in most instances, retain its original alignment. Rocks formed thousands of millions of years ago thus "remember" the location of the magnetic poles at the time of their formation and are said to possess fossil magnetism, or **paleomagnetism**.

Polar Wandering. A study of lava flows conducted in Europe in the 1950s led to an amazing discovery. The magnetic alignment in the iron-rich minerals in lava flows of different ages was found to vary widely. A plot of the apparent positions of the magnetic north pole revealed that, during the past 500 million years, the location of the pole had gradually wandered from a spot near Hawaii northward through eastern Siberia and finally to its present site (Figure 7.17A). This was clear evidence that either the magnetic poles had migrated through time, an idea known as **polar wandering**, or that the lava flows had moved—in other words, the continents had drifted.

Although the magnetic poles are known to move, studies of the magnetic field indicated that the average positions of the magnetic poles correspond closely to the positions of the geographic poles. This is consistent with our knowledge of Earth's magnetic field, which is generated in part by the rotation of Earth about its axis. If the geographic poles do not wander appreciably, which we believe is true, neither can the magnetic poles. Therefore, a more acceptable explanation for the apparent polar wandering is provided by the plate tectonics theory. *If the magnetic poles remain stationary, their apparent movement was produced by the drifting of the continents.*

Further evidence for plate tectonics came a few years later when polar wandering curves were constructed for North America and Europe (Figure 7.17A). To nearly everyone's surprise, the curves for North America and Europe had similar paths, except that they were separated by about 24 degrees of longitude. When these rocks solidified, could there have been two magnetic north poles which migrated parallel to each other? This is very unlikely. The differences in these migration paths, however, can be reconciled if the two presently separated continents are placed next to one another, as we now believe they were prior to the opening of the Atlantic Ocean (Figure 7.17B).

Magnetic Reversals and Seafloor Spreading. Another discovery came when geophysicists learned that Earth's magnetic field periodically reverses polarity; that is, the north magnetic pole becomes the south magnetic pole, and vice versa. A rock solidifying during one of the periods of reverse polarity will be magnetized with the polarity opposite that of rocks being formed today.

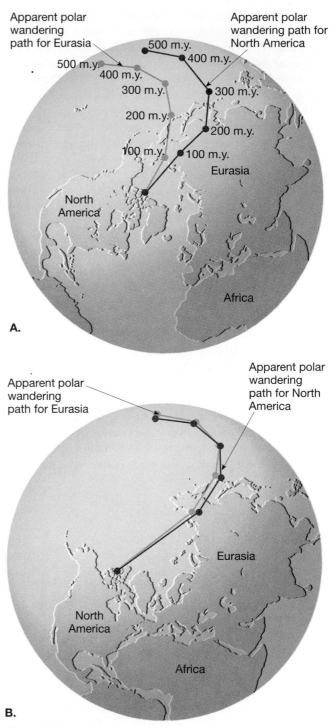

Figure 7.17

Simplified apparent polar wandering paths as established from North American and Eurasian paleomagnetic data. **A.** The more westerly path determined from North American data is thought to have been caused by the westward drift of North America by about 24 degrees from Eurasia. **B.** The positions of the wandering paths when the landmasses are reassembled in their predrift locations.

When rocks exhibit the same magnetism as the present magnetic field, they are said to possess **normal polarity**, while those rocks exhibiting the opposite magnetism are said to have **reverse polarity**. Evidence for magnetic reversals was obtained from lavas and sediments from around the world. Once the concept of magnetic reversals was confirmed, researchers set out to establish a time scale for polarity reversals. There are many areas where volcanic activity has occurred sporadically for periods of millions of years (Figure 7.18). The task was to measure the directions of paleomagnetism in numerous lava flows of various ages. These data were collected from several places and were used to determine the dates when the polarity of Earth's magnetic field changed. Figure 7.19 shows the time scale of the polarity reversals established for the last few million years.

A significant relationship was uncovered between the magnetic reversals and the seafloor spreading hypothesis. Very sensitive instruments called *magnetometers* were towed by research vessels across a segment of the ocean floor located off the west coast of the United States. Here workers from the Scripps Institute of Oceanography discovered alternating strips of high- and low-intensity magnetism that trended in roughly a north-south direction. This relatively simple pattern of magnetic variation defied explanation until 1963, when it was tied to the concept of seafloor spreading. The strips of high-intensity magnetism are regions where the paleomagnetism of the ocean crust is of the normal type . Consequently, these positively magnetized rocks *enhance* the existing magnetic field. Conversely, the low-intensity strips represent regions where the ocean crust is polarized in the reverse direction and, therefore, *weaken* the existing magnetic field. But how do parallel strips of normally and reversely magnetized rock become distributed across the ocean floor?

As new basalt is added to the ocean floor at the oceanic ridges, it becomes magnetized according to the existing magnetic field (Figure 7.20). Since new rock is added in approximately equal amounts to the trailing edges of both plates, we should expect strips of equal size and polarity to parallel both sides of the ocean ridges, as shown in Figure 7.20C. This explanation of the alternating strips of normal and reverse polarity, which lay as mirror images across the ocean ridges, was the strongest evidence so far presented in support of the concept of seafloor spreading.

Now that the dates of the most recent magnetic reversals have been established, the rate at which spreading occurs at the various ridges can be determined accurately. In the Pacific Ocean, for example, the magnetic strips are much wider for corresponding time intervals than those of the Atlantic Ocean. Hence, we conclude that a faster spreading rate exists for the spreading center of the Pacific as compared to the Atlantic. When we apply absolute dates to these magnetic events, we find that the spreading rate for the North Atlantic Ridge is only 1 or 2 centimeters per year (note that each side spreads at this rate). The rate is somewhat faster for the South Atlantic. The spreading rates

Figure 7.18

Schematic illustration of paleomagnetism preserved in lava flows of various ages. Data such as these from various locales were used to establish the time scale of polarity reversals shown in Figure 7.19.

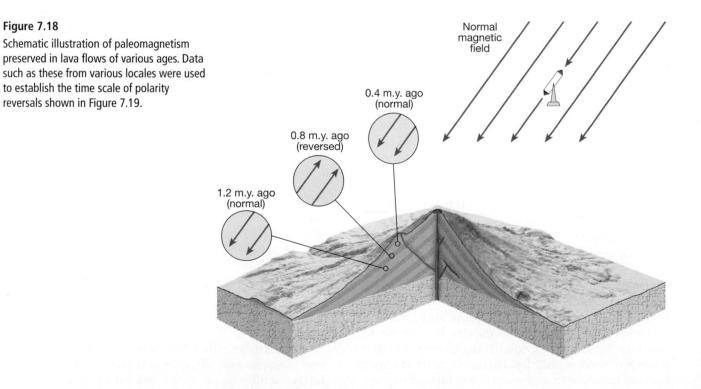

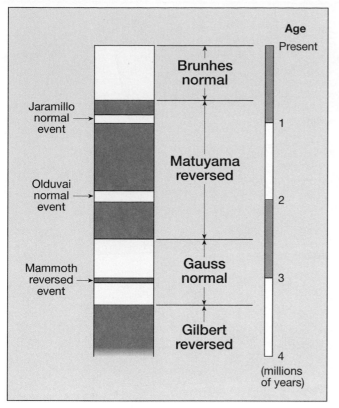

Figure 7.19
Time scale of Earth's magnetic field in the recent past. This time scale was developed by establishing the magnetic polarity for lava flows of known age. (Data from Allen Cox and G. B. Dalrymple)

for the East Pacific Rise generally range between 3 and 8 centimeters per year, with a maximum rate of about 10 centimeters per year in one segment. Thus, we have a magnetic tape recorder that records changes in Earth's magnetic field. This recorder also permits us to determine the rate of seafloor spreading.

Evidence: Earthquake Patterns

By 1968, the basic outline of global tectonics was firmly established. In this same year, three seismologists at Lamont-Doherty Earth Observatory published papers demonstrating how successfully the new plate tectonics model accounted for the global distribution of earthquakes (Figure 7.21). In particular, these scientists were able to account for the close association between deep-focus earthquakes and ocean trenches. Furthermore, the absence of deep-focus earthquakes along the oceanic ridge system was also shown to be consistent with the new theory.

The close association between plate boundaries and earthquakes can be seen by comparing the distribution of earthquakes shown in Figure 7.21 with the map of plate boundaries in Figure 7.8 and trenches in Figure 7.14. In trench regions where dense slabs of lithosphere plunge into the mantle, this association is especially striking. When the depths of earthquake foci and their locations within the trench systems are plotted, an interesting pattern emerges. Figure 7.22, which shows the distribution of earthquakes in the vicinity of the Japan trench, is an example. Here most shallow-focus earthquakes occur within, or adjacent to, the trench, whereas intermediate- and deep-focus earthquakes occur toward the mainland.

In the plate tectonics model, deep-ocean trenches are produced where cold, dense slabs of oceanic lithosphere plunge into the mantle. Shallow-focus earthquakes are produced as the descending plate interacts with the overriding lithosphere. As the slab descends further into the asthenosphere, deeper-focus earthquakes are generated (Figure 7.22). Because the earthquakes occur within the rigid subducting plate rather than in the "plastic" mantle, they provide a method for tracking the plate's descent. Very few earthquakes have been recorded below 700 kilometers (435 miles), possibly because the slab has been heated sufficiently to lose its rigidity.

Evidence: Ocean Drilling

Some of the most convincing evidence confirming the plate tectonics theory has come from drilling directly into ocean-floor sediment. From 1968 until 1983, the source of these important data was the Deep Sea Drilling Project, an international program sponsored by several major oceanographic institutions and the National Science Foundation. A new drilling ship was built. The *Glomar Challenger* represented a significant technological breakthrough, because this ship could lower drill pipe thousands of meters to the ocean floor and then drill hundreds of meters into the sediments and underlying basaltic crust. In the South Atlantic, at several sites, holes were drilled through the entire thickness of sediments to the basaltic rock below. An important objective was to gather samples of sediment from just above the igneous crust as a means of dating the sea floor at each site. (Radiometric dates of the ocean crust itself are unreliable because seawater alters basalt.)

When the oldest sediment from each drill site was plotted against its distance from the ridge crest, it was revealed that the age of the sediment increased with increasing distance from the ridge. This finding agreed with the seafloor spreading hypothesis, which predicted that the youngest oceanic crust would be found at the ridge crest, and that the oldest oceanic crust would be at the continental margins.

Figure 7.20

As new basalt is added to the ocean floor at the mid-ocean ridges, it is magnetized according to Earth's existing magnetic field. Hence, it behaves much like a tape recorder as it records each reversal of the planet's magnetic field.

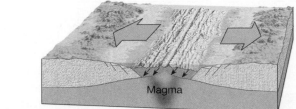

A. Period of normal magnetism

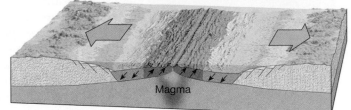

B. Period of reverse magnetism

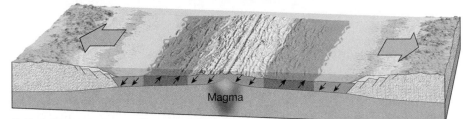

C. Period of normal magnetism

The data from the Deep Sea Drilling Project also reinforced the idea that the ocean basins are geologically youthful, because no sediment with an age in excess of 160 million years was found. By comparison, some continental crust has been dated at 3.9 billion years.

During its 15 years of operation, the *Glomar Challenger* drilled 1092 holes and obtained more than 96 kilometers (60 miles) of invaluable core samples. The Ocean Drilling Program has succeeded the Deep Sea Drilling Project and, like its predecessor, it is a major international program. A

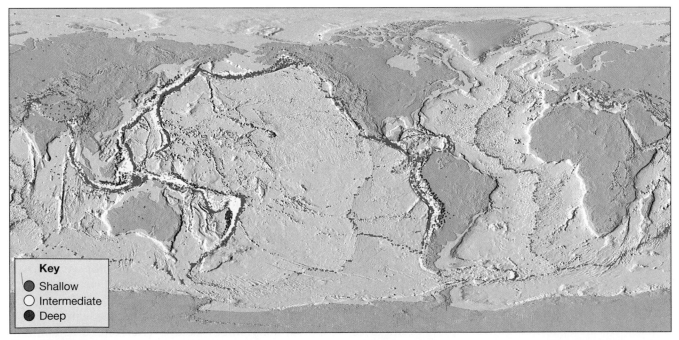

Key
- 🔵 Shallow
- ⚪ Intermediate
- ⚫ Deep

Figure 7.21

Distribution of shallow-, intermediate-, and deep-focus earthquakes. Note that deep-focus earthquakes only occur in association with subduction zones. (Data from NOAA)

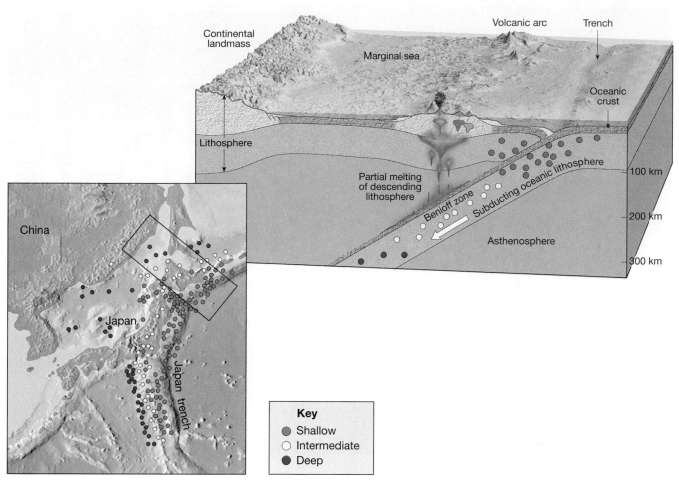

Figure 7.22

Distribution of earthquake foci in the vicinity of the Japan trench. Note that intermediate- and deep-focus earthquakes occur only within the sinking slab of oceanic lithosphere. (Data from NOAA)

more technologically advanced drilling ship, the *JOIDES Resolution*, now continues the work of the *Glomar Challenger* (see Figure 7.23).

Evidence: Hot Spots

Mapping of seafloor volcanoes called seamounts in the Pacific revealed a chain of volcanic structures extending from the Hawaiian Islands to Midway Island and then continuing northward toward the Aleutian trench (Figure 7.24). Radiometric dates of volcanoes in this chain revealed that the volcanoes increase in age with increasing distance from Hawaii. Suiko Seamount, which is located near the Aleutian trench, is 65 million years old, Midway Island is 27 million years old, and the island of Hawaii built up from the sea floor less than one million years ago (Figure 7.24).

Researchers have proposed that a rising plume of mantle material is located below the island of Hawaii. Melting of this hot rock as it enters the low-pressure en-

vironment near the surface generates a volcanic area or **hot spot**. Presumably, as the Pacific plate moved over the hot spot, successive volcanic mountains have been built. The age of each volcano indicates the time when it was situated over the relatively stationary mantle plume. This pattern is shown in Figure 7.24. Kauai is the oldest of the large islands in the Hawaiian chain. Five million years ago, when it was positioned over the hot spot, Kauai was the only Hawaiian Island in existence (Figure 7.24). Visible evidence of the age of Kauai can be seen by examining its extinct volcanoes, which have been eroded into jagged peaks and vast canyons. By contrast, the south slopes of the relatively youthful island of Hawaii consist of fresh lava flows, and two of Hawaii's volcanoes, Mauna Loa and Kilauea, remain active.

This evidence supports the fact that the plates do indeed move relative to Earth's interior. The hot spot "tracks" also trace the direction of plate motion. Notice, for example, in Figure 7.24 that the Hawaiian Island–Emperor Seamount chain bends. This particular

Figure 7.23

The *JOIDES Resolution*, the drilling ship of the Ocean Drilling Program. This modern drilling ship has replaced the *Glomar Challenger* in the important work of sampling the floors of the world's oceans. JOIDES is an acronym for Joint Oceanographic Institutions for Deep Earth Sampling. (Photo courtesy of Ocean Drilling Program)

bend in the trace occurred about 40 million years ago when the motion of the Pacific plate changed from nearly due north to a northwesterly path.

Pangaea: Before and After

Robert Dietz and John Holden have projected the gross details of the migrations of individual continents over the past 500 million years. By extrapolating plate motion back in time using such evidence as the orientation of volcanic structures left behind on moving plates, the distribution and movements of transform faults, and paleomagnetism, Dietz and Holden were able to reconstruct Pangaea (Figure 7.25A). The use of radiometric dating helped them establish the time frame for the formation and eventual breakup of Pangaea, and the relatively stationary positions of hot spots through time helped to fix the locations of the continents.

Breakup of Pangaea

The fragmentation of Pangaea began about 200 million years ago. Figure 7.25 illustrates the breakup and subsequent paths taken by the landmasses involved. As we can readily see in Figure 7.25B, two major rifts initiated the breakup. The rift zone between North America and Africa generated numerous outpourings of Jurassic-age basalts which are presently visible along the eastern seaboard of the United States. Radiometric dating of these basalts indicates that rifting occurred between 200 and 165 million years ago. This date can be used as the birth date of this section of the North Atlantic. The rift that formed in the southern landmass of Gondwanaland developed a Y-shaped fracture which sent India on a northward journey and simultaneously separated South America-Africa from Australia-Antarctica.

Figure 7.25C illustrates the position of the continents 135 million years ago, about the time Africa and South America began splitting apart to form the South

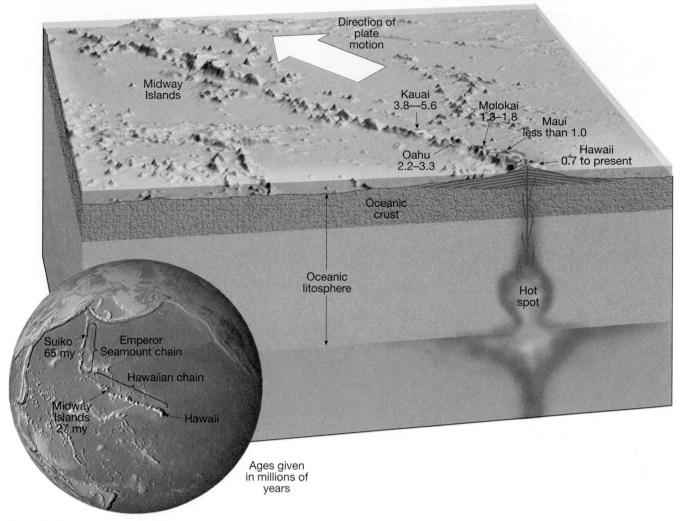

Figure 7.24
The chain of islands and seamounts that extends from Hawaii to the Aleutian trench results from the movement of the Pacific plate over an apparently stationary hot spot. Radiometric dating of the Hawaiian Islands shows that the volcanic activity decreases in age toward the island of Hawaii.

Atlantic. India can be seen halfway into its journey to Asia, while the southern portion of the North Atlantic has widened greatly. By the beginning of the Cenozoic, about 65 million years ago, Madagascar had separated from Africa, and the South Atlantic had emerged as a full-fledged ocean (Figure 7.25D). At this juncture, India had drifted over a hot spot that generated numerous fluid basalt flows across a region in western India now called the Deccan Plateau. These lava flows are very similar to those that make up the Columbia Plateau in the Pacific Northwest.

The current map (Figure 7.25E) shows India in contact with Asia, an event that began about 45 million years ago and created the highest mountains on Earth, the Himalayas, along with the Tibetan Highlands. It is interesting to note that the *average* height of Tibet is 5000 meters, higher than any spot in the contiguous United States. India's continued northward migration is believed to cause the numerous and often destructive earthquakes which plague that part of the world.

By comparing Figures 7.25D and 7.25E, we can see that the separation of Greenland from Eurasia was a recent event in geologic history. Also notice the recent formation of the Baja Peninsula along with the Gulf of California. This event is thought to have occurred less than 10 million years ago.

Before Pangaea

Prior to the formation of Pangaea, the landmasses had probably gone through several episodes of fragmentation similar to what we see happening today. Also like today, these ancient continents moved away from each other only to collide again at some other location. During the period between 500 and 225 million years ago,

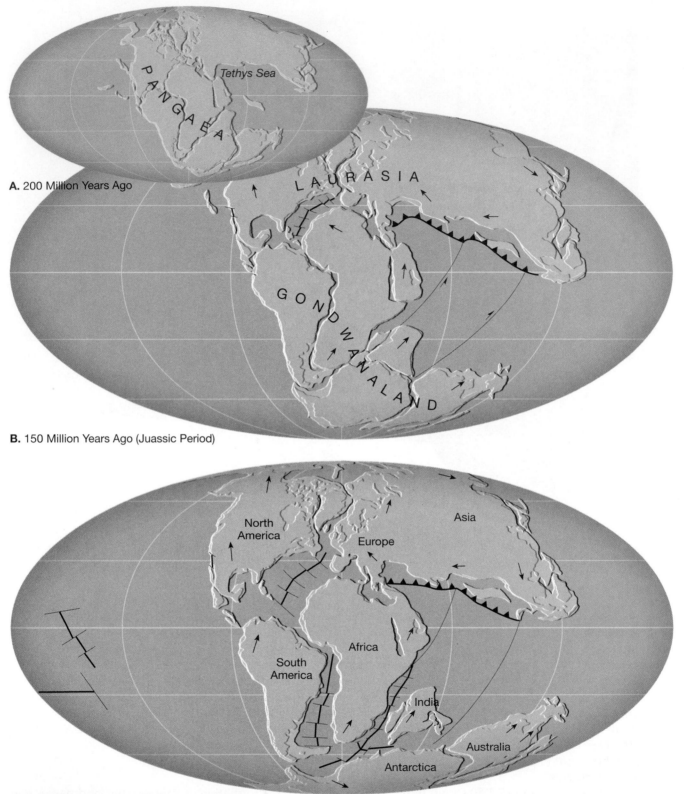

A. 200 Million Years Ago

B. 150 Million Years Ago (Juassic Period)

C. 100 Million Years Ago (Cretaceous Period)

Figure 7.25

Several views of the breakup of Pangaea over a period of 200 million years. (After D. Walsh and C. Scotese)

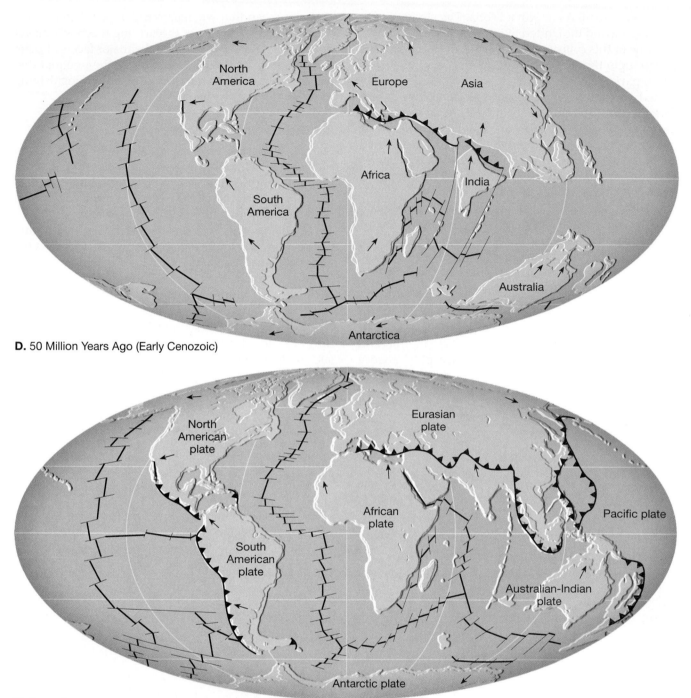

D. 50 Million Years Ago (Early Cenozoic)

E. Present

the fragments of an earlier dispersal began collecting to form the continent of Pangaea. Evidence of these earlier continental collisions include the Ural Mountains of Russia and the Appalachian Mountains, which flank the east coast of North America.

Available evidence indicates that about 500 million years ago the northern continent of Laurasia was fragmented into three major sections—North America,

northern Europe (southern Europe was part of Africa), and Siberia—with each section separated by a sizable ocean. The southern continent of Gondwanaland probably was intact and lay near the South Pole. The first collision is believed to have occurred as North America and Europe closed the pre-North Atlantic. This activity resulted in the formation of the northern Appalachians. Parts of the floor of the former ocean can be seen today high above sea level in Nova Scotia. The sliver of eastern Canada and the United States which lies seaward of the zone of this collision is truly a gift from Europe. It is also thought that before North America and Europe collided, part of Scotland, Ireland, and Norway were attached to the North American plate. While North America and Europe were joining, Siberia was closing the gap between itself and Europe, which lay farther to the west. This closing culminated about 300 to 350 million years ago in the formation of the Ural Mountains. The consolidation of these landmasses completed the northern continent of Laurasia.

During the next 50 million years the northern and southern landmasses converged, producing the supercontinent of Pangaea. At this time (about 250 to 300 million years ago), Africa and North America collided to produce the southern Appalachians (see Box 7.2).

The Driving Mechanism

The plate tectonics theory *describes* plate motion and the *effects* of this motion. Because we accept the theory does not mean that we *understand the forces* that move the plates. In fact, none of the driving mechanisms yet proposed can account for *all* of the major facets of plate motion. Nevertheless, it is clear that the unequal distribution of heat within Earth is the underlying driving force for plate movement.

Convection Current Hypothesis

One of the first models to explain the movements of plates was originally proposed as a possible driving mechanism for continental drift. Adapted to plate tectonics, this hypothesis suggests that large convection currents within the mantle drive plate motion (Figure 7.26A). The warm, less dense material of the mantle rises very slowly in the regions of oceanic ridges. As the material spreads laterally, it might drag the lithosphere along. Eventually, the material cools and begins to sink back into the mantle, where it is reheated. Partly because of its simplicity, this proposal had wide appeal. However, modern research techniques reveal that the flow of material in the mantle is far more complex than such simple convection cells.

Box 7.2

Plate Tectonics into the Future

Two geologists, Robert Dietz and John Holden, extrapolated present-day plate movements into the future. Figure 7.B illustrates where they envision Earth's landmasses will be 50 million years from now if present plate movements persist for this time span.

In North America we see that the Baja Peninsula and the portion of southern California that lies west of the San Andreas fault will have slid past the North American plate. If this northward migration takes place, Los Angeles and San Francisco will pass each other in about 10 million years, and in about 60 million years, Los Angeles will begin to descend into the Aleutian trench.

Significant changes are seen in Africa, where a new sea emerges as East Africa parts company with the mainland. In addition, Africa will have moved slowly into Europe, perhaps initiating the next major mountain-building stage on our dynamic planet. Meanwhile, the Arabian Peninsula continues to diverge from Africa, allowing the Red Sea to widen and closing the Persian Gulf.

In other parts of the world, Australia is now astride the equator and, along with New Guinea, is on a collision course with Asia. Meanwhile, North and South America are beginning to separate, while the Atlantic and Indian oceans continue to grow at the expense of the Pacific Ocean.

These projections into the future, although interesting, must be viewed with caution because many assumptions must be correct for these events to unfold as just described. Nevertheless, changes in the shapes and positions of continents that are equally profound will undoubtedly occur for millions of years to come.

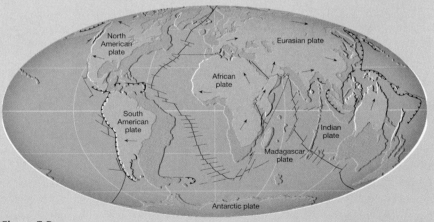

Figure 7.B
The world as it may look 50 million years from now. (From "The Breakup of Pangaea," Robert S. Dietz and John C. Holden. Copyright 1970 by Scientific American, Inc. All rights reserved)

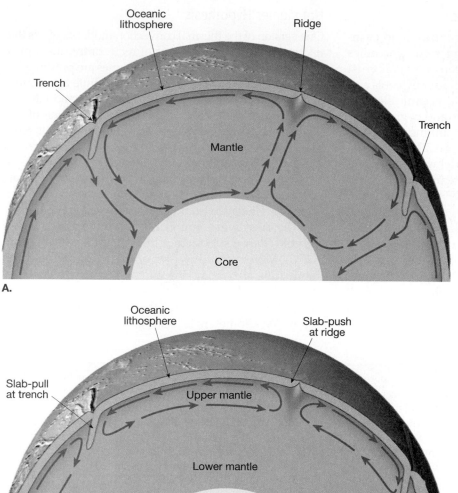

A.

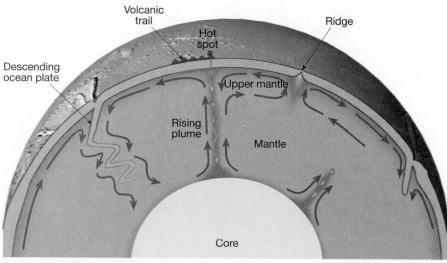

B.

Figure 7.26
Proposed models of the driving force for plate tectonics. **A.** Large convection cells in the mantle may carry the lithosphere in a conveyor-belt fashion. **B.** Slab-pull results because a subducting slab is more dense than the underlying material. Slab-push is a form of gravity sliding caused by the elevated position of lithosphere at a ridge crest. **C.** The hot plume model suggests that all upward convection is confined to a few narrow plumes, while the downward limbs of these convection cells are the cold, dense subducting oceanic plates.

C.

Slab-Push and Slab-Pull Hypotheses

Many other mechanisms that may contribute to or influence plate motion have been suggested. One relies on the fact that, as a newly formed slab of oceanic crust moves away from the ridge crest, it gradually cools and becomes denser. Eventually, the cold oceanic slab becomes denser than the asthenosphere and begins to descend. When this occurs, the dense sinking slab pulls the trailing lithosphere along. This so-called slab-pull hypothesis is similar to another model which suggests that the elevated position of an oceanic ridge could cause the lithosphere to slide under the influence of gravity. However, some ridge systems are subdued, which would reduce the effectiveness of the slab-push model. Further, some ocean basins, notably the Atlantic, lack subduction zones; thus, the slab-pull mechanism cannot adequately explain the spreading occurring at all ridges. Nevertheless, the slab-push and slab-pull phenomena appear to be active in some ridge-trench systems (Figure 7.26B).

Hot Plumes Hypothesis

One version of the thermal convection model suggests that relatively narrow, hot plumes of rock contribute to plate motion (Figure 7.25C). These hot plumes are presumed to extend upward from the vicinity of the mantle-core boundary. Upon reaching the lithosphere, they spread laterally and facilitate plate motion away from the zone of upwelling. These mantle plumes reveal themselves as long-lived volcanic areas (hot spots) in such places as Iceland. A dozen or so hot spots have been identified along ridge systems where they may contribute to plate divergence. Recall, however, that many hot spots, including the one that generated the Hawaiian Islands, are not located in ridge areas.

The downward limbs of these convection cells are the cold, dense subducting lithospheric plates (Figure 7.25C). Moreover, advocates of this view suggest that subducting slabs may descend all the way to the core-mantle boundary. However, convincing seismic evidence for the existence of these sheetlike structures below 700 kilometers is lacking.

Box 7.3

The Earth System: Plate Tectonics and the Rock Cycle

When the rock cycle was originally presented in Chapter 2, it allowed you to view many of the interrelationships among different parts of the Earth System. Now a basic knowledge of plate tectonics allows you to develop a more complete understanding of the processes that produce and alter Earth's materials.

Figure 7.C presents the rock cycle in terms of the plate tectonics model. Here, weathered material from elevated landmasses is transported to the continental margins, where it is deposited in layers that collectively are thousands of meters thick. Once lithified, these sediments create a thick wedge of sedimentary rocks flanking the continents.

Eventually the relatively quiet activity of sedimentation along a continental margin is interrupted when the region becomes a convergent plate boundary. When this occurs, oceanic lithosphere adjacent to the continent begins to inch downward into the asthenosphere beneath the continent. Along active continental margins such as this, convergence deforms the strata into linear belts of metamorphic rocks.

As the oceanic plate descends, some of the overlying sediments and sedimentary rocks that were not crumpled into mountains are carried downward into the hot asthenosphere, where they too undergo metamorphism. Eventually some of these metamorphic rocks are transported to depths where the conditions are conducive to melting. The newly formed magma then migrates upward through the overlying lithosphere to produce igneous rocks. Some will crystallize prior to reaching the surface and the remainder will erupt and solidify at the surface. After erosion eventually exposes the igneous rocks at the surface, they are immediately attacked by the processes of weathering. Thus, the rock cycle is ready to begin anew.

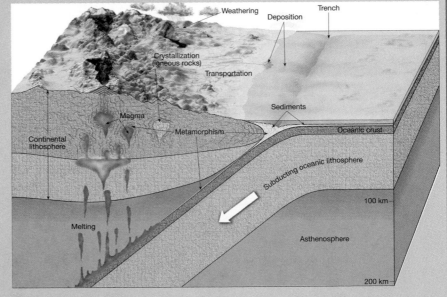

Figure 7.C
The rock cycle as it relates to the plate tectonics model.

Although there is still much to be learned about the mechanisms that cause plates to move, some facts are clear. The unequal distribution of heat in Earth generates some type of thermal convection in the mantle which ultimately drives plate motion. Whether the upwelling is mainly in the form of rising limbs of convection currents or cylindrical plumes of various sizes and shapes is yet to be determined. Furthermore, the descending lithospheric plates are active components of downwelling, and they serve to transport cold material into the mantle.

Review Questions

1. Who is credited with developing the continental drift hypothesis?

2. What was probably the first evidence that led some to suspect the continents were once connected?

3. What was Pangaea?

4. List the evidence that Wegener and his supporters gathered to substantiate the continental drift hypothesis.

5. Explain why the discovery of the fossil remains of *Mesosaurus* in both South America and Africa, but nowhere else, supports the continental drift hypothesis.

6. Early in this century, what was the prevailing view of how land animals migrated across vast expanses of ocean?

7. How did Wegener account for the existence of glaciers in the southern landmasses, while at the same time areas in North America, Europe, and Siberia supported lush tropical swamps?

8. On what basis were plate boundaries first established?

9. What are the three major types of plate boundaries? Describe the relative plate motion at each of these boundaries.

10. What is seafloor spreading? Where is active seafloor spreading occurring today?

11. What is a subduction zone? With what type of plate boundary is it associated?

12. Where is lithosphere being consumed? Why must the production and destruction of lithosphere be going on at approximately the same rate?

13. Briefly describe how the Himalaya Mountains formed.

14. Differentiate between transform faults and the other two types of plate boundaries.

15. Some predict that California will sink into the ocean. Is this idea consistent with the theory of plate tectonics?

16. Define the term *paleomagnetism*.

17. How does the continental drift hypothesis account for the apparent wandering of Earth's magnetic poles?

18. Describe the distribution of earthquake epicenters and foci depths as they relate to oceanic trench systems.

19. What is the age of the oldest sediments recovered by deep-ocean drilling? How do the ages of these sediments compare to the ages of the oldest continental rocks?

20. How do hot spots and the plate tectonics theory account for the fact that the Hawaiian Islands vary in age?

21. With what type of plate boundary are the following places or features associated (be as specific as possible): Himalayas, Aleutian Islands, Red Sea, Andes Mountains, San Andreas fault, Iceland, Japan, Mount St. Helens?

Key Terms

continental drift (p. 181)
convergent boundary (p. 188)
deep-ocean trench (p. 192)
divergent boundary (p. 188)

hot spot (p. 201)
island arc (p. 193)
normal polarity (p. 197)
paleomagnetism (p. 197)
Pangaea (p. 181)
plate (p. 185)
plate tectonics (p. 185)

polar wandering (p. 197)
reverse polarity (p. 198)
rift (rift valley) (p. 190)
seafloor spreading (p. 188)
subduction zone (p. 192)
transform boundary (p. 188)
volcanic arc (p. 192)

CHAPTER 8

Igneous Activity

Oregon's Mount Hood. (Photo by John M. Roberts/The Stock Market)

The significance of igneous activity may not be obvious at first glance. However, since volcanoes extrude molten rock that formed at great depth, they provide the only windows we have for direct observation of processes that occur many kilometers below Earth's surface. Furthermore, the gases emitted during volcanic eruptions are thought to be the material from which the atmosphere and oceans evolved. Either of these facts is reason enough for igneous activity to warrant our attention. All volcanic eruptions are spectacular, but why are some destructive and others quiescent? Is the entire island of Hawaii a volcano as high as Mount Everest resting on the ocean floor? Do the large volcanoes of Washington and Oregon present a threat to human lives? This chapter considers these and other questions as we explore the formation and movement of magma.

On Sunday, May 18, 1980, the largest volcanic eruption to occur in North America in historic times transformed a picturesque volcano into a decapitated remnant (Figure 8.1). On this date in southwestern Washington State, Mount St. Helens erupted with tremendous force. The blast blew out the entire north flank of the volcano, leaving a gaping hole. In one brief moment, a prominent volcano whose summit had been more than 2900 meters (9500 feet) above sea level was lowered by more than 400 meters (1350 feet).

Figure 8.1
Before and after photographs show the transformation of Mount St. Helens caused by the May 18, 1980 eruption. (Photos courtesy of U.S. Geological Survey)

Figure 8.2
Forest lands destroyed by the lateral blast of Mount St. Helens on May 18, 1980. (Photo by Wolfgang Kaehler)

The event devastated a wide swath of timber-rich land on the north side of the mountain. Trees within a 400-square-kilometer area lay intertwined and flattened, stripped of their branches and appearing from the air like toothpicks strewn about (Figure 8.2). The accompanying mudflows carried ash, trees, and water-saturated rock debris 29 kilometers down the Toutle River. The eruption claimed 59 lives, some dying from the intense heat and the suffocating cloud of ash and gases.

The events leading to the May 18th eruption began about two months earlier, on the 20th of March, as a series of minor earth tremors centered beneath the awakening mountain (Figure 8.3A). The first volcanic activity took place on March 27th when a small amount of ash and steam rose from the summit. Over the next several weeks, sporadic eruptions of varied intensity occurred.

Prior to the main eruption, the primary concern had been the potential hazard of mudflows. These moving lobes of saturated debris were created when ice and snow were melted by heat from the magma within the volcano. The only sign of a potentially hazardous eruption was a bulge on the volcano's north flank (Figure 8.3B). Careful monitoring of this dome-shaped structure indicated a very slow but steady growth rate of a few meters per day. Geologists monitoring the activity suggested that if the growth rate of the bulge changed appreciably, an eruption might quickly follow. Unfortunately, no such variation was detected prior to the eruption. In fact, the seismic activity decreased during the two days preceding the huge blast.

"Vancouver, Vancouver, this it!" was the only warning to precede the unleashing of tremendous quantities of pent-up gases. The trigger was an Earth tremor with a rating of 5.1 on the Richter scale. The vibrations sent the north slope of the cone plummeting into the Toutle River, effectively removing the overburden which had trapped the magma below (Figure 8.3C). With the pressure reduced, the water-rich magma is thought to have ruptured like an over-heated steam boiler. Since the

eruption originated in the vicinity of the bulge, which was several hundred meters below the summit, the main impact of the eruption was directed laterally rather than vertically. Had the full force of the eruption been upward, far less destruction would have occurred.

Mount St. Helens is one of 15 large volcanoes and innumerable smaller ones that comprise the Cascade Range, which extends from British Columbia to northern California. Eight of the largest volcanoes have been active in the past few hundred years. The last eruptive phase of Mount St. Helens ended in 1857. Of the remaining seven "active" volcanoes, the most likely to erupt again are Mount Baker, Mount Shasta, Lassen Peak, Mount Hood, and Mount Rainier.

Not all volcanic eruptions are as violent as the 1980 Mount St. Helens event. Some volcanoes, such as Hawaii's Kilauea volcano, generate relatively quiet outpourings of fluid lavas (Figure 8.4). These "gentle" eruptions are not without some fiery displays; occasionally fountains of incandescent lava spray hundreds of meters into the air. Such events, however, are typically short-lived and harmless, and the lava generally falls back into a lava pool.

Testimony to the quiet nature of Kilauea's eruptions is the fact that the Hawaiian Volcanoes Observatory has operated on its summit since 1912. This, despite the fact that Kilauea has had more than 50 eruptive phases since record keeping began in 1823. Further, the longest and largest of Kilauea's rift eruptions began in 1983 and remains active at the time of this writing, although it has received only modest media attention.

Why do volcanoes like Mount St. Helens erupt explosively, whereas others like Kilauea are relatively quiet? Why do volcanoes occur in chains like the Aleutian Islands or the Cascade Range? Why do some volcanoes form on the ocean floor, while others occur on the continents? This chapter will deal with these and other questions as we explore the formation and movement of magma.

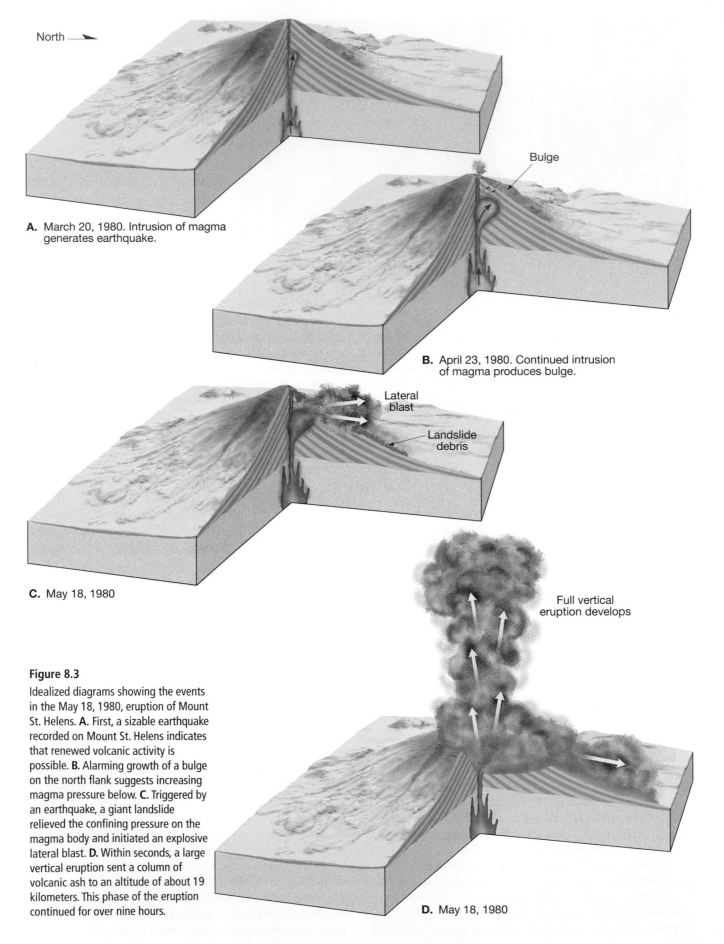

North ➤

Bulge

A. March 20, 1980. Intrusion of magma generates earthquake.

B. April 23, 1980. Continued intrusion of magma produces bulge.

Lateral blast

Landslide debris

C. May 18, 1980

Full vertical eruption develops

D. May 18, 1980

Figure 8.3

Idealized diagrams showing the events in the May 18, 1980, eruption of Mount St. Helens. **A.** First, a sizable earthquake recorded on Mount St. Helens indicates that renewed volcanic activity is possible. **B.** Alarming growth of a bulge on the north flank suggests increasing magma pressure below. **C.** Triggered by an earthquake, a giant landslide relieved the confining pressure on the magma body and initiated an explosive lateral blast. **D.** Within seconds, a large vertical eruption sent a column of volcanic ash to an altitude of about 19 kilometers. This phase of the eruption continued for over nine hours.

Figure 8.4
Fluid basaltic lava erupting from Kilauea
Volcano, Hawaii. (Photo by Douglas Peebles)

The Nature of Volcanic Eruptions

Volcanic activity is commonly perceived as a process that produces a picturesque, cone-shaped structure which periodically erupts in a violent manner. Although some eruptions are very explosive, many others are not. What determines whether a volcano extrudes magma violently or "gently"? The primary factors include the magma's *composition*, its *temperature*, and the amount of *dissolved gases* it contains. To varying degrees these factors affect the magma's **viscosity**. The more viscous ("thicker") the material, the greater its resistance to flow. For example, syrup is more viscous than water.

Factors Affecting Viscosity

The effect of temperature on viscosity is easily visualized. Just as heating syrup makes it more fluid (less viscous), the mobility of lava is strongly influenced by temperature changes. As a lava flow cools and begins to congeal, its mobility decreases and eventually the flow halts.

But more significant to volcanic behavior is the chemical composition of magmas. This was discussed in Chapter 2 along with the classification of igneous rocks. Recall that a major difference among various igneous rocks is their silica (SiO_2) content. The same is true of the magmas from which rocks form. Magmas that produce basaltic rocks contain about 50 percent silica, whereas magmas that produce granitic rocks contain over 70 percent silica (Table 8.1).

Table 8.1 Variations in properties among magmas of differing compositions.

Property	Basaltic magma	Andesitic magma	Granitic magma
Silica content	Least (about 50%)	Intermediate (about 60%)	Most (about 70%)
Viscosity	Least ("thinnest")	Intermediate	Greatest ("thickest")
Tendency to form lavas	Highest	Intermediate	Least
Tendency to form pyroclastics	Least	Intermediate	Greatest
Melting temperature	Highest	Intermediate	Lowest

Note that *a magma's viscosity is directly related to its silica content.* In general, the more silica in magma, the greater is its viscosity. The flow of magma is impeded because silicate structures link together into long chains, even before crystallization begins. Consequently, because of high silica content, granitic lavas are very viscous and tend to form comparatively short, thick flows. By contrast, basaltic lavas, which contain less silica, tend to be more fluid and have been known to travel distances of 150 kilometers (90 miles) or more before congealing (Figure 8.4).

In Hawaiian eruptions, the magmas are hot and basaltic, so they are extruded with ease. By contrast, highly viscous granitic magmas are more difficult to force through a vent. On occasion, the vent may become plugged with viscous magma, which results in a buildup of gases and a great pressure increase, so a potentially explosive eruption may result. However, a viscous magma is not explosive by itself. It is the gas content that puts the "bang" into a violent eruption.

Importance of Dissolved Gases in Magma

Dissolved gases in magma provide the force that extrudes molten rock from the vent. These gases are mostly water vapor and carbon dioxide. As magma moves into a near-surface environment, such as within a volcano, the confining pressure in the uppermost portion of the magma body is greatly reduced. This reduced confining pressure allows dissolved gases to be released suddenly, just as opening a soda bottle allows dissolved carbon dioxide gas to bubble out of the soda.

At high temperatures and low, near-surface pressures, these gases will expand to occupy hundreds of times their original volume. Very fluid basaltic magmas allow the expanding gases to bubble upward and escape from the vent with relative ease. As they escape, the gases will often propel incandescent lava hundreds of meters into the air, producing lava fountains. Although spectacular, such fountains are mostly harmless and not generally associated with major explosive events that cause great loss of life and property. Rather, eruptions of fluid basaltic lavas, such as those that occur in Hawaii, are relatively quiet.

At the other extreme, highly viscous magmas impede the upward migration of expanding gases. The gases collect in bubbles and pockets that increase in size and pressure until they explosively eject the semimolten rock from the volcano. The result is a Mount St. Helens or Mt. Pinatubo.

To summarize, the viscosity of magma, plus the quantity of dissolved gases and the ease with which they can escape, determines the nature of a volcanic eruption. We can now understand the "gentle" volcanic eruptions of hot, fluid lavas in Hawaii and the explosive, violent, dangerous eruptions of viscous lavas in some volcanoes.

What Is Extruded During Eruptions?

Lava appears to be the primary material extruded from a volcano, but this is not always the case. Just as often, explosive eruptions eject huge quantities of broken rock, lava "bombs," fine ash, and dust. Moreover, all volcanic eruptions emit large amounts of gas. In this section we will examine each of these materials associated with a volcanic eruption.

Lava Flows

Due to their low silica content, basaltic lavas are usually very fluid. They flow in thin, broad sheets or streamlike ribbons. On the island of Hawaii, such lavas have been clocked at speeds of 30 kilometers (20 miles) per hour down steep slopes. These velocities are rare, however, and flow rates of 10 to 300 meters per hour are more common. In contrast, the movement of silica-rich lava is on occasion too slow to be perceptible.

When fluid basaltic lavas of the Hawaiian type congeal, they commonly form a relatively smooth skin that wrinkles as the still-molten subsurface lava continues to advance (Figure 8.5A). These are known as **pahoehoe flows** (pronounced *pah-hoy-hoy*) and resemble the twisted braids in ropes.

Another common type of basaltic lava has a surface of rough, jagged blocks with dangerously sharp edges and spiny projections (Figure 8.5B). The name **aa** (pronounced *ah-ah*) is given to these flows. Active aa flows are relatively cool and thick and, depending on the slope, advance at rates from 5 to 50 meters per hour. Further, escaping gases fragment the cool surface and produce numerous voids and sharp spines in the congealing lava. As the molten interior advances, the outer crust is broken further, giving the flow the appearance of an advancing mass of lava rubble.

Gases

Magmas contain varied amounts of dissolved gases held in the molten rock by confining pressure, just as carbon dioxide is held in soft drinks. As with soft drinks, as soon as the pressure is reduced, the gases begin to escape. Obtaining gas samples from an erupting volcano is often difficult and dangerous, so geologists usually only estimate the amount of gas originally contained within the magma.

The gaseous portion of most magmas is believed to make up 1 to 5 percent of the total weight, with most of this being water vapor. Although the percentage may be small, the actual quantity of emitted gas can exceed thousands of tons per day.

The composition of volcanic gases is important to scientists because these gases are thought to be the original source of the water for the oceans. Further, volcanic eruptions have contributed significantly to gases that compose the atmosphere. Analysis of samples taken during Hawaiian eruptions indicates that the gases are about 70 percent water vapor, 15 percent carbon dioxide, 5 percent nitrogen, 5 percent sulfur, and lesser amounts of chlorine, hydrogen, and argon. Sulfur compounds are easily recognized by their pungent odor and because they readily form sulfuric acid. Volcanoes are a natural source of air pollution, and can cause serious problems.

Pyroclastics

When basaltic lava is extruded, the dissolved gases escape quite freely and continually. As stated earlier, these gases propel incandescent blobs of lava to great heights, thereby producing spectacular lava fountains. Some ejected material may land near the vent and build a cone structure, whereas smaller particles will be carried great distances by the wind.

The gases in highly viscous magmas, on the other hand, become superheated, and upon release they expand a thousandfold as they blow pulverized rock, lava, and glass fragments from the vent (see Box 8.1). The particles produced by these processes are called **pyroclastics** (meaning "fire fragments"). These ejected lava fragments range in size from very fine dust and sand-sized volcanic ash, to large pieces (Figure 8.6).

The fine *ash* particles are produced when the extruded lava contains so many gas bubbles that it resembles the froth flowing from a newly opened bottle of champagne.

A.

B.

Figure 8.5
A. Typical pahoehoe (ropy) lava flow, Kilauea Hawaii. **B.** Typical slow-moving aa flow. (Photos by J. D. Griggs)

Box 8.1

The Year Without a Summer

The graph in Figure 8.A allows us to compare the volume of volcanic debris extruded during some well-known eruptions, beginning with Mount Vesuvius in A.D. 79. The eruption of the volcano named *Tambora* is clearly the largest of the modern era. Between April 7 and 12, 1815, this nearly 4000-meter Indonesian structure violently ejected an estimated 30 cubic kilometers of volcanic debris. That is 30 times more ash than was emitted during the eruption of Mount St. Helens in May 1980.

Although the Tambora eruption occurred in an isolated part of the world, circulation patterns high in the atmosphere spread its influence far and wide. The impacts of the volcanic dust and gases on climate are believed to have been widespread in the Northern Hemisphere. According to one researcher, "The extreme cold that prevailed during the spring and summer of 1816 in some regions of the world represents one of the most unusual climatic episodes that has occurred since the advent of instrumental weather observations."[*] The effects were especially severe in New England, where 1816 has come to be known as the "year without a summer."

[*]*American Weather Stories* (Washington, D.C.: National Oceanic and Atmospheric Administration, 1976), p. 43.

From May through September 1816, an unprecedented series of cold spells affected the northeastern United States and adjacent portions of Canada. The result was a late spring, a cold summer, and an early fall. There were heavy snows in June and frosts in July and August. Across the Northeast, crops were reported killed by the cold. Temperatures in New England averaged up to 3.5 °C below normal in June and between 1 °C and 2 °C below normal in August. Although temperatures this cold had occurred before, there has never been such a protracted span of cold since records have been kept. The reduced averages may seem modest, but they took place in a region where even a small drop in minimum temperatures can mean severe frost.

New England and adjacent Canada were not the only areas to experience a "year without a summer." Patrick Hughes writes, "Although the New England farmer considered it a local tragedy, the abnormal weather was widespread throughout the Northern Hemisphere. In England it was almost as cold as in the United States, and 1816 was a famine year there, as it was in France and Germany."[†]

The unusual meteorological events of 1816 which followed the massive 1815 eruption of Tambora are regarded by many as a spectacular example of the influence of explosive volcanism on climate. Although the effects were rela-

[†]*American Weather Stories*, p. 43.

tively short-lived, this geological event had a significant atmospheric and human impact.

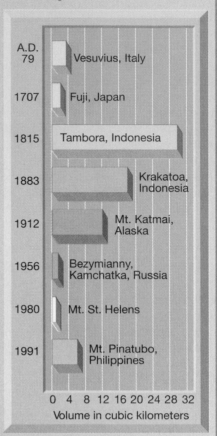

Figure 8.A

Approximate volume of volcanic debris emitted during some well-known eruptions. The 1815 eruption of Tambora, the largest-known eruption in historic time, ejected over 30 times more ash than did Mount St. Helens in 1980.

As the hot gases expand explosively, the lava is disseminated into very fine glassy fragments. When the hot ash falls, the glassy shards often fuse to form *welded tuff*. Sheets of this material, as well as ash deposits that later consolidated, cover vast portions of the western United States. Sometimes the frothlike lava is ejected in larger pieces called *pumice*. This material has so many voids (air spaces) that it is often light enough to float in water.

Pyroclastics the size of walnuts, called *lapilli* ("little stones"), and pea-sized particles called *cinders* are also very common. Cinders contain numerous voids and form when ejected lava blocks are pulverized by escaping gases. Particles larger than lapilli are called *blocks* when they are made of hardened lava and *bombs* when they are ejected as incandescent lava. Since volcanic bombs are semimolten upon ejection, they often take on a streamlined shape, as shown in Figure 8.7.

Volcano Types

Successive eruptions from a central vent result in a mountainous accumulation of material known as a **volcano**. Located at the summit of many volcanoes is a steep-walled depression called a **crater**. The crater is connected to a magma chamber via a pipelike conduit, or **vent**. Some volcanoes have unusually large summit depressions that exceed one kilometer in diameter and are known as **calderas**.

When fluid lava leaves a conduit, it is often stored in the crater or caldera until it overflows. On the other hand, viscous lava forms a plug in the pipe. It rises slowly or is blown out, often enlarging the crater. However, lava does not always issue from a central crater. Sometimes it is easier for the magma or escaping gases to push through fissures on the volcano's flanks. Mount

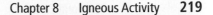

Figure 8.6
Pyroclastics and lava ejected from a flank eruption at Kilauea volcano, Hawaii. (Photo by Paul Chesley/Tony Stone Images)

Figure 8.7
Volcanic bomb. Ejected lava fragments take on a streamlined shape as they sail through the air. (Photo by GeoScience/PH)

Etna in Italy, for example, has more than 200 secondary vents. Some of these emit only gases and are appropriately called *fumaroles*.

The eruptive history of each volcano is unique. Consequently, all volcanoes are somewhat different in form and size. Nevertheless, volcanologists recognize three general eruptive patterns and characteristic forms: shield volcanoes, cinder cones, and composite cones (see Box 8.2).

Shield Volcanoes

When fluid lava is extruded, the volcano takes the shape of a broad, slightly domed structure called a **shield volcano** (Figure 8.8). They are so called because they roughly resemble the shape of a warrior's shield. Shield volcanoes are built primarily of basaltic lava flows and contain only a small percentage of pyroclastic material.

Mauna Loa, probably the largest volcano on Earth, is one of the five shield volcanoes that together make up the island of Hawaii. Its base rests on the ocean floor

5000 meters below sea level, and its summit is 4170 meters (13,680 feet) above the water. Nearly one million years and numerous eruptive cycles built this truly gigantic pile of volcanic rock. Many other volcanic structures, including Midway Island and the Galapagos Islands, have been built in a similar manner from the ocean's depths.

Perhaps the most active and intensively studied shield volcano is Kilauea, located on the island of Hawaii on the southeastern flank of the larger volcano Mauna Loa. Kilauea has erupted more than 50 times in recorded history and is still active today. Several months before an eruptive phase, Kilauea's summit inflates as magma rises from its source 60 kilometers or more below the surface. This molten rock gradually works its way upward and accumulates in smaller reservoirs 3 to 5 kilometers below the summit. For up to 24 hours in advance of each eruption, swarms of small earthquakes warn of the impending activity.

The longest and largest rift eruption ever recorded at Kilauea began in 1983 and continues to the present day. The eruption began along a 6.5-kilometer (4-mile) fissure in an inaccessible forested area east of the summit caldera (Figure 8.9). Between 1983 and 1986, nearly 50 eruptive phases were observed. Each consisted of a period of high lava fountaining which lasted a few hours to a few days, followed by an inactive period that averaged one month. After a few months, the eruptions localized in one spot along the fissure, building a major spatter cone 250 meters (820 feet) high.

Box 8.2

Volcano Sizes

In the section on the three basic volcano types (shield, cinder cone, composite cone), we often mentioned their sizes. It is useful to compare the three types side-by-side, as in Figure 8.B.

Smallest are the cinder cones (Figure 8.B). Recall that Parícutin rose from a Mexican cornfield during a period of just nine years. Because their eruption history is short, most cinder cones do not exceed 300 meters (1000 feet) in height.

In contrast to cinder cones, composite cones and shield volcanoes have recurring eruptions that can span a million years or longer. As shown in Figure 8.B, Mount Rainier, one of the largest composite cones in the Cascade Range, dwarfs a relatively large cinder cone, Sunset Crater in Arizona. Sunset Crater formed over a few years, but Mount Rainier gradually formed over the past 700,000 years.

Mauna Loa, one of five shield volcanoes that compose the island of Hawaii, is considered the world's largest active volcano (Figure 8.B). This massive pile of basaltic lava has an estimated volume of 40,000 cubic kilometers that was extruded over approximately one million years. From its base on the floor of the Pacific Ocean to its summit, Mauna Loa approaches 9 kilometers in height.

Despite its enormous size, Mauna Loa is not the largest known volcano in the solar system. Olympus Mons, a huge shield volcano on Mars, is 25 kilometers high and 600 kilometers wide (see Chapter 20).

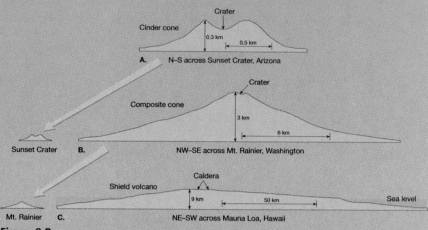

Figure 8.B
Profiles of volcanic landforms. **A.** Profile of Sunset Crater, Arizona, a typical steep-sided cinder cone. **B.** Sunset Crater compared with Mt. Rainier, Washington, a large composite cone in the Cascade Range. **C.** Mt. Rainier compared with Mauna Loa, Hawaii, the largest shield volcano in the Hawaiian chain.

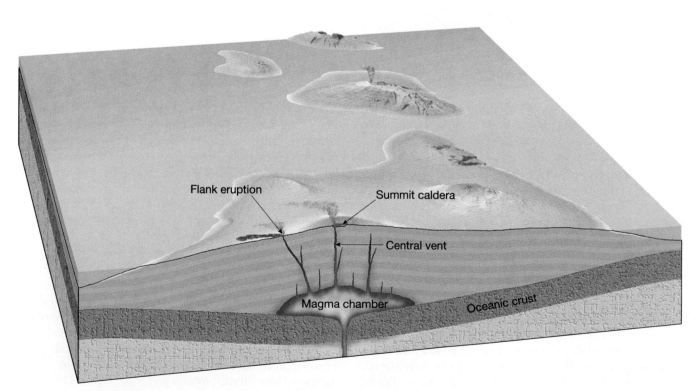

Figure 8.8
Shield volcanoes are built primarily of fluid basaltic lava flows and contain only a small percentage of pyroclastic materials. These broad, slightly domed structures, exemplified by the Hawaiian Islands, are the largest volcanoes on Earth.

Figure 8.9
Fissure eruption ("Curtain of Fire") at Kilauea, Hawaii in 1984. (Photo by Bern Pedit)

Lava falling back from these fountains formed aa flows that migrated a few hundred meters per hour. Although inhabitants in a nearby settlement had time to be evacuated, many of their homes were burned and buried in the aa.

By the summer of 1986, the eruptions shifted three kilometers downslope from the cone. Here, smooth-surfaced pahoehoe lava formed a lava lake. Eventually, it overflowed, and the fast-moving pahoehoe destroyed nearly a hundred rural homes, covered a major roadway, and eventually flowed into the sea. Lava has been intermittently pouring into the ocean since that time, adding new land to the island.

Eruptions of the type just described are typical of shield volcanoes and have been occurring sporadically on the Island of Hawaii for nearly one million years. The result is the formation of mountains with summits over 4000 meters (13,000 feet) above sea level, and if we take into account the portion below sea level, they are the tallest mountains on Earth, over 9000 meters (30,000 feet).

Cinder Cones

As the name suggests, **cinder cones** are built from ejected lava fragments. Loose pyroclastic material has a high angle of repose (between 30 and 40 degrees), the steepest angle at which the material remains stable. Thus, volcanoes of this type have very steep slopes (Figure 8.10). Cinder cones are rather small, usually less than 300 meters (1000 feet) high, and often form as parasitic cones on or near larger volcanoes. In addition, they frequently occur in groups.

One of the very few volcanoes whose formation has been observed by geologists from beginning to end is a cinder cone called Parícutin. In 1943, about 200 miles west of Mexico City, the volcano Parícutin was born (Figure 8.11). It erupted in a cornfield owned by Dionisio Pulido, who witnessed the event as he prepared the field for planting.

For two weeks prior to the first eruption, numerous earth tremors caused apprehension in the village of Parícutin about 3.5 kilometers away. Then sulfurous smoke began billowing from a small hole that had been in the cornfield for as long as Señor Pulido could remember. During the night, hot, glowing rock fragments thrown into the air from the hole produced a spectacular fireworks display. In one day the cone grew to 40 meters (130 feet) and by the fifth day it was over 100 meters (330 feet) high. Explosive eruptions threw hot fragments 1000 meters (3300 feet) above the crater rim. Larger fragments fell near the crater, some remaining incandescent as they rolled down the slope. These built an aesthetically pleasing cone, while finer ash fell over a much larger area, burning and eventually covering the village of Parícutin. Within two years the cone attained its final height of about 400 meters (1300 feet).

The first lava flow came from a fissure that opened just north of the cone, but after a few months flows began to emerge from the base of the cone itself. In June 1944, a clinkery aa flow 10 meters thick moved over much of the village of San Juan Parangaricutiro, leaving only the church steeple exposed (Figure 8.11). After nine years, the activity ceased almost as quickly as it had begun. Today, Parícutin is just another one of the numerous quiet cinder cones dotting the landscape in this region of Mexico. Like the others, it will probably not erupt again.

Composite Cones

Earth's most picturesque volcanoes are composite cones. Most active composite cones are in a narrow zone that encircles the Pacific Ocean, appropriately named the *Ring of Fire*. Found in this region are Fujiyama (Mt. Fuji) in Japan, Mount Mayon in the Philippines, and the picturesque volcanoes of the Cascade Range in the northwestern United States, including Mount St. Helens, Mount Rainier, and Mount Shasta (Figure 8.12).

Figure 8.10
SP Crater, a cinder cone north of Flagstaff, Arizona. Note the road across the lava flow for scale. (Photo by Michael Collier)

Pyroclastic material

Crater

Central vent filled
with rock fragments

A **composite cone** or **stratovolcano** is a large, nearly symmetrical structure composed of interbedded lava flows and pyroclastic deposits, emitted mainly from a central vent. Just as shield volcanoes owe their shape to the highly fluid nature of the extruded lavas, so too do composite cones reflect the nature of the erupted material.

Composite cones are produced when relatively viscous lavas of andesitic composition are extruded. A composite cone may extrude viscous lava for long periods. Then, suddenly, the eruptive style changes and the vol-

cano violently ejects pyroclastic material. Most of the ejected pyroclastic material falls near the summit, building a steep-sided mound of cinders. In time, this debris becomes covered by lava. Occasionally, both activities occur simultaneously, and the resulting structure consists of alternating layers of lava and pyroclastics. Two of the most perfect cones, Mount Mayon in the Philippines and Fujiyama in Japan, exhibit the classic form of the stratovolcano with its steep summit area and more gently sloping flanks.

Figure 8.11
The village of San Juan Parangaricutiro engulfed by lava from Parícutin, shown in the background. Only the church towers remain. (Photo by Tad Nichols/Peter L. Kresan Photography)

Composite cones represent the most violent type of volcanic activity. An eruption can be unexpected and devastating, as was the A.D. 79 eruption of the Italian volcano we now call Vesuvius. Prior to this eruption, Vesuvius had been dormant for centuries. Although minor earthquakes probably warned of the events to follow, Vesuvius was covered with dense vegetation and hardly looked threatening. On August 24, however, the tranquility ended, and in the next three days the city of Pompeii (near Naples) and more than 2000 of its 20,000 residents were buried. They remained so for nearly 17 centuries, until the city was rediscovered and excavated.

Nuée Ardente. Although the destruction of Pompeii was catastrophic, even more devastating eruptions occur when hot gases infused with incandescent ash are ejected. This produces a fiery cloud called a **nuée ardente**. Also referred to as *glowing avalanches*, these turbulent steam clouds and companion ash flows race down steep volcanic slopes at speeds that can approach 200 kilometers (125 miles) per hour (Figure 8.13). The ground-hugging portions of glowing avalanches are rich in particulate matter that is suspended by hot, buoyant gases. Thus, these flows, which can include larger rock fragments in addition to ash, travel downslope in a nearly frictionless environment cushioned by expanding volcanic gases. This fact explains why some nuée ardente deposits extend more than 100 kilometers (60 miles) from their source.

Figure 8.12
Mount Shasta, California, one of the largest composite cones in the Cascade Range. (Photo by John S. Shelton)

Crater   Pyroclastic layers   Lava flows

Figure 8.13
Nuée ardente races down the slope of Mount St. Helens on August 7, 1980, at speeds in excess of 100 kilometers (60 miles) per hour. (Photo by Peter W. Lipman, U.S. Geological Survey)

Figure 8.14
St. Pierre as it appeared shortly after the eruption of Mount Pelée, 1902. (Reproduced from the collection of the Library of Congress)

In 1902, a nuée ardente from Mount Pelée, a small volcano on the Caribbean island of Martinique, destroyed the port town of St. Pierre. The destruction happened in moments and was so devastating that almost all of St. Pierre's 28,000 inhabitants were killed. Only a prisoner protected in a dungeon, a shoemaker, and a few people on ships in the harbor were spared (Figure 8.14).

When we compare the destruction of St. Pierre with that of Pompeii, you can see several differences. Pompeii was buried by an event lasting three days, whereas St. Pierre was destroyed in moments and its remains were mantled by only a thin layer of volcanic debris. Also, the structures of Pompeii remained intact, except for roofs that collapsed under the weight of the ash. In St. Pierre, masonry walls nearly one meter thick were knocked over like dominoes; large trees were uprooted and cannons were torn from their mounts. Clearly, volcanic hazards vary from one volcano to another.

Lahars. In addition to their violent eruptions, volcanoes are potential hazards in other ways. In particular, large composite cones often generate a type of mudflow known as **lahar**. These destructive events occur when deposits of volcanic ash and debris become saturated with water and flow down steep volcanic slopes, generally following stream valleys. Some lahars are produced when rainfall saturates volcanic deposits, whereas others are triggered as large volumes of ice and snow melt during an eruption.

When Mount St. Helens erupted in May 1980, several lahars formed. The flows and accompanying floods raced down the valleys of the north and south forks of

the Toutle River at speeds in excess of 30 kilometers per hour. Water levels reached 4 meters above flood stage, and nearly all the homes and bridges along the river were destroyed or severely damaged (Figure 8.15). Fortunately, the affected area was not densely populated. This was not the case in November 1985, when Nevado del Ruiz, a volcano in the Andes, erupted and generated a lahar that killed nearly 20,000 people (see the section entitled "Mudflow" in Chapter 3).

Other Volcanic Landforms

The most obvious volcanic landform is a cone. But other distinctive landforms are associated with volcanic activity.

Calderas

As noted, most volcanoes have a steep-walled *crater*. A crater is called a *caldera* when it exceeds one kilometer in diameter. Calderas are usually circular, with rather

Figure 8.16
Crater Lake in Oregon occupies a caldera about 6 miles in diameter. (Photo by Greg Vaughn/Tom Stack and Associates)

flat floors and steep walls. Several large calderas exist; the largest in the United States, LaGanita in Colorado, is 28 kilometers (17 miles) across.

Most calderas probably form when the summit of a volcanic structure collapses into the partially emptied magma chamber below. Crater Lake in Oregon, 8-10 kilometers (5-6 miles) wide and 1175 meters (over 3800 feet) deep, is located in such a depression (Figure 8.16). The formation of Crater Lake began about 7000 years ago when the volcano, later to be named Mount Mazama, put forth a violent ash eruption that extruded an estimated 50-70 cubic kilometers of volcanic material (Figure 8.17). With the loss of support, 1500 meters (4900 feet) of this once-prominent 3600-meter cone collapsed. After the collapse, rainwater filled the caldera. Later volcanic activity built a small cinder cone called Wizard Island, which today provides a mute reminder of past activity.

At least 138 calderas that exceed 5 kilometers in diameter are known to exist. One of the largest of these is about 100 kilometers (60 miles) across. Unlike Crater Lake, these calderas are so large and irregular that many remained undetected until high-quality aerial, or satellite photos of the region became available. One example, the Valles Caldera located west of Los Alamos, New Mexico, is about 16 kilometers across and 1 kilometer deep. These features are usually associated with large granitic intrusive bodies located near the surface. Following a violent eruption that may emit tens of cubic kilometers of pyroclastic debris, the roof of the partially emptied magma chamber apparently collapses. Other examples of large calderas believed to have formed in a similar manner are California's Long Valley Caldera and Yellowstone Caldera in Wyoming.

Not all calderas are produced following explosive eruptions. For example, the summits of Hawaii's active shield volcanoes, Mauna Loa and Kilauea, have large calderas that formed as a result of relatively quiet activity. These calderas, which measure 3-5 kilometers (2-3 miles) across and nearly 200 meters (650 feet) deep, formed by collapse as magma slowly drained from the summit magma chambers during flank eruptions.

Volcanic Necks

Volcanoes, like all land areas, are continually being lowered by weathering and erosion. Cinder cones are easily eroded, because they are composed of unconsolidated materials. As erosion progresses, the rock occupying the vent is often more resistant and may remain standing above the surrounding terrain long after most of the cone has vanished. Ship Rock, New Mexico, is such a feature, called a **volcanic neck** (Figure 8.18). This structure, higher than many skyscrapers, is but one of many such landforms that protrude conspicuously from the red desert landscapes of the Southwest. Ultimately, even these resistant necks will succumb to relentless erosion over geologic time.

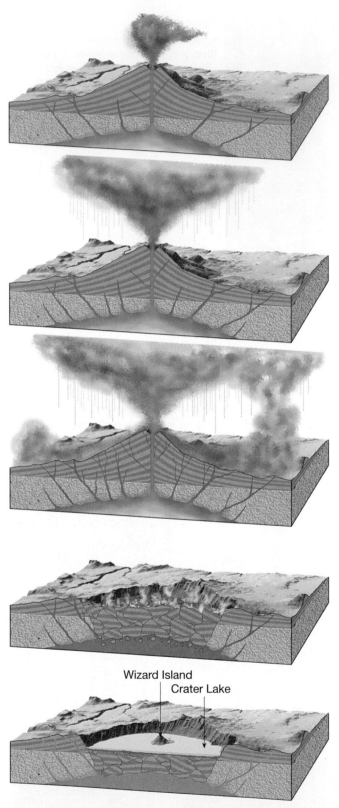

Figure 8.17

Sequence of events that formed Crater Lake, Oregon. About 7000 years ago, the summit of former Mount Mazama collapsed following a violent eruption which partly emptied the magma chamber. Subsequent eruptions produced the cinder cone called Wizard Island. Rainfall and groundwater contributed to form the lake. (After H. Williams, *The Ancient Volcanoes of Oregon*, p. 47. Courtesy of the University of Oregon)

Figure 8.18
Ship Rock, New Mexico, is a volcanic neck. This structure which stands over 420 meters (1380 feet) high, consists of igneous rock that crystallized in the vent of a volcano that has long since been eroded away. The tabular structure in the background is a dike that served to feed lava flows along the flanks of the once active volcano. (Photo by Michael Collier)

Fissure Eruptions and Lava Plateaus

We think of volcanic eruptions as occurring from a central vent and building a cone or mountain, but by far the greatest volume of volcanic material is extruded from fractures in the crust called **fissures**. Rather than building a cone, these long, narrow cracks pour forth lava, blanketing a wide area. The extensive Columbia Plateau in the northwestern United States was formed this way (Figure 8.19). Here, numerous fissure eruptions extruded very fluid basaltic lava. Successive flows, some 50 meters thick, buried the existing landscape as they built a lava plateau, nearly a mile thick in places. The fluidity is evident, because some lava remained molten long enough to flow 150 kilometers (90 miles) from its source. The term **flood basalts** appropriately describes these flows (Figure 8.20).

Pyroclastic Flows

When silica-rich magma is extruded, **pyroclastic flows** of ash and pumice fragments usually result. They are propelled away from the vent at high speeds and may blanket extensive areas before coming to rest. Once deposited, the pyroclastic materials may closely resemble lava flows.

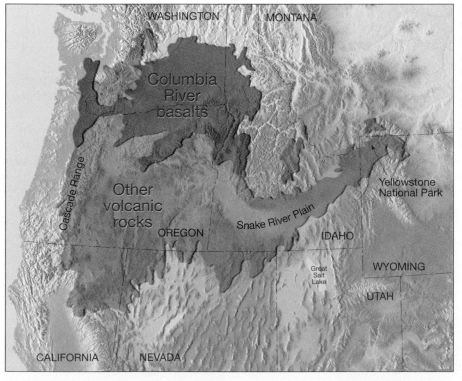

Figure 8.19
Volcanic areas in the northwestern United States. The Columbia River basalts cover an area of nearly 200,000 square kilometers (80,000 square miles). Activity here began about 17 million years ago as lava began to pour out of large fissures, eventually producing a basalt plateau with an average thickness of more than 1 kilometer. (After U.S. Geological Survey)

Figure 8.20
Basalt flows near Idaho Falls. (Photo by John S. Shelton)

Extensive pyroclastic flow deposits exist in many parts of the world, most associated with large calderas. Perhaps best known is the Yellowstone Plateau in northwestern Wyoming. Here a large magma body, rich in silica, still exists a few kilometers below the surface. Several times over the past two million years, fracturing of the rocks overlying the magma chamber has resulted in huge outpourings, accompanied by the formation of calderas. In Yellowstone National Park, numerous buried fossil forests have been discovered. During volcanic activity, a forest would develop on a newly formed volcanic surface, only to be covered by ash from the next eruptive phase. Fortunately, no eruption of this type has occurred in historic times.

Intrusive Igneous Activity

Volcanic eruptions can be among the most violent and spectacular events in nature and thus receive much scientific attention. Yet most magma is emplaced at depth, intruding into existing rocks. An understanding of this intrusive igneous activity is therefore as important to geologists as the study of volcanic events. In mythology, Pluto was lord of the underworld. Thus, underground igneous rock bodies are called *plutons*.

Figure 8.21 shows intrusive igneous bodies (plutons) that form when magma crystallizes within Earth's crust. Notice that some of these structures have a tabular or sheetlike shape, while others are bulky masses. Some cut across existing structures, such as layers of sedimentary rocks, while others form when magma is injected between sedimentary layers. Due to these differences, plutons are generally classified according to their shape as either *tabular* (sheetlike) or *massive*, and by their orientation with respect to the host rock. Intrusive igneous bodies are said to be *discordant* if they cut across existing sedimentary beds, and *concordant* if they form parallel to the existing sedimentary beds.

Dikes and Sills

Dikes are sheetlike bodies produced when magma is injected into fractures that cut across rock layers. They range in thickness from less than one centimeter to more than one kilometer. The largest are hundreds of kilometers long. Dikes are often oriented vertically, forming in pathways followed by molten rock that fed ancient lava flows. Frequently dikes are more resistant to weathering than the surrounding rock. When exposed, such dikes have the appearance of a wall.

Sills are tabular plutons formed when magma is injected along sedimentary bedding surfaces (Figures 8.21 and 8.22). Horizontal sills are the most common, although all orientations, even vertical, are known to exist where the strata have been tilted. Because of their relatively uniform thickness and large extent, sills must form from very fluid basaltic magma. The emplacement of a sill requires that the overlying sedimentary rock be lifted to a height equal to the thickness of the sill. Consequently, sills form only at relatively shallow depths where the pressure exerted by the weight of overlying strata is relatively low.

One of the largest and best-known sills in the United States is the Palisades Sill, which is exposed along the Hudson River near New York City. This resistant, 300-meter-thick sill has formed an imposing cliff easily seen from across the river.

In many respects, sills closely resemble buried lava flows. Both are tabular and often exhibit **columnar jointing** (Figure 8.23). Further, because sills generally form in near-surface environments and may be only a few meters thick, the emplaced magma is often chilled quickly enough to generate a fine-grained texture.

When attempts are made to reconstruct the geologic history of a region, it becomes important to differentiate between sills and buried lava flows. Fortunately, under close examination these two structures can be readily distinguished. The upper portion of a buried

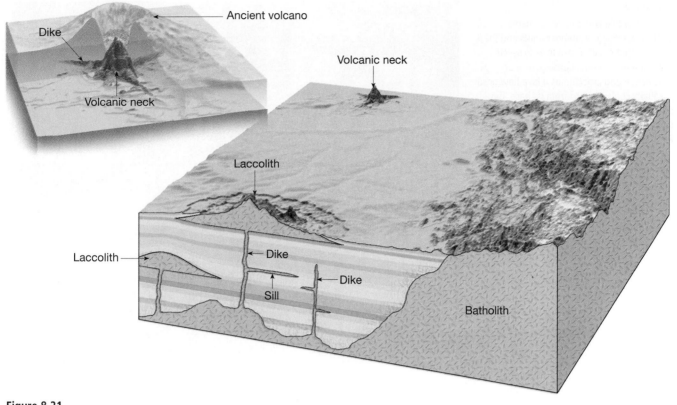

Figure 8.21
Basic intrusive (underground) igneous structures.

lava flow usually contains voids produced by entrapped gas bubbles, and only the rocks beneath a lava flow show evidence of metamorphic alteration. Sills, on the other hand, form when magma has been forcefully intruded between sedimentary layers. Thus, fragments of the overlying wall rock occur only in sills, since lava flows are extruded before the overlying strata are deposited. Further "baked" zones above and below are trademarks of a sill.

Laccoliths and Batholiths

Laccoliths form the same way as sills, but from more viscous magma, which collects as a lens-shaped mass that arches the overlying strata upward. Consequently, a laccolith can occasionally be detected because of the dome it creates at the surface (Figure 8.21).

Batholiths are by far the largest intrusive igneous bodies. The Idaho batholith, for example, encompasses more than 40,000 square kilometers. Indirect evidence from gravitational studies indicates that batholiths are also very thick, possibly extending tens of kilometers into the crust. Geologists define a batholith as an intrusive body with a surface exposure of more than 100 square kilometers (40 square miles). Similar but smaller intrusive bodies are termed *stocks*. Many stocks appear to be portions of batholiths that are not yet fully exposed.

Batholiths frequently form the cores of mountain systems. Here, uplift and erosion have removed the surrounding rock to expose the resistant igneous body (Figure 8.21). Some of the highest mountain peaks, such as

Figure 8.22
Salt River Canyon, Arizona. The dark, essential horizontal band is a sill of basaltic composition that intruded into horizontal layers of sedimentary rock. (Photo by E. J. Tarbuck)

Figure 8.23
Devil's Post Pile National Monument, California, exhibits columnar joints and the columns that result. These five- to seven-sided columns are the consequence of contraction and cracking as a lava flow or sill gradually cools upward from its base. The small photo inset is a view looking down on these columns. (Courtesy of the National Park Service, U.S. Department of the Interior)

Mount Whitney in the Sierra Nevada, are carved from such a granitic mass. Large batholiths form over millions of years. The intrusive activity that created the Sierra Nevada batholith, for example, occurred nearly continuously over a 130-million-year span (Figure 8.24).

Large expanses of granitic rock are also exposed in the stable interiors of the continents, such as the Canadian Shield of North America. These relatively flat outcrops are believed to be the remnants of ancient mountains that erosion has long since leveled.

Igneous Activity and Plate Tectonics

The origin of magma has been controversial in geology, almost from the beginning of the science. How do magmas of different compositions form? Why do volcanoes in the deep-ocean basins primarily extrude basaltic lava, whereas those adjacent to oceanic trenches extrude mainly andesitic lava? Why does an area of igneous activity, commonly called the *Ring of Fire*, surround the Pacific Ocean? New insights from the theory of plate tectonics are providing answers. We will first examine the origin of magma and then look at the global distribution of volcanic activity as viewed by the plate tectonics model.

Origin of Magma

Based on available scientific evidence, the crust and mantle are composed primarily of solid rock. Further, although the outer core is in a fluid state, this material is very dense and remains deep within Earth. What then is the source of magma that produces volcanic activity?

Temperature and Magma Generation. Geologists conclude that magma must originate from essentially solid rock located in the crust and mantle. The most obvious way to generate magma from solid rock is to raise its temperature.

Figure 8.24
Half Dome in Yosemite National Park, California. This feature is just a tiny portion of the Sierra Nevada batholith, a huge structure that extends for approximately 400 kilometers. (Photo by Marc Muench/David Muench Photography, Inc.)

In a near-surface environment, silica-rich rocks of granitic composition begin to melt around 750°C, whereas basaltic rocks must reach temperatures above 1000°C before melting commences.

What is the source of heat sufficient to melt this rock? One source is the heat liberated during the decay of radioactive elements found in the mantle and crust. Workers in underground mines have long recognized that temperatures increase as they descend to greater depths.

This gradual increase in temperature with depth is thought to contribute to magma production in two important ways. First, at deep-ocean trenches, slabs of cool oceanic lithosphere descend into the hot mantle. Here heat supplied by the surrounding rocks is thought to be sufficient to melt the subducting oceanic crust and produce basaltic magma. Second, once a hot magma body forms as just described, it could migrate upward to the base of the crust, and intrude granitic rocks. Because granitic rocks have melting temperatures well below those of basalt, heat derived from the hotter basaltic magma could melt the already warm crustal rocks. The volcanic activity that produced the vast aa flows in Yellowstone National Park is believed to have resulted from

such activity. Here basaltic magma from the mantle transported heat to the crust, where the melting of silica-rich rocks generated explosive outflows of pyroclastics.

Role of Pressure. If temperature were the only factor that determined whether or not rock melts, Earth would be a molten ball covered with a thin, solid outer shell. This, of course, is not the case. The reason is that pressure also increases with depth.

In general, an increase in the confining pressure increases a rock's melting temperature. Reducing confining pressure lowers a rock's melting temperature. Consequently, a drop in confining pressure can lower the melting temperature of rock sufficiently to trigger melting. This occurs when rock ascends, thereby moving into zones of lower pressures.

Partial Melting. One important difference exists between the melting of a substance that consists of a single compound, such as ice, and the melting of igneous rocks, which are mixtures of several different minerals. Whereas ice melts at a definite temperature, most igneous rocks melt over a temperature range of a few hundred degrees. (Recall Bowen's Reaction Series in Chapter 2.) As a rock is heated, the first melt to form will contain a higher percentage of the low-melting-temperature minerals than the original rock. Should melting continue, the composition of the melt will steadily approach the overall composition of the rock from which it is derived. Most often, however, melting is not complete. This process, known as **partial melting**, produces most, if not all, magma.

Thus, an important consequence of partial melting is the production of a melt with a higher silica content than the parent rock. Recall that basaltic rocks have a relatively low silica content and that granitic rocks have a much higher silica content. Consequently, magmas generated by partial melting are more granitic than the parent material from which they formed. As we shall see, this will help us understand the global distribution of the different types of volcanic activity.

Distribution of Igneous Activity

Most of Earth's more than 600 active volcanoes are near convergent plate margins. Further, extensive volcanic activity occurs along spreading centers of the oceanic ridge system, hidden from easy view by the world ocean. In this section we will examine three zones of volcanic activity and relate all of them to global tectonic activity. These active areas are along the oceanic ridges, adjacent to ocean trenches, and within the plates themselves (Figure 8.25).

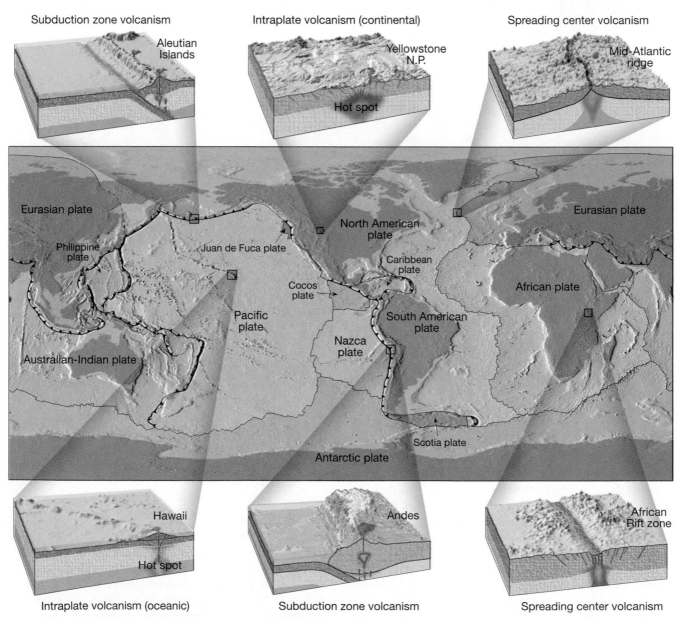

Figure 8.25

Three zones of volcanism. Two of these zones are plate boundaries, and the third includes areas within the plates.

Spreading Center Volcanism.   The greatest volume of volcanic rock is produced along the oceanic ridge system, where seafloor spreading is active (Figure 8.25). As the rigid lithosphere pulls apart, the pressure on the underlying rocks is lessened. This reduced pressure, in turn, lowers the melting temperature of the mantle rocks. Partial melting of these rocks (primarily peridotite) generates large quantities of basaltic magma that moves upward to fill the newly formed cracks.

Some of the molten basalt reaches the ocean floor, where it produces extensive lava flows or occasionally grows into a volcanic pile. Sometimes this activity produces a volcanic cone that rises above sea level, as the island of Surtsey did near the coast of Iceland in 1963 (Figure 8.26). Numerous submerged volcanic cones also dot the flanks of the ridge system and the adjacent deep-ocean floor. Many of these formed along the ridge crests and were moved away as new oceanic crust was created by the process of seafloor spreading.

Subduction Zone Volcanism.   Recall that ocean trenches are sites where slabs of oceanic crust are bent and move downward into the upper mantle (Figure 8.25). When the descending oceanic plate reaches a depth of about 100 kilometers (60 miles), partial melting of the water-rich ocean crust and the overlying mantle rocks takes place. The partial melting of these materials is thought to produce basaltic and andesitic magmas.

After a sufficient quantity of magma has accumulated, the molten material slowly migrates upward toward the surface because it is less dense than the surrounding rock. When subduction zone volcanism occurs in the ocean, a chain of volcanoes called an *island arc* is produced. Examples are numerous in the Pacific and include the Aleutian Islands (Figure 8.25), Tonga Islands, and Mariana Islands.

When subduction occurs beneath continental crust, the magma generated is often altered before it solidifies. Assimilation of crustal fragments into the ascending magma body can create a melt with an andesitic to granitic composition. The volcanoes of the Andes Mountains, from which andesite gets its name, are examples of this mechanism at work (Figure 8.25).

Many volcanoes border the Pacific Basin. Because of this pattern, the region often is called the *Ring of Fire*. Volcanism here is associated with subduction and partial melting of the Pacific sea floor. As oceanic plates sink, they carry with them sediments and oceanic crust containing abundant water. Because water reduces the melting point of rock, it aids the melting process.

The presence of water contributes to the high gas content (water vapor) and explosive nature of volcanoes that make up the *Ring of Fire* (see Box 8.3.) The volcanoes of the Cascade Range in the northwestern United States, including Mount St. Helens, Mount Rainier, and Mount Shasta, are all of this type (Figure 8.27).

Intraplate Volcanism.   The processes that trigger volcanic activity within a rigid plate are difficult to establish. Consider this: activity in the Yellowstone region produced silica-rich lava, pumice, and ash flows, whereas extensive basaltic flows have formed the Columbia Plateau. Yet, these rocks of greatly varying compositions actually overlie one another in several locations.

Because basaltic volcanism occurs on the continents as well as within the ocean basins, it may have a common source. Most likely the source is the partial melting of mantle rocks. One proposal suggests that some intraplate basaltic magma comes from rising plumes of hot mantle material. These hot plumes, which may extend to the core-mantle boundary, produce **hot spots**, volcanic regions a few hundred kilometers across. Most hot spots

Figure 8.26
The new volcanic island of Surtsey emerged from the ocean just south of Iceland in 1963. (Photo by P. Vauthey/Sygma)

appear to have persisted for a few tens of millions of years. A hot spot is believed to be located beneath the island of Hawaii and one may have produced the large outpourings of lava that compose the Columbia Plateau.

Generally, lavas and ash of granitic composition are extruded from vents located inland from the continental margins. This suggests that remelting of the continental crust may form these silica-rich magmas. But what mechanism causes large quantities of continental material to melt? One proposal suggests that a thick segment of continental crust occasionally becomes situated over a rising plume of hot mantle material. Rather than producing vast outpourings of basaltic lava, as oc-

curs at oceanic sites such as Hawaii, magma from the rising plume is emplaced at depth. Here the incorporation and melting of surrounding rock forms a secondary, silica-rich magma, which slowly migrates upward. Continued hot spot activity supplies heat to the rising mass, thereby aiding its ascent. This may explain the activity in the Yellowstone region (Figure 8.25).

Although the theory of plate tectonics has answered many questions that plagued volcanologists for decades, new questions have arisen: Why does seafloor spreading occur in some areas and not in others? How do hot spots originate? These and other questions are the subject of continuing scientific research.

Box 8.3

Volcanic Activity and Climate

The idea that explosive volcanic eruptions may cause changes in Earth's climate was first proposed many years ago and is still regarded as a plausible explanation for some aspects of climatic variability. Explosive eruptions emit huge quantities of gases and fine-grained debris into the atmosphere. The greatest eruptions are sufficiently powerful to inject material high into the atmosphere, where it spreads around the globe and remains for many months or even years. The basic premise is that this suspended volcanic material will filter out a portion of the incoming solar radiation, which, in turn, will lower temperatures in the lower atmosphere. Two hundred years ago, Benjamin Franklin used this same idea to argue that material from the eruption of a large Icelandic volcano could have reflected sunlight back to space and therefore could have been responsible for unusually cold weather during the winter of 1783–1784.

Perhaps the most notable cool period linked to a volcanic event is the "year without a summer" that followed the 1815 eruption of Tambora in Indonesia. (see Box 8.1). Similar, although apparently less dramatic, effects were associated with other explosive volcanoes, including Krakatoa in 1883 and Mount Agung in 1963. In recent years, three major volcanic events have provided considerable data and insight regarding the impact

of volcanoes on global temperatures. The eruptions of Washington State's Mount St. Helens in 1980, the Mexican volcano El Chichón in 1982, and the Philippine's Mount Pinatubo in 1991 have given scientists an opportunity to study the atmospheric effects of volcanic eruptions with the aid of more sophisticated technology than was available in the past. Satellite images and remote sensing instruments allowed scientists to monitor closely the effects of the clouds of gases and ash that these volcanoes emitted.

When Mount St. Helens erupted, there was almost immediate speculation about the possible effects of this event on our climate. Can such eruptions cause our climate to change? There is no doubt that the large quantity of volcanic ash emitted by the explosive eruption had significant local and regional effects for a short period. Still, studies indicate that any longer-term lowering of hemispheric temperatures was negligible. That is, the cooling was so slight, probably less than 0.1°C, that it could not be distinguished from other natural temperature fluctuations.

Two years of monitoring and studies following the El Chichón eruption indicated that its cooling effect on global mean temperatures was on the order of 0.3°C to 0.5°C. Because the eruption of El Chichón was less explosive than the Mount St. Helens blast, why did it have a greater impact on global temperatures? The reason was that the material emitted by Mount St. Helens was largely fine ash that settled out in a relatively short time. El Chichón, on the other hand, emitted far greater quanti-

ties of sulfur-rich gases (an estimated 40 times more) than did Mount St. Helens. Theses gases combined with water vapor in the stratosphere to produce a dense cloud of tiny sulfuric acid droplets. Such clouds take several years to settle out completely and are capable of decreasing the troposphere's mean temperature because the droplets both absorb solar radiation and scatter it back to space. It now appears that volcanic clouds that remain in the stratosphere for a year or more are composed largely of sulfuric acid droplets and not of dust as once thought. Thus, the volume of the fine debris emitted during an explosive event is not the best criterion for predicting the global atmospheric effects of an eruption.

The Philippines volcano, Mt. Pinatubo, erupted explosively in June 1991, injecting 25 to 30 million tons of sulfur dioxide into the stratosphere. The event provided scientists with an opportunity to study the climatic impact of a major volcanic eruption using NASA's spaceborne Earth Radiation Budget Experiment. During the next year, the haze of tiny particles acted to lower global temperatures by 0.5°C.

It may be true that the impact on global temperatures of an eruption like that of El Chichón is relatively minor, but many scientists agree that the cooling produced could alter the general pattern of atmospheric circulation for a limited period. Such a change, in turn, could have an effect on the weather in some regions. Predicting or even identifying specific regional effects still presents a considerable challenge to scientists, however.

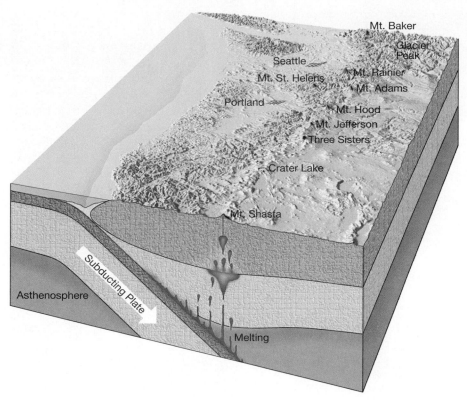

Figure 8.27
Locations of several of the larger composite cones that comprise the Cascade Range.

Review Questions

1. What triggered the May 18, 1980, eruption of Mount St. Helens?

2. Name five volcanoes in the western United States that geologists believe will erupt again.

3. What is the difference between magma and lava?

4. What three factors determine the nature of a volcanic eruption? What role does each play?

5. Why is a volcano fed by highly viscous magma likely to be a greater threat than a volcano supplied with very fluid magma?

6. Describe pahoehoe and aa lava.

7. List the main gases released during a volcanic eruption.

8. Analysis of samples taken during Hawaiian eruptions on the island of Hawaii indicate that _____ was the most abundant gas released.

9. Describe each type of pyroclastic material.

10. Compare and contrast the main types of volcanoes (size, shape, eruptive style, and so forth).

11. Name one example of each of the three types of volcanoes.

12. Compare the formation of Hawaii with that of Parícutin.

13. How is a caldera different from a crater?

14. Describe the formation of Crater Lake. Compare it to the caldera formed during the eruption of Kilauea.

15. Briefly describe the mechanism by which explosive volcanic eruptions are thought to influence Earth's climate.

16. What is Ship Rock, New Mexico, and how did it form?

17. Describe each of the four intrusive features discussed in the text.

18. Why might a laccolith be detected at Earth's surface before being exposed by erosion?

19. What is the largest of all intrusive igneous bodies? Is it tabular or massive? Concordant or discordant (see Figure 8.21)?

20. Explain how most magma is thought to originate.

21. Spreading center volcanism is associated with which rock type? What causes rocks to melt in regions of spreading center volcanism?

22. What is the *Ring of Fire*?

23. Are volcanic eruptions in the *Ring of Fire* generally quiet or violent? Name a volcano that would support your answer.

24. The Hawaiian Islands and Yellowstone are thought to be associated with which of the three zones of volcanism?

25. Volcanic islands in the deep ocean are composed primarily of what igneous rock type?

26. Explain why andesitic and granitic rocks are confined to the continents and oceanic margins but are absent from deep-ocean basins.

Key Terms

aa flow (p. 217)
batholith (p. 229)
caldera (p. 218)
cinder cone (p. 221)
columnar joints (p. 228)
composite cone (stratovolcano) (p. 222)

crater (p. 218)
dike (p. 228)
fissure eruption (p. 227)
flood basalt (p. 227)
hot spot (p. 233)
laccolith (p. 229)
lahar (p. 224)
nuée ardente (p. 223)
pahoehoe flow (p. 217)

partial melting (p. 232)
pyroclastic flow (p. 227)
pyroclastics (p. 217)
shield volcano (p. 219)
sill (p. 228)
vent (p. 218)
viscosity (p. 215)
volcanic neck (p. 226)
volcano (p. 218)

CHAPTER 9
Mountain Building

Mount Williamson, East Face of the Sierra Nevada, California. (Photo by Carr Clifton)

Mountains provide some of the most spectacular scenery on our planet. This splendor has been captured by poets, painters, and songwriters alike. Geologists believe that at some time all continental regions were mountainous masses and have concluded that the continents grow by the addition of mountains to their flanks. Consequently, by unraveling the secrets of mountain formation, geologists will have taken a major step in understanding the evolution of Earth. If continents do indeed grow by adding mountains to their flanks, how do geologists explain the existence of mountains (the Urals, for example) that are located in the interior of a landmass? In order to answer this and other related questions, this chapter attempts to piece together the sequence of events that is believed to generate these lofty structures.

Mountains often are spectacular features that rise several hundred meters or more above the surrounding terrain (Figure 9.1). Some occur as isolated masses; the volcanic cone Kilimanjaro, for example, stands almost 6000 meters (20,000 feet) above sea level, overlooking the grasslands of East Africa. Other peaks are parts of extensive mountain belts, such as the American Cordillera, which runs continuously from the tip of South America through Alaska, including the Andes and the Rocky Mountains. Chains such as the Himalayas consist of youthful, towering peaks that are still rising, whereas others, including the Appalachian Mountains in the eastern United States, are much older and have been eroded much lower than their original lofty height.

The name for the processes that collectively produce a mountain system is **orogenesis**, from the Greek *oros* ("mountain") and *genesis* ("to come into being"). Mountain systems show evidence of enormous forces that have bent (folded), broken (faulted), and generally deformed large sections of Earth's crust.

Figure 9.1
Mount Patterson in Alberta's Banff National Park is part of the Canadian Rockies. (Photo by Carr Clifton)

The first encompassing explanation of orogenesis came a little over three decades ago as part of the plate tectonics theory. The idea of crustal plates in slow motion, sometimes converging with great force, has opened many new and exciting avenues to geologists. Before examining mountain building according to the plate tectonics model, however, it will be advantageous to view some basic principles: the processes of crustal uplifting and rock deformation.

Crustal Uplift

The fossilized shells of marine invertebrates are often found at high elevations in mountains, an indication that the sedimentary rock in the mountain was once below sea level. This is convincing evidence that some dramatic changes have occurred between the time these animals died and the discovery of their fossilized remains!

Evidence for crustal uplift such as this is common in rocks, which form the "geologic record." Many examples exist along the coast of the western United States. When a coastal area remains undisturbed for an extended period, wave action cuts a gently sloping platform. In parts of California, ancient wave-cut platforms can now be found as terraces, hundreds of meters above sea level (Figure 9.2). Each terrace represents a period when that area was at sea level. Such evidence of crustal uplift is easy to find, but unfortunately, the *reasons* for uplift are not so easily determined.

Isostasy

We know that the force of gravity must play an important role in determining the elevation of the land. In particular, the less dense crust is believed to float on top of the denser and more easily deformed rocks of the mantle. This concept—a floating crust in gravitational balance—is called **isostasy**.

Perhaps the easiest way to grasp the concept of isostasy is to envision a series of wooden blocks of different heights floating in water, as shown in Figure 9.3. Note that the taller wooden blocks float higher in water than do the shorter blocks. In a similar manner, mountain belts stand higher above the surface and have "roots" that extend deeper into the supporting material below (Figure 9.4A). The average thickness of the continental crust is about 35 kilometers, but crustal thicknesses for some mountain chains exceeds 70 kilometers. Thus, mountains exist where unusually thick blocks of light crustal material are supported from below by the dense rocks of the mantle.

Carrying this idea one step further, the crust beneath the oceans must be thinner than that of the continents, because its elevation is lower. Not only is this true, but oceanic rocks also have a greater density than continental rocks, another factor contributing to their lower position.

If the concept of isostasy is correctly applied to mountains, we should expect that adding weight to the crust makes it respond by subsiding, and when that weight is removed, crustal uplifting occurs. (Visualize what happens to a ship as cargo is being loaded—the ship sinks deeper—and unloaded—the ship rises.)

Figure 9.2
Former wave-cut platforms now exist as a series of elevated terraces on the west side of San Clemente Island off the southern California coast. Once at sea level, the highest terraces are now about 400 meters (1320 feet) above it. (Photo by John S. Shelton)

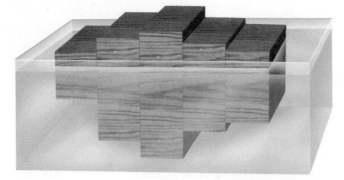

Figure 9.3

This drawing illustrates how wooden blocks of different thicknesses float in water. In a similar manner, thick sections of crustal material float higher than thinner crustal slabs.

Isostatic Adjustment

Evidence for this type of movement exists, strongly supporting the concept of **isostatic adjustment**. A classic example is provided by Ice Age glaciers. When continental ice sheets occupied portions of North America during the Pleistocene epoch, the added weight of the 3-kilometer-thick masses of ice caused downwarping of Earth's crust by hundreds of meters. In the 8000 to 10,000 years since the last ice sheets vanished, uplifting of as much as 330 meters has occurred in the Hudson Bay region of Canada, where the thickest ice had accumulated.

As the foregoing examples illustrate, isostatic adjustment can account for considerable *vertical* crustal movement. Thus, we can understand why, as erosion slowly lowers the summits of mountains, the crust grad-

Figure 9.4

This sequence illustrates how the combined effect of erosion and isostatic adjustment results in a thinning of the crust in mountainous regions. **A.** When mountains are young, the continental crust is thickest. **B.** As erosion lowers the mountains, the crust rises in response to the reduced load. **C.** Erosion and uplift continue until the mountains reach "normal" crustal thickness.

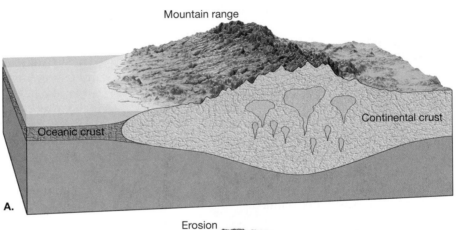

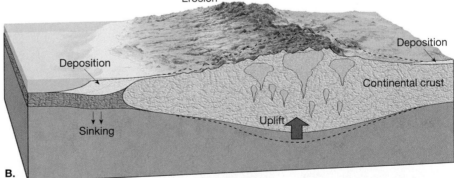

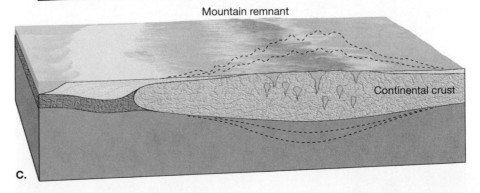

ually rises in response to the reduced load. The processes of uplifting and erosion will continue together until the mountain block reaches "normal" crustal thickness (Figure 9.4). When this occurs, the mountains will have been eroded to near sea level and the deeply buried portions of the mountains will become exposed at the surface. In addition, as the mountains succumb to erosion, sediment derived from the shrinking masses is deposited on the adjacent continental margin, causing the margin to subside (Figure 9.4B, C).

To summarize, we have learned that Earth's major mountain chains consist of unusually thick sections of deformed crustal material that have been elevated above the surrounding terrain. The support for these massive features comes from the buoyancy of light crustal material supported by denser mantle rocks. As erosion lowers the peaks, isostatic adjustment gradually raises the mountains in response. Eventually, the deepest portions of the mountains rise to the shallower depths of the surrounding crust. The question still to be answered is: How do these thickened sections of Earth's crust come into existence in the first place? We attempt to answer this question later in the chapter.

Rock Deformation

When rocks are subjected to stresses greater than their own strength, they begin to deform, usually by folding or fracturing (Figure 9.5). It is easy to visualize how individual rocks break. But how are large rock units, many meters thick and many kilometers long, bent into

intricate folds without being appreciably broken during the process? In an attempt to answer this, geologists used the laboratory. Here rocks were subjected to stresses under conditions that simulated those existing at various depths within the crust.

Although each rock type deforms somewhat differently, the general characteristics of rock deformation were determined from these experiments. Geologists discovered that when stress is applied, rocks first respond by deforming elastically. Changes resulting from *elastic deformation* are reversible; that is, like a rubber band, the rock will return to nearly its original size and shape when the stress is removed.

However, once the *elastic limit* is surpassed, rocks either deform plastically or fracture. *Plastic deformation* results in permanent changes; that is, the size and shape of a rock unit are altered through folding and flowing. Laboratory experiments confirmed that, at the high temperatures and pressures that are typical deep in Earth, most rocks deform plastically once their elastic limit is surpassed.

Rocks tested under surface conditions also deform elastically, but once they exceed their elastic limit, most behave like a brittle solid and fracture. Recall that the energy for most earthquakes comes from stored elastic energy that is released as rock fractures and snaps back to its original shape.

One factor that researchers cannot duplicate in the laboratory is *geologic time*. We know that if stress is applied quickly, as with a hammer, rocks tend to fracture. On the other hand, identical rocks may deform plastically if stress is applied over an extended period. For

Figure 9.5
Deformed Paleozoic strata west of Teakettle Junction in Death Valley, California. (Photo by Michael Collier)

example, marble benches have been known to sag under their own weight over a period of a hundred years or so. In nature, even modest forces applied over long periods surely play an important role in the deformation of rock.

Folds

During mountain building, flat-lying sedimentary and volcanic rocks are often bent into a series of wavelike undulations called **folds** (Figure 9.6). Folds in sedimentary strata are much like those that would form if you pushed on one edge of a carpet, rumpling it. In nature, folds vary widely in size and shape. Some folds are broad features in which rock units hundreds of meters thick are slightly warped. Others are very tight, microscopic structures found in metamorphic rocks. Size differences notwithstanding, most folds are the result of *compressional stresses* that result in the shortening and thickening of the crust. (Occasionally, folds are found singly, but most often they occur as a series of undulations.)

The two most common types of folds are anticlines and synclines (Figure 9.6). An **anticline** is most commonly formed by the upfolding, or arching, of rock layers. Anticlines are sometimes spectacularly displayed where highways have been cut through deformed strata. Often found in association with anticlines are downfolds, or troughs, called **synclines**. Anticlines and synclines vary in symmetry. They may be *symmetrical*, *asymmetrical*, or, if one limb has been tilted beyond the vertical, *overturned* (Figure 9.6).

Folds do not continue on forever; rather, their ends die out, much like wrinkles in cloth. These folds are said to be *plunging*, since the axis of the fold plunges to the ground (Figure 9.7). Figure 9.8A shows an example of a plunging anticline and the pattern produced when erosion removes the upper layers of this structure and ex-

Figure 9.7

Sheep Mountain, a doubly plunging anticline. Note that erosion has cut the flanking sedimentary beds into low ridges that make a "V" pointing in the direction of plunge. (Photo by John S. Shelton)

poses its interior. Note that the outcrop pattern of an anticline points in the direction it is plunging, while the opposite is true for synclines.

A good example of all this—the topography that results when erosional forces wear down folded sedimentary strata for millions of years—is the Valley and Ridge Province of the Appalachians. Here, resistant sandstone beds remain as imposing ridges, separated by valleys that have been cut into more easily eroded shale or limestone beds.

Although most folds are caused by horizontal compression, which squeezes strata, some are a consequence of *vertical* displacement. When *upwarping* produces a circular or somewhat elongated structure, the feature is

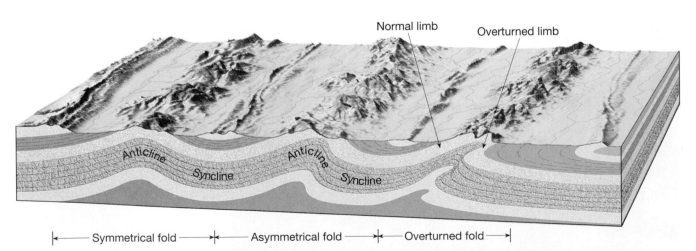

Normal limb   Overturned limb

Anticline   Syncline   Anticline   Syncline

|← Symmetrical fold →|← Asymmetrical fold →|← Overturned fold →|

Figure 9.6

Block diagram of principal types of folded strata. The upfolded or arched structures are anticlines. The downfolds or troughs are synclines. Notice that the limb of an anticline is also the limb of the adjacent syncline.

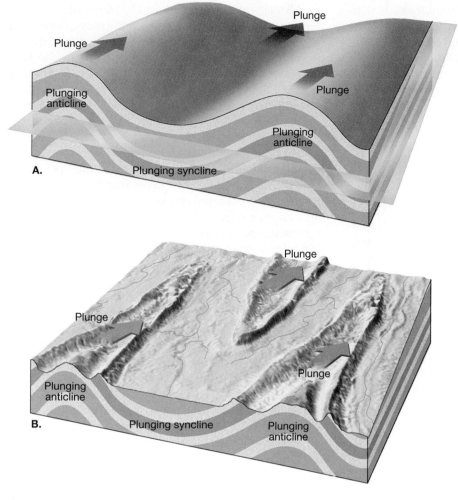

Figure 9.8
Plunging folds. **A.** Idealized view of plunging folds in which a horizontal surface has been added. **B.** View of plunging folds as they might appear after extensive erosion. Notice that in a plunging anticline the outcrop pattern "points" in the direction of the plunge, while the opposite is true of plunging synclines.

called a **dome** (Figure 9.9A). The Black Hills of western South Dakota are a domal structure. Erosion has stripped away the upwarped sedimentary beds, exposing older igneous and metamorphic rocks in the center (see Figure 9.18).

Structures of similar shape, except *downwarped*, are termed **basins** (Figure 9.9B). Several basins exist in the United States. The basins of Michigan and Illinois, each roughly the size of its state, have very gently sloping beds, similar to vast saucers.

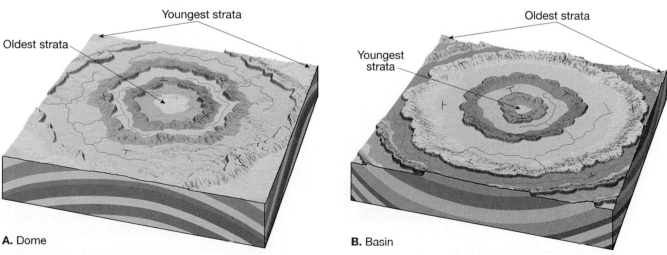

A. Dome

B. Basin

Figure 9.9
Gentle upwarping and downwarping of crustal rocks produce domes (**A**) and basins (**B**). Erosion of these structures results in an outcrop pattern that is roughly circular or elongated.

Because large basins contain sedimentary beds sloping at such low angles, they are usually identified by the age of the rocks composing them. The youngest rocks are found near the center and the oldest rocks are at the flanks. This is just the opposite order of a domal structure such as the Black Hills, where the oldest rocks form the core.

Faults and Joints

Faults. **Faults** are fractures in the crust along which appreciable displacement has occurred, on a scale from centimeters to kilometers. Occasionally, small faults can be recognized in road cuts where sedimentary beds have been offset a few meters, as shown in Figure 9.10. Faults of this scale usually occur as single discrete breaks. By contrast, large faults, like the infamous San Andreas fault in California, have displacements of hundreds of kilometers. These large faults consist of multiple interconnecting fault surfaces, forming *fault zones* that can be several kilometers wide. Often they are easier to identify from high-altitude images than at ground level (see Figure 9.A in Box 9.1).

Figure 9.10
Faulting caused the vertical displacement of these sedimentary beds in southern Nevada. Arrows show relative motion of rock units. (Photo by E. J. Tarbuck)

Movement along a fault can be horizontal, vertical, or a combination. Faults in which the movement is primarily *vertical* are called **dip-slip faults**, because the displacement is along the inclination (*dip*) of the fault plane. Movement along dip-slip faults can be either up or down the fault plane—that is, rocks on one side of the fault can be either pushed up or dropped down with respect to the other side. Thus, two types of dip-slip faults are recognized. To distinguish between the two, we call the rock that is higher than the fault surface the *hanging wall*, and the rock below the fault surface the *footwall*. This nomenclature arose from prospectors and miners who excavated shafts along fault zones because such zones are frequently sites of ore deposition. During these operations, the miners would walk on the rocks below the fault plane (hence, footwall), and hang their lanterns on the rocks above (hence, hanging wall).

Dip-slip faults are classified as **normal faults** when the hanging wall moves downward relative to the footwall (Figure 9.11A). Conversely, **reverse faults** occur when the hanging wall moves upward relative to the footwall (Figure 9.11B). Reverse faults having a very low angle are also called **thrust faults** (Figure 9.11C). In mountainous regions such as the Alps and the Appalachians, thrust faults have displaced rock as far as 50 kilometers over adjacent strata. Thrust faults of this type result from strong compressional stresses.

Faults in which the dominant displacement is along the *strike*, or trend, of the fault are called **strike-slip faults** (Figure 9.11D).Many large strike-slip faults are associated with plate boundaries and are called *transform faults*. Transform faults have nearly vertical dips and connect large structures, such as segments of an ocean ridge. The San Andreas fault in California is a well-known transform fault in which the displacement is several hundred kilometers (see Box 9.1). When faults have both vertical and horizontal movement, they are called **oblique-slip faults**.

Tensional and Compressional Forces. Fault motion assists the geologist in determining the direction of forces at work within Earth. Normal faults indicate *tensional* stresses that pull the crust apart (Figure 9.11A). This "pulling apart" can be accomplished either by uplifting that causes the surface to stretch and break, or by horizontal forces that actually rip the crust apart.

Normal faulting is known to occur at spreading centers, where plates are diverging. Here a central block called a **graben**, bounded by normal faults, drops as the plates separate (Figure 9.12). These grabens produce an elongated valley bounded by upfaulted structures called **horsts**. The Great Rift Valley of East Africa consists of several large grabens, above which tilted horsts produce a linear mountainous topography (see Figure 7.11). This valley, nearly 6000 kilometers (3700 miles) long, contains

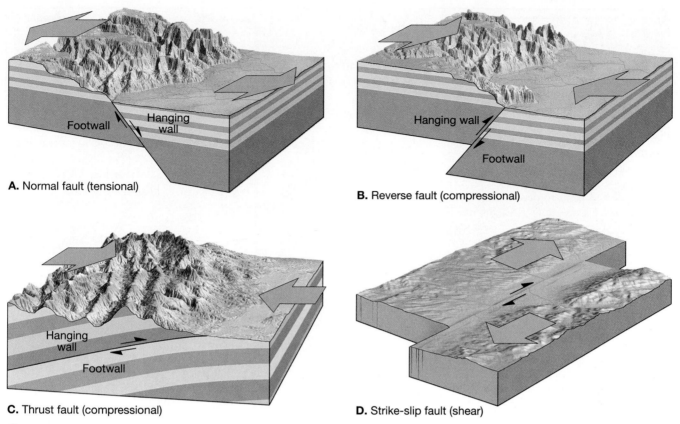

Figure 9.11
Block diagrams of four types of faults. **A.** Normal fault. **B.** Reverse fault. **C.** Thrust fault. **D.** Strike-slip fault.

A. Normal fault (tensional)

B. Reverse fault (compressional)

C. Thrust fault (compressional)

D. Strike-slip fault (shear)

the discovery sites of some of the earliest human fossils. Other rift valleys include the Rhine Valley in Germany and the valley of the Dead Sea in the Middle East.

In reverse and thrust faulting, the sections of crust are displaced toward one another, so geologists conclude that *compressional forces* are at work (Figure 9.11B, C). The primary regions of such activity are convergent zones where plates are colliding. Compressional forces generally produce folds as well as faults and result in a general thickening and shortening of the crust. As you shall see, most orogenesis occurs in compressional situations.

Joints. Among the most common rock structures are fractures called **joints**. Unlike faults, joints are fractures along which *no appreciable displacement* has occurred. Although some joints have random orientation, most occur in roughly parallel groups.

We have already considered two types of joints. Earlier we learned that *columnar joints* form when igneous rocks cool and develop shrinkage fractures, producing elongated, pillarlike columns (see Figure 8.23). Also recall that *sheeting* produces a pattern of gently curved joints that develop more or less parallel to the surface of large exposed igneous bodies such as batholiths. Here

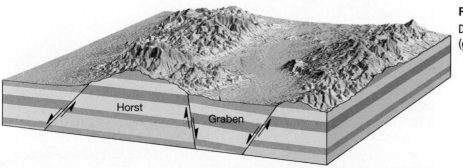

Horst

Graben

Figure 9.12
Diagrammatic sketch of downfaulted block (graben) and upfaulted block (horst).

Box 9.1

The San Andreas Fault System

The San Andreas, the best-known and largest fault system in North America, first attracted wide attention after the great 1906 San Francisco earthquake and fire. Following this devastating event, geologic studies demonstrated that a displacement of as much as 5 meters along the fault had been responsible for the earthquake. It is now known that this dramatic event is just one of many thousands of earthquakes that have resulted from repeated movements along the San Andreas throughout its 29-million-year history.

Where is the San Andreas fault system located? As shown in Figure 9.A, it trends in a northwesterly direction for nearly 1300 kilometers (780 miles) through much of western California. At its southern end, the San Andreas connects with a spreading center located in the Gulf of California. In the north, the fault enters the Pacific Ocean at Point Arena, where it is thought to continue its northwesterly trend, eventually joining the Mendocino fracture zone. In the central section, the San Andreas is relatively simple and straight. However, at its two extremities, several branches spread from the main trace, so that in some areas the fault zone exceeds 100 kilometers (60 miles) in width.

Over much of its extent, a linear trough reveals the presence of the San Andreas fault. When the system is viewed from the air, linear scars, offset stream channels, and elongated ponds mark the trace in a striking manner. On the ground, however, surface expressions of the faults are much more difficult to detect. Some of the most distinctive landforms include long, straight escarpments, narrow ridges, and sag ponds formed by settling of blocks within the fault zone. Further, many stream channels characteristically bend sharply to the right where they cross the fault (Figure 9.B).

With the advent of the theory of plate tectonics, geologists began to realize the significance of this great fault system. The San Andreas fault is a transform boundary separating two crustal plates that move very slowly. The Pacific plate, located to the west, moves northwestward relative to the North American plate, causing earthquakes along the fault (Table 9.A).

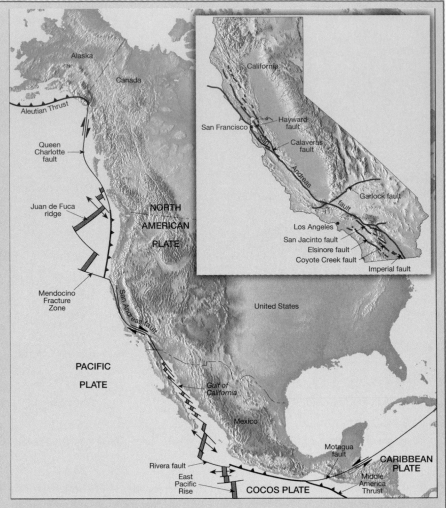

Figure 9.A

Map showing the extent of the San Andreas fault system. Inset shows only a few of the many splinter faults that are part of this great fault system.

The San Andreas is undoubtedly the most studied of any fault system in the world. Although many questions remain unanswered, geologists have learned that each fault segment exhibits somewhat different behavior. Some portions of the

(continues)

Figure 9.B

Aerial view showing the displacement of stream channels along the trace of the San Andreas fault. (Photo by Michael Collier)

San Andreas exhibit a slow creep with little noticeable seismic activity. Other segments regularly slip, producing small earthquakes, while still other segments seem to store elastic energy for hundreds of years and rupture in great earthquakes. This knowledge is useful when assigning earthquake hazard potential to a given segment of the fault zone.

Because of the great length and complexity of the San Andreas fault, it is more appropriately referred to as a "fault system." This major fault system consists primarily of the San Andreas fault and several major branches including the Hayward and Calaveras faults of central California and the San Jacinto and Elsinore faults of southern California (see Figure 9.A). These major segments, plus a vast number of smaller faults that include the Imperial fault, San Fernando fault, and the Santa Monica fault, collectively accommodate the relative motion between the North American and Pacific plates.

Since the great San Francisco earthquake of 1906, when as much as 5 meters of displacement occurred, geologists have attempted to establish the cumulative displacement along this fault over its 29-million-year history. By matching rock units across the fault, geologists have determined that the total accumulated displacement from earthquakes and creep exceeds 560 kilometers (340 miles).

Table 9.A Major earthquakes on the San Andreas Fault system.

Date	Location	Magnitude	Remarks
1812	Wrightwood, CA	7	Church at San Juan Capistrano collapsed, killing 40 worshipers.
1812	Santa Barbara channel	7	Churches and other buildings wrecked in and around Santa Barbara.
1838	San Francisco peninsula	7	At one time thought to have been comparable to the great earthquake of 1906.
1857	Fort Tejon, CA	8.25	One of the greatest U.S. earthquakes. Occurred near Los Angeles, then a city of 4000.
1868	Hayward, CA	7	Rupture of the Hayward fault caused extensive damage in San Francisco Bay area.
1906	San Francisco, CA	8.25	The great San Francisco earthquake. As much as 80 percent of the damage caused by fire.
1940	Imperial Valley	7.1	Displacement on the newly discovered Imperial fault.
1952	Kern County	7.7	Rupture of the White Wolf fault. Largest earthquake in California since 1906. Sixty million dollars in damages and 12 people killed.
1971	San Fernando Valley	6.5	One-half billion dollars in damages and 58 lives claimed.
1989	Santa Cruz Mountains	7.1	Loma Prieta earthquake. Six billion dollars in damages, 62 lives lost, and 3757 people injured.
1994	Northridge (Los Angeles area)	6.9	Over 15 billion dollars in damages, 51 lives lost, and over 5000 injured.

the jointing is thought to result from the gradual expansion that occurs as erosion removes the pressure of the overlying load (see Figure 3.4).

In contrast to the situations just described, most joints are produced when rocks are deformed by the stresses associated with crustal movements during mountain building. Extensive joint patterns can also develop in response to relatively subtle and often barely perceptible regional upwarping and downwarping of the crust.

Many rock units are broken by two or even three sets of intersecting joints that divide the rock into numerous regularly shaped blocks. These joint sets often exert a strong influence on other geologic processes. For example, chemical weathering tends to concentrate along joints, and in many areas, groundwater movement and the resulting solution activity in soluble rocks is controlled by the joint pattern (Figure 9.13). Moreover, a system of joints can influence the path taken by a stream. The rectangular drainage pattern shown in Figure 4.18C is such a case.

Now that you have some knowledge of crustal uplift and rock deformation, we will examine different types of mountains and how they come to be.

Mountain Types

No two mountain ranges are exactly alike, but they can be classified according to their structural characteristics. Using this approach, four main categories result: (1) fault-block mountains, (2) upwarped mountains, (3) folded mountains (complex mountains), and (4) volcanic mountains.

Mountain ranges of the same structural type commonly occur in close proximity, forming a *mountain system*. For example, nearly the entire state of Nevada is composed of numerous elongated fault-block mountains that are separated by faulted basins (Figure 9.14). Further, within any mountainous belt, such as that portion of the American Cordillera in the western United States, mountain ranges representing each of the four groups can be found.

Figure 9.13
Chemical weathering is enhanced along joints in granitic rocks near the top of Lembert Dome, Yosemite National Park. (Photo by E. J. Tarbuck)

In addition to these basic varieties, some regions exhibit mountainous topography that was produced *without appreciable crustal deformation*. For example, plateaus (areas of high-standing rocks that are essentially horizontal) can become deeply dissected by erosion into rugged, mountainlike terrains. Although these highlands have the topographical expression of mountains, they lack the structures associated with orogenesis.

In the following sections we will examine three of the four basic mountain types: fault-block, upwarped, and, the most common, folded mountains. (Volcanic mountains are treated in detail in Chapter 5.)

Fault-Block Mountains

Most orogenesis occurs in compressional environments. Evidence for this is the predominance of large thrust faults and folded strata in mountainous areas. However, some mountain building is associated with tensional stresses (Figure 9.15). The mountains that form under such circumstances, termed **fault-block mountains**, are bounded on at least one side by normal faults of high to moderate angle.

Some fault-block mountains form in response to broad crustal uplifting, which causes elongation (stretching) and faulting. Such a situation is exemplified by the fault blocks that rise above the rift valleys of East Africa. Other fault-block mountains appear to develop where isolated blocks are tilted, or lifted nearly vertically, high above adjacent undeformed valleys.

The *Basin and Range Province*, a region that encompasses Nevada and portions of Utah, New Mexico, Arizona, and California, contains excellent examples of fault-block mountains. Here the crust has been broken into hundreds of blocks, giving rise to nearly parallel mountain ranges, averaging about 80 kilometers in length, which rise precipitously above the adjacent sediment-laden basins (Figure 9.15).

No single explanation for the development of the Basin and Range is universally accepted. It is clear, however, that important geological events preceded the period of stretching and faulting. During this earlier time span, a nearly horizontal oceanic plate was subducted eastward, beneath western North America (Figure 9.16A). This caused compressional stresses in the overlying plate, thickening the crust as far inland as the Rocky Mountains of Colorado.

Then, about 30 million years ago, the western portion of the North American plate began to override a spreading center in the Pacific basin. This attached the southwestern portion of the United States to the Pacific plate, which was moving northwesterly. This movement continues today and is most pronounced along the San Andreas fault, where portions of southern California and Mexico's Baja Peninsula have migrated several hundred kilometers toward the northwest.

In addition to creating movement along the San Andreas fault system, this change in the direction of plate motion also may have stretched and fractured the crust in the region now occupied by the Basin and Range Province, generating its fault-block topography.

Some geologists disagree with part of the preceding scenario. Instead, they suggest that extension in the Basin and Range was triggered when the horizontal oceanic slab that had been thrust under the continent began to sink (Figure 9.16B). Sinking of the oceanic slab produced an upwelling of magma from the asthenosphere. The buoyancy of the magma caused upwarping and tensional fracturing in the overlying crust. This event was associated with volcanism and east-west stretching of the crust.

Regardless of which explanation is correct, if either, there is general agreement that the fault-block mountains in this region resulted from tensional forces that stretched the crust by nearly 150 kilometers.

Other fault-block mountains in the United States are the Teton Range of Wyoming (see Figure 9.17) and the Sierra Nevada of California. Both are faulted along

Box 9.2

Oil Traps

Unlike the altered organic matter from which they formed, petroleum and natural gas are mobile. These fluids are gradually squeezed from the compacting, mud-rich layers where they originated and may migrate hundreds of kilometers into permeable beds such as sandstone, where openings between sediment grains are larger. The rock layers containing the oil and gas are saturated with water, and, since they are not as dense as water, oil and gas migrate upward through the water-filled pore spaces of the enclosing rocks. Unless something acts to halt this upward migration, the fluids will eventually reach the surface, at which point the volatile components will evaporate.

A geologic environment that allows for economically significant amounts of oil and gas to accumulate is termed an *oil trap*. Although, as we shall see, several geologic structures may act as oil traps, all have two basic conditions in common: a porous permeable *reservoir rock* that will yield petroleum and natural gas in sufficient quantities to make drilling worthwhile; and a *cap rock*, such as shale, that is virtually impermeable to oil and gas. The cap rock keeps the upwardly mobile oil and gas from escaping at the surface.

Figure 9.C illustrates some common oil and natural gas traps. One of the simplest traps is an *anticline*, an uparched series of sedimentary strata (Part A). As the strata are folded, the rising oil and gas collect at the apex of the fold. Because of its lower density, the natural gas collects above the oil. Both rest upon the denser water that saturates the reservoir rock. One of the world's largest oil fields, El Nala in Saudi Arabia, is the result of an anticlinal trap, as is the famous Teapot Dome in Wyoming. *Fault*

traps form when strata are displaced in such a manner as to bring a dipping reservoir rock opposite an impermeable bed, as shown in Part B. In this case the upward migration of the oil and gas is halted at the fault zone. In the Gulf coastal plain region of the United States important accumulations of oil occur in association with *salt domes*. In such areas which are characterized by thick accumulations of sedimentary strata, layers of rock salt occurring at great depth have been forced to rise in columns by the pressure of overlying strata. Since oil and gas migrate to the highest level possible, they accumulate in the upturned sandstone beds adjacent to the salt column (Part C). Yet another important geologic circumstance that may lead to significant accumulations of oil and gas is termed a *stratigraphic trap*. These oil-bearing structures result primarily from the original pattern of sedimentation rather than structural deformation. The stratigraphic trap illustrated in Part D of the figure exists because a sloping bed of sandstone thins to the point of disappearance.

When the lid created by the cap rock is punctured by drilling, the oil and natural gas, which are under pressure, migrate from the pore spaces of the reservoir rock to the drill hole. On rare occasions the fluid pressure may force oil up the drill hole to the surface. Usually, however, a pump is required to lift the oil out.

A drill hole is not the only means by which oil and gas can escape from a trap. Traps can be broken by natural forces as well. For example, Earth movements may create fractures that allow the hydrocarbon fluids to escape, or surface erosion may breach a trap with similar results. The older the rock strata, the greater the chance that deformation or erosion has affected a trap. Indeed, not all ages of rock yield oil and gas in the same proportions. The greatest production comes from the youngest rocks, those of Cenozoic age. Older Mesozoic rocks produce considerably less, followed by even smaller yields from the still older Paleozoic strata. There is virtually no oil produced from the most ancient rocks, those of Precambrian age.

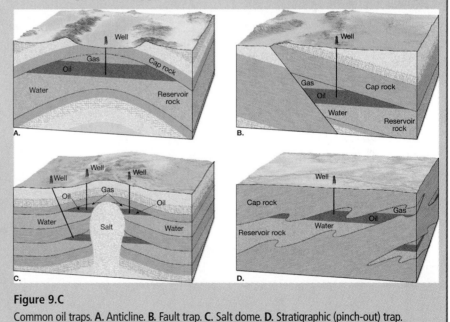

Figure 9.C

Common oil traps. **A.** Anticline. **B.** Fault trap. **C.** Salt dome. **D.** Stratigraphic (pinch-out) trap.

their eastern flanks, which were uplifted as the blocks tilted downward to the west. Looking west from Jackson Hole, Wyoming, and Owens Valley, California, the eastern fronts of these ranges (the Tetons and the Sierra Nevada, respectively) rise over 2 kilometers, making them two of the most precipitous mountain fronts in the United States (Figure 9.17).

Folded Mountains

The largest and most complex mountain systems are **folded mountains** (also called **complex mountains**). Although folding is often more conspicuous, faulting, metamorphism, and igneous activity are always present in varying degrees. Most major mountain belts, including the Alps,

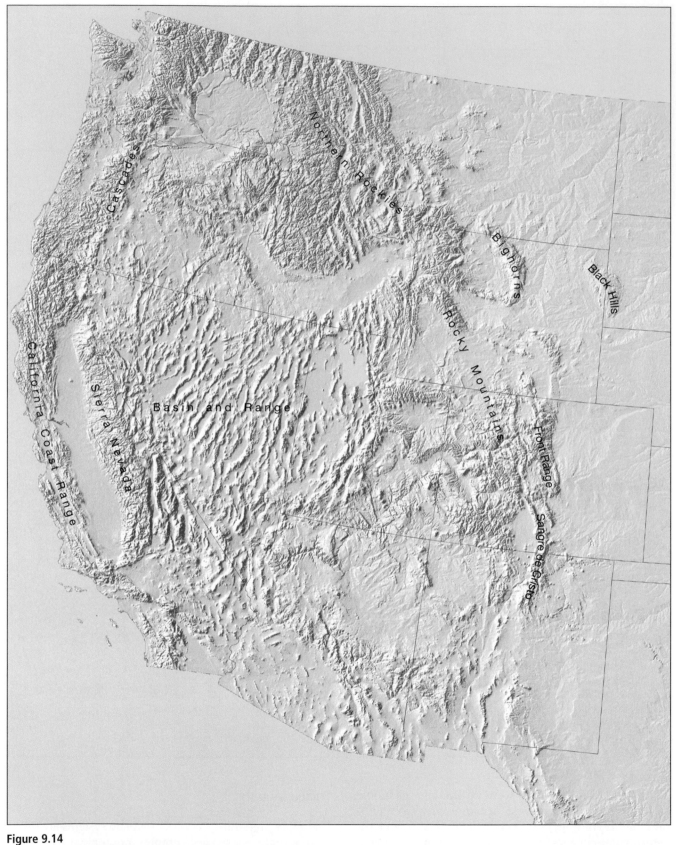

Figure 9.14
Map of major mountainous landforms of the western United States. (After Thelin and Pike, USGS)

Urals, Himalayas, Appalachians, and northern Rockies, are of this type (see Box 9.3). Folded mountains represent the world's major mountain systems, so the process of mountain building is usually described in terms of their formation. Thus the next section is devoted to the evolution of these majestic and always complex mountain systems.

Figure 9.15

Normal faulting in the Basin and Range Province. Here, tensional stresses have elongated and fractured the crust into numerous blocks. Movement along these fractures has tilted the blocks, producing parallel mountain ranges called fault-block mountains. (Photo by Michael Collier)

Figure 9.16

One proposal for the formation of the Basin and Range province. Top: Nearly horizontal subduction of an oceanic plate produced compressional stresses which generally thickened the crust in the Basin and Range. Bottom: Sinking of this oceanic slab allowed for the upwelling of hot material from the asthenosphere. The buoyancy of the warm material caused upwarping and tensional fracturing in the crust above. This event was associated with volcanism and east-west extension of the crust by nearly 150 kilometers.

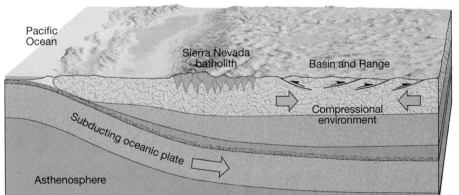

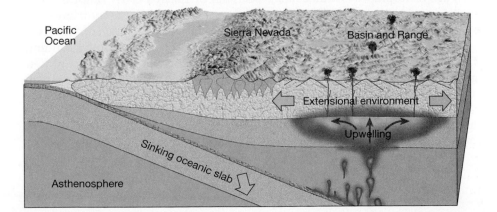

Figure 9.17
The Teton Range in Grand Teton National Park, Wyoming is an example of fault-block mountains. (Photo by Carr Clifton)

Upwarped Mountains

Upwarped mountains are caused by a broad arching of the crust or, in some instances, because of great vertical displacement along a high-angle fault. The Black Hills in western South Dakota and the Adirondack Mountains in upstate New York are examples of arching. They consist of older igneous and metamorphic bedrock that was once eroded flat and subsequently mantled with sediment. As these regions were upwarped, erosion removed the veneer of sedimentary strata, leaving a core of igneous and metamorphic rocks standing above the surrounding terrain (Figure 9.18).

Another example of upwarped mountains is that portion of the Rockies that extends from southern Montana through Colorado and into New Mexico.

Mountain Building

Mountains are spectacular features that often rise several hundred meters or more above the surrounding terrain. All mountain systems show evidence of enormous forces that have folded, faulted, and generally deformed large sections of Earth's crust (Figure 9.19). Although this folding and faulting have contributed to the majestic structure of mountains, much of the credit for their beauty must be given to erosion by running water and glacial ice, which sculpture these uplifted masses in an unending effort to lower them to sea level. **Mountain building**, however, refers to the processes that uplift the mountains.

Mountain building has operated during the recent geologic past in several locations around the world. These relatively young mountainous belts include: the

Box 9.3

The Rocky Mountains

The portion of the Rocky Mountains that extends from southern Montana to New Mexico was produced by a period of deformation known as the *Laramide Orogeny*. This event, which created some of the most picturesque scenery in the United States, peaked about 60 million years ago (Figure 9.D). The mountain ranges generated during the Laramide Orogeny include the Front Range of Colorado, the Sangre de Cristo of New Mexico and Colorado, and the Bighorns of Wyoming.

These mountains are structurally much different from the northern Rockies, which include the Canadian Rockies and those portions of the Rockies found in Idaho, western Wyoming, and western Montana. Whereas the latter ranges are folded mountains, composed of thick sequences of sedimentary rocks that were deformed by folding and low-angle thrust faulting, the middle and southern Rockies are upwarped mountains pushed almost vertically upward as part of broad upwarping of the crust.

In general, these mountains consist of ancient basement rocks overlain by relatively thin layers of younger strata. Since the time of deformation, much of the sedimentary cover has been eroded from the highest portions of the uplifted blocks, exposing their igneous and metamorphic cores. Examples include a number of granitic outcrops that project as steep summits, such as Pikes Peak and Longs Peak in Colorado's Front Range. In many areas, remnants of the sedimentary strata that once covered this region are visible as prominent angular ridges called *hogbacks* flanking the crystalline cores of the mountain ranges.

Although the Rockies have been extensively studied for over a century, there is still a good deal of debate regarding the mechanisms that led to uplift. According to the plate tectonics model, most mountain belts are produced along continental margins in association with convergent plate boundaries. Here, crustal buckling forces generate thick sequences of folded, faulted, and metamorphosed strata that are often intruded by massive igneous bodies.

However, this model does not fit the middle and southern Rocky Mountains, which consist of elongated, uplifted blocks of Precambrian basement rocks separated by sediment-filled basins.

One hypothesis for the formation of the middle and southern Rocky Mountains has been gaining widespread support. According to this proposal, a subducted plate of oceanic lithosphere remained nearly horizontal as it pushed eastward under North America as far inland as the Black Hills of South Dakota (see Figure 9.16). As the subducted slab scraped beneath North America, compressional stresses shortened and thickened the rocks in the lower crust. Furthermore, this event locally uplifted blocks of ancient basement rocks along high-angle faults to produce the Rockies and intermountain basins.

Thus, the Laramide Orogeny may be associated with a special type of convergent plate boundary. Whereas plate subduction along a typical Andean-type plate boundary occurs at a steep angle, the middle and southern Rocky Mountains may have been produced by the nearly horizontal subduction of an oceanic plate.

Figure 9.D
The spectacular Maroon Bells are part of the Colorado Rockies. (Photo by Larry Ulrich/Tony Stone Images)

American Cordillera, which runs along the western margin of the Americas from Cape Horn to Alaska (Rockies, Andes); the Alpine-Himalayan chain, which extends from the Mediterranean through Iran to northern India and into Indochina; and the mountainous terrains of the western Pacific, which include mature island arcs such as Japan, the Philippines, and Sumatra. Most of these young mountain belts have come into existence within the last 100 million years. Some, including the Himalayas, began their growth as recently as 45 million years ago.

In addition to these recently formed mountains, several chains of much older mountains exist. Although these structures are deeply eroded and topographically less prominent, they clearly possess the same structural

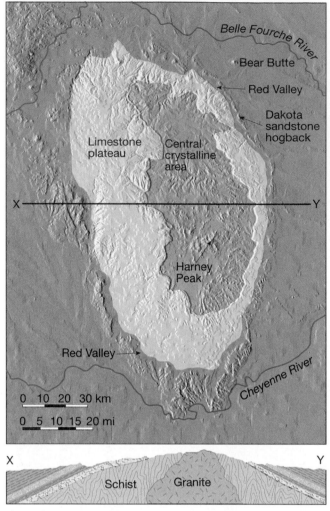

Figure 9.18

The Black Hills of South Dakota, a large domal structure with resistant igneous and metamorphic rocks exposed in the core. (After Arthur N. Strahler, *Introduction to Physical Geography*, 3rd ed., New York: John Wiley & Sons, 1973. Reprinted by permission.)

features found in younger mountains. Typical of this older group are the Appalachians in the eastern United States and the Urals in Russia.

The first encompassing explanation of mountain building came a little over three decades ago as part of plate tectonic theory. It is now known that Earth's major mountain systems formed along convergent plate boundaries. Recall that convergence can occur between one oceanic and one continental plate, between two oceanic plates, or between two continental plates. We will consider these sites of mountain building in the following sections.

Mountain Building at Convergent Boundaries

To unravel the events that produce mountains, many studies have been conducted along active subduction zones where plates are converging. At most modern-day subduction zones, volcanic arcs are forming. This situation is typified by Alaska's Aleutian Islands and by the Andes Mountains of western South America. Although all volcanic arcs are similar, *Aleutian-type subduction zones* occur where *two oceanic plates* converge (Figure 9.20). *Andean-type subduction zones* are situated where *oceanic crust is being thrust beneath a continental mass*. Consequently, the events that are generating these volcanic arcs followed different evolutionary paths. Further, although the development of a volcanic arc does result in the formation of mountainous topography, this activity is viewed as just one of the phases in the development of a major mountain belt.

Mountain Building Where Oceanic and Continental Crust Converge. Mountain building along continental margins involves the convergence of an oceanic plate and a plate whose leading edge contains continental

Figure 9.19

Highly deformed sedimentary strata in the Rocky Mountains of British Columbia. These sedimentary rocks are continental shelf deposits that were displaced toward the interior of Canada by low-angle thrust faults. (Photo by John Montagne)

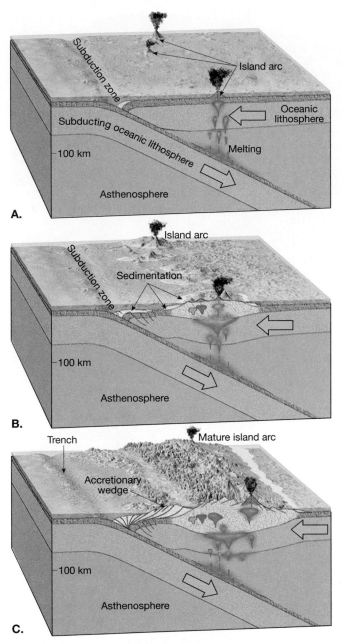

Figure 9.20
The development of a mature volcanic island arc by the convergence of two oceanic plates. These Aleutian-type subduction zones produce a mountainous topography that is similar to the volcanic arc that comprises the Aleutian islands of Alaska.

crust. Exemplified by the Andes Mountains, this *Andean type* of convergence generates structures resembling those of a developing volcanic island arc.

The first stage in the development of an Andean-type mountain belt occurs prior to the formation of the subduction zone. During this period the continental margin is *passive*. It is not a plate boundary but a part of the same plate as the adjoining oceanic crust. The East

Coast of the United States is a present-day example of a passive continental margin. Here, as at other passive continental margins surrounding the Atlantic, deposition of sediment along the continental margin is producing a thick wedge of sandstones, limestones, and shales (Figure 9.21A).

At some point the continental margin becomes active. A subduction zone forms and the deformation process begins (Figure 9.21B). A good place to examine such an active continental margin is the western coast of South America. Here, an oceanic plate is being subducted eastward beneath the South American plate along the Peru-Chile trench (see Figure 7.8).

In an idealized Andean-type subduction, convergence of the continental block and the subducting oceanic plate leads to deformation and metamorphism of the continental margin. Once the oceanic plate descends to about 100 kilometers, partial melting generates magma that slowly migrates upward, intruding and further deforming these strata (Figure 9.21B). During the development of the volcanic arc, sediment derived from the land as well as that scraped from the subducting plate becomes plastered against the landward side of the trench. This chaotic accumulation of sedimentary and metamorphic rocks, including occasional scraps of ocean crust, is called an **accretionary wedge** (Figure 9.21B). Prolonged subduction can build an accretionary wedge that is large enough to stand above sea level (Figure 9.21C).

Andean-type mountain belts, like island arcs, are composed of two roughly parallel zones. The landward segment is the volcanic arc, made up of volcanoes and large bodies of intrusive igneous rock, intermixed with high-temperature metamorphic rocks. The seaward segment is the accretionary wedge. It consists of folded, faulted, and metamorphosed sediments and volcanic debris (Figure 9.21C).

One of the best examples of an inactive Andean-type mountain belt is found in the western United States. It includes the Sierra Nevada and the Coast Ranges in California (Figure 9.22). These parallel mountain belts were produced by the subduction of a portion of the Pacific basin under the western edge of the North American plate. The Sierra Nevada batholith is a remnant of a portion of the volcanic arc that was produced by several surges of magma over tens of millions of years. Subsequent uplifting and erosion have removed most evidence of past volcanic activity and exposed a core of crystalline metamorphic and igneous rocks.

In the trench region, sediments scraped from the subducting plate, plus those provided by the eroding volcanic arc, were intensely folded and faulted into the accretionary wedge, which presently constitutes part of California's Coast Ranges. Uplifting of the Coast Ranges took place only recently, as evidenced by the unconsolidated sediments that still mantle portions of these highlands.

Figure 9.21
Mountain building along an Andean-type subduction zone. **A.** Passive continental margin with extensive wedge of sediments. **B.** Plate convergence generates a subduction zone, and partial melting produces a developing volcanic arc. **C.** Continued convergence and igneous activity further deform and thicken the crust, elevating the mountain belt, while the accretionary wedge grows.

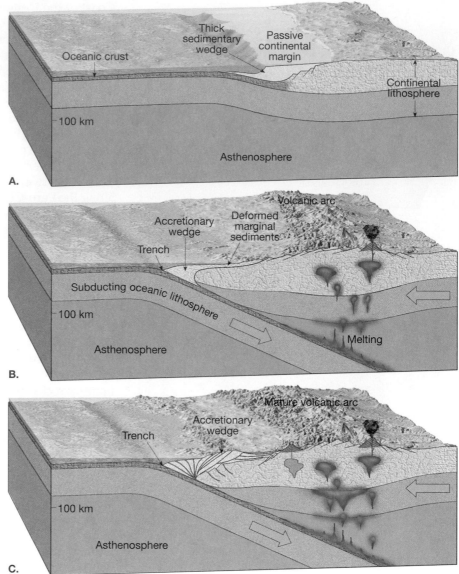

Mountain Building Where Continents Converge.   So far, we have discussed the formation of mountain belts where the leading edge of just one of the two converging plates contained continental crust. However, it is possible for two converging plates to be carrying continental crust. Because continental lithosphere is too buoyant to undergo any appreciable subduction, a collision between the continental fragments eventually results (Figure 9.23).

An example of such a collision occurred about 45 million years ago when India collided with the Eurasian plate. India was once part of Antarctica but split from that continent and moved a few thousand kilometers due north. The result of this collision was the formation of the spectacular Himalaya Mountains and the Tibetan Highlands.

Although most of the oceanic crust that separated these landmasses prior to the collision was subducted, some was caught up in the squeeze, along with sediment

that lay offshore (Figure 9.23). Today it is elevated high above sea level. Geologists think that, following such a collision, the subducted oceanic plate decouples from the rigid continental plate and continues its descent into the mantle.

A similar but much older collision is believed to have taken place when the European continent collided with the Asian continent to produce the Ural Mountains, which extend north-south through Russia. Prior to the discovery of plate tectonics, geologists had difficulty explaining the existence of mountain ranges, like the Urals, which are located deep within continental interiors. How could thousands of meters of marine sediment be deposited and then become highly deformed while situated in the middle of a large landmass?

Other mountain ranges showing evidence of continental collisions are the Alps and the Appalachians. The Appalachians resulted from collisions between North

Figure 9.22
Map of the California Coast Ranges and the Sierra Nevada.

America, Europe, and northern Africa. Although they have since separated, these landmasses were juxtaposed as part of the supercontinent Pangaea less than 200 million years ago. Detailed studies in the southern Appalachians indicate that the formation of this mountain belt was more complex than once thought. Rather than forming during a single continental collision, the Appalachians resulted from several distinct episodes of mountain building that occurred over a period of nearly 300 million years.

Mountain Building and Continental Accretion

Plate tectonics theory originally suggested two mechanisms for mountain building: (1) continental collisions were proposed to explain the formation of such mountainous terrains as the Appalachians, Himalayas, and Urals; and (2) subduction of oceanic lithosphere was thought to be the underlying tectonic process for many circum-Pacific mountain chains, as typified by the Andes. Recent investigations, however, indicate a third mechanism of mountain building: *Smaller crustal fragments col-*

lide and accrete to continental margins. Through this process of collision and accretion, many of the mountainous regions rimming the Pacific have been generated.

What is the nature of these small crustal fragments, and where do they come from? Researchers suggest that, prior to their accretion to a continental block, some of the fragments may have been microcontinents similar to the present-day island of Madagascar near Africa. Others were island arcs like Japan, the Philippines, and the Aleutian Islands. Still others may have been submerged crustal fragments that extended high above the ocean floor. It is believed that these fragments originated as submerged continental fragments, extinct volcanic island arcs, or submerged volcanic chains associated with hot spot activity.

Accretion of Foreign Terranes. The widely accepted view today is that, as oceanic plates move, they carry the embedded oceanic plateaus or microcontinents to a subduction zone. Here the upper portions of these thickened zones are peeled from the descending plate and thrust in relatively thin sheets onto the adjacent continental block. This newly added material increases the width of the continent. The material may later be overridden and displaced further inland by colliding with other fragments.

Geologists refer to these accreted crustal blocks as terranes. Simply, the term **terrane** designates any crustal fragment whose geologic history is distinct from that of the adjoining terranes. (Do not confuse the term *terrane* with the word *terrain*, which indicates the topography or lay of the land.) Terranes come in varied shapes and sizes; some are no larger than volcanic islands. Others, such as the one composing the entire Indian subcontinent, are very large.

Accretion and Mountain Building. The idea that mountain building occurs in association with the accretion of small crustal fragments to a continental mass arose principally from studies conducted in the North American Cordillera (Figure 9.24). Some mountainous areas, principally those of Alaska and British Columbia, contain fossil and magnetic evidence that these strata formed much nearer the equator.

It is now assumed that many other terranes found in the North American Cordillera were once scattered throughout the eastern Pacific, much as we find island arcs and oceanic plateaus distributed in the western Pacific today. Over the last 200 million years, these fragments migrated toward and collided with the west coast of North America (Figure 9.24). Apparently, this activity resulted in the piecemeal addition of fragments to the entire Pacific Coast, from the Baja Peninsula to northern Alaska. In a like manner, many modern microcontinents will eventually be accreted to active continental margins, thus resulting in the formation of new mountainous belts.

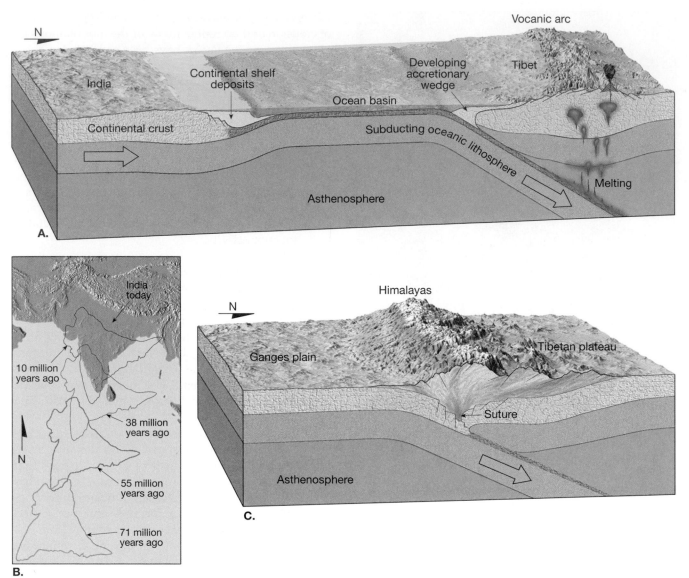

Figure 9.23
Simplified diagrams showing the northward migration and collision of India with the Eurasian plate. **A.** Converging plates generated a subduction zone, while partial melting of the subducting oceanic slab produced a volcanic arc. Sediments scraped from the subducting plate were added to the accretionary wedge. **B.** Position of India in relation to Eurasia at various times.(Modified after Peter Molnar) **C.** Eventually the two landmasses collided, deforming and elevating the accretionary wedge and continental shelf deposits. In addition, slices of the Indian crust were thrust up onto the Indian plate.

Review Questions

1. State the two lines of evidence that support the concept of crustal uplift. Can you think of another?

2. What happens to a floating object when weight is added? Subtracted? How do these principles apply to changes in the elevations of mountains? What term is applied to the adjustment that causes crustal uplift of this type?

3. List one line of evidence that supports the idea of a floating crust.

4. What conditions favor rock deformation by folding? By faulting?

5. The San Andreas fault is an excellent example of a _____ fault.

6. Compare the movement of normal and reverse faults. What type of force produces each?

7. At which of the three types of plate boundaries does normal faulting predominate? Reverse faulting? Strike-slip faulting?

8. Describe a horst and a graben. Explain how a graben valley forms and name one.

9. Compare and contrast anticlines and synclines. Domes and basins. Anticlines and domes.

10. Although we classify many mountains as folded, why might this description be misleading?

11. What type of faults are associated with fault-block mountains?

12. During the formation of fault-block mountains, are the forces that act upon the region compressional or tensional?

13. There are four types of mountain ranges. Name the type of mountain range exemplified by each of the following:
   a. Black Hills   b. Basin and Range   c. Adirondacks
   d. Cascades   e. Appalachians   f. Tetons
   g. Bighorns   h. Himalayas   i. Front Range

14. What do we call the site where sediments are deposited along the margin of the continent where they have a good chance of being squeezed into a mountain range?

15. Which type of plate boundary is most directly associated with mountain building?

16. Describe an accretionary wedge and explain its formation.

17. Suppose a sliver of oceanic crust was discovered 1000 kilometers into the interior of a continent. Would this support or refute the theory of plate tectonics? Why?

18. Why might it have been difficult for geologists to conclude that the Appalachian Mountains were formed by plate collision if examples like the Himalayas did not exist?

19. How does the plate tectonics theory explain the existence of fossil marine life on top of the Ural Mountains?

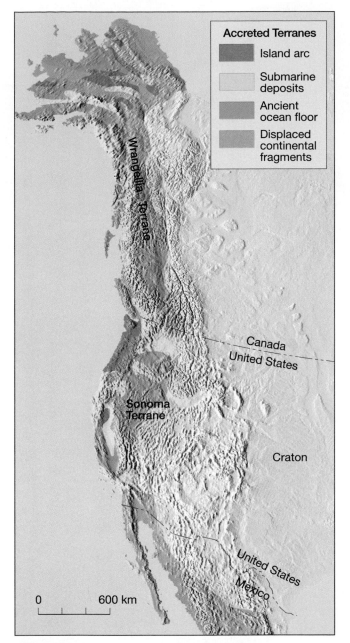

Figure 9.24
Map showing terranes thought to have been added to western North America during the past 200 million years. (Redrawn after D. R. Hutchinson and others)

Accreted Terranes: Island arc, Submarine deposits, Ancient ocean floor, Displaced continental fragments

Wrangellia Terrane, Canada/United States, Sonoma Terrane, Craton, United States/Mexico, 0 600 km

Key Terms

accretionary wedge (p. 255)
anticline (p. 242)
basin (p. 243)
dip-slip fault (p. 244)
dome (p. 243)
fault (p. 244)
fault-block mountains (p. 248)

fold (p. 242)
folded mountains (complex mountains) (p. 249)
graben (p. 244)
horst (p. 244)
isostasy (p. 239)
isostatic adjustment (p. 240)
joint (p. 245)
mountain building (p. 252)

normal fault (p. 244)
oblique-slip fault (p. 244)
orogenesis (p. 238)
reverse fault (p. 244)
strike-slip fault (p. 244)
syncline (p. 242)
terrane (p. 257)
thrust fault (p. 244)
upwarped mountains (p. 252)

CHAPTER 10
Geologic Time

Sedimentary strata representing millions of years of Earth history exposed at Junction Butte, Canyonlands National Park, Utah. (Photo by Carr Clifton)

In the late 1700s, James Hutton recognized that Earth is very old. But how old? For many years there was no reliable method to determine the age of Earth or the dates of various events in the geologic past. Rather, a geologic time scale was developed that showed the sequence of events based on relative dating principles. What are these principles? What part do fossils play? With the discovery of radioactivity and radiometric dating techniques, geologists now can assign fairly accurate dates to many of the events in Earth history. What is radioactivity? Why is it a good "clock" for dating the geologic past? In this chapter we shall answer these questions.

In 1869, John Wesley Powell, who was later to head the U.S. Geological Survey, led a pioneering expedition down the Colorado River and through the Grand Canyon. Writing about the rock layers that were exposed by the downcutting of the river, Powell said that ". . .the canyons of this region would be a Book of Revelations in the rock-leaved Bible of geology." He was undoubtedly impressed with the millions of years of Earth history exposed along the walls of the Grand Canyon (Figure 10.1).

Powell realized that the evidence for an ancient Earth is concealed in its rocks. Like the pages in a long and complicated history book, rocks record the geological events and changing life-forms of the past. The book, however, is not complete. Many pages, especially in the early chapters, are missing. Others are tattered, torn, or smudged. Yet enough of the book remains to allow much of the story to be deciphered.

Interpreting Earth history is a prime goal of the science of geology. Like a modern-day sleuth, the geologist must interpret clues found preserved in the rocks. By studying rocks, especially sedimentary rocks, and the features they contain, geologists can unravel the complexities of the past.

Geological events by themselves, however, have little meaning until they are put into a time perspective. Studying history, whether it be the Civil War or the Age of Dinosaurs, requires a calendar. Among geology's major contributions is a calendar called the *geologic time scale* and the discovery that Earth history is exceedingly long.

Figure 10.1
The Grand Canyon of the Colorado River as viewed from West Rim Drive. (Photo by Carr Clifton)

A Short History of Geology

In the mid-1600s, James Ussher, Anglican Archbishop of Armagh, Primate of all Ireland, published a work that had immediate and profound influence on people's view of Earth's age. A respected scholar of the Bible, Ussher constructed a chronology of human and Earth history in which he determined that Earth was only a few thousands of years old, having been created in 4004 B.C. Ussher's treatise earned widespread acceptance among scientific and religious leaders alike, and his chronology was soon printed in the margins of the Bible itself.

During the seventeenth and eighteenth centuries the doctrine of **catastrophism** strongly influenced people's thinking about Earth. Briefly stated, catastrophists believed that Earth's landscape had been developed primarily by great catastrophes. Features such as mountains and canyons, which today we know take great periods of time to form, were explained as having been produced by sudden and often worldwide disasters produced by unknowable causes that no longer operate. This philosophy was an attempt to fit the rate of Earth processes to the prevailing ideas on the age of Earth.

Birth of Modern Geology

Modern geology, began in the late 1700s when James Hutton, a Scottish physician and gentleman farmer, published his *Theory of the Earth* (Figure 10.2). In this work, Hutton put forth a fundamental principle that is a pillar of geology today: **uniformitarianism**. It simply states that *the physical, chemical, and biological laws that operate today have also operated in the geologic past*. This means that the forces and processes that we observe presently shaping our planet have been at work for a very long time. Thus, to understand ancient rocks, we must first understand present-day processes and their results. This idea is commonly stated by saying "the present is the key to the past."

Prior to Hutton's *Theory of the Earth*, no one had effectively demonstrated that geological processes occur over extremely long periods of time. However, Hutton persuasively argued that weak, slow-acting processes could, over long spans of time, produce effects just as great as those resulting from sudden catastrophic events. Unlike his predecessors, Hutton cited verifiable observations to support his ideas.

Unfortunately, Hutton's literary style was cumbersome and difficult, so his work was not widely read nor easily understood. It is the English geologist Sir Charles Lyell who is given the most credit for advancing the basic principles of modern geology. Between 1830 and 1872, Lyell produced eleven editions of his great work, *Principles of Geology*. As was customary in those days, Lyell's book had a lengthy subtitle that outlined the

Figure 10.2
James Hutton, the eighteenth-century Scottish geologist who is often called the "founder of modern geology." (The Natural History Museum, London)

main theme of the work: *Being an Attempt to Explain the Former Changes of the Earth's Surface, by Reference to Causes Now in Operation*.

Lyell painstakingly illustrated the uniformity of nature through time. He was able to show more convincingly than his predecessors that geologic processes observed today can be assumed to have operated in the past. Although the doctrine of uniformitarianism did not originate with Lyell, he was most successful in interpreting and publicizing it for society at large.

Geology Today

Today the basic tenets of uniformitarianism are just as viable as in Lyell's day. Indeed, we realize more strongly than ever that the present gives us insight into the past and that the physical, chemical, and biological laws that govern geological processes remain unchanging through time. However, we also understand that the doctrine should not be taken too literally. To say that geological processes in the past were the same as those occurring today is not to suggest that they always had the same relative importance or that they operated at precisely the same rate. Although the same processes have prevailed through time, their rates have undoubtedly varied.

The acceptance of uniformitarianism meant the acceptance of a very long history for Earth. Although Earth's processes vary in intensity, they still take a very long time to create or destroy major landscape features.

For example, rocks containing fossils of organisms that lived in the sea more than 15 million years ago are now part of mountains that stand 3000 meters (10,000 feet) above sea level. This means that the mountains were uplifted 3000 meters in about 15 million years, which works out to a rate of only 0.2 millimeter per year! Rates of erosion are equally slow. Estimates indicate that the North American continent is being lowered by erosion at just 3 centimeters per 1000 years. Thus, as you can see, it takes tens of millions of years for nature to build mountains and wear them down again.

But even these time spans are relatively short on the time scale of Earth history, for the rock record contains evidence that shows Earth has experienced many cycles of mountain building and erosion. Concerning the ever-changing nature of Earth through great expanses of geologic time, Hutton stated: "We find no vestige of a beginning, no prospect of an end."

It is important to remember that, although many features of our physical landscape may seem to be unchanging over our lifetimes, they are nevertheless changing, but on time scales of hundreds, thousands, or even many millions of years.

Relative Dating—Key Principles

During the late 1800s and early 1900s, a number of attempts were made to determine the age of Earth. Although some of the methods appeared promising at the time, none proved to be reliable. What these scientists were seeking was an **absolute date**. Such dates pinpoint the time in history when something took place—for example, the extinction of the dinosaurs about 66 million years ago. Today our understanding of radioactivity allows us to accurately determine absolute dates for rock units that represent important events in Earth's distant past. We will study radioactivity later in this chapter. Prior to the discovery of radioactivity, geologists had no accurate and dependable method of absolute dating and had to rely solely on relative dating.

Relative dating means placing rocks in their proper *sequence of formation*—which formed first, second, third, and so on. Relative dating cannot tell us how long ago something took place, only that it followed one event and preceded another. The relative dating techniques that were developed are valuable and still widely used. Absolute dating methods did not replace these techniques; they simply supplemented them. To establish a relative time scale, a few simple principles or rules had to be discovered and applied. Although they may seem obvious to us today, they were major breakthroughs in thinking at the time, and their discovery and acceptance was an important scientific achievement.

Law of Superposition

Nicolaus Steno, a Danish anatomist, geologist, and priest (1636–1686), is credited with being the first to recognize a sequence of historical events in an outcrop of sedimentary rock layers. Working in the mountains of western Italy, Steno applied a very simple rule that has come to be the most basic principle of relative dating—the **law of superposition**. The law simply states that in an undeformed sequence of sedimentary rocks, each bed is older than the one above it and younger than the one below. Although it may seem obvious that a rock layer could not be deposited unless it had something older beneath it for support, it was not until 1669 that Steno clearly stated the principle. This rule also applies to other surface-deposited materials such as lava flows and beds of ash from volcanic eruptions. Applying the law of superposition to the beds exposed in the upper portion of the Grand Canyon (Figure 10.3), you can easily place the layers in their proper order. Among those that are shown, the sedimentary rocks in the Supai Group must be the oldest, followed in order by the Hermit Shale, Coconino Sandstone, Toroweap Formation, and Kaibab Limestone.

Principle of Original Horizontality

Steno is also credited with recognizing the **principle of original horizontality**. This principle simply states that most layers of sediment are deposited in a horizontal position. Thus, if we observe rock layers that are flat, it means they have not been disturbed, and still have their *original* horizontality (Figure 10.4A). But if they are folded or inclined at a steep angle, they must have been moved into that position by crustal disturbances sometime after their deposition (Figure 10.4B).

Principle of Cross-Cutting Relationships

When a fault cuts through other rocks, or when magma intrudes and crystallizes, we can assume that the fault or intrusion is younger than the rocks affected. For example, in Figure 10.5, the faults and dikes clearly must have occurred *after* the sedimentary layers were deposited.

This is the **principle of cross-cutting relationships**. By applying the cross-cutting principle you can see that fault A occurred *after* the sandstone layer was deposited, because it "broke" the layer. However, fault A occurred *before* the conglomerate was laid down, because that layer is unbroken.

A.   **B.**

Figure 10.3
Applying the law of superposition to these layers exposed in the upper portion of the Grand Canyon, the Supai Group is oldest and the Kaibab Limestone is youngest. (Photo by E. J. Tarbuck)

We can also state that dike B and its associated sill are older than dike A, because dike A cuts the sill. In the same manner, we know that the batholith was emplaced after movement occurred along fault B, but before dike B was formed. This is true because the batholith cuts across fault B and dike B cuts across the batholith.

Inclusions

Sometimes inclusions can aid the relative dating process. **Inclusions** are pieces of one rock unit that are contained within another. The basic principle is logical and straightforward. The rock mass adjacent to the one

A.   **B.**

Figure 10.4
A. The principle of original horizontality states that most layers of sediment are deposited in a nearly horizontal position. **B.** When we see folded rock layers such as these, we can assume they must have been moved into that position by crustal disturbances *after* their deposition. (Photos by E. J. Tarbuck)

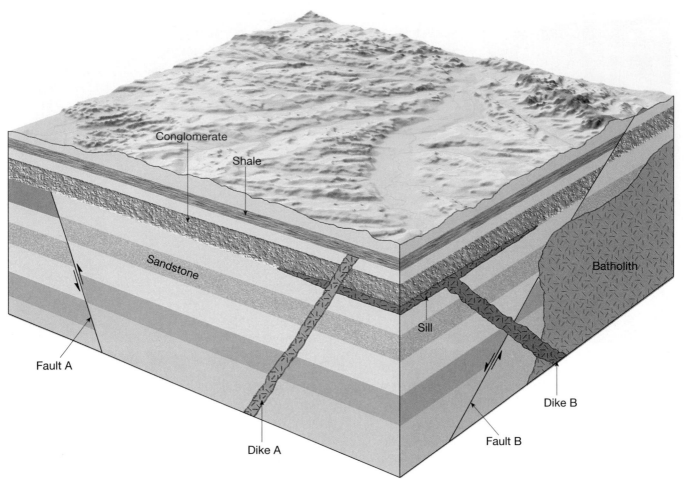

Figure 10.5
Cross-cutting relationships are an important principle used in relative dating. An intrusive rock body is younger than the rocks it intrudes. A fault is younger than the rock layers it cuts.

containing the inclusions must have been there first in order to provide the rock fragments. Therefore, the rock mass containing inclusions is the younger of the two. Figure 10.6 provides an example. Here the inclusions of granite in the adjacent sedimentary layer indicate the sedimentary layer was deposited on top of a weathered mass of granite rather than being intruded from below by a mass of magma that later crystallized.

Unconformities

When we observe layers of rock that have been deposited essentially without interruption, we call them **conformable**. Many areas exhibit conformable beds representing certain spans of geologic time. However, no place on Earth has a complete set of conformable strata.

Throughout Earth history, the deposition of sediment has been interrupted again and again. All such breaks in the rock record are termed unconformities.

An **unconformity** represents a long period during which deposition ceased, erosion removed previously formed rocks, and then deposition resumed. In each case Earth's crust has undergone uplift and erosion, followed by subsidence and renewed sedimentation. Unconformities are important features because they represent significant geologic events in Earth history. Moreover, their recognition helps us identify what intervals of time are not represented by strata and thus are missing from the geologic record.

The rocks exposed in the Grand Canyon of the Colorado River represent a tremendous span of geologic history. It is a wonderful place to take a trip through time. The canyon's colorful strata record a long history of sedimentation in a variety of environments—advancing seas, rivers and deltas, tidal flats, and sand dunes. But the record is not continuous. Unconformities represent vast amounts of time that have not been recorded in the canyon's layers. Figure 10.7 is a geologic cross-section of

unconformity indicates that during the pause in deposition, a period of deformation (folding or tilting) and erosion occurred (Figure 10.8).

Disconformity.   When contrasted with angular unconformities, **disconformities** are more common, but usually far less conspicuous because the strata on either side are essentially parallel. Many disconformities are difficult to identify because the rocks above and below are similar and there is little evidence of erosion. Such a break often resembles an ordinary bedding plane. Other disconformities are easier to identify because the ancient erosion surface is cut deeply into the older rocks below.

Nonconformity.   The third basic type of unconformity is a **nonconformity**. Here the break separates older metamorphic or intrusive igneous rocks from younger sedimentary strata (Figure 10.6). Just as angular unconformities and disconformities imply crustal movements, so too do nonconformities. Intrusive igneous masses and metamorphic rocks originate far below the surface. Thus, for a nonconformity to develop, there must be a period of uplift and the erosion of overlying rocks. Once exposed at the surface, the igneous or metamorphic rocks are subjected to weathering and erosion prior to subsidence and the renewal of sedimentation.

Using Relative Dating Principles

If you apply the principles of relative dating to the hypothetical geologic cross-section in Figure 10.9, you can place in proper sequence the rocks and the events they represent. The statements within the figure summarize the logic used to interpret the cross-section.

In this example, we establish a relative time scale for the rocks and events in the area of the cross-section. Remember that this method gives us no idea of how many years of Earth history are represented, for we have no absolute dates. Nor do we know how this area compares to any other.

Correlation of Rock Layers

To develop a geologic time scale that applies to the whole Earth, rocks of similar age in different regions must be matched up. Such a task is referred to as **correlation**. Within a limited area correlating the rocks of one locality with those of another may be done simply by walking along the outcropping edges. However, this may not be possible when the rocks are mostly concealed by soil and vegetation. Correlation over short distances is often achieved by noting the position of a distinctive rock layer in a sequence of strata. Or, a layer may be identified in another location if it is composed of very distinctive or uncommon minerals.

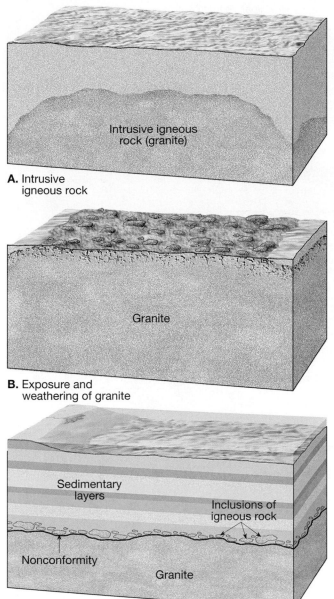

A. Intrusive igneous rock

B. Exposure and weathering of granite

C. Deposition of sedimentary layers

Figure 10.6

Since pieces of granite (*inclusions*) are contained within the overlying sedimentary bed, we know the granite must be older. When older intrusive igneous rocks are overlain by younger sedimentary layers, a type of unconformity termed a *nonconformity* is said to exist.

the Grand Canyon. Refer to it as you read about the three basic types of unconformities: angular unconformities, disconformities, and nonconformities.

Angular Unconformity.   Perhaps the most easily recognized unconformity is an **angular unconformity**. It consists of tilted or folded sedimentary rocks that are overlain by younger, more flat-lying strata. An angular

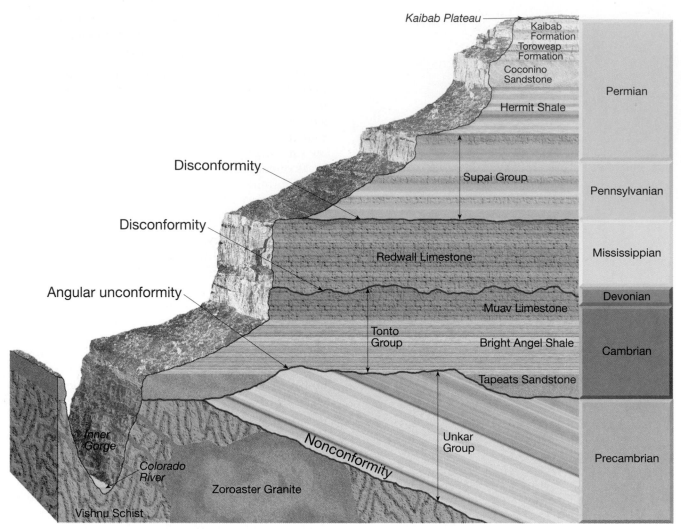

Figure 10.7
This cross-section through the Grand Canyon illustrates the three basic types of unconformities. An angular unconformity can be seen between the tilted Precambrian Unkar Group and the Cambrian Tapeats Sandstone. Two disconformities are marked, above and below the Redwall Limestone. A nonconformity occurs between the igneous and metamorphic rocks of the Inner Gorge and the sedimentary strata of the Unkar Group.

By correlating the rocks from one place to another, a more comprehensive view of the geologic history of a region is possible. Figure 10.10, for example, shows the correlation of strata at three sites on the Colorado Plateau. No single locale exhibits the entire sequence, but correlation reveals a more complete picture of the sedimentary rock record.

Many geologic studies involve relatively small areas. Such studies are important in their own right, but their full value is realized only when the rocks are correlated with those of other regions. Although the methods just described are sufficient to trace a rock formation over relatively short distances, they are not adequate for matching rocks that are separated by great distances.

When correlation between widely separated areas or between continents is the objective, the geologist must rely on fossils.

Fossils: Evidence of Past Life

Fossils, the remains or traces of prehistoric life, are important inclusions in sediment and sedimentary rocks. They are important tools for interpreting the geologic past. Knowing the nature of the life forms that existed at a particular time helps researchers understand past environmental conditions. Further, fossils are important time indicators and play a key role in correlating rocks of similar ages that are from different places.

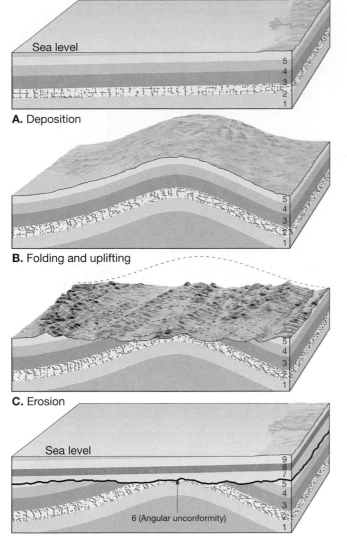

A. Deposition

B. Folding and uplifting

C. Erosion

D. Subsidence and renewed deposition

Figure 10.8

Formation of an angular unconformity. An angular unconformity represents a period during which deformation and erosion occurred.

Types of Fossils

Fossils are of many types. The remains of relatively recent organisms may not have been altered at all. Such objects as teeth, bones, and shells are common examples. Far less common are entire animals, flesh included, that have been preserved because of rather unusual circumstances. Remains of prehistoric elephants called mammoths that were frozen in the Arctic tundra of Siberia and Alaska are examples, as are the mummified remains of sloths preserved in a dry cave in Nevada.

Given enough time, the remains of an organism are likely to be modified. Often fossils become *petrified* (literally, "turned into stone"), meaning that the small in-

ternal cavities and pores of the original structure are filled with precipitated mineral matter (Figure 10.11A). In other instances *replacement* may occur. Here the cell walls and other solid material are removed and replaced with mineral matter. Sometimes the microscopic details of the replaced structure are faithfully retained.

Molds and casts constitute another common class of fossils. When a shell or other structure is buried in sediment and then dissolved by underground water, a *mold* is created. The mold faithfully reflects only the shape and surface marking of the organism; it does not reveal any information concerning its internal structure. If these hollow spaces are subsequently filled with mineral matter, *casts* are created (Figure 10.11B).

A type of fossilization called *carbonziation* is particularly effective in preserving leaves and delicate animal forms. It occurs when fine sediment encases the remains of an organism. As time passes, pressure squeezes out the liquid and gaseous components and leaves behind a thin residue of carbon (Figure 10.11C). Black shales deposited as organic-rich mud in oxygen-poor environments often contain abundant carbonized remains. If the film of carbon is lost from a fossil preserved in fine-grained sediment, a replica of the surface, called an *impression*, may still show considerable detail (Figure 10.11D).

Delicate organisms, such as insects, are difficult to preserve and consequently are relatively rare in the fossil record. Not only must they be protected from decay, they must not be subjected to any pressure that would crush them. One way in which some insects have been preserved is in *amber*, the hardened resin of ancient trees. The fly in Figure 10.11E was preserved after being trapped in a drop of sticky resin. Resin sealed off the insect from the atmosphere and protected the remains from damage by water and air. As the resin hardened, a protective, pressure-resistant case was formed.

In addition to the fossils already mentioned, there are numerous other types, many of them only traces of prehistoric life. Examples of such indirect evidence include:

1. Tracks—animal footprints made in soft sediment that was later lithified (Figure 10.11F).

2. Burrows—tubes in sediment, wood, or rock made by an animal. These holes may later become filled with mineral matter and preserved. Some of the oldest-known fossils are believed to be worm burrows.

3. Coprolites—fossil dung and stomach contents that can provide useful information pertaining to food habits of organisms.

4. Gastroliths—highly polished stomach stones that were used in the grinding of food by some extinct reptiles.

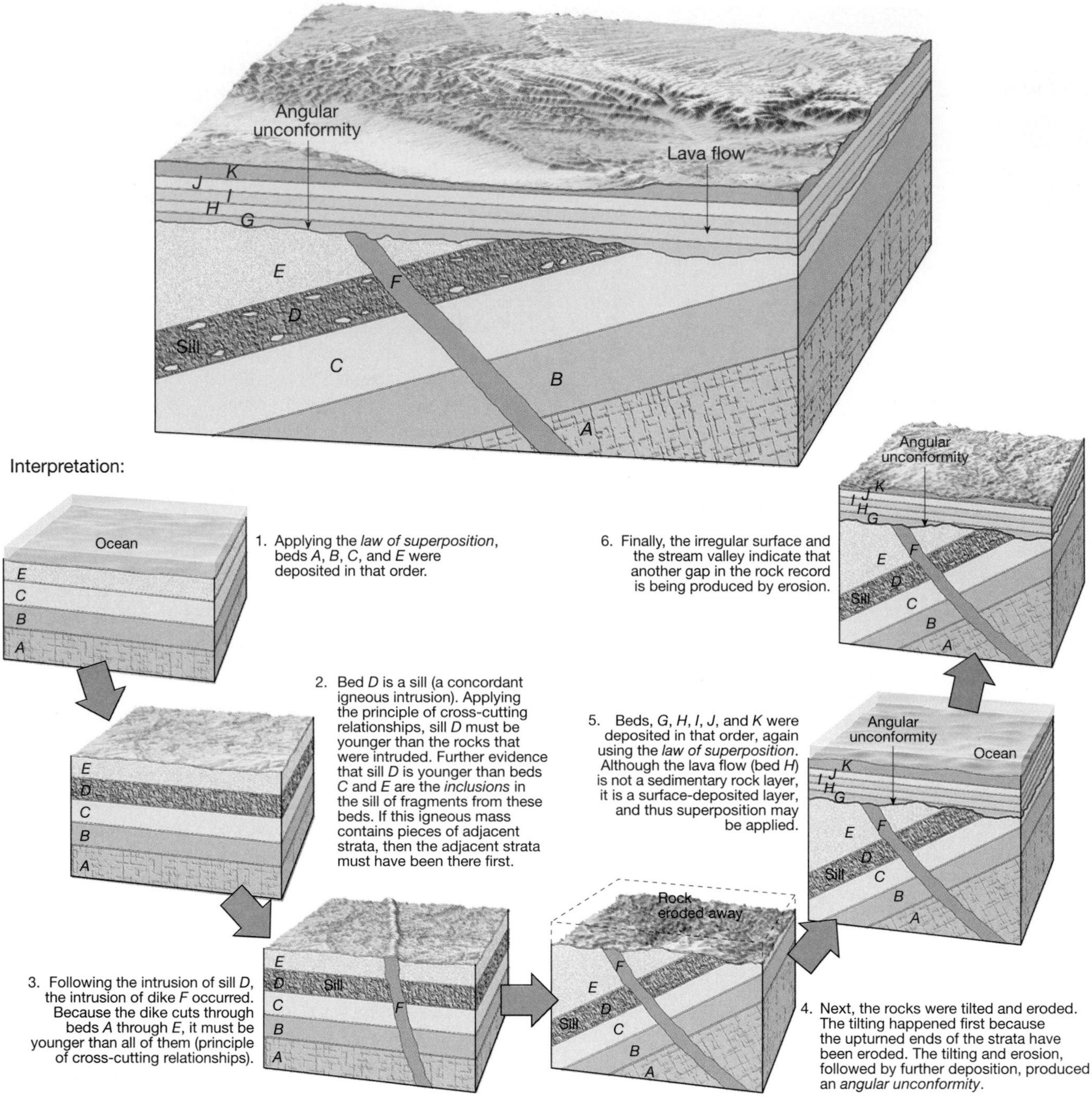
Angular unconformity
Lava flow
K
J
I
H
G
E
F
D
Sill
C
B
A
Interpretation:
Ocean
1. Applying the *law of superposition*, beds *A*, *B*, *C*, and *E* were deposited in that order.
2. Bed *D* is a sill (a concordant igneous intrusion). Applying the principle of cross-cutting relationships, sill *D* must be younger than the rocks that were intruded. Further evidence that sill *D* is younger than beds *C* and *E* are the *inclusions* in the sill of fragments from these beds. If this igneous mass contains pieces of adjacent strata, then the adjacent strata must have been there first.
3. Following the intrusion of sill *D*, the intrusion of dike *F* occurred. Because the dike cuts through beds *A* through *E*, it must be younger than all of them (principle of cross-cutting relationships).
Rock eroded away
4. Next, the rocks were tilted and eroded. The tilting happened first because the upturned ends of the strata have been eroded. The tilting and erosion, followed by further deposition, produced an *angular unconformity*.
5. Beds, *G*, *H*, *I*, *J*, and *K* were deposited in that order, again using the *law of superposition*. Although the lava flow (bed *H*) is not a sedimentary rock layer, it is a surface-deposited layer, and thus superposition may be applied.
6. Finally, the irregular surface and the stream valley indicate that another gap in the rock record is being produced by erosion.

Figure 10.9
Geologic cross-section of a hypothetical region.

Conditions Favoring Preservation

Only a tiny fraction of the organisms that have lived during the geologic past have been preserved as fossils. Normally the remains of an animal or plant are destroyed. Under what circumstances are they preserved? Two special conditions appear to be necessary: rapid burial and the possession of hard parts.

When an organism perishes, its soft parts usually are quickly eaten by scavengers or decomposed by bacteria. Occasionally, however, the remains are buried by sediment. When this occurs, the remains are protected from the environment where destructive processes operate. Rapid burial therefore is an important condition favoring preservation.

In addition, animals and plants have a much better chance of being preserved as part of the fossil record if they have hard parts. Although traces and imprints of soft-bodied animals such as jellyfish, worms, and insects exist, they are rare. Flesh usually decays so rapidly that preservation is exceedingly unlikely. Hard parts such as shells, bones, and teeth predominate in the record of past life.

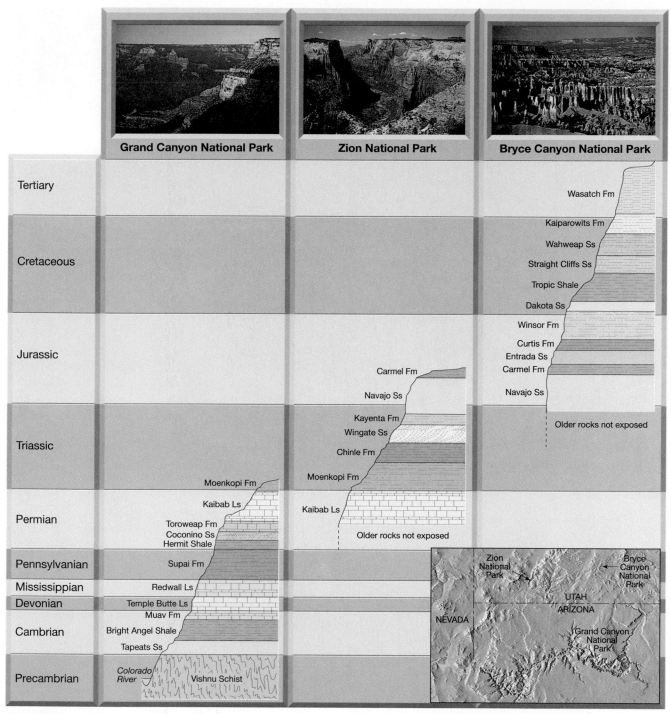

Grand Canyon National Park **Zion National Park** **Bryce Canyon National Park**

Figure 10.10

Correlation of strata at three locations on the Colorado Plateau reveals the total extent of sedimentary rocks in the region. (After U.S. Geological Survey)

Because preservation is contingent on special conditions, the record of life in the geologic past is biased. The fossil record of those organisms with hard parts that lived in areas of sedimentation is quite abundant. However, we get only an occasional glimpse of the vast array of other life-forms that did not meet the special conditions favoring preservation.

Fossils and Correlation

The existence of fossils had been known for centuries, yet it was not until the late 1700s and early 1800s that their significance as geologic tools was made evident. During this period an English engineer and canal builder, William Smith, discovered that each rock for-

Figure 10.11
There are many types of fossilization. Six examples are shown here. **A.** Petrified wood in Petrified Forest National Park, Arizona. **B.** Natural casts of shelled invertebrates. **C.** A fossil bee preserved as a thin carbon film. **D.** Impressions are common fossils and often show considerable detail. **E.** Insect in amber. **F.** Dinosaur footprint in fine-grained limestone near Tuba City, Arizona. (Photo A by David Muench; Photos B, D, and F by E. J. Tarbuck; Photo C courtesy of the National Park Service; Photo E by Breck P. Kent)

mation in the canals he worked on contained fossils unlike those in the beds either above or below. Further, he noted that sedimentary strata in widely separated areas could be identified—and correlated—by their distinctive fossil content.

Based on Smith's classic observations and the findings of many geologists who followed, one of the most important and basic principles in historical geology was formulated: *Fossil organisms succeed one another in a definite and determinable order, and therefore any time period can be recognized by its fossil content.* This has come to be known as the **principle of fossil succession**. In other words, when fossils are arranged according to their age by applying the law of superposition to the rocks in which they are found, they do not present a random or haphazard picture. To the contrary, fossils show progressive change documenting the evolution of life through time.

For example, an Age of Trilobites is recognized quite early in the fossil record. Then, in succession, paleontologists recognize an Age of Fishes, an Age of Coal Swamps, an Age of Reptiles, and an Age of Mammals. These "ages" pertain to groups that were especially plentiful and characteristic during particular time periods. Within each of the "ages," there are many subdivisions based, for example, on certain species of trilobites, and certain types of fish, reptiles, and so on. This same succession of dominant organisms, never out of order, is found on every major landmass.

Once fossils were recognized as time indicators, they became the most useful means of correlating rocks of similar age in different regions. Geologists pay particular attention to certain fossils called **index fossils**. These fossils are widespread geographically and are limited to a short span of geologic time, so their presence provides an important method of matching rocks of the same age. Rock formations, however, do not always contain a specific index fossil. In such situations, groups of fossils are used to establish the age of the bed. Figure 10.12 illustrates how a group of fossils can be used to date rocks more precisely than could be accomplished by the use of only one of the fossils.

In addition to being important and often essential tools for correlation, fossils are important environmental indicators. Although much can be deduced about past environments by studying the nature and characteristics of sedimentary rocks, a close examination of any fossils present can usually provide a great deal more information. For example, when the remains of certain clam shells are found in limestone, the geologist can assume that the region was once covered by a shallow sea, because that is where clams live today. Also, by using what we know of living organisms, we can conclude that fossil animals with thick shells capable of

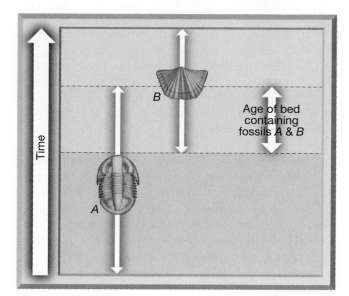

Figure 10.12
Overlapping ranges of fossils help date rocks more exactly than using a single fossil.

withstanding pounding and surging waves must have inhabited shorelines. On the other hand, animals with thin, delicate shells probably indicate deep, calm offshore waters. Hence, by looking closely at the types of fossils, the approximate position of an ancient shoreline may be identified.

Further, fossils can indicate the former temperature of the water. Certain present-day corals require warm and shallow tropical seas like those around Florida and the Bahamas. When similar corals are found in ancient limestones, they indicate that a Florida-like marine environment must have existed when they were alive. These are just a few examples of how fossils can help unravel the complex story of Earth history.

Absolute Dating with Radioactivity

In addition to establishing relative dates by using the principles described in the preceding sections, it is also possible to obtain reliable absolute dates for events in the geologic past. For example, we know that Earth is about 4.6 billion years old and that the dinosaurs became extinct about 66 million years ago. Dates that are expressed in millions and billions of years truly stretch our imagination because our personal calendars involve time measured in hours, weeks, and years. Nevertheless, the vast expanse of geologic time is a reality and it is radiometric dating that allows us to accurately measure it. In this section you will learn about radioactivity and its application in radiometric dating.

Recall from Chapter 1 that each atom has a *nucleus* containing protons and neutrons, and that the nucleus is orbited by electrons. *Electrons* have a negative electrical charge and *protons* have a positive charge. A *neutron* is actually a proton and an electron combined, so it has no charge (it is neutral).

The *atomic number* (each element's identifying number) is the number of protons in the nucleus. Every element has a different number of protons, and thus a different atomic number (hydrogen = 1, carbon = 6, oxygen = 8, uranium = 92, etc.) Atoms of the same element always have the same number of protons, so the atomic number stays constant.

Practically all of an atom's mass (99.9%) is in the nucleus, indicating that electrons have virtually no mass at all. So, by adding the protons and neutrons in an atom's nucleus, we derive the atom's *mass number*. The number of neutrons can vary, and these variants, or *isotopes*, have different numbers.

To summarize with an example, uranium's nucleus always has 92 protons, so its atomic number always is 92. But its neutron population varies, so uranium has three isotopes: uranium-234 (mass of protons + neutrons = 234), uranium-235, and uranium-238. All three isotopes are mixed in nature. They look the same and behave the same in chemical reactions.

Radioactivity

The forces that bind protons and neutrons together in the nucleus usually are strong. However, in some isotopes, the nuclei are unstable because the forces binding protons and neutrons together are not strong enough. As a result, the nuclei spontaneously break apart (decay), a process called **radioactivity**.

What happens when unstable nuclei break apart? Three common types of radioactive decay are illustrated in Figure 10.13 and are summarized as follows:

1. Alpha particles (α particles) may be emitted from the nucleus. An alpha particle is composed of 2 protons and 2 neutrons. Consequently, the emission of an alpha particle means (a) the mass number of the isotope is reduced by 4, and (b) the atomic number is decreased by 2.

2. When a beta particle (β particle), or electron, is given off from a nucleus, the mass number remains unchanged, because electrons have practically no mass. However, because the

electron has come from a neutron (remember, a neutron is a combination of a proton and an electron), the nucleus contains one more proton than before. Therefore, the atomic number increases by 1.

3. Sometimes an electron is captured by the nucleus. The electron combines with a proton and forms an additional neutron. As in the last example, the mass number remains unchanged. However, since the nucleus now contains one less proton, the atomic number decreases by 1.

An unstable (radioactive) isotope of an element is called the *parent*. The isotopes resulting from the decay of the parent are the *daughter products*. Figure 10.14 provides an example of radioactive decay. Here it can be seen that when the radioactive parent, uranium-238 (atomic number 92, mass number 238), decays, it follows a number of steps, emitting 8 alpha particles and 6 beta particles before finally becoming the stable daughter product lead-206 (atomic number 82, mass number 206). One of the unstable daughter products produced during this decay series is radon. Box10.1 examines the hazards associated with this radioactive gas.

Certainly among the most important results of the discovery of radioactivity is that it provided a reliable means of calculating the ages of rocks and minerals that contain particular radioactive isotopes. The procedure is called **radiometric dating**. Why is radiometric dating reliable? Because the rates of decay for many isotopes have been precisely measured and do not vary under the physical conditions that exist in Earth's outer layers. Therefore, each radioactive isotope used for dating has been decaying at a fixed rate since the formation of the rocks in which it occurs, and the products of decay have been accumulating at a corresponding rate. For example, when uranium is incorporated into a mineral that crystallizes from magma, there is no lead (the stable daughter product) from previous decay. The radiometric "clock" starts at this point. As the uranium in this newly formed mineral disintegrates, atoms of the daughter product are trapped and measurable amounts of lead eventually accumulate.

Half-Life

The time required for one-half of the nuclei in a sample to decay is called the **half-life** of the isotope. Half-life is a common way of expressing the rate of radioactive disintegration. Figure 10.15 illustrates what occurs when a radioactive parent decays directly into its stable daughter product. When the quantities of parent and daugh-

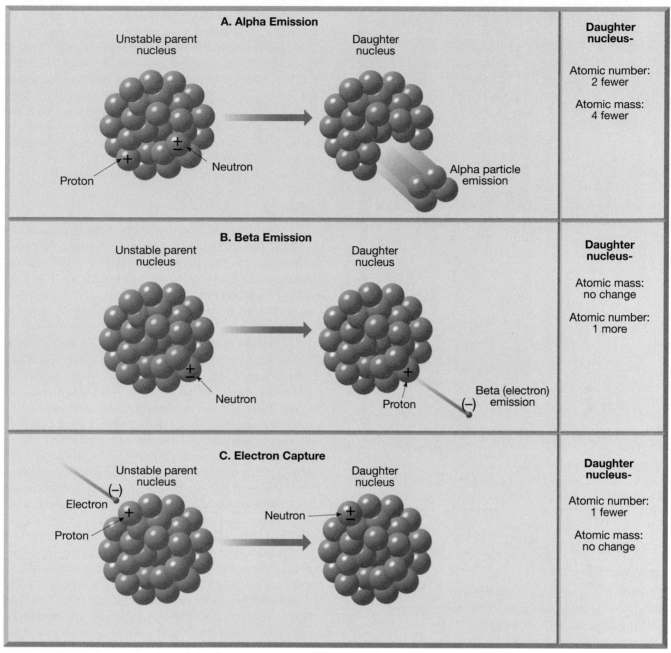

Figure 10.13
Common types of radioactive decay. Notice that in each case the number of protons (atomic number) in the nucleus changes, thus producing a different element.

ter are equal (ratio 1:1), we know that one half-life has transpired. When one-quarter of the original parent atoms remain and three-quarters have decayed to the daughter product, the parent/daughter ratio is 1:3 and we know that two half-lives have passed. After three half-lives, the ratio of parent atoms to daughter atoms is 1:7 (one parent for every seven daughter atoms).

If the half-life of a radioactive isotope is known and the parent/daughter ratio can be measured, the age of the sample can be calculated. For example, assume that the half-life of a hypothetical unstable isotope is one million years and the parent/daughter ratio in a sample is 1:15. Such a ratio indicates that four half-lives have passed and that the sample must be four million years old.

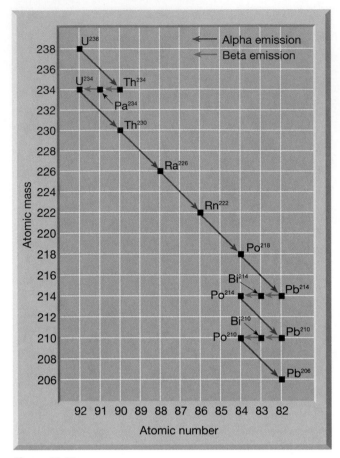

Figure 10.14
The most common isotope of uranium (U-238) is an example of a radioactive decay series. Before the stable end product (Pb-206) is reached, many different isotopes are produced as intermediate steps.

Radiometric Dating

Notice that the *percentage* of radioactive atoms that decay during one half-life is always the same: 50 percent. However, the *actual number* of atoms that decay with the passing of each half-life continually decreases. Thus, as the percentage of radioactive parent atoms declines, the proportion of stable daughter atoms rises, with the increase in daughter atoms just matching the drop in parent atoms. This fact is the key to radiometric dating.

Of the many radioactive isotopes that exist in nature, five have proven particularly important in providing radiometric ages for ancient rocks (Table 10.1). Rubidium-87, uranium-238, and uranium-235 are used for dating rocks that are millions of years old, but potassium-40 is more versatile. Although the half-life of potassium-40 is 1.3 billion years, recent analytical techniques have made possible the detection of tiny amounts of its stable daughter product, argon-40, in some rocks that are younger than 100,000 years. Another important reason for its frequent use is that potassium is abundant in many common minerals, particularly micas and feldspars.

It is important to realize that an accurate radiometric date can be obtained only if the mineral remained a closed system during the entire period since its formation. A correct date is not possible unless there was neither the addition nor loss of parent or daughter isotopes. This is not always the case. In fact, an important limitation of the potassium-argon method arises from the fact that argon is a gas, and it may leak from minerals, throwing off measurements. Cross-checking of samples, using two different radiometric methods, is done where possible to ensure accurate age determinations.

Carbon-14 Dating

To date very recent events, carbon-14 is used. Carbon-14 is the radioactive isotope of carbon. Because the half-life of carbon-14 is only 5730 years, it can be used for dating events from the historic past as well as those from recent geologic history. In some cases carbon-14 can be used to date events as far back as 75,000 years.

Carbon-14 is continuously produced in the upper atmosphere as a consequence of cosmic ray bombardment. Cosmic rays, which are high-energy particles, shatter the nuclei of gas atoms, releasing neutrons. Some of the neutrons are absorbed by nitrogen atoms (atomic number 7) causing their nuclei to emit a proton. As a result, the atomic number decreases by 1 (to 6), and a different element, carbon-14, is created (Figure 10.16A). This isotope of carbon quickly becomes incorporated into carbon dioxide, which circulates in the atmosphere and is absorbed by living matter. As a result, all organisms contain a small amount of carbon-14, including yourself.

While an organism is alive, the decaying radiocarbon is continually replaced, and the proportions of carbon-14 and carbon-12 remain constant. Carbon-12 is the stable and most common isotope of carbon. However, when any plant or animal dies, the amount of carbon-14 gradually decreases as it decays to nitrogen-14 by beta emission (Figure 10.16B). By comparing the proportions of carbon-14 and carbon-12 in a sample, radiocarbon dates can be determined.

Although carbon-14 is useful in dating only the last small fraction of geologic time, it has become a very valuable tool for anthropologists, archeologists, and historians, as well as for geologists who study very recent Earth history. In fact, the development of radiocarbon dating was considered so important that the chemist who discovered this application, Willard F. Libby, received a Nobel prize.

Importance of Radiometric Dating

Bear in mind that, although the basic principle of radiometric dating is simple, the actual procedure is quite complex. The analysis which determines the quantities of

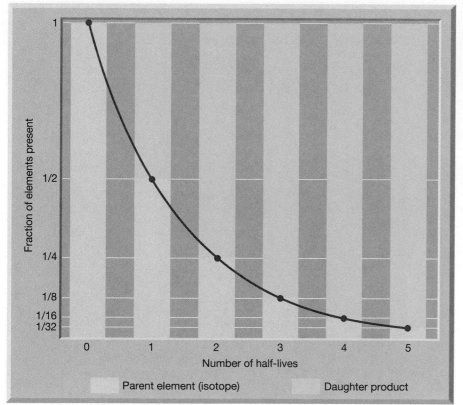

Figure 10.15
The radioactive decay curve shows change that is exponential. Half of the radioactive parent remains after one half-life. After a second half-life one-quarter of the parent remains, and so forth.

parent and daughter must be painstakingly precise. In addition, some radioactive materials do not decay directly into the stable daughter product. As you saw in Figure 10.14, uranium-238 produces thirteen intermediate unstable daughter products before the fourteenth and final daughter product, the stable isotope lead-206, is produced.

Radiometric dating methods have produced literally thousands of dates for events in Earth history. Rocks from several localities have been dated at more than three billion years, and geologists realize that still-older rocks exist. For example, a granite from South Africa has been dated at 3.2 billion years—and it contains inclusions of quartzite that must be older. Quartzite itself is a metamorphic rock which originally was the sedimentary rock sandstone. Sandstone, in turn, is the product of the lithification of sediments produced by the weathering of existing rocks. Thus we have a positive indication that much older rocks existed.

Radiometric dating has vindicated the ideas of Hutton, Darwin, and others who inferred that geologic time must be immense. Indeed, modern dating methods have proven that there has been enough time for the processes we observe to have accomplished tremendous tasks.

The Geologic Time Scale

Geologists have divided the whole of geologic history into units of varying magnitude. Together they comprise the **geologic time scale** of Earth history (Figure 10.17). The major units of the time scale were delineated during the 1800s, principally by workers in western Europe and Great Britain. Because absolute dating was unavailable at that time, the entire time scale was created using methods of relative dating. It has been only in this century that radiometric dating permitted absolute dates to be added.

Radioactive Parent	Stable Daughter Product	Currently Accepted Half-life Values
Uranium-238	Lead-206	4.5 billion years
Uranium-235	Lead-207	713 million years
Thorium-232	Lead-208	14.1 billion years
Rubidium-87	Strontium-87	47.0 billion years
Potassium-40	Argon-40	1.3 billion years

Table 10.1 Radioactive isotopes frequently used in radiometric dating.

Box 10.1

Radon

Richard L. Hoffman*

Radioactivity is defined as the spontaneous emission of atomic particles and/or electromagnetic waves from unstable atomic nuclei. For example, in a sample of uranium-238, unstable nuclei decay and produce a variety of radioactive progeny or "daughter" products as well as energetic forms of radiation (Table 10.A). One of its radioactive decay products is *radon*—a colorless, odorless, invisible gas.

Radon gained public attention in 1984 when a worker in a Pennsylvania nuclear power plant set off radiation alarms as he entered the plant at the start of his workday. It was discovered that his clothing and hair were contaminated with large amounts of radon decay products. Investigation revealed that he lived in a home whose basement radon level was 2800 times the average level in indoor air. The home was located along a geologic formation known as the Reading Prong—a mass of uranium-bearing black shale that runs from near Reading, Pennsylvania, to near Trenton, New Jersey.

Originating in the radio-decay of traces of uranium and thorium found in almost all soils, radon isotopes (Rn-222 and Rn-220) are continually renewed in an ongoing natural process. Geologists estimate that the top six feet of soil from an average acre of land contain about 50 pounds of uranium (about 2 to 3 parts per million). Some types of rocks contain more than average levels of uranium. These include light-colored volcanic rocks, granites, dark shales, and sedimentary and metamorphic rocks that contain phosphates. Because uranium is relatively plentiful and has a half-life of about 4.5 billion years, radon and its progeny will be with us forever.

Radon has a half-life of about 4 days. Its decay products (except lead-206) are all radioactive solids that adhere to dust particles, many of which are ultimately inhaled. And, too, during prolonged exposure to a radon-contaminated environment, it is certain that some radio-decay will occur while the gas is in the lungs, thereby placing the radioactive radon progeny in direct contact with delicate lung tissue. The radiation hazard of radon is not trivial.

Table 10.A Decay Products of Uranium-238

Some Decay Products of Uranium-238	Decay Particle Produced	Half-Life
Uranium-238	alpha	4.5 billion years
Radium-226	alpha	1600 years
Radon-222	**alpha**	**3.82 days**
Polonium-218	alpha	3.1 minutes
Lead-214	beta	26.8 minutes
Bismuth-214	beta	19.7 minutes
Polonium-214	alpha	1.6×10^{-4} second
Lead-210	beta	20.4 years
Bismuth-210	beta	5.0 days
Polonium-210	alpha	138 days
Lead-206	none	stable

- Steadily accumulating evidence indicates radon to be a significant cause of lung cancer second only to smoking.

Radon Measurement

Radon levels are reported in units of "picoCuries per Liter," (pCi/L)—a measured of radioactive disintegrations per time, per volume of air. One Curie is a radioactive decay rate of 3.7×10^{-10} disintegrations per second. This represents a very large number of disintegrations in a short time. Thus, environmental scientists use a smaller unit, the picoCurie, 3.7×10^{-2} nuclear disintegrations, occurring per second. This is equivalent to about 2.22 disintegrations (counts) per minute per liter of air. A house with a radon level of 4.0 pCI/L has about 8 to 9 atoms of radon decaying every minute in every liter of air. Outdoor air averages about 0.2 pCi/L.

- The EPA suggests that indoor radon levels be kept below 4.0 pCi/L.
- EPA risk estimates are conservative—they are based on an assumption that one would spend 75 percent of a 70-year time span (about 52 years) in the contaminated space.

*Dr. Hoffman is Professor of Chemistry, Illinois Central College.

The Entry Mechanism

Certainly the amount of uranium in the soil near a structure has a significant bearing on the structure's potential for a radon problem. Even so, there must be both a pathway for entry and a driving force to cause the gas to move indoors from its point of origin in the soil. Mitigation techniques and modern construction practices are specifically designed to interfere with both the access route and the driving forces which enable radon to enter a structure.

Once radon is produced in the soil, it diffuses throughout the tiny spaces between soil particles. Some ultimately percolates to the soil surface where it dissipates into the air. Its radio-decay becomes part of the terrestrial background radiation to which we are all exposed. That portion of the gas which decays while still in Earth poses no significant hazard because the solid radio-decay products adhere to soil particles. As a consequence, that radiation becomes a minor part of the natural background. Radon need not diffuse very far from its point of origin in order to enter the building because it is being continuously generated in the soil immediately adjacent to the structure's basement.

(continues)

(Box 10.1, continued)

Radon enters through pores, holes, cracks, and/or any other discontinuities in the basement floors and/or walls. Radon is fairly soluble in water; some enters the structure via the sub-slab drainage system. The density of radon gas is greater than that of air. Any which enters below-ground-level spaces in buildings tends to remain there during the remainder of its relatively short decay cycle.

A difference in interior/exterior atmospheric pressure is the major driving force that causes radon to enter a structure.

High winds around and over a house can draw air from inside. Likewise, air is drawn out of a house through aboveground leaks that permit warmed air to flow up and out through those openings. Thus, soil gas enters the basement through any openings that may be found there. Inflow is augmented by ventilation fans (attic, bathroom, kitchen range, and the like) designed to exhaust fumes or bring in fresh air.

Prognosis

The source of radon is as enduring as its generation mechanism within Earth; radon will never go away. However, cost-effective mitigation strategies are available to reduce radon levels to acceptable levels. In most cases radon levels in structures can be reduced without great expense.

The geologic time scale subdivides the 4.6-billion-year history of Earth into many different units and provides a meaningful time frame within which the events of the geologic past are arranged. As shown in Figure 10.17, **eons** represent the greatest expanses of time. The eon that began about 570 million years ago is the **Phanerozoic**, meaning *visible life*. It is an appropriate description because the rocks and deposits of the Phanerozoic eon contain abundant fossils that document major evolutionary trends.

Another glance at the time scale reveals that the Phanerozoic eon is divided into **eras**. The three eras within the Phanerozoic are the **Paleozoic** ("ancient life"), the **Mesozoic** ("middle life"), and the **Cenozoic** ("recent life"). As the names imply, the eras are bounded by profound worldwide changes in life-forms. Each era is subdivided into **periods**. The Paleozoic has seven, the Mesozoic three, and the Cenozoic two. Each of these twelve periods is characterized by a somewhat less profound change in life-forms as compared with the eras. Finally, periods are divided into still smaller units called **epochs**. As you can see in Figure 10.17, seven epochs have been named for the periods of the Cenozoic. The epochs of other periods, however, are not usually referred to by specific names. Instead, the terms *early, middle,* and *late* are generally applied to the epochs of these earlier periods.

Notice that the detail of the geologic time scale does not begin until about 570 million years ago, the date for the beginning of the Cambrian period. The more than four billion years prior to the Cambrian is divided into three eons, the *Hadean*, the *Archean*, and the *Proterozoic*. It is also common for this vast expanse of time to simply be referred to as the **Precambrian**. Although it represents more than 85 percent of Earth history, the Precambrian is not divided into nearly as many smaller time units as the Phanerozoic eon.

The quantity of information geologists have deciphered about Earth's past is somewhat analogous to the detail of human history. The farther back we go, the less

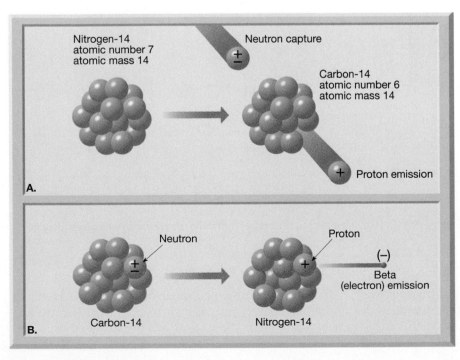

Figure 10.16
A. Production and B. decay of carbon-14. These sketches represent the nuclei of the respective atoms.

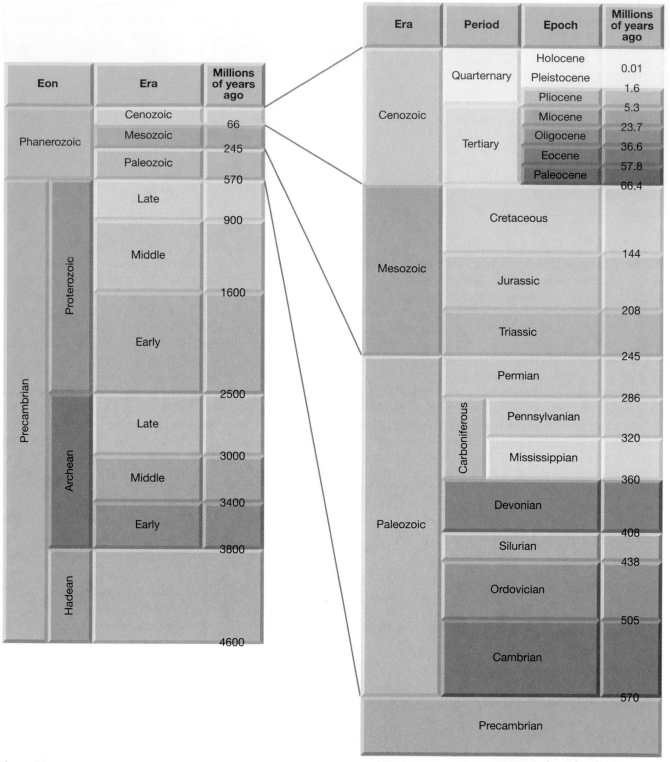

Figure 10.17

The geologic time scale. The absolute dates were added long after the time scale had been established using relative dating techniques. (Data from Geological Society of America)

we know. Certainly more data and information exist about the past ten years than for the first decade of the twentieth century; the events of the nineteenth century have been documented much better than the events of the first century A.D.; and so on. So it is with Earth history. The more recent past has the freshest, least disturbed, and most observable record. The farther back in time the geologist goes, the more fragmented the record and clues become.

Difficulties in Dating the Geologic Time Scale

Although reasonably accurate absolute dates have been worked out for the periods of the geologic time scale (see Figure 10.17), the task is not without difficulty. The primary problem in assigning absolute dates to units of time is the fact that not all rocks can be dated radiometrically. Recall that for a radiometric date to be useful, all minerals in the rock must have formed at approximately the same time. For this reason, radioactive isotopes can be used to determine when minerals in an igneous rock crystallized and when pressure and heat created new minerals in a metamorphic rock.

However, samples of sedimentary rock can only rarely be dated directly by radiometric means. A sedimentary rock may include particles that contain radioactive isotopes, but the rock's age cannot be accurately determined because the grains composing the rock are not the same age as the rock in which they occur. Rather, the sediments have been weathered from rocks of diverse ages.

Radiometric dates obtained from metamorphic rocks may also be difficult to interpret, because the age of a particular mineral in a metamorphic rock does not necessarily represent the time when the rock initially formed. Instead, the date may indicate any one of a number of subsequent metamorphic phases.

If samples of sedimentary rocks rarely yield reliable radiometric ages, how can absolute dates be assigned to sedimentary layers? Usually the geologist must relate them to datable igneous masses, as in Figure 10.18. In this example, radiometric dating has determined the ages of the volcanic ash bed within the Morrison Formation and the dike cutting the Mancos Shale and Mesaverde Formation. The sedimentary beds below the ash are obviously older than the ash, and all the layers above the ash are younger (principle of superposition). The dike is younger than the Mancos Shale and the Mesaverde Formation but older than the Wasatch Formation because the dike does not intrude the Tertiary rocks (cross-cutting relationships).

From this kind of evidence, geologists estimate that a part of the Morrison Formation was deposited about 160 million years ago, as indicated by the ash bed. Further, they conclude that the Tertiary period began after the intrusion of the dike, 66 million years ago. This is one example of literally thousands that illustrates how datable materials are used to *bracket* the various episodes in Earth history within specific time periods. It shows the necessity of combining laboratory dating methods with field observations of rocks.

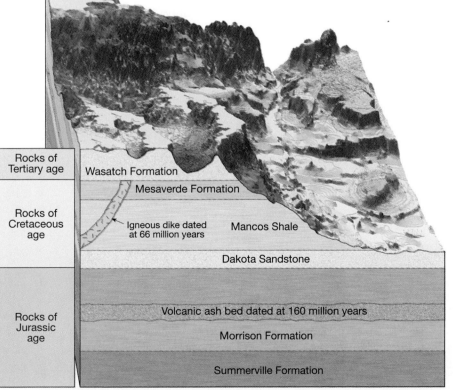

Figure 10.18

Absolute dates for sedimentary layers are usually determined by examining their relationship to igneous rocks. (After U.S. Geological Survey)

Rocks of Tertiary age

Rocks of Cretaceous age

Rocks of Jurassic age

Wasatch Formation

Mesaverde Formation

Igneous dike dated at 66 million years

Mancos Shale

Dakota Sandstone

Volcanic ash bed dated at 160 million years

Morrison Formation

Summerville Formation

Review Questions

1. Contrast the philosophies of catastrophism and uniformitarianism. How did the proponents of each perceive the age of Earth?

2. Distinguish between absolute and relative dating.

3. What is the law of superposition? How are cross-cutting relationships used in relative dating?

4. When you observe an outcrop of steeply inclined sedimentary layers, what principle allows you to assume that the beds became tilted *after* they were deposited?

5. Refer to Figure 10.5 and answer the following questions:
   (a) Is fault A older or younger than the sandstone layer?
   (b) Is dike A older or younger than the sandstone layer?
   (c) Was the conglomerate deposited before or after fault A?
   (d) Was the conglomerate deposited before or after fault B?
   (e) Which fault is older, A or B?
   (f) Is dike A older or younger than the batholith?

6. A mass of granite is in contact with a layer of sandstone. Using a principle described in this chapter, explain how you might determine whether the sandstone was deposited on top of the granite or the granite was intruded from below after the sandstone was deposited.

7. Distinguish among angular unconformity, disconformity, and nonconformity.

8. What is meant by the term *correlation*?

9. List and briefly describe a least five different types of fossils.

10. List two conditions that improve an organism's chances of being preserved as a fossil.

11. Why are fossils such useful tools in correlation?

12. In addition to being important aids in dating and correlating rocks, how else are fossils helpful in geologic investigations?

13. If a radioactive isotope of thorium (atomic number 90, mass number 232) emits 6 alpha particles and 4 beta particles during the course of radioactive decay, what are the atomic number and mass number of the stable daughter product?

14. Why is radiometric dating the most reliable method of dating the geologic past?

15. Assume that a hypothetical radioactive isotope has a half-life of 10,000 years. If the ratio of radioactive parent to stable daughter product is 1:3, how old is the rock containing the radioactive material?

16. To make calculations easier, let us round the age of Earth to five billion years.
   (a) What fraction of geologic time is represented by recorded history (assume 5000 years for the length of recorded history)?
   (b) The first abundant fossil evidence does not appear until the beginning of the Cambrian period (approximately 600 million years ago). What percentage of geologic time is represented by abundant fossil evidence?

17. What subdivisions make up the geologic time scale? What is the primary basis for differentiating the eras?

18. Briefly describe the difficulties in assigning absolute dates to layers of sedimentary rock.

solar system led most astronomers to conclude that its members formed at essentially the same time and from the same primordial material. This proposal, known as the **nebular hypothesis** , suggests that the bodies of our solar system condensed from an enormous cloud. It was composed mostly of hydrogen and helium, with only a small percentage of all other heavier elements.

About five billion years ago, this huge cloud of minute rocky fragments and gases began to contract under its own gravitational influence (Figure 11.2A). The contracting material somehow began to rotate. Like a spinning ice skater pulling in her arms, the cloud rotated faster and faster as it contracted. This rotation in turn caused the nebular cloud to flatten into a disk (Figure 11.2B). Within the rotating disk, smaller condensations formed nuclei form which the planets eventually coalesced. However, the great concentration of material was pulled toward the center of this rotating mass. As it packed inward upon itself, it gravitationally heated, forming the hot *protosun* (sun in the making).

After the protosun formed, the temperature out in the rotating disk dropped significantly. This cooling caused substances with high melting points to condense into small particles, perhaps the size of sand grains. Iron and nickel solidified first. Next to condense were the elements of which rocky substances are composed. As these fragments collided over a few tens of millions of years, they accreted into the planets (Figure 11.2 C, D). In the same manner, but on a lesser scale, the processes of condensation and accretion acted to form the moons and other small bodies of the solar system.

As the *protoplanets* (planets in the making) accumulated more and more debris, the space within the solar system began to clear. This removal of debris allowed sunlight to reach planetary surfaces unimpeded and to heat them. The resulting high surface temperatures of the inner planets (Mercury, Venus, Earth, Mars), coupled with their comparatively weak gravitational fields, meant that Earth and its neighbors were unable to retain appreciable amounts of the lighter components of the primordial cloud. These light materials, which included hydrogen, helium, ammonia, methane, and water, vaporized from their surfaces and were eventually whisked from the inner solar system by streams of solar particles called the *solar winds*.

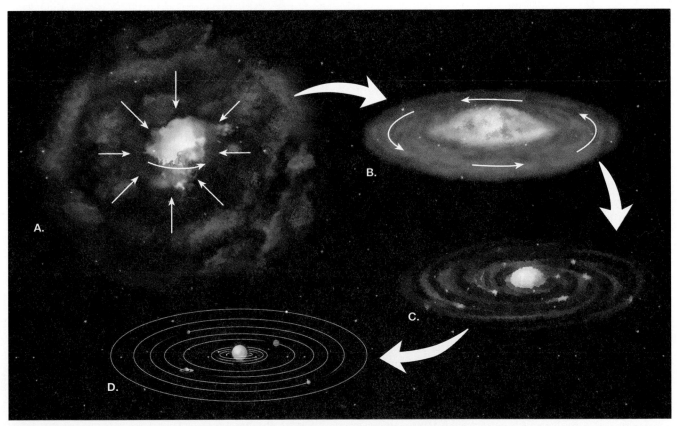

Figure 11.2

Nebular hypothesis. **A.** The nebula, a huge rotating cloud of dust and gases, began to contract. **B.** Most of the material was gravitationally swept toward the center, producing the sun. However, due to rotational motion, some dust and gases remained orbiting the central body as a flattened disk. **C.** The planets began to accrete from the material that was orbiting within the flattened disk. **D.** In time, most of the remaining debris either coalesced to form the nine planets and their moons, or was swept into space by the solar wind.

At distances beyond Mars, temperatures were much cooler. Consequently, the large outer planets (Jupiter, Saturn, Uranus, and Neptune) accumulated huge amounts of hydrogen and other light materials from the primordial cloud. The accumulation of these gaseous substances is thought to account for the comparatively large sizes and low densities of the outer planets.

Shortly after Earth formed, the decay of radioactive elements, coupled with heat released by colliding particles, produced at least some melting of the interior. Melting allowed the denser elements, principally iron and nickel, to sink to Earth's center, while the lighter rocky components floated outward, toward the surface. The sorting of material by density, which began early in Earth's history, is believed to be occurring still, but on a much smaller scale. As a result of this differentiation, Earth's interior is not homogeneous. Rather, it consists of shells or spheres of materials that have different properties.

An important consequence of this period of differentiation is that gaseous materials were allowed to escape from Earth's interior, just as gases are given off today during volcanic eruptions. By this process, an atmosphere gradually evolved, composed chiefly of gasses expelled from within the planet.

Earth's Atmosphere Evolves

Today, the air you breathe is a stable mixture of 79 percent nitrogen, 20 percent oxygen, about 1 percent argon (an inert gas), and trace gases like carbon dioxide and water vapor. But our planet's original atmosphere, several billion years ago, was far different.

Earth's very earliest atmosphere probably was swept into space by the *solar wind* , a vast stream of particles emitted by the sun. As Earth slowly cooled, a more enduring atmosphere formed. The molten surface solidified into a crust, and gases that had been dissolved in the molten rock were gradually released, a process called **outgassing**. Outgassing continues today from hundreds of active volcanoes worldwide. Thus, geologists hypothesize that Earth's original atmosphere was made up of gases similar to those released in volcanic emissions today: water vapor, carbon dioxide, nitrogen, and several trace gases.

As the planet continued to cool, the water vapor condensed to form clouds, and great rains commenced. At first, the water evaporated in the hot air before reaching the ground, or quickly boiled away upon contacting the surface, just like water sprayed on a hot grill. This accelerated the cooling of Earth's crust. When the surface had cooled below water's boiling point (100°C or 212°F), torrential rains slowly filled low areas, forming the oceans. This reduced not only the water vapor in the air, but the amount of carbon dioxide as well, for it became dissolved in the water. What remained was a nitrogen-rich atmosphere.

If Earth's primitive atmosphere resulted from volcanic outgassing, we have a problem, because volcanoes do not emit free oxygen. Where did the very significant percentage of oxygen in our present atmosphere (20 percent) come from?

The major source of oxygen is green plants. Put another way, *life itself* has strongly influenced the composition of our present atmosphere. Plants did not just adapt to their environment; they actually influenced it, dramatically altering the composition of the entire planet's atmosphere by using carbon dioxide and releasing oxygen. This is a good example of how Earth operates as a giant system in which living things interact with their environment.

How did plants come to alter the atmosphere? The key is the way in which plants create their own food. They employ *photosynthesis*, in which they use light energy to synthesize food sugars from carbon dioxide and water. The process releases a waste gas, oxygen. Those of us in the animal kingdom rely on oxygen to metabolize our food, and we in turn exhale carbon dioxide as a waste gas. The plants use this carbon dioxide for more photosynthesis, and so on, in a continuing system.

The first life-forms on Earth, probably bacteria, did not need oxygen. Their life processes were geared to the earlier, oxygenless atmosphere. Even today, many *anaerobic* bacteria thrive in environments that lack free oxygen. Later, primitive plants evolved that used photosynthesis and released oxygen. Slowly, the oxygen content of Earth's atmosphere increased. The Precambrian rock record suggests that much of the first free oxygen did not remain free because it combined with (oxidized) other substances dissolved in water, especially iron. Iron has tremendous affinity for oxygen, and the two elements combine to form iron oxides (rust) at any opportunity.

Then, once the available iron satisfied its need for oxygen, substantial quantities of oxygen accumulated in the atmosphere. By the beginning of the Paleozoic era, about four billion years into Earth's existence (after seven-eighths of Earth's history had transpired), the fossil record reveals abundant ocean-dwelling organisms that require oxygen to live. Hence, the composition of Earth's atmosphere has evolved together with its life-forms, from an oxygenless envelope to today's oxygen-rich environment.

With this background on the evolution of the solar system, sun, Earth, and atmosphere, we will look at Earth's history.

Precambrian Time: Vast and Enigmatic

The Precambrian encompasses immense geological time, from Earth's distant beginnings 4.6 billion years ago until the start of the Cambrian period, some four billion years later. Thus, the Precambrian spans about 87 percent of Earth's history. Our knowledge of this ancient time is sketchy, for much of the early rock record has been obscured by the very Earth processes you have been studying, especially plate tectonics, erosion, and deposition.

Untangling the long, complex Precambrian rock record is a formidable task, and we are far from done. Most Precambrian rocks are devoid of fossils, which hinders correlation of rocks. Rocks of this great age are metamorphosed and deformed, extensively eroded, and obscured by overlying strata. Consequently, this least-understood span of Earth's history has not been successfully divided into briefer time units, as have later intervals. Indeed, Precambrian history is written in scattered, speculative episodes, like a long book with many missing chapters.

Precambrian Rocks

Looking upon Earth from the Space Shuttle, astronauts see plenty of ocean (71 percent) and much less land area (29 percent). Over large expanses of the continents, the orbiting space scientists gaze upon many Paleozoic, Mesozoic, and Cenozoic rock surfaces, but fewer Precambrian surfaces. This demonstrates the law of super-

position: Precambrian rocks in these regions are buried from view beneath varying thicknesses of more recent rocks. Here, Precambrian rocks peek through the surface where younger strata are extensively eroded, as in the Grand Canyon and in some mountain ranges. However, on each continent, large "core areas" of Precambrian rocks dominate the surface, mostly as deformed metamorphic rocks. These areas are called **shields** because they roughly resemble a warrior's shield in shape.

Figure 11.3 shows these shield areas of Precambrian rocks worldwide. In North America (including Greenland), the Canadian Shield encompasses 7.2 million square kilometers (2.8 million square miles), the equivalent of about ten states of Texas put together.

Much of what we know about Precambrian rocks comes from mining the ores that some contain. The mining of iron, nickel, gold, silver, copper, chromium, uranium, and diamonds has provided Precambrian rock samples for study, and surveys to locate valuable ore deposits have revealed much about the rocks.

Noteworthy are extensive iron ore deposits. Rocks from the middle Precambrian (1.2–2.5 billion years ago) contain most of Earth's iron ore, mainly as the mineral hematite (Fe_2O_3). These iron-rich sedimentary rocks probably represent the time when oxygen became sufficiently abundant to react with iron dissolved in shallow lakes and seas. Later, after much of the iron was oxidized and deposited on lake and sea bottoms, formation of these iron-rich deposits declined and oxygen levels in the ocean and atmosphere began to increase. Because

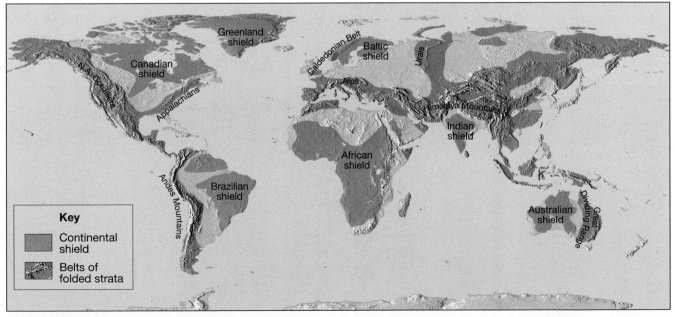

Figure 11.3
Today's continents look very different from the Precambrian. Remnants of Precambrian rocks are the continental shields, composed largely of metamorphic rocks.

most of Earth's free oxygen results from plant photosynthesis, the formation of extensive Precambrian iron ore deposits is linked to life in the sea.

Notably absent in the Precambrian are fossil fuels (coal, oil, natural gas). The reason is clear—a virtual absence of land plants to form coal swamps and of certain animals to form petroleum. Fossil fuels are from a later time.

Precambrian Fossils

A century ago, the earliest fossils known dated from the Cambrian period, about 570 million years ago. None were known from the Precambrian. This created a major problem for science at that time: How could complex organisms like trilobites abruptly appear in the geologic record? The answer, or course, was that there were fossils in Precambrian rocks, but they were rare, small, obscure, and simply had not been discovered yet. Today, our knowledge of Precambrian life, although far from complete, is quite extensive.

Precambrian fossils are disappointing if you are expecting to see fascinating plants and large animals, for these had not yet evolved. Instead, the most common Precambrian fossils are **stromatolites**. These are distinctively layered mounds or columns of calcium carbonate (Figure 11.4). Stromatolites are not the remains of actual organisms, but are material deposited by algae. They are indirect evidence of algae because they closely resemble similar deposits made by modern algae.

Stromatolites did not become common until the middle Precambrian, around 2 billion years ago. Stromatolites are large, but most actual organisms preserved in Precambrian rocks are microscopic. Well-preserved remains of many tiny organisms have been discovered, extending the record of life back beyond 3.5 billion years.

Many of these most ancient fossils are preserved in *chert*, a hard, dense chemical sedimentary rock. Chert must be very thinly sliced and studied under powerful microscopes to observe bacteria and algae fossils within it.

Microfossils have been found at several locations worldwide. Two notable areas are in southern Africa, where the rocks date to more than 3.1 billion years, and in the Gunflint Chert (named for its use in flintlock rifles) of Lake Superior, which dates to 1.7 billion years. In both places, bacteria and blue-green algae have been discovered. The fossils are of the most primitive organisms, *prokaryotes*. Their cells lack organized nuclei, and they reproduce asexually.

More advanced organisms, *eukaryotes*, have cells that contain nuclei. Eukaryotes are among billion-year-old fossils discovered at Bitter Springs in Australia, such as green algae. Unlike prokaryotes, eukaryotes reproduce sexually, which means that genetic material is exchanged between organisms. This reproductive mode permits greatly increased genetic variation. Thus, development of eukaryotes may have dramatically increased the rate of evolutionary change.

Plant fossils date from the middle Precambrian, but animal fossils came a bit later, in the late Precambrian. Many of these fossils are *trace fossils*, meaning that they are not of the animals themselves, but of their activities, such as trails and worm holes. Areas in Australia and Newfoundland have yielded hundreds of fossil impressions of soft-bodied creatures. Most, if not all, of the Precambrian fauna lacked shells, which would develop as protective armor during the Paleozoic.

As the Precambrian came to a close, the fossil record disclosed diverse and complete multicelled organisms. This set the stage for more complex plants and animals to evolve at the dawn of the Paleozoic era.

A.   **B.**

Figure 11.4

Stromatolites are among the most common Precambrian fossils. **A.** Precambrian fossil stromatolites composed of calcium carbonate deposited by algae in the Helena formation, Glacier National Park. (Photo by Ken M. Johns/Photo Researchers, Inc.) **B.** Modern stromatolites growing in shallow saline area, western Australia. (Photo by Bill Bachman/Photo Researchers, Inc.)

Paleozoic Era: Life Explodes

Following the long Precambrian, the most recent 570 million years of Earth history are divided into three eras: Paleozoic, Mesozoic, and Cenozoic. The Paleozoic era encompasses more than 345 million years and is by far the longest of the three. Seven periods make up the Paleozoic era (Figure 11.5).

Before the Paleozoic, life-forms possessed no hard parts—shells, scales, bones, or teeth. The beginning of the Paleozoic is marked by the appearance of the first life-forms with hard parts (Figure 11.6). Hard parts greatly enhanced their chance of being preserved as part of the fossil record. Therefore, our knowledge of life's diversification improves greatly from the Paleozoic onward. This diversity is demonstrated in Figure 11.7.

Era	Period	Epoch		Development of Plants and Animals
Cenozoic	Quarternary	Holocene	0.01	Humans develop
		Pleistocene	1.6	
	Tertiary	Pliocene	5.3	"Age of Mammals"
		Miocene	23.7	
		Oligocene	36.6	
		Eocene	57.8	
		Paleocene	66.4	Extinction of dinosaurs and many other species
Mesozoic	Cretaceous	"Age of Reptiles"		First flowering plants
		144		First birds
	Jurassic			
		208		Dinosaurs dominant
	Triassic			
		245		
Paleozoic	Permian	"Age of Amphibians"		Extinction of trilobites and many other marine animals
		286		First reptiles
	Carboniferous — Pennsylvanian			
		320		Large coal swamps
	Carboniferous — Mississippian			Amphibians abundant
		360		
	Devonian	"Age of Fishes"		First insect fossils
				Fishes dominant
		408		
	Silurian			First land plants
		438		
	Ordovician	"Age of Invertebrates"		First fishes
		505		Trilobites dominant
	Cambrian			
				First organisms with shells
		570		
Precambrian—comprises about 87% of geologic time				First multicelled organisms
				First one-celled organisms
				Origin of the earth
		4600		

Figure 11.5

The geologic time scale. Numbers on the time scale represent time in millions of years before the present. These dates were added long after the time scale had been established using relative dating techniques. The Precambrian accounts for about 87 percent of geologic time. (Data from Geological Society of America)

A.

B.

Figure 11.6

Fossils of common Paleozoic life-forms. **A.** Natural cast of a trilobite. Trilobites dominated the Paleozoic ocean, scavenging food from the bottom. **B.** Extinct coiled cephalpods. Like their modern descendants, these were highly developed marine organisms.

Abundant Paleozoic fossils have allowed geologists to construct a far more detailed time scale for the last one-eighth of geologic time than for the preceding seven-eighths, the Precambrian. Moreover, because every organism is associated with a particular environment, the greatly improved fossil record provided invaluable information for deciphering ancient environments. To facilitate our brief tour of the Paleozoic, we divide it into Early Paleozoic (Cambrian, Ordovician, Silurian periods) and Late Paleozoic (Devonian, Mississippian, Pennsylvanian, Permian periods).

Early Paleozoic History

The early Paleozoic consists of a 162-million-year span that embraces the Cambrian, Ordovician and Silurian periods. Anyone approaching Earth from space at this time would have seen the familiar blue planet with plentiful white clouds, but the arrangement of continents would have looked very different than today (Figure 11.8). At this time, the vast southern continent of Gondwanaland encompassed five continents (South America, Africa, Australia, Antarctica, India, and perhaps China). Evidence of an extensive continental glaciation places western Africa near the South Pole!

Landmasses that were not part of Gondwanaland existed as five separate units and some scattered fragments. Although the exact position of these "northern"

continents is uncertain, ancestral North America and Europe are thought to have been near the equator and separated by a narrow sea, as shown on the map.

As the Paleozoic opened, North America was a land with no living things, plant or animal. There were no Appalachian or Rocky Mountains; rather, the continent was largely a barren lowland. Several times during the Cambrian and Ordovician periods, shallow seas moved inland and then receded from the interior of the continent. Deposits of clean sandstones, used today to make glass, mark the edge of these shallow seas in the midcontinent.

Early in the Paleozoic, a mountain-building event affected eastern North America from the present-day central Appalachians to Newfoundland. The mountains produced during this event, the Taconic orogeny, have since eroded away, leaving behind deformed strata and a large volume of detrital sedimentary rocks that were derived from the weathering of these mountains.

During the Silurian period, much of North America was once again inundated by shallow seas. This time large barrier reefs restricted circulation between shallow marine basins and the open ocean. Water in these basins evaporated, causing deposition of large quantities of rock salt and gypsum. Today these thick *evaporite beds* are important resources for the chemical, rubber, plasterboard, and photographic industries in Ohio, Michigan, and western New York State.

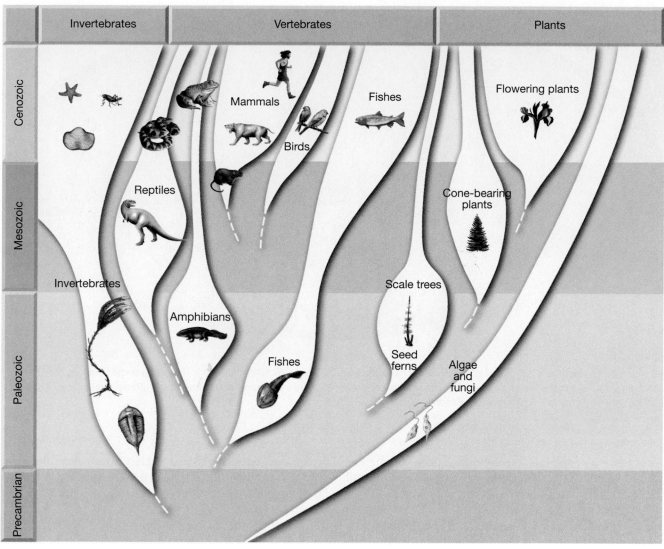

Figure 11.7

This chart indicates the times of appearance and relative abundance of major groups of organisms. The wider the band, the more dominant the group.

Early Paleozoic Life

Life in early Paleozoic time was restricted to the seas. Vertebrates had not yet evolved, so life consisted of several invertebrate groups (shown in Figure 11.9). The Cambrian period was the golden age of *trilobites*. More than 600 genera of these mud-burrowing scavengers flourished worldwide. By Ordovician times, *brachiopods* outnumbered the trilobites. Brachiopods are among the most widespread Paleozoic fossils and, except for one modern group, are now extinct. Although the adults lived attached to the sea floor, the young larvae were free-swimming. This mobility accounts for the group's wide geographic distribution.

The Ordovician also marked the appearance of abundant *cephalopods*, mobile, highly developed mollusks that became the major predators of their time. The

descedents of cephalopods include the modern squid, octopus, and nautilus. Cephalopods were the first truly large organisms on Earth (Figure 11.9). Whereas the largest trilobites seldom exceeded 30 centimeters (12 inches) in length and the biggest brachiopods were no more than about 20 centimeters (8 inches) across, one species of cephalopod reached a length of nearly 10 meters (30 feet).

The beginning of the Cambrian period marks an important event in animal evolution. For the first time, organisms appeared that secreted material which formed *hard parts*, such as shells. Why several diverse life-forms began to develop hard parts about the same time remains unanswered. One proposal suggests that, because an external skeleton provides protection from predators, hard parts evolved for survival. Yet, the fossil record does not seem to support this hypothesis. Organisms with hard

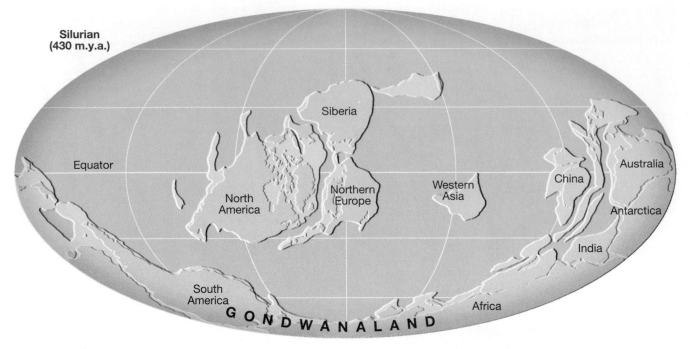

**Silurian
(430 m.y.a.)**

Siberia

Equator

North
America

Northern
Europe

Western
Asia

China

Australia

Antarctica

India

South
America

Africa

GONDWANALAND

Figure 11.8

Reconstruction of Earth as it may have appeared in early Paleozoic time. The southern continents were joined into a single landmass called Gondwanaland. Four of the five landmasses that would later join to form the northern continent of Laurasia lay scattered roughly along the equator. (After C. Scotese, R. K. Bambach, C. Barton, R. VanderVoo, and A. Ziegler)

parts were plentiful in the Cambrian period, whereas predators such as cephalopods were not abundant until the Ordovician period, some 70 million years later.

Whatever the answer, hard parts clearly served many useful purposes and aided adaptations to new ways of life. Sponges, for example, developed a network of fine interwoven silica spicules that allowed them to grow larger and more erect, capable of extending above the surface in search for food. Mollusks (clams and snails) secreted external shells of calcium carbonate that protected them and allowed body organs to function in a more controlled environment. The successful trilobites developed an exoskeleton of a protein called *chitin*, which permitted them to burrow through soft sediment in search of food (Figure 11.10).

Late Paleozoic History

The late Paleozoic consists of four periods—the Devonian, Mississippian, Pennsylvanian, and Permian—that span about 160 million years. Tectonic forces reorganized Earth's landmasses during this time, culminating with the formation of the supercontinent *Pangaea* (Figure 11.11).

Forming Pangaea, the Supercontinent. As ancestral North America collided with Africa, the narrow sea that separated these landmasses began to close slowly (compare Figure 11.11B and 11.11C). Strong compressional forces from this collision deformed the rocks to produce the original northern Appalachian Mountains of eastern North America.

During the fusion of North America and Africa, the other northern continents began to converge (Figure 11.11). By the Permian period, this newly formed landmass had collided with western Asia and the Siberian landmass along the line of the Ural Mountains. Through this union, the northern continent of *Laurasia* was born, encompassing present-day North America, Europe, western Asia, Siberia, and perhaps China.

As Laurasia was forming, Gondwanaland migrated northward. By the Pennsylvanian period, Gondwanaland collided with Laurasia, forming a mountainous belt through central Europe. Simultaneously, a collision between the African fragment of Gondwanaland and the southeastern edge of North America rumpled up the southern Appalachian Mountains.

By the close of the Paleozoic, all the continents had fused into the supercontinent of Pangaea (Figure 11.11). With only a single vast continent, the world's climate became very seasonal, having extremes far greater than those we experience today. As illustrated in Box 11.1, these altered climatic conditions caused one of the most dramatic biological declines in all of Earth history.

Figure 11.9
During the Ordovician period (505-438 million years ago), the shallow waters of an inland sea over central North America contained an abundance of marine invertebrates. Shown in this reconstruction are straight-shelled cephalopods, trilobites, brachiopods, snails, and corals. (© The Field Museum, Neg. #GEO80820c, Chicago)

Late Paleozoic Life

During most of the late Paleozoic, organisms diversified dramatically. Some 400 million years ago, plants that had adapted to survive at the water's edge began to move inland, becoming *land plants*. These earliest land plants were leafless, vertical spikes about the size of your index finger. However, by the end of the Devonian, 40 million years later, the fossil record indicates the existence of forests with trees tens of meters high.

In the oceans, armor-plated fishes that had evolved during the Ordovician continued to adapt. Their armor plates thinned to lightweight scales that increased their speed and mobility (Figure 11.12). Other fishes evolved during the Devonian, including primitive sharks that had a skeleton made of cartilage and bony fishes, the groups to which virtually all modern fishes belong. Because of this, the Devonian period is often called the "age of fishes."

By late Devonian time, two groups of bony fishes, the lung fish and the lobe-finned fish, became adapted to land environments. Not unlike their modern relatives, these fishes had primitive lungs that supplemented their breathing through gills. It is believed that the lobe-finned fish occupied tidal flats or small ponds and that in times of drought they may have used their bone fins to "walk" from dried-up pools in search of other ponds. Through time, the lobe-finned fish began to rely more on their lungs and less on their gills. By late Devonian time, they had evolved into true air-breathing amphibians with fishlike heads and tails. It should be noted that insects had already invaded the land.

Modern amphibians, like frogs, toads, and salamanders, are small and occupy limited biological niches. But conditions during the remainder of the Paleozoic were ideal for these newcomers to the land. Plants and insects, which were their main diet, already were very abundant and large. Having only minimal competition from other land dwellers, the amphibians rapidly diversified. Some groups took on roles and forms that were more similar to modern reptiles, such as crocodiles, than to modern amphibians (Figure 11.13).

Figure 11.10
These trilobites date from the Devonian period, 408 to 360 million years ago. (Photo by Phil Degginger/Tony Stone Images)

By the Pennsylvanian period, large tropical swamps extended across North America, Europe, and Siberia (Figure 11.14). Trees approached 30 meters (100 feet), with trunks over one meter across. The coal deposits that fueled the Industrial Revolution, and which provide a substantial portion of our electric power today, originated in these vast swamps. Further, it was in the lush coal swamp environment of the late Paleozoic that the amphibians evolved quickly into a variety of species.

Mesozoic Era: Age of the Dinosaurs

Spanning nearly 180 million years, the Mesozoic era is divided into three periods: the Triassic, Jurassic, and Cretaceous. The Mesozoic era witnessed the beginning of the breakup of the supercontinent Pangaea. Also in this era, organisms that had survived the great Permian extinction began to diversify in spectacular ways (see Box 11.1). On land, dinosaurs became dominant and remained unchallenged for over 100 million years. Because of this, the Mesozoic Era is often called the "age of dinosaurs."

Early geologists recognized a profound difference between the fossils in Permian strata and those in younger Triassic rocks, as if someone had drawn a bold line to separate the two time periods. Clearly one-half of the fossil groups that occurred in late Paleozoic rocks were missing in Mesozoic rocks. On this basis, it was decided to separate the Paleozoic and Mesozoic at the Permian-Triassic boundary.

Mesozoic History

The Mesozoic era began with much of the world's land above sea level. In fact, in North America no period exhibits a more meager marine sedimentary record than the Triassic period. Of the exposed Triassic strata, most are red sandstones and mudstones that lack fossils and contain features indicating a terrestrial environment.

As the second period opened, the Jurassic, the sea invaded western North America. Adjacent to this shallow sea, extensive continental sediments were deposited on what is now the Colorado Plateau. The most prominent is the Navajo Sandstone, a windblown, white quartz sandstone that in places approaches a thickness of 300 meters (1000 feet). These massive dunes indicate that a major desert occupied much of the American Southwest during early Jurassic times.

A well-known Jurassic deposit is the Morrison Formation, within which is preserved the world's richest storehouse of dinosaur fossils. Included are fossilized bones of huge dinosaurs such as *Apatosaurus* (formerly *Brontosaurus*), *Brachiosaurus*, and *Stegosaurus*.

As the Jurassic period gave way to the Cretaceous, shallow seas once again invaded much of western North America, the Atlantic, and Gulf coastal regions. This created great swamps like those of the Paleozoic era, forming Cretaceous coal deposits that are very important economically in the western United States and Canada. For example, on the Crow Indian reservation in Montana, there exists nearly 20 billion tons of high-quality coal of Cretaceous age.

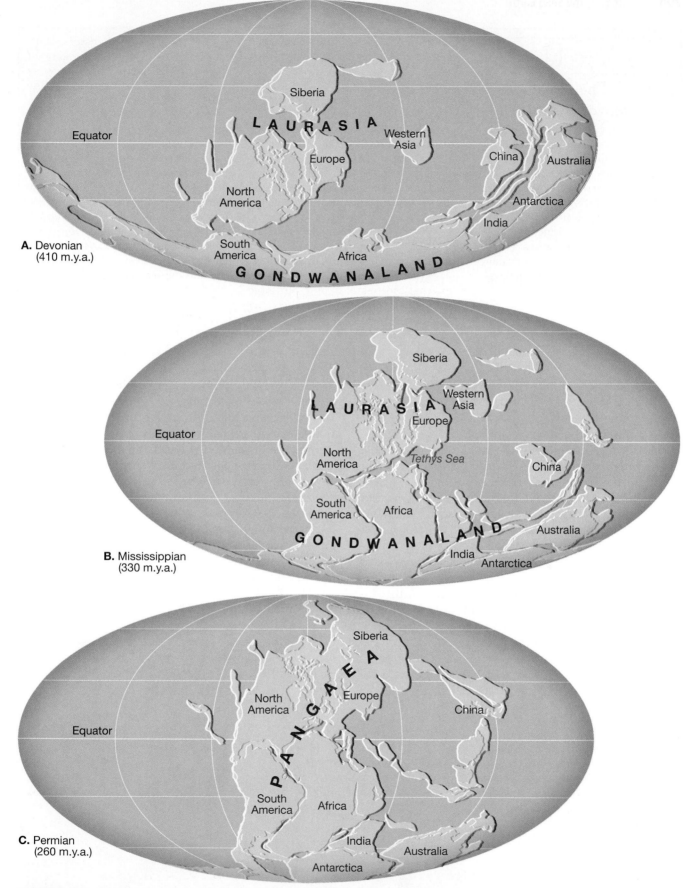

Figure 11.11

During the late Paleozoic, plate movements were joining together the major landmasses to produce the supercontinent of Pangaea. (After C. Scotese, R. K. Bambach, C. Barton, R. VanderVoo, and A. Ziegler)

Box 11.1

The Great Paleozoic Extinction

The Paleozoic ended with the Permian period, a time when Earth's major landmasses joined to form the supercontinent Pangaea. This redistribution of land and water and changes in the elevations of landmasses brought pronounced changes in world climates. Broad areas of the northern continents became elevated above sea level and the climate grew drier. These changes apparently triggered extinctions of many species on land and sea.

By the close of the Permian, 75 percent of the amphibian families had disappeared, and plants had declined in number and variety. Although many amphibian groups became extinct, their descendants, the reptiles, would become the most successful and advanced animals on Earth. Marine life was not spared. At least 80 percent, and perhaps as much as 95 percent, of marine life disappeared. Many marine invertebrates that had been dominant during the Paleozoic, including all the remaining trilobites as well as some types of corals and brachiopods, failed to adapt to the widespread environmental changes.

The late Paleozoic extinction was the greatest of at least five mass extinctions to occur over the past 600 million years. Each extinction wreaked havoc with the existing biosphere, wiping out large numbers of species. In each case, however, the survivors formed new biological communities that were more diverse than their predecessors. Thus, mass extinctions actually invigorated life on Earth, as the few hardy survivors eventually filled more niches than the ones left by the victims.

The cause of the great Paleozoic extinction is uncertain. The climate changes from the formation of Pangaea and the associated drop in sea level undoubtedly stressed many species. In addition, at least two million cubic kilometers of lava flowed across Siberia to produce what is called the Siberian Traps. Perhaps debris from these eruptions blocked incoming sunlight, or perhaps enough sulfuric acid was emitted to make the seas virtually uninhabitable. Whatever caused the late Paleozoic extinction, it is clear that, without it, a very different population of organisms would today inhabit this planet.

A major event of the Mesozoic era was the breakup of Pangaea (Figure 11.15). A rift developed between what is now the eastern United States and western Africa, marking the birth of the Atlantic Ocean. It also represents the beginning of the breakup of Pangaea, a process that continued for 200 million years, through the Mesozoic and into the Cenozoic.

As Pangaea fragmented, the westward-moving North American plate began to override the Pacific plate. This tectonic event began a continuous wave of deformation that moved inland along the entire western margin of the continent. By Jurassic times, subduction of the Pacific plate had begun to produce the chaotic mixture of rocks that exists today in the Coast Ranges of California. Further inland, igneous activity was widespread, and for nearly 60 million years, huge masses of magma rose to within a few miles of the surface, where they cooled and solidified. The remnants of this intrusive activity included the granitic rocks of the Sierra Nevada, the Idaho batholith, and the Coast Range batholith of British Columbia.

Tectonic activity that began in the Jurassic continued throughout the Cretaceous, ultimately forming the vast mountains of western North America (Figure 11.16). Compressional forces moved huge rock units in a shingle-like fashion toward the east. Throughout much of the western margin of North America, older rocks were thrust eastward over younger strata, for a distance exceeding 150 kilometers (90 miles).

Toward the end of the Mesozoic, the middle and southern ranges of the Rocky Mountains formed. This mountain-building event, called the Laramide Orogeny, resulted when tectonic forces uplifted large blocks of Precambrian rocks in Colorado and Wyoming.

Mesozoic Life

As the Mesozoic era dawned, its life forms were survivors of the great Paleozoic extinction. These survivors diversified in many new ways to fill the biological voids created at the close of the Paleozoic. On land, conditions favored those that could adapt to drier climates. Among

Figure 11.12

These placoderms or "plate-skinned" fish were abundant during the Devonian (408–360 million years ago). (Drawing after A. S. Romer)

Figure 11.13
Permian scene (286-245 million years ago) shows the large amphibian *Eryops*, which exceeded 2 meters (6 feet) in length. Eryops was a carnivore, as you can see from the teeth.

plants, the gymnosperms were one such group. Unlike the first plants to invade the land, the seed-bearing gymnosperms did not depend on free-standing water for fertilization. Consequently, these plants were not restricted to a life near the water's edge.

The gymnosperms quickly became the dominant trees of the Mesozoic. They included the cycads, the conifers, and the ginkgoes. The cycads resembled a large pineapple plant. The ginkgoes had pan-shaped leaves, much like their modern relatives. Largest were the conifers, whose modern descendants include the pines, firs, and junipers. The best-known fossil occurrence of these ancient trees is in northern Arizona's Petrified Forest National Park. Here, huge petrified logs lie exposed at the surface, having been weathered from rocks of the Triassic Chinle Formation (Figure 11.17).

The Shelled Egg.   Among the animals, reptiles readily adapted to the drier Mesozoic environment. They were the first true terrestrial animals. Unlike amphibians, reptiles have shell-covered eggs that can be laid on land.

The elimination of a water-dwelling stage (like the tadpole stage in frogs) was an important evolutionary step. Of interest is the fact that the watery fluid within the reptilian egg closely resembles seawater in chemical composition. Because the reptile embryo develops in this watery environment, the shelled egg has been characterized as a "private aquarium" in which the embryos of these land vertebrates spend their water-dwelling stage of life.

Dinosaurs Dominate.   With the perfection of the shelled egg, reptiles quickly became the dominant land animals. They continued this dominance for more than 160 million years. Most awesome of the Mesozoic reptiles were the dinosaurs. Some of the huge dinosaurs were carnivorous (*Tyrannosaurus*), whereas others were herbivorous (like the ponderous *Apatosaurus* formerly *Brontosaurus*). The extremely long neck of *Apatosaurus* may have been an adaptation for feeding on tall conifer trees. However, not all dinosaurs were large. In fact, certain small forms closely resembled modern fleet-

Figure 11.14
Restoration of a Pennsylvania coal swamp (320-286 million years ago). Shown are scale trees (left), seed ferns (lower left), and scouring rushes (right). Also note the large dragonfly. (© The Field Museum, Neg. #GEO85637, Chicago. Photographer John Weinstein.)

footed lizards. Further, evidence indicates that some dinosaurs, unlike their present-day reptile relatives, were warm blooded.

The reptiles made one of the most spectacular adaptive radiations in all of Earth history (Figure 11.18). One group, the pterosaurs, took to the air. These "dragons of the sky" possessed huge membranous wings that allowed them rudimentary flight (Figure 11.19). Another group of reptiles, exemplified by the fossil *Archaeopteryx*, led to more successful flyers: the birds. Whereas some reptiles took to the skies, others returned to the sea, including the fish-eating plesiosaurs and ichthyosaurs. These reptiles became proficient swimmers, but retained their reptilian teeth and breathed by means of lungs.

At the close of the Mesozoic, many reptile groups became extinct. Only a few types survived to recent times, including the turtles, snakes, crocodiles, and lizards. The huge land-dwelling dinosaurs, the marine plesiosaurs, and the flying pterosaurs all are known only through the fossil record. What caused this great extinction? (See Box 11.2.)

Cenozoic Era: Age of Mammals

The Cenozoic era, or "era of recent life," encompasses the past 66 million years of Earth history. It is the "post-dinosaur" era, the time of mammals, including humans. It is during this span that the physical landscapes and life-forms of our modern world came into being. The Cenozoic era represents a much smaller fraction of geologic time than either the Paleozoic or the Mesozoic. Although shorter, it nevertheless possesses a rich history, because the completeness of the geologic record improves as time approaches the present. The rock formations of this time span are more widespread and less disturbed than those of any preceding time.

The Cenozoic era is divided into two periods of very unequal duration, the Tertiary period and the Quaternary period. The Tertiary period includes five epochs and embraces about 64 million years, practically all of

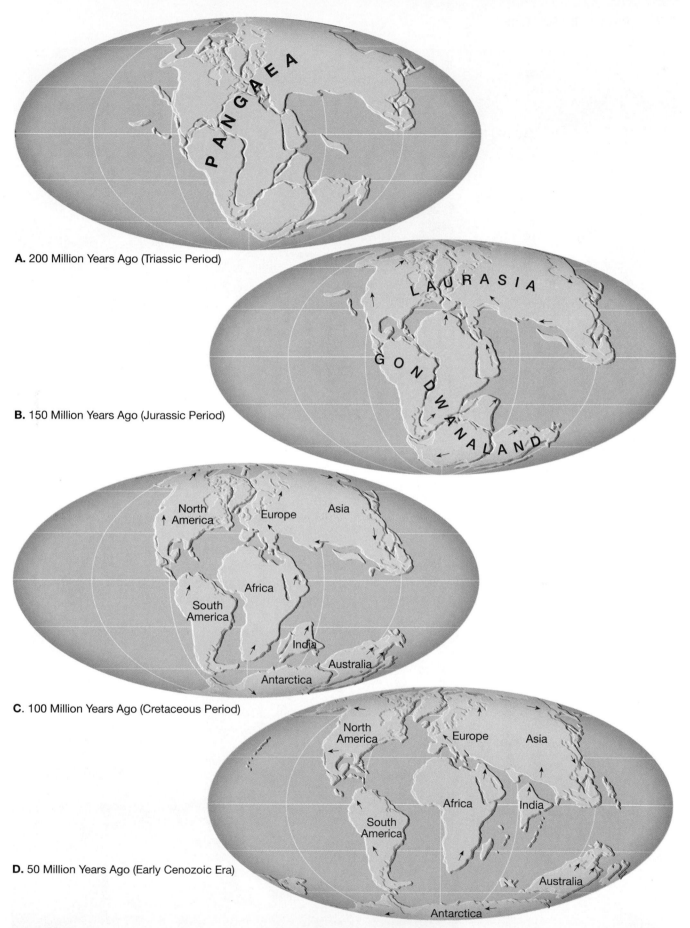

A. 200 Million Years Ago (Triassic Period)

B. 150 Million Years Ago (Jurassic Period)

C. 100 Million Years Ago (Cretaceous Period)

D. 50 Million Years Ago (Early Cenozoic Era)

Figure 11.15
The breakup of the supercontinent of Pangaea began and continued throughout the Mesozoic era. (After D. Walsh and C. Scotese)

Figure 11.16
Uplifted and deformed sedimentary strata in the Canadian Rockies near Lake Louise, Alberta, Canada. (Photo by Carr Clifton/Minden Pictures)

the Cenozoic era. The Quaternary period consists of two epochs that represent only the last two million years of geologic time.

Cenozoic North America

Most of North America was above sea level throughout the Cenozoic era. However, the eastern and western margins of the continent experienced markedly contrasting events, because of their different relationships with plate boundaries. The Atlantic and Gulf coastal regions, far removed from an active plate boundary, were tectonically stable. Western North America, on the other hand, was the leading edge of the North American plate. As a result, plate interactions during the Cenozoic gave rise to many events of mountain building, volcanism, and earthquakes in the West.

Figure 11.17
Petrified logs that have been weathered out of the Triassic Chinle Formation in the Petrified Forest of Arizona. (Photo by David Muench)

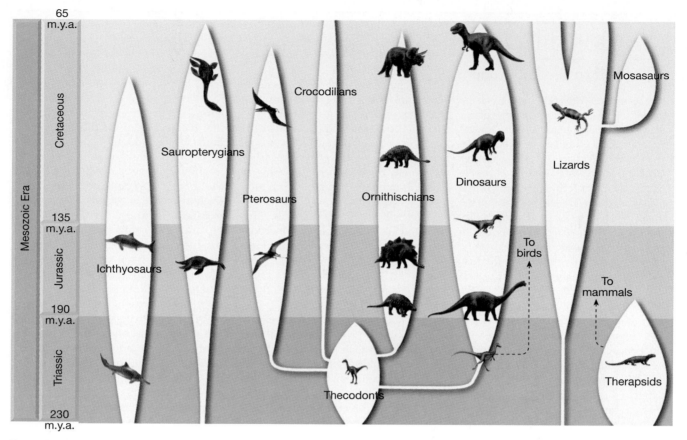

Figure 11.18

Adaptive radiation of major reptile groups during the Mesozoic era—"the golden age of reptiles." Dinosaurs, pterosaurs (flying reptiles), and large marine reptiles all became extinct by the end of the Mesozoic. (After Robert W. Tope)

Eastern North America. The stable continental margin of eastern North America was the site of abundant marine sedimentation. The most extensive deposition

Figure 11.19

Fossils of the great flying reptile Pteranodon have been recovered from Cretaceous chalk deposits located in Kansas. Although Pteranodon had a wingspan of 7 meters (22 feet), flying reptiles with twice this wingspan have been discovered in strata of west Texas.

surrounded the Gulf of Mexico, from the Yucatan Peninsula to Florida. Here, the great buildup of sediment caused the crust to downwarp and produced numerous faults. In many instances, the faults created traps in which oil and natural gas accumulated. Today, these and other petroleum traps are the most economically important resource in Cenozoic strata of the Gulf Coast, as evidenced by the Gulf's well-known offshore drilling platforms.

By early Cenozoic time, most of the original Appalachians had been eroded to a low plain. Then, by the mid-Cenozoic, isostatic adjustments raised the region once again, rejuvenating its rivers. Streams eroded with renewed vigor, gradually sculpturing the surface into its present-day topography (Figure 11.20). The sediments from all of this erosion were deposited along the eastern margin of the continent, where they attained a thickness of many kilometers. Today, portions of the strata deposited during the Cenozoic are exposed as the gently sloping Atlantic and Gulf coastal plains.

Western North America. In the West, the Laramide orogeny that built the Rocky Mountains was coming to an end. As erosion lowered the mountains, the basins between uplifted ranges filled with sediments. Eastward, a great wedge of sediment from the eroding Rockies was building, creating the Great Plains.

Box 11.2

Farewell to the Dinosaurs

The boundaries between divisions on the geologic time scale represent times of significant geological and/or biological change. About 66 million years ago, more than half of all plant and animal species died out. The event marks the end of the Mesozoic era and the beginning of the Cenozoic era. It represents the end of the era in which dinosaurs and other reptiles dominated the landscape and the beginning of the era when mammals became very important (Figure 11.A). Because the Cretaceous (abbreviated K to avoid confusion with other "C" periods) is the last period of the Mesozoic, and the Tertiary (abbreviated T) is the first period of the Cenozoic, the time of this mass extinction is called the Cretaceous-Tertiary or KT boundary.

The dinosaurs met their demise around the KT boundary, along with many other animal and plant groups, both terrestrial and marine. Equally important is the fact that many species survived. Human beings are descended from these survivors. Perhaps this fact explains why an event that occurred 66 million years ago has captured the interest of so many people.

The extinction of the great reptiles is generally attributed to this group's inability to adapt to some radical change in environmental conditions. What event could have triggered the sudden extinction of the dinosaurs—the most successful group of land animals ever to have lived?

One hypothesis is that about 66 million years ago, a large asteroid or comet, perhaps 10 kilometers in diameter, collided with Earth. The impact would have produced a dust cloud thousands of times larger than that released during most volcanic eruptions. For many months the dust-laden atmosphere would have greatly restricted the amount of sunlight that penetrated to Earth's surface. Without sunlight for photosynthesis, plants would die and plant-eaters would starve, causing delicate food chains to collapse. It is further hypothesized that large dinosaurs would be affected more adversely than would smaller life-forms. In addition, acidic rains and global fires may have added to the environmental disaster. It is estimated that when the sunlight returned, more than half of the species on Earth, including numerous marine organisms, had become extinct.

What evidence supports this catastrophic collision 66 million years ago? First a thin layer of sediment nearly 1 centimeter thick has been discovered at the KT boundary worldwide. This sediment contains a high level of the element *iridium*, which is rare in Earth's crust but is found in similar elevated proportions in stony meteorites. Could this layer represent the scattered remains of an asteroid that was responsible for the environmental changes that led to the demise of many reptile groups?

Second, this period of mass extinction appears to have affected all land animals larger than dogs. Supporters of this catastrophic-event scenario suggest that small, ratlike mammals could survive a breakdown of food chains lasting perhaps several months. Large animals, they contend, could not survive such an event.

Other scientists disagree. They claim that what appears to be a mass extinction over a short period of time actually occurred over a much broader time span. Based on an examination of the fossil record at the Cretaceous-Tertiary boundary, these geologists concluded that the decline of the dinosaurs was much more gradual.

Those who disagree with the impact hypothesis have suggested that volcanism could account for the mass extinction. In fact, enormous outpourings of basaltic lavas occurred in India approximately 66 million years ago, forming the region known as the Deccan traps. Although much more extensive, the Deccan traps resemble the flood basalts of the Columbia Plateau in the Pacific Northwest of the United States.

Whichever hypothesis is correct, the fact remains that one of the more successful groups ever to inhabit our planet, the dinosaurs, died out at the close of the Mesozoic. The decline of these reptiles provided vacancies in habitants for the mammals. Although small and inconspicuous during the Mesozoic, mammals today dominate the Cenozoic.

Figure 11.A
Dinosaurs dominated the Mesozoic landscape until their extinction at the close of the Cretaceous period. This skeleton of *Triceratops* is on display in London's Natural History Museum. (Photo by Natural History Museum, London)

Beginning in the Miocene epoch, a broad region from northern Nevada into Mexico experienced crustal movements that formed more than 150 fault-block mountain ranges. They rise abruptly above the adjacent basins, creating the Basin and Range Province (see Chapter 9).

As the Basin and Range Province was forming, the

entire western interior of the continent was gradually uplifted. This uplift re-elevated the Rockies and rejuvenated many of the West's major rivers. As the rivers became entrenched, many spectacular gorges were formed, including the Grand Canyon of the Colorado River, the Grand Canyon of the Snake River, and the Black Canyon of the Gunnison. The present topography of the Rocky Mountains is in large measure the result of this late Tertiary uplift and the subsequent excavation of the early Tertiary basin deposits by rejuvenated streams (Figure 11.21).

Volcanic activity was common in the West during much of the Cenozoic time. Beginning in the Miocene epoch, great volumes of fluid basaltic lava flowed from fissures in portions of present-day Washington, Oregon, and Idaho. These eruptions built the extensive (1.3 million square kilometers) Columbia Plateau. Immediately west of the Columbia Plateau, volcanic activity was quite different. Here, thicker magmas with a higher silica content erupted explosively, creating a chain of stratovolcanoes from northern California to the Canadian border, some of which remain active—like Mount St. Helens (Figure 11.22).

A final episode of folding occurred in the West in late Tertiary time, creating the Coast Ranges that stretch along the Pacific Coast. Meanwhile, the Sierra Nevada became fault-block mountains as they were uplifted along their eastern flank, creating the imposing mountain front we know today.

As the Tertiary period drew to a close, the effects of mountain building, volcanic activity, isostatic adjustments, and extensive erosion and sedimentation had created a physical landscape very similar to the configuration of today. All that remained of Cenozoic time was the final two-million-year episode called the Quaternary period. During this most recent (and current) phase of Earth history, in which humans evolved, the action of glacial ice and other erosional agents added the finishing touches.

Cenozoic Life

Mammals replaced reptiles as the dominant land animals in the Cenozoic. Angiosperms (flowering plants with covered seeds) replaced gymnosperms as the dominant land plants. Marine invertebrates took on a modern look. Microscopic animals called *foraminifera* became

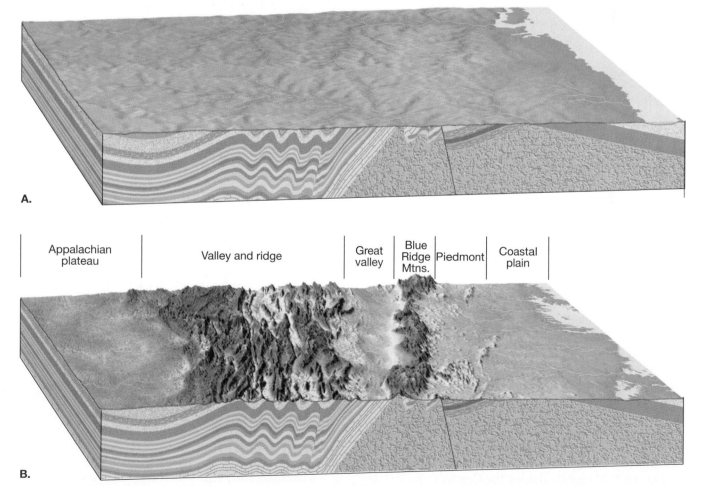

Figure 11.20
The formation of the modern Appalachian Mountains. **A.** The original Appalachians eroded to a low plain. **B.** The recent upwarping and erosion of the Appalachians began nearly 30 million years ago to produce the present topography.

Figure 11.21
Wyoming's spectacular Teton Range is part of the Rocky Mountains. They were formed during the late Tertiary uplift about 10 million years ago. (Photo by Tom Till)

especially important. Today foraminifera are among the most intensely studied of all fossils because their widespread occurrence makes them invaluable in correlating Tertiary sediments. Tertiary strata are very important to the modern world, for they yield more oil than rocks of any other age.

The Cenozoic is often called the "age of mammals," because these animals came to dominate land life. It could also be called the "age of flowering plants," for the angiosperms enjoy a similar status in the plant world. As a result of advances in seed fertilization and dispersal, angiosperms experienced rapid development and expansion as the Mesozoic drew to a close. Thus, as the Cenozoic era began, angiosperms were already the dominant land plants.

Development of the flowering plants, in turn, strongly influenced the evolution of both birds and mammals. Birds that feed on seeds and fruits, for example, evolved rapidly during the Cenozoic in close association with the flowering plants. During the middle Tertiary, grasses developed rapidly and spread over the plains. This fostered the emergence of herbivorous (plant-eating) mammals that were mainly grazers. In turn, the development and spread of grazing animals established the setting for the evolution of the carnivorous mammals that preyed upon them.

Mammals Replace Reptiles.   Back in the Mesozoic, an important evolutionary event was the appearance of primitive mammals in the late Triassic time, about the same time that the dinosaurs emerged. Yet throughout the period of dinosaur dominance, mammals remained in the background as small and inconspicuous animals. By the close of the Mesozoic era, dinosaurs and other reptiles no longer dominated the land. It was only after these large reptiles became extinct that mammals came into their own as the dominant land animals. The transition is a major example in the fossil record of the replacement of one large group by another.

Mammals are distinct from reptiles in important respects. Mammalian young are born live and mammals maintain a steady body temperature, that is, they are

Figure 11.22
Mount Hood, Oregon. This dormant volcano is one of several large composite cones that comprise the Cascade Range. (Photo by John M. Roberts/The Stock Market)

"warm blooded." This latter adaptation allowed mammals to lead more active and diversified lives than reptiles because they could survive in cold regions and search for food during any season or time of day. Other mammalian adaptations included the development of insulating body hair and more efficient heart and lungs.

It is worth noting that perfect boundaries cannot be drawn between mammalian and reptilian traits. One minor group of animals, the monotremes, still lay eggs. The two species in this group, the duck-billed platypus and the spiny anteater, are found only in Australia. Moreover, although modern reptiles are "cold-blooded," some paleontologists believe that dinosaurs may have been "warm blooded."

With the demise of most Mesozoic reptiles, Cenozoic mammals diversified rapidly. The many forms that exist today evolved from small primitive mammals that were characterized by short legs, flat five-toed feet, and small brains. Their development and specialization took four principle directions: (1) increase in size, (2) increase in brain capacity, (3) specialization of teeth to better accommodate a particular diet, and (4) specialization of limbs to better equip the animal for life in a particular environment.

Marsupials, Placentals, and Diversity.   Following the reptilian extinctions at the close of the Mesozoic, two groups of mammals, the marsupials and the placentals, evolved and expanded to dominate the Cenozoic. The groups differ principally in their modes of reproduction. Young marsupials are born live but at a very early stage of development. After birth, the tiny and immature young crawl into the mother's external stomach pouch to complete their development. Examples are kangaroos, opossums, and koala bears.

Placental mammals, on the other hand, develop within the mother's body for a much longer period, so that birth occurs after the young are relatively mature and independent. Most mammals are placental, including humans.

Today marsupials are found primarily in Australia, where they went through a separate evolutionary expansion during the Cenozoic, largely isolated from placental mammals.

In South America, both primitive marsupials and placentals coexisted before that landmass became completely isolated during the break-up of Pangaea. Evolution and specialization of both groups continued undisturbed for approximately 40 million years until the close of the Pliocene epoch, when the Central American land bridge emerged, connecting the two American continents. Then an invasion of advanced carnivores from North America brought the extinction of many hoofed mammals that had persisted in South America for millions of years. The marsupials, except for opossums, also could not compete and became extinct. Both Australia and South America provide excellent examples of how isolation caused by the separation of continents increased the diversity of animals in the world.

Large Mammals and Extinction.   As we have seen, mammals diversified quite rapidly during the Cenozoic era. One tendency was for some groups to became very large. For example, by the Oligocene epoch a hornless rhinoceros that stood nearly 5 meters high (16 feet) had evolved. It is the largest land mammal known to have existed. As time approached the present, many other types evolved to a large size as well; more, in fact, than now exist. Many of these large forms were common as recently as 11,000 years ago. However, a wave of late Pleistocene extinctions rapidly eliminated these animals from the landscape. In North America, the mastodon and mammoth, both huge relatives of the elephant, became extinct. In addition, saber-toothed cats, giant beavers, large ground sloths, horses, camels, giant bison, and others died out (Figure 11.23). In Europe, late Pleistocene extinctions included woolly rhinos, large cave bears, and the Irish elk. The reason for this recent wave of large animal extinctions puzzles scientists. These animals had survived several major glacial advances and interglacial periods, so it is difficult to ascribe these extinctions to climatic change. Some scientists believe that early humans hastened the decline of these mammals by selectively hunting large forms. Although this hypothesis is preferred by many, it is not yet accepted by all.

Figure 11.23
Skeleton of an extinct sabre-toothed cat that lived about 15,000 years ago. It was about the size of a modern lion. (Photo by Lois Ellen Frank/Westlight)

Review Questions

1. Briefly describe the events believed to have led to the formation of the solar system.
2. What is the major source of free oxygen in Earth's atmosphere?
3. Explain why Precambrian history is more difficult to decipher than more recent geological history.
4. Match the following words and phrases to the most appropriate time span. Select among the following: Precambrian, early Paleozoic, late Paleozoic, Mesozoic, Cenozoic.

   (a) Pangaea came into existence.
   (b) Bacteria and blue-green algae preserved in chert.
   (c) The era that encompasses the least amount of time.
   (d) Shields.
   (e) "Age of dinosaurs."
   (f) Formation of the original northern Appalachian mountains.
   (g) Mastodons and mammoths.
   (h) Extensive deposits of rock salt.
   (i) Triassic, Jurassic, and Cretaceous.
   (j) Coal swamps extended across North America, Europe, and Siberia.
   (k) Gulf Coast oil deposits formed.

(l) Formation of most of the world's major iron ore deposits.
(m) Massive sand dunes covered large portions of the Colorado Plateau region.
(n) The "age of fishes" occurred during this span.
(o) Cambrian, Ordovician, and Silurian.
(p) Pangaea began to break apart.
(q) "Age of mammals."
(r) Animals with hard parts first appeared in abundance.
(s) Gymnosperms were the dominant trees.
(t) Columbia Plateau formed.
(u) Stromatolites are among its more common fossils.
(v) "Golden age of trilobites" occurred during this span.
(w) Fault-block mountains form in the Basin and Range Region

5. briefly discuss two proposals that attempt to explain why several groups developed hard parts at the beginning of the Cambrian period. Do these proposals appear to provide a satisfactory explanation? Why or why not?

6. List some differences between amphibians and reptiles. List differences between reptiles and mammals.

7. Describe one hypothesis that attempts to explain the extinction of the dinosaurs. Cite a major objection to this proposal.

8. Contrast the eastern and western margins of North America during the Cenozoic era in terms of their relationships to plate boundaries.

Key Terms

nebular hypothesis (p. 287)
outgassing (p. 288)
shields (p. 289)
stromatolites (p. 290)

PART 2
The Oceans

Waimea Bay, Oahu, Hawaii. (Photo by Douglas Peebles)

CHAPTER 12
Ocean Waters and the Ocean Floor

A coral reef in the tropical Pacific Ocean. (Photo by Nancy Sefton/Photo Researchers, Inc.)

Iow deep is the ocean? How much of Earth is covered by the global sea? These are basic questions, but they went unanswered for thousands of years. People's interest in the oceans undoubtedly dates back to ancient times, but it was not until rather recently that these seemingly simple questions began to be answered. The beginning of this chapter deals with these answers. Then, after examining the composition of seawater and the resources extracted from it, the remainder of the chapter provides a look at Earth beneath the sea. Suppose that all of the water were drained from the ocean. What would we see? Plains? Mountains? Canyons? Plateaus? Indeed, the ocean conceals all of these features, and more. In fact, the topography of the ocean floor is as varied as that of any continent. And what about the carpet of sediment that covers much of the sea floor? Where did it come from and what can we learn by examining it?

Calling Earth the "water planet" is certainly appropriate, because nearly 71 percent of its surface is covered by the global ocean. Although the ocean comprises a much greater percentage of Earth's surface than the continents, it was only in the relatively recent past that the ocean became an important focus of study. Recently there has been a virtual explosion of data about the oceans, and with it, oceanography has grown dramatically. **Oceanography** is a composite science that draws on the methods and knowledge of biology, chemistry, physics, and geology to study all aspects of the world ocean.

The Vast World Ocean

A glance at a globe or a view of Earth from space reveals a planet dominated by the world ocean (Figure 12.1). Indeed, it is for this reason that Earth is often referred to as the *blue planet*. The area of Earth is about 510 million square kilometers (197 million square miles). Of this total, approximately 360 million square kilometers (140 million square miles), or 71 percent, is represented by oceans and marginal seas (meaning seas around the ocean's margin, like the Mediterranean Sea and Caribbean Sea). Continents and islands comprise the remaining 29 percent, or 150 million square kilometers (58 million square miles).

By studying a globe or world map, it is readily apparent that the continents and oceans are not evenly divided between the Northern and Southern hemispheres (Figure 12.1). When we compute the percentages of land and water in the Northern Hemisphere, we find that nearly 61 percent of the surface is water, while about 39 percent is land. In the Southern Hemisphere, on the other hand, almost 81 percent of the surface is water, and only 19 percent is land. It is no wonder then that the Northern Hemisphere is called the *land hemisphere*, and the Southern Hemisphere the *water hemisphere*.

Figure 12.2 shows the distribution of land and water in the Northern and Southern hemispheres. Between latitudes 45 degrees north and 70 degrees north, there is actually more land than water, whereas between 40 degrees south and 65 degrees south there is almost no land to interrupt the oceanic and atmospheric circulation.

The volume of the ocean basins is many times greater than the volume of the continents above sea level. In fact, the volume of all land above sea level is only $\frac{1}{18}$ that of the ocean.

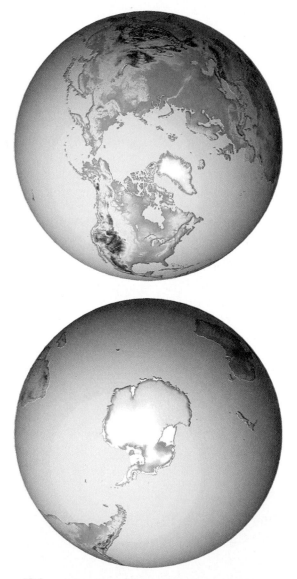

Figure 12.1

These views of Earth show the uneven distribution of land and water between the Northern and Southern hemispheres. Almost 81 percent of the Southern Hemisphere is covered by the oceans—20 percent more than the Northern Hemisphere.

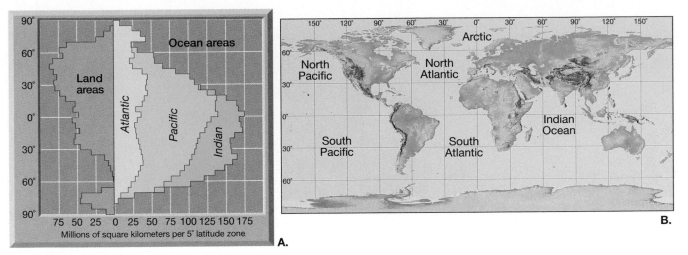

Figure 12.2
Distribution of land and water. **A.** The graph shows the amount of land and water in each 5-degree latitude belt. **B.** The world map provides a more familiar view.

An obvious difference between continents and the ocean basins is their relative levels. The average elevation of the continents above sea level is about 840 meters (2750 feet), whereas the average depth of the oceans is more than 4.5 times this figure—3800 meters (12,500 feet). If Earth's solid mass were perfectly smooth (level) and spherical, the oceans would cover it to a uniform depth of more than 2000 meters (1.2 miles).

A comparison of the three major oceans reveals that the Pacific is by far the largest; it is nearly as large as the Atlantic and Indian oceans combined (Figure 12.2). The Pacific contains slightly more than half of the water in the world ocean, and because it includes few shallow seas along its margins, it has the greatest average depth—3940 meters (12,900 feet).

The Atlantic, bounded by almost parallel continental margins, is a relatively narrow ocean when compared to the Pacific. When the Arctic Ocean is included, the Atlantic has the greatest north–south extent and connects the two polar regions. Because the Atlantic has many shallow adjacent seas, including the Caribbean, Gulf of Mexico, Baltic, and Mediterranean, as well as wide continental shelves along its borders, it is the shallowest of the three oceans, with an average depth of 3310 meters (10,860 feet).

As Figure 12.2 shows, the Indian Ocean covers less area than the Pacific or the Atlantic. Unlike the others, it is largely a Southern Hemisphere water body.

Composition of Seawater

Seawater is a complex solution of salts, consisting of about 3.5 percent (by weight) dissolved mineral substances. Although the percentage of salts may seem small, the actual quantity is huge. If all of the water were evaporated from the oceans, a layer of salt approaching 60 meters (200 feet) thick would cover the entire ocean floor.

Salinity is the proportion of dissolved salts to pure water. In many things, we express proportion in parts-per-hundred, or percent, %. But because the proportion of salts in seawater is a small number, scientists express salinity in parts-per-thousand, noted ‰. Thus, the average salinity of the ocean is about 35‰.

The principal elements that contribute to the ocean's salinity are shown in Figure 12.3. If we made our own seawater, we could come reasonably close by following the recipe shown in Table 12.1. From this table it is evident that most of the salt is sodium chloride (common table salt). Sodium chloride together with the next four most abundant salts comprise 99 percent of the salt in the sea. Although only eight elements make up these five most abundant salts, seawater contains more than 70 of Earth's other naturally occurring elements. Despite their presence in minute quantities, many of these elements are very important in maintaining the necessary chemical environment for life in the sea.

The relative abundances of the major components in sea salt are essentially constant, no matter where the ocean is sampled. Variations in salinity, therefore, are primarily a consequence of changes in the water content of the solution. As a result, high salinities are found where evaporation is high, as is the case in the dry subtropics. Conversely, where heavy precipitation dilutes ocean waters, as in the mid-latitudes and near the equator, lower salinities prevail (Figure 12.4).

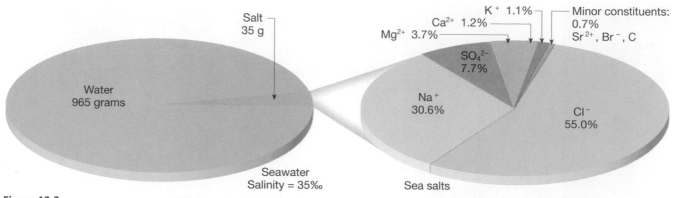

Figure 12.3

Relative proportions of water and dissolved salts in seawater. Ions shown by chemical symbol are chlorine (Cl), sodium (Na), sulfate (SO₄), magnesium (Mg), calcium (Ca), potassium (K), strontium (Sr), bromine (Br), and carbon (C).

Whereas salinity variations in the open ocean normally range from 33‰ to 37‰, some marginal seas demonstrate extraordinary extremes. For example, in the restricted waters of the Persian Gulf and the Red Sea, where evaporation far exceeds precipitation, salinity may exceed 42‰. Conversely, very low salinities occur where large quantities of freshwater are supplied by rivers and precipitation. Such is the case for the Baltic Sea, where salinity is often below 10‰.

What are the primary sources for the vast quantities of salts in the ocean? Chemical weathering of rocks on the continents is one source. These soluble materials are delivered to the oceans by streams at an estimated rate of more than 2.5 billion tons annually. The second major source of elements found in ocean water is Earth's interior. Through volcanic eruptions, large quantities of water and dissolved gases have been emitted during much of geologic time. This process, called **outgassing**, is thought to be the principal source of water in the oceans as well as in the atmosphere. Certain elements, notably chlorine, bromine, sulfur, and boron, are much more abundant in the ocean than in Earth's crust. Because they result from outgassing, their abundance in the sea tends to confirm the hypothesis that volcanic action is largely responsible for the present oceans.

Table 12.1 Recipe for artificial seawater.

MIX:	
Sodium chloride (NaCl)	23.48 grams
Magnesium chloride (MgCl₂)	4.98
Sodium sulfate (Na₂SO₄)	3.92
Calcium chloride (CaCl₂)	1.10
Potassium chloride (KCl)	0.66
Sodium bicarbonate (NaHCO₃)	0.192
Potassium bromide (KBr)	0.096
Hydrogen borate (H₃BO₃)	0.026
Strontium chloride (SrCl₂)	0.024
Sodium fluoride (NaF)	0.003

ADD:
Water (H₂O) to form 1000 grams of solution.

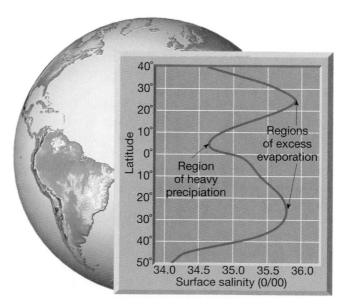

Figure 12.4

The subtropics are dry. On the continents, deserts such as the Sahara occur in this latitude belt. Over the oceans, high evaporation rates remove more water than is replaced by the meager rainfall. Thus, surface salinities are high here. By contrast, near the equator rainfall is heavy, therefore surface salinities are reduced here.

WORLD OCEAN FLOOR
BY BRUCE C. HEEZEN AND MARIE THARP

Based on Research and Exploration Initiated and Supported by the

UNITED STATES NAVY · OFFICE OF NAVAL RESEARCH

1977

Figure 12.11

This map and profile show the major topographic divisions of the North Atlantic.

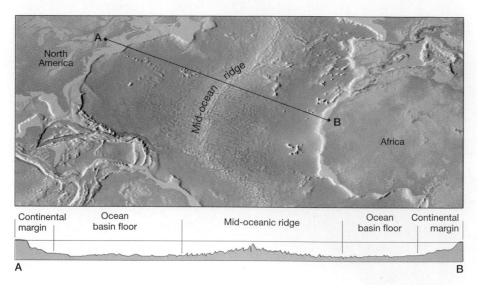

Oceanographers studying the topography of the ocean basins have delineated three major units: *continental margins*, the *ocean basin floor*, and *mid-ocean ridges*. The map in Figure 12.11 shows these provinces for the North Atlantic, and the profile at the bottom of the illustration shows the varied topography. Such profiles usually have their vertical dimension exaggerated many times to make topographic features more conspicuous. Because of this, the slopes in the profile of the sea floor in Figure 12.11 appear to be much steeper than they actually are.

Although information about the ocean floor is substantial and growing every day, scientists have nevertheless only scratched the surface. Ironically, in many respects we know more about the details of the surface of Venus than of the floor of the ocean! For example, in 1991 and 1992 the *Magellan* spacecraft used special radar to scan and map more than 90 percent of the surface of Venus. By contrast, in the 500 years since Ferdinand Magellan circumnavigated the world's oceans, oceanographers have scanned *less than 10 percent* of the deep ocean floor at the resolution provided by the *Magellan* spacecraft.

Why have details about the ocean floor come so slowly? To the present day, efforts to accurately map the sea floor are made from ships equipped with echo sounding equipment because radar waves (which are used to map our neighboring planets) do not travel far through seawater. In addition, ships move across the ocean at a mere 18 kilometers per hour rather than orbiting a planet at 19,500 kilometers per hour.

Continental Margins

We will now look at the three major topographic regions of the sea floor (continental margin, ocean basin floor, and mid-ocean ridge). To keep your perspective during this discussion, refer to Figures 12.10, 12.11 and 12.12.

Continental Shelf

The zones that collectively make up the **continental margin** include the continental shelf, continental slope, and continental rise (Figure 12.12). The first of these parts,

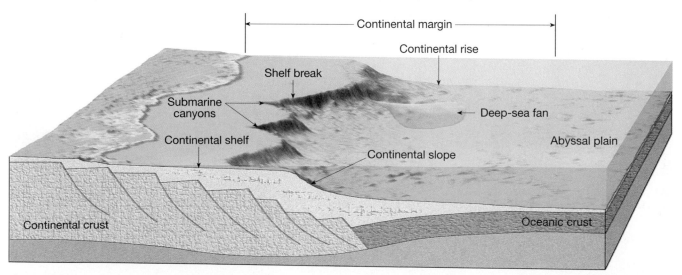

Figure 12.12

Schematic view showing the parts of the continental margin.

the **continental shelf**, is a gently sloping submerged surface extending from the shoreline toward the deep-ocean basin. It is a flooded extension of the continents.

The continental shelf varies greatly in width, as you can see in Figures 12.10 and 12.11. Almost nonexistent along some continents, the shelf may extend seaward as far as 1500 kilometers (900 miles) along others. On the average, the continental shelf is about 80 kilometers (50 miles) wide and 130 meters (423 feet) deep at its seaward edge, the *shelf break*. The average inclination of the continental shelf is less than one-tenth of one degree, a drop of only about 2 meters per kilometer (10 feet per mile). The slope is so slight that it appears horizontal.

The continental shelves represent 7.5 percent of the total ocean area, which is equivalent to about 18 percent of Earth's total land area. These areas have taken on increased economic and political significance because they contain important mineral deposits, including large reservoirs of petroleum and natural gas, as well as huge sand and gravel deposits. The waters of the continental shelf also contain many important fishing grounds that are significant sources of food.

When compared with many parts of the deep-ocean floor, the surface of the continental shelf is relatively featureless. This is not to say that the shelves are completely smooth. In many regions, long valleys run from the coastline into deeper waters. Many of these *shelf valleys* are seaward extensions of river valleys on the adjacent landmass. Such valleys were carved by streams during the Pleistocene epoch (Ice Age).

During this time, great quantities of water were stored in vast ice sheets on the continent. The result was that sea level dropped by 100 meters (330 feet) or more, exposing large portions of the continental shelves (see Figure 5.B, p. 136). Because of the lower sea level, rivers extended their courses, and land-dwelling plants and animals moved onto the newly exposed portions of the continents. Today these areas are again covered by the sea and inhabited by marine organisms. Scientists dredging along the eastern coast of North America have discovered the remains of numerous land dwellers, including mammoths, mastodons, and horses. Bottom sampling has also revealed that freshwater peat bogs existed, adding to the evidence that the continental shelves were once above sea level.

Continental Slope and Rise

Marking the seaward edge of the continental shelf is the **continental slope** (Figure 12.12). It leads into deep water and has a steep gradient compared to the continental shelf. While the gradient varies from place to place, it has an average drop of about 70 meters per kilometer (370 feet per mile). The continental slope is the true edge of the continent.

In regions where trenches do not exist, the steep continental slope merges into a more gradual incline known as the **continental rise** (Figure 12.12). Here the gradient lessens to between 4 and 8 meters per kilometer (20 to 40 feet per mile). Whereas the width of the continental slope averages about 20 kilometers (12 miles), the continental rise may extend for hundreds of kilometers. This feature consists of a thick accumulation of sediment that moved downslope from the continental shelf to the deep-ocean floor. The sediments are delivered to the base of the continental slope by *turbidity currents* that follow submarine canyons (Figure 12.12). (We will discuss these shortly.) When these muddy currents emerge from the mouth of a canyon onto the relatively flat ocean floor, they deposit sediment that forms a **deep-sea fan**. Deep-sea fans have the same basic shape as alluvial fans, which form at the foot of steep mountain slopes on land. As fans from adjacent submarine canyons grow, they coalesce to produce the continuous apron of sediment at the base of the continental slope.

Along some mountainous coasts, the continental slope does not grade gently into the continental rise, but descends abruptly into a deep-ocean trench that parallels the landmass. In such cases, the shelf is very narrow or does not exist at all. The side of the trench and the continental slope are essentially the same feature and grade into the adjacent mountains that tower thousands of meters above sea level. These narrow continental margins are primarily located around the Pacific Ocean where plates are converging. The stress between colliding plates results in deformation of the continental margin. An example of this activity is found along the west coast of South America (see Figure 12.10). Here the vertical distance from the peaks of the Andes Mountains to the floor of the deep Peru-Chile trench bordering the continent exceeds 12,000 meters.

Submarine Canyons and Turbidity Currents

About 100 kilometers south of San Francisco, deep beneath the waters of California's Monterey Bay, is Monterey Canyon. In places the canyon walls of this 100-kilometer-long chasm gradually drop 2300 meters, nearly 400 meters more than the highest cliff in the Grand Canyon. Monterey Canyon is one of many impressive examples of deep, steep-sided valleys known as **submarine canyons** that originate on the continental slope and may extend to depths as great as 3 kilometers (Figure 12.13). Although some of these canyons appear to be the seaward extensions of river valleys, many others do not line up in this manner. Furthermore, since these canyons extend to depths far below the maximum lowering of the sea level during the Ice Age, we cannot attribute their formation to stream erosion. These features must be created by some process that operates far below the ocean surface. Most available information seems to favor the view that submarine canyons have been excavated by turbidity currents (see Box 12.1)

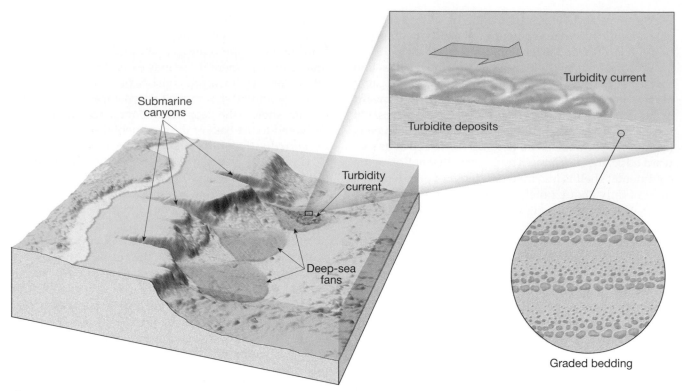

Figure 12.13
Turbidity currents move downslope, eroding the continental marin to enlarge submarine canyons. These sediment-laden density currents eventually lose momentum and deposit their loads of sediment as deep-sea fans. Beds deposited by these currrents are called *turbidites*. Each event produces a single bed characterized by a decrease in sediment size from bottom to top, a feature known as a *graded bed*.

Turbidity currents are downslope movements of dense, sediment-laden water. They are created when sand and mud on the continental shelf and slope are dislodged and thrown into suspension. Since the mud-choked water is denser than normal sea water, it flows downslope, eroding and accumulating more sediment as it goes. The erosional work repeatedly carried on by these muddy torrents is thought to be the major force in the excavation of most submarine canyons.

Turbidity currents usually originate along the continental slope and continue across the continental rise, still cutting channels. Eventually they lose momentum and come to rest along the ocean basin floor (Figure 12.13). As these currents slow, suspended sediments begin to settle out. First, the coarser sand is dropped, followed by successively finer accumulations of silt and then clay. Consequently, these deposits, called **turbidites**, are characterized by a decrease in sediment grain size from bottom to top, a phenomenon known as **graded bedding** (Figure 12.13).

Although there is still more to be learned about the complex workings of turbidity currents, it has been well established that they are a very important mechanism of sediment transport in the ocean. By the action of turbidity currents, submarine canyons are created and sediments are carried to the deep-ocean floor.

The Ocean Basin Floor

Between the continental margin and the mid-ocean ridge system lies the ocean basin floor (see Figure 12.10). The size of this region—almost 30 percent of Earth's surface—is roughly comparable to the percentage of the surface that presently projects above the sea as land. On the ocean basin floor, we find dramatically deep grooves, which are *ocean trenches*; remarkably flat regions, known as *abyssal plains*; and steep-sided volcanic peaks, called *seamounts*. We will now examine each of these features.

Deep-Ocean Trenches

Deep-ocean trenches are long, narrow troughs that are the deepest parts of the ocean. Several in the western Pacific approach or exceed depths of 10,000 meters (33,000 feet). A portion of one, the Challenger Deep in the Mariana trench, is more than 11,000 meters (36,000 feet) below sea level.

Although deep-ocean trenches represent only a very small portion of the ocean floor area, they are very significant geologically. Trenches are the sites where moving crustal plates plunge back into the mantle (see Figure 7.11, p. 190). In addition to the earthquakes created as one plate descends beneath another, melting of the sub-

Box 12.1

Evidence for the Existence of Turbidity Currents

For many years the existence of turbidity currents in the ocean was a matter of considerable debate among marine geologists. Not until the 1950s did the speculation begin to subside. Two lines of evidence helped establish turbidity currents as important mechanisms of submarine erosion and sediment transportation.

The first important evidence came from records of a rather severe earthquake that took place off the coast of Newfoundland in 1929 and resulted in the breakage of 13 transatlantic telephone and telegraph cables. At the time, it was presumed that the tremor had caused the multiple breaks. However, when the data were analyzed, it appeared that this was not the case. After plotting the locations of the breaks on a map, researchers saw that all the breaks had occurred along the steep continental slope and the gentler

continental rise. Since the time of each break was known from information provided by automatic recorders, a pattern of what had happened could be deduced. The breaks high up on the continental slope took place first, almost concurrently with the earthquake. The other breaks happened in succession, the last occurring 13 hours later, some 720 kilometers (450 miles) from the source of the quake (Figure 12.A).

The breaks downslope had obviously taken place too long after the tremor to have been caused by the shock of the earthquake. The existence of a turbidity current, triggered by the quake, thus appeared as a plausible alternative. As the avalanche of sediment-choked water raced downslope, it snapped the cables in its path. Investigators calculated that the current reached speeds approaching 80 kilometers (50 miles) per hour on the continental slope and about 24 kilometers (15 miles) per hour on the more gently sloping continental rise.

A second compelling line of evidence relating turbidity currents to submarine erosion and transportation of sediment came from the examination of deep-sea

sediment samples. These cores show that extensive graded beds of sand, silt, and clay exist in the quiet waters of the deep ocean. Some samples also include fragments of plants and animals that live only in the shallower waters of the continental shelves. No mechanism other than turbidity currents has been identified to satisfactorily explain the existence of these deposits.

More recently, turbidity currents have been measured directly. In one study, instruments were deployed in Bute Inlet, a deep fiord along the coast of British Columbia in western Canada. The inlet served as a natural laboratory for the study of turbidity current dynamics. The year-long program found turbidity currents to be common events.

Although the majority were low-velocity flows that covered limited distances, occasional faster-moving, large-scale currents were also measured. These larger events moved significant amounts of sand down the gently sloping floor of the fiord for distances of 40 to 50 kilometers. The dense currents of sediment in this investigation were linked to submarine landslides in a delta at the head of the fiord.

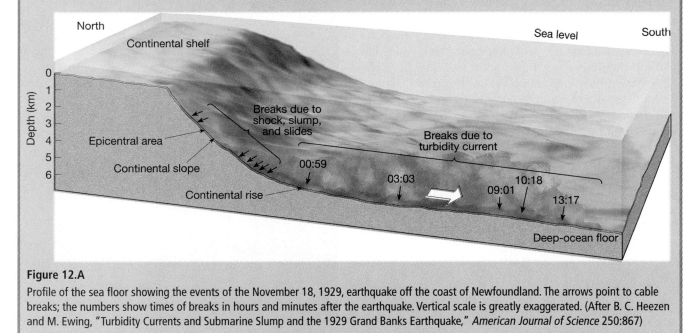

Figure 12.A
Profile of the sea floor showing the events of the November 18, 1929, earthquake off the coast of Newfoundland. The arrows point to cable breaks; the numbers show times of breaks in hours and minutes after the earthquake. Vertical scale is greatly exaggerated. (After B. C. Heezen and M. Ewing, "Turbidity Currents and Submarine Slump and the 1929 Grand Banks Earthquake," *American Journal of Science* 250:867)

ducting plate is also associated with trench regions. Thus, trenches in the open ocean are often paralleled by volcanic island arcs. Moreover, volcanic mountains, such as those making up a portion of the Andes, are situated adjacent to trenches that parallel the continental margins.

Abyssal Plains

Abyssal plains are incredibly flat features; in fact, these regions are likely the most level places on Earth. The abyssal plain found off the coast of Argentina, for example, has less than 3 meters (10 feet) of relief over a distance exceeding 1300 kilometers (800 miles). The monotonous topography of abyssal plains is occasionally interrupted by the protruding summit of a buried volcanic structure.

By employing seismic profilers, instruments whose signals penetrate far below the ocean floor, researchers have shown that abyssal plains consist of thick accumulations of sediment that were deposited atop the low, rough portions of the ocean floor. The nature of the sediment indicates that these plains consist primarily of sediments transported far out to sea by turbidity currents. These deposits are interbedded with sediments composed of minute clay-sized particles that continuously settle onto the ocean floor.

Abyssal plains occur in all of the oceans. However, they are more widespread where there are no deep-ocean trenches adjacent to the continents. Because the Atlantic Ocean has fewer trenches to act as traps for the sediments carried down the continental slope, it has more extensive abyssal plains than the Pacific.

Seamounts

Dotting the ocean floors are isolated volcanic peaks called **seamounts** that may rise hundreds of meters above the seafloor. Although these steep-sided conical peaks are found on the floors of all the oceans, the greatest number have been identified in the Pacific (see Figure 12.10).

Some, like the Emperor Seamount chain that stretches from the Hawaiian Islands to the Aleutian Trench, form in association with volcanic hot spots. Others are born near oceanic ridges, divergent plate boundaries where the plates of the lithosphere move apart (see Chapter 7). If the volcano grows large enough before being carried from the zone that nourishes it, the structure emerges as an island. Examples in the Atlantic include the Azores, Ascension, Tristan da Cunha, and St. Helena.

During the time they exist as islands, some of these volcanoes are eroded to near sea level by running water and wave action. Over a span of millions of years the islands gradually sink as the moving plate slowly carries them away from the elevated oceanic ridge or hot spot where they originated. These submerged, flat-topped seamounts are called **guyots**.

Mid-Ocean Ridges

Our knowledge of **mid-ocean ridges**, the sites of seafloor spreading, comes from soundings taken of the ocean floor, core samples obtained from deep-sea drilling, and visual inspection using deep-diving submersibles (Figure 12.14). We also have first-hand inspection of slices of ocean floor that have been shoved up onto dry land by plate movements. Ocean ridge systems are characterized by an elevated position, extensive faulting, and numerous volcanic structures that have developed on the newly formed crust. This is quite evident as you follow the path of mid-ocean ridges in Figure 12.10.

Mid-ocean ridges exist in all major oceans and represent more than 20 percent of Earth's surface. They are certainly the most prominent topographic features in the oceans, for they form an almost continuous mountain range, which extends for about 65,000 kilometers (40,000 miles) in a manner similar to the seam on a baseball. Although ocean ridges stand above the adjacent deep-ocean basins, they are much different from the mountains found on the continents. Rather than containing thick sequences of folded and faulted sedimentary rocks, oceanic ridges consist of layer upon layer of basaltic rocks that have been faulted and uplifted. The term *ridge* may also be misleading, for these features are not narrow, but have widths from 500 to 5000 kilometers. In places they may occupy as much as one-half of the total ocean floor area.

Ridges are broad features, yet much of the geologic activity occurs along a relatively narrow region on the ridge crest. This belt, called the **rift zone**, is the site where magma from the asthenosphere moves upward to create new slivers of oceanic crust. Active rift zones have frequent but generally weak earthquakes and more rapid heat flow than other crustal segments. Here vertical dis-

Figure 12.14

The deep-diving submersible *Alvin* is 7.6 meters long, weighs 16 tons, has a cruising speed of 1 knot, and can reach depths as great as 4000 meters. A pilot and two scientific observers are along during a normal 6- to 10-hour dive. Vessels such as this are important tools for detecting and studying the fine-scale features of the ocean floor. (Photo by Rod Catanach @ Woods Hole Oceangraphic Institution)

placement of large slabs of oceanic crust caused by faulting and the growth of volcanic piles contribute to the characteristically rugged topography of the oceanic ridge system. Further, the rocks along the ridge axis appear very fresh and are nearly void of sediment. Away from the ridge axis the topography becomes more subdued and the thickness of the sediments as well as the depth of the water increase. Gradually the ridge system grades into the less rugged sediment-laden regions of the deep-ocean floor.

During seafloor spreading, new material is added about equally to the two diverging plates. Hence, we would expect new ocean floor to grow symmetrically on both sides of a centrally located ridge. Indeed, the ridge systems of the Atlantic and Indian oceans are located near the middles of these water bodies and as a consequence are named mid-ocean ridges (Figure 12.10). However, the East Pacific Rise is situated far from the center of the Pacific Ocean. Despite uniform spreading along the East Pacific Rise, much of the Pacific basin that once lay east of this spreading center has been overridden by the westward migration of the American plate.

Partly because of its accessibility to both American and European scientists, the Mid-Atlantic Ridge has been studied more thoroughly than other ridge systems. The Mid-Atlantic Ridge is a gigantic submerged mountain range standing 2500–3000 meters (8200–9800 feet) above the adjacent floor of the deep-ocean basins. In a few places, such as Iceland, the ridge actually extends above sea level (Figure 12.10). Throughout most of its length, however, this divergent plate boundary lies 2500 meters below sea level.

Another prominent feature of the Mid-Atlantic Ridge is a deep linear valley extending along the ridge axis. In places, this rift valley is deeper than the Grand Canyon of the Colorado River and two or three times as wide. The name *rift valley* has been applied to this feature because it is strikingly similar to continental rift valleys such as those in East Africa. An examination of Figure 12.10 reveals that this central rift is broken into sections that are offset by transform faults.

Coral Reefs and Atolls

Coral reefs are among the most picturesque features in the ocean. They are constructed primarily from the skeletal remains and secretions of corals and certain algae, built up over thousands of years. The term *coral reef* is somewhat misleading in that it makes no mention of the skeletons of many other small animals and plants found inside the branching framework built by the corals, or the fact that limy secretions of algae help bind the entire structure together.

Coral thrive in the warm water of the Pacific and Indian oceans, so most reefs are in these oceans, although a few occur elsewhere. Reef-building corals grow best in waters with an average annual temperature of about 24°C (75°F). They can survive neither sudden temperature changes nor prolonged exposure to temperatures below 18°C (65°F). In addition, these reef-builders require clear sunlit water. For this reason, the limiting depth of active reef growth is about 45 meters (150 feet).

In 1831, the naturalist Charles Darwin set out aboard the British ship H.M.S. *Beagle* on its famous five-year expedition that circumnavigated the globe. One outcome of Darwin's studies during the voyage was his hypothesis on the formation of coral islands, called **atolls**. As Figure 12.15 illustrates, atolls consist of a continuous or broken ring of coral reef surrounding a central lagoon.

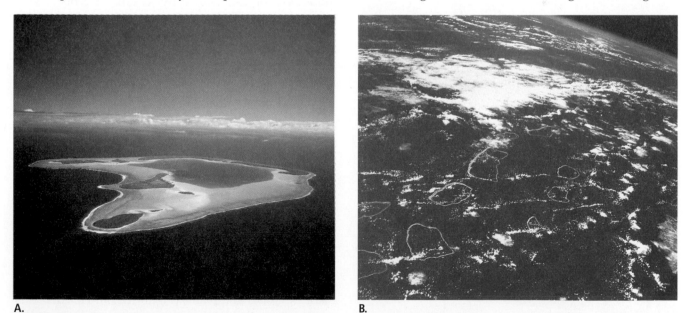

A.   **B.**

Figure 12.15

A. An aerial view of Tetiaroa Atoll in the Pacific. The light blue waters of the relatively shallow lagoon contrast with the dark blue color of the deep ocean surrounding the atoll. (Photo by Douglas Peebles Photography) **B.** View from space of a group of atolls in the Pacific Ocean. (Courtesy of NASA)

Atolls present a paradox: how can corals, which require warm, shallow, sunlit water no deeper than 45 meters, create structures that reach thousands of meters to the floor of the ocean? The essence of Darwin's hypothesis was that coral reefs form on the flanks of sinking volcanic islands. As an island slowly sinks, the corals continue to build the reef complex upward (Figure 12.16).

Thus atolls, like guyots, are thought to owe their existence to the gradual sinking of oceanic crust. Drilling on atolls has revealed that volcanic rock does indeed underlie the thick coral reef structure. This finding confirmed Darwin's explanation.

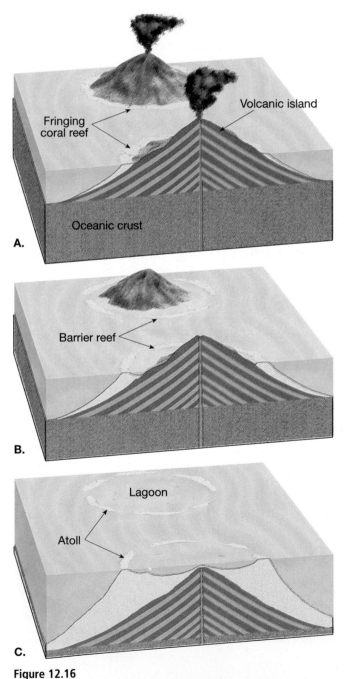

Figure 12.16
Formation of a coral atoll caused by the gradual sinking of oceanic crust.

Seafloor Sediments

Except for a few areas, such as near the crests of mid-ocean ridges, the ocean floor is mantled with sediment. Part of this material has been deposited by turbidity currents, and the rest has slowly settled to the bottom from above. The thickness of this carpet of debris varies greatly. In some trenches, which act as traps for sediments originating on the continental margin, accumulations may approach 10 kilometers (6 miles). In general, however, sediment accumulations are considerably less. In the Pacific Ocean, uncompacted sediment measures about 600 meters (2000 feet) or less, whereas on the floor of the Atlantic, the thickness varies from 500 to 1000 meters (1500 to 3000 feet).

Although deposits of sand-sized particles are found on the deep-ocean floor, mud is the most common sediment covering this region. Muds also predominate on the continental shelf and slope, but the sediments in these areas are coarser overall because of greater quantities of sand.

Seafloor sediments can be classified according to their origin into three broad categories: (1) terrigenous ("derived from land") sediment, (2) biogenous ("derived from organisms") sediment, and (3) hydrogenous ("derived from water") sediment. Although we discuss each category separately, remember that all seafloor sediments are mixtures. No body of sediment comes entirely from a single source.

Terrigenous Sediment

Terrigenous sediment consists primarily of mineral grains that were weathered from continental rocks and transported to the ocean. The sand-sized particles settle near shore. Because the very smallest particles take years to settle to the ocean floor, they may be carried for thousands of kilometers by ocean currents. As a consequence, virtually every area of the ocean receives some terrigenous sediment. The rate at which this sediment accumulates on the deep-ocean floor is indeed very slow. From 5000 to 50,000 years are necessary for a 1-centimeter layer to form. Conversely, on the continental margins near the mouths of large rivers, terrigenous sediment accumulates rapidly. In the Gulf of Mexico, for example, the sediment has reached a depth of many kilometers.

Since fine particles remain suspended in the water for a very long time, there is ample opportunity for chemical reactions to occur. Because of this, the colors of deep-sea sediments are often red or brown. This results when iron in the particle or in the water reacts with dissolved oxygen in the water and produces a coating of iron oxide (rust).

Biogenous Sediment

Biogenous sediment consists of shells and skeletons of marine animals and plants (Figure 12.17). This debris is produced mostly by microscopic organisms living in the sunlit waters near the ocean surface. The remains continually "rain" down upon the sea floor (see Box 12.2).

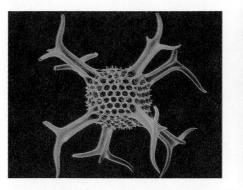
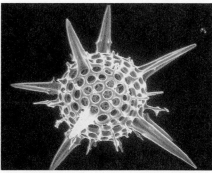

Figure 12.17
Microscopic radiolarian hard parts are examples of biogenous sediments. These photomicrographs have been enlarged *hundreds* of times. (Photo by Manfred Kage/Peter Arnold, Inc.)

The most common biogenous sediments are known as *calcareous* ($CaCO_3$) *oozes*, and as the name implies, they have the consistency of thick mud. These sediments are produced by organisms that inhabit warm surface waters. When calcareous hard parts slowly sink through a cool layer of water, they begin to dissolve.

Box 12.2

Seafloor Sediments and Climatic Change

Reliable climate records go back only a couple of hundred years, at best. How do scientists learn about climates and climatic changes prior to that time? The obvious answer is that they must reconstruct past climates from *indirect evidence*; that is, they must examine and analyze phenomena that respond to and reflect changing atmospheric conditions. An interesting and important technique for analyzing Earth's climatic history is the study of sediments from the ocean floor.

Although seafloor sediments are of many types, most contain the remains of organisms that once lived near the sea surface (the ocean-atmosphere interface). When such near-surface organisms die, their shells slowly settle to the ocean floor where they become part of the sedimentary record. One reason that seafloor sediments are useful recorders of worldwide climatic changes is that the numbers and types of organisms living near the sea surface change as the climate changes. This principle is explained by Richard Foster Flint as follows:

...we would expect that in any area of the ocean/atmosphere interface the average annual temperature of the surface water of the ocean would approximate that of the contiguous atmosphere. The temperature equilibrium established between surface seawater and the air above it should mean

that...changes in climate should be reflected in changes in organisms living near the surface of the deep sea...When we recall that the seafloor sediments in vast areas of the ocean consist mainly of shells of pelagic foraminifers, and that these animals are sensitive to variations in water temperature, the connection between such sediments and climatic change become obvious.*

Thus, in seeking to understand climatic change as well as other environmental transformations, scientists are tapping the huge reservoir of data in seafloor sediments. The sediment cores

*Glacial and Quaternary Geology (New York: Wiley, 1971), p. 718.

gathered by drilling ships and other research vessels have provided invaluable data that have significantly expanded our knowledge and understanding of past climates (Figure 12.B).

One notable example of the importance of seafloor sediments to our understanding of climate change relates to unraveling the fluctuating atmospheric conditions of the Ice Age. The record of temperature changes contained in cores of sediment from the ocean floor have proven critical to our present understanding of this recent span of Earth history.†

†For more on this topic, see "Causes of Glaciation" in Chapter 5.

Figure 12.B
The *JOIDES Resolution*, drilling ship of the Ocean Drilling Program. The seafloor sediments recovered by this and other research vessels provide scientists with data that allow them to reconstruct past climates. (Photo courtesy of the Ocean Drilling Program)

This results because cold seawater contains more carbon dioxide and is thus more acidic than warm water. In seawater deeper than about 4500 meters (15,000 feet), calcareous shells will completely dissolve before they reach bottom. Consequently, calcareous ooze does not accumulate where depths are great.

Other biogenous sediments include *siliceous* (SiO_2) *oozes*. They are composed primarily of skeletons of diatoms (single-celled algae) and radiolarians (single-celled animals). Other sediments are derived from the bones, teeth, and scales of fish and other marine organisms.

Hydrogenous Sediment

Hydrogenous sediment consists of minerals that crystallize directly from seawater through various chemical reactions. For example, some limestones are formed when calcium carbonate precipitates directly from the water; however, most limestone is composed of biogenous sediment.

One type of hydrogenous sediment is significant for its economic potential. **Manganese nodules** are rounded blackish lumps composed of a complex mixture of minerals that form very slowly on the floor of the ocean basins (Figure 12.18). In fact, their formation rate is one of the slowest chemical reactions known.

Although manganese nodules may contain more than 20 percent manganese, the interest in them as a potential resource lies in the fact that other more valuable metals may be concentrated in them. In addition to manganese, nodules may contain significant iron, copper, nickel, and cobalt. All regions containing nodules, however, are not equally good potential sites for mining. Possible mining locations must have abundant nodules and contain the economically optimum mix of cobalt, copper, and nickel. Sites meeting these criteria are relatively limited. Furthermore, before such areas prove to be valuable commercial sources for these metals, the technical problems of extracting nodules from the floor

Figure 12.18

Manganese nodules photographed at a depth of 2909 fathoms (5323 meters) beneath the *Robert Conrad* south of Tahiti. (Photo courtesy of Lawrence Sullivan, Lamont-Doherty Earth Observatory)

of the deep-ocean basins must be solved. Also the political and legal ramifications of mining the sea bed must be worked out.

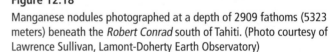

Review Questions

1. How does the area covered by the oceans compare with that of the continents? Describe the distribution of land and water on Earth.

2. Answer the following questions about the Atlantic, Pacific, and Indian oceans:
   (a) Which is the largest in area? Which is smallest?
   (b) Which has the greatest north-south extent?
   (c) Which is shallowest? Explain why its average depth is least.

3. How does the average depth of the ocean compare to the average elevation of the continents?

4. What is meant by *salinity*? What is the average salinity of the ocean?

5. Why do variations in salinity occur? Give some specific examples to illustrate your answer.

6. What is the origin of the minerals dissolved in seawater?

7. List three commercial products that are extracted from seawater.

8. Why is desalination not likely to be a significant source of water for agriculture in the foreseeable future?

9. Assuming that the average speed of sound waves in water is 1500 meters per second, determine the water depth if the signal sent out by an echo sounder requires 6 seconds to strike bottom and return to the recorder (see Figure 12.9).

10. List the three major subdivisions of the continental margin. Which subdivision is considered a flooded extension of the continent? Which has the steepest slope?

11. How does the continental margin along the west coast of South America differ from the continental margin along the east coast of North America?

12. Defend or rebut the statement "Most submarine canyons found on the continental slope and rise were formed during the Ice Age when rivers extended their valleys seaward."

13. What are turbidites? What is meant by the term *graded bedding*?

14. Discuss the evidence that helped confirm the existence of turbidity currents in the ocean and establish them as significant mechanisms of erosion and sediment transport. (See Box 12.1.)

15. Why are abyssal plains more extensive on the floor of the Atlantic than on the floor of the Pacific?

16. How are mid-ocean ridges and deep-ocean trenches related to seafloor spreading?

17. What is an atoll? Describe Darwin's proposal on the origin of atolls. Was it ever confirmed?

18. Distinguish among the three basic types of seafloor sediment.

19. If you were to examine recently deposited biogenous sediment taken from a depth in excess of 4500 meters (15,000 feet), would it more likely be rich in calcareous materials or siliceous materials? Explain.

20. Why are seafloor sediments useful in studying climates of the past? (See Box 12.2.)

Key Terms

abyssal plain (p. 325)
atoll (p. 327)
biogenous sediment (p. 328)
continental margin (p. 322)
continental rise (p. 323)
continental shelf (p. 323)
continental slope (p. 323)
coral reef (p. 327)

deep-ocean trench (p. 324)
deep-sea fan (p. 323)
desalination (p. 317)
echo sounder (p. 319)
graded bedding (p. 324)
guyot (p. 326)
halocline (p. 318)
hydrogenous sediment (p. 330)
manganese nodule (p. 330)
mid-ocean ridge (p. 326)

oceanography (p. 314)
outgassing (p. 316)
rift zone (p. 326)
salinity (p. 315)
seamount (p. 326)
submarine canyon (p. 323)
terrigenous sediment (p. 328)
thermocline (p. 318)
turbidite (p. 324)
turbidity current (p. 324)

CHAPTER 13
The Restless Ocean

Waves crashing at Montana de Oro State Park, California. (Photo by David Muench)

The restless waters of the ocean are constantly in motion. Winds generate surface currents, the moon and sun produce tides, and density differences create deep-ocean circulation. Further, waves carry the energy from storms to distant shores, where their impact erodes the land. This chapter examines the movements of ocean waters and their importance.

Ocean circulation involves surface currents and movements of deep-water masses. Surface waters are set in motion by the wind (Figure 13.1). Some of these currents are short-lived and affect only small areas. Such water movements are responses to local or seasonal influences. Other surface currents are relatively permanent phenomena that extend over large portions of the oceans. These major horizontal movements of surface waters are closely related to the general atmospheric circulation, which in turn is driven by the unequal heating of Earth by the sun.

In contrast to the largely horizontal movements of surface currents, the deep-ocean circulation has a significant vertical component and accounts for the thorough mixing of deep-water masses. This component of ocean circulation is a response to density differences among water masses and reflects contrasts in temperature and salinity.

Surface Currents

There is a river in the ocean. In the severest droughts it never fails, and in the mightiest floods...it never overflows. Its banks and its bottoms are of cold water, while its current is of warm. The Gulf of Mexico is its fountain, and its mouth is in the Arctic Sea. It is the Gulf Stream. There is in the world no other such majestic flow of waters.

This quote from Matthew Fontaine Maury's 1855 book, *The Physical Geography of the Sea*, describes perhaps the best-known and most-studied of the surface ocean currents, the Gulf Stream. This current is just a portion of a huge, slowly moving, circular whorl, or **gyre**. Figure 13.2 shows the entire gyre, which fills the North Atlantic, and other major gyres of the world ocean.

Figure 13.1
Wind is not only responsible for creating waves, but it also provides the force that drives the ocean's surface circulation. (Photo by Arnulf Husmo/Tony Stone Worldwide)

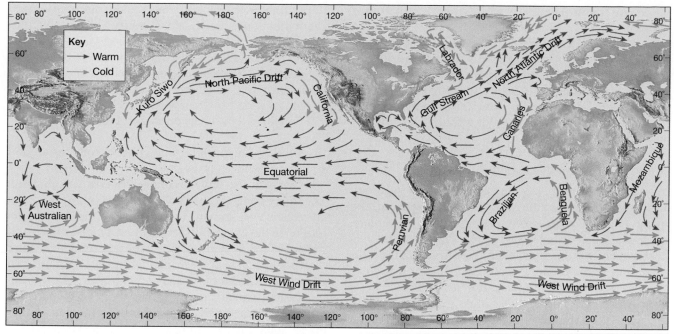

Figure 13.2
Average position and extent of the principal surface ocean currents.

Ocean Circulation Patterns

Where the atmosphere and ocean are in contact, energy is passed from moving air to the water through friction. The drag exerted by winds blowing steadily across the ocean causes the surface layer of water to move. Thus, winds are the primary driving force of surface currents. This means there is a clear relationship between the oceanic circulation and the general atmospheric circulation. A comparison of Figures 13.2 and 16.13 (p. 430) illustrates this.

A further clue to the influence of winds on ocean circulation is provided by the currents in the northern Indian Ocean. Here, there are seasonal wind shifts known as the summer and winter monsoons. When the winds change direction, the surface currents also reverse direction.

Although wind is the force that generates surface currents, other factors also influence the movement of ocean waters. The most significant of these is the **Coriolis effect**. Due to Earth's rotation, currents are deflected to the right in the Northern Hemisphere and to the left in the Southern Hemisphere. (The Coriolis effect will be explained in Chapter 16.) As a consequence, the direction of surface currents does not coincide precisely with the wind direction. Rather, they are altered by Earth's rotation.

North and south of the equator are two westward-moving currents, the Equatorial Currents (Figure 13.2). They derive their energy principally from the trade winds that blow from the northeast and southeast, respectively, toward the equator. The equatorial currents can be thought of as the "backbone" of the system of ocean currents. Because of the Coriolis effect, these currents are deflected poleward to form clockwise gyres in the Northern Hemisphere and counterclockwise gyres in the Southern Hemisphere. These nearly circular patterns are found in each major ocean basin (Figure 13.2).

In the North Atlantic, the equatorial current is deflected northward through the Caribbean, where it becomes the Gulf Stream (Figure 13.2). As the Gulf Stream moves along the East Coast of the United States, it is strengthened by the prevailing westerly winds and is deflected to the east (to the right) between the Carolinas and New England. As it continues northeastward beyond the Grand Banks, it gradually widens and slows until it becomes a vast, slowly moving current known as the North Atlantic Drift (Figure 13.2).

As the North Atlantic Drift approaches Western Europe, it splits, with part moving northward past Great Britain, Norway, and Iceland, carrying heat to these otherwise chilly areas. The other part is deflected southward as the cool Canaries Current. As the Canaries Current moves southward, it eventually merges into the North Equatorial Current, completing the gyre.

The clockwise circulation of the North Atlantic leaves a large central area which has no well-defined currents. This zone of calmer waters is known as the Sargasso Sea, named for the large quantities of *Sargassum*, a type of seaweed, encountered there.

In the South Atlantic, surface ocean circulation is very much the same as in the North Atlantic. And, circulation in the North and South Pacific generally parallels that of the North and South Atlantic (Figure 13.2).

Around the ice-covered continent of Antarctica, with no large landmasses in the way, cold surface waters circulate continuously. They move in response to prevailing westerly winds, giving this circulation the name West Wind Drift.

Ocean Currents and Upwelling

In addition to producing surface currents, winds may also cause *vertical* water movements. **Upwelling**, the rising of cold water from deeper layers to replace warmer surface water, is a common wind-induced vertical movement. It is most characteristic along the eastern shores of the oceans, most notably along the coasts of California, South America, and West Africa.

Upwelling occurs in these areas when winds blow toward the equator and parallel to the coast. Due to the Coriolis effect, the surface water movement is deflected away from the shore. As the surface layer moves away from the coast, it is replaced by water that "upwells" from below the surface. This slow upward flow from depths of 50 to 300 meters (165 to 1000 feet) brings water that is cooler than the original surface water and creates a characteristic zone of lower temperatures near the shore.

For swimmers who are accustomed to the warm waters along the mid-Atlantic shore of the United States, a dip in the Pacific off the coast of central California can be a chilling surprise. In August, when temperatures in the Atlantic are 21°C (70°F) or higher, central California's surf is only about 15°C (60°F). Coastal upwelling also brings to the ocean surface greater concentrations of dissolved nutrients, such as nitrates and phosphates. These nutrient-enriched waters from below promote the growth of plankton, which in turn supports extensive populations of fish.

The Importance of Ocean Currents

Anyone who navigates the oceans needs to be aware of currents. By understanding the direction and strength of ocean currents, sailors soon realize that their voyage time can be reduced if they travel with a current, or increased if they travel against a current. For example, the first good chart of the Gulf Stream, which was printed by Benjamin Franklin, was developed as a result of this need to know (Figure 13.3). P.L. Richardson describes the situation as follows:

> While he was in London as Deputy Postmaster General for the American colonies, Franklin was consulted on the question of why the mail packets took a fortnight longer to sail to America than the merchant ships. In October 1768 Franklin discussed this problem with his cousin Timothy Folger, a Nantucket ship captain then visiting London. Folger told him the packet captains were ignorant of the Gulf Stream and frequently sailed in this current...Folger sketched the Gulf Stream on a chart and added written notes on how to avoid the Gulf Stream and Franklin had the chart printed in 1769 or 1770.[*]

Even today, the Franklin-Folger chart remains a good summary of the average path of the Gulf Stream.

Currents also have an important effect on climates. The moderating effect of poleward-moving warm ocean currents is well known. The North Atlantic Drift, an extension of the warm Gulf Stream, keeps Great Britain and much of northwestern Europe warmer during the winter than one would expect for their latitudes, which are similar to the latitudes of Alaska and Labrador. The prevailing westerly winds carry the moderating effects far inland. For example, Berlin (52 degrees north latitude) has an average January temperature similar to that experienced at New York City, which lies 12 degrees farther south.

In contrast to warm ocean currents whose effects are felt most in the middle latitudes in winter, the influence of cold currents is most pronounced in the tropics or during summer months in the middle latitudes. Cold currents, such as the Benguela Current along Africa and the Peruvian Current (see Figure 13.2), moderate the tropical heat.

In addition to influencing temperatures of adjacent land areas, cold currents have other climatic influences. For example, where tropical deserts exist along the west coasts of continents, cold ocean currents have a dramatic impact. The principal west coast deserts are the Atacama in Peru and Chile, and the Namib in southern Africa. The aridity along these coasts is intensified because the lower air is chilled by cold offshore waters. When this occurs, the air becomes very stable and resists the upward movement necessary to create precipitation-producing clouds. In addition, the presence of cold currents causes temperatures to approach and often reach the dew point, the temperature at which water vapor condenses. As a result, these areas are characterized by high relative humidities and much fog. Thus, not all tropical deserts are hot with low humidities and clear skies. Rather, the presence of cold currents transforms some tropical deserts into relatively cool, damp places that are often shrouded in fog.

Ocean currents also play a major role in maintaining Earth's heat balance. They accomplish this task by transferring heat from the tropics, where there is an excess of heat, to the polar regions, where a deficit exists

---

[*]P.L. Richardson, "Benjamin Franklin and Timothy Folger's First Printed Chart of the Gulf Stream," *Science* 207 (4431): 643.

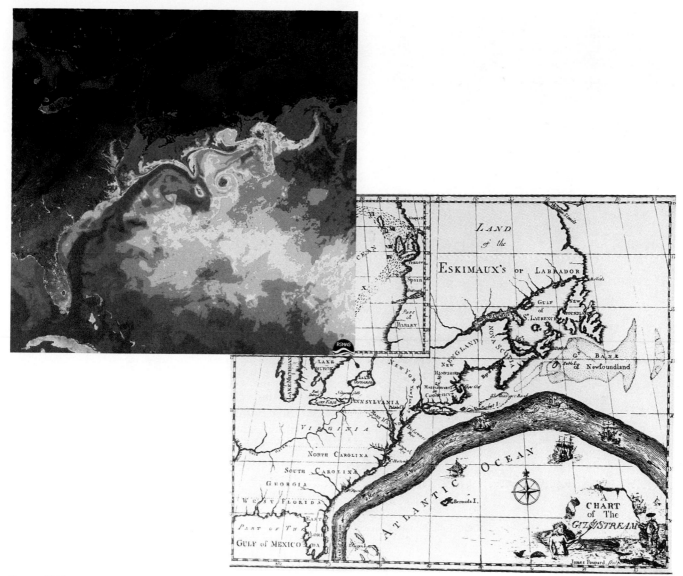

Figure 13.3

To aid ships crossing the Atlantic, Benjamin Franklin, with the assistance of his cousin, Timothy Folger, produced the first detailed chart of the Gulf Stream about 1769 or 1770. (Courtesy of NOAA) Today satellite images provide views of the Gulf Stream's complexities. Reds and yellows denote warmer waters. The current transports heat from lower latitudes toward the north pole. Meanders of the stream pinch off to form eddies that may move about the ocean for up to two years before dissipating. (Courtesy of O. Brown, R. Evans, and M. Carle/University of Miami Rosenstiel School of Marine and Atmospheric Science)

(Figure 13.3). Ocean water movement accounts for about a quarter of this heat transport, and winds transport the remaining three-quarters.

Deep-Ocean Circulation

Ocean circulation is not confined to surface currents. Significant deep-water movements occur as well. Unlike the wind-induced movements of surface and near-surface waters, deep-ocean circulation is governed by gravity and driven by density differences.

Two factors—temperature and salinity—are most significant in creating a dense mass of water. Water that is cold and salty is denser than warmer, less salty water. Consequently, deep-ocean circulation is called **thermohaline circulation** (*thermo*-"heat," *haline*-"salt"). After sinking from the surface of the ocean, waters will not reappear at the surface for an average of 500–2000 years.

The basic idea of thermohaline circulation is easily understood. Water at the surface of the ocean is made colder (by heat loss to the atmosphere) and/or more salty (by removal of water by evaporation or freezing) and becomes more dense. This denser water then sinks

Figure 13.4
Sea ice near the Antarctic Peninsula. When seawater freezes, sea salts do not become part of the ice. Consequently, the salt content of the remaining seawater becomes more concentrated, which makes it denser and prone to sink. (Photo by Wolfgang Kaehler)

toward the ocean bottom and displaces lighter (less dense) water, which moves back toward the zone where the denser water formed. In this manner, cold, dense waters flow away from their source near the poles and are replaced by warmer waters from lower latitudes.

Arctic and Antarctic waters represent the two major regions where dense water masses are created. Antarctic waters are chilled during the winter. The temperatures here are low enough to form sea ice, and since sea salts do not get frozen into the ice, the remaining water becomes saltier. The result is the densest water in all of the oceans (Figure 13.4). This cold saline brine slowly sinks to the sea floor, where it moves through the ocean basins in sluggish currents.

Centuries later, subsurface waters return to the sea surface. The return flow is accomplished by slow, upward movements of deep waters. Much of this return flow seems to be concentrated in upwelling along the equator and in coastal areas. Because water movements in the deep ocean are difficult to measure directly, relatively little is known about the details of deep-ocean circulation.

Tides

Tides are daily changes in the elevation of the ocean surface. Their rhythmic rise and fall along coastlines have been known since antiquity. Other than waves, they are the easiest ocean movements to observe. An exceptional example of extreme daily tides is shown in Figure 13.5.

Although known for centuries, tides were not explained satisfactorily until Sir Isaac Newton applied the law of gravitation to them. Newton showed that there is a mutual attractive force between two bodies, as between Earth and the moon. Because the atmosphere and the ocean both are fluids, and free to move, both are deformed by this force. Hence, ocean tides result from the gravitational attraction exerted upon Earth by the moon, and to a lesser extent by the sun.

Figure 13.5
High tide and low tide on Nova Scotia's Minas Basin in the Bay of Fundy. The areas exposed during low tide and flooded during high tide are called *tidal flats*. Tidal flats here are extensive. (Courtesy of Nova Scotia Dept. of Tourism)

Causes of Tides

To illustrate how tides are produced, we will assume Earth is a rotating sphere covered to a uniform depth with water (Figure 13.6). It is easy to see how the moon's gravitational force can cause the water to bulge on the side of Earth nearest the moon. In addition, however, an equally large tidal bulge is produced on the side of Earth directly opposite the moon.

Both tidal bulges are caused, as Newton discovered, by the pull of gravity. Gravity is inversely proportional to the square of the distance between two objects, meaning simply that it quickly weakens with distance. In this case, the two objects are the moon and Earth. Because the force of gravity decreases with distance, the moon's gravitational pull on Earth is slightly greater on the near side of Earth than on the far side. The result of this differential pulling is to stretch (elongate) the solid Earth, very slightly. In contrast, the world ocean, which is mobile, is deformed quite dramatically by this effect to produce the two opposing tidal bulges.

Because the position of the moon changes only moderately in a single day, the tidal bulges remain in place while Earth rotates "through" them. For this reason, if you stand on the seashore for 24 hours, Earth will rotate you through alternating areas of deeper and shallower water. As you are carried into regions of deeper water, the tide rises, and as you are carried away, the tide falls. Therefore, during one day you experience two high tides and two low tides.

Further, the tidal bulges migrate as the moon revolves around Earth every 29 days. As a result, the tides, like the time of moonrise, shift about 50 minutes later each day. After 29 days, the cycle is complete and a new one begins.

There may be an inequality between the high tides during a given day. Depending on the moon's position, the tidal bulges may be inclined to the equator as in Figure 13.6. This figure illustrates that the first high tide experienced by an observer in the Northern Hemisphere is considerably lower than the high tide half a day later. On the other hand, a Southern Hemisphere observer would experience the opposite effect.

The sun also influences the tides. It is far larger than the moon, but because it is so far away, the effect is considerably less than that of the moon. In fact, the tide-generating potential of the sun is slightly less than half that of the moon.

Near the times of new and full moons, the sun and moon are aligned and their forces are added together (Figure 13.7A). Accordingly, the combined gravity of

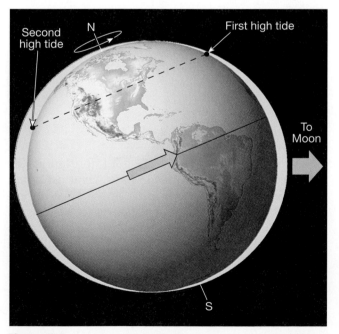

Figure 13.6

Tides on an Earth that is covered to a uniform depth with water. Depending on the moon's position, tidal bulges may be inclined to the equator. In this situation an observer will experience two unequal high tides.

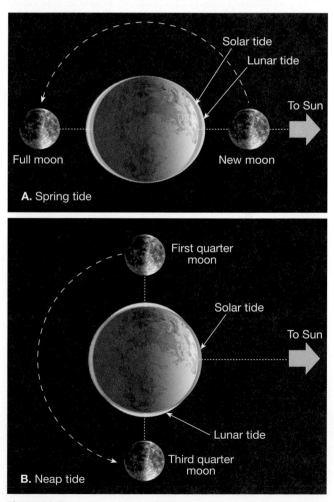

Figure 13.7

Relationship of the moon and sun to Earth during **A.** spring tides and **B.** neap tides.

these two tide-producing bodies cause higher tidal bulges (high tides) and lower tidal troughs (low tides). These are called the **spring tides**. Spring tides create the largest daily tidal range, that is, the largest variation between high and low tides. Conversely, at about the time of the first and third quarters of the moon, the gravitational forces of the moon and sun act on Earth at right angles, and each partially offsets the influence of the other (Figure 13.7B). As a result, the daily tidal range is less. These are called **neap tides**.

Types of Tides

So far, we have explained the basic causes and patterns of tides. But keep in mind these theoretical considerations cannot be used to predict either the height or the time of actual tides at a particular place. The shape of coastlines and the configuration of ocean basins greatly influence the tides. Consequently, tides at various locations respond differently to the tide-producing forces. This being the case, the nature of the tide at any place can be determined most accurately by actual observation. The predictions in tidal tables and the tidal data on nautical charts are based on such observations.

Tides are classified as one of three types according to the tidal pattern occurring at a particular locale. The *semidiurnal* type fits the twice-daily pattern described earlier. There are two high and two low tides each tidal day, with a relatively small difference in the high and low water heights. Tides along the Atlantic coast of the United States are representative of the semidiurnal type, which is illustrated by the tidal curve for Boston in Figure 13.8A. *Diurnal* tides are characterized by a single high and low water height each tidal day (Figure 13.8B). Tides of this type occur along the northern shore of the Gulf of Mexico, among other locations. More common than the diurnal tide, the *mixed* type of tide is characterized by a large inequality in high water heights, low water heights, or both (Figure 13.8C). In this case, there are usually two high and two low waters each day. Such tides are prevalent along the Pacific coast of the United States and in many other parts of the world.

Tidal Currents

Tidal current is the term used to describe the *horizontal* flow of water accompanying the rise and fall of the tide. These water movements induced by tidal forces can be important in some coastal areas. Tidal currents that advance into the coastal zone as the tide rises are called *flood currents*. As the tide falls, seaward-moving water generates *ebb currents*. Periods of little or no current, called *slack water*, separate flood and ebb. The areas affected by these alternating tidal currents are called **tidal flats** (see Figure 13.5). Depending on the nature of the coastal zone, tidal flats vary from narrow strips seaward of the beach to zones that may extend for several kilometers.

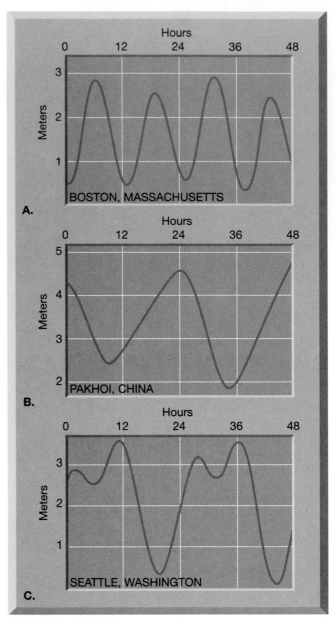

Figure 13.8

Types of tides. **A.** Semidiurnal. **B.** Diurnal. **C.** Mixed.

Although tidal currents are not important in the open sea, they can be rapid in bays, river estuaries, straits, and other narrow places. Off the coast of Brittany in France, for example, tidal currents that accompany a high tide of 12 meters (40 feet) may attain a speed of 20 kilometers (12 miles) per hour. While tidal currents are not generally considered to be major agents of erosion and sediment transport, notable exceptions occur where tides move through narrow inlets. Here they scour the small entrances to many good harbors that would otherwise be blocked.

Sometimes deposits called **tidal deltas** are created by tidal currents (Figure 13.9). They may develop either as *flood deltas* landward of an inlet or as *ebb deltas* on the seaward side of an inlet. Because wave activity and

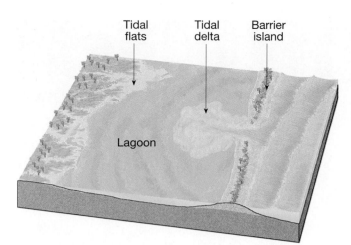

Tidal flats   Tidal delta   Barrier island

Lagoon

Figure 13.9
Because this tidal delta is forming in the relatively quiet waters on the landward side of a barrier island, it is termed a flood delta. As a rapidly moving tidal current emerges from the inlet, it slows and deposits sediment. The shapes of tidal deltas range from irregular to typical delta shaped.

longshore currents are reduced on the sheltered landward side, flood deltas are more common and actually more prominent. They form after the tidal current moves rapidly through an inlet. As the current emerges into more open waters from the narrow passage, it slows and deposits its load of sediment(see Box 13.1).

Waves Modify the Shoreline

The waters of the ocean are constantly in motion. The restless nature of the water is most noticeable along the shore—the dynamic interface between land and sea. Here we can observe the rhythmic rise and fall of tides and see waves constantly rolling in and breaking. Sometimes the waves are low and gentle. At other times, they pound the shore with an awesome fury.

Although it may not be obvious to occasional visitors, the shoreline is constantly being modified by waves. For example, south of Boston at Cape Cod, Massachusetts, wave activity is causing cliffs of poorly consolidated glacial sediment to retreat at nearly 1 meter per year (Figure 13.10A). By contrast, north of San Francisco at Point Reyes, California, the durable bedrock cliffs are far less susceptible to wave attack and therefore are retreating much more slowly (Figure 13.10B). Along both coasts, wave activity is moving sediment near the shore and building narrow sand bars that protrude across some bays.

The nature of present-day shorelines is not just the result of the relentless attack of the land by the sea. The shore is a complex zone whose unique character results from many geologic processes. For example, practically all coastal areas were affected by the worldwide rise in sea level that accompanied the melting of glaciers at the close of the Pleistocene epoch. As the sea edged

landward, the shoreline became superimposed upon landscapes that resulted from such diverse processes as stream erosion, glaciation, volcanic activity, and the forces of mountain building.

Wind-generated waves provide most of the energy that shapes and modifies shorelines. Where the land and sea meet, waves that may have traveled unimpeded for hundreds of kilometers suddenly encounter a barrier that will not allow them to advance farther and that must absorb their energy. Stated another way, the shore is where a practically irresistible force confronts an almost immovable object. The conflict that results is never-ending and sometimes dramatic.

Characteristics of Waves

The undulations of the water surface, called waves, derive their energy and motion from the wind. If a breeze of less than 3 kilometers (2 miles) per hour starts to blow across still water, small wavelets appear almost instantly. When the breeze dies, the ripples disappear as suddenly as they formed. However, if the wind exceeds 3 kilometers per hour, more stable waves gradually form and progress with the wind.

All waves have the characteristics illustrated in Figure 13.11. The tops of the waves are the *crests*, which are separated by *troughs*. The vertical distance between trough and crest is called the **wave height**, and the horizontal distance separating successive crests is the **wavelength**. The **wave period** is the time interval between the passage of successive crests at a stationary point. The height, length, and period that are eventually achieved by a wave depend on three factors: (1) wind speed; (2) length of time the wind has blown; and (3) **fetch**, the distance that the wind has traveled across open water. As the quantity of energy transferred from the wind to the water increases, the height and steepness of the waves increase as well. Eventually, a critical point is reached and ocean breakers called *whitecaps* form.

When the wind stops or changes direction, or the waves leave the stormy area where they were created, they continue on without relation to local winds. The waves also undergo a gradual change to *swells* that are lower in height and longer in length and may carry a storm's energy to distant shores. Because many independent wave systems exist at the same time, the sea surface acquires a complex and irregular pattern. Hence, the sea waves we watch from the shore are usually a mixture of swells from faraway storms and waves created by local winds.

Types of Waves

It is important to realize that in the open sea the motion of the wave is different from the motion of the water particles within it. It is the wave form or shape that moves forward, not the water itself, despite the way it

A.

Figure 13.10

A. This satellite image includes the familiar outline of Cape Cod. Boston is in the upper left corner. The two large islands off the south shore of Cape Cod are Martha's Vineyard (left) and Nantucket (right). Although the work of waves constantly modifies this coastal landscape, shoreline processes are not responsible for creating it. Rather, the present size and shape of Cape Cod result from the positioning of moraines and other glacial materials deposited during the Pleistocene epoch. (Photo courtesy of Earth Satellite Corporation) **B. (opposite)** High-altitude image of the Point Reyes area north of San Francisco, California. The 5.5-kilometer-long south-facing cliffs at Point Reyes (lower left corner) are exposed to the full force of the waves from the Pacific Ocean. Nevertheless, this promontory retreats slowly because the bedrock from which it formed is very resistant. (Photo courtesy of USDA-ASCS)

looks. Waves of *energy* simply move through the nearly stationary medium of the water. Each water particle moves in a nearly circular path during the passage of a wave (Figure 13.12). As a wave passes, a water particle returns almost to its original position. When water is part of the wave crest, it moves in the same direction as the advancing wave form. When it is in the trough, the water actually moves backward. This is demonstrated by observing the behavior of a floating object as a wave passes. The toy boat in Figure 13.12 merely seems to bob up and down and sway slightly to and fro without advancing appreciably from its original position. (The wind does drag the water forward slightly, causing the surface circulation of the oceans.) For this reason, waves in the open sea are called **waves of oscillation**.

The energy contributed by the wind to the water is transmitted not only along the surface of the sea but downward as well. However, beneath the surface, the circular

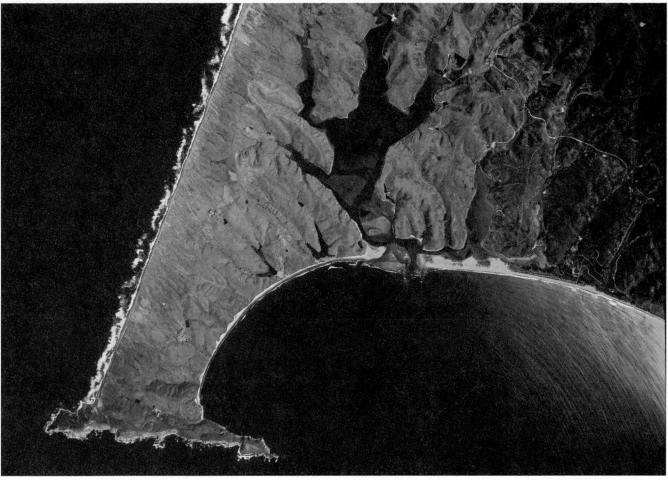

B.

motion rapidly diminishes until, at a depth equal to about one-half the wavelength, the movement of water particles becomes negligible. This is shown by the rapidly diminishing diameters of water-particle orbits in Figure 13.11.

As long as a wave is in deep water it progresses smoothly across the ocean (Figure 13.13, right). However, when a wave approaches the shore, the water becomes shallower and influences wave behavior. The

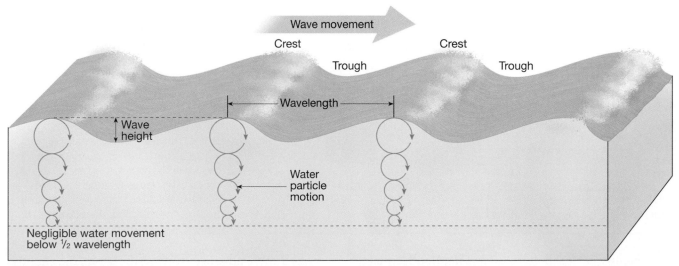

Figure 13.11

This diagram illustrates the basic parts of a wave as well as the movement of water particles with the passage of a wave. Negligible water movement occurs below a depth equal to one-half the wavelength (the level of the dashed line).

Box 13.1

Tidal Power

Several methods of generating electrical energy from the oceans have been proposed, but the ocean's energy potential remains largely untapped. The development of tidal power is the principal example of energy production from the ocean.

Tides have been used as a source of power for centuries. Beginning in the twelfth century, water wheels driven by the tides were used to power grist-mills and sawmills. During the seventeenth and eighteenth centuries, much of Boston's flour was produced at a tidal mill. Today, far greater energy demands must be met, and more sophisticated ways of using the force created by the perpetual rise and fall of the ocean must be employed.

Tidal power is harnessed by constructing a dam across the mouth of a bay or an estuary in a coastal area having a large tidal range (Figure 13.A). The narrow opening between the bay and the open ocean magnifies the variations in water level that occur as the tides rise and fall. The strong in-and-out flow that results at such a site is then used to drive turbines and electrical generators.

Tidal energy utilization is exemplified by the tidal power plant at the mouth of the Rance River in France. By far the largest yet constructed, this plant went into operation in 1966 and produces enough power to satisfy the needs of Brittany and also contribute to the demands of other regions. Much smaller experimental facilities have been built near Murmansk in Russia and near Taliang in China, and on the Bay of Fundy in the Canadian province of Nova Scotia (Figure 13.B). The United States has not yet tapped its tidal power potential, although a site at Passamaquoddy Bay in Maine, where the tidal range approaches 15 meters (50 feet), has been under review for many years.

Along most of the world's coasts it is not possible to harness tidal energy. If the tidal range is less than 8 meters (25 feet) or if narrow, enclosed bays are absent, tidal power development is uneconomical. For this reason, the tides will never provide a very high portion of our ever-increasing electrical energy requirements. Nevertheless, the development of tidal power may be worth pursuing at feasible sites because electricity produced by the tides consumes no exhaustible fuels and creates no noxious wastes.

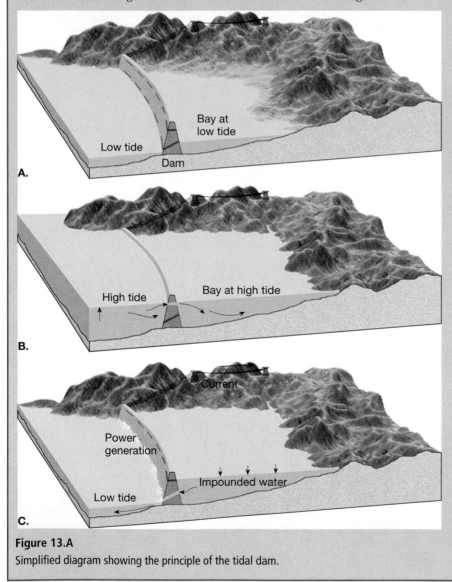

Figure 13.A
Simplified diagram showing the principle of the tidal dam.

Figure 13.B
The first tidal power generating station in North America opened in 1984 at Annapolis Royal, Nova Scotia. This pilot plant on Canada's Bay of Fundy is being used to assess the merits and drawbacks of building a larger facility. (Photo by Stephen J. Krasemann/Photo Researchers, Inc.)

wave begins to "feel bottom" at a water depth equal to about one-half its wavelength. Such depths interfere with water movement at the base of the wave and slow its advance (Figure 13.13, center).

As a wave advances toward the shore, the slightly faster seaward waves catch up, decreasing the wavelength. As the speed and length of·the wave diminish, the energy must go somewhere, and the only way to go

called *swash*, moves up the slope of the beach. When the energy of the swash has been expended, the water flows back down the beach toward the surf zone as *backwash*.

Wave Erosion

During periods of calm weather, wave action is at a minimum. However, just as streams do most of their work during floods, so do waves perform most of their work during storms. The impact of high, storm-induced waves against the shore can be awesome in its violence. Each breaking wave may hurl thousands of tons of water against the land, sometimes causing the ground literally to tremble. The pressures exerted by Atlantic Ocean waves, for example, average nearly 10,000 kilograms per square meter (more than 2000 pounds per square foot) in winter. The force during storms is even greater.

It is no wonder that cracks and crevices are quickly opened in cliffs, seawalls, breakwaters, and anything else that is subjected to these enormous shocks. Water is forced into every opening, causing air in the cracks to become highly compressed by the thrust of crashing waves. When the wave subsides, the air expands rapidly, dislodging rock fragments and enlarging and extending preexisting fractures (Figure 13.14).

In addition to the erosion caused by wave impact and pressure, **abrasion**, the sawing and grinding action of the water armed with rock fragments, is also important. In fact, abrasion is probably more intense in the surf zone than in any other environment. Smooth and rounded stones and pebbles along the shore are obvious reminders of the relentless grinding action of rock against rock in the surf zone. Further, such fragments are used as "tools" by the waves as they cut horizontally into the land (Figure 13.15).

Wave Refraction

The bending of waves, called **wave refraction**, plays an important part in shoreline processes. It affects the distribution of energy along the shore, and thus strongly influences where and to what degree erosion, sediment transport, and deposition will take place.

Waves seldom approach the shore straight on. Rather, most waves move toward the shore at an angle. When they reach the shallow water of a smoothly sloping bottom, however, they are bent and tend to become parallel to the shore. Such bending occurs because the part of the wave nearest the shore touches bottom and slows first, while the end that is still in deep water continues forward at its full speed. The net result is a wave front that may approach nearly parallel to the shore regardless of the original direction of the wave.

Due to refraction, wave impact is concentrated against the sides and ends of headlands projecting into the water, while wave attack is weakened in bays. This

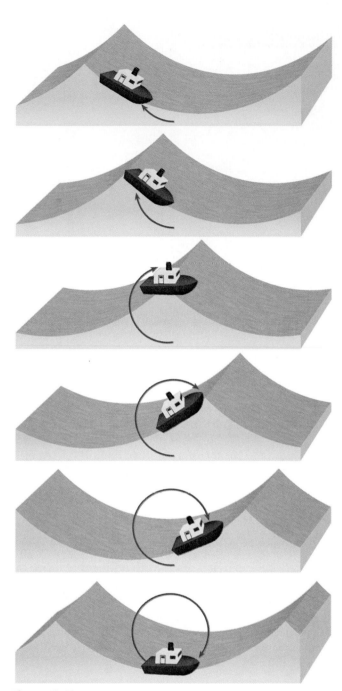

Figure 13.12

The movements of the toy boat show that the wave form advances, but the water does not advance appreciably from its original position. In this sequence, the wave moves from left to right as the boat (and the water in which it is floating) rotates in an imaginary circle. The boat moves slightly to the left up the front of the approaching wave, then after reaching the crest, slides to the right down the back of the wave.

is up, so the wave steadily grows higher. Finally a critical point is reached when the steep wave front collapses or *breaks* (Figure 13.13, left). What had been a wave of oscillation now becomes a **wave of translation** in which the water advances up the shore.

The turbulent water created by breaking waves is called **surf**. On the landward margin of the surf zone, the turbulent sheet of water from collapsing breakers,

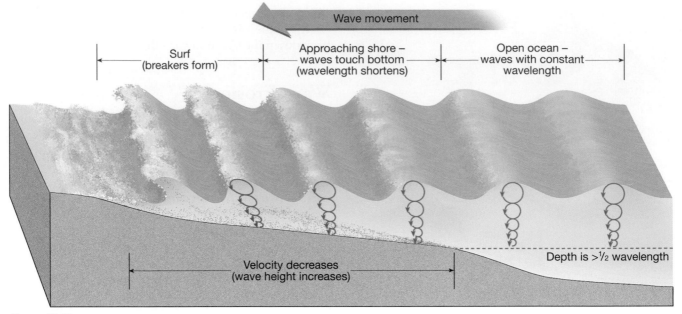

Figure 13.13
Changes that occur when a wave moves onto shore.

differential wave attack along irregular coastlines is illustrated in Figure 13.16. Because the waves reach the shallow water in front of the headland sooner than they do in adjacent bays, they are bent more nearly parallel to the protruding land and strike it from all three sides. By contrast, refraction in the bays causes waves to diverge and expend less energy. In these zones of weakened wave activity, sediments can accumulate and form sheltered sandy beaches. Over a long period, erosion of the headlands and deposition in the bays will straighten an irregular shoreline.

Moving Sand Along the Beach

Although waves are refracted, most still reach the shore at an angle, however slight. Consequently, the uprush of water from each breaking wave (the swash) is not head-on, but oblique. However, the backwash is in the direction of the slope of the beach. The effect of this pattern of water movement is to transport particles of sediment in a zigzag pattern along the beach (Figure 13.17). This movement, called **beach drift**, can transport sand and pebbles hundreds or even thousands of meters each day.

Oblique waves also produce currents within the surf zone that flow parallel to the shore. Since the water here is turbulent, these **longshore currents** easily move the fine suspended sand, and roll larger sand and gravel along the bottom. When the sediment transported by longshore currents is added to the quantity moved by beach drift, the total amount can be very large. At Sandy

Hook, New Jersey, for example, the quantity of sand transported along the shore over a 48-year period averaged almost 750,000 tons per year. For a 10-year period at Oxnard, California, more than 1.5 million tons of sediment moved along the shore each year.

There should be little wonder that beaches have been characterized as "rivers of sand." At any point along a beach there is likely to be more sediment that was derived elsewhere than material eroded from the land immediately behind it. It is also worth noting that much of the sediment composing beaches is not wave-eroded debris. Rather, in many areas sediment-laden rivers that discharge into the ocean are the major source of material. For that reason, if it were not for beach drift and longshore currents, many beaches would be nearly sandless.

Shoreline Features

Shoreline features vary, depending on the rocks of the shore, currents, wave intensity, and whether the coast is stable, sinking, or rising. This section summarizes many shoreline features, and several are shown in Figure 13.22.

Wave-Cut Cliffs and Platforms

Whether along the rugged and irregular New England coast or along the steep shorelines of the West Coast, the effects of wave erosion are often easily seen. **Wave-cut cliffs**, as their name implies, originate by the cut-

Figure 13.14
When waves break against the shore, the force of the water can be powerful and the erosional work that is accomplished can be great. (Photo by H. Richard Johnston/Tony Stone Worldwide)

ting action of the surf against the base of coastal land. As erosion progresses, rocks overhanging the notch at the base of the cliff crumble into the surf and the cliff retreats. A relatively flat, benchlike surface, the **wave-cut platform**, is left behind by the receding cliff (Figure 13.18). The platform broadens as wave attack continues. Some debris produced by the breaking waves remains along the water's edge as part of the beach, while the remainder is transported farther seaward.

Arches, Stacks, Spits, and Bars

As explained, headlands that extend into the sea are vigorously attacked by waves because of refraction. The surf erodes the rock selectively, wearing away the softer or more highly fractured rock at the fastest rate. At first, sea caves may form. When two caves on opposite sides of a headland unite, a **sea arch** results (Figure 13.19). Finally the arch falls in, leaving an isolated remnant, or **sea stack**, on the wave-cut platform (Figure 13.19). Eventually it too will be consumed by the action of the waves.

Where beach drift and longshore currents are active, several features related to the movement of sediment along the shore may develop. **Spits** are elongated ridges of sand that project from the land into the mouth of an adjacent bay. Often the end in the water hooks landward in response to wave-generated currents (Figure 13.20). The term **baymouth bar** is applied to a sand bar that completely crosses a bay, sealing it off from the open ocean (Figure 13.20). Such a feature tends to form across bays where currents are weak, allowing a spit to

Figure 13.15

Cliff undercut by wave erosion along the Oregon coast. (Photo by E. J. Tarbuck)

extend to the other side. A **tombolo**, a ridge of sand that connects an island to the mainland or to another island, forms in much the same manner as does a spit.

Barrier Islands

The Atlantic and Gulf Coastal Plains are relatively flat and slope gently seaward. The shore zone is characterized by **barrier islands**. These low ridges of sand parallel the coast at distances from 3 to 30 kilometers offshore. From Cape Cod, Massachusetts, to Padre Island, Texas, nearly 300 barrier islands rim the coast (Figure 13.21).

Most barrier islands are from 1 to 5 kilometers wide and between 15 and 30 kilometers long. The highest features are sand dunes, which usually reach heights of 5 to 10 meters. The lagoons that separate these narrow islands from the shore are zones of relatively quiet water that allow small craft traveling between New York and northern Florida to avoid the rough waters of the North Atlantic.

Barrier islands probably originate in several ways. Some originate as spits that were subsequently severed from the mainland by wave erosion or by the general rise in sea level following the last episode of glaciation. Others are created when turbulent waters in the line of breakers heap up sand that has been scoured from the bottom. Also some barrier islands may be former sand dune ridges that originated along the shore during the

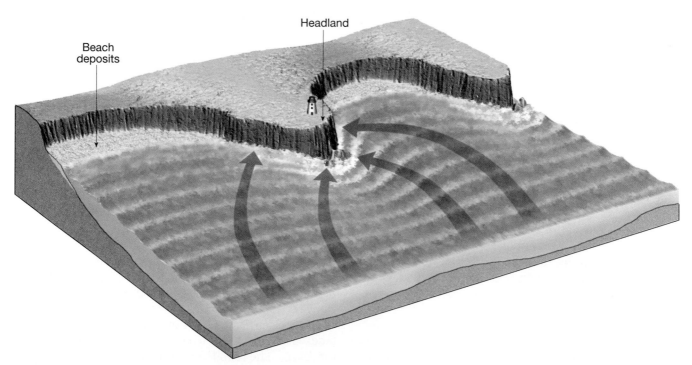

Figure 13.16

Because of wave refraction, the greatest erosional power is concentrated on the headlands. In the bays, the force of the waves is much weaker.

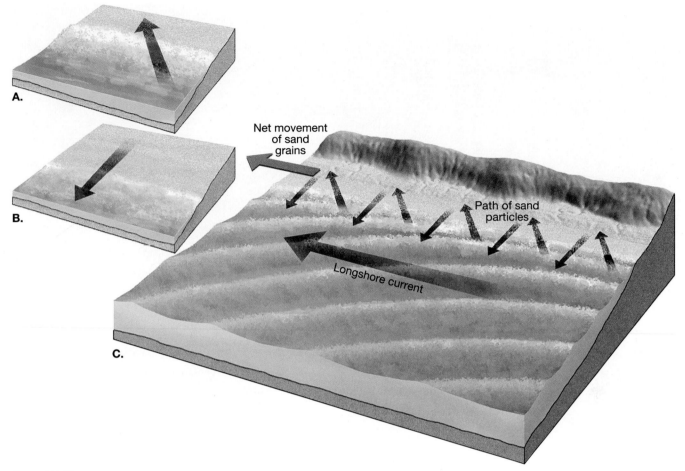

Figure 13.17
Beach drift and longshore currents are created by obliquely breaking waves. These processes transport large quantities of material along the beach and in the surf zone.

last glacial period, when sea level was lower. As the ice sheets melted, sea level rose and flooded the area behind the beach-dune complex.

The Evolving Shore

There is little question that a shoreline continually undergoes modification regardless of its initial configuration. At first most coastlines are irregular, although the degree of and reason for the irregularity may differ considerably from place to place. Along a coastline that is characterized by varied geology, the pounding surf may at first increase its irregularity because the waves will erode the weaker rocks more easily than the stronger ones. Be that as it may, it is commonly agreed that if a shoreline remains stable, marine erosion and deposition will eventually produce a straighter, more regular coast. Figure 13.22 illustrates the evolution of an initially irregular coast. As waves erode the headlands, creating cliffs and a wave-cut platform, sediment is carried along the shore. Some material is de-

posited in the bays, while other debris is formed into spits and baymouth bars. At the same time rivers fill the bays with sediment. Ultimately, a smooth coast results.

Shoreline Erosion Problems

Today the coastal zone teems with human activity. Unfortunately, people often treat the shoreline as if it were a stable platform on which structures can be built safely. This attitude jeopardizes both people and the shoreline. Many coastal landforms are relatively fragile, short-lived features that are easily damaged by development. And, as anyone who has endured a tropical storm knows, the shoreline is not always a safe place to live.

Compared with other natural hazards such as earthquakes, volcanic eruptions, and landslides, shoreline erosion appears to be a more continuous and predictable process that causes relatively modest damage to limited areas. In reality, the shoreline is a dynamic

Figure 13.18
Elevated wave-cut platform along the California coast north of San Francisco. A new platform is being created at the base of the cliff. (Photo by John S. Shelton)

Figure 13.19
Sea stack (left) and sea arches along the rugged coastline of Washington State. (Photo by David Muench)

Figure 13.20
High-altitude image of a well-developed spit and baymouth bar along the coast of Martha's Vineyard, Massachusetts. Also notice the tidal delta in the lagoon adjacent to the inlet through the baymouth bar. (Photo courtesy of USDA-ASCS)

place that can change rapidly in response to natural forces. Exceptional storms are capable of eroding beaches and cliffs at rates that far exceed the long-term average. Such bursts of accelerated erosion not only have a significant impact on the natural evolution of a coast, but can also have a profound impact on people who reside in the coastal zone (Figure 13.23). Erosion along our coasts causes significant property damage. Large sums are spent annually not only to repair damage, but

also to control erosion. Already a problem at many sites, shoreline erosion is certain to become increasingly serious as extensive coastal development continues.

Although the same processes cause change along every coast, not all coasts respond in the same way. Interactions among different processes and the relative importance of each process depend on local factors. These factors include: (1) the proximity of a coast to sediment-laden rivers, (2) the degree of tectonic activity, (3)

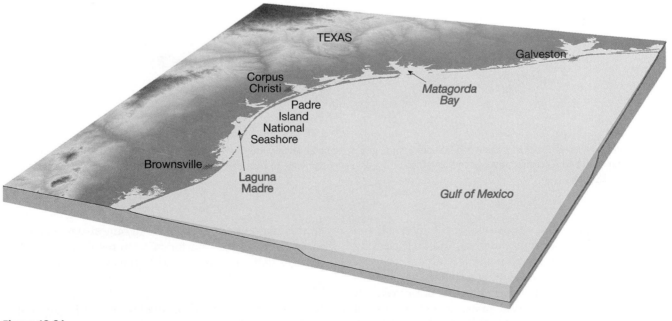

Figure 13.21
The islands along the south Texas coast are excellent examples of the nearly 300 barrier islands that rim the Atlantic and Gulf coasts.

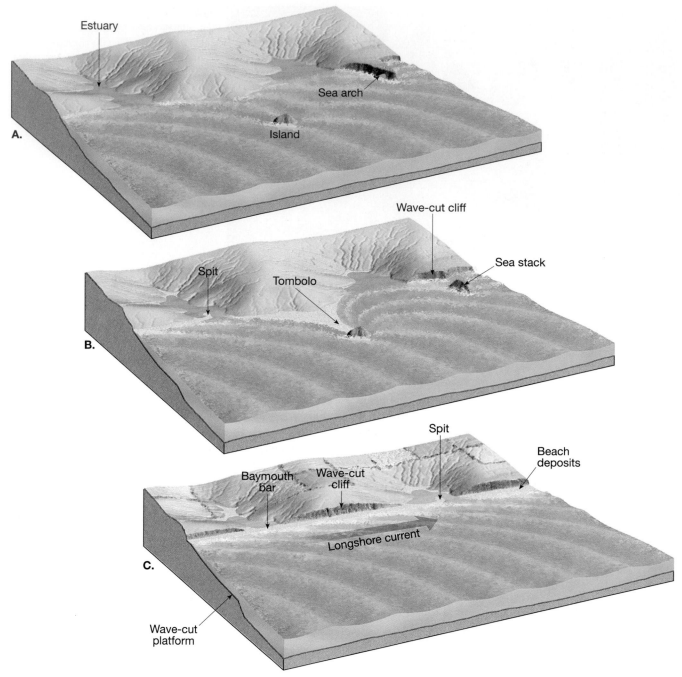

Figure 13.22

These diagrams illustrate the changes that can take place through time along an initially irregular coastline that remains relatively stable. The coastline shown in part **A** gradually evolves to **B** and then **C**. The diagrams also serve to illustrate many of the features described in the section on the shoreline features.

the topography and composition of the land, (4) prevailing winds and weather patterns, and (5) the configuration of the coastline and nearshore areas.

During this century, growing affluence and increasing demands for recreation have brought unprecedented development to many coastal areas. As the number and value of buildings has increased, so too have efforts to protect property from storm waves. Also, controlling the

natural migration of sand is an ongoing struggle in many coastal areas. Such interference can result in unwanted changes that are difficult and expensive to correct.

Groins

To maintain or widen beaches that are losing sand, **groins** are sometimes constructed. A groin is a barrier built at a right angle to the beach to trap sand that is

Figure 13.23
This destruction occurred when Hurricane Emily hit North Carolina's Outer Banks on August 31, 1993. (Photo courtesy of Raleigh News and Observer)

moving parallel to the shore (Figure 13.24). These structures often do their job so effectively that the longshore current beyond the groin becomes sand-starved. As a result, the current erodes sand from the beach on the leeward side of the groin.

To offset this effect, property owners downcurrent from the structure may erect a groin on their property. In this manner, the number of groins multiplies. An example of such proliferation is the shoreline of New Jersey, where hundreds of these structures have been built. Since it has been shown that groins often do not provide a satisfactory solution, they are no longer the preferred method of keeping beach erosion in check.

Figure 13.24
A series of groins at Ship Bottom, New Jersey. Since groins trap sand on the upcurrent side, the movement of sand along this coast, caused by the longshore current, must be toward the bottom of the photograph. (Photo by John S. Shelton)

Breakwaters and Seawalls

In some coastal areas a **breakwater** may be constructed parallel to the shoreline. The purpose of such a structure is to protect boats from the force of large breaking waves by creating a quiet water zone near the shore. However, when this is done, the reduced wave activity along the shore behind the structure may allow sand to accumulate. If this happens, the marina will eventually fill with sand while the downstream beach erodes and retreats. At Santa Monica, California, where the building of a breakwater created such a problem, the city had to install a dredge to remove sand from the protected quiet water zone and deposit it down the beach where longshore currents and beach drift could recirculate the sand (Figure 13.25).

As development has moved ever closer to the beach, seawalls are sometimes built to defend property from the force of breaking waves. **Seawalls** are simply massive barriers intended to prevent waves from reaching the areas behind the wall. Waves expend much of their energy as they move across an open beach. Seawalls cut this process short by reflecting the force of unspent waves seaward. As a consequence, the beach to the seaward side of the seawall experiences significant erosion and may, in some instances, be eliminated entirely. Once the width of the beach is reduced, the seawall is subjected to even greater pounding by the waves. Eventually this battering will cause the wall to fail, and a larger, more expensive wall must be built to take its place.

The wisdom of building temporary protective structures along shorelines is increasingly questioned. The feelings of many coastal scientists are expressed in the following excerpt from a position paper that grew out of a conference on America's Eroding Shoreline:

It is now clear that halting the receding shoreline with protective structures benefits only a few and seriously degrades or destroys the natural beach and the value it holds for the majority. Protective structures divert the ocean's energy temporarily from private properties, but usually refocus the energy on the adjacent natural beaches. Many interrupt the natural sand flow in coastal currents, robbing many beaches of vital sand replacement.[*]

Beach Nourishment

Beach nourishment represents another approach to stabilizing shoreline sands. As the term implies, this practice simply involves the addition of large quantities of sand to the beach system. By building the beaches seaward, beach quality and storm protection are both improved. Beach nourishment, however, is not a permanent solution to the problem of shrinking beaches. The same processes that removed the sand in the first place will eventually remove the replacement sand as well. In addition, beach nourishment is very expensive. When beach nourishment was used to renew 24 kilometers of Miami Beach, the cost was $64 million. Here the restoration must be redone every 10 to 12 years.

In some instances, beach nourishment can lead to unwanted environmental effects. For example, beach replenishment at Waikiki Beach, Hawaii, involved replacing coarse calcareous sand with softer, muddier calcareous sand. Destruction of the soft beach sand by breaking waves increased the water's turbidity and killed offshore coral reefs. At Miami Beach, increased turbidity also damaged local coral communities.

---

[*]"Strategy for Beach Preservation Proposed," *Geotimes* 30 (No. 12, December 1985): 15.

Figure 13.25
Aerial view of a breakwater at Santa Monica, California. The breakwater appears as a faint line in the water behind which many boats are anchored. The construction of the breakwater disrupted longshore transport and caused the seaward growth of the beach. (Photo by John S. Shelton)

So far, two basic responses to shoreline erosion problems have been considered: (1) the building of structures, such as groins and seawalls, to hold the shoreline in place, and (2) the addition of sand to replenish eroding beaches. However, a third option is also available, and that is to relocate buildings away from the beach. For many areas, the wisdom of adding new sand to the beaches and/or building stronger and stronger seawalls is indeed questionable. When this is the case, abandonment and relocation are the alternatives.

Contrasting the Atlantic and Pacific Coasts

The shoreline along the Pacific Coast of the United States is strikingly different from that characterizing the Atlantic and Gulf coast regions. Some of the differences are related to plate tectonics. The West Coast represents the leading edge of the North American plate, and because of this, it experiences active uplift and deformation. By contrast, the East Coast is a tectonically quiet region that is far from any active plate margin. Because of this basic geological difference, the nature of shoreline erosion problems along America's opposite coasts is different.

Atlantic and Gulf Coasts.   During this century growing affluence and increasing demand for recreation have brought unprecedented development to many coastal areas. Much of this development has occurred on barrier islands. Typically, barrier islands consist of a wide beach that is backed by dunes and separated from the mainland by marshy lagoons. The broad expanses of sand and exposure to the ocean have made barrier islands exceedingly attractive sites for development. Unfortunately, development has taken place more rapidly than our understanding of barrier island dynamics.

Because barrier islands face the open ocean, they receive the full force of major storms that strike the coast. When a storm occurs, the barriers absorb the energy of the waves primarily through the movement of sand. Frank Lowenstein describes this process and the dilemma that results:

Waves may move sand from the beach to offshore areas or, conversely, into the dunes; they may erode the dunes, depositing sand onto the beach or carrying it out to sea; or they may carry sand from the beach and the dunes into the marshes behind the barrier, a process known as overwash. The common factor is movement. Just as a flexible reed may survive a wind that destroys an oak tree, so the barriers survive hurricanes and nor'easters not through unyielding strength but by giving before the storm.

This picture changes when a barrier is developed for homes or a resort. Storm waves that previously rushed harmlessly through gaps between the dunes now encounter buildings and roadways. Moreover, since the dynamic nature of the barriers is readily perceived only during storms, homeowners tend to attribute damage to a particular storm, rather than to the basic mobility of coastal barriers. With their homes or investments at stake, local residents are more likely to seek to hold the sand in place and the waves at bay than to admit that development was improperly placed to begin with.[*]

Pacific Coast.   In contrast to the broad, gently sloping coastal plains of the East, much of the Pacific Coast is characterized by relatively narrow beaches that are backed by steep cliffs and mountain ranges. Recall that America's western margin is a more rugged and tectonically active region than the East. Since uplift continues, the apparent rise in sea level in the West is not so readily apparent. Nevertheless, like the shoreline erosion problems facing the East's barrier islands, West Coast difficulties also stem largely from the alteration of a natural system by people.

A major problem facing the Pacific shoreline, and especially portions of southern California, is a significant narrowing of many beaches. The bulk of the sand on many of these beaches is supplied by rivers that transport it from the mountainous regions to the coast. Over the years this natural flow of material to the coast has been interrupted by dams built for irrigation and flood control. The reservoirs effectively trap the sand that would otherwise nourish the beach environment. When the beaches were wider, they served to protect the cliffs behind them from the force of storm waves. Now, however, the waves move across the narrowed beaches without losing much energy and cause more rapid erosion of the sea cliffs.

Although the retreat of the cliffs provides material to replace some of the sand impounded behind dams, it also endangers homes and roads built on the bluffs. In addition, development atop the cliffs aggravates the problem. Urbanization increases runoff which, if not carefully controlled, can result in serious bluff erosion. Watering lawns and gardens adds significant quantities of water to the slope. This water percolates downward toward the base of the cliff, where it may emerge in small seeps. This action reduces the slope's stability and facilitates mass wasting.

Shoreline erosion along the Pacific Coast varies considerably from one year to the next, largely because of the sporadic occurrence of storms. As a consequence, when the infrequent but serious episodes of erosion occur, the damage is often blamed on the unusual storms and not on coastal development or the sediment-trapping dams that may be great distances away. If, as predicted, sea level rises at an increasing rate in the years to come, increased shoreline erosion and sea cliff retreat should be expected along many parts of the Pacific Coast (see Box 13.2).

---

[*]"Beaches or Bedrooms—The Choice as Sea Level Rises," *Oceanus* 28 (3) (Fall 1985): 22.

Emergent and Submergent Coasts

The great variety of present-day shorelines suggests that they are complex areas. Indeed, to understand the nature of any particular coastal area, many factors must be considered, including rock types, size and direction of waves, number of storms, tidal range, and submarine profile. Moreover, recent tectonic events and changes in sea level must also be taken into account. These many variables make shoreline classification difficult.

Many geologists classify coasts based on changes that have occurred with respect to sea level. This commonly used classification divides coasts into two very general categories: emergent and submergent. **Emergent coasts** develop either because an area experiences

---

Box 13.2

Is Global Warming Causing Sea Level to Rise?

The shifting dynamic nature of barrier islands and the ineffectiveness of most shoreline protection measures are now relatively well-established facts. Unfortunately, recent research, which indicates that sea level is rising, has compounded this already distressing situation. Studies indicate that sea level has risen between 10 and 15 centimeters over the past century. Furthermore, some investigators predict an accelerated sea-level rise in the years to come—30 cm or more by the middle of the next century. Although such a vertical change may seem modest, many coastal geologists believe that any given rise in sea level along the gently sloping Atlantic and Gulf coasts will cause from 10 to 1000 times as much horizontal shoreline retreat (Figure 13.C).

The idea that sea level will continue to rise in the coming decades is linked to the results of climatic studies that predict a global warming trend. Such predictions are based on the now well-established fact that the carbon dioxide (CO_2) content of the atmosphere has been rising at an accelerating rate for more than a century. The CO_2 is added primarily as a by-product of the combustion of ever-increasing quantities of fossil fuels. If we assume that the use of fossil fuels will continue to rise at projected rates, current estimates indicate that the atmosphere's CO_2 content will grow by an additional 40 percent or more by some time in the second half of the next century. The importance of CO_2 lies in the fact that it traps a portion of the radiation emitted by Earth and thereby keeps the air near the surface warmer than it would be without CO_2. Because CO_2 is an important heat-absorbing gas, an increase in the air's CO_2 content is believed to contribute to higher atmospheric temperatures. In addition, other trace gases generated by human activities also play a role.

How is a warmer atmosphere related to a global rise in sea level? First, higher temperatures can cause glacial ice to melt. About one-half of the 10- to 15-centimeter rise in sea level over the past century is attributed to the melting of small glaciers and ice caps. Second, a warmer atmosphere causes an increase in ocean volume through thermal expansion. That is, higher air temperatures raise the temperature of the upper layers of the ocean, which, in turn, causes the water to expand and sea level to rise. It is also believed that a warmer ocean may spur storm development. Of course, an increase in storm activity would compound an already serious problem in many coastal areas.

Since rising sea level is a gradual phenomenon, it may be overlooked by coastal residents as a significant contributor to shoreline erosion problems. Rather, the blame is assigned to other forces, especially storm activity. Although a given storm may be the immediate cause, the magnitude of its destruction may result from the relatively small sea level rise that allowed the storm's power to cross a much greater land area.

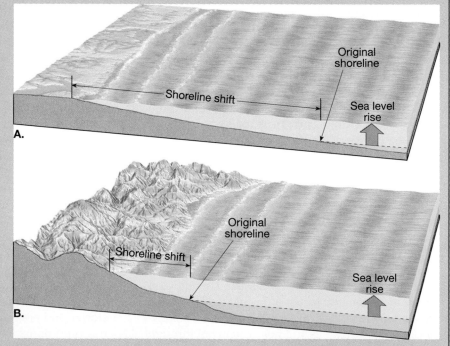

Figure 13.C

The slope of a shoreline is critical to determine the degree to which sea-level changes will affect it. **A.** When the slope is gentle, small changes in sea level cause a substantial shift. **B.** The same sea-level rise along a steep coast results in only a small shoreline shift.

uplift or as a result of a drop in sea level. Conversely, **submergent coasts** are created when sea level rises or the land adjacent to the sea subsides.

In some areas the coast is clearly emergent because rising land or a falling water level expose wave-cut cliffs and platforms above sea level. Excellent examples include portions of coastal California where uplift has occurred in the recent geological past. The elevated wave-cut platform shown in Figure 13.18 illustrates this. In the case of the Palos Verdes Hills, south of Los Angeles, seven different terrace levels exist, indicating seven episodes of uplift. The ever-persistent sea is now cutting a new platform at the base of the cliff. If uplift follows, it too will become an elevated marine terrace.

Other examples of emergent coasts include regions that were once buried beneath great ice sheets. When glaciers were present, their weight depressed the crust and when the ice melted, the crust began to gradually spring back. As a result, prehistoric shoreline features today are found high above sea level. The Hudson Bay region of Canada is one such area, portions of which are still rising at a rate of more than one centimeter per year.

In contrast to the preceding examples, other coastal areas show definite signs of submergence. The shoreline of a coast that has been submerged in the relatively recent past is often highly irregular because the sea typically floods the lower reaches of river valleys flowing into the ocean. The ridges separating the valleys, however, remain above sea level and project into the sea as headlands. These drowned river mouths, which are called **estuaries**, characterize many coasts today. Along the Atlantic coast, the Chesapeake and Delaware bays are examples of estuaries created by submergence (Figure 13.26). The picturesque coast of Maine, particularly in the vicinity of Acadia National Park, is another excellent example of an area that was flooded by the post-glacial rise in sea level and transformed into a highly irregular submerged coastline.

Keep in mind that most coasts have a complicated geologic history. With respect to sea level, many have at various times emerged and then submerged again. Each time they retain some of the features created during the previous situation.

Figure 13.26
Estuaries along the East Coast of the United States. The lower portions of many river valleys were submerged by the rise in sea level that followed the end of the Ice Age. Chesapeake Bay and Delaware Bay are especially prominent examples.

Review Questions

1. What is the primary driving force of surface ocean currents? How does the Coriolis effect influence these currents?

2. Describe the process of coastal upwelling.

3. How do ocean currents influence climate?

4. What is the driving force of deep-ocean circulation?

5. Discuss the origin of ocean tides.

6. Explain why an observer can experience two unequal high tides during a day (see Figure 13.6).

7. How does the sun influence tides?

8. How do semidiurnal, diurnal, and mixed tides differ?

9. Distinguish between flood current and ebb current.

10. What advantages does tidal power production offer? Is it likely that tides will provide a significant proportion of the world's electrical energy needs in the foreseeable future? (See Box 13.1.)

11. List three factors that determine the height, length, and period of a wave.

12. Describe the motion of a water particle as a wave passes (see Figure 13.12).

13. Explain what happens when a wave breaks.

14. How do waves cause erosion?

15. What is wave refraction? What is the effect of this process along irregular coastlines?

16. Why are beaches often called "rivers of sand"?

17. Describe the formation of the following features: wave-cut cliff, wave-cut platform, sea stack, spit, baymouth bar, tombolo.

18. Examine the satellite image of Cape Cod in Figure 13.10A. Locate some examples of spits and baymouth bars.

19. List three ways that barrier islands may form.

20. For what purpose is a groin built? Why might the building of one groin lead to the building of others?

21. How was the beach at Santa Monica, California, affected when a breakwater was constructed (see Figure 13.25)?

22. How might a seawall increase beach erosion?

23. It is believed that global temperatures will be increasing in the decades to come. How can a warmer atmosphere lead to rise in sea level? (See Box 13.2.)

24. Relate the damming of rivers to the shrinking of beaches at some locations along the West Coast of the United States.

25. What observable features would lead you to classify a coastal area as emergent?

26. Are estuaries associated with submergent or emergent coasts? Why?

Key Terms

abrasion (p. 345)
barrier island (p. 348)
baymouth bar (p. 347)
beach drift (p. 346)
beach nourishment (p. 355)
breakwater (p. 354)
Coriolis effect (p. 335)
emergent coast (p. 356)
estuary (p. 357)
fetch (p. 341)

groin (p. 352)
gyre (p. 334)
longshore current (p. 346)
neap tide (p. 340)
sea arch (p. 347)
sea stack (p. 347)
seawall (p. 354)
spit (p. 347)
spring tide (p. 340)
submergent coast (p. 357)
surf (p. 345)
thermohaline circulation (p. 337)
tidal current (p. 340)

tidal delta (p. 340)
tidal flat (p. 340)
tide (p. 338)
tombolo (p. 348)
upwelling (p. 336)
wave-cut cliff (p. 346)
wave-cut platform (p. 347)
wave height (p. 341)
wavelength (p. 341)
wave of oscillation (p. 342)
wave of translation (p. 345)
wave period (p. 341)
wave refraction (p. 345)

PART 3
The Atmosphere

Satellite image of a storm system over the British Isles.
(Photo by European Space Agency/Science Photo Library/Photo
Researchers, Inc.)

CHAPTER 14

Composition, Structure, and Temperature

Fall scene, Ottawa National Forest, Michigan. (Photo by Carr Clifton)

The study of the atmosphere—Earth's gaseous envelope—goes back thousands of years. Present knowledge suggests that no other planet has the exact mixture of gases or the heat and moisture conditions necessary to sustain life as we know it. The gases that make up Earth's atmosphere and the controls to which they are subject are vital to our existence. In this chapter we begin our examination of the ocean of air in which we all must live. We shall attempt to answer a number of basic questions. What is the composition of the atmosphere? At what point do we leave the atmosphere and enter outer space? What causes the seasons? How is air heated? What factors control temperature variations over the globe?

Weather influences our everyday activities, our jobs, and our health and comfort (Figure 14.1). Many of us pay little attention to the weather unless we are inconvenienced by it or when it adds to our enjoyment outdoors. Nevertheless, there are few other aspects of our physical environment that affect our lives more than the phenomena we collectively call the weather.

The United States has the greatest variety of weather of any country in the world. Beyond its direct impact on the lives of individuals, the weather has a strong effect on the world economy by influencing agriculture, energy use, water resources, transportation, and industry.

Although the atmosphere influences our lives a great deal, it is also important to realize that the reverse is true: People influence the atmosphere and its behavior. There are, and will continue to be, significant political and scientific decisions that must be made involving these impacts. Important examples are air pollution control and the effects of human industry on global climate and the atmosphere's protective ozone layer (see Box 14.1). We all need increased awareness and understanding of our atmosphere and its behavior.

Weather and Climate

If you were to search through the library for books about the atmosphere, many of the titles would contain the term *weather*, whereas others would have the word

Figure 14.1
New York City following a major blizzard that hit on January 7-8, 1996. The storm paralyzed much of the Northeast. Probably no other aspect of our physical environment affects the daily lives of people more than the weather. (Photo by Porter Gifford/Gamma Liaison)

climate. What is the difference between these two terms? **Weather** is a word used to denote the state of the atmosphere at a particular place for a *short period of time*. Weather is constantly changing—hourly, daily, and seasonally. **Climate**, on the other hand, might best be described as an aggregate or composite of weather. Stated another way, the climate of a place or region is a generalization of the weather conditions over a *long period of time*. Therefore, a climatic description is possible only after weather records have been kept for many years.

Although weather and climate are not identical, the nature of both is expressed in terms of the same **elements**, those quantities or properties that are measured regularly. The most important are (1) air temperature, (2) humidity, (3) type and amount of cloudiness, (4) type and amount of precipitation, (5) air pressure, and (6) the speed and direction of the wind. These elements are the major variables from which weather patterns and climate types are deciphered. Although we shall study these elements separately at first, keep in mind that they are very much interrelated. A change in any one of the elements will often bring about changes in the others.

Composition of the Atmosphere

In the days of Aristotle all things were thought to be a combination of four fundamental substances—fire, air, earth (soil), and water. Today we know that matter is much more complex. For example, Earth is composed of minerals, which, in turn, are made of numerous elements and compounds. Furthermore, **air**, which appears to be a unique substance, is really a *mixture* of many discrete gases, each with its own physical properties, in which varying quantities of tiny solid and liquid particles are suspended.

Major Components

The composition of air is not constant; it varies from time to time and from place to place. If the water vapor, dust, and other variable components were removed from the atmosphere, we would find that its makeup is very stable worldwide up to an altitude of about 80 kilometers (50 miles).

Clean, dry air is composed almost entirely of two gases—78 percent nitrogen and 21 percent oxygen (Figure 14.2). Although these gases are the most plentiful components of air and are of great significance to life on Earth, they are of minor importance in affecting weather phenomena. The remaining 1 percent of dry air is mostly the inert gas argon (0.93 percent) plus tiny quantities of a number of other gases. Carbon dioxide, although present in only minute amounts (0.035 percent),

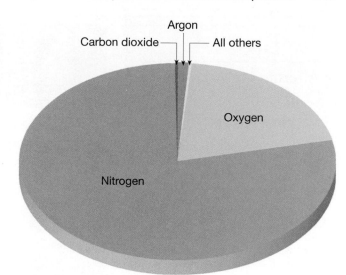

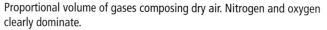

Figure 14.2

Proportional volume of gases composing dry air. Nitrogen and oxygen clearly dominate.

is nevertheless an important constituent of air, because it has the ability to absorb heat energy radiated by Earth and thus helps keep the atmosphere warm.

Variable Components

Air includes many gases and particles that vary significantly from time to time and place to place. Important examples include water vapor, dust particles, and ozone. Although usually present in small percentages, they can have significant effects on weather and climate.

Water Vapor. The amount of water vapor in the air varies considerably, from practically none at all up to about 4 percent by volume. Why is such a small fraction of the atmosphere so significant? Certainly the fact that water vapor is the source of all clouds and precipitation would be enough to explain its importance. However, water vapor has other roles. Like carbon dioxide, it has the ability to absorb heat energy given off by Earth as well as some solar energy. It is therefore important when we examine the heating of the atmosphere.

When water changes from one state to another (see Figure 15.2, p. 389), it absorbs or releases heat. This energy is termed *latent heat*, which means "hidden" heat. As we shall see in later chapters, water vapor in the atmosphere transports this latent heat from one region to another, and it is the energy source that helps drive many storms.

Dust. Most of us probably think of dust as small, barely visible bits of dirt. Nevertheless, from a meteorological standpoint, dust is much more than that. It includes

Box 14.1

Altering the Atmosphere's Composition: Sources and Types of Air Pollution

Air pollutants are airborne particles and gases that occur in concentrations that endanger the health and well-being of organisms or disrupt the orderly functioning of the environment. One category of pollutants, the *primary pollutants*, are emitted directly from identifiable sources. They pollute the air immediately upon being emitted. Figure 14.A shows the major primary pollutants and the sources that produce them. When the sources are examined, the significance of the transportation category is obvious. It accounts for nearly half of our air pollution (by weight). In addition to highway vehicles, this category includes trains, ships, and airplanes. Still, the tens of millions of cars and trucks on U.S. roads are, without a doubt, the greatest contributors (Figure 14.B).

Sometimes the direct impact of primary pollutants on human health and the environment is less severe than the effects of the secondary pollutants they form. *Secondary pollutants* are not emitted directly into the air, but form in the atmosphere when reactions take place

Figure 14.B

This crowded freeway reminds us that the transportation category accounts for nearly half of the air pollution (by weight) in the United States. (Photo by Will and Deni McIntyre/Photo Researchers, Inc.)

among primary pollutants. The chemicals that make up smog are important examples, as is the sulfuric acid that falls as acid precipitation. After the primary pollutant, sulfur dioxide, is emitted into the atmosphere, it combines with oxygen to produce sulfur trioxide, which then combines with water to create this irritating and corrosive acid (see Box 3.1).

Many reactions that produce secondary pollutants are triggered by strong sunlight and so are called *photochemical reactions*. One common example occurs when nitrogen oxides absorb solar radiation, initiating a chain of complex reactions. When certain volatile organic compounds are present, the result is the formation of a number of undesirable secondary products that are very reactive, irritating, and toxic. Collectively, this noxious mixture of gases and particles is called *photochemical smog*.

Air quality standards have not yet been met in a large number of places. Nevertheless, the United States has made significant progress in reducing air pollution. In the early 1990s annual emissions of the five primary pollutants shown in Figure 14.A totaled about 125 million metric tons. By contrast, in 1970, when the first Clean Air Act became law, the yearly output of the same five pollutants totaled 194 million metric tons. This improvement in air quality (a 35 percent reduction in primary pollutants) has been achieved during a time when urban areas have grown dramatically. Still, air pollution is a continuing threat to our health and welfare.

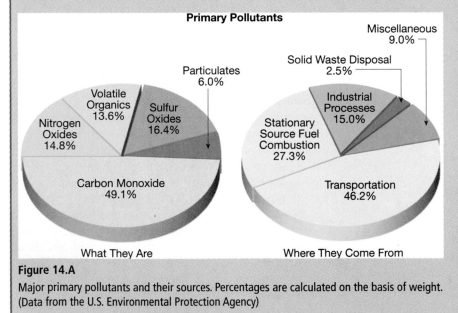

Figure 14.A

Major primary pollutants and their sources. Percentages are calculated on the basis of weight. (Data from the U.S. Environmental Protection Agency)

many microscopic particles that are invisible to the naked eye, among them organic materials like pollen, spores, and seeds. Dust particles are most numerous in the lower atmosphere near their primary source, Earth's surface. Still, the upper atmosphere is not free of them, because some dust is carried to great heights by rising currents of air. Other dust particles are provided to the upper atmosphere by meteors that disintegrate as they pass through Earth's envelope of air.

Some particles are important because they act as surfaces upon which water vapor may condense. This function is essential to the formation of clouds and fog. In addition, dust may absorb or reflect incoming solar radiation. Thus, when the dust content of the atmosphere is high, as it may be following an explosive volcanic eruption, the amount of sunlight reaching Earth's surface can be measurably reduced. Finally, dust in the air contributes to a phenomenon we all have observed—the red and orange colors of sunrise and sunset (Figure 14.3).

Ozone.   Another important component of the atmosphere is *ozone*. It is a form of oxygen that combines three oxygen atoms into each molecule (O_3). Ozone is not the same as oxygen we breathe, which has two atoms per molecule (O_2). There is very little of this gas in the atmosphere and its distribution is not uniform. In the lowest portion of the atmosphere, ozone represents less than one part in 100 million. It is concentrated well above the surface in a layer called the *stratosphere*, between 10 and 50 kilometers (6 and 31 miles).

In this altitude range, oxygen molecules (O_2) are split into single atoms of oxygen (O) when they absorb ultraviolet radiation emitted by the sun. Ozone is then created when a single atom of oxygen (O) and a molecule of oxygen (O_2) collide. This must happen in the presence of a third, neutral molecule that acts as a *catalyst* by allowing the reaction to take place without itself being consumed in the process. Ozone is concentrated

in the 10- to 50-kilometer height range because a crucial balance exists there: The ultraviolet radiation from the sun is sufficient to produce single atoms of oxygen, and there are enough gas molecules to bring about the required collisions.

The presence of the ozone layer in our atmosphere is crucial to those of us who dwell on Earth. The reason is that ozone absorbs the potentially harmful ultraviolet (UV) radiation from the sun. If ozone did not filter a great deal of the ultraviolet radiation, and if the sun's UV rays reached the surface of Earth undiminished, our planet would be uninhabitable for most life as we know it. Thus, anything that reduces the amount of ozone in the atmosphere could affect the well-being of life on Earth. Just such a problem exists and is described in Box 14.2.

Height and Structure of the Atmosphere

To say that the atmosphere begins at Earth's surface and extends upward is obvious. However, where does the atmosphere end and outer space begin? There is no sharp boundary; the atmosphere rapidly thins as you travel away from Earth, until there are too few gas molecules to detect.

Pressure Changes

To understand the vertical extent of the atmosphere, let us examine the changes in the atmospheric pressure with height. Atmospheric pressure is simply the weight

Figure 14.3
Dust in the air can cause sunsets to be especially colorful. (Photo by Marc Muench/David Muench Photography, Inc.)

Box 14.2

The Ozone Problem

Although stratospheric ozone is 10–50 kilometers up, it is vulnerable to human activities. Chemicals we produce are breaking up ozone molecules in the stratosphere, weakening our shield against UV rays. This loss of ozone is a serious global-scale environmental problem. Measurements over the past two decades confirm that ozone-depletion is occurring worldwide and is especially pronounced above Earth's poles. You can see this effect over the South Pole in Figure 14.C.

Over the past half century, people have unintentionally placed the ozone layer in jeopardy by polluting the atmosphere. The offending chemicals are known as *chlorofluorocarbons* (CFCs for short). Over the decades many uses were developed for CFCs: coolants for air conditioning and refrigeration equipment, cleaning solvents for electronic components and computer chips, and propellants for aerosol sprays. CFCs are also used to produce certain plastic foams for products such as cups and home insulation.

Because CFCs are practically inert (that is, not chemically active) in the lower atmosphere, a portion of these gases gradually makes its way to the ozone layer, where sunlight separates the chemicals into their constituent atoms. The chlorine atoms released this way break up some of the ozone molecules.

Because ozone filters out most of the UV radiation from the sun, a decrease in its concentration permits more of these harmful wavelengths to reach Earth's surface. The most serious threat to human health is an increased risk of skin cancer. An increase in damaging UV radiation also can impair the human immune system as well as promote cataracts, a clouding of the eye lens that reduces vision and may cause blindness if not treated.

In response to this problem, an international agreement known as the *Montreal Protocol* was developed under the sponsorship of the United Nations. If ozone-depleting CFCs are eliminated by the Montreal Protocol, it is estimated that the ozone layer will gradually recover and return to former levels by the middle of the next century. Positive action has been taken, yet the problem will remain for decades to come.

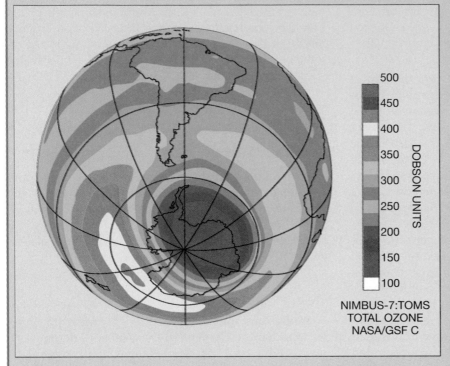

500
450
400
350
300
250
200
150
100

DOBSON UNITS

NIMBUS-7:TOMS
TOTAL OZONE
NASA/GSF C

Figure 14.C

This image shows ozone distribution for the Southern Hemisphere for the month of October 1992. The area of greatest depletion, called the ozone "hole," forms over Antarctica during the Southern Hemisphere spring and is shown in shades of pink and dark red. In October 1992, ozone levels reached a minimum of about 120 Dobson units, far below the 220 Dobson units typically seen over Antarctica before the hole forms. The image was produced with data acquired by the Total Ozone Mapping Spectrometer (TOMS), an instrument on board NASA's NIMBUS-7 satellite. (Courtesy of Arlin J. Krueger, NASA/Goddard Space Flight Center)

of the air above. At sea level, the average pressure is slightly more than 1000 millibars. This corresponds to a weight of slightly more than 1 kilogram per square centimeter (14.7 pounds per square inch). Obviously the pressure at higher altitudes is less (Figure 14.4).

One-half of the atmosphere lies below an altitude of 5.6 kilometers (3.5 miles). At about 16 kilometers (10 miles), 90 percent of the atmosphere has been traversed, and above 100 kilometers (62 miles), only 0.00003 percent of all the gases composing the atmosphere remain. Even so, traces of our atmosphere extend far beyond this altitude, gradually merging with the emptiness of space.

Temperature Changes

By the early twentieth century, much had been learned about the lower atmosphere. The upper atmosphere was partly known from indirect methods. Data from balloons and kites had revealed that the air temperature dropped with increasing height above Earth's surface. This phenomenon is felt by anyone who has climbed a high mountain, and is obvious in pictures of snowcapped mountaintops rising above snow-free lowlands (Figure 14.5). We divide the atmosphere vertically into four layers on the basis of temperature (Figure 14.6).

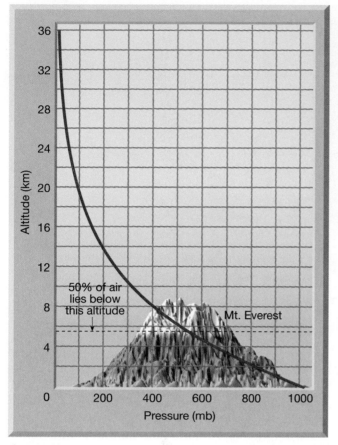

Figure 14.4
Atmospheric pressure variation with altitude. The rate of pressure decrease with an increase in altitude is not constant. Rather, pressure decreases rapidly near Earth's surface and more gradually at greater heights.

Figure 14.5
Temperatures drop with an increase in altitude in the troposphere. Therefore, it is possible to have snow on a mountaintop and warmer, snow-free lowlands below. Mount Kerkeslin, Jasper National Park, Alberta, Canada. (Photo by Carr Clifton/Minden Pictures)

Troposphere.   The bottom layer in which we live, where temperature decreases with an increase in altitude, is the **troposphere**. The term literally means the region where air "turns over," a reference to the turbulent weather in this lowermost zone. The troposphere is the chief focus of meteorologists, because it is in this layer that essentially all important weather phenomena occur.

The temperature decrease in the troposphere is called the **environmental lapse rate**. Although its average value is 6.5°C per kilometer (3.5°F per 1000 feet), a figure known as the *normal lapse rate*, its value is quite variable. Thus, to determine the actual environmental lapse rate for any particular time and place, as well as gather information about vertical changes in pressure, wind, and humidity, radiosondes are used. The radiosonde is an instrument package that is attached to a balloon and transmits data by radio as it ascends through the atmosphere (Figure 14.7).

The thickness of the troposphere is not the same everywhere; it varies with latitude and the season. On the average, the temperature drop continues to a height of about 12 kilometers (7.4 miles). The outer boundary of the troposphere is the *tropopause*.

Stratosphere.   Beyond the tropopause is the **stratosphere**. In the stratosphere, the temperature remains constant to a height of about 20 kilometers and then begins a gradual increase that continues until the *stratopause*, at a height of about 50 kilometers above Earth's surface. Below the tropopause, atmospheric properties like temperature and humidity are readily transferred by large-scale turbulence and mixing. Above the tropopause, in the stratosphere, they are not. The reason for the increased temperatures in the stratosphere is that the atmosphere's ozone is concentrated in this layer. (Recall that ozone absorbs ultraviolet radiation from the sun.) As a consequence, the stratosphere is heated.

Mesosphere.   In the third layer, the **mesosphere**, temperatures again decrease with height until, at the *mesopause*, approximately 80 kilometers (50 miles) above the surface, the temperature approaches –90°C.

Figure 14.6

Thermal structure of the atmosphere to a height of about 110 kilometers (68 miles).

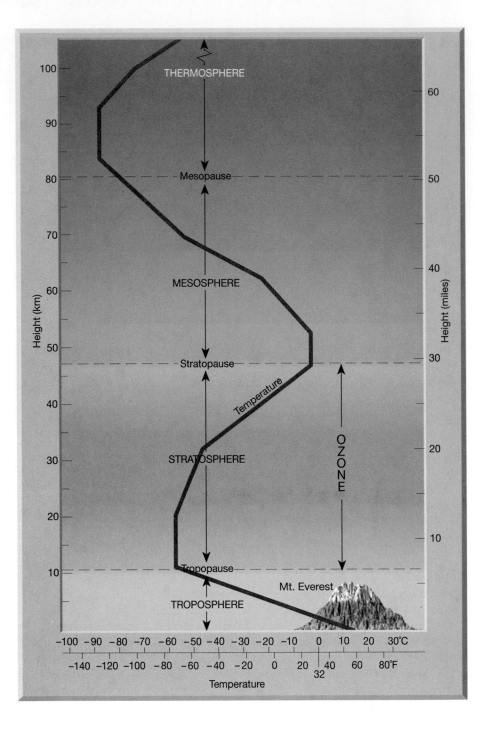

Thermosphere. The fourth layer extends outward from the mesopause and has no well-defined upper limit. It is the **thermosphere**, a layer that contains only a minute fraction of the atmosphere's mass. In the extremely rarefied air of this outermost layer, temperatures again increase due to the absorption of very short-wave, high-energy solar radiation by atoms of oxygen and nitrogen.

Temperatures rise to extremely high values of more than 1000°C in the thermosphere. But such temperatures are not comparable to those experienced near Earth's surface. Temperature is defined in terms of the average speed at which molecules move. Since the gases of the thermosphere are moving at very high speeds, the temperature is very high. But the gases are so sparse that collectively, they possess only an insignificant quantity of heat. For this reason, the temperature of a satellite orbiting Earth in the thermosphere is determined chiefly by the amount of solar radiation it absorbs and not by the high temperature of the almost nonexistent surrounding air. If an astronaut inside were to expose his or her hand, it would not feel hot.

Figure 14.7
Atmospheric soundings using radiosondes supply data on vertical changes in temperature, pressure, and humidity. (Courtesy of NOAA)

Earth-Sun Relationships

Always remember that nearly all of the energy that drives Earth's variable weather and climate comes from the sun. Earth intercepts only a minute percentage of the energy given off by the sun—less than one two-billionth. This may seem to be an insignificant amount until we realize that it is several hundred thousand times the electrical-generating capacity of the United States.

If solar energy were distributed uniformly over Earth's surface, our planet would have more uniform temperatures. But solar energy is not distributed evenly over Earth's surface. It is this unequal heating that drives the ocean's currents and creates the winds, which, in turn, transport heat from the tropics to the poles in an unending attempt to reach an energy bal-

ance. If the sun were to be "turned off," global winds would quickly subside. Yet as long as the sun shines, the winds will blow and the phenomena we know as weather will persist.

Energy from the sun is the most important control of our weather and climate. To understand atmospheric processes, we must understand what causes the variations in the amount of solar energy reaching Earth. The principal causes are variations in sun angle and exposure time.

Earth's Rotation and Revolution

Earth has two principal motions—rotation and revolution. **Rotation** is the spinning of Earth about its axis. The axis is an imaginary line running through the poles. Our planet rotates once every 24 hours, producing the daily cycle of daylight and darkness. At any moment, half of Earth is experiencing daylight, and the other half darkness. The line separating the dark half of Earth from the lighted half is called the **circle of illumination**.

Revolution refers to the movement of Earth in its orbit around the sun. Hundreds of years ago, most people believed that Earth was stationary in space and that the sun and stars revolved around our planet. Today, we know that Earth is traveling at more than 107,000 kilometers per hour in its orbit about the sun.

Seasons

We know that it is colder in winter than in summer. But why? Length of daylight certainly accounts for some of the difference. Long summer days expose us to more solar radiation, whereas short winter days expose us to less. But an even bigger factor is the angle of the sun above the horizon.

During mid-summer the sun is high above the horizon as it makes its daily journey across the sky. But as summer gives way to autumn, the noon sun appears lower and lower in the sky. What we observe here is the annual shifting of the solar angle or *altitude* of the sun.

The seasonal variation in the altitude of the sun affects the amount of energy received at Earth's surface in two ways. First, when the sun is high in the sky, the solar rays are most concentrated (you can see this in Figure 14.8A). The lower the angle, the more spread out and less intense is the solar radiation that reaches the surface (Figure 14.8B, C).

Second, and of lesser importance, the angle of the sun determines the amount of atmosphere the rays must penetrate (Figure 14.9). When the sun is directly overhead, the rays pass through a thickness of only 1 atmosphere, whereas rays entering at a 30-degree angle travel through twice this amount, and 5-degree rays travel through a thickness roughly equal to 11 atmospheres. The longer the path, the greater are the chances for absorption, reflection, and scattering by the atmosphere, all of which reduce the intensity at the surface.

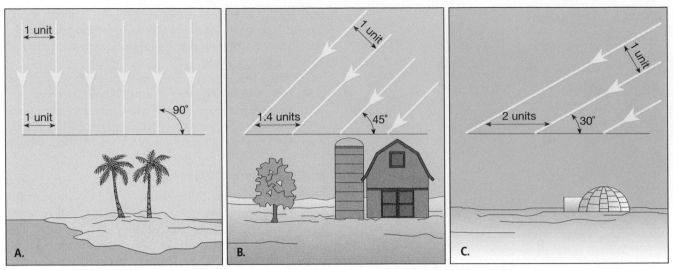

Figure 14.8

Changes in the sun angle cause variations in the amount of solar energy reaching Earth's surface. The higher the angle, the more intense the solar radiation. Notice that one unit of solar energy striking at a 30° angle (**C**) spreads over twice as much area as that striking at a 90° angle (**A**).

The same effects account for the fact that the midday sun can be literally blinding, while the setting sun can be a sight to behold.

If Earth were a flat surface, oriented at a right angle to the sun, all places would receive the same amount of radiation. But Earth is spherical. Hence, on any given day only places located at a particular latitude receive vertical (90-degree) rays from the sun. As we move either north or south of this location, the sun's rays strike at an ever-decreasing angle. Thus, the nearer a place is to the latitude receiving vertical rays of the sun, the higher will be its noon sun (Figure 14.9).

Causes of Seasons. What causes the yearly fluctuations in the sun angle and length of daylight? They occur because Earth's orientation to the sun continually

Figure 14.9

Rays striking at a low angle must travel through a greater thickness of the atmosphere than rays striking at a higher angle. Thus, the rays striking at a low angle are subject to greater depletion by reflection and absorption.

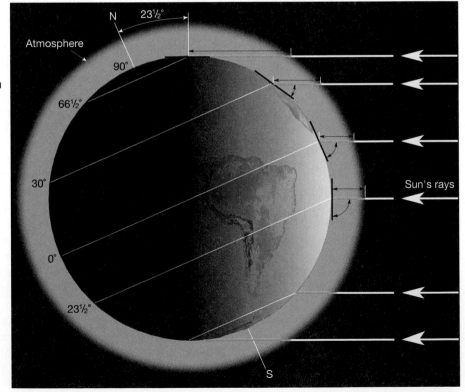

changes as it travels along its orbit. Earth's axis is not perpendicular to the plane of its orbit around the sun, but instead is tilted $23\frac{1}{2}$ degrees from the perpendicular. This is termed the **inclination of the axis**, and as we see, if the axis were not inclined, we would have no seasonal changes. In addition, because the axis remains pointed in the same direction (toward the North Star) as Earth journeys around the sun, the orientation of Earth's axis to the sun's rays is constantly changing (Figure 14.10).

On one day each year the axis is such that the Northern Hemisphere is "leaning" $23\frac{1}{2}$ degrees *toward* the sun. Six months later, when Earth has moved to the opposite side of its orbit, the Northern Hemisphere "leans" $23\frac{1}{2}$ degrees *away* from the sun. On days between these extremes, Earth's axis "leans" at amounts less than $23\frac{1}{2}$ degrees to the rays of the sun. This change in orientation causes the vertical rays of the sun to make a yearly migration from $23\frac{1}{2}$ degrees north of the equator to $23\frac{1}{2}$ degrees south of the equator.

This migration in turn causes the altitude angle of the noon sun to vary by as much as 47 degrees ($23\frac{1}{2}$ + $23\frac{1}{2}$) during the year at places located poleward of latitude $23\frac{1}{2}$ degrees. For example, a mid-latitude city like New York (about 40 degrees north latitude) has a maximum noon sun angle of $73\frac{1}{2}$ degrees when the sun's vertical rays reach their farthest northward location and a minimum noon sun angle of $26\frac{1}{2}$ degrees six months later.

Solstices and Equinoxes.   Historically, four days each year have been given special significance based on the annual migration of the direct rays of the sun and its importance to the yearly weather cycle. On June 21 or 22, Earth is in a position such that the north end of its axis is tilted $23\frac{1}{2}$ degrees toward the sun (Figure 14.11A). At this time the vertical rays of the sun strike $23\frac{1}{2}$ degrees north latitude ($23\frac{1}{2}$ degrees north of the equator), a latitude known as the **Tropic of Cancer**. For people in the Northern Hemisphere, June 21 or 22 is known as the **summer solstice**.

Six months later, on about December 21 or 22, Earth is in the opposite position, with the sun's vertical rays striking at $23\frac{1}{2}$ degrees south latitude (Figure 14.11C). This parallel is known as the **Tropic of Capricorn**. For those in the Northern Hemisphere, December 21 and 22 is the **winter solstice**. However, at the same time in the Southern Hemisphere, people are experiencing just the opposite—the summer solstice.

Midway between the solstices are the equinoxes. September 22 or 23 is the date of the **autumnal equinox** in the Northern Hemisphere, and March 21 or 22 is the date

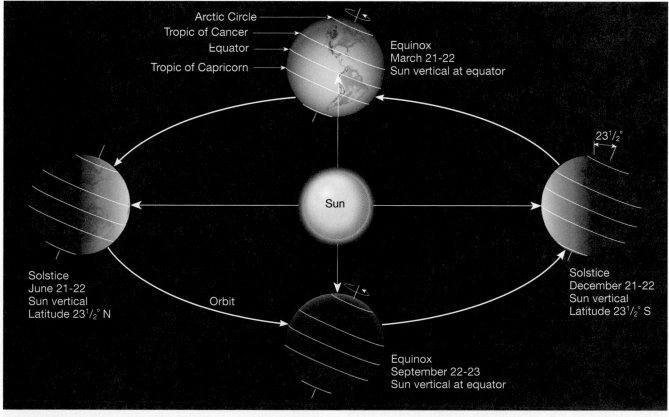

Figure 14.10
Earth–sun relationships.

Figure 14.11
Characteristics of the solstices and equinoxes.

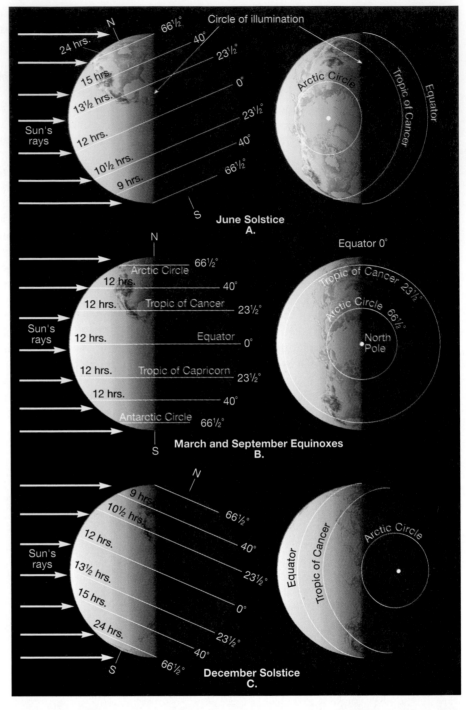

of the **spring equinox**. On these dates, the vertical rays of the sun strike the equator (0 degrees latitude) because Earth is in such a position in its orbit that the axis is tilted neither toward nor away from the sun (Figure 14.11B).

The length of daylight versus darkness is also determined by Earth's position in orbit. The length of daylight on June 21, the summer solstice in the Northern Hemisphere, is greater than the length of night. This fact can be established from Figure 14.11A by comparing the fraction of a given latitude that is on the "day" side of the circle of illumination with the fraction on the "night" side. The opposite is true for the winter solstice, when the nights are longer than the days. Again for comparison let us consider New York City, which has 15 hours of daylight June 21 and only 9 hours on December 21 (you can see this in Figure 14.11 and Table 14.1).

Table 14.1 Length of daylight.

Latitude (degrees)	Summer Solstice	Winter Solstice	Equinoxes
0	12 h	12 h	12 h
10	12 h 35 min	11 h 25 min	12
20	13 12	10 48	12
30	13 56	10 04	12
40	14 52	9 08	12
50	16 18	7 42	12
60	18 27	5 33	12
70	24 h (for 2 mo)	0 00	12
80	24 h (for 4 mo)	0 00	12
90	24 h (for 6 mo)	0 00	12

As a review of the characteristics of the summer solstice for the Northern Hemisphere, examine Figure 14.11A and Table 14.1 and consider the following facts:

1. The solstice occurs on June 21 or 22.
2. The vertical rays of the sun are striking the Tropic of Cancer ($23\frac{1}{2}$ degrees north latitude).
3. Locations in the Northern Hemisphere are experiencing their greatest length of daylight. (Opposite for the Southern Hemisphere.)
4. Locations north of the Tropic of Cancer are experiencing their highest noon sun angles. (Opposite for places south of the Tropic of Capricorn.)
5. The farther you are north of the equator, the longer the period of daylight, until the Arctic Circle is reached, where daylight lasts for 24 hours. (Opposite for the Southern Hemisphere.)

Also note from Table 14.1 that on June 21 the farther you are north of the equator the longer is the period of daylight, until the Arctic Circle is reached, where the length of daylight is 24 hours (Figure 14.12).

During an equinox (meaning "equal night"), the length of daylight is 12 hours everywhere on Earth, because the circle of illumination passes directly through the poles, dividing the latitudes in half.

The facts about the winter solstice are just the opposite. It should now be apparent why a mid-latitude location is warmest in the summer, for it is then that days are longest and sun's altitude is highest.

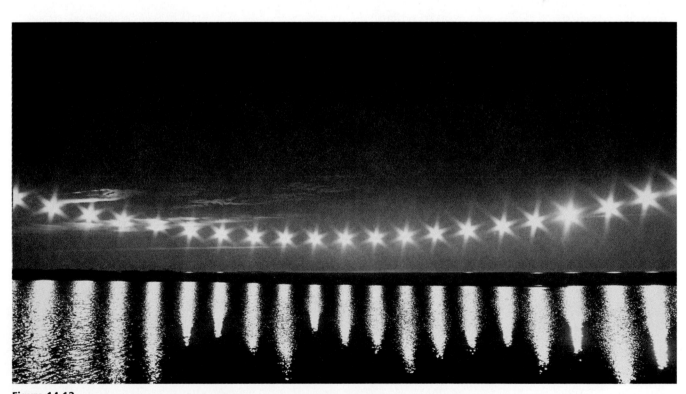

Figure 14.12

Multiple exposures of the midnight sun in late June or July in high northern latitudes—Alaska, Scandinavia, Northern Canada, etc. (Photo by Brian Stablyk/Tony Stone Images)

In summary, seasonal variations in the amount of solar energy reaching places on Earth's surface are caused by the migrating vertical rays of the sun and the resulting variations in sun angle and length of daylight. All places at the same latitude have identical sun angles and lengths of daylight. If the Earth–sun relationships just described were the only controls of temperature, we would expect these places to have identical temperatures as well. Obviously this is not the case. Although the altitude of the sun is the main control of temperature, it is not the only control, as we shall see.

Box 14.3

Solar Energy

A high percentage of the world's energy needs are derived from fossil fuels, primarily oil, coal, and natural gas. Of course, as world population soars, consumption will climb. Thus, reserves will eventually be in short supply. In the meantime, the environmental impact of burning huge quantities of fossil fuels will undoubtedly have an adverse effect on the environment. How can a growing demand for energy be met without radically altering the planet we inhabit? Although no clear answer has yet been formulated, the need to rely more heavily on alternate energy sources such as solar power must be considered.

The term *solar energy* generally refers to the direct use of the sun's rays to supply energy for the needs of people. The simplest, and perhaps most widely used, solar collectors are south-facing windows. As sunlight passes through the glass, energy is absorbed by objects in the room. These objects, in turn, radiate heat that warms the air in the room. In the United States, the use of south-facing windows, along with better-insulated and more airtight construction, is widely practiced, substantially reducing heating costs.

More elaborate systems used for home heating involve an active solar collector. These roof-mounted devices are normally large, blackened boxes that are covered with glass. The heat collected in the system can be transferred by the movement of air or by fluids that are circulated through a network of tubes. Solar collectors are also used successfully to heat water for domestic and commercial needs. In Israel, for example, about 20 percent of all homes are equipped with some type of solar device.

Although solar energy is free, the necessary equipment and its installation are not. The initial costs of setting up a system, including a supplemental heating unit, can be substantial. Nevertheless, over the long term, solar energy is economical in most parts of the United States and will become even more cost effective as the prices of other fuels increase.

Research is currently underway to improve the technologies for concentrating sunlight. One method being examined uses mirrors that track the sun and keep its rays focused on a receiving tower. A prototype facility, with an array of 2000 mirrors, has been constructed near Barstow, California (Figure 14.D). Solar energy focused on the tower heats water in pressurized panels to over 500°C. The superheated water is then transferred to turbines, which turn electrical generators. This pilot project has demonstrated that solar collectors have the potential of becoming commercially feasible.

Another type of collector uses photovoltaic (solar) cells that convert the sun's energy directly into electricity. A large experimental facility that uses photovoltaic cells is located near Hesperia, California, and supplies electricity to customers of Southern California Edison.

Recently, small rooftop photovoltaic systems have begun appearing in rural households of some Third World countries, including the Dominican Republic, Sri Lanka, and Zimbabwe. These units are about the size of an open briefcase and use a battery to store electricity that is generated during the daylight hours. In the tropics, these small photovoltaic systems are capable of running a television or radio, plus a few light bulbs for 3 to 4 hours. Although much cheaper than building conventional electric generators, these units are still too expensive for poor families. Consequently, an estimated two billion people in developing countries still lack electricity.

Figure 14.D

Solar One, a solar installation used to generate electricity in the Mojave Desert near Barstow, California. (Photo by Thomas Braise/The Stock Market)

Mechanisms of Heat Transfer

All forms of matter, whether solid, liquid, or gas, are composed of atoms or molecules that are in constant motion. Because of this vibratory motion, all matter is said to contain *thermal energy*. Whenever a substance is heated, its atoms move faster and faster. This leads to an increase in thermal energy. It is the average motion of the atoms or molecules in objects that we sense when we determine how hot or cold something is. We commonly describe the "hotness" or "coldness" of an object using the measurement called *temperature*. More specifically, temperature is a measure of the average motion of the atoms or molecules within a substance.

Although closely related, temperature and heat are different concepts (recall the situation in the thermosphere). Whereas temperature is a measure of the average motion of molecules, heat is the energy that flows because of temperature differences. In all situations, *heat is transferred from warmer to cooler objects*. Thus, if two objects of different temperature are in contact, the warmer object will become cooler and the cooler object will become warmer until they both reach the same temperature.

Three mechanisms of heat transfer are recognized: conduction, convection, and radiation. We experience all three in everyday life, as you shall see.

Conduction and Convection

Conduction is familiar to all of us. Anyone who has touched a metal spoon that was left in a hot pan has discovered that heat was conducted through the spoon. **Conduction** *is the transfer of heat through matter by molecular activity*. The energy of molecules is transferred through collisions from one molecule to another, with the heat flowing from the higher temperature to the lower temperature.

The ability of substances to conduct heat varies considerably. Metals are good conductors, as those of us who have touched hot metal have quickly learned (Figure 14.13). Air, on the other hand, is a very poor conductor of heat. Consequently, conduction is important only between Earth's surface and the air directly in contact with the surface. As a means of heat transfer for the atmosphere as a whole, conduction is the least significant.

Heat gained by the lowest layer of the atmosphere from radiation or conduction is most often transferred by convection. **Convection** *is the transfer of heat by the movement of a mass or substance from one place to another*. It can take place only in liquids and gases (Figure 14.13). Convective motions in the atmosphere transport heat from equatorial regions to the poles and from the surface upward. The term *advection* is usually reserved for horizontal convective motions such as winds, while *convection* is used to describe vertical motions in the atmosphere.

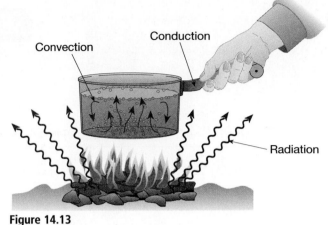

Figure 14.13
Illustration of conduction, convection, and radiation.

We will look at radiation, the third mechanism, in a moment. But first, look at Figure 14.13 which summarizes the mechanisms of heat transfer. The heat produced by the campfire passes through the pan by conduction, warming the water at the bottom of the pan. Convection currents carry the warmed water throughout the container, heating the remaining water. Meanwhile, the camper is warmed by radiation from the fire and the pan. Further, because metals are good conductors, the camper's hand is likely to be burned if a pot holder is not used. In most situations involving heat transfer, conduction, convection, and radiation occur simultaneously.

Radiation

Radiation is the only mechanism of heat transfer that can transmit heat through the relative emptiness of space. Thus, the vast majority of energy coming to and leaving Earth must be in this form (see Box 14.3).

From our everyday experience we know that the sun emits light and heat as well as the ultraviolet rays that cause suntan. Although these forms of energy comprise a major portion of the total energy that radiates from the sun, they are only part of a large array of energy called **radiation** or **electromagnetic radiation**. This array, or spectrum, of electromagnetic energy is shown in Figure 14.14. All radiation, whether X rays, radio waves, or heat waves, transmit energy through the vacuum of space at 300,000 kilometers (186,000 miles) per second and only slightly slower through our atmosphere.

Nineteenth-century physicists were so puzzled by the seemingly impossible phenomenon of energy traveling through the vacuum of space, without a medium to transmit it, that they assumed that a material, which they named ether, existed between the sun and Earth. This medium was thought to transmit radiant energy in much the same way that air transmits sound waves. Of course, this was incorrect. Today we know that, like gravity, radiation requires no material to transmit it.

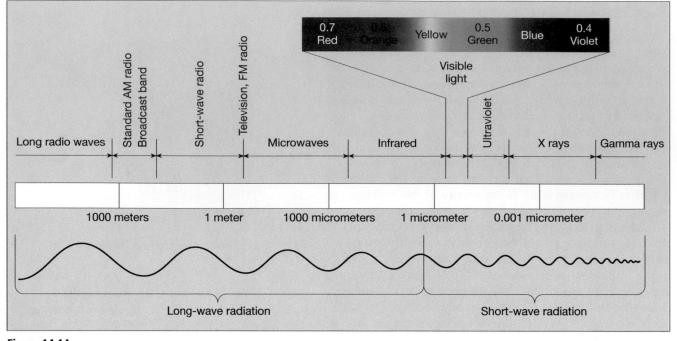

Figure 14.14

The electromagnetic spectrum.

In some respects, the transmission of radiant energy parallels the motion of the gentle swells in the open ocean. Like ocean swells, electromagnetic waves come in various sizes. For our purpose, the most important difference among electromagnetic waves is their wavelength, or the distance from one crest to the next. Radio waves have the longest wavelengths, ranging to tens of kilometers, whereas gamma waves are the very shortest, being less than one billionth of a centimeter long.

Visible light, as the name implies, is the only portion of the spectrum we can see. We often refer to visible light as "white" light since it appears "white" in color. However, it is easy to show that white light is really a mixture of colors, each corresponding to a particular wavelength (Figure 14.15). Using a prism, we can divide white light into a color array rainbow. Figure 14.14 shows that violet has the shortest wavelength—0.4 micrometer—and red has the longest—0.7 micrometer.

Located adjacent to red, and having a longer wavelength, is **infrared** radiation, which we cannot see but which we can detect as heat. The closest invisible waves to violet are called **ultraviolet** rays and are responsible for sunburn after an intense exposure to the sun. Although we divide radiant energy into groups based on our ability to perceive them, all forms of radiation are basically the same. When any form of radiant energy is absorbed by an object, the result is an increase in molecular motion, which causes a corresponding increase in temperature.

To better understand how the atmosphere is heated, it is useful to have a general understanding of the basic laws governing radiation.

1. All objects, at whatever temperature, emit radiant energy. Hence, not only hot objects like the sun but also Earth, including its polar ice caps, continually emit energy.

2. Hotter objects radiate more total energy per unit area than do colder objects.

3. The hotter the radiating body, the shorter the wavelength of maximum radiation. The sun, with a surface temperature of 6000 K, radiates maximum energy at 0.5 micrometer, which is in the visible range. The maximum radiation for Earth occurs at a wavelength of 10 micrometers, well within the infrared (heat) range. Because the maximum Earth radiation is roughly 20 times longer than the maximum solar radiation, terrestrial radiation is often called long-wave radiation, and solar radiation is called short-wave radiation.

4. Objects that are good absorbers of radiation are good emitters as well. Earth's surface and the sun approach being perfect radiators because they absorb and radiate with nearly 100 percent efficiency for their respective temperatures. On the other hand, *gases are selective absorbers and radiators.* Thus the atmosphere, which is nearly transparent to (does not absorb) certain wavelengths of radiation, is nearly opaque (a good absorber) to others. Our experience tells us that the atmosphere is transparent to visible light; hence, it readily reaches Earth's surface. This is not the case for the longer wavelength radiation emitted by Earth.

Figure 14.15
Visible light consists of an array of colors we commonly call the "colors of the rainbow." Rainbows are relatively common optical phenomena produced by the bending and reflection of light by drops of water. (Photo by Tony Stone Worldwide)

Paths Taken by Incoming Solar Radiation

Although the atmosphere is largely transparent to incoming solar radiation, only about 25 percent penetrates directly to Earth's surface without some sort of interference by the atmosphere. The remainder is either *absorbed* by the atmosphere, *scattered* about until it reaches Earth's surface or returns to space, or is *reflected* back to space (Figure 14.16). What determines whether radiation will be absorbed, scattered, or reflected outward? It depends greatly on the wavelength of the energy being transmitted, as well as on the nature of the intervening material.

Scattering

Although solar radiation travels in a straight line, the gases and dust particles in the atmosphere can redirect this energy, a process called *scattering*. Some of the light is backscattered into space, while the remainder continues Earthward, where it interacts with other molecules that scatter it further by changing the direction of the radiation, but not its wavelength.

About 30 percent of the solar energy reaching the outer atmosphere is reflected back to space. Included in this figure is the amount sent skyward by backscattering. This energy is lost to Earth and does not play a role in heating the atmosphere.

Albedo

The fraction of the total radiation that is reflected by a surface is called its **albedo**. Thus, the albedo for Earth as a whole (the planetary albedo) is 30 percent. However, the albedo from place to place as well as from time to time in the same locale varies considerably, depending on the amount of cloud cover and particulate matter in the air, as well as on the angle of the sun's rays

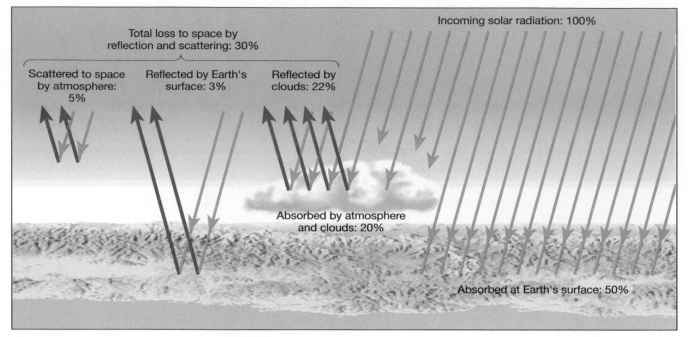

Figure 14.16

Distribution of incoming solar radiation by percentage. More solar energy is absorbed by Earth's surface than by the atmosphere. Consequently, the air is not heated directly by the sun but is heated chiefly by reradiation from Earth's surface.

and the nature of the surface. A lower angle means that more atmosphere must be penetrated, thus making the "obstacle course" longer, and therefore the loss of solar radiation greater (see Figure 14.9). Table 14.2 gives the albedo for various surfaces and clouds. Note that the angle at which the sun's rays strike a water surface greatly affects its albedo.

Table 14.2 Albedo of various surfaces.

Surface	Percent Reflected
Clouds, stratus	
<150 meters thick	25-63
150-300 meters thick	45-75
300-600 meters thick	59-84
Average of all types	
and thicknesses	50-55
Concrete	17-27
Crops, green	5-25
Forest, green	5-10
Meadows, green	5-25
Ploughed field, moist	14-17
Road, blacktop	5-10
Sand, white	30-60
Snow, fresh-fallen	80-90
Snow, old	45-70
Soil, dark	5-15
Soil, light (or desert)	25-30
Water	8*

* Typical albedo value for a water surface. The albedo of a water surface varies greatly depending upon the sun angle. If the sun angle is greater than 30 degrees, the albedo is less than 5 percent. When the sun is near the horizon (sun angle less than 3 degrees), the albedo is more than 60 percent.

Absorption

As stated earlier, gases are selective absorbers, meaning that they absorb strongly in some wavelengths, moderately in others, and only slightly in still others. When a gas molecule absorbs light waves, the energy is transformed into internal molecular motion, which is detectable as a rise in temperature. Nitrogen, the most abundant constituent in the atmosphere, is a poor absorber of all types of incoming radiation. Oxygen and ozone are efficient absorbers of ultraviolet radiation. Oxygen removes most of the shorter ultraviolet radiation high in the atmosphere, and ozone absorbs most of the remaining ultraviolet rays in the stratosphere. The absorption of UV radiation in the stratosphere accounts for the high temperatures experienced there. The only other significant absorber of incoming solar radiation is water vapor, which along with oxygen and ozone accounts for most of the solar radiation absorbed within the atmosphere.

For the atmosphere as a whole, none of the gases are effective absorbers of visible radiation. This explains why most visible radiation reaches Earth's surface and why we say that the atmosphere is transparent to incoming solar radiation. Thus, the atmosphere does not acquire the bulk of its energy directly from the sun. Rather, it is heated chiefly by energy that is first absorbed by Earth's surface and then reradiated to the sky.

Heating the Atmosphere: The Greenhouse Effect

Approximately 50 percent of the solar energy that strikes the top of the atmosphere reaches Earth's surface and is absorbed. Most of this energy is then reradiated skyward. Since Earth has a much lower surface temperature than the sun, the radiation that it emits has longer wavelengths than solar radiation.

The atmosphere as a whole is an efficient absorber of the longer wavelengths emitted by Earth (*terrestrial radiation*). Water vapor and carbon dioxide are the principal absorbing gases. Water vapor absorbs roughly five times more terrestrial radiation than do all other gases combined and accounts for the warm temperatures found in the lower troposphere, where it is most highly concentrated. Because the atmosphere is quite transparent to shorter-wavelength solar radiation and more readily absorbs longer-wavelength terrestrial radiation, the atmosphere is heated from the ground up rather than vice versa. This explains the general drop in temperature with increasing altitude experienced in the troposphere. The farther from the "radiator," the colder it becomes.

When the gases in the atmosphere absorb terrestrial radiation, they warm, but they eventually radiate this energy away. Some travels skyward, where it may be reabsorbed by other gas molecules, a possibility less likely with increasing height because the concentration of water vapor decreases with altitude. The remainder travels Earthward and is again absorbed by Earth. For this reason, Earth's surface is continually being supplied with heat from the atmosphere as well as from the sun. Without these absorptive gases in our atmosphere, Earth would not be a suitable habitat for humans and numerous other life-forms.

This very important phenomenon has been termed the **greenhouse effect** because it was once thought that greenhouses were heated in a similar manner (Figure 14.17). The gases of our atmosphere, especially water vapor and carbon dioxide, act very much like the glass in the greenhouse. They allow shorter-wavelength solar radiation to enter, where it is absorbed by the objects inside. These objects in turn reradiate the heat, but at longer wavelengths, to which glass is nearly opaque. The heat therefore is trapped in the greenhouse. However, a more important factor in keeping a greenhouse warm is the fact that the greenhouse itself prevents mixing of air inside with cooler air outside. Nevertheless, the term *greenhouse effect* is still used.

Temperature Measurement and Data

Changes in air temperature are probably noticed by people more often than changes in any other element of weather. At a weather station, the temperature is read on a regular basis from instruments mounted in an instrument shelter (Figure 14.18). The shelter protects the instruments from direct sunlight and allows a free flow of air. In addition to a standard mercury thermometer, the shelter is likely to contain a thermograph to continuously record temperature and a set of maximum–minimum thermometers. As their name implies, these thermometers record the highest and lowest temperatures during a measurement period, usually 24 hours.

The daily maximum and minimum temperatures are the bases for many of the temperature data compiled by meteorologists:

1. By adding the maximum and minimum temperatures and then dividing by two, the *daily mean temperature* is calculated.

2. The *daily range* of temperature is computed by finding the difference between the maximum and minimum temperatures for a given day.

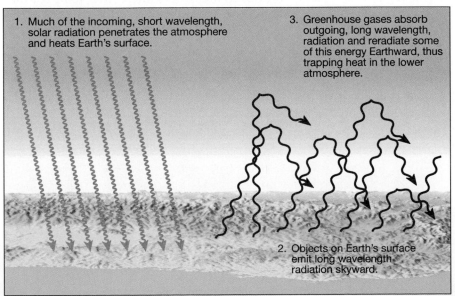

1. Much of the incoming, short wavelength, solar radiation penetrates the atmosphere and heats Earth's surface.

3. Greenhouse gases absorb outgoing, long wavelength, radiation and reradiate some of this energy Earthward, thus trapping heat in the lower atmosphere.

2. Objects on Earth's surface emit long wavelength radiation skyward.

Figure 14.17

The heating of the atmosphere. Most of the solar radiation that is not reflected back to space passes through the atmosphere and is absorbed at Earth's surface. Earth's surface, in turn, emits longer wavelength radiation. A portion of this energy is absorbed by certain gases in the atmosphere.

Figure 14.18
Standard instrument shelter. A shelter protects instruments from direct sunlight and allows for the free flow of air. (Courtesy of Qualimetrics, Inc.)

Mean temperatures are particularly useful for making comparisons, whether on a daily, monthly, or annual basis. It is quite common to hear a weather reporter state, "Last month was the hottest July on record," or "Today Chicago was ten degrees warmer than Miami." Temperature ranges are also useful statistics, because they give an indication of extremes.

Many different applications of temperature data have been devised. One familiar category of applications relates to human perceptions of temperature. People who listen to weather reports are probably acquainted with indices that attempt to portray levels of human comfort. Such indices are based on the fact that the sensation of temperature that the human body feels is often different from the actual temperature of the air as recorded by a thermometer. The human body is a heat engine that is continually releasing energy, and anything that influences the rate of heat loss from the body also influences the sensation of temperature that the body feels, thereby affecting human comfort. Several factors play a part in controlling the thermal comfort of the human body, and certainly air temperature is a major factor. Other environmental conditions, such as relative humidity, wind, and solar radiation, are also significant. For example, many of us know that a cold, windy winter day can feel much colder than the air temperature would seem to indicate (Table 14.3).

Controls of Temperature

A "temperature control" is any factor that causes temperature to vary from place to place and from time to time. Earlier in this chapter we examined the single greatest cause for temperature variations—differences in the receipt of solar radiation. Because variations in sun angle and length of daylight are a function of latitude, they are responsible for warm temperatures in the tropics and colder temperatures at more poleward locations.

3. The *monthly mean* is calculated by adding together the daily means for each day of the month and dividing by the number of days in the month.

4. The *annual mean* is an average of the twelve monthly means.

5. The *annual temperature range* is computed by finding the difference between the highest and lowest monthly means.

Table 14.3 Windchill equivalent temperature (°C).

Actual Temperature (°C)	Wind Speed (km per hr)								
	0	10	20	30	40	50	60	70	80
12	12	9	5	3	1	0	−0	−1	−1
8	8	5	0	−3	−5	−6	−7	−7	−8
4	4	0	−5	−8	−11	−12	−13	−14	−14
0	0	−4	−10	−14	−17	−18	−19	−20	−21
−4	−4	−8	−15	−20	−23	−25	−26	−27	−27
−8	−8	−13	−21	−25	−29	−31	−32	−33	−34
−12	−12	−17	−26	−31	−35	−37	−39	−40	−40
−16	−16	−22	−31	−37	−41	−43	−45	−46	−47
−20	−20	−26	−36	−43	−47	−49	−51	−52	−53
−24	−24	−31	−42	−48	−53	−56	−58	−59	−60
−28	−28	−35	−47	−54	−59	−62	−64	−65	−66
−32	−32	−40	−52	−60	−65	−68	−70	−72	−73
−36	−36	−44	−57	−65	−71	−74	−77	−78	−79
−40	−40	−49	−63	−71	−77	−80	−83	−85	−86

Source: NOAA, National Weather Service.

Review Questions

1. Distinguish between weather and climate.

2. List the basic elements of weather and climate.

3. What are the two major components of clean, dry air? Are they important meteorologically?

4. Why are water vapor and dust important constituents of our atmosphere?

5. What source is responsible for the most pollution? (See Figure 14.A.)

6. What is the difference between primary and secondary pollutants?

7. **(a)** Why is ozone important to life on Earth?

   (b) What is the most serious threat to human health of a decrease in the stratosphere's ozone?

8. What is the troposphere and why do some call it the "weather sphere"?

9. Why do temperatures rise in the stratosphere?

10. Briefly explain the primary cause of the seasons.

11. After examining Table 14.1, write a general statement that relates the season, the latitude, and the length of daylight.

12. Describe the relationship between the temperature of a radiating body and the wavelengths it emits.

13. Distinguish among the three basic mechanisms of heat transfer.

14. Figure 14.16 illustrates what happens to incoming solar radiation. The percentages shown, however, are only global averages. In particular, the amount of solar radiation reflected (albedo) may vary considerably. What factors might cause variations in albedo?

15. How does Earth's atmosphere act as a "greenhouse"?

16. Briefly describe two methods by which solar energy might be used to produce electricity.

17. How are the following temperature data computed: daily mean, daily range, monthly mean, annual mean, annual range?

18. Referring to Table 14.3 determine the wind-chill equivalent temperature if the wind speed were 10 kilometers per hour and the air temperature were –8°C. What if the air temperature remained unchanged but the wind speed were 30 kilometers per hour?

19. Quito, Ecuador, is located on the equator and is not a coastal city. It has an average annual temperature of only 13°C (55°F). What is the likely cause for this low average temperature?

20. In what ways can geographic position be considered a control of temperature?

21. How does cloud cover influence the maximum temperature on an overcast day? How is the nighttime minimum influenced by clouds?

22. Yakutsk is located in Siberia at about 60 degrees north latitude. This Russian city has one of the highest average annual temperature ranges in the world: 62.2°C (112°F). Explain the reasons for the very high annual temperature range.

Key Terms

air (p. 363)
albedo (p. 377)
autumnal equinox (p. 371)
circle of illumination (p. 369)
climate (p. 363)
conduction (p. 375)
convection (p. 375)
electromagnetic radiation (p. 375)

element (of weather and climate) (p. 363)
environmental lapse rate (p. 367)
greenhouse effect (p. 379)
inclination of the axis (p. 371)
infrared (p. 376)
isotherm (p. 383)
mesosphere (p. 367)
radiation (p. 375)
revolution (p. 369)
rotation (p. 369)
spring equinox (p. 372)

stratosphere (p. 367)
summer solstice (p. 371)
thermosphere (p. 368)
Tropic of Cancer (p. 371)
Tropic of Capricorn (p. 371)
troposphere (p. 367)
ultraviolet (p. 376)
visible light (p. 376)
weather (p. 363)
winter solstice (p. 371)

CHAPTER 15
Moisture

Pine trees covered by rime in California's Sierra Nevada Range. (Photo by Carr Clifton)

As you observe day-to-day weather changes, you witness the powerful role of water in the air. This raises questions: What is humidity and how is it measured? Why do clouds form on some occasions but not on others? What processes produce clouds and precipitation? Why are some clouds thin, white, and harmless, while others are towering, gray, and ominous? What is the difference between sleet and hail? What is fog? Are all fogs alike? This chapter investigates the remarkable phenomena associated with moisture in the air.

Scientists agree: when it comes to understanding atmospheric processes, *water vapor is the most important gas in the atmosphere*. Water vapor constitutes only a small fraction of the gases in the atmosphere, varying from nearly 0 to about 4 percent by volume. But the importance of water in the air vastly exceeds what these small percentages would indicate.

Recall from Chapter 14 that water vapor is an important heat-absorbing gas. This property is critical to the heating of the atmosphere. Water vapor is also the source of all condensation and precipitation. Clouds and fog, as well as rain, snow, sleet, and hail, are among the most conspicuous and observable of weather phenomena (Figure 15.1).

We all expect rain to fall when certain cloud types are present, yet most of us have no idea of the complex processes that must take place in a cloud to produce rain. The formation of a single average raindrop requires water from nearly one million microscopic cloud droplets. What mechanisms foster the creation of raindrops? The wide variation in the amount of precipitation from place to place, as well as local differences from time to time, have a significant impact on the nature of the physical landscape and on our lifestyles.

Changes of State

Water vapor is an odorless, colorless gas that mixes freely with the other gases of the atmosphere. Oxygen and nitrogen, the two gases that comprise 99 percent of the atmosphere, change state (solid to liquid to gas) at extremely low temperatures (–200°C or so). But water vapor changes state (solid to liquid to gas) at the temperatures and pressures experienced near Earth's surface.

It is because of these changes of state at "ordinary" temperatures that water leaves the oceans as a gas and returns again as a liquid. The processes that involve such changes of state require that *heat* be absorbed or released (Figure 15.2). This heat energy is measured in calories. One **calorie** is the amount of heat required to raise the temperature of 1 gram of water 1°C. Thus, when 10 calories of heat are added to 1 gram of water, a 10°C temperature rise occurs.

Importance of Latent Heat

Under certain conditions, heat may be added to a substance without an accompanying temperature change. This may sound impossible, but consider what occurs during a change in state. To illustrate, let us apply heat to a container of ice water. Its temperature is 0°C. We keep adding heat, but the temperature remains constant until all the ice is melted. *Then* the temperature begins to rise.

Where did the heat go during melting? In this case, the energy was used to disrupt the internal crystalline structure of the ice cubes, freeing the water molecules to move around—in other words, to melt. *Because this heat energy is not associated with a temperature change,* it is referred to as **latent heat** (latent means hidden). This energy is not available as heat until the liquid returns to the solid state, when the heat is released.

Figure 15.1

Clouds are among the most observable of weather phenomena. (Photo by Michio Hoshino/Minden Pictures)

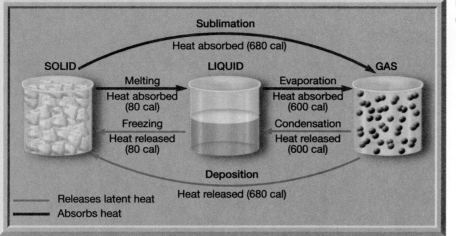

Figure 15.2
Changes of state.

As we shall examine later, latent heat plays a crucial role in many atmospheric processes. In particular, the release of latent heat aids the formation of the towering clouds often seen on warm summer days, and it is the major source of energy for violent thunderstorms, tropical storms, and hurricanes.

Liquid/Gas Change of State.   The process of converting a liquid to a gas is termed **evaporation**. It takes approximately 600 calories of energy to convert 1 gram of water to water vapor (Figure 15.2). The energy absorbed by the water molecules during evaporation is used solely to give them the motion needed to escape the surface of the liquid and become a gas. This energy is referred to as *latent heat of vaporization.* During the process of evaporation, it is the higher-temperature (faster-moving) molecules that escape the surface. As a result, the average molecular motion (temperature) of the remaining water is reduced, causing a cooling effect—hence the common expression, "evaporation is a cooling process." You have undoubtedly experienced this cooling effect when stepping dripping wet from a swimming pool or bathtub.

Condensation is the process whereby water vapor changes to the liquid state. For condensation to occur, the water molecules must release their stored heat energy (*latent heat of condensation*), equal to what was absorbed during evaporation. This released energy plays an important role in producing violent weather and can transfer great quantities of heat from tropical oceans to more poleward locations. When condensation occurs in the atmosphere, it results in the formation of fog and clouds.

Solid/Liquid Change of State.   **Melting** is the process by which a solid is changed to a liquid. It requires absorption of approximately 80 calories of heat per gram of water (*latent heat of melting*). **Freezing,** the reverse process, releases these 80 calories per gram as *latent heat of fusion.*

Solid/Gas Change of State.   The last of the processes illustrated in Figure 15.2 are sublimation and deposition. **Sublimation** is the conversion of a solid directly to a gas, without passing through the liquid state. You may have observed this change in watching the sublimation of "dry ice" (frozen carbon dioxide) into white, wispy vapor, sometimes used to generate "fog" or "smoke" in theatrical productions. Water ice sublimates too, which is why unused ice cubes gradually shrink in a freezer.

The term **deposition** is used to denote the reverse process, the conversion of a vapor directly to a solid. This change occurs, for example, when water vapor is deposited as ice on cold objects such as grass or windows (Figure 15.3). These deposits are termed *white frost* or *hoar frost* but are generally just called *frost*. A household example of deposition is the "frost" that accumulates in a freezer compartment. As shown in Figure 15.2, sublimation and deposition involve an amount of energy equal to the total of the other two processes.

Humidity: Water Vapor in the Air

Humidity is the general term for the amount of water vapor in air. There are different ways of looking at humidity; we will examine two—specific humidity and relative humidity.

But, before we consider these humidity measures, you need to understand the concept of **saturation**. Imagine a closed jar half full of water and half full of dry air, both at the same temperature. As the water begins to evaporate from the water surface, a small increase in pressure can be detected in the air above. This increase is the result of the motion of the water vapor molecules that were added to the air through evaporation. In the open atmosphere, this pressure is termed

Figure 15.3
Frost on a window pane is an example of deposition. (Photo by D. Cavagnaro/DRK Photo)

vapor pressure and is defined as that part of the total atmospheric pressure that can be attributed to the water vapor content.

In the closed container, as more and more molecules escape from the water surface, the steadily increasing vapor pressure in the air above forces more and more of these molecules to return to the liquid. Eventually the number of vapor molecules returning to the surface will balance the number leaving. At that point, the air is said to be *saturated*, or filled to capacity. However, if we add heat to the container, increasing the temperature of the water and air, more water will evaporate before a balance is reached. Consequently, at higher temperatures, more moisture is required for saturation. Stated anoth-

Table 15.1 Amount of water vapor (grams) required to saturate a kilogram of air at various temperatures.

Temperature		Grams of water vapor per kg of air
(°C)	(°F)	
−40	−40	0.1
−30	−22	0.3
−20	−4	0.75
−10	14	2
0	32	3.5
5	41	5
10	50	7
15	59	10
20	68	14
25	77	20
30	86	26.5
35	95	35
40	104	47

er way, the water vapor capacity of air is *temperature dependent*, with warm air having a much greater capacity than cold air. The amount of water vapor required for saturation at various temperatures varies surprisingly, as shown in Table 15.1.

Specific Humidity

Not all air is saturated, of course. Thus, we need ways to express how humid a parcel of air is. One method specifies the amount of water vapor contained in a unit of air. **Specific humidity** is the weight of water vapor per weight of a chosen mass of air, including the water vapor. Since it is measured in units of weight (usually in grams per kilogram), specific humidity is not affected by changes in pressure or temperature. *Specific humidity is a measure of the actual quantity of water vapor in a given mass of air.*

Relative Humidity

The most familiar, and perhaps the most misunderstood, term used to describe the moisture content of air is relative humidity. Stated in an admittedly oversimplified manner, **relative humidity** *is the ratio of the air's actual water vapor content to its potential water vapor capacity at a given temperature.* To illustrate, we see from Table 15.1 that at 25°C air is saturated when it contains 20 grams of water vapor per kilogram of air. Thus, if the air contains only 10 grams per kilogram on a 25°C day, the relative humidity is expressed as 10/20 or 50 percent. Further, if air with a temperature of 25°C had a water vapor content of 20 grams per kilogram, the relative humidity would be expressed as 20/20 or 100 percent. On those occasions when the relative humidity reaches 100 percent, the air is saturated.

Because relative humidity is a comparison of the air's water vapor content (specific humidity) and its capacity, relative humidity can be changed in either of two ways. First, if moisture is added to or subtracted from air, its relative humidity will change. This is shown clearly in Figure 15.4.

The second condition that affects relative humidity is air temperature. Examine Figure 15.5 carefully, and note in Part A that, when air at 20°C contains 7 grams of water vapor per kilogram of air, it has a relative humidity of 50 percent. When the air is cooled from 20°C to 10°C as shown in Part B, it still contains 7 grams of water vapor, but the relative humidity increases from 50 percent to 100 percent. We can conclude from this that, when the water vapor content remains constant, *a decrease in temperature results in an increase in relative humidity.*

What happens when the air is cooled further, below the temperature at which saturation occurs? Part C of Figure 15.5 illustrates this situation. Notice from Table 15.1 that, when the flask is cooled to 0°C, the air is saturated at 3.5 grams of water vapor per kilogram of air. Because this flask originally contained 7 grams of water vapor, 3.5 grams of water vapor have to go somewhere, because there is no room for them in the cooler air. They will condense to form liquid droplets that collect on the

walls of the container. In the meantime, the relative humidity of the air inside remains at 100 percent. This brings up an important concept. When air aloft is cooled below its saturation point, some of the water vapor condenses to form clouds. Because clouds are made of liquid droplets, this moisture is no longer part of the *water vapor* content of the air.

We can generalize the effects of temperature on relative humidity as follows. When the water vapor content of air remains at a constant level, a decrease in air temperature results in an increase in relative humidity, and an increase in temperature causes a decrease in relative humidity. In Figure 15.6 the variations in temperature and relative humidity during a typical day demonstrate the relationship just described (see Box 15.1).

Dew Point

Another important idea related to relative humidity is the dew-point temperature. **Dew point** is the *temperature to which air would have to be cooled to reach saturation.* Note that in Figure 15.5 unsaturated air at 20°C is cooled to 10°C before saturation occurs. Therefore, 10°C would be the dew-point temperature for this air. If this same parcel of air were cooled further, the air's capacity would be exceeded, and the excess vapor would

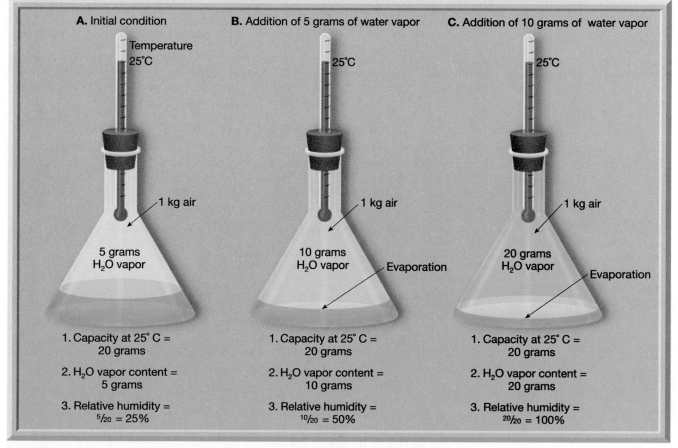

Figure 15.4

When the temperature remains constant, relative humidity will increase as water vapor is added to the air. Here the water vapor capacity remains constant at 20 grams per kilogram, while the relative humidity rises from 25 percent to 100 percent as the water vapor content (specific humidity) increases.

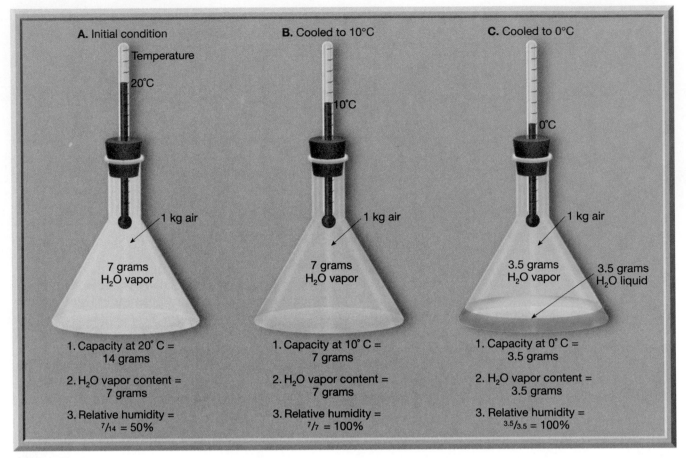

Figure 15.5

When the water vapor content remains constant, the relative humidity can be changed by increasing or decreasing the air temperature. In this example, when the temperature of the air in the flask was lowered from 20°C to 10°C, the relative humidity increased from 50 percent to 100 percent. Thus, 10°C is the dew point. Further cooling (from 10°C to 0°C) causes one-half of the water vapor to condense because colder air holds less moisture. In nature, cooling of air below its dew point generally causes condensation in the form of clouds, dew, or fog.

condense, typically as dew, fog, or clouds. The term dew point stems from the fact that during nighttime hours, objects near the ground often become coated with dew

(Figure 15.7). Contrary to popular belief, white frost is not frozen dew. Rather, white frost forms on occasions when the dew point of the air is 0°C or below. Thus,

Figure 15.6

Typical daily variations in temperature and relative humidity during a spring day at Washington, D.C. When temperature increases, relative humidity drops (see mid-afternoon) and vice versa.

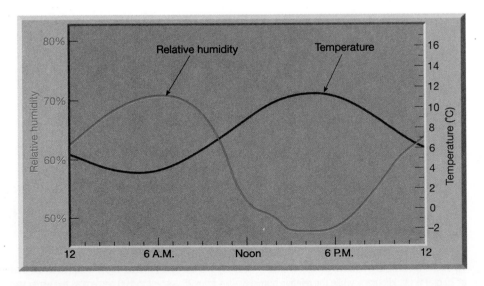

Does air with higher relative humidity always contain more water vapor than air with low relative humidity? At the *same temperature*, yes. However, if temperatures are different, there can be less water in "humid" air than in "dry" air. To illustrate, let us compare a typical January day at Winnipeg, Manitoba, to one at Phoenix, Arizona. On this hypothetical day the temperature in Winnipeg is a frigid –10°C and the relative humidity is 100 percent. By referring to Table 15.1, we can see that saturated –10°C air has a water vapor content of 2 grams per kilogram. By contrast, the desert air at Phoenix on this January day is a warm 25°C and the relative humidity is just 20 percent. A look at Table 15.1 reveals that 25°C air has a capacity of 20 grams per kilogram. Therefore, with a relative humidity of 20 percent, the air at Phoenix has a water vapor content of 4 grams per kilogram (20 grams times 20 percent). Consequently, the "dry" air at Phoenix actually contains twice the water vapor as the "humid" air at Winnipeg.

It now should be clear why places that are very cold are also very dry. The low water vapor content of frigid air helps to explain why many arctic areas receive only meager precipitation and are really polar deserts. This also explains why you may experience dry skin and chapped lips during the winter months. The water vapor content of cold air is low, even when compared to some hot arid regions.

frost forms when water vapor changes directly from a gas into a solid (ice) without entering the liquid state. This process, called deposition, produces delicate patterns of ice crystals that frequently decorate windows on frigid winter days (see Figure 15.3).

Measuring Humidity

Relative humidity is commonly measured using a psychrometer or a hygrometer. A **psychrometer** consists of two identical thermometers mounted side by side (Figure 15.8). One thermometer, the *dry-bulb*, gives the present air temperature. The other, called the *wet-bulb* thermometer, has a thin muslin wick tied around the end (see end of thermometers in the photo).

To use the psychrometer, the cloth sleeve is saturated with water and a continuous current of air is passed over the wick. This is done either by swinging the instrument freely in the air or by fanning air past it. As a consequence, water evaporates from the wick, and the heat absorbed by the evaporating water makes the

Figure 15.7
Dew on a spider web. (Photo by Wolfgang Kaehler)

Figure 15.8
Sling psychrometer. This instrument is used to determine both relative humidity and dew point. The dry-bulb thermometer gives the current air temperature. The web-bulb thermometer is covered with a cloth wick that is dipped in water. The thermometers are spun until the temperature of the wet-bulb thermometer stops declining. Then the thermometers are read and the data used in conjunction with Tables 15.2 and 15.3. (Photo by E. J. Tarbuck)

temperature of the wet bulb drop. The loss of heat that was required to evaporate water from the wet bulb lowers the thermometer reading.

The amount of cooling that takes place is directly proportional to the dryness of the air. The drier the air, the more moisture evaporates. The more heat the evaporating water absorbs, the greater the cooling. Therefore, the larger the difference that is observed between the thermometer readings, the lower the relative humidity; the smaller the difference, the higher the relative humidity. If the air is saturated, no evaporation will occur, and the two thermometers will have identical readings.

To determine the precise relative humidity from the thermometer readings, a standard table is used (Table 15.2). With the same information, but using a different table (Table 15.3), the dew-point temperature may also be calculated.

The second commonly used instrument for measuring relative humidity, the **hygrometer**, can be read directly, without the use of tables. The hair hygrometer operates on the principle that hair or certain synthetic fibers change their length in proportion to changes in the relative humidity, lengthening as relative humidity increases and shrinking as the relative humidity drops. The tension of a bundle of hairs is linked mechanically to an indicator that is calibrated between 0 and 100 percent. Thus we need only glance at the dial to determine the relative humidity. Unfortunately, the hair hygrometer is less accurate than the psychrometer. Further, it requires frequent calibration and is slow in responding to changes in humidity, especially at low temperatures.

A different type of hygrometer is used in remote-sensing instrument packages such as radiosondes that transmit upper-air observations back to ground stations.

Table 15.2 Relative humidity (percent).[*]

Dry bulb (°C)	Depression of Wet-Bulb Temperature (Dry-Bulb Temperature Minus Wet-Bulb Temperature = Depression of the Wet Bulb)																					
	1	2	3	4	5	6	7	8	9	10	11	12	13	14	15	16	17	18	19	20	21	22
−20	28																					
−18	40																					
−16	48	0																				
−14	55	11																				
−12	61	23																				
−10	66	33	0																			
−8	71	41	13																			
−6	73	48	20	0																		
−4	77	54	32	11																		
−2	79	58	37	20	1																	
0	81	63	45	28	11																	
2	83	67	51	36	20	6																
4	85	70	56	42	27	14																
6	86	72	59	46	35	22	10	0														
8	87	74	62	51	39	28	17	6														
10	88	76	65	54	43	33	24	13	4													
12	88	78	67	57	48	38	28	19	10	2												
14	89	79	69	60	50	41	33	25	16	8	1											
16	90	80	71	62	54	45	37	29	21	14	7	1										
18	91	81	72	64	56	48	40	33	26	19	12	6	0									
20	91	82	74	66	58	51	44	36	30	23	17	11	5									
22	92	83	75	68	60	53	46	40	33	27	21	15	10	4	0							
24	92	84	76	69	62	55	49	42	36	30	25	20	14	9	4	0						
26	92	85	77	70	64	57	51	45	39	34	28	23	18	13	9	5						
28	93	86	78	71	65	59	53	47	42	36	31	26	21	17	12	8	4					
30	93	86	79	72	66	61	55	49	44	39	34	29	25	20	16	12	8	4				
32	93	86	80	73	68	62	56	51	46	41	36	32	27	22	19	14	11	8	4			
34	93	86	81	74	69	63	58	52	48	43	38	34	30	26	22	18	14	11	8	5		
36	94	87	81	75	69	64	59	54	50	44	40	36	32	28	24	21	17	13	10	7	4	
38	94	87	82	76	70	66	60	55	51	46	42	38	34	30	26	23	20	16	13	10	7	5
40	94	89	82	76	71	67	61	57	52	48	44	40	36	33	29	25	22	19	16	13	10	7

Relative Humidity Values

[*] To determine the relative humidity, find the air (dry-bulb) temperature on the vertical axis (far left) and the depression of the wet bulb on the horizontal axis (top). Where the two meet, the relative humidity is found. For example, when the dry-bulb temperature is 20°C and a wet-bulb temperature is 14°C, then the depression of the wet-bulb is 6°C (20°C − 14°C). From Table 15.2, the relative humidity is 51 percent and from Table 15.3, the dew point is 10°C.

The electric hygrometer contains an electrical conductor coated with a moisture-absorbing chemical. It works on the principle that the passage of current varies as the relative humidity varies.

The Basis of Cloud Formation: Adiabatic Cooling

Up to this point, we have considered basic properties of water vapor and how its variability is measured. We are now ready to examine some of the important roles that water vapor plays in weather, especially the formation of clouds.

Fog and Dew versus Cloud Formation

Recall that condensation occurs when water vapor changes to a liquid. This condensation may form dew, fog, or clouds. Although these three forms are different, all require saturated air to develop. As indicated earlier, saturation occurs either when water vapor is added to the air or, more commonly, when the air is cooled to its dew point. Near Earth's surface heat is readily exchanged between the ground and the air above. During evening hours, the surface radiates heat away, and the surface and adjacent air cool rapidly. This "radiation cooling" accounts for the formation of dew and some types of fog. At higher altitudes, however, clouds often form during the warmest part of the day, not during the cool evening or night. Consequently, some other mechanism must operate to form clouds.

Adiabatic Temperature Change

The process that is responsible for most cloud formation is easily visualized if you have ever pumped up a bicycle tire and noticed that the pump barrel became quite warm. The heat you felt was the consequence of the work you did on the air to compress it. When energy is used to

Table 15.3 Dew-point temperature (°C).[*]

Dry bulb (°C)	Dry-Bulb Temperature Minus Wet-Bulb Temperature = Depression of the Wet Bulb																					
	1	2	3	4	5	6	7	8	9	10	11	12	13	14	15	16	17	18	19	20	21	22
-20	-33																					
-18	-28																					
-16	-24																					
-14	-21	-36																				
-12	-18	-28																				
-10	-14	-22																				
-8	-12	-18	-29																			
-6	-10	-14	-22																			
-4	-7	-22	-17	-29																		
-2	-5	-8	-13	-20																		
0	-3	-6	-9	-15	-24																	
2	-1	-3	-6	-11	-17																	
4	1	-1	-4	-7	-11	-19																
6	4	1	-1	-4	-7	-13	-21															
8	6	3	1	-2	-5	-9	-14															
10	8	6	4	1	-2	-5	-9	-14	-28													
12	10	8	6	4	1	-2	-5	-9	-16													
14	12	11	9	6	4	1	-2	-5	-10	-17												
16	14	13	11	9	7	4	1	-1	-6	-10	-17											
18	16	15	13	11	9	7	4	2	-2	-5	-10	-19										
20	19	17	15	14	12	10	7	4	2	-2	-5	-10	-19									
22	21	19	17	16	14	12	10	8	5	3	-1	-5	-10	-19								
24	23	21	20	18	16	14	12	10	8	6	2	-1	-5	-10	-18							
26	25	23	22	20	18	17	15	13	11	9	6	3	0	-4	-9	-18						
28	27	25	24	22	21	19	17	16	14	11	9	7	4	1	-3	-9	-16					
30	29	27	26	24	23	21	19	18	16	14	12	10	8	5	1	-2	-8	-15				
32	31	29	28	27	25	24	22	21	19	17	15	13	11	8	5	2	-2	-7	-14			
34	33	31	30	29	27	26	24	23	21	20	18	16	14	12	9	6	3	-1	-5	-12	-29	
36	35	33	32	31	29	28	27	25	24	22	20	19	17	15	13	10	7	4	0	-4	-10	
38	37	35	34	33	32	30	29	28	26	25	23	21	19	17	15	13	11	8	5	1	-3	-9
40	39	37	36	35	34	32	31	30	28	27	25	24	22	20	18	16	14	12	9	6	2	-2

Dew point temperatures

[*] See footnote to Table 15.2.

compress air, the motion of the gas molecules increases and therefore the temperature of the air rises. Conversely, air that is allowed to escape from a bicycle tire *expands and cools*. This results because the expanding air pushes (does work on) the surrounding air and must cool by an amount equivalent to the energy expended.

You may have experienced the cooling effect of an expanding gas while applying a spray deodorant. As the compressed gas propellant in the aerosol can is released, it quickly expands and cools. This drop in temperature occurs *even though heat is neither added nor subtracted.* Such variations are known as **adiabatic temperature changes** and result when air is compressed or allowed to expand. In summary, when air is allowed to expand, it cools, and when it is compressed, it warms.

Dry Adiabatic Rate. As you travel from Earth's surface upward through the atmosphere, the atmospheric pressure rapidly diminishes, because there are fewer and fewer gas molecules. Thus, any time a parcel of air moves upward, it passes through regions of successively lower pressure. As a result, the ascending air expands. As it expands, it cools adiabatically. Unsaturated air cools at the rather constant rate of 10°C for every 1000 meters of ascent (1°C per 100 meters).

Conversely, descending air comes under increasingly higher pressures, compresses, and is heated 10°C for every 1000 meters of descent. This rate of cooling or heating applies only to *unsaturated air* and is known as the **dry adiabatic rate**.

Wet Adiabatic Rate. If air rises high enough, it will cool sufficiently to reach the dew point and cause condensation. From this point on along its ascent, *latent heat of condensation* stored in the water vapor will be liberated. Although the air will continue to cool after condensation begins, the released latent heat works against the adiabatic process, thereby reducing the rate at which the air cools. This slower rate of cooling caused by the addition of latent heat is called the **wet adiabatic rate** of cooling. Because the amount of latent heat released depends on the quantity of moisture present in the air, the wet adiabatic rate varies from 5°C per 1000 meters for air with a high moisture content to 9°C per 1000 meters for dry air.

Figure 15.9 illustrates the role of adiabatic cooling in the formation of clouds. Note that from the surface up to the condensation level the air cools at the dry adiabatic rate. The wet adiabatic rate commences at the condensation level.

Stability of Air

As you have learned, if air rises, it will cool adiabatically and eventually produce clouds. But why does air rise on some occasions and not on others? Why do the size of clouds and the amount of precipitation vary so much when air does rise? The answers to these questions are closely related to the stability of the air.

Imagine, if you will, a large bubble of air with a thin flexible cover that allows it to expand but prevents it from mixing with the surrounding air. If the imaginary bubble were forced to rise, its temperature would *decrease because of expansion*. By comparing the bubble's temperature to that of the surrounding air, we can determine the stability of the bubble.

Figure 15.9

Rising air cools at the dry adiabatic rate of 10° per 1000 meters, until the air reaches the dew point and condensation (cloud formation) begins. As air continues to rise, the latent heat released by condensation reduces the rate of cooling. The wet adiabatic rate is therefore always less than the dry adiabatic rate.

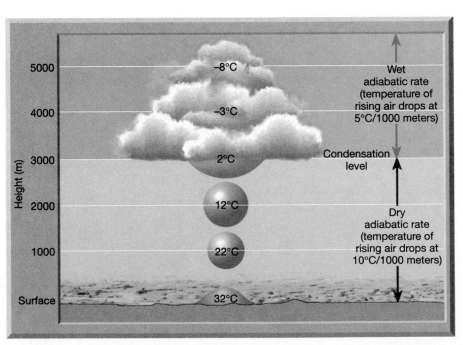

Figure 15.10
As long as air is warmer than its surroundings, it will rise. Hot air balloons rise up through the atmosphere for this reason. (Photo by Bachmann/Photo Researchers, Inc.)

If the bubble's temperature is lower than that of its environment, it will be denser, and if allowed to move freely, it would sink to its original position. Air of this type is termed *stable air* because it resists vertical displacement.

On the other hand, if our imaginary bubble were warmer, and therefore less dense than the surrounding air, it would continue to rise until it reached an altitude having the same temperature, much as a hot-air balloon rises as long as it is less dense than the surrounding air (Figure 15.10). This type of air is called *unstable air*.

Determining Stability

The stability of air is determined by examining the temperature of the atmosphere at various heights. Recall from Chapter 14 that this measure is termed the *environmental lapse rate*. The environmental lapse rate is the temperature of the atmosphere as determined from observations made by radiosondes and aircraft. It is important not to confuse this with *adiabatic temperature changes*, which are changes in temperature caused by expansion or compression as a parcel of air rises or descends.

Stable Air. For illustration, we will examine a situation where the environmental lapse rate is 5°C per 1000 meters (Figure 15.11). Under this condition, when air at the surface has a temperature of 25°C, the air at 1000 meters will be 5 degrees cooler, or 20°C, the air at 2000 meters will have a temperature of 15°C, and so forth. At first glance it appears that the air at the surface is less dense than the air at 1000 meters, because it is 5 degrees warmer. However, if the air near the surface is unsaturated and were to rise to 1000 meters, it would expand and cool at the dry adiabatic rate of 10°C per

1000 meters. Therefore, upon reaching 1000 meters, its temperature would have dropped 10°C. Being 5 degrees cooler than its environment, it would be denser and

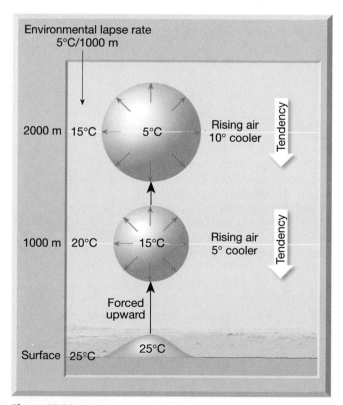

Figure 15.11

In a stable atmosphere, as an unsaturated parcel of air is lifted, it expands and cools at the dry adiabatic rate of 10°C per 1000 meters. Because the temperature of the rising parcel of air is lower than that of the surrounding environment, it will be heavier and, if allowed to do so, will sink to its original position.

tend to sink to its original position. Hence, we say that the air near the surface is potentially cooler than the air aloft and therefore will not rise on its own. The air just described is *stable* and resists vertical movement.

Stated quantitatively, **absolute stability** prevails when the environmental lapse rate is less than the wet adiabatic rate. Figure 15.12 depicts this situation using an environmental lapse rate of 5°C per 1000 meters and a wet adiabatic rate of 6°C per 1000 meters. Note that at 1000 meters the temperature of the surrounding air is 15°C, while the rising parcel of air has cooled to 10°C and is therefore the denser air. Even if this stable air were to be forced above the condensation level, it would remain cooler and denser than its environment, and thus it would tend to return to the surface.

The most stable conditions occur when the temperature in a layer of air actually increases with altitude. When such a reversal occurs, a *temperature inversion* is said to exist. Temperature inversions frequently occur on clear nights as a result of radiation cooling of Earth's surface. Under these conditions, an inversion is created because the ground and the air immediately above will cool more rapidly than the air aloft. When warm air overlies cooler air, it acts as a lid and prevents appreciable vertical mixing. Because of this, temperature inversions are responsible for trapping pollutants in a narrow zone near Earth's surface. This idea is explored more fully in Box 15.2.

Unstable Air. At the other extreme, air is said to exhibit **absolute instability** when the environmental lapse rate is greater than the dry adiabatic rate. As shown in Figure 15.13, the ascending parcel of air is always warmer than its environment and will continue to rise because of its own buoyancy. However, absolute instability is generally limited to near Earth's surface. On hot, sunny days the air above some surfaces, like paved shopping centers, is heated more than the air over adjacent surfaces. These invisible pockets of more intensely heated air, being less dense than the air aloft, will rise like a hot-air balloon. This phenomenon produces the small fluffy clouds we associate with fair weather. Occasionally, when the surface air is considerably warmer than the air aloft, clouds with considerable vertical development can form, as shown in Figure 15.13. Clouds of this type can be associated with rain showers.

Conditional Instability. A more common type of atmospheric instability is called **conditional instability**. This situation prevails when moist air has an environmental lapse rate between the dry and wet adiabatic rates (between 5 and 10°C per 1000 meters). Simply, the atmosphere is said to be conditionally unstable when it is *stable* for an *unsaturated* parcel of air, but *unstable* for a *saturated* parcel of air. Notice in Figure 15.14 that the rising parcel of air is cooler than the surrounding air for the first 4000 meters. With the addition of latent heat

Figure 15.12
Absolute stability prevails when the environmental lapse rate is less than the wet adiabatic rate. The rising parcel of air is therefore always cooler and denser than the surrounding air. When stable air is forced to rise, it spreads out, producing flat, layered clouds.

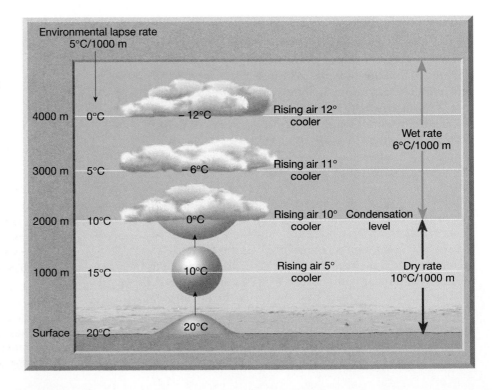

Box 15.2

Atmospheric Stability Influences Air Pollution

Air quality is not just a function of the quantity and types of pollutants emitted into the air, but it is also closely linked to the atmosphere's ability to disperse these noxious substances. Perhaps you have heard the following well-known phrase: "The solution to pollution is dilution." To a significant degree this is true. If the air into which the pollution is released is not dispersed, the air will become more toxic. Two of the most important atmospheric conditions affecting the dispersion of pollutants are the strength of the wind and the stability of the air. These factors are critical because they determine how rapidly pollutants are diluted by mixing with the surrounding air after leaving the source.

The way in which wind speed influences the concentration of pollutants is straightforward. When winds are weak or calm, the concentration of pollutants is higher than when winds are strong. High wind speeds mix polluted air into a greater volume of surrounding air and therefore cause the pollution to be more diluted. When winds are light, there is less turbulence and mixing so the concentration of pollutants is higher.

Whereas wind speed governs the amount of air into which pollutants are initially mixed, atmospheric stability determines the extent to which vertical motions will mix the pollution with cleaner air above. The distance between Earth's surface and the height to which vertical air movements extend is termed the *mixing depth*. Generally, the greater the mixing depth, the better the air quality. When the mixing depth is several kilometers, pollutants are mixed through a large volume of cleaner air and dilute rapidly. When the mixing depth is shallow, pollutants are confined to a much smaller volume of air and concentrations can reach unhealthy levels. When air is stable, vertical motions are suppressed and mixing depths are small. Conversely, an unstable atmosphere promotes vertical air movements and greater mixing depths. Since heating of Earth's surface by the sun enhances convectional

movements, mixing depths are usually greater during the afternoon hours. For the same reason, mixing depths during the summer months are typically greater than during the winter months.

Temperature inversion represents a situation in which the atmosphere is very stable and the mixing depth is significantly restricted. Warm air overlying cooler air acts as a lid and prevents upward movement, leaving the pollutants trapped in a relatively narrow zone near the ground (Figure 15.A).

Inversions are generally classified into one of two categories—those that

Figure 15.A

Many air pollution episodes are linked to the occurrence of temperature inversions. This event occurred in Cubatao, Brazil. (Photo by Ted Spiegel/Black Star)

form near the ground and those that form aloft. A *surface inversion* develops close to the ground on clear and relatively calm nights. It forms because the ground is a more effective radiator than the air above. This being the case, radiation from the ground to the clear night sky causes more rapid cooling at the surface than higher in the atmosphere. The result is that the air close to the ground is cooled more than the air above, yielding a temperature profile similar to the one shown in Figure 15.B.

(continues)

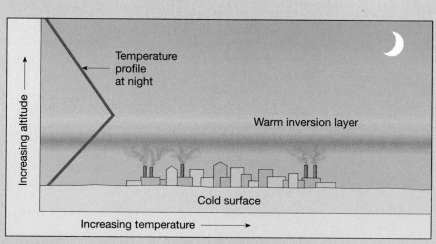

Figure 15.B

Generalized temperature profile for a surface inversion.

(Box 15.2, continued)

After sunrise the ground is heated and the inversion disappears. Although surface inversions are usually shallow, they may be quite thick in regions where the land surface is uneven. Because cold air is denser (heavier) than warm air, the chilled air near the surface gradually drains from the uplands and slopes into adjacent lowlands and valleys. As might be expected, these thicker surface inversions will not dissipate as quickly after sunrise.

Many extensive and long-lived air pollution episodes are linked to temperature inversions that develop in association with the sinking air that characterizes slow-moving centers of high pressure (Figure 15.C). As the air sinks to lower altitudes, it is compressed and so its temperature rises. Because turbulence is almost always present near the ground, this lower-most portion of the atmosphere is generally prevented from participating in the general subsidence. Thus an inversion develops aloft between the lower turbulent zone and the subsiding warmed layers above.

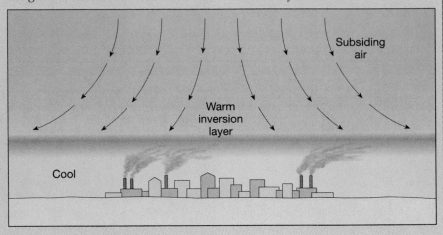

Figure 15.C
Inversions aloft frequently develop in association with slow-moving centers of high pressure where the air aloft subsides and warms by compression. The turbulent surface zone does not subside as much. Thus, an inversion often forms between the lower turbulent zone and the subsiding layers above.

above the lifting condensation level, the parcel becomes warmer than the surrounding air. From this point along its ascent the parcel will continue to rise because of its own buoyancy, without an outside force. Thus, conditional instability depends on whether or not the rising air is saturated. The word *conditional* is used because the air must be forced upward, such as over mountainous terrain, before it becomes unstable and rises because of its own buoyancy.

Summary. In summary, the stability of air is determined by examining the temperature of the atmosphere at various heights. In simple terms, a column of air is deemed unstable when the air near the bottom of this layer is significantly warmer (less dense) than the air aloft, indicating a steep environmental lapse rate. Under these conditions the air actually turns over, since the warm air below rises and displaces the colder air aloft. Conversely, the air is considered to be stable when

Figure 15.13
Absolute instability illustrated using an environmental lapse rate of 12°C per 1000 meters. The rising air is always warmer and therefore lighter than the surrounding air.

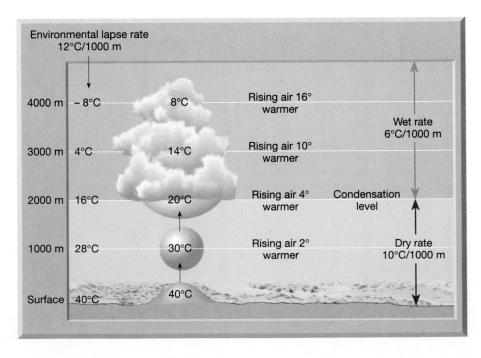

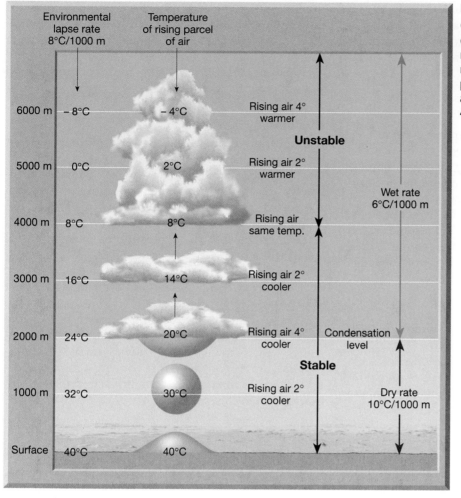

Figure 15.14
Conditional instability illustrated using an environmental lapse rate of 8°C per 1000 meters, which lies between the dry adiabatic rate and the wet adiabatic rate. The rising parcel of air is cooler than the surrounding air below 4000 meters and warmer above 4000 meters.

the temperature drops gradually with increasing altitude. The most stable conditions occur during a temperature inversion when the temperature actually increases with height. Under these conditions, there is very little vertical air movement.

Stability and Daily Weather

From the previous discussion we can conclude that stable air resists vertical movement, whereas unstable air ascends freely because of its own buoyancy. But how do these facts manifest themselves in our daily weather?

Because stable air resists upward movement, we might conclude that clouds will not form when stable conditions prevail in the atmosphere. Although this seems reasonable, processes exist that *force* air aloft. These will be discussed in the following section, "Processes That Lift Air." On occasions when stable air is forced aloft, the clouds that form are widespread and have little vertical thickness when compared to their horizontal dimension, and precipitation, if any, is light to moderate.

By contrast, clouds associated with unstable air are towering and are usually accompanied by heavy precipitation (Figure 15.15). For this reason, we can con-

clude that on a dreary, overcast day with light drizzle, stable air has been forced aloft. On the other hand, during a day when cauliflower-shaped clouds appear to be growing as if bubbles of hot air are surging upward, we can be fairly certain that the ascending air is unstable.

Figure 15.15
Cauliflower-shaped clouds provide evidence of unstable conditions in the atmosphere. (Photo by Jim Brandenburg/Minden Pictures)

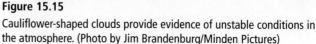

In summary, the role of stability in determining our daily weather is very important. To a large degree, stability determines the type of clouds that develop and whether precipitation will come as a gentle shower or a heavy downpour.

Processes That Lift Air

Earlier you learned that on some summer days, surface heating can be sufficient to cause pockets of air to be warmed more than the surrounding air. If the temperature of this air is sufficiently high and ample moisture is present, these unstable parcels will rise to produce clouds and occasionally precipitation. By contrast, air that is stable will not rise on its own. It requires some mechanism to initiate the vertical movement. Three such mechanisms are orographic lifting, frontal wedging, and convergence.

Orographic Lifting

Orographic lifting occurs when elevated terrains, such as mountains, act as barriers to flowing air (Figure 15.16). As air is pushed up a mountain slope, adiabatic cooling often generates clouds and copious precipitation. In fact, many of the rainiest places in the world are located on windward mountain slopes, such as the Olympic Mountains in Washington State.

In addition to providing the lift to render air unstable, mountains remove additional moisture in other ways. By slowing the horizontal flow of air, they retard the passage of storm systems. Moreover, the irregular topography of mountains enhances differential heating and instability. These combined effects account for the generally higher precipitation associated with mountainous regions compared with surrounding lowlands.

By the time air reaches the leeward side of a mountain, much of its moisture has been lost. And, if the air descends, it becomes compressed and warms adiabatically, making condensation and precipitation even less likely. As shown in Figure 15.16, the result can be a **rainshadow desert**. The Great Basin desert of the western United States lies only a few hundred kilometers from the Pacific but is effectively cut off from the ocean's moisture by the imposing Sierra Nevada range (Figure 15.17). The Gobi desert of Mongolia, the Takla Makan of China, and the Patagonia desert of Argentina are other examples of rainshadow deserts found on the leeward sides of mountains.

Frontal Wedging

If orographic lifting were the only mechanism that forced air aloft, the relatively flat central portion of North America would be an expansive desert. Fortunately, this is not the case. **Frontal wedging** occurs when cool air acts as a barrier over which warmer, less dense air rises. Figure 15.18 illustrates frontal wedging of both stable and unstable air. As this figure shows, forceful lifting is important in producing clouds. The stability of the air, however, determines to a great extent the type of clouds formed and the amount of precipitation that may be expected.

These weather-producing fronts are part of the storm systems called *middle-latitude cyclones*. These cyclones, which show up as the "lows" you see on weather maps, are responsible for producing a high proportion of the precipitation in the middle latitudes. We will examine them closely in Chapter 17.

Convergence

Whenever air masses flow together, **convergence** is said to occur. Such flow results in general upward movement, because when air converges, it has to go somewhere, so the height of the air column increases. Consequently, air within the column must move upward.

The Florida peninsula provides an excellent example of the role that convergence can play in initiating cloud development and precipitation. On warm days, the airflow is from the ocean to the land along both coasts of Florida. This leads to general convergence over the peninsula. Such a pattern of air movement and the uplift that results is aided by intense solar heating. Rising parcels of humid air expand and cool adiabatically, the water vapor condenses to form clouds, and

Figure 15.16

Orographic lifting. The name comes from *oro-*, which means "mountain."

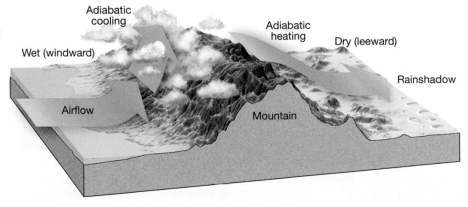

Adiabatic cooling

Wet (windward)

Airflow

Adiabatic heating

Dry (leeward)

Rainshadow

Mountain

Figure 15.17
Rainshadow desert. The arid conditions in Death Valley can be partially attributed to the adjacent mountains, which effectively remove the moisture from air originating over the Pacific. (Photo by Jack Dykinga)

the result is the greatest frequency of midafternoon thunderstorms in the United States. More important, convergence as a mechanism of forceful lifting is a major contributor to the stormy weather associated with middle-latitude cyclones and hurricanes, discussed in Chapter 17.

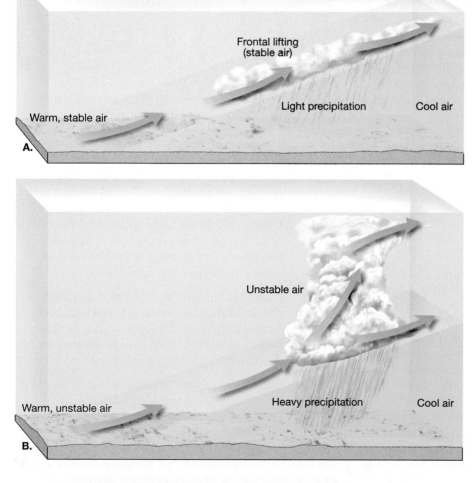

Figure 15.18
A. When stable air is lifted, layered clouds usually result. **B.** When warm, unstable air is forced to rise over cooler air, "towering" clouds develop.

Condensation and Cloud Formation

To review briefly, condensation occurs when water vapor in the air changes to a liquid. The result of this process may be dew, fog, or clouds. For any of these forms of condensation to occur, the air must be saturated. Saturation occurs most commonly when air is cooled to its dew point, or less often when water vapor is added to the air.

Generally, there must be a surface on which the water vapor may condense. When dew occurs, objects at or near the ground serve this purpose, like grass and car windows. But when condensation occurs in the air above the ground, tiny bits of particulate matter, known as **condensation nuclei**, serve as surfaces for water vapor condensation. These nuclei are very important, for in their absence, a relative humidity well in excess of 100 percent is needed to produce clouds.

Condensation nuclei such as microscopic dust, smoke, and salt particles (from the ocean) are profuse in the lower atmosphere. Because of this abundance of particles, relative humidity rarely exceeds 101 percent. Some particles, such as ocean salt, are particularly good nuclei because they absorb water. These particles are termed **hygroscopic** ("water-seeking") **nuclei**. When condensation takes place, the initial growth rate of cloud droplets is rapid. It diminishes quickly because the excess water vapor is quickly absorbed by the numerous competing particles. This results in the formation of a cloud consisting of millions upon millions of tiny water droplets, all so fine that they remain suspended in air. When cloud formation occurs at below-freezing temperatures, tiny ice crystals form. Thus, a cloud may consist of water droplets, ice crystals, or both.

The slow growth of cloud droplets by additional condensation and the immense size difference between cloud droplets and raindrops suggest that condensation alone is not responsible for the formation of drops large enough to fall as rain. We will first examine clouds and then return to the questions of how precipitation forms.

Types of Clouds

Clouds are among the most conspicuous and observable aspects of the atmosphere and its weather. **Clouds** are a form of condensation best described as *visible aggregates of minute droplets of water or tiny crystals of ice.* In addition to being prominent and sometimes spectacular features in the sky, clouds are of continual interest to meteorologists, because they provide a visible indication of what is going on in the atmosphere. Anyone who observes clouds with the hope of recognizing different types often finds that there is a bewildering variety of these familiar white and gray masses streaming across the sky. Still, once one comes to know the basic classification scheme for clouds, most of the confusion vanishes.

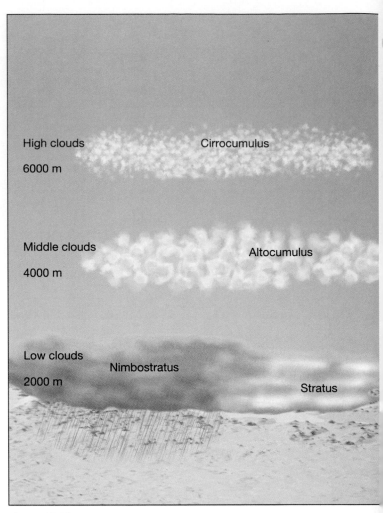

Figure 15.19

Classification of clouds according to height and form. (After Ward's Natural Science Establishment, Inc., Rochester, N.Y.)

Clouds are classified on the basis of their *form* and *height* (Figure 15.19). Three basic forms are recognized: cirrus, cumulus, and stratus.

- **Cirrus** clouds are high, white, and thin. They can occur as patches or as delicate veil-like sheets or extended wispy fibers that often have a feathery appearance.

- **Cumulus** clouds consist of globular individual cloud masses. Normally they exhibit a flat base and have the appearance of rising domes or towers. Such clouds are frequently described as having a cauliflower structure.

- **Stratus** clouds are best described as sheets or layers that cover much or all of the sky. While there may be minor breaks, there are no distinct individual cloud units.

All other clouds reflect one of these three basic forms or are combinations or modifications of them.

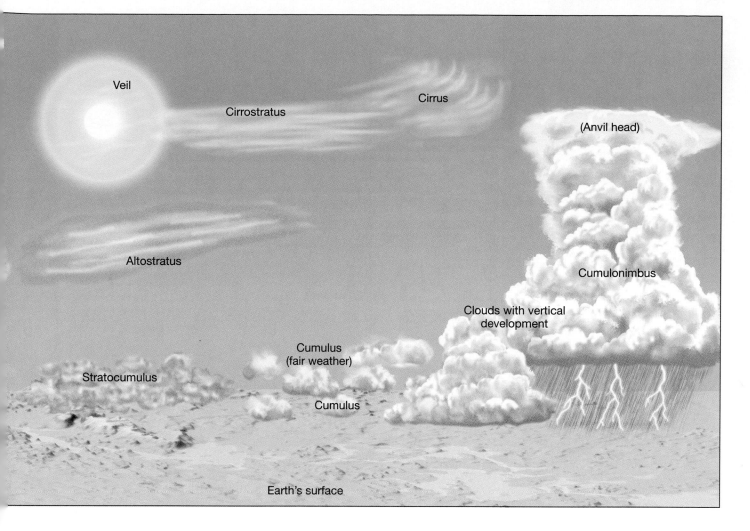

Three levels of cloud heights are recognized: high, middle, and low (Figure 15.19). **High clouds** normally have bases above 6000 meters (20,000 feet), **middle clouds** generally occupy heights from 2000 to 6000 meters (6500 to 20,000 feet), and **low clouds** form below 2000 meters (6,500 feet). The altitudes listed for each height category are not hard and fast. There is some seasonal as well as latitudinal variation. For example, at high latitudes or during cold winter months in the midlatitudes, high clouds are often found at lower altitudes.

High Clouds.   Three cloud types make up the family of high clouds (above 6000 meters): cirrus, cirrostratus, and cirrocumulus. *Cirrus* clouds are thin and delicate and sometimes appear as hooked filaments called "mares' tails" (Figure 15.20A). As the names suggest, *cirrocumulus* clouds consist of fluffy masses (Figure 15.20B), whereas *cirrostratus* clouds are flat layers (Figure 15.20C). Because of the low temperatures and small quantities of water vapor present at high altitudes, all high clouds are thin and white and are made up of ice crystals. Further, these clouds are not considered precipitation makers. However, when cirrus clouds are followed by cirrocumulus clouds and increased sky coverage, they may warn of impending stormy weather.

Middle Clouds.   Clouds that appear in the middle range (2000 to 6000 meters) have the prefix *alto* as part of their name. *Altocumulus* clouds are composed of globular masses that differ from cirrocumulus clouds in that they are larger and denser (Figure 15.20D). *Altostratus* clouds create a uniform white to grayish sheet covering the sky with the sun or moon visible as a bright spot (Figure 15.20E). Infrequent light snow or drizzle may accompany these clouds.

Low Clouds.   There are three members in the family of low clouds: stratus, stratocumulus, and nimbostratus. *Stratus* are a uniform foglike layer of clouds that frequently covers much of the sky. On occasions these clouds may produce light precipitation. When stratus clouds develop a scalloped bottom that appears as long parallel rolls or broken globular patches, they are called *stratocumulus* clouds.

Nimbostratus clouds derive their name from the Latin *nimbus*, which means "rainy cloud," and *stratus*, which means "to cover with a layer" (Figure 15.20F). As the name suggests, nimbostratus clouds are one of the chief precipitation producers. Nimbostratus clouds form in association with stable conditions. We might not expect clouds to grow or persist in stable air, yet

A. Cirrus

B. Cirrocumulus

C. Cirrostratus

D. Altocumulus

Figure 15.20

These photos depict common forms of several different cloud types. (Photos A, B, D, E, F, and G by E. J. Tarbuck. Photo C by A. and J. Verkaik/The Stock Market. Photo H by Joe Towers/The Stock Market)

cloud growth of this type is common when air is forced to rise, as occurs along a mountain range, a front, or near the center of a cyclone where converging winds cause air to ascend. Such forced ascent of stable air leads to the formation of a stratified cloud layer that is large horizontally compared to its depth.

E. Altostratus

F. Nimbostratus

G. Cumulus

H. Cumulonimbus

Clouds of Vertical Development. Some clouds do not fit into any one of the three height categories mentioned. Such clouds have their bases in the low height range but often extend upward into the middle or high altitudes. Consequently, clouds in this category are called **clouds of vertical development**. They are all related to one another and are associated with unstable air. Although *cumulus* clouds are often connected with "fair" weather (Figure 15.20G), they may grow dramatically under the proper circumstances. Once upward movement is triggered, acceleration is powerful and clouds with great vertical extent form. The end result is often a towering cloud, called a *cumulonimbus,* that may produce rain showers or a thunderstorm (Figure 15.20H).

Definite weather patterns can often be associated with particular clouds or certain combinations of cloud types, so it is important to become familiar with cloud descriptions and characteristics. Table 15.4 lists the ten basic cloud types that are recognized internationally and gives some characteristics of each.

Fog

Fog is generally considered to be an atmospheric hazard. When it is light, visibility is reduced to 2 or 3 kilometers. However, when it is dense, visibility may be cut to a few tens of meters or less, making travel by any mode not only difficult but often dangerous. Officially, visibility must be reduced to 1 kilometer or less before fog is reported. While this figure is arbitrary, it does provide an objective criterion for comparing fog frequencies at different locations.

Fog is defined as a *cloud with its base at or very near the ground.* Physically, there is no basic difference between a fog and a cloud; their appearance and structure are the same. The essential difference is the method and place of formation. Whereas clouds result when air rises and cools adiabatically, most fogs are the consequence of radiation cooling or the movement of air over a cold surface. (The exception is upslope fog.) In other circumstances, fogs form when enough water vapor is added to the air to bring about saturation (evaporation fogs). We will examine both of these types.

Fogs Caused by Cooling

When warm, moist air moves over a cool surface, the result may be a blanket of fog called **advection fog** (Figure 15.21). Examples of such fogs are very common. The

Table 15.4 Cloud types and characteristics.

Cloud Family and Height	Cloud Type	Characteristics
High clouds— above 6000 meters (20,000 feet)	Cirrus	Thin, delicate, fibrous, ice-crystal clouds. Sometimes appear as hooked filaments called "mares' tails." (Figure 15.20A)
	Cirrocumulus	Thin, white, ice-crystal clouds in the form of ripples, waves, or globular masses all in a row. May produce a "mackerel sky." Least common of the high clouds. (Figure 15.20B)
	Cirrostratus	Thin sheet of white, ice-crystal clouds that may give the sky a milky look. Sometimes produce halos around the sun or moon. (Figure 15.20C)
Middle clouds— 2000–6000 meters (6500–20,000 feet)	Altocumulus	White to gray clouds often composed of separate globules; "sheep-back" clouds. (Figure 15.20D)
	Altostratus	Stratified veil of clouds that are generally thin and may produce very light precipitation. When thin, the sun or moon may be visible as a "bright spot," but no halos are produced. (Figure 15.20E)
Low clouds— below 2000 meters (6500 feet)	Stratocumulus	Soft, gray clouds in globular patches or rolls. Rolls may join together to make a continuous cloud.
	Stratus	Low uniform layer resembling fog but not resting on the ground. May produce drizzle.
	Nimbostratus	Amorphous layer of dark gray clouds. One of the chief precipitation-producing clouds. (Figure 15.20F)
Clouds of vertical development— 500–18,000 meters (1600–60,000 feet)	Cumulus	Dense, billowy clouds often characterized by flat bases. May occur as isolated clouds or closely packed. (Figure 15.20G)
	Cumulonimbus	Towering cloud sometimes spreading out on top to form an "anvil head." Associated with heavy rainfall, thunder, lightning, hail, and tornadoes. (Figure 15.20H)

Figure 15.21
Advection fog rolling into San Francisco Bay. (Photo by Ed Pritchard/Tony Stone Worldwide)

foggiest location in the United States, and perhaps in the world, is Cape Disappointment, Washington. The name is indeed appropriate, because this station averages 2552 hours of fog each year (there are about 8760 hours in a year). The fog experienced at Cape Disappointment, and at other West Coast locations, is produced when warm, moist air from the Pacific Ocean moves over the cold California Current and is then carried onto shore by the prevailing winds. Advection fogs are also quite common in the winter season when warm air from the Gulf of Mexico blows across cold, often snow-covered surfaces of the Midwest and East.

Radiation fog forms on cool, clear, calm nights, when Earth's surface cools rapidly by radiation. As the night progresses, a thin layer of air in contact with the ground is cooled below its dew point. As the air cools and becomes denser, it drains into low areas, resulting in "pockets" of fog. The largest "pockets" are often river valleys, where thick accumulations may occur (Figure 15.22).

As its name implies, **upslope fog** is created when relatively humid air moves up a gradually sloping plain or, in some cases, up the steep slopes of a mountain. As a result of this upward movement, air expands and cools adiabatically. If the dew point is reached, an extensive layer of fog may form. In the United States, the Great Plains offers an excellent example. When humid easterly or southeasterly winds blow westward from the Mississippi River toward the Rocky Mountains, the air gradually rises, resulting in an adiabatic temperature decrease of about 13°C. When the difference between the air temperature and dew point of westward-moving air is less than 13°C, an extensive fog can result in the western plains.

Evaporation Fogs

When the saturation of air in contact with the ground occurs primarily because of the addition of water vapor, the resulting fogs may be called *evaporation fogs*. Two types of evaporation fogs are recognized: steam fog and frontal or precipitation fog.

When cool air moves over warm water, enough moisture may evaporate from the water surface to produce saturation. As the rising water vapor meets the cold air, it immediately recondenses and rises with the air that is being warmed from below. Since the water has a steaming appearance, the phenomenon is called **steam fog** (Figure 15.23). Steam fog is fairly common over lakes and rivers in the fall and early winter, when the water may still be relatively warm and the air is rather crisp. Steam fog is often shallow because as the steam rises it reevaporates in the unsaturated air above.

Figure 15.22

Radiation fog in the Illinois River valley at Peoria, Illinois. (Photo by David Zalaznik/ Peoria Journal Star)

When frontal wedging occurs, warm air is lifted over colder air. If the resulting clouds yield rain, and the cold air below is near the dew point, enough rain will evaporate to produce fog. A fog formed in this manner is called **frontal fog**, or **precipitation fog**. The result is a more or less continuous zone of condensed water droplets reaching from the ground up through the clouds.

In summary, both steam fog and frontal fog result from the addition of moisture to a layer of air. As you saw, the air is usually cool or cold and already near saturation. Because air's capacity to hold water vapor at low temperatures is small, only a relatively modest amount of evaporation is necessary to produce saturated conditions and fog.

How Precipitation Forms

All clouds contain water, but why do some produce precipitation while others drift placidly overhead? This simple question perplexed meteorologists for many years.

First, cloud droplets are very small, averaging less than 10 micrometers in diameter (for comparison, a human hair is about 75 micrometers in diameter). Because of their small size, cloud droplets fall incredibly slowly. Theoretically, an average cloud droplet falling from a cloud base at 1000 meters would require about 48 hours to reach the ground. Of course, it would never complete its journey. Even falling through humid air, a cloud droplet would evaporate before it fell a few meters below the cloud base. In addition, clouds are made up of many billions of these droplets, all competing for the available water vapor; thus, their continued growth via condensation is very slow.

A raindrop large enough to reach the ground without evaporating contains roughly a million times more water than a cloud droplet. Therefore, for precipitation to form, millions of cloud droplets must somehow coalesce (join together) into drops large enough to sustain themselves during their descent. Two mechanisms have been proposed to explain this phenomenon: the Bergeron process and the collision-coalescence process.

Figure 15.23

Early morning steam fog rising from the Blackwater River in West Virginia. (Photo by Carr Clifton)

Bergeron Process

The **Bergeron process** is named after its discoverer, Swedish meteorologist Tor Bergeron. The process relies on two interesting properties of water: supercooling and supersaturating.

Supercooling.   Cloud droplets do not freeze at 0°C as expected. In fact, pure water suspended in air does not freeze until it reaches a temperature of nearly –40°C (–40°F). Water in the liquid state below 0°C is referred to as **supercooled**. Supercooled water will readily freeze if sufficiently agitated. This explains why airplanes collect ice when they pass through a cloud composed of supercooled droplets.

In addition, supercooled droplets will freeze upon contact with solid particles that have a crystal form closely resembling that of ice. Such materials are termed **freezing nuclei**. The need for freezing nuclei to initiate the freezing process is similar to the requirement for condensing nuclei in the process of condensation. However, in contrast to condensation nuclei, freezing nuclei are very sparse in the atmosphere and do not generally become active until the temperature reaches –10°C or less. Only at temperatures well below freezing will ice crystals begin to form in clouds, and even at that, they will be few and far between. Once ice crystals form, they are in direct competition with the supercooled droplets for the available water vapor.

Supersaturation.   When air is saturated (100 percent relative humidity) with respect to water, it is supersaturated (relative humidity is greater than 100 percent) with respect to ice. Table 15.5 shows that at –10°C, when the relative humidity is 100 percent with respect to water, the relative humidity with respect to ice is nearly 110 percent. Thus, ice crystals cannot coexist with water droplets, because the air always "appears" supersaturated to the ice crystals. So the ice crystals begin to consume the "excess" water vapor, which lowers the relative humidity near the surrounding droplets. In turn, the water droplets evaporate to replenish the diminishing water vapor, thereby providing a continual source of vapor for the growth of the ice crystals (Figure 15.24).

Table 15.5 Relative humidity with respect to ice when relative humidity with respect to water is 100 percent.

Temperature (°C)	Relative Humidity with Respect to:	
	Water (%)	Ice (%)
0	100	100
–5	100	105
–10	100	110
–15	100	116
–20	100	121

Because the level of supersaturation with respect to ice can be quite great, the growth of ice crystals is generally rapid enough to generate crystals large enough to fall. During their descent, these ice crystals enlarge as they intercept cloud drops, which freeze upon them. Air movement will sometimes break up these delicate crystals and the fragments will serve as freezing nuclei. A chain reaction develops, producing many ice crystals, which by accretion form into large crystals called snowflakes. When the surface temperature is above 4°C (39°F), snowflakes usually melt before they reach the ground and continue their descent as rain. Even a summer rain may have begun as a snowstorm in the clouds overhead.

Collision-Coalescence Process

A few decades ago, meteorologists believed that the Bergeron process was responsible for the formation of most precipitation. However, it was discovered that copious rainfall can be associated with clouds located well below the freezing level (*warm clouds*), particularly in the tropics. This led to the proposal of a second mechanism thought to produce precipitation—the **collision-coalescence process**.

Research has shown that clouds composed entirely of liquid droplets must contain droplets larger than 20 micrometers if precipitation is to form. These large droplets form when "giant" condensation nuclei are present, or when hygroscopic particles such as sea salt exist. Hygroscopic particles begin to remove water vapor from the air at relative humidities under 100 percent and can grow quite large. Because the rate at which drops fall is size-dependent, these "giant" droplets fall most rapidly. As they plummet, they collide with smaller, slower droplets and coalesce. Growing larger in the process, they fall even more rapidly (or in an updraft, rise more slowly), increasing their chances of collision and rate of growth (Figure 15.25). After a great many such collisions they are large enough to fall to the surface without completely evaporating. Because of the number of collisions required for growth to raindrop size, droplets in clouds that have great vertical thickness and abundant moisture have a better chance of reaching the required size. Updrafts also aid this process for they allow the droplets to traverse the cloud repeatedly.

Raindrops can grow to a maximum size of 5 millimeters, at which point they fall at the rate of 33 kilometers (20 miles) per hour. At this size and speed, the water's surface tension, which holds the drop together, is overcome by the drag imposed by the air, which in turn pulls the drops apart. The resulting breakup of a large raindrop produces numerous smaller drops which begin anew the task of sweeping up cloud droplets. Drops that are less than 0.5 millimeter upon reaching the ground are termed *drizzle* and require about ten minutes to fall from a cloud 1000 meters (3300 feet) overhead.

Figure 15.24
The Bergeron process. Ice crystals grow at the expense of cloud droplets until they are large enough to fall. The size of these particles has been greatly exaggerated.

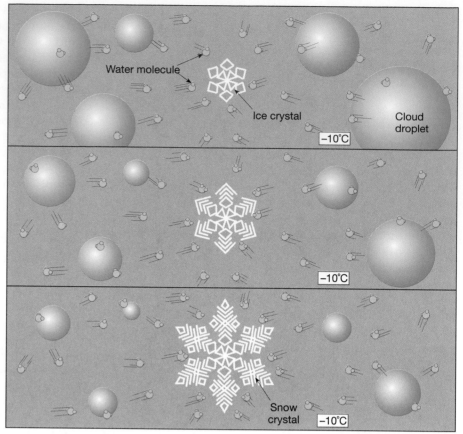

The collision-coalescence process is not quite as simple as described. First, as the large droplets descend, they produce an air stream around them similar to that produced by a fast-moving automobile. If an automobile is driven at night and we use the bugs that are often out to be analogous to the cloud droplets, it is easy to visualize how most cloud droplets are swept aside. The larger the cloud droplet (or bug), the better its chance of colliding with the giant droplet (or car). Second, collision does not guarantee coalescence. Experiments indicate that atmospheric electricity may hold these droplets together once they collide. If a droplet with a negative charge should collide with a positively charged droplet, electrical attraction of opposite charges may bind them together.

Forms of Precipitation

Much of the world's precipitation begins as snow crystals or other solid forms such as hail or graupel. Entering the warmer air below the cloud, these ice particles often melt and reach the ground as raindrops. In some parts of the world, particularly the subtropics, precipitation often forms in clouds that are warmer than 0°C (32°F). These rains frequently occur over the ocean where cloud condensation nuclei are not plentiful and those that exist vary in size. Under such conditions, cloud droplets can grow rapidly by the collision-coalescence process to produce copious amounts of rain.

Because atmospheric conditions vary greatly from place to place as well as seasonally, several different forms of precipitation are possible. Rain and snow are the most common and familiar, but the other forms of precipitation listed in Table 15.6 are important as well. The occurrence of sleet, glaze, and hail are often associated with important weather events. Although limited in occurrence and sporadic in both time and space, these forms, especially glaze and hail, can cause considerable damage.

Rain and Drizzle

We are familiar with rain. In meteorology, the term **rain** is restricted to drops of water that fall from a cloud and have a diameter of at least 0.5 millimeter. Most rain originates either in nimbostratus clouds or in towering cumulonimbus clouds that are capable of producing unusually heavy rainfalls known as *cloudbursts.* Raindrops rarely exceed about 5 millimeters in diameter. Larger drops do not survive because surface tension, which holds the drops together, is exceeded by the frictional drag of the air. Consequently, large raindrops regularly break apart into smaller ones.

Fine, uniform drops of water having a diameter less than 0.5 millimeter are called *drizzle.* Drizzle can be so fine that the tiny drops appear to float and their impact is almost imperceptible. Drizzle and small raindrops

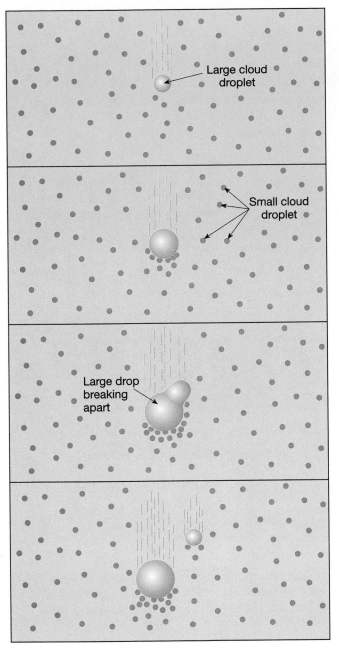

Figure 15.25
The collision-coalescence process. Because large cloud droplets fall more rapidly than smaller droplets, they are able to sweep up the smaller ones in their path and grow. Most cloud droplets are so small that the motion of the air keeps them suspended. Even if these small cloud droplets were to fall, they would evaporate long before reaching the surface.

generally are produced in stratus or nimbostratus clouds where precipitation may be continuous for several hours, or on rare occasions for days.

Snow

Snow is precipitation in the form of ice crystals (snowflakes) or, more often, aggregates of crystals. The size, shape, and concentration of snowflakes depend to a great extent on the temperature at which they form.

Recall that at very low temperatures, the moisture content of air is small. The result is the generation of very light, fluffy snow made up of individual six-sided ice crystals. This is the "powder" that downhill skiers talk so much about. By contrast, at temperatures warmer than about –5°C, the ice crystals join together into larger clumps consisting of tangled aggregates of crystals. Snowfalls composed of these composite snowflakes are generally heavy and have a high moisture content, which makes them ideal for making snowballs.

Sleet and Glaze

Sleet is a wintertime phenomenon and refers to the fall of small particles of ice that are clear to translucent. For sleet to be produced, a layer of air with temperatures above freezing must overlie a subfreezing layer near the ground. When raindrops, which are often melted snow, leave the warmer air and encounter the colder air below, they freeze and reach the ground as small pellets of ice the size of the raindrops from which they formed.

On some occasions, when the vertical distribution of temperatures is similar to that associated with the formation of sleet, freezing rain or **glaze** results instead. In such situations, the subfreezing air near the ground is not thick enough to allow the raindrops to freeze. The raindrops, however, do become supercooled as they fall through the cold air and turn to ice upon colliding with solid objects. The result can be a thick coating of ice having sufficient weight to break tree limbs, down power lines, and make walking or driving extremely hazardous (Figure 15.26).

Hail

Hail is precipitation in the form of hard, rounded pellets or irregular lumps of ice. Moreover, large hailstones often consist of a series of nearly concentric shells of differing densities and degrees of opaqueness (Figure 15.27). Most hailstones have diameters between 1 centimeter (pea size) and 5 centimeters (golf ball size), although some can be as big as an orange or larger. Occasionally, hailstones weighing a pound or more have been reported. Many of these were probably composites of several stones frozen together.

The heaviest authenticated hailstone on record fell on Coffeyville, Kansas, September 3, 1970 (Figure 15.27). With a 14-centimeter (5.5-inch) diameter, this "giant" weighed 766 grams (1.67 pounds). It is estimated that this stone hit the ground at a speed in excess of 160 kilometers (100 miles) per hour.

The destructive effects of large hailstones are well known, especially to farmers whose crops have been devastated in a few minutes and to people whose windows and roofs have been damaged (Figure 15.28). In the United States, hail damage each year is significant. In 1994 alone, there was more than $450 million in property damage attributed to hail.

Table 15.6 Forms of precipitation.

Type	Appropriate Size	State of Matter	Description
Mist	0.005 to 0.05 mm	Liquid	Droplets large enough to be felt on the face when air is moving 1 meter/second. Associated with stratus clouds.
Drizzle	Less than 0.5 mm	Liquid	Small uniform drops that fall from stratus clouds, generally for several hours.
Rain	0.5 to 5 mm	Liquid	Generally produced by nimbostratus or cumulonimbus clouds. When heavy, it can show high variability from one place to another.
Sleet	0.5 to 5 mm	Solid	Small, spherical to lumpy ice particles that form when raindrops freeze while falling through a layer of subfreezing air. Because the ice particles are small, damage, if any, is generally minor. Sleet can make travel hazardous.
Glaze	Layers 1 mm to 2 cm thick	Solid	Produced when supercooled raindrops freeze on contact with solid objects. Glaze can form a thick coating of ice having sufficient weight to seriously damage trees and power lines.
Rime	Variable accumulations	Solid	Deposits usually consisting of ice feathers that point into the wind. These delicate, frostlike accumulations form as supercooled cloud or fog droplets encounter objects and freeze on contact.
Snow	1 mm to 2 cm	Solid	The crystalline nature of snow allows it to assume may shapes including six-sided crystals, plates, and needles. Produced in supercooled clouds where water vapor is deposited as ice crystals that remain frozen during their descent.
Hail	5 mm to 10 cm or larger	Solid	Precipitation in the form of hard, rounded pellets or irregular lumps of ice. Produced in large convective, cumulonimbus clouds, where frozen ice particles and supercooled water coexist.
Gaupel	2 to 5 mm	Solid	Sometimes called soft hail, graupel forms as rime collects on snow crystals to produce irregular masses of "soft" ice. Because these particles are softer than hailstones, they normally flatten out upon impact.

Hail is produced only in large cumulonimbus clouds where updrafts can sometimes reach speeds approaching 160 kilometers (100 miles) per hour, and where there is an abundant supply of supercooled water. Hailstones begin as small embryonic ice pellets that grow by collecting supercooled water droplets as they fall through the cloud. If they encounter a strong updraft, they may be carried upward again and begin the downward journey anew. Each trip through the supercooled portion of the cloud may be represented by an additional layer of ice. Hailstones may also form from a single descent through an updraft. Either way, the process continues until the hailstone encounters a downdraft or grows too heavy to remain suspended by the thunderstorm's updraft.

Rime

Rime is a deposit of ice crystals formed by the freezing of supercooled fog or cloud droplets on objects whose surface temperature is below freezing. When rime forms on trees, it adorns them with its characteristic ice feathers, which can be spectacular to behold (Figure 15.29). In these situations, objects such as pine needles act as freezing nuclei, causing the supercooled droplets to freeze on contact. On occasions when the wind is blowing, only the windward surfaces of objects will accumulate the layer of rime.

Measuring Precipitation

The most common form of precipitation, rain, is the easiest to measure. Any open container having a consistent cross-section throughout can be a rain gauge. In general practice, however, more sophisticated devices are used so that small amounts of rainfall can be measured more accurately, and to reduce losses from evaporation. The *standard rain gauge* (Figure 15.30) has a diameter of about 20 centimeters (8 inches) at the top. Once the water is caught, a funnel conducts the rain into a cylindrical measuring tube that has a cross-sectional area only one-tenth as large as the receiver. Consequently, rainfall depth is magnified ten times, which allows for accurate measurements to the nearest 0.025 centimeter (0.01 inch). The narrow opening also minimizes evaporation. When the amount of rain is less than 0.025 centimeter, it is reported as a *trace* of precipitation.

Figure 15.26
Glaze forms when supercooled raindrops freeze on contact with objects. (Photo by Annie Griffiths Belt/DRK Photo)

Figure 15.28
Hail damage to a corn field in central Illinois. (Courtesy of the National Center for Atmospheric Research)

In addition to the standard rain gauge, several types of recording gauges are routinely used. These instruments record not only the amount of rain but also its time of occurrence and intensity (amount per unit of time).

No matter which rain gauge is used, proper exposure is critical. Errors arise when the gauge is shielded from obliquely falling rain by buildings,

Figure 15.27
A cross-section of the Coffeyville hailstone. This largest recorded hailstone fell over Kansas in 1970 and weighed 766 grams (1.67 pounds). (Courtesy of the National Center for Atmospheric Research)

Figure 15.29
Rime consists of delicate ice crystals that form when supercooled fog or cloud droplets freeze on contact with objects. Here, rime has formed on the needles of a pine tree. (Photo by Carr Clifton)

trees, or other high objects. Hence, the instrument should be at least as far away from such obstructions as the objects are high. Another cause of error is the wind. It has been shown that with increasing wind and turbulence, it becomes more difficult to collect a representative quantity of rain.

When snow records are kept, two measurements are normally taken: depth and water equivalent. Usually the depth of snow is measured with a calibrated stick. The actual measurement is simple, but choosing a *representative* spot often poses a dilemma. Even when winds are light or moderate, snow drifts freely. As a rule, it is best to take several measurements in an open place away from trees and obstructions and then average them. To obtain the water equivalent, samples may be melted and then weighed or measured as rain.

The quantity of water in a given volume of snow is not constant. A general ratio of 10 units of snow to 1 unit of water is often used when exact information is not available, but the actual water content of snow may deviate widely from this figure. It may take as much as 30 centimeters of light, fluffy dry snow or as little as 4 centimeters of wet snow to produce 1 centimeter of water.

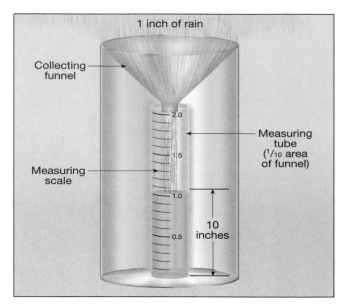

Figure 15.30
Precipitation measurement. The standard rain gauge allows for accurate rainfall measurement to the nearest 0.025 centimeter (0.01 inch). Because the cross-sectional area of the measuring tube is only one-tenth as large as the collector, rainfall is magnified ten times.

Review Questions

1. Summarize the processes by which water changes from one state to another. Indicate whether heat energy is absorbed or liberated.
2. After studying Table 15.1, write a generalization relating temperature and the capacity of air to hold water vapor.
3. How do relative and specific humidity differ?
4. Referring to Figure 15.6, answer the following questions:
   (a) During a typical day, when is the relative humidity highest? Lowest?
   (b) At what time of day would dew most likely form? Write a generalization relating changes in air temperature to changes in relative humidity.
5. If the temperature remains unchanged and the specific humidity decreases, how will relative humidity change?
6. On a cold winter day when the temperature is −10°C and the relative humidity is 50 percent, what is the specific humidity (refer to Table 15.1)? What is the specific humidity for a day when the temperature is 20°C and the relative humidity is 50 percent?
7. Explain the principle of the sling psychrometer and the hair hygrometer.
8. Using the standard tables (Tables 15.2 and 15.3), determine the relative humidity and dew-point temperature if the dry-bulb thermometer reads 16°C and the wet-bulb thermometer reads 12°C. How would the relative humidity and dew point change if the wet-bulb thermometer read 8°C?
9. On a warm summer day when the relative humidity is high, it may seem even warmer than the thermometer indicates. Why do we feel so uncomfortable on a "muggy" day?
10. Why does air cool when it rises through the atmosphere?
11. Explain the difference between environmental lapse rate and adiabatic cooling.
12. If unsaturated air at 23°C were to rise, what would its temperature be at 500 meters? If the dew-point temperature at the condensation level were 13°C, at what altitude would clouds begin to form?
13. Why does the adiabatic rate of cooling change when condensation begins? Why is the wet adiabatic rate not a constant figure?
14. The contents of an aerosol can are under very high pressure. When you push the nozzle on such a can, the spray feels cold. Explain.
15. How do orographic lifting and frontal wedging act to force air to rise?
16. Explain why the Great Basin area of the western United States is so dry. What term is applied to such a situation?
17. How does stable air differ from unstable air? Describe the general nature of the clouds and precipitation expected with each.
18. How do temperature inversions influence air pollution? (See Box 15.2.)
19. What is the function of condensation nuclei in cloud formation? The function of the dew point?
20. As you drink an ice-cold beverage on a warm day, the outside of the glass or bottle becomes wet. Explain.
21. What is the basis for the classification of clouds?
22. Why are high clouds always thin?
23. Which cloud types are associated with the following characteristics: thunder, halos, precipitation, hail, mackerel sky, lightning, mares' tails?
24. List five types of fog and discuss the details of their formation.
25. What is the difference between precipitation and condensation?
26. List the forms of precipitation and the circumstances of their formation.
27. Sometimes, when rainfall is light, the amount is reported as a trace. When this occurs, how much (or little) rain has fallen?
28. Using Box 15.3 as a guide, calculate the precipitation probability if there is a 70 percent chance that a storm will move into the forecast area and that it will affect 60 percent of the area.

Box 15.3

Precipitation Probability Forecasts

Many of us are satisfied with a precipitation forecast that simply tells us whether rain is likely or not (Figure 15.D). In traditional precipitation forecasts like these, the inherent uncertainty is expressed using qualifiers such as "chance" and "likely." However, such vague terminology is subject to a wide range of misinterpretations by everyone. Moreover, compared with the use of probabilities to express level of uncertainty, traditional forecast terminology is imprecise and often ambiguous. To overcome these shortcomings, the National Weather Service instituted *precipitation probability forecasts* in 1965.

Probability refers to the chance that an event will occur and is represented as a number between 0 (the probability of an impossible one) and 1 (the probability of an inevitable one). Probability can also be expressed as a percentage, so that a 0.3 chance of an event occurring is expressed as 30 percent. This expresses the idea of probability in terms that people quickly understand.

In forecasting, probability is the percentage chance that at least 0.01 inch (0.025 cm) of precipitation will fall at *any point* in the area during the time period covered by the forecast. Thus, a 70 percent probability indicates a 7 in 10 chance of precipitation, and a 3 in 10 chance of no measurable precipitation at any location in the forecast area. In general, these forecasts cover 12-hour periods and moderate-sized metropolitan areas.

Unfortunately, most people interpret this forecast to mean that there is a 70 percent chance of precipitation somewhere in the forecast area, and a 30 percent chance that it will not occur anywhere in that area. This is not the case.

In actuality, this forecast states that at any point in the forecast area (for example, your home), there is a 70 percent chance of measurable precipitation.

Further, the chance of a shower occurring at a specific site is the product of two quantities: the probability that a precipitation-producing storm will develop or move into the area and the percent of the area that the storm is expected to cover.

For example, a forecaster can have a high degree of confidence that a storm will move into an area (say, 80 percent), but determine that only 40 percent of the area will be affected.

Under these conditions, the forecaster will call for a 30 percent chance of precipitation ($0.80 \times 0.40 = 0.32$, or about 30 percent). Although precipitation is nearly certain, the chance that it will affect you, wherever you are in the forecast area, is only 3 in 10. Consequently, in the summer, when storms are frequently isolated or scattered, the probability that your immediate area will have rain tends to be smaller than during the winter months when storms are more widespread.

Figure 15.D
The forecast must have been accurate; these people were obviously prepared for the rainy weather. (Photo by Charles O'Rear/Westlight.)

Key Terms

absolute instability (p. 398)
absolute stability (p. 398)
adiabatic temperature change (p. 396)
advection fog (p. 408)
Bergeron process (p. 411)
calorie (p. 388)
cirrus (p. 404)
cloud (p. 404)
cloud of vertical development (p. 407)
collision-coalescence process (p. 411)
condensation (p. 389)
condensation nuclei (p. 404)
conditional instability (p. 398)
convergence (p. 402)
cumulus (p. 404)
deposition (p. 389)

dew point (p. 391)
dry adiabatic rate (p. 396)
evaporation (p. 389)
fog (p. 408)
freezing (p. 389)
freezing nuclei (p. 411)
frontal fog (p. 410)
frontal wedging (p. 402)
glaze (p. 413)
hail (p. 413)
high cloud (p. 405)
humidity (p. 389)
hygrometer (p. 394)
hygroscopic nuclei (p. 404)
latent heat (p. 388)
low cloud (p. 405)
melting (p. 389)
middle cloud (p. 405)
orographic lifting (p. 402)

precipitation fog (p. 410)
psychrometer (p. 393)
radiation fog (p. 409)
rain (p. 412)
rainshadow desert (p. 402)
relative humidity (p. 390)
rime (p. 414)
saturation (p. 389)
sleet (p. 413)
snow (p. 413)
specific humidity (p. 390)
steam fog (p. 409)
stratus (p. 404)
sublimation (p. 389)
supercooled (p. 411)
upslope fog (p. 409)
vapor pressure (p. 390)
wet adiabatic rate (p. 396)

CHAPTER 16

Air Pressure and Wind

Wind turbines, California. (Photo by ML Sinibaldi/The Stock Market)

We have already dealt with the two elements of weather and climate that generally are of the greatest interest to people—temperature and moisture. In this chapter we shall investigate the remaining two elements—pressure and wind—that may not seem as important but indeed are. It is the wind that often brings changes in temperature and moisture conditions, and it is pressure differences that drive the wind. Moreover, we will also see that an understanding of atmospheric pressure patterns is a key factor in understanding global and regional precipitation distribution. Among the questions we will try to answer are: How can the weight of the air be measured? What are the factors that control the winds? What is a prevailing wind? Why are "highs" and "lows" always shown on the weather map? How can the weather be predicted by checking the barometer?

Of the various elements of weather and climate, changes in air pressure are the least noticeable. When listening to a weather report, we are generally interested in moisture conditions (humidity and precipitation), temperature, and perhaps wind. It is the rare individual, however, who wonders about air pressure. Although the hour-to-hour and day-to-day variations in air pressure are generally not noticed by people, they are very important in producing changes in our weather. Variations in air pressure from place to place cause the movement of air we call wind and are a significant factor in weather forecasting (Figure 16.1). As we shall see, air pressure is tied very closely to the other elements of weather in a cause-and-effect relationship.

Measuring Air Pressure

In Chapter 14 you saw that air has weight; at sea level, it exerts a pressure of 1 kilogram per square centimeter (14.7 pounds per square inch). The air pressure at a particular place is simply the force exerted by the weight of the air above. With an increase in altitude, there is less air, so the weight of the air, and thus the pressure, decreases. It decreases rapidly at first, then much more slowly (see Figure 14.4, p. 367, and Table 16.1).

When meteorologists measure atmospheric pressure, they employ a unit called the *millibar*. Standard sea level pressure is 1013.2 millibars. Although the millibar has been the unit of measure on all U.S. weather maps since January 1940, the media use "inches of mercury" to describe atmospheric pressure. In the United States, the National Weather Service converts millibar values to inches of mercury for public and aviation use.

"Inches of mercury" is easy to understand. The use of mercury for measuring air pressure dates from 1643, when Torricelli, a student of the famous Italian scientist Galileo, invented the **mercury barometer**. Torricelli correctly described the atmosphere as a vast ocean of air that exerts pressure on us and all objects about us. To measure this force he filled a glass tube, which was closed at one end, with mercury. The tube was then inverted into a dish of mercury (Figure 16.2A). Torricelli found that the mercury flowed out of the tube until the weight of the column was balanced by the pressure that the atmosphere exerted on the surface of the mercury in the dish. In other words, the weight of mercury in the column equaled the weight of the same diameter column of air that extended from the ground to the top of the atmosphere.

Torricelli noted that when air pressure increased, the mercury in the tube rose. Conversely, when air pressure decreased, so did the height of the column of mercury. The length of the column of mercury, therefore, became the measure of the air pressure. With some refinements the mercurial barometer invented by Torricelli is still the standard pressure-measuring instrument used today. Standard atmospheric pressure at sea level equals 29.92 inches of mercury.

Figure 16.1

Wind, the horizontal movement of air, is created by differences in air pressure at Earth's surface. (Photo by Pal Hermansen/Tony Stone Images)

Figure 16.1 Pressure changes with altitude.

Altitude (kilometers)	Altitude (miles)	Pressure (millibars)
0	0	1013
1.0	0.6	899
2.0	1.2	795
3.0	1.9	701
4.0	2.5	617
5.0	3.1	540
10.0	6.2	265
20.0	12.4	55
30.0	18.6	12
40.0	24.8	3

The need for a smaller and more portable instrument for measuring air pressure led to the development of the **aneroid** ("without liquid") **barometer**. Based on a different principle from the mercury barometer, this instrument consists of partially evacuated metal chambers that have a spring inside to keep them from collapsing. The metal chambers, being very sensitive to air pressure variations, change shape, compressing as the pressure increases and expanding as the pressure decreases. Aneroids are often used in making **barographs**, instruments that continuously record pressure changes (Figure 16.2B).

Factors Affecting Wind

In Chapter 15 we examined the upward movement of air and its role in cloud formation. As important as vertical motion is, far more air moves horizontally, the phenomenon we call **wind**. What causes wind? Simply stated, wind is the result of horizontal differences in air pressure. *Air flows from areas of higher pressure to areas of lower pressure.* You may have experienced this when opening a vacuum-packed can of coffee. The noise you hear is caused by air rushing from the higher pressure outside the can to the lower pressure inside. Wind is nature's attempt to balance such inequalities in air pressure. Because unequal heating of Earth's surface generates these pressure differences, *solar energy is the ultimate driving force of wind.*

If Earth did not rotate, and if there were no friction between moving air and Earth's surface, air would flow in a straight line from areas of higher pressure to areas of lower pressure. But because Earth does rotate and friction exists, wind is controlled by the following combination of forces: (1) the pressure gradient force, (2) Coriolis effect, and (3) friction. We will now examine each of these factors.

Pressure Gradient Force

Pressure differences create wind, and the greater these differences, the greater the wind speed. Over Earth's surface, variations in air pressure are determined from

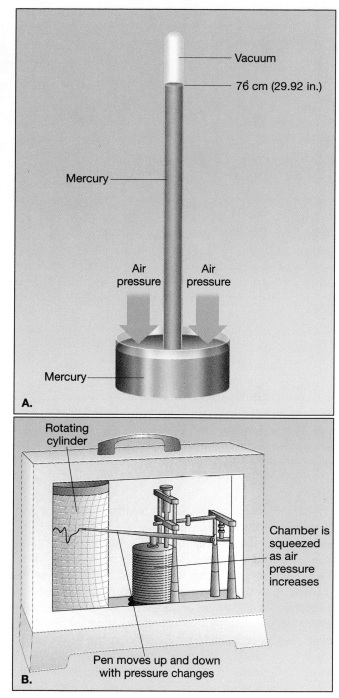

Figure 16.2

A. Simple mercury barometer. The weight of the column of mercury is balanced by the pressure exerted on the dish of mercury by the air above. If the pressure decreases, the column of mercury falls; if the pressure increases, the column rises. **B.** An aneroid barograph makes a continuous record of pressure changes. One important advantage of the aneroid barometer is that it is easily adapted to a recording mechanism.

barometric readings taken at hundreds of weather stations. These pressure data are shown on a weather map using **isobars**, lines that connect places of equal air pressure (Figure 16.3). The spacing of isobars indicates the amount of pressure change occurring over a given distance and is expressed as the **pressure gradient**.

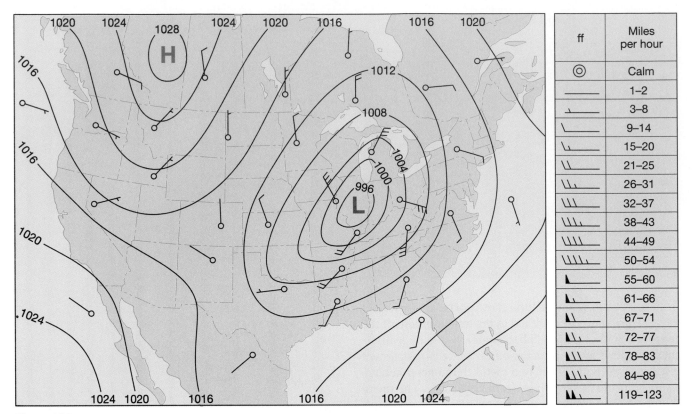

ff	Miles per hour
◎	Calm
——	1–2
⌐	3–8
⌐	9–14
⌐⌐	15–20
⌐⌐	21–25
⌐⌐⌐	26–31
⌐⌐⌐	32–37
⌐⌐⌐⌐	38–43
⌐⌐⌐⌐	44–49
⌐⌐⌐⌐⌐	50–54
◣	55–60
◣	61–66
◣⌐	67–71
◣⌐⌐	72–77
◣⌐⌐	78–83
◣⌐⌐⌐	84–89
◣◣⌐	119–123

Figure 16.3

The black lines are isobars which connect places of equal barometric pressure. They show the distribution of pressure on weather maps. The lines usually curve and often join around cells of high and low pressure. The flags indicate the expected airflow surrounding cells of high and low pressure and are plotted like flags flying with the wind. Wind speed is indicated by flags and "feathers" as shown along the right-hand side of this drawing.

You might find it easier to visualize a pressure gradient if you think of it like the slope of a hill. A steep pressure gradient, like a steep hill, causes greater acceleration of an air parcel than does a weak pressure gradient. Thus, the relationship between wind speed and the pressure gradient is straightforward: *Closely spaced isobars indicate a steep pressure gradient and high winds, whereas widely spaced isobars indicate a weak pressure gradient and light winds.* Figure 16.3 illustrates the relationship between the spacing of isobars and wind speed. Notice that wind speeds are greater in Ohio, Kentucky, Michigan, and Illinois, where isobars are more closely spaced, than in the western states where isobars are more widely spaced.

The pressure gradient is the driving force of wind, and it has both magnitude and direction. Its magnitude is determined from the spacing of isobars. The direction of force is always from areas of higher pressure to areas of lower pressure and at right angles to the isobars. Once the air starts to move, the Coriolis effect and friction come into play, but then only to modify the movement, not to produce it.

Coriolis Effect

Figure 16.3 shows the typical air movements associated with high- and low-pressure systems. As expected, the air moves out of the regions of higher pressure and into the regions of lower pressure. However, the wind does not cross the isobars at right angles as the pressure gradient force directs it to do. The direction deviates as a result of Earth's rotation. This has been named the **Coriolis effect** after the French scientist who first thoroughly described it. All free-moving objects or fluids, including the wind, are deflected to the *right* of their path of motion in the Northern Hemisphere and to the *left* in the Southern Hemisphere.

The reason for this deflection can be illustrated by imagining the path of a rocket launched from the North Pole toward a target located on the equator (Figure 16.4). If the rocket took an hour to reach its target, during its flight Earth would have rotated 15 degrees to the east. To someone standing on Earth it would look as if the rocket veered off its path and hit Earth 15 degrees west of its target. The true path of the rocket is straight and would appear so to someone out in space looking at Earth. It was Earth turning under the rocket that gave it its *apparent* deflection.

Note that the rocket was deflected to the right of its path of motion because of the counterclockwise rotation of the Northern Hemisphere. In the Southern Hemisphere, the effect is reversed. Clockwise rotation produces a similar deflection, but to the *left* of the path of motion. The same deflection is experienced by wind regardless of the direction it is moving.

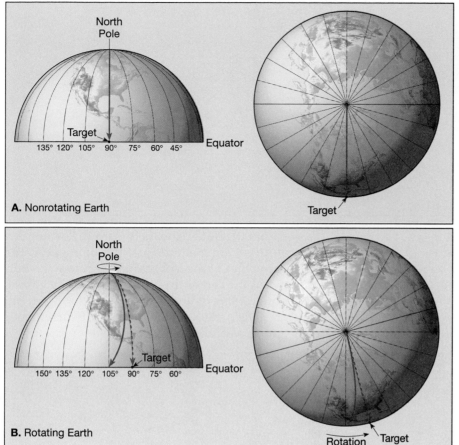

Figure 16.4
The Coriolis effect illustrated using a 1-hour flight of a rocket traveling from the North Pole to a location on the equator. **A.** On a nonrotating Earth, the rocket would travel straight to its target. **B.** However, Earth rotates 15° each hour. Thus, although the rocket travels in a straight line, when we plot the path of the rocket on Earth's surface, it follows a curved path that veers to the right of the target.

We attribute the apparent shift in wind direction to the Coriolis effect. This deflection: (1) is always directed at right angles to the direction of airflow; (2) affects only wind direction, not wind speed; (3) is affected by wind speed (the stronger the wind, the greater the deflection); and (4) is strongest at the poles and weakens equatorward, becoming nonexistent at the equator.

Friction with Earth's Surface

The frictional effect on wind is important only within the first few kilometers from Earth's surface. Friction acts to slow air movement and, as a consequence, alters wind direction. To illustrate friction's effect on wind direction, let us look at a situation in which it has no role. Above the friction layer, the pressure gradient force and Coriolis effect work together to direct the flow of air. Under these conditions, the pressure gradient force causes air to start moving across the isobars. As soon as the air starts to move, the Coriolis effect acts at right angles to this motion. The faster the wind speed, the greater the deflection.

Eventually, the Coriolis effect will balance the pressure gradient force and the wind will blow parallel to the isobars (Figure 16.5). Upper-air winds generally take this path and are called **geostrophic winds**. Because of the lack of friction with Earth's surface, geostrophic winds travel at higher speeds than do surface winds. This can be observed in Figure 16.6 by noting the wind flags, many of which indicate winds of 50 to 100 miles per hour.

The most prominent features of upper-level flow are the **jet streams**. First encountered by high-flying bombers during World War II, these fast-moving "rivers" of air travel between 120 and 240 kilometers (75 and 150 miles) per hour in a west-to-east direction. One such stream is situated over the polar front, which is the zone separating cool polar air from warm subtropical air.

Below 600 meters (2000 feet), friction complicates the airflow just described. Recall that the Coriolis effect is proportional to wind speed. Friction lowers the wind speed, so it reduces the Coriolis effect. Because the pressure gradient force is not affected by wind speed, it wins the tug of war shown in Figure 16.7. The result is a movement of air at an angle across the isobars toward the area of lower pressure.

The roughness of the terrain determines the angle of airflow across the isobars. Over the smooth ocean surface, friction is low and the angle is small. Over rugged terrain, where friction is higher, the angle that air makes as it flows across the isobars can be as great as 45 degrees.

In summary, upper airflow is nearly parallel to the isobars, while the effect of friction causes the surface winds to move more slowly and cross the isobars at an angle.

Figure 16.5

The geostrophic wind. Upper-level winds are deflected by the Coriolis effect until the Coriolis effect just balances the pressure gradient force. Above 600 meters (1970 feet), where friction is negligible, these winds will flow nearly parallel to the isobars and are called geostrophic winds.

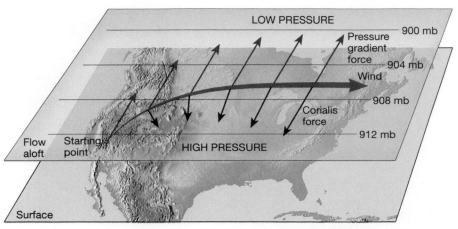

Cyclones and Anticyclones

Among the most common features on any weather map are areas designated as pressure centers (see Box 16.1). **Cyclones**, or lows, are centers of low pressure, and **anticyclones**, or highs, are high-pressure centers. As Figure 16.8 illustrates, the pressure decreases from the outer isobars toward the center in a cyclone. In an anticyclone, just the opposite is the case—the values of the isobars increase from the outside toward the center. By knowing just a few basic facts about centers of high and low pressure, you can greatly increase your understanding of current and forthcoming weather.

ff	Miles per hour
◎	Calm
——	1–2
⌐	3–8
⌐	9–14
⌐	15–20
⌐	21–25
⌐	26–31
⌐	32–37
⌐	38–43
⌐	44–49
⌐	50–54
◣	55–60
◣	61–66
◣	67–71
◣	72–77
◣	78–83
◣	84–89
◣◣	119–123

A. Upper-level weather chart

B. Representation of upper-level chart

Figure 16.6

Upper-air winds. This map shows the direction and speed for the upper-air wind for a particular day. Note that the airflow is nearly parallel to the contours. These isolines are height contours for the 500-millibar level.

Box 16.1

Scales of Atmospheric Motion

The time and space scales commonly used to describe atmospheric motions are provided in Table 16.A. The largest scale wind pattern is called a *macroscale wind system* and is exemplified by long waves in the westerlies. The planetary-scale flow patterns extend around the entire globe and often remain essentially unchanged for weeks at a time. A somewhat smaller type of macroscale circulation is commonly called the *synoptic scale*, or *weather map scale*, and consists mainly of individual traveling cyclones and anticyclones (Figure 16.A). The airflow in these less-than-global-sized macroscale circulations consists primarily of horizontal flow with only modest amounts of vertical motion. In contrast, *mesoscale* winds influence smaller areas and exhibit extensive vertical flow, which can be rapid, as in a developing thunderstorm or tornado (Figure 16.B). The smallest scale of air motion is referred to as *microscale*. These small, often chaotic winds normally last seconds, or at most are measured in minutes. Examples include simple gusts, which hurl debris into the air, and well-developed vortices such as dust devils (Figure 16.C).

Although it is common practice to divide atmospheric motions according to scale, it is important to remember that this flow is complex—much like a

turbulent river with smaller eddies within large eddies within still larger eddies. Also, like the flow of a turbulent river, each scale of motion is related to the other. Let us examine the flow associated with the hurricanes that form over the North Atlantic, for example. When we view one of these tropical cyclones on satellite images, the storm appears as a large whirling cloud migrating slowly across the ocean (see Figure 17.22, p. 460). From this perspective (synoptic scale), the general counterclockwise rotation can

be seen easily and the storm's path, which persists for a week or so, can be

(continues)

Figure 16.A

Traveling cyclonic storms, such as this hurricane, are examples of macroscale circulation. (Photo courtesy of the National Hurricane Center)

Figure 16.B

This tornado is an example of mesoscale circulation. (Photo by A. and J. Verkaik/The Stock Market)

Table 16.A Time and space scales for atmospheric motion.

Name of Scale	Time Scale	Size Scale	Examples
Macroscale			
Planetary Scale	Weeks to years	1000 to 40,000 km	Waves in the Westerlies
Synoptic Scale	Days to weeks	100 to 500 km	Cyclones, anticyclones, and hurricanes
Mesoscale	Minutes to days	$1/2$ to 100 km	Land-sea breezes, thunderstorms, and tornadoes
Microscale	Seconds to minutes	$<1/2$ km	Turbulence, dust devils, and gusts

followed easily. When we average the winds of a hurricane, we find that they have net motion from east to west, thereby indicating that these large eddies are embedded in a still larger flow (the general circulation) that is moving westward across the tropical portion of the North Atlantic. When we examine a hurricane more closely by flying an airplane through it, some of the small-scale aspects of the storm become noticeable. As the plane approaches the outer edge of the system, it becomes evident that the large rotating cloud that we saw in the satellite images is made of numerous towering cumulonimbus clouds. Each of these mesoscale phenomena (thunderstorms) lasts for only a few hours and must be continually replaced by new ones if the hurricane is to persist. When we fly into these storms, we quickly realize that the individual clouds are also made up of even smaller-scale turbulences. The small thermals of rising air that compose these clouds can make for a rough trip. Thus a typical hurricane exhibits several different scales of motion, including many mesoscale thunderstorms, which, in turn, consist of numerous microscale turbulences, whereas the counterclockwise circulation of the hurricane is itself part of the larger general circulation in the tropics.

Figure 16.C
Gusts are examples of microscale winds. (Photo by E. J. Tarbuck)

Cyclonic and Anticyclonic Winds

From the preceding section, you learned that the two most significant factors that affect wind are the pressure gradient force and the Coriolis effect. Winds move from higher pressure to lower pressure and are deflected to the right or left by Earth's rotation. When these controls of airflow are applied to pressure centers in the Northern Hemisphere, the result is that winds blow inward and counterclockwise around a low. Around a high, they blow outward and clockwise (Figure 16.8).

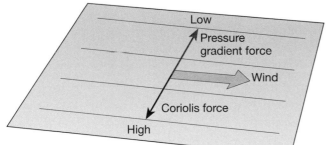

A. Upper-level wind (no friction)

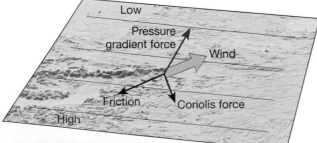

B. Surface wind (effect of friction)

Figure 16.7
Comparison between upper-level winds and surface winds showing the effects of friction on airflow. Friction slows surface wind speed, which weakens the Coriolis effect, causing the winds to cross the isobars and move toward the lower pressure.

In the Southern Hemisphere the Coriolis effect deflects the winds to the left, and therefore winds around a low are blowing clockwise, and winds around a high are moving counterclockwise. In either hemisphere, friction causes a net inflow (**convergence**) around a cyclone and a net outflow (**divergence**) around an anticyclone.

Weather Generalizations About Highs and Lows

Rising air is associated with cloud formation and precipitation, whereas subsidence produces clear skies. In this section we will learn how the movement of air can itself create pressure change and hence generate winds. After doing so, we will examine the relation between horizontal and vertical flow, and their effects on the weather.

Let us first consider the situation around a surface low-pressure system where the air is spiraling inward. Here the net inward transport of air causes a shrinking of the area occupied by the air mass, a process that is termed *horizontal convergence*. Whenever air converges horizontally, it must pile up, that is, increase in height to allow for the decreased area it now occupies. This generates a "taller" and therefore heavier air column. Yet a surface low can exist only as long as the column of air exerts less pressure than that occurring in surrounding regions. We seem to have encountered a paradox—a low-pressure center causes a net accumulation of air, which increases its pressure. Consequently, a surface cyclone should quickly eradicate itself in a manner not unlike what happens when a vacuum-packed can is opened.

It should be apparent that for a surface low to exist for very long, compensation must occur at some layer aloft. For example, surface convergence could be maintained if divergence (spreading out) aloft occurred at a rate equal to the inflow below. Figure 16.9 shows the relationship between surface convergence (inflow) and divergence (outflow) aloft that is needed to maintain a

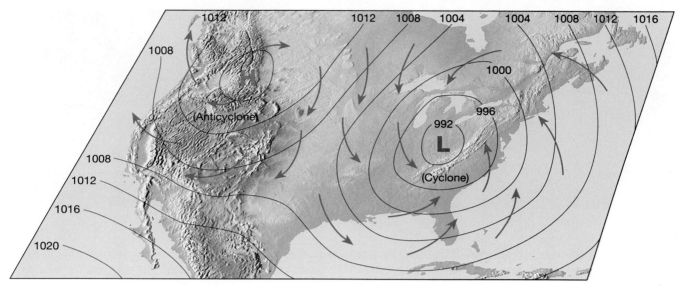

Figure 16.8

Cyclonic and anticyclonic winds in the Northern Hemisphere. Arrows show that winds blow into and counterclockwise around a low. By contrast, around a high, winds blow outward and clockwise.

low-pressure center. You can see that surface convergence around a cyclone causes a net upward movement.

The rate of this vertical movement is quite slow, generally less than 1 kilometer per day. Nevertheless, because rising air cools adiabatically, a low-pressure system often brings cloudiness and precipitation (Figure 16.10A). On occasion, divergence aloft may even exceed surface convergence, resulting in intensified surface inflow (faster winds) and increased vertical motion. For this reason, divergence aloft can intensify these storm centers as well as maintain them.

Diverging surface winds are associated with high-pressure centers. The outflow near the surface is accompanied by convergence aloft and general subsidence of the air column (Figure 16.9). Because descending air is compressed and warmed, cloud formation and precipitation are unlikely in an anticyclone, and "fair" weather can usually be expected with the approach of a high (Figure 16.10B).

For reasons that should now be obvious, it has been common practice to print on household barometers the words "stormy" at the low-pressure end and "fair" on

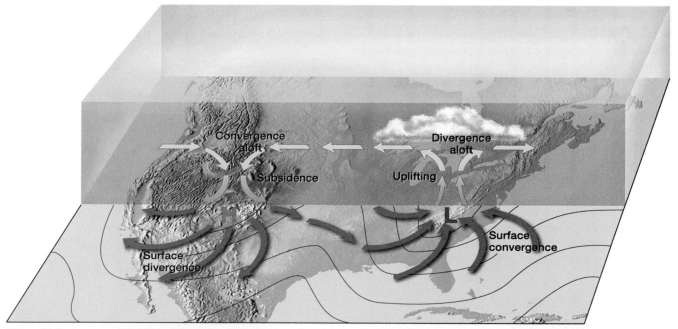

Figure 16.9

Airflow associated with surface cyclones and anticyclones. A low, or cyclone, has converging surface winds and rising air causing cloudy conditions. A high, or anticyclone, has diverging surface winds and descending air, which leads to clear skies and fair weather.

A.

B.

Figure 16.10
These two photographs illustrate the basic weather generalizations associated with pressure centers. **A.** Centers of low pressure are frequently associated with cloudy conditions and precipitation. (Photo by Cindy Charles/Liaison International) **B.** By contrast, clear skies and "fair" weather may be expected when an area is under the influence of high pressure. (Photo by Brenda Tharp/Photo Researchers, Inc.)

the high-pressure end. By noting whether the pressure is rising, falling, or steady, we have a good indication of what the forthcoming weather will be. Such a determination, called the **pressure**, or **barometric tendency**, is a very useful aid in short-range weather prediction.

The generalizations relating cyclones and anticyclones to weather conditions are the basis for several weather proverbs, including the one that follows (*glass* refers to the barometer):

When the glass falls low, prepare for a blow;
When it rises high, let all your kites fly.

You should now be better able to understand why television weather reporters emphasize the positions and projected paths of cyclones and anticyclones. The "villain" on these weather programs is always the low-pressure center, which produces "bad" weather in any season. Lows move in roughly a west-to-east direction across the United States and require a few days to more than a week for the journey. Their paths can be somewhat erratic; thus, accurate prediction of their migration is difficult, although essential, for short-range forecasting. Meteorologists must also determine if the flow aloft will intensify an embryo storm or act to suppress its development. Because of the close tie between conditions at the surface and those aloft, a great deal of emphasis has been placed on the importance and understanding of the total atmospheric circulation, particularly in the mid-latitudes. We will now examine the workings of Earth's general atmospheric circulation, and then again consider the structure of the cyclone in light of this knowledge.

General Circulation of the Atmosphere

As noted, the underlying cause of wind is unequal heating of Earth's surface. In tropical regions, more solar radiation is received than is radiated back to space. In polar regions the opposite is true—less solar energy is received than is lost. Attempting to balance these differences, the atmosphere acts as a giant heat transfer system, moving warm air poleward and cool air equatorward. On a smaller scale, but for the same reason, ocean currents also contribute to this global heat transfer. The general circulation is very complex, and there is a great deal that has yet to be explained. We can, however, develop a general understanding by first considering the circulation that would occur on a nonrotating Earth having a uniform surface. We will then modify this system to fit observed patterns.

Circulation on a Nonrotating Earth

On a hypothetical nonrotating planet with a smooth surface of either all land or all water, two large thermally produced cells would form (Figure 16.11). The heated equatorial air would rise until it reached the

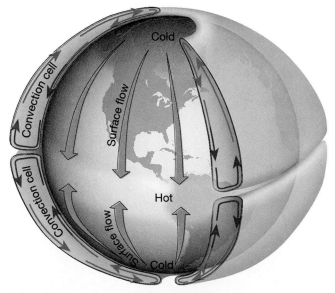

Figure 16.11
Global circulation on a nonrotating Earth. A simple convection system is produced by unequal heating of the atmosphere on a nonrotating Earth.

tropopause, which, acting like a lid, would deflect the air poleward. Eventually, this upper-level airflow would reach the poles, sink, spread out in all directions at the surface, and move back toward the equator. Once there, it would be reheated and start its journey over again. This hypothetical circulation system has upper-level air flowing poleward and surface air flowing equatorward.

If we add the effect of rotation, this simple convection system will break down into smaller cells. Figure 16.12 illustrates the three pairs of cells proposed to carry on the task of heat redistribution on a rotating planet. The polar and tropical cells retain the characteristics of the thermally generated convection described earlier. The nature of the mid-latitude circulation is more complex and will be discussed in more detail in a later section.

Idealized Global Circulation

Near the equator, the rising air is associated with the pressure zone known as the **equatorial low**—a region marked by abundant precipitation. As the upper-level flow from the equatorial low reaches 20-30 degrees latitude, north or south, it sinks back toward the surface. This subsidence and associated adiabatic heating produce the hot, arid regions in this latitude range. The center of this zone of subsiding dry air is the **subtropical high**, which encircles the globe near 30 degrees latitude, north and south (Figure 16.12). Located here are extensive arid and semiarid regions. The great deserts of Australia, Arabia, and North Africa exist because of the stable dry condition caused by the subtropical highs.

At the surface, airflow is outward from the center of the subtropical high. Some of the air travels equatorward and is deflected by the Coriolis effect, producing

the reliable **trade winds**. The remainder travels poleward and is also deflected, generating the prevailing **westerlies** of the mid-latitudes. As the westerlies move poleward, they encounter the cool **polar easterlies** in the region of the **subpolar low**. The interaction of these warm and cool winds produces the stormy belt known as the **polar front**. The source region for the variable polar easterlies is the **polar high**. Here, cold polar air is subsiding and spreading equatorward.

In summary, this simplified global circulation is dominated by four pressure zones. The subtropical and polar highs are areas of dry subsiding air that flows outward at the surface, producing the prevailing winds. The low-pressure zones of the equatorial and subpolar regions are associated with inward and upward airflow accompanied by clouds and precipitation.

Influence of Continents

Up to this point, we have described the surface pressure and associated winds as continuous belts around Earth. However, the only truly continuous pressure belt is the subpolar low in the Southern Hemisphere. Here the ocean is uninterrupted by landmasses. At other latitudes, particularly in the Northern Hemisphere where landmasses break up the ocean surface, large seasonal temperature differences disrupt the pattern. Figure 16.13 shows the resulting pressure and wind patterns for January and July. The circulation over the oceans is dominated by semipermanent cells of high pressure in the subtropics and cells of low pressure over the subpolar regions. The subtropical highs are responsible for the trade winds and westerlies, as mentioned earlier.

Figure 16.12
Idealized global circulation.

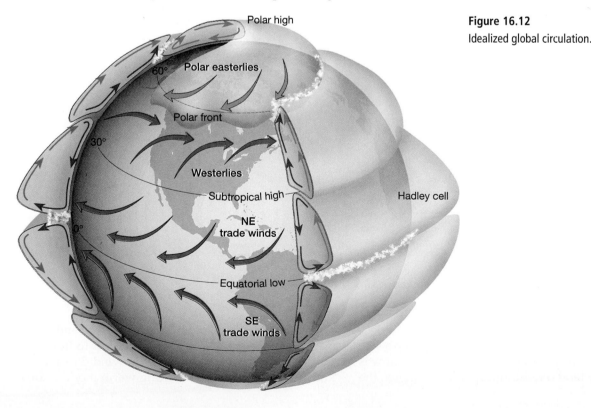

Polar high

Polar easterlies

60°

Polar front

30°

Westerlies

Subtropical high

Hadley cell

NE trade winds

0°

Equatorial low

SE trade winds

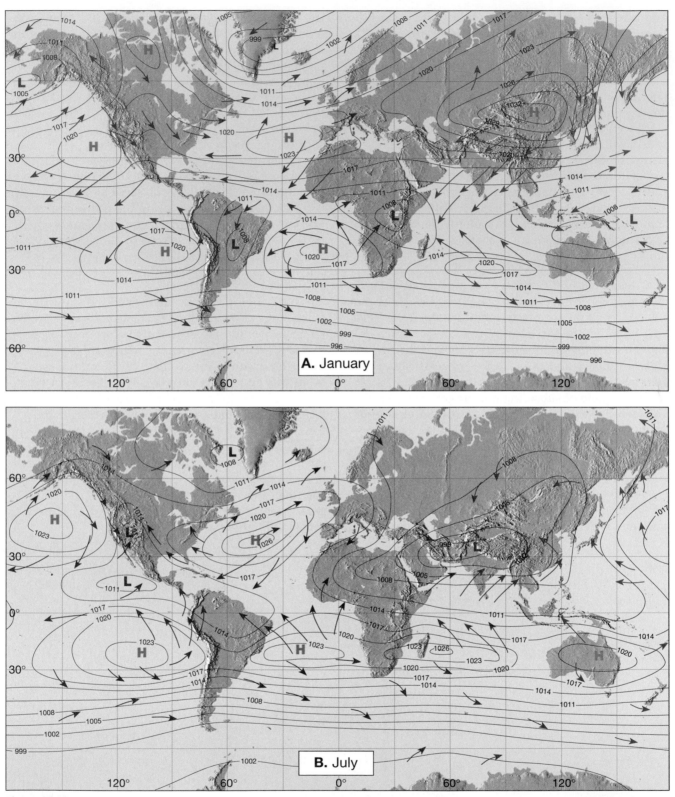

Figure 16.13

Average surface barometric pressure in millibars for **A.** January and **B.** July, with associated winds.

The large landmasses, on the other hand, particularly Asia, become cold in the winter and develop a seasonal high-pressure system from which surface flow is directed off the land (Figure 16.13). In the summer, the opposite occurs; the landmasses are heated and develop a low-pressure cell, which permits air to flow onto the land. These seasonal changes in wind direction are known as the **monsoons**. During warm months, areas

CHAPTER 17
Weather Patterns and Severe Storms

Thunderstorm over northern Arizona. (Photo by Richard Kaylin/Tony Stone Images)

Tornadoes and hurricanes rank among nature's most destructive forces. Each spring, newspapers report the death and destruction left in the wake of a "band" of tornadoes. During late summer and fall we hear occasional reports about hurricanes in the news. Storms with names like Andrew, Camille, Gilbert, and Hugo have made front-page headlines. Thunderstorms, although less intense and far more common than tornadoes and hurricanes, will also be part of our discussion in this chapter on severe weather disturbances. Before looking at violent weather, however, we shall study those atmospheric phenomena that most often affect our day-to-day weather: air masses, fronts, and traveling middle-latitude cyclones. Here we shall see the interplay of the elements of weather discussed in Chapters 14, 15, and 16.

Air Masses

For many people who live in the middle latitudes, including much of the United States, summer heat waves and winter cold spells are familiar experiences. In the first instance, several days of high temperatures and oppressive humidities may finally end when a series of thunderstorms pass through the area, followed by a few days of relatively cool relief. By contrast, the clear skies that often accompany a span of frigid subzero days may be replaced by thick stratus clouds and a period of snow as temperatures rise to levels that seem mild when compared to those that existed just a day earlier. In both examples, what was experienced was a period of generally constant weather conditions followed by a relatively short period of change, and then the subsequent reestablishment of a new set of weather conditions that remained for perhaps several days before changing again.

The weather patterns just described result from movements of large bodies of air, called air masses. An **air mass**, as the term implies, is an immense body of air, typically 1600 kilometers (1000 miles) or more across, and perhaps several kilometers thick, that is characterized by a similarity of temperature and moisture at any given altitude. When this air moves out of its region of origin, it will carry these temperatures and moisture conditions with it, eventually affecting a large portion of a continent.

The horizontal uniformity of an air mass is not perfect, of course. Because air masses extend over large areas, small differences in temperature and humidity from place to place are to be expected. Nevertheless, the differences observed within an air mass are small in comparison to the rapid changes experienced across boundaries between air masses. Because it may take several days for an air mass to traverse an area, the region under its influence will probably experience fairly constant weather, a situation called **air-mass weather**. Certainly there may be some day-to-day variations, but the events will be quite unlike those in adjacent air masses that have different characteristics. For this reason the boundary between two adjoining air masses that have contrasting characteristics, called a *front*, marks a change in weather.

Source Regions

When a portion of the lower atmosphere moves slowly or stagnates over a relatively uniform surface, the air will assume the distinguishing features of that area, particularly with regard to temperature and moisture conditions (Figure 17.1).

A.

B.

Figure 17.1

An air mass forms when air stagnates or moves slowly over a relatively uniform surface. It assumes the distinguishing features of that region, especially temperature and moisture conditions. **A.** Air masses that form over the ocean are humid. (Photo by Wolfgang Kaehler) **B.** Those that form over land are dryer. (Photo by Craig J. Brown/Liaison International)

The area where an air mass acquires its characteristic properties of temperature and moisture is called its **source region**. The source regions that produce air masses which influence North America are shown in Figure 17.2.

Air masses are classified according to their source region. **Polar (P) air masses** originate in high latitudes toward Earth's poles, whereas those that form in low latitudes are called **tropical (T) air masses**. The designation *polar* or *tropical* gives an indication of the temperature characteristics of the air masses. *Polar* indicates cold and *tropical* indicates warm.

In addition, air masses are classified according to the nature of the surface in the source region. **Continental (c) air masses** form over land and **maritime (m) air masses** originate over water. The designation *continental* or *maritime* thus suggests the moisture characteristics of the air mass. Continental air is likely to be dry, and maritime air, humid.

The four basic types of air masses according to this scheme of classification are continental polar (cP), continental tropical (cT), maritime polar (mP), and maritime tropical (mT).

Weather Associated with Air Masses

Continental polar and maritime tropical air masses influence the weather of North America most, especially east of the Rocky Mountains. Continental polar air masses originate in northern Canada, interior Alaska, and the Arctic—areas that are uniformly cold and dry in winter and cool and dry in summer. In winter, an invasion of continental polar air brings the clear skies and cold temperatures we associate with a cold wave as it moves southward through Canada into the United States. In summer, this air mass may bring a few days of cooling relief.

Although cP air masses are not normally associated with heavy precipitation, those that cross the Great Lakes in winter pick up moisture from the lakes and sometimes bring snow to the leeward shores. These localized storms often form when the surface weather map indicated no apparent cause for a snowstorm to occur. These are known as **lake-effect snows** and they make Buffalo and Rochester, New York, among the snowiest cities in the United States (see Box 17.1).

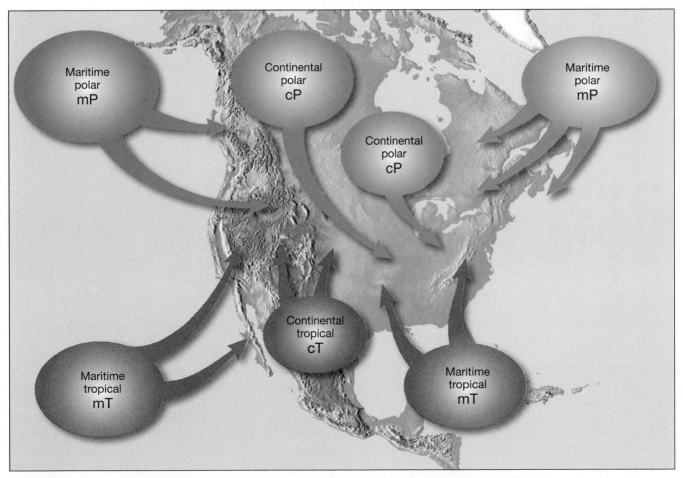

Figure 17.2

Air masses are classified on the basis of their source region. The designation continental (c) or maritime (m) gives an indication of moisture content, whereas polar (P) and tropical (T) indicate temperature conditions.

Box 17.1

Lake-Effect Snow

Every winter a unique and interesting weather phenomenon takes place along the downwind shores of the Great Lakes. Periodically brief, heavy snow showers issue from dark clouds that move onshore from the lakes. Seldom do to storms move more than about 40 kilometers inland from the shore before the snows come to an end. These highly localized storms occurring along the leeward shores of the Great Lakes create what are known as lake-effect snows.

Lake-effect storms account for a high percentage of the snowfall in many areas adjacent to the lakes. The strips of land that are most frequently affected, called *snowbelts*, are shown in Figure 17A. When snowfalls statistics for large cities were examined for a recent 10-year span, the two snowiest metropolitan areas in the United States turned out to be Buffalo and Rochester in upstate New York. Buffalo, on the eastern shore of Lake Erie, had more than 2700 centimeters (nearly 90 feet) during the period, while Rochester, situated on the south side of Lake Ontario, recorded slightly less (2635 centimeters). A comparison of average snowfalls totals at Thunder Bay in Ontario, Canada, on the north shore of Lake Superior, and Marquette, Michigan, along the southern shore, provides another excellent example of lake-effect snow. Amounts are in centimeters.

	Oct.	Nov.	Dec.	Jan.
Thunder Bay, Ontario	3.0	14.9	19.0	22.6
Marquette Michigan	5.3	37.6	56.4	53.1

Because Marquette is situated on the leeward shore of the lake, it receives substantial lake-effect snow and therefore has a much higher snowfall total than does Thunder Bay.

What causes lake-effect snow? The answer is closely linked to the differential heating of water and land (Chapter 14) and to the concept of atmospheric instability (Chapter 15). During the summer months water bodies, including the Great Lakes, absorb huge quantities of energy from the sun and from the warm air that passes over them. Although these water bodies do not reach particularly high temperatures, they nevertheless represent huge reservoirs of heat. The surrounding land, on the other hand, cannot store heat nearly as effectively. Consequently, during autumn and winter, the temperature of

the land drops quickly while water bodies lose their heat more gradually and cool slowly. From late November through late January, the contrasts in average temperatures between water and land range from about 8°C in the southern Great Lakes to 17°C farther north. However, the temperature differences can be much greater (perhaps 25°C) when a very cold cP air mass pushes southward across the lakes. When such a dramatic temperature contrast exists, the lakes interact with the air to produce major lake-effect storms. As a cP air mass moves across one of the Great Lakes, the air acquires large quantities of heat and moisture from the relatively warm lake surface. By the time it reaches the opposite shore, the air mass is humid and unstable and heavy snow showers are likely.

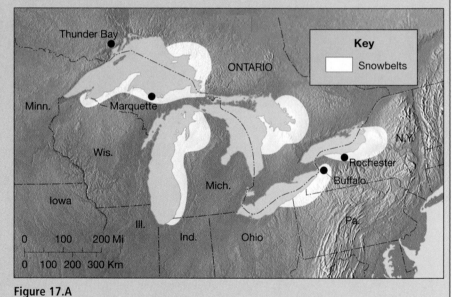

Figure 17.A

The snowbelts of the Great Lakes region are the zones that most frequently experience lake-effect snowstorms. As continental polar air crosses the lakes in winter, it acquires moisture and is made unstable because of warming from below. Snow showers on the lee side of the lakes are often the consequence of this air-mass modification.

Maritime tropical air masses affecting North America most often originate in the Gulf of Mexico, the Caribbean Sea, or the adjacent Atlantic Ocean. As we might expect, these air masses are warm, moisture-laden, and usually unstable. Maritime tropical air is the source of much, if not most, of the precipitation received in the eastern two-thirds of the United States. In summer, when an mT air mass invades the central and eastern United States, and occasionally southern Canada, it brings the high temperatures and oppressive humidity typically associated with its source region.

Of the two remaining air masses, maritime polar and continental tropical, the latter has the least influence on the weather of North America. Hot, dry continental tropical air, originating in the Southwest and Mexico during the summer, seldom affects the weather outside its source region.

During the winter, maritime polar air masses coming from the North Pacific often originate as continental polar air masses in Siberia. As they move across the Pacific, they warm and gradually accumulate moisture. Upon entering North America, these air masses drop much of their moisture because of orographic lifting in the western mountains. Maritime polar air also originates off the coast of eastern Canada and occasionally influences the weather of the northeastern United States. When New England is on the northern or northwestern edge of a passing low, the counterclockwise cyclonic winds draw in maritime polar air. The result is a storm characterized by snow and cold temperatures, known locally as a *nor'easter*.

Fronts

Fronts are boundaries that separate different air masses, one warmer than the other and often higher in moisture content. Fronts can form between any two contrasting air masses. Considering the vast size of the air masses involved, fronts are relatively narrow, being 15- to 200-kilometer-wide bands of discontinuity. On the scale of a weather map, they are generally narrow enough to be represented by a broad line (as in Figure 17.8).

Above the ground, the frontal surface slopes at a low angle so that warmer air overlies cooler air (Figure 17.3). In the ideal case, the air masses on both sides of the front move in the same direction and at the same speed. Under this condition, the front acts simply as a barrier, moving along between the two air masses, and through which they cannot penetrate. Generally, however, the air mass on one side of a front is moving faster in the direction perpendicular to the front than the air mass on the other side. Thus, one air mass actively advances into another and "clashes"

with it. In fact, the boundaries were tagged *fronts* during World War I by Norwegian meteorologists, who visualized them as analogous to battle lines between two armies.

As one air mass moves into another, some mixing does occur along the frontal surface, but for the most part the air masses retain their distinct identities as one is displaced upward over the other. No matter which air mass is advancing, *it is always the warmer, less dense air that is forced aloft*, while *the cooler, denser air acts as the wedge upon which lifting takes place*.

Warm Fronts

When the surface position of a front moves so that warm air occupies territory formerly covered by cooler air, it is called a **warm front** (Figure 17.3). On a weather map, the surface position of a warm front is shown by a line with semicircles extending into the cooler air (Figure 17.3). East of the Rockies, warm tropical air often enters the United States from the Gulf of Mexico and overruns receding cool air. As the cold wedge retreats, friction slows the advance of the surface position of the front more so than its position aloft; for this reason, the boundary separating these air masses acquires a small slope. The average slope of a warm front is about 1:200, which means that if you are 200 kilometers ahead of the surface location of a warm front, you will find the frontal surface at a height of 1 kilometer.

As warm air ascends the retreating wedge of cold air, it cools adiabatically to produce clouds and frequently, precipitation. The sequence of clouds shown in Figure 17.3 typically precedes a warm front. The first

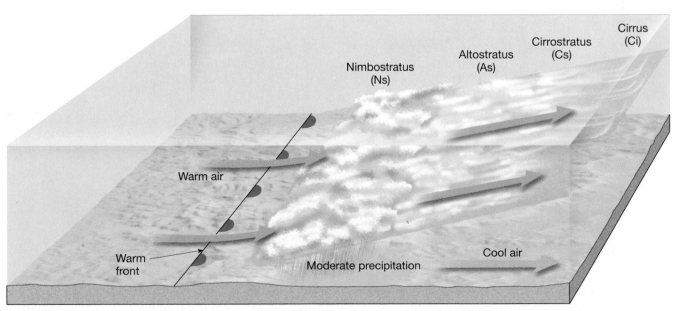

Figure 17.3

Warm front produced as warm air glides up over a cold air mass. Precipitation is moderate and occurs within a few hundred kilometers of the surface front.

sign of the approach of a warm front is the appearance of cirrus clouds overhead (Figure 17.4). These high clouds form 1000 kilometers or more ahead of the surface front where the overrunning warm air has ascended high up the wedge of cold air.

As the front nears, cirrus clouds grade into cirrostratus, which blend into denser sheets of altostratus. About 300 kilometers ahead of the front, thicker stratus and nimbostratus clouds appear and rain or snow begins. Because of their slow rate of advance and very low slope, warm fronts usually produce light-to-moderate precipitation over a large area for an extended period. Warm fronts, however, are occasionally associated with cumulonimbus clouds and thunderstorms. This occurs when the overrunning air is unstable and the temperatures on opposite sides of the front contrast sharply. At the other extreme, a warm front associated with a dry air mass could pass unnoticed at the surface.

A gradual increase in temperature occurs with the passage of a warm front. The increase is most noticeable when there is a large temperature difference between the adjacent air masses. The moisture content and stability of the encroaching warm air mass largely determine when clear skies will return. During the summer, cumulus, and occasionally cumulonimbus, clouds are embedded in the warm unstable air mass that follows the front. Precipitation from these clouds is usually sporadic and not extensive.

Cold Fronts

When cold air is actively advancing into a region occupied by warmer air, the boundary is called a **cold front** (Figure 17.5). As with warm fronts, friction

Figure 17.4

Cirrus clouds are high, thin clouds that are often the first sign that a warm front is approaching. As warm air is wedged aloft, clouds form at increasingly greater heights. (Photo by E. J. Tarbuck)

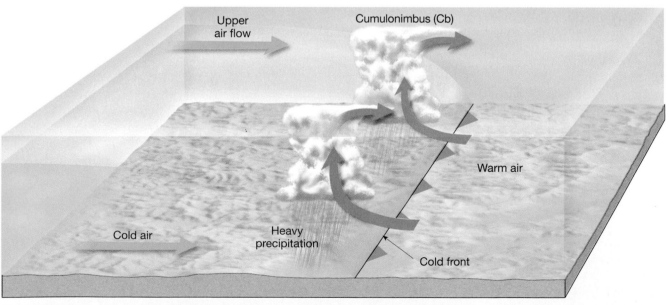

Figure 17.5

Fast-moving cold front and cumulonimbus clouds. Often thunderstorms occur if the warm air is unstable.

tends to slow the surface position of a cold front more so than its position aloft. However, because of the relative positions of the adjacent air masses, the cold front steepens as it moves. On the average, cold fronts are about twice as steep as warm fronts, having a slope of perhaps 1:100. In addition, cold fronts advance more rapidly than warm fronts. These two differences—rate of movement and steepness of slope—largely account for the more violent nature of cold-front weather.

The forceful lifting of air along a cold front is often so rapid that the latent heat released when water vapor condenses will increase the air's buoyancy appreciably. The heavy downpours and vigorous wind gusts associated with mature cumulonimbus clouds frequently result. Because a cold front produces roughly the same amount of lifting as a warm front, but over a shorter distance, the intensity of precipitation is greater, but the duration is shorter.

As a cold front approaches, generally from the west or northwest, towering clouds often can be seen in the distance. Near the front, a dark band of ominous clouds foretells the ensuing weather. Usually a marked temperature drop and a wind shift from the south to west or northwest accompany the passage of the front. The sometimes violent weather and sharp temperature contrast along the cold front are symbolized on a weather map by a line with triangle-shaped points that extend into the warmer air mass (Figure 17.5).

The weather behind a cold front is dominated by a subsiding and relatively cold air mass. Thus, clearing begins relatively soon after the front passes. Although the compression of air by subsidence causes some adiabatic heating, the effect on surface temperatures is minor. In winter, the clear skies that follow the passing of a cold front further reduce surface temperatures because of the high rate of radiation cooling that occurs during the night. If the continental polar air mass, which most frequently accompanies a cold front, moves into a relatively warm and humid area, surface heating can produce shallow convection. This, in turn, may generate low cumulus or stratocumulus clouds behind the front.

Stationary Fronts and Occluded Fronts

Occasionally the flow on both sides of a front is almost parallel to the position of the front. The surface position of the front does not move and it is therefore named a **stationary front**. On a weather map, stationary fronts are shown with triangular points on one side of the front and semicircles on the other. At times some overrunning occurs along a stationary front. This can result in an extended period of relatively widespread cloudiness, light rain, or light snow.

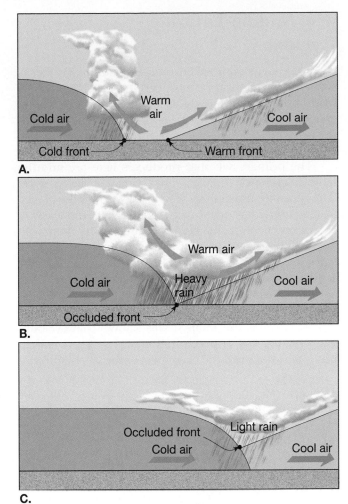

Figure 17.6
Stages in the formation of an occluded front.

Another commonly occurring front is the **occluded front**. Here an active cold front overtakes a warm front (Figure 17.6). As the advancing cold air wedges the warm front upward, a new front emerges between the advancing cold air and the air over which the warm front is gliding. The weather of an occluded front often is complex. Most of the precipitation is associated with the warm air being forced aloft. However, when conditions are suitable, the newly formed front can produce precipitation of its own.

A word of caution is in order concerning the weather associated with various fronts. Although the preceding discussion will help you recognize the weather patterns associated with fronts, remember that these descriptions are generalizations. The weather generated along any individual front may or may not conform fully to this idealized picture. Fronts, like all aspects of nature, never lend themselves to classification as nicely as we would like.

The Middle-Latitude Cyclone

So far we have examined the basic elements of weather as well as the dynamics of atmospheric motions. We are now ready to apply our knowledge of these diverse phenomena to an understanding of day-to-day weather patterns in the middle latitudes. For our purposes, *middle latitudes* refers to the region between southern Florida and Alaska.

The primary weather producers here are **middle-latitude cyclones**. They are large centers of low pressure that generally travel from west to east across the planet. Lasting from a few days to more than a week, these weather systems have a counterclockwise circulation pattern with an air flow inward toward their centers. Most middle-latitude cyclones also have a cold front extending from the central area of low pressure, and frequently a warm front as well. Convergence and forceful lifting initiate cloud development and frequently cause abundant precipitation.

As early as the 1800s it was known that cyclones were the bearers of precipitation and severe weather. But it was not until the early part of the twentieth century that a model was developed to explain how cyclones form. It was formulated by a group of Norwegian scientists and published in 1918. The model was created primarily from near-surface observations.

Years later, as data from the middle and upper troposphere and from satellite images became available, some modifications were necessary. Yet this model is still an accepted working tool in interpreting the weather. It provides a visual picture of the dynamic atmosphere as it generates a storm. If you keep this model in mind when you observe changes in the weather, the changes will no longer come as a surprise. You should begin to see some order in what had appeared to be disorder, and you might even occasionally "predict" the impending weather.

Life Cycle

Middle-latitude cyclones form along fronts where they change in a somewhat predictable way. Figure 17.7 is a schematic representation of the stages in the development of a "typical" middle-latitude cyclone. As the figure shows, cyclones originate along a front where air

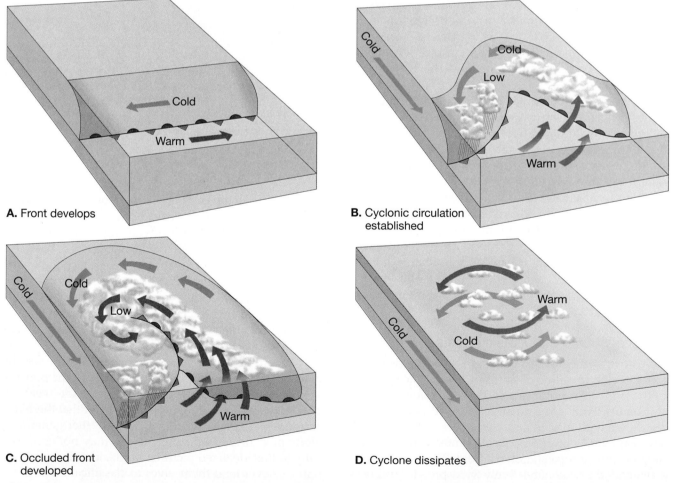

Figure 17.7

The life cycle of a hypothetical middle-latitude cyclone. **A.** Front develops. **B.** Cyclonic circulation is developed. **C.** Occluded front is fully developed. **D.** Cyclone dissipates.

masses of different densities (temperatures) are moving parallel to the front in opposite directions. In the classic model, continental polar air associated with the polar easterlies is north of the front, and maritime tropical air of the westerlies is south of the front. The result of this opposing airflow is counterclockwise (cyclonic) rotation. (To better visualize this effect, place a pencil between the palms of your hands. Now move your right hand ahead of your left hand and notice that your pencil rotates in a counterclockwise fashion.) Under the correct conditions, the frontal surface will take on a wave shape. The waves are usually several hundred kilometers long. Some dampen out, whereas others become unstable and grow in amplitude. The latter ones change in shape with time much as a gentle ocean swell does as it moves into shallow water and becomes a tall breaking wave.

Once a small wave forms, warm air moves poleward, while the surrounding cold air moves equatorward. This change causes a readjustment in pressure that results in nearly circular isobars with the low pressure centered at the apex of the wave. The creation of the low encourages the inflow (convergence) of air and general vertical lifting, particularly where warm air is overrunning colder air.

You can see in Figure 17.8 that the air in the warm sector is flowing from the southwest toward colder air flowing from the southeast. Since the warm air is moving faster than the cold air in a direction perpendicular to this front, we can conclude that warm air is invading a region formerly occupied by cold air; hence, this must be a warm front. Similar reasoning indicates that in the rear of the cyclonic disturbance, cold air is underrunning the air of the warm section, generating a cold front there.

Generally, the position of the cold front advances faster than the warm front and begins to close the warm sector as shown in Figure 17.7. This process, called *occlusion*, results in an occluded front with the displaced warm sector located aloft. The cyclone enters maturity (maximum intensity) when it reaches this stage in its development. A steep pressure gradient and strong winds develop as lifting continues. Eventually all of the warm sector is forced aloft and cold air surrounds the cyclone at low levels (Figure 17.7D). Once the sloping discontinuity (front) between the air masses no longer exists, the pressure gradient weakens. At this point, the cyclone has exhausted its source of energy, and the storm comes to an end.

Idealized Weather

The middle-latitude cyclone model provides a useful tool for examining the weather patterns of the middle latitudes. Figure 17.8 illustrates the distribution of clouds and thus the regions of possible precipitation associated with a mature system. Compare this drawing to the satellite image of a cyclone shown in Figure 17.9.

Guided by the westerlies aloft, cyclones generally move eastward across the United States, so we can expect the first signs of their arrival in the west. However, often in the region of the Mississippi valley, cyclones begin a more northeasterly path and occasionally move directly northward. A mid-latitude cyclone typically requires two to four days to pass over a given region. During that short time, rather abrupt changes in atmospheric conditions may be experienced. This is particularly true in the spring, when the largest temperature contrasts occur across the mid-latitudes.

Using Figure 17.8 as a guide, we will now consider these weather producers and what we should expect from them as they pass an area during the spring. To facilitate our discussion, Figure 17.8 includes two profiles along lines *A-E* and *F-G*.

- First, imagine the change in weather as you move along profile *A-E*. At point *A* the sighting of high cirrus clouds would be the first sign of the approaching cyclone. These high clouds can precede the surface front by 1000 kilometers or more and they generally will be accompanied by falling pressure. As the warm front advances, a lowering and thickening of the cloud deck is noticed.

- Usually within 12 to 24 hours after the first sighting of cirrus clouds, light precipitation begins (point *B*). As the front nears, the rate of precipitation increases, a rise in temperature is noticed, and winds begin to change from east or southeast to south or southwest.

- With the passage of the warm front, the area is under the influence of the maritime tropical air mass of the warm sector (point *C*). Usually the region affected by this sector of the cyclone experiences warm temperatures, south or southwest winds, and generally clear skies, although fair-weather cumulus or altocumulus are not uncommon here.

- The rather pleasant weather of the warm sector passes quickly and is replaced by gusty winds and precipitation generated along the cold front. The approach of a rapidly advancing cold front is marked by a wall of dark clouds (point *D*). Severe weather accompanied by heavy precipitation, hail, and an occasional tornado is a definite possibility at this time of year. The passage of the cold front is easily detected by a wind shift; the southerly flow is replaced by winds from the west to northwest and by a pronounced drop in temperature. Also, rising pressure hints of the subsiding cool, dry air behind the front.

- Once the front passes, the skies clear quickly as the cooler air invades the region (point *E*). Often a day or two of almost cloudless deep-blue skies can be expected unless another cyclone is edging into the region.

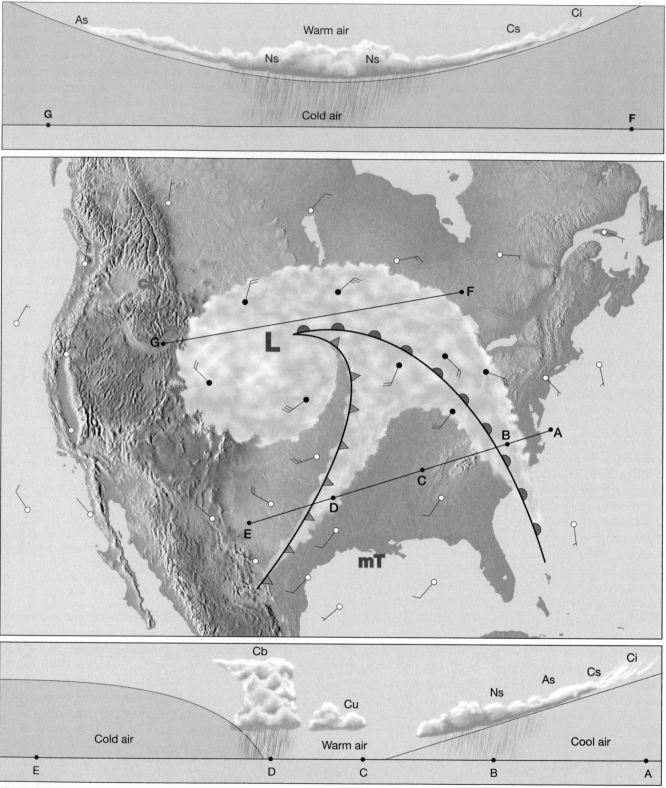

Figure 17.8
Distribution of clouds associated with an idealized middle-latitude cyclone.

A very different set of weather conditions will prevail in those regions north of the storm's center along profile *F-G*. Often the storm reaches its greatest intensity in this zone and the area along profile *F-G* receives the brunt of the storm's fury. Here temperatures remain cold during the passage of the system and heavy snow, sleet, and/or freezing rain may develop during the winter months.

Figure 17.9
This false-color infrared satellite image shows a very strong cyclonic storm that struck March 13–15, 1993, and brought blizzard conditions to a region that stretched from Alabama to eastern Canada. It was responsible for at least 220 deaths and was called by many the "storm of the century." (Photo courtesy of NESDIS/NOAA)

The Role of Airflow Aloft

When the earliest studies of cyclones were made, little was known about the nature of the airflow in the middle and upper troposphere. Since then, a close relationship between surface disturbances and the flow aloft has been established. Airflow aloft plays an important role in maintaining cyclonic and anticyclonic circulation. In fact, more often than not, these rotating surface wind systems are actually generated by upper-level flow.

Recall that the airflow about a surface low is inward, a fact that leads to mass convergence, or coming together (Figure 17.10). The resulting accumulation of air must be accompanied by a corresponding increase in surface pressure. Consequently, we might expect a low-pressure system to "fill" rapidly and be eliminated, just as the vacuum in a coffee can is quickly dissipated when we open it. However, this does not occur. On the contrary, cyclones often exist for a week or longer. For this to happen, surface convergence must be offset by a mass outflow at some level aloft (Figure 17.10). As long as divergence (spreading out) aloft is equal to, or greater than, the surface inflow, the low pressure and its accompanying convergence can be sustained.

Because cyclones are bearers of stormy weather, they have received far more attention than anticyclones. Nevertheless, a close relation exists, which makes it difficult to separate any discussion of these two types of pressure systems. The surface air that feeds a cyclone, for example, generally originates as air flowing out of an anticyclone. Consequently, cyclones and anticyclones

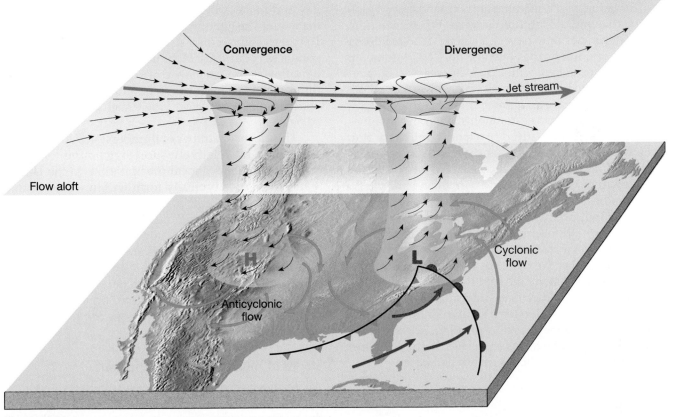

Figure 17.10
Idealized diagram depicting the support that divergence and convergence aloft provide to cyclonic and anticyclonic circulation at the surface. Divergence aloft initiates upward air movement, reduced surface pressure, and cyclonic flow. On the other hand, convergence along the jet stream results in general subsidence of the air column, increased surface pressure, and anticyclonic surface winds.

typically are found adjacent to one another. Like the cyclone, an anticyclone depends on the flow far above to maintain its circulation. In this instance, divergence at the surface is balanced by convergence aloft and general subsidence of the air column (Figure 17.10).

What's in a Name?

So far we have examined the middle-latitude cyclones that play such an important role in causing day-to-day weather changes. Yet the use of the term *cyclone* is often confusing. To many people, the term implies only an intense storm, such as a hurricane or a tornado. When a hurricane unleashes its fury on India or Bangladesh, for example, it is usually reported in the media as a cyclone (the term denoting a hurricane in that part of the world). Similarly, tornadoes are occasionally referred to as cyclones. This custom is particularly common in portions of the Great Plains of the United States. Recall that in the *Wizard of Oz*, Dorothy's house was carried from her Kansas farm to the land of Oz by a cyclone. Indeed, the nickname for the athletic teams at Iowa State University is the Cyclones. Although hurricanes and tornadoes are, in fact, cyclones, the vast majority of cyclones are not hurricanes or tornadoes. The term *cyclone* simply refers to the circulation around any low-pressure center, no matter how large or intense it is.

Tornadoes and hurricanes are both smaller and more violent than middle-latitude cyclones. Whereas middle-latitude cyclones may have a diameter of 1600 kilometers or more, hurricanes average only 600 kilometers across, and tornadoes, with a diameter of just $1/4$ kilometer, are much too small to show up on a weather map.

The thunderstorm, a much more familiar weather event, hardly needs to be distinguished from tornadoes, hurricanes, and mid-latitude cyclones. Unlike the flow of air about these latter storms, the circulation associated with thunderstorms is characterized by strong up-and-down movements. Winds in the vicinity of a thunderstorm do not follow the inward spiral of a cyclone, but they are typically variable and gusty.

Although thunderstorms form "on their own" away from cyclonic storms, they also form in conjunction with cyclones. For instance, thunderstorms are frequently spawned along the cold front of a mid-latitude cyclone, where on rare occasions a tornado may descend from the thunderstorm's cumulonimbus tower. Hurricanes also generate widespread thunderstorm activity. Thus, thunderstorms are related in some manner to all three types of cyclones mentioned here.

Thunderstorms

This is the first of three severe weather types we will examine in this chapter. Sections on tornadoes and hurricanes follow this look at thunderstorms.

Severe weather has a fascination that everyday weather phenomena cannot provide. The lightning display and booming thunder generated by a severe thunderstorm can be a spectacular event that elicits both awe and fear (Figure 17.11). Of course, hurricanes and tornadoes also attract a great deal of much-deserved attention. A single tornado outbreak or hurricane can cause billions of dollars in property damage as well as many deaths (see Box 17.2).

Thunderstorm Occurrence

Most everyone has observed a small-scale phenomenon that is caused by the vertical motion of warm, unstable air. Perhaps on a hot day you have seen a small dust devil that formed over an open field and whirled its dusty load to great heights. Or maybe you have noticed a bird glide skyward effortlessly upon an invisible thermal of hot air. These examples illustrate the dynamic thermal instability that occurs during the development of a **thunderstorm**. Thunderstorm activity is associated with cumulonimbus clouds that generate heavy rainfall, thunder, lightning, and occasionally hail.

At any given time there are an estimated 2000 thunderstorms in progress on Earth. As we would expect, the greatest proportion occur in the tropics where warmth, plentiful moisture, and instability are always present. About 45,000 thunderstorms take place each day and more than 16 million occur annually around the world. The lightning from these storms strikes Earth 100 times each second. Annually, the United States experiences about 100,000 thunderstorms and millions of lightning strikes. A glance at Figure 17.12 shows that thunderstorms are most frequent in Florida and the eastern Gulf Coast region, where such activity is recorded between 70 and 100 days each year.

The greatest number of thunderstorms occur in association with relatively short-lived cumulonimbus clouds that produce local precipitation. In the United States these cells typically form within warm, humid (maritime tropical) air masses that originate over the Gulf of Mexico and migrate northward. Occasionally, thunderstorms grow very large and remain active for hours. These *severe thunderstorms* produce frequent lightning and are accompanied by locally damaging winds or hail. Most severe thunderstorms in the middle latitudes form along or ahead of cold fronts. Here, forceful lifting of unstable mT air masses triggers thunderstorm development.

Stages of Thunderstorm Development

All thunderstorms require warm, moist air, which, when lifted, will release sufficient latent heat to provide the buoyancy necessary to maintain its upward flight. Although this instability and associated buoyancy are triggered by a number of different processes, all thunderstorms have a similar life history.

Figure 17.11
Cumulonimbus clouds are associated with lightning, thunderstorms, and other forms of severe weather. (Photo by Richard Kaylin/Tony Stone Images)

Because instability and buoyancy are enhanced by high surface temperatures, thunderstorms are most common in the afternoon and early evening. However, surface heating alone is not sufficient for the growth of towering cumulonimbus clouds. A solitary cell of rising hot air produced by surface heating could, at best, produce a small cumulus cloud, which would evaporate within 10–15 minutes.

The development of 12,000-meter (or on rare occasions 18,000-meter) cumulonimbus towers requires a continual supply of moist air (Figure 17.13). Each new surge of warm air rises higher than the last, adding to the height of the cloud (Figure 17.14). These updrafts must occasionally reach speeds over 100 kilometers (60 miles) per hour to accommodate the size of hailstones they are capable of carrying upward. Usually within an hour the amount and size of precipitation that has accumulated is too much for the updrafts to support, and in one part of the cloud downdrafts develop, releasing heavy precipitation. This represents the most active stage of the thunderstorm. Gusty winds, lightning, heavy precipitation, and sometimes hail are experienced.

Box 17.2

The Deadliest Weather

Tornadoes and hurricanes are nature's most awesome storms. Because of this status, they are logically the focus of much well-deserved attention. Yet, surprisingly, these dreaded events are not responsible for the greatest number of weather-related deaths. That distinction is reserved for lightning and flash floods.

Figure 17.B and Table 17.A provide data on weather-related deaths in the United States for the 1940-1992 time span. During this period, more than 8,300 people were killed by lightning, making it the leading weather killer. Not everyone who is struck by lightning is killed. In fact, fatalities represent only about 20 percent of lightning casualties. The risks from lightning are actually greater than the official statistics because some deaths and injuries go unreported.

Lightning as a safety threat receives less attention than the other causes of weather-related deaths. Tornadoes, hurricanes, and flash floods usually receive considerably more publicity because they often result in multiple deaths, many injuries, and widespread damage. By contrast, lightning usually takes its victims one or two at a time.

For the 20-year span ending in 1991, lightning ranked second to floods (primarily flash floods) as a cause of weather-related deaths (Figure 17.C). Slow-moving intense thunderstorms are most often responsible for triggering the sudden overflow of stream channels. Dam failures and the sudden breakup of ice jams can also release large volumes of water in a short time. Such floods rise rapidly with little or no warning and diminish almost as quickly.

Table 17.A Average annual severe weather fatalities by decade, 1940–1992.

Years	Lightning	Tornado	Flood	Hurricane
1940–1949	337	154	144	22
1950–1959	184	135	79	87
1960–1969	133	94	121	59
1970–1979	98	99	182	21
1980–1989	72	52	110	12
1990–1992	73	40	102	14

Source: National Weather Service

Flash floods are particularly common in mountainous terrain, where steep slopes can quickly channel runoff into narrow valleys. The hazard is most acute when the soil is already nearly saturated from earlier rains or consists of impermeable materials. The disastrous nature of flash floods is illustrated by the Big Thompson River flood of July 31, 1976, in Colorado (Figure 17.D). During a 4-hour period more than 30 centimeters of rain fell on portions of the small area drained by the river. This amounted to nearly three-quarters of the average yearly total. The flash flood in the narrow canyon lasted only a few hours, but cost 139 people their lives. Damages were estimated at more than $35 million. Despite efforts to improve observa-

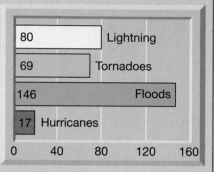

Figure 17.C

Annual average number of weather-related deaths from 1972 through 1991. (Source: National Weather Service)

tions and warning procedures, flash floods remain an elusive natural killer.

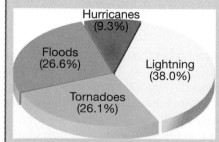

Figure 17.B

Weather-related deaths from 1940 through 1991. (Source: National Weather Service)

Figure 17.D

Effects of the July, 1976, flash flood in Colorado's Big Thompson Canyon. (Photo courtesy of U.S. Geological Survey)

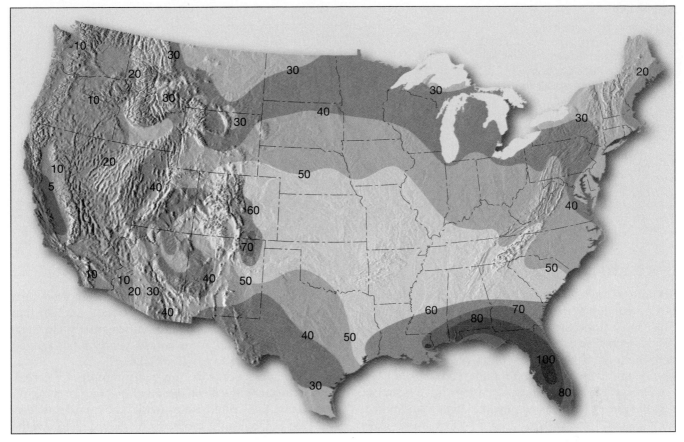

Figure 17.12

Average number of days per year with thunderstorms. Because of its close proximity to the source region for warm, humid, and unstable air masses, the Gulf Coast receives much of its precipitation from thunderstorms. (Source: Environmental Data Service, NOAA)

Eventually downdrafts dominate throughout the cloud. The cooling effect of falling precipitation coupled with the influx of colder air aloft mark the end of the thunderstorm activity. The life span of a cumulonimbus cell within a thunderstorm complex is only about an hour, but as the storm moves, fresh supplies of warm, water-laden air generate new cells to replace those that are dissipating.

Figure 17.13

This developing cumulonimbus cloud became a towering August thunderstorm over the high plains near Boulder, Colorado. (Photo by Henry Lansford)

Figure 17.14

Stages in the development of a thunderstorm. During the cumulous stage, strong updrafts act to build the storm. The mature stage is marked by heavy precipitation and cool downdrafts in part of the storm. When the warm updrafts disappear completely, precipitation becomes light, and the cloud begins to evaporate.

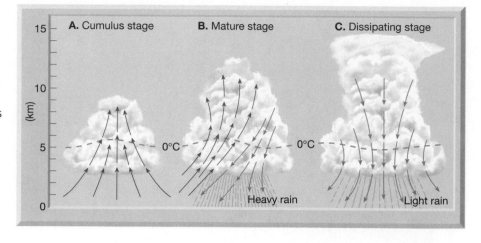

Tornadoes

Tornadoes are local storms of short duration that must be ranked high among nature's most destructive forces. Their sporadic occurrence and violent winds cause many deaths each year. Tornadoes are violent windstorms that take the form of a rotating column of air that extends downward from a cumulonimbus cloud (Figure 17.15).

Pressures within some tornadoes have been estimated to be as much as 10 percent lower than immediately outside the storm. Drawn by the much lower pressure in the center of the storm, air near the ground rushes into the tornado from all directions. As the air streams inward, it is spiraled upward around the core until it eventually merges with the airflow of the parent thunderstorm deep in the cumulonimbus tower. Because of the tremendous pressure gradient associated with a strong tornado, maximum winds can sometimes approach 480 kilometers (300 miles) per hour.

Tornado Occurrence and Development

An average of about 780 tornadoes are reported each year in the United States. Still, the actual number that occur from one year to the next varies greatly. During a recent 40-year span, for example, yearly totals ranged from a low of 421 to a high of 1126.

Tornadoes occur during every month of the year. April through June is the period of greatest tornado frequency in the United States, while the number is lowest during December and January (Figure 17.16). Of the more than 21,000 confirmed tornadoes reported over the contiguous 48 states during the period depicted in Figure 17.16, an average of almost five per day occurred during May. At the other extreme, a tornado was reported only every other day in January.

Tornadoes form in association with severe thunderstorms that produce high winds, heavy rainfall, and often damaging hail. Fortunately, less than 1 percent of

Figure 17.15

A tornado is a violently rotating column of air in contact with the ground. The air column is visible when it contains condensation or dust and debris. Often the appearance is the result of both. When the column of air is aloft and does not produce damage, the visible portion is properly called a funnel cloud. (Photo by A. and J. Verkaik/The Stock Market)

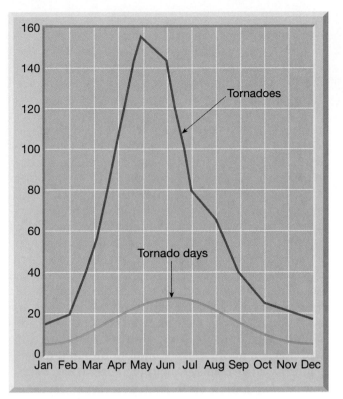

Figure 17.16
Average number of tornadoes and tornado days each month in the United States for a 27-year period. (After NOAA)

all thunderstorms produce tornadoes. Although weather scientists are still not sure what triggers tornado formation, it has become apparent that they are products of the interaction between strong updrafts in the thunderstorm and winds in the troposphere. In spite of recent advances in modeling the many variables that eventually produce a strong tornado, our knowledge is still limited. Nevertheless, the general atmospheric conditions that are most likely to develop into tornado activity are known.

Severe thunderstorms—and hence tornadoes—are most often spawned along the cold front of a middle-latitude cyclone. Throughout spring, air masses associated with middle-latitude cyclones are most likely to have greatly contrasting conditions. Continental polar air from the Canadian Arctic may still be very cold and dry, whereas maritime tropical air from the Gulf of Mexico is warm, humid, and unstable. The greater the contrast when these air masses meet, the more intense the storm. Because these two contrasting air masses are more likely to meet in the central United States, it is not surprising that this region generates more tornadoes than any other area of the country or, in fact, the world. Figure 17.17, which depicts tornado incidence in the United States based on data for a 27-year period, readily substantiates this fact.

An average tornado has a diameter of between 150 and 600 meters (500 and 2000 feet), travels across the landscape at approximately 45 kilometers (30 miles) per hour, and cuts a path about 10 kilometers (6 miles) long.[*] Since tornadoes usually occur slightly ahead of a cold front, in the zone of southwest winds, most move toward the northeast. The Illinois example demonstrates this movement (Figure 17.18). Figure 17.18 also shows that many tornadoes do not fit the description of the "average" tornado. Some have had paths a great deal longer than 10 kilometers and have traveled not at 45 kilometers per hour, but at speeds in excess of 100 kilometers (60 miles) per hour. Furthermore, there have been tornadoes that have had diameters of up to 1.6 kilometers (1 mile)—more than four times the "average" size.

Tornado Destruction

The potential for tornado destruction depends largely on the strength of the winds generated by the storm. One commonly used guide to tornado intensity is the *Fujita intensity scale*, or simply the *F-scale* (Table 17.1). Because tornado winds cannot be measured directly, a rating on the F-scale is determined by assessing the worst damage produced by a storm.

Tornadoes take many lives each year, sometimes hundreds in a single day. When tornadoes struck an area stretching from Canada to Georgia on April 3, 1974, the death toll exceeded 300, the worst in half a century. Most tornadoes, however, do not result in a loss of life. In one statistical study that examined a 29-year period, there were 689 tornadoes that resulted in deaths. This figure represents slightly less than 4 percent of the total 19,312 reported storms.

Although the percentage of tornadoes that result in death is small, each tornado is potentially lethal. When tornado fatalities and storm intensities are compared, the results are quite interesting: The majority (63 percent) of tornadoes are weak (F0 and F1) and the number of storms decreases as tornado intensity increases. The distribution of tornado fatalities, however, is just the opposite. Although only 2 percent of tornadoes are classified as violent (F4 and F5), they account for nearly 70 percent of the deaths. If there is some question as to the causes of tornadoes, there certainly is no question about the destructive effects of these violent storms (Figure 17.19).

Tornado Warnings

Because tornadoes are small and relatively short-lived phenomena, they are among the most difficult weather features to forecast precisely. When conditions appear

---

[*]The 10-kilometer figure applies to documented tornadoes. Because many small tornadoes go undocumented, the real average path of all tornadoes is unknown, but shorter than 10 kilometers.

favorable for tornado formation, a **tornado watch** is issued for areas covering about 65,000 square kilometers (25,000 square miles). Between 35 and 40 percent of the predictions are correct; that is, one or more tornadoes do occur somewhere in the specified region. The incorrect forecasts are about evenly divided between cases in which no tornadoes are sighted and cases when tornadoes occur outside, but near, the watch area.

Whereas a tornado watch is designed to alert people to the possibility of tornadoes, a **tornado warning** is issued when a funnel cloud has actually been sighted or is indicated by radar. It warns of high probability of imminent danger.

When severe weather threatens, radar screens are monitored for very intense echoes, which in turn are associated with heavy precipitation and the greater likelihood of hail, strong winds, and tornadoes. In addition, the echo from a tornadic storm sometimes displays a hook-shaped appendage. If the direction and approximate speed of the storm are known, the storm's most probable path can be estimated. Because tornadoes often move erratically, the warning area is fan-shaped downwind from the point where the tornado has been spotted.

Since the late 1960s, warnings have been given for most major tornadoes. It is believed that such warnings substantially reduce the number of deaths and serious injuries that might otherwise occur.

The tornado warning system that has been used throughout the United States for many years relies heavily on visual sightings by a few trained observers as well as the general public. Unfortunately, such a system is prone to incomplete coverage and mistakes. The errors are most likely to occur at night when tornadoes may go unnoticed or harmless clouds may be mistaken for funnel clouds. Hence, there may be a lack of adequate warning on the one hand, or unnecessary warnings on the other.

Many of the difficulties that have limited the accuracy of tornado warnings are being reduced or eliminated by an advancement in radar technology called **Doppler radar.** Doppler radar not only performs the same tasks as conventional radar, but also has the ability to detect motion directly. Doppler radar can detect the initial formation and subsequent development of a **mesocyclone**, an intense rotating wind system in the lower part of a thunderstorm that frequently precedes tornado development. Almost all mesocyclones produce damaging hail, severe winds, or tornadoes. Those that produce tornadoes (about 60 percent) can be distinguished by their stronger wind speeds and their sharper gradients of wind speeds.

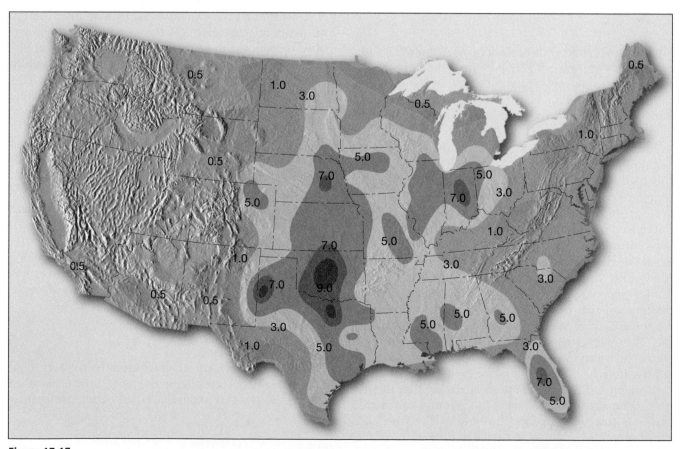

Figure 17.17

Average annual tornado incidence per 10,000 square miles (26,000 square kilometers) for a 27-year period.

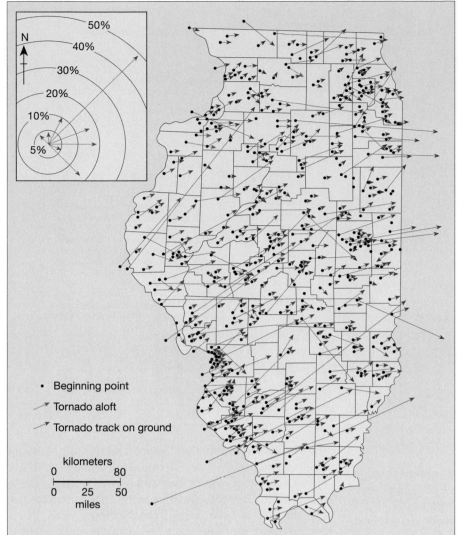

Figure 17.18
Paths of Illinois tornadoes (1916–1969). Because most tornadoes occur slightly ahead of a cold front, in the zone of southwest winds, they tend to move toward the northeast. Tornadoes in Illinois verify this. Over 80 percent exhibited directions of movement toward the northeast through east. (After John W. Wilson and Stanley A. Changnon, Jr., *Illinois Tornadoes*, Illinois State Water Survey Circular 103, 1971, pp. 10, 24)

- Beginning point
- Tornado aloft
- Tornado track on ground

kilometers
0   80

0   25   50
miles

As a practical tool for tornado detection, Doppler radar has significant advantages over a system that uses observers and conventional radar. Recognizing these advantages, the National Weather Service is replacing its conventional weather radars with more advanced systems that use Doppler principles of direct wind speed and direction measurements. The new radars are part of the *Next Generation Weather Radar* (NEXRAD) program aimed at improving the forecasting of severe storms, tornadoes, and flash floods.

Hurricanes

Most of us view the weather in the tropics with favor. Places like Hawaii and the islands of the Caribbean are known for their lack of significant day-to-day variations. Warm breezes, steady temperatures, and rains that come as heavy but brief tropical showers are expected. It is ironic that these relatively tranquil regions sometimes produce the most violent storms on Earth.

The whirling tropical cyclones that on occasion have wind speeds attaining 300 kilometers (185 miles) per hour are known in the United States as **hurricanes** and are the greatest storms on Earth. Out at sea, they can generate 15-meter (50-foot) waves capable of inflicting destruction hundreds of kilometers from their source. Should a hurricane smash onto land, strong winds coupled with extensive flooding can impose billions of dollars in damage and great loss of life (Figure 17.20). The vast majority of hurricane-related deaths and damage are caused by relatively infrequent, yet powerful, storms. Hurricane Andrew was one such storm and is described in Box 17.3.

Hurricanes are becoming a growing threat because more and more people are living and working along and near coasts. In 1990, 50 percent of the U.S. population lived within 75 kilometers (45 miles) of a coast. This number is projected to increase to 75 percent by the year 2010. The concentration of such large numbers of people near the shoreline means that hurricanes and other large storms place millions at risk. Moreover, the potential costs of property damage are incredible.

Table 17.1 Fujita intensity scale.

Scale	Wind Speed		Expected Damage
	Km/Hr	Mi/Hr	
F0	<116	<72	**Light damage.** Damage to chimneys and billboards; broken branches; shallow-rooted trees pushed over.
F1	116–180	72–112	**Moderate damage.** The lower limit is near the beginning of hurricane wind speed. Surface peeled off roofs; mobile homes pushed off foundations or overturned; moving autos pushed off the road.
F2	181–253	113–157	**Considerable damage.** Roofs torn off frame houses; mobile homes demolished; boxcars pushed over; large trees snapped or uprooted; light-object missiles generated.
F3	254–332	158–206	**Severe damage.** Roofs and some walls torn off well-constructed houses; trains overturned; most trees in forest uprooted; heavy cars lifted off ground and thrown.
F4	333–419	207–260	**Devastating damage.** Well-constructed houses leveled; structures with weak foundations blown some distance; cars thrown and large missiles generated.
F5	>419	>260	**Incredible damage.** Strong frame houses lifted off foundations and carried considerable distance to disintegrate; automobile-sized missiles fly through the air farther than 100 m; trees debarked; incredible phenomena occur.

Profile of a Hurricane

Hurricanes form in all tropical waters (except those of the South Atlantic and eastern South Pacific) between the latitudes of 5 degrees and 20 degrees. They are known by different names in various parts of the world. In the western Pacific, they are called *typhoons* and in the Indian Ocean, *cyclones*. The North Pacific has the greatest number of storms, averaging 20 per year. Fortunately for those living in the coastal regions of the southern and eastern United States, fewer than 5 hurricanes, on the average, develop each year in the warm sector of the North Atlantic.

Although many tropical disturbances develop each year, only a few reach hurricane status. By international agreement, a hurricane has wind speeds in excess of 119 kilometers (74 miles) per hour and a rotary circulation. Hurricanes average 600 kilometers (375 miles) in diameter and often extend 12,000 meters (40,000 feet) above the ocean surface. From the outer edge to the center, the barometric pressure has on occasion dropped 60 millibars, from 1010 millibars to 950 millibars. The lowest pressures ever recorded in the Western Hemisphere are associated with these storms. A steep pressure gradient generates the rapid, inward-spiraling winds of a hurricane.

As the inward rush nears the core of the storm, it turns upward and ascends in a ring of cumulonimbus towers. This doughnut-shaped wall of intense convective activity surrounding the center of the storm is called the **eye wall**. It is here that the greatest wind speeds and heaviest rainfall occur. Surrounding the eye wall are curved bands of clouds that trail away in a spiral fashion. Near the top of the hurricane the airflow is outward, carrying the rising air away from the storm center, thereby providing room for more inward flow at the surface.

At the very center of the storm is the **eye** of the hurricane (Figure 17.21). This well-known feature is a zone about 20 kilometers (12.5 miles) in diameter where precipitation ceases and winds subside. It offers a brief but deceptive break from the extreme weather in the enormous curving wall clouds that surround it. The air within the eye gradually descends and heats by compression, making it the warmest part of the storm. Although many people believe that the eye is characterized by clear blue skies, such is usually not the case because the subsidence in the eye is seldom strong enough to produce cloudless conditions. Although the sky appears much brighter in this region, scattered clouds at various levels are common.

Hurricane Formation and Decay

A hurricane is a heat engine that is fueled by the latent heat liberated when huge quantities of water vapor condense. The amount of energy produced by a typical hurricane in just a single day is truly immense—roughly equivalent to the entire electrical energy production of the United States in a year. The release of latent heat warms the air and provides buoyancy for its upward flight. The result is to reduce the pressure near the surface, which encourages a more rapid inward flow of air. To get this engine started, a large quantity of warm, moisture-laden air is required, and a continual supply is needed to keep it going.

Hurricanes develop most often in the late summer when water temperatures have reached 27°C (80°F) or higher and thus are able to provide the necessary heat and moisture to the air. This ocean-water temperature requirement is thought to account for the fact that hurricanes do not form over the relatively cool waters of the South Atlantic and the eastern South Pacific. For the same reason, few hurricanes form poleward of 20 degrees of latitude. Although water temperatures are sufficiently high, hurricanes do not form within 5 degrees of the equator, because the Coriolis effect is too weak to initiate the necessary rotary motion.

The initial stage of a tropical cyclone's life cycle is not well understood, due in part to the complexity of these storms. Another important factor is the lack of observations in regions where storms form. However, we do know that smaller tropical cyclones initiate the process. These initial disturbances are regions of low-level convergence and lifting.

Many tropical disturbances of this type occur each year, but only a few develop into full-fledged hurricanes. By international agreement, lesser tropical cyclones are placed in different categories, based on wind strength. When a cyclone's strongest winds do not exceed 61 kilometers (38 miles) per hour, it is called a **tropical depression**; when winds are between 61 and 119 kilometers (38 and 74 miles) per hour, the cyclone is termed a **tropical storm**. Each year between 80 and 100 tropical storms develop, and of this total, half or more become hurricanes.

Hurricanes diminish in intensity whenever they (1) move over ocean waters that cannot supply warm, moist tropical air; (2) move onto land; or (3) reach a location where the large-scale flow aloft is unfavorable. Whenever a hurricane moves onto land, it loses its punch rapidly. The most important reason for this rapid demise is the fact that the storm's source of warm, moist air is cut off. When an adequate supply of water vapor does not exist, condensation and the release of latent heat must diminish. In addition, friction from the increased roughness of the land surface rapidly slows surface wind speeds. This factor causes the winds to move more directly into the center of the low, thus helping to eliminate the large pressure differences.

Figure 17.19
This home was completely destroyed by a tornado. (Photo by Michael Meinhardt/Sygma)

Box 17.3

Hurricane Andrew*

Ever since 1965 when Hurricane Betsy struck, the residents of South Florida's vulnerable east coast have waited for the next big one. In the intervening years, many powerful storms skirted the region and caused destruction elsewhere. In August 1992, the inevitable became reality.

On August 16, 1992, a tropical depression formed out in the Atlantic, closer to the west coast of Africa than to the United States. The next day the National Hurricane Center declared that the depression had reached the status of a tropical storm and christened it *Andrew*. For the next week Andrew made its way westward, achieving hurricane status on August 22 (Figure 17.E). By Sunday, the, the spiraling storm was on a collision course with South Florida.

Hurricane Andrew made landfall south of Key Biscayne, Florida, during the dark, early morning hours of August 24 (Figure 17.F). Maximum sustained surface winds were about 230 kilometers (145 miles) per hour, with gusts exceeding 280 kilometers (175 miles) per hour. With a central pressure of 922 millibars (27.23 inches), Andrew was a category 4 storm on the Saffir-Simpson scale (see Table 17.2). It was close to being a rare category 5. Andrew had the third lowest pressure this century for a hurricane making landfall in the United States.

*This material was prepared with the assistance of Gregory Ballinger, Miami-Dade Community College.

Figure 17.F

Color radar image of Hurricane Andrew at 4:35a.m. EDT, August 24, 1992. The picture is from the last full sweep of the National Weather Service's Miami radar located at the National Hurricane Center (NHC) before the radar was destroyed by the storm. The digitized radar imagery shows the eye centered over Elliott Key just before landfall at Homestead Air Force Based (HAFB). As Andrew traveled due west, the heaviest damage occurred in those areas affected by the eye wall (doughnut-shaped region with echoes greater than 42 dBZ). The weather radar measures the power from the portion of the radar beam scattered back by raindrops and ice particles. The colors associated with higher dBZ (i.e., red) correspond to areas with larger amounts of rain, which typically are also regions of stronger winds. Areas with high dBZ in the center of the eye are because of ground clutter from islands. (Ground clutter is the reflection of the radar beam by terrain, large structures, and rough water.) (Courtesy of Hurricane Research Division/NOAA)

It took Hurricane Andrew just hours to cut its destructive path across Florida. Property damage from Andrew was primarily wind damage along a 40-kilometer-wide swath of destruction that was centered a few kilometers north of the town of Homestead. Fortunately, the highly developed coastline of Miami Beach was more than 27 kilometers from the eye of the storm and so did not experience sustained hurricane-force winds. Two other aspects of the storm made it less damaging than it might otherwise have been: The storm surge was small and there was little rainwater flooding because the hurricane did not linger but advanced rapidly across the region.

The storm's destructive accomplishments were awesome. In one area, the wind carried 6-meter-long steel and concrete beams, with roofs still attached, more than 50 meters. Cars and boats were tossed about like toys, tens

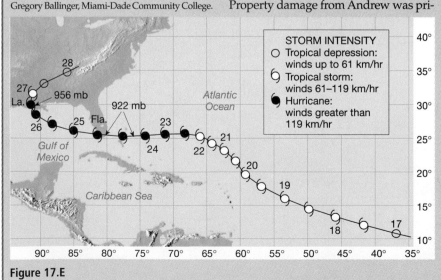

Figure 17.E

Positions for Hurricane Andrew, August 17–28, 1992. (National Hurricane Center)

(continues)

Hurricane Destruction

A location only a few hundred kilometers from a hurricane—just one day's striking distance away—may experience clear skies and virtually no wind. Prior to the age of weather satellites, such a situation made the job of warning people of impending storms very difficult.

The worst natural disaster in U.S. history resulted from a hurricane that struck an unprepared Galveston, Texas, on September 8, 1900. The strength of the storm, coupled with the lack of adequate warning, caught the population by surprise and cost 6000 people in the city their lives. At least 2000 more were killed elsewhere. Fortunately, hurricanes are no longer the unheralded killers they once were. Once a storm develops cyclonic flow and the spiraling bands of clouds characteristic of a hurricane, it receives continuous monitoring. For ex-

ample, when hurricanes Hugo (1989) and Andrew (1992) formed, satellites were able to identify and track the storms long before they made landfall (Figure 17.22). In the United States, early warning systems have greatly reduced the number of deaths caused by hurricanes. At the same time, however, there has been an astronomical rise in the amount of property damage. The primary reason, of course, has been the extensive building of homes and businesses in coastal areas.

Although the amount of damage caused by a hurricane depends on several factors, including the size and population density of the area affected and the shape of the ocean bottom near the shore, certainly the most significant factor is the strength of the storm itself. By studying past storms, a scale has been established to rank the relative intensity of hurricanes (Table 17.2). As Table 17.2 indicates, a *category 5* storm is the worst possible, whereas a *category 1*

Figure 17.20
Hurricane Andrew devastated portions of South Florida in August 1992. (Photo by Allan Tannenbaum/Sygma)

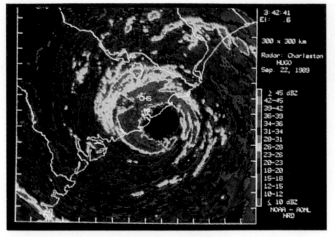

Figure 17.21
Radar image of Hurricane Hugo moving over the coast of South Carolina on September 22, 1989. Colors show rainfall intensity. Red is the most intense and blue is less intense. No rain is falling in the black areas. The rainless "hole" in the center is the eye. As expected, the heaviest rains are occurring in the eye wall. (Courtesy of Peter Dodge, NOAA, Hurricane Research Division)

Figure 17.22
A satellite image of Hurricane Hugo as the storm moved onto the South Carolina coast. The eye is clearly visible in the center of the storm. The eye wall, the most intense part of the storm, is a doughnut-shaped wall of cumulonimbus development surrounding the eye. Satellites are essential tools for detecting and tracking hurricanes. (Courtesy of National Hurricane Center/NOAA)

hurricane is least severe. During the hurricane season it is common to hear scientists and reporters alike use the numbers from the *Saffir-Simpson Hurricane Scale*. The famous Galveston hurricane just mentioned, with winds in excess of 209 kilometers (130 miles) per hour and a pressure of 931 millibars, would be placed in category 4. Storms that fall into category 5 are rare. Hurricane Camille, a 1969 storm that caused catastrophic damage along the coast of Mississippi, is one well-known example.

Damage caused by hurricanes can be divided into three categories: (1) wind damage, (2) storm surge, and (3) inland freshwater flooding.

Wind Damage.   For some structures, the force of the wind is sufficient to cause total destruction. This was demonstrated in South Florida in 1992. The billions of dollars in property damages from Hurricane Andrew were largely the result of strong winds. However, wind damage is not necessarily responsible for a hurricane's greatest destructiveness.

Storm Surge.   The most devastating damage in the coastal zone is caused by the storm surge. It not only accounts for a large share of coastal property losses, but is also responsible for 90 percent of all hurricane-caused deaths. A **storm surge** is a dome of water 65 to 80 kilometers (40 to 50 miles) wide that sweeps across the coast near the point where the eye makes landfall. Ignoring wave activity, the storm surge is the height of the water above normal tide level. Thus, a storm surge commonly adds 2 to 3 meters (6 to 10 feet) to normal tide heights—to say nothing of tremendous wave activity superimposed atop the surge.

We can easily imagine the damage this surge of water can inflict on low-lying coastal areas. In the delta region of Bangladesh, for example, the land is mostly less than 2 meters above sea level. When a storm surge superimposed upon normal high tide inundated that area on November 13, 1970, the official death toll was 200,000; unofficial estimates ran to 500,000. This was one of the worst disasters of modern times.

Table 17.2 Saffir-Simpson hurricane scale.

Scale Number (category)	Central Pressure (millibars)	Winds (km/hr)	Storm Surge (meters)	Damage
1	≥980	119–153	1.2–1.5	Minimal
2	965–979	154–177	1.6–2.4	Moderate
3	945–964	178–209	2.5–3.6	Extensive
4	920–944	210–250	3.7–5.4	Extreme
5	<920	>250	>5.4	Catastrophic

Inland Flooding. The torrential rains that accompany most hurricanes represent a third significant threat—flooding. Whereas the effects of storm surge and strong winds are concentrated in coastal areas, heavy rains may affect places hundreds of kilometers from the coast for several days after the storm has lost its hurricane-force winds.

Hurricanes weaken rapidly as they move inland, yet the remnants of the storm can still yield 15 to 30 centimeters (6 to 12 inches) or more of rain as they move inland. A good example of such destruction was Hurricane Agnes (1972). Although this was just a category 1 storm on the Saffir-Simpson scale, it was one of the costliest hurricanes of the century, creating more than $2 billion in damage and taking 122 lives. Most destruction was attributed to flooding caused by an inordinate amount of rainfall.

To summarize, extensive damage and loss of life in the coastal zone can result from storm surge, torrential rains, and strong winds. When loss of life occurs, it is commonly caused by the storm surge, which can devastate entire barrier islands and low-lying land along the coast. Although wind damage is usually not as catastrophic as the storm surge, it affects a much larger area. Where building codes are inadequate, economic losses can be especially severe. Because hurricanes weaken as they move inland, most wind damage occurs within 200 kilometers of the coast. Far from the coast a weakening storm can produce extensive flooding long after the winds have diminished below hurricane levels. Sometimes the damage from inland flooding exceeds storm-surge destruction.

Review Questions

1. Describe the weather associated with a continental polar air mass in the winter and in the summer. When would this air mass be most welcome in the United States?

2. What are the characteristics of a maritime tropical air mass? Where are the source regions for the maritime tropical air masses that affect North America? Where are the source regions for the maritime polar air masses?

3. Why are snowfall totals in the Great Lakes region highest on the leeward (downwind) shores of the lakes? What term is applied to these heavy snows? (See Box 17.1.)

4. Describe the weather along a cold front where very warm, moist air is being displaced.

5. The formation of an occluded front marks the beginning of the end of a middle-latitude cyclone. Why is this true?

6. For each of the weather elements that follow, describe the changes that an observer experiences when a middle-latitude cyclone passes with its center north of the observer: wind direction, pressure tendency, cloud type, cloud cover, precipitation, temperature.

7. Describe the weather conditions an observer would experience if the center of a middle-latitude cyclone passed to the south.

8. Briefly explain how the flow aloft aids the formation of cyclones at the surface.

9. Compare the wind speeds and sizes of middle-latitude cyclones, tornadoes, and hurricanes. How are thunderstorms related to each?

10. What is the primary requirement for the formation of thunderstorms?

11. Based on your answer to Question 10, where would you expect thunderstorms to be most common on Earth? In the United States?

12. Why do tornadoes have such high wind speeds?

13. What general atmospheric conditions are most conducive to the formation of tornadoes?

14. When is the "tornado season"? That is, during what months is tornado activity most pronounced?

15. Distinguish between a tornado watch and a tornado warning.

16. What advantage does Doppler radar have over conventional radar?

17. During what time of year do most of the hurricanes that affect North America form? Why is hurricane formation favored at this time?

18. Which has stronger winds, a tropical storm or a tropical depression?

19. Why does the intensity of a hurricane diminish rapidly when it moves onto land?

20. Hurricane damage can be divided into three broad categories. Name them. Which category is responsible for the highest percentage of hurricane-related deaths?

21. Great damage and significant loss of life can take place a day or more after a hurricane has moved ashore and weakened. When this occurs, what is the likely cause?

22. A hurricane has slower wind speeds than a tornado, yet it inflicts more total damage. How might this be explained?

23. Although tornadoes and hurricanes are dangerous storms, they are not responsible for the greatest number of weather-related deaths. What weather phenomena cause more deaths than tornadoes and hurricanes? (See Box 17.2.)

Key Terms

air mass (p. 438)
air-mass weather (p. 438)
cold front (p. 442)
continental (c) air mass (p. 439)
Doppler radar (p. 454)
eye (p. 456)
eye wall (p. 456)
front (p. 441)
hurricane (p. 451)
lake-effect snow (p. 439)
maritime (m) air mass (p. 439)
mesocyclone (p. 454)
middle-latitude cyclone (p. 444)
occluded front (p. 443)
polar (P) air mass (p. 439)
source region (p. 439)
stationary front (p. 443)
storm surge (p. 460)
thunderstorm (p. 448)
tornado (p. 452)
tornado warning (p. 454)
tornado watch (p. 454)
tropical (T) air mass (p. 439)
tropical depression (p. 457)
tropical storm (p. 457)
warm front (p. 441)

CHAPTER 18
Climate

Expedition to Nepal's Mount Everest. (Photo by Pascal Tournaire/Sygma)

Anyone who has the opportunity to travel around the world will find such an incredible variety of climates that it is hard to believe they could all occur on the same planet. The broad diversity of climates around the globe is an important focus of this chapter. Climate strongly influences the nature of plant and animal life, the soil, and many external geological processes. Climate influences people as well.

Although climate has a significant impact on people, we are learning that people also have a strong influence on climate. The latter portion of this chapter examines the ways in which humans may be changing global climate.

The Climate System

The focus of this chapter is *climate*. In Chapter 14, where this term was introduced, climate was characterized as being an aggregate of weather. It consists not only of average atmospheric values, but also involves the variability of elements and information on the occurrence of extreme events.

To understand and appreciate climate, it is important to realize that climate involves more than just the atmosphere. Indeed, we must recognize that there is a **climate system**. In addition to the atmosphere, it consists of the hydrosphere, solid Earth, biosphere, and cryosphere (Figure 18.1). The *cryosphere* refers to the ice and snow that exist at Earth's surface. Powered by the energy from the sun, the climate system involves the exchanges of energy and moisture that occur among the five parts. These exchanges link the atmosphere to the other parts of the system to produce an integrated and extremely complex interactive unit.

The atmosphere is the central component of the complex, connected, and interactive global environmental system upon which all life depends. Climate may be broadly defined as the long-term behavior of this environmental system. To understand fully and to predict changes in the atmospheric component of the climate system, one must understand the sun, oceans, ice sheets, solid Earth and all forms of life[*].

[*] The American Meteorological Society and the University Corporation for Atmospheric Research. "Weather and the Nation's Well-Being." *Bulletin of the American Meteorological Society*, 73, no. 12 (December 1991) 2038.

Figure 18.1
The climate system involves the complex interactions that occur among the atmosphere, hydrosphere, solid Earth, cryosphere, and biosphere. All of the components are represented in this scene in Wyoming. (Photo by E. J. Tarbuck)

The climate system provides a framework for the study of climate. The interactions and exchanges among the parts of the climate system create a complex network that links the five spheres. Changes to the system do not occur in isolation. Rather, when one part of this interactive unit changes, the other components also react. This well-established relationship is demonstrated often when we study climate and climate change.

World Climates

Previous chapters have already presented the spatial and seasonal variations of the major elements of weather and climate. Chapter 14 examined the controls of temperature and the world distribution of temperature. In Chapter 16, you studied the general circulation of the atmosphere and the global distribution of precipitation. You are now ready to investigate the *combined* effects of these variations in different parts of the world. The varied nature of Earth's surface and the many interactions that occur between atmospheric processes give every location on our planet a distinctive, even unique, climate. Our intention, however, is not to describe the unique climatic character of countless different locales. Instead, the purpose is to introduce the major climatic regions of the world. The discussion examines large areas and uses particular places only to illustrate the characteristics of these major climatic regions.

Temperature and precipitation are the most important elements in a climatic description because they have the greatest influence on people and their activities and also have an important impact on the distribution of such phenomena as vegetation and soils. Nevertheless, other factors are also important for a complete climatic description. When possible, some of these factors are introduced into our discussion of world climates. (See Box 18.1.)

Climate Classification

The distribution of the major atmospheric elements is, to say the least, complex. Because of the many differences from place to place as well as from time to time at a particular locale, it is unlikely that any two sites on Earth experience exactly the same weather conditions. The fact that the number of places on Earth is virtually infinite makes it readily apparent that the number of different climates must be extremely large.

Of course, having a great diversity of information to investigate is not unique to the study of the atmosphere; it is a problem that is basic to all science. To cope with such variety, it is not only desirable, but essential to devise some means of classifying the vast array of data to be studied. By establishing groups consisting of items that have certain important characteristics in common, order and simplicity are introduced. Bringing order to large quantities of information not only aids comprehension and understanding, but it also facilitates analysis and explanation.

During the twentieth century, many climate-classification schemes have been devised. It should be remembered that the classification of climates (or of anything else) is not a natural phenomenon but the product of human ingenuity. The value of any particular classification is determined largely by its intended use. A system designed for one purpose is not necessarily applicable to another.

In this chapter, we use a classification devised by Russian-born German climatologist Wladimir Köppen (1846-1940). As a tool for presenting the general world pattern of climates, the **Köppen classification** has been the best-known and most used system for more than 70 years. It is widely accepted for many reasons. For one, it uses only easily obtained data: mean monthly and annual values of temperature and precipitation. Furthermore, the criteria are unambiguous, relatively simple to apply, and divide the world into climatic regions in a realistic way.

Köppen believed that the distribution of natural vegetation was an excellent expression of the totality of climate. Consequently, the boundaries he chose were largely based on the limits of certain plant associations. Five principal groups were recognized; each group was designated by a capital letter as follows:

A   Humid tropical. Winterless climates; all months having a mean temperature above 18°C.

B   Dry. Climates where evaporation exceeds precipitation; there is a constant water deficiency.

C   Humid middle-latitude. Mild winters; the average temperature of the coldest month is below 18°C but above -3°C.

D   Humid middle-latitude. Severe winters; the average temperature of the coldest month is below -3°C and the warmest monthly mean exceeds 10°C.

E   Polar. Summerless climates; the average temperature of the warmest month is below 10°C.

Notice that four of the major groups (A, C, D, E) are defined on the basis of temperature characteristics, and the fifth, the B group, has precipitation as its primary criterion. Each of the five groups is further subdivided by using the criteria and symbols presented in Table 18.1. In addition, the world distribution of climates according to the Köppen* classification is shown in Figure 18.2. You will be referred to this figure several times as Earth's climates are discussed in the following pages.

---

* When classifying climatic data using Table 18.1, you should first determine whether the data meet the criteria for the E climates. If the station is not a polar climate, proceed to the criteria for B climates. If your data do not fit into either the E or B groups, check the data against the criteria for A, C, and D climates, in that order.

Box 18.1

The World's Hottest and Coldest Places

Maximum Temperature Records

Most people living in the United States have experienced temperatures of 38°C (100°F) or more. When statistics for the 50 states are examined, we find that *every* state has a maximum temperature of 38°C or higher. Even Alaska has recorded a temperature this high. The record was set June 27, 1915, at Fort Yukon, a town along the Arctic Circle in the interior of the state. With one exception, every other state has a maximum temperature record *in excess* of 38°C. Surprisingly, the state that ties Alaska for the "lowest high" is Hawaii. Panaloa, on the south coast of the big island, recorded 38°C on April 27, 1931. Although humid tropical and subtropical places like Hawaii are known for being warm throughout the year, they seldom experience maximum temperatures that surpass the low- to mid-30's Celsius (90s Fahrenheit).

The highest accepted temperature record for the United States as well as the entire Western Hemisphere is 57°C (134°F). This long-standing record was set at Death Valley, California , on July 10, 1913. Summer temperatures at Death Valley are consistently among the highest in the Western Hemisphere. During June, July, and August, temperatures exceeding 49°C (120°F) are to be expected. Fortunately, Death Valley has no summertime residents.

Why are summer temperatures at Death Valley so high? In addition to having the lowest elevation in the Western Hemisphere (53 meters below sea level), Death Valley is a desert. Although it is only about 300 kilometers form the Pacific Ocean, mountains cut off the valley from the ocean's moderating influence and moisture. Clear skies allow a maximum of sunshine to strike the dry, barren surface. Because no energy is used to evaporate moisture as occurs in humid regions, all of the energy is available to heat the ground. In addition, subsiding air that warms by compression as it descends is also common to the region and contributes to its high maximum temperatures.

When maximum temperature records for other continents are examined (Table 18.A), we find situations similar to that in North America. That is, the highest temperatures have been recorded at locations that are arid or nearly so. The nearly 59°C reading at Azizia, Libya, in North Africa's Sahara Desert, is the world record. Of course, Antarctica's highest recorded temperature of 15°C (59°F) does not come close to the others.

Minimum Temperature Records

Table 18.B presents minimum temperature records for each of the continents plus Greenland. The stations are ranked, with the lowest minimum at the top of the list. Remember that many othe places have no doubt experienced equally low or even lower temperatures; they just were not officially recorded.

The temperature controls that make these places so frigid are predictable, and should come as no surprise. We should expect cold temperatures during winter in high-latitude places that lack marine influence. Moreover, stations located on ice sheets and glaciers should be especially cold, as should stations positioned high in the mountains. Clearly *all* of these criteria apply to Vostok in Antarctica and to Greenland's Northice station. Snag in Canada's Yukon Territory holds the record for North America, so no U.S. locations appear on the list. Nevertheless, Prospect Creek, located north of the Arctic Circle in the Endicott Mountains of Alaska, came close to the North American record on January 23, 1971, when the temperature plunged to -62°C (-80°F). In the lower 48 states, the record of -57°C (-70°F) was set in the mountains of Rogers Pass, Montana on January 20, 1954.

Table 18.A Maximum temperature records.

Area	Highest (°C)	Highest (°F)	Place	Elevation (Meters)	Date
Africa	58	136	Azizia, Libya	114	Sept.13, 1922
North America	57	134	Death Valley, Calif.	−53	July 10, 1913
Asia	54	129	Tirat Tsvi, Israel	−217	June 21, 1942
Australia	53	128	Cloncurry, Queensland	187	Jan. 16, 1889
Europe	50	122	Seville, Spain	8	Aug. 4, 1881
South America	49	120	Rivadavia, Argentina	203	Dec. 11, 1905
Oceania	42	108	Tuguegarao, Philippines	22	April 29, 1912
Antarctica	15	59	Vanda Station	8	Jan. 5, 1974

Table 18.B Minimum temperature records.

Area	Lowest (°C)	Lowest (°F)	Place	Elevation (Meters)	Date
Antarctica	−89	−129	Vostok	3366	Aug. 24, 1960
Asia	−68	−90	Oymykon, Russia	788	Feb. 6, 1933
Greenland	−66	−87	Northice	2307	Jan. 9, 1954
North America	−63	−81	Snag Yukon, Canada	578	Feb. 3, 1947
Europe	−55	−67	Ust'Shchugor, Russia	84	Jan.*
South America	−33	−27	Sarmiento, Argentina	264	June 1, 1907
Africa	−24	−11	Ifrane, Morocco	1609	Feb. 11, 1935
Australia	−22	−8	Charlotte Pass, NSW	†	July 22, 1947**

* Exact date unknown; lowest in 15-year period.
** And earlier dates.
† Elevation unknown.

Table 18.1 Köppen system of climatic classification.

Letter Symbol			
1st	2nd	3rd	
A			Average temperature of the coldest month is 18°C or higher
	f		Every month has 6 cm of precipitation or more
	m		Short dry season; precipitation in driest month less than 6 cm but equal to or greater than 10 − R/25 (R is annual rainfall in cm)
	w		Well-defined winter dry season; precipitation in driest month less than 10 − R/25
	s		Well-defined summer dry season (rare)
B			Potential evaporation exceeds precipitation. The dry-humid boundary is defined by the following formulas: (Note: R is the average annual precipitation in cm and T is the average annual temperature in °C) R < 2T + 28 when 70% or more of rain falls in warmer 6 months R < 2T when 70% or more of rain falls in cooler 6 months R < 2T + 14 when neither half year has 70% or more of rain
	S		Steppe The BS–BW boundary is ½ the dry-humid boundary
	W		Desert
		h	Average annual temperature is 18°C or greater
		k	Average annual temperature is less than 18°C
C			Average temperature of the coldest month is under 18°C and above −3°C
	w		At least ten times as much precipitation in a summer month as in the driest winter month
	s		At least three times as much precipitation in a winter month as in the driest summer month; precipitation in driest summer month less than 4 cm
	f		Criteria for w and s cannot be met
		a	Warmest month is over 22°C; at least 4 months over 10°C
		b	No month above 22°C; at least 4 months over 10°C
		c	One to 3 months above 10°C
D			Average temperature of coldest month is −3°C or below; average temperature of warmest month is greater than 10°C
	s		Same as under C
	w		Same as under C
	f		Same as under C
		a	Same as under C
		b	Same as under C
		c	Same as under C
		d	Average temperature of the coldest month is −38°C or below
E			Average temperature of the warmest month is below 10°C
	T		Average temperature of the warmest month is greater than 0°C and less than 10°C
	F		Average temperature of the warmest month is 0°C or below

Humid Tropical (A) Climates

Within the A group of climates, two main types are recognized—wet tropical climates (Af and Am) and tropical wet and dry (AW).

The Wet Tropics

The constantly high temperatures and year-round rainfall in the wet tropics combine to produce the most luxuriant vegetation found in any climatic realm: the **tropical rain forest** (Figure 18.3).

The environment of the wet tropics characterizes almost 10 percent of Earth's land area. An examination of Figure 18.2 shows that Af and Am climates form a discontinuous belt astride the equator that typically extends 5 to 10° into each hemisphere. The poleward margins are most often marked by diminishing rainfall, but occasionally decreasing temperatures mark the boundary. Because of the general decrease in temperature with height in the troposphere, this climate region is restricted to elevations below 1000 meters. Consequently, the major interruptions near the equator are principally cooler highland areas.

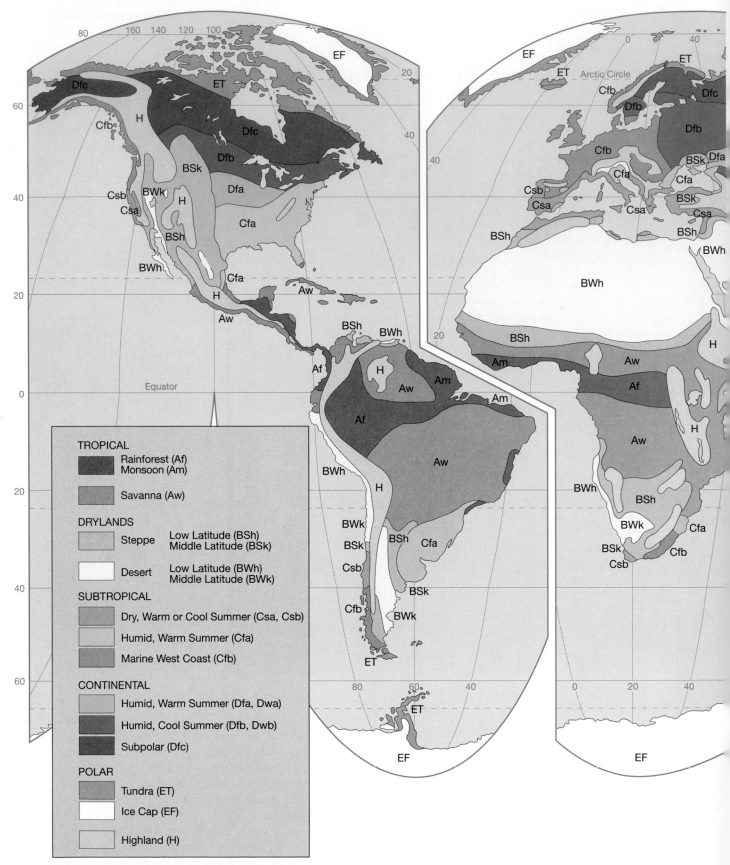

Figure 18.2
Climates of the world based on the Köppen classification.

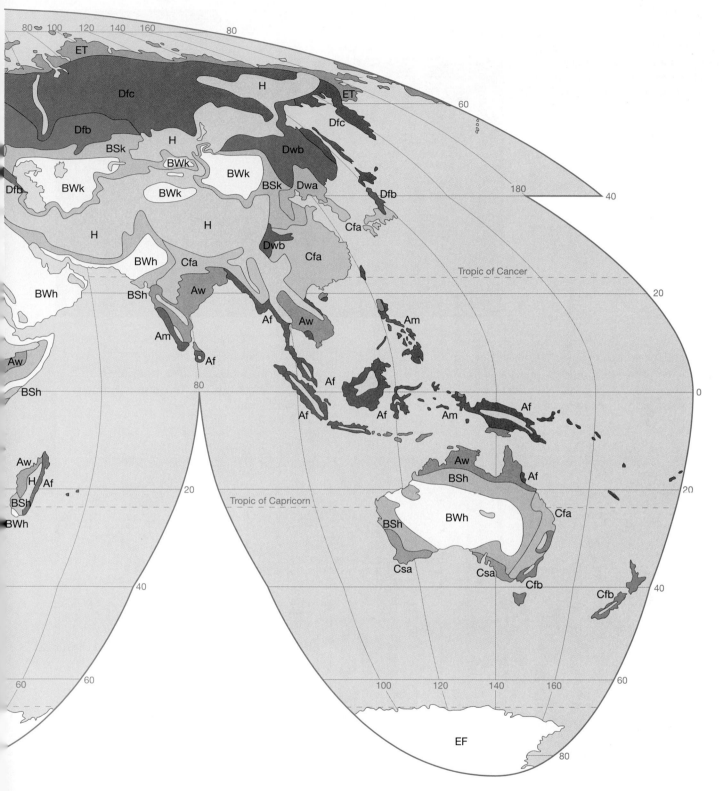

Figure 18.3
Unexcelled in luxuriance and characterized by hundreds of different species per square kilometer, the tropical rain forest is a broadleaf evergreen forest that dominates the wet tropics. (Photo by Wolfgang Kaehler)

Data for some representative stations in the wet tropics are shown in Figures 18.4A and B. A brief examination reveals the most obvious features that characterize the climate in these areas.

1. Temperatures usually average 25°C or more each month. Consequently, not only is the annual mean high, but the annual range is also very small.

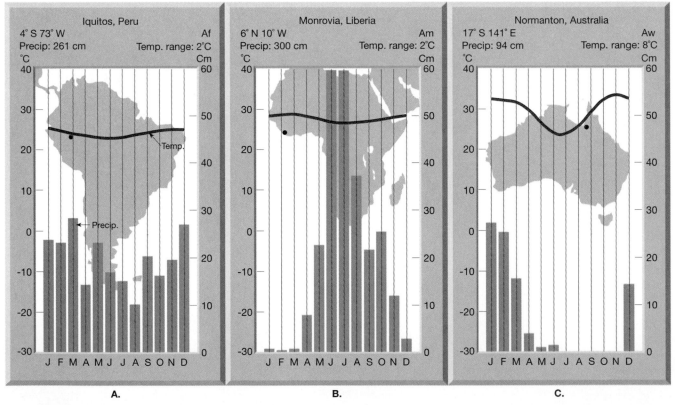

Figure 18.4
By comparing these three climatic diagrams, the primary differences among the *A* climates can be seen. **A.** Iquitos, the Af station, is wet throughout the year. **B.** Monrovia, the Am station, has a short, dry season. **C.** As is true for all Aw stations, Normanton has an extended dry season and a higher annual temperature range than the others.

2. The total precipitation for the year is high, often exceeding 200 centimeters.

3. Although rainfall is not evenly distributed throughout the year, tropical rain forest stations are generally wet in all months. If a dry season exists, it is a very short one.

Because places with an Af or Am designation lie near the equator, the reason for the uniform temperature rhythm experienced in such locales is clear: The intensity of solar radiation is consistently high. The vertical rays of the sun are always relatively close and changes in the length of daylight throughout the year are slight; therefore, seasonal temperature variations are minimal.

The region is strongly influenced by the equatorial low. Its converging trade winds and the accompanying ascent of warm, humid, unstable air, produce conditions that are ideal for the formation of precipitation.

Tropical Wet and Dry

In the latitude zone poleward of the wet tropics and equatorward of the subtropical deserts lies a transitional climatic region called **tropical wet and dry**. Here the rain forest gives way to the *savanna*, a tropical grassland with scattered drought-tolerant trees (Figure 18.5). Because temperature characteristics among all A climates are quite similar, the primary factor that distinguishes the Aw climate from Af and Am is precipitation. Although the overall amount of precipitation in the tropical wet and dry realm is often considerably less than in the wet tropics, the most distinctive feature of this climate is not the annual rainfall total but the markedly seasonal character of the rainfall. The climatic diagram for Normanton, Australia (Figure 18.4C) clearly illustrates this trait. As the equatorial low advances poleward in summer, the rainy season commences and features weather patterns typical of the wet tropics. Later, with the retreat of the equatorial low, the subtropical high advances into the region and brings with it intense drought conditions. In some Aw regions such as India, Southeast Asia, and portions of Australia, the alternating periods of rainfall and drought are associated with a pronounced monsoon circulation (see Chapter 16).

Dry (B) Climates

It is important to realize that the concept of dryness is a relative one and refers to any situation in which a water deficiency exists. Climatologists define a dry climate as one in which the yearly precipitation is not as great as the potential loss of water by evaporation. Thus,

Figure 18.5
This tropical savanna in Kenya, with its stunted, drought-resistant trees, probably resulted from seasonal burnings carried out by native human populations. (Photo by Michele Burgess/The Stock Market)

dryness is not only related to annual rainfall totals, but is also a function of evaporation, which in turn is closely dependent upon temperature.

To establish the boundary between dry and humid climates, the Köppen classification uses formulas that involve three variables: average annual precipitation, average annual temperature, and seasonal distribution of precipitation. The use of average annual temperature reflects its importance as an index of evaporation. The amount of rainfall defining the humid-dry boundary increases as the annual mean temperature increases. The use of seasonal precipitation as a variable is also related to this idea. If rain is concentrated in the warmest months, loss to evaporation is greater than if the precipitation were concentrated in the cooler months.

Within the regions defined by a general water deficiency are two climatic types: **arid** or **desert** (BW) and **semiarid** or **steppe** (BS) (Figure 18.6). These two groups have many features in common; their differences are primarily a matter of degree The semiarid is a marginal and more humid variant of the arid and represents a transition zone that surrounds the desert and separates it from the bordering humid climates.

Low-Latitude Deserts and Steppes

The heart of low-latitude dry climates lies in the vicinities of the Tropics of Cancer and Capricorn. A glance at Figure 18.6 shows a virtually unbroken desert environment stretching for more than 9300 kilometers from the Atlantic coast of North Africa to the dry lands of north-western India. In addition to this single great expanse, the Northern Hemisphere contains another much smaller area of subtropical desert and steppe in northern Mexico and the southwestern United States. In the Southern Hemisphere, dry climates dominate Australia. Almost 40 percent of the continent is desert, and much of the remainder is steppe. In addition, arid and semiarid areas are found in southern Africa and make a limited appearance in coastal Chile and Peru.

The existence of this dry subtropical realm is primarily the result of the prevailing global distribution of air pressure and winds. Earth's low-latitude deserts and steppes coincide with the subtropical high-pressure belts (see Figures 16.12, p. 429, and 16.13, p. 430). Here, air is subsiding. When air sinks, it is compressed and warmed. Such conditions are just opposite of what is needed for cloud formation and precipitation. Therefore, clear skies, a maximum of sunshine, and drought are to be expected. The climate diagrams for Cairo, Egypt, and Monterrey, Mexico (Figure 18.7A, B) illustrate the characteristics of low-latitude dry climates.

Middle-Latitude Deserts and Steppes

Unlike their low-latitude counterparts, middle-latitude deserts and steppes are not controlled by the subsiding air masses associated with high pressure. Instead, these dry lands exist principally because of their positions in the deep interiors of large landmasses far removed from the oceans, which are the ultimate source of moisture for cloud formation and precipitation. In addition, the

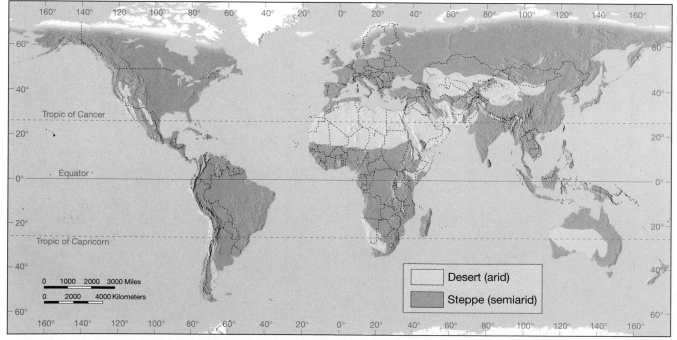

Figure 18.6

Arid and semiarid climates cover about 30 percent of Earth's land surface. No other climatic group covers so large an area.

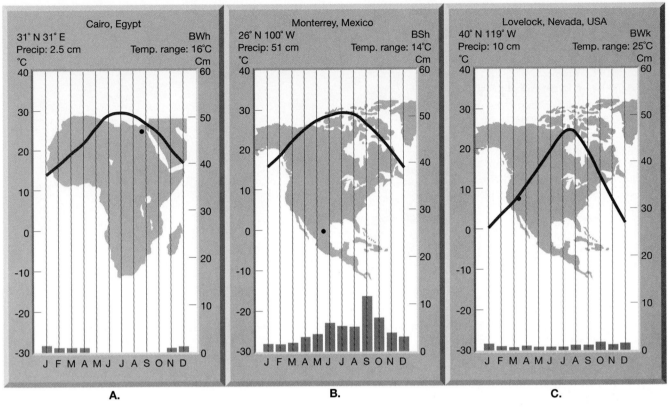

Figure 18.7

Climatic diagrams for representative arid and semiarid stations. Stations **A.** and **B.** are in the subtropics, whereas **C.** is in the middle-latitudes. Cairo and Lovelock are classified as deserts; Monterey is a steppe. Lovelock, Nevada may also be called a rainshadow desert.

presence of high mountains across the paths of prevailing winds further acts to separate these areas from water-bearing, maritime air masses.

Windward sides of mountains are often wet. As prevailing winds meet mountain barriers, the air is forced to ascend, producing clouds and precipitation (see Box 18.2). By contrast, the leeward sides of mountains are usually much drier and are often arid enough to be referred to as **rainshadow deserts** (see Figure 15.16, p. 402). Because many middle-latitude deserts occupy sites on the leeward sides of the mountains, they can also be classified as rainshadow deserts (Figure 18.7C). In North America, the Coast Ranges, Sierra Nevada, and Cascades are the foremost mountain barriers. In Asia, the great Himalayan chain prevents the summertime monsoon flow of moist Indian Ocean air from reaching the interior. Because the Southern Hemisphere lacks extensive land areas in the middle latitudes, only a small area of desert and steppe is found in this latitude range, existing primarily in the rainshadow of the towering Andes.

In the case of middle-latitude deserts, we have an example of the impact of tectonic processes on climate. Rainshadow deserts exist by virtue of the mountains produced when plates collide. Without such mountain-building episodes, wetter climates would prevail where many dry regions exist today.

Humid Middle-Latitude Climates with Mild Winters (C Climates)

Although the term *subtropical* is often used for the C climates, it can be misleading. While many areas with C climates do indeed possess some near-tropical characteristics, other regions do not. For example, we would be stretching the use of the term *subtropical* to describe the climates of coastal Alaska and Norway, which belong to the C group. Within the C group of climates, several subgroups are recognized.

Humid Subtropics

Located on the eastern sides of the continents, in the 25-to 40-degree latitude range, the **humid subtropical climate** dominates the southeastern United States, as well as other similarly situated areas around the world (Figures 18.2 and 18.8A). In the summer, the humid subtropics experience hot, sultry weather of the type one expects to find in the rainy tropics. Daytime temperatures are generally high, and since both specific and relative humidities are high, the night brings little relief. An afternoon or evening thunderstorm is also possible, for these areas experience such storms on an average of 40 to 100 days each year, the majority during the summer months.

Box 18.2

Precipitation Records and Mountainous Terrain

Many of the rainiest places in the world are located on windward mountain slopes. Typically these rainy areas are situated such that the mountains act as a barrier to the general circulation. Thus, the prevailing winds are forced to ascend the sloping terrain, thereby generating clouds and often abundant precipitation. A station at Mt. Waialeale, Hawaii, for example, records the highest average annual rainfall in the world, some 1234 centimeters (486 inches). The station is located on the windward (northeast) coast of the island of Kauai at an elevation of 1569 meters (5148 feet). Incredibly, only 31 kilometers (19 miles) away, lies the popular sunspot Barking Sands, with an annual precipitation that averages less than 50 centimeters (20 inches).

The greatest recorded rainfall for a twelve-month period occurred at Cherrapunji, India where an astounding 2647 centimeters (1042 inches), over 86 feet, fell. Cherrapunji, which is located at an elevation of 1293 meters (4309 feet), lies just north of the Bay of Bengal in an ideal location to receive the full effect of India's wet, summer monsoon. Most all of this rainfall occurred in the summer, particularly during

Figure 18.A

Mountain snows are important sources of water. (Photo by Philippe Ledru/Sygma)

the month of July, when a record of 930 centimeters (366 inches) fell. For comparison, 10 times more rain fell in a single month at Cherrapunji than falls in an average year at Chicago, Illinois.

Because mountains can be sites of abundant precipitation, they are frequently very important sources of water. This is particularly true for arid regions such as the southwestern United States. Here the snow pack that accumulates high in the mountains during the winter is a major source of

water for the summer season when precipitation is light and demand is high (Figure 18.A). Reservoirs in the Sierra Nevada, for example, accumulate and store spring runoff, which is then delivered to cities such as Los Angeles by way of an extensive network of canals. The record for greatest annual snowfall in the United States goes to Paradise Ranger Station, located on Mount Rainier, where 2850 centimeters (1122 inches) of snow fell during the winter of 1971–72.

As summer turns to autumn, the humid subtropics lose their similarity to the rainy tropics. Although winters are mild, frosts are common in the higher-latitude Cfa areas and occasionally plague the tropical margins as well. The winter precipitation is also different in character from the summer. Some is in the form of snow, and most is generated along fronts of the frequent middle-latitude cyclones that sweep over these regions.

Marine West Coast

Situated on the western (windward) side of continents, from about 40 to 65 degrees north and south latitude, is a climatic region dominated by the onshore flow of oceanic air. In North America, the **marine west coast climate** extends from near the U.S.–Canadian border northward as a narrow belt into southern Alaska (Figure 18.8B). The largest area of Cfb climate is found in Europe, for here there is no mountain barrier blocking the movement of cool maritime air from the North Atlantic.

The prevalence of maritime air masses means that mild winters and cool summers are the rule, as is an ample amount of rainfall throughout the year. Although there is no pronounced dry period, there is a drop in monthly precipitation totals during the summer. The reason for the reduced summer rainfall is the poleward migration of the oceanic subtropical highs. Although the areas of marine west coast climate are situated too far poleward to be dominated by these dry anticyclones, their influence is sufficient to cause a decrease in warm season rainfall.

Dry-Summer Subtropics

The **dry-summer subtropical climate** is typically located along the west sides of continents between latitudes 30 and 45°. Situated between the marine west coast climate on the poleward side and the subtropical steppes on the equatorward side, this climatic region is best described as transitional in character. It is unique because it is the only humid climate that has a strong winter rainfall maximum, a feature that reflects its intermediate position (Figure 18.8C). In summer, the region is dominated by stable conditions associated with the oceanic subtropical highs. In winter, as the wind and pressure systems follow the sun equatorward, it is within range of the cyclonic storms of the polar front. Thus, during the course of a year, these areas alternate between becoming a part of the dry tropics and an extension of the humid middle latitudes. Although middle-latitude changeability characterizes the winter, tropical constancy describes the summer.

As was the case for the marine west coast climate, mountain ranges limit the dry-summer subtropics to a relatively narrow coastal zone in both North and South America. Because Australia and southern Africa barely reach to the latitudes where dry-summer climates exist, the development of this climatic type is limited on these continents as well. Consequently, because of the arrangement of the continents, and of their mountain ranges, inland development occurs only in the Mediterranean basin. Here the zone of subsidence extends far to the east in summer; in winter, the sea is a major route of cyclonic disturbances. Because the dry-summer climate is particularly extensive in this region, the name *Mediterranean climate* is often used as a synonym (Figure 18.9).

Humid Middle-Latitude Climates with Severe Winters (D Climates)

The C climates that were just described characteristically have mild winters. By contrast, D climates experience severe winters. Two types of D climates are recognized, the humid continental and the subarctic. Climatic diagrams of representative locations are shown in Figure 18.10. The D climates are land-controlled climates, the result of broad continents in the middle latitudes. Because continentality is a basic feature, D climates are absent in the Southern Hemisphere where the middle-latitude zone is dominated by the oceans.

Humid Continental

The **humid continental climate** is confined to the central and eastern portions of North America and Eurasia in the latitude range between approximately 40 and 50°N latitude. It may at first seen unusual that a continental climate should extend eastward to the margins of the ocean. However, because the prevailing atmospheric circulation is from the west, deep and persistent incursions of maritime air from the east are not likely to occur.

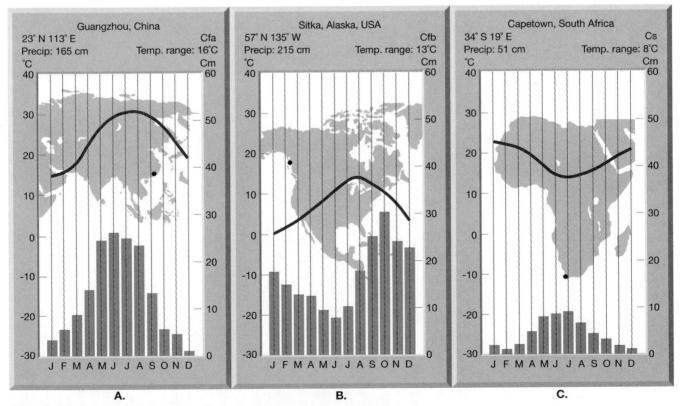

Figure 18.8

Each of these climatic diagrams represents one of the three main types of C climates: **A.** humid subtropical, **B.** marine west coast, and **C.** dry-summer subtropical.

Figure 18.9

A vineyard in southern France. The dry-summer subtropical climate is especially well developed here and in other regions adjacent to the Mediterranean Sea. (Photo by Michael Busselle/Tony Stone Images)

Both winter and summer temperatures in the humid continental climate may be characterized as relatively severe. Consequently, annual temperature ranges are high throughout the climate.

Precipitation is generally greater in summer than in winter. Precipitation totals generally decrease toward the interior of the continents as well as from south to north, primarily because of increasing distance from the sources of mT air. Furthermore, the more northerly stations are also influenced for a greater part of the year by drier polar air masses.

Wintertime precipitation in humid continental climates is chiefly associated with the passage of fronts connected with traveling middle-latitude cyclones. Part of this precipitation is in the form of snow, the proportion increasing with latitude. Although precipitation is often considerably less during the cold season, it is usually more conspicuous than the greater amounts that fall during summer. An obvious reason is that snow remains on the ground, often for extended periods, and rain, of course, does not.

Figure 18.10

D climates are associated with the interiors of large landmasses in the mid-to-high latitudes of the Northern Hemisphere. Although winters can be harsh in Chicago's humid continental (Dfa) climate, the subarctic environment (Dfc) of Moose Factory is more extreme.

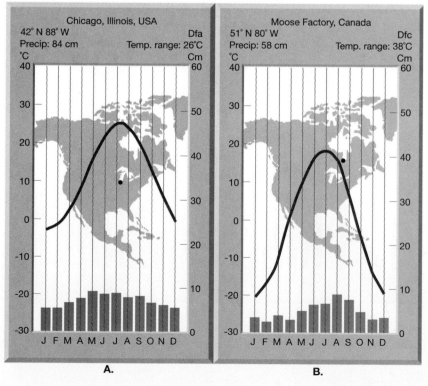

A.

B.

Subarctic

Situated north of the humid continental climate and south of the polar tundra is an extensive **subarctic climate** region covering broad, uninterrupted expanses from western Alaska to Newfoundland in North America and from Norway to the Pacific coast of Russia in Eurasia. It is often referred to as the *taiga* climate, for its extent closely corresponds to the northern coniferous forest region of the same name. Although scrawny, the spruce, fir, larch, and birch trees in the taiga represent the largest stretch of continuous forest on the surface of Earth.

Here in the source regions of continental polar air masses, the outstanding feature is certainly the dominance of winter. Not only is it long, but temperatures are also bitterly cold. Winter minimum temperatures are among the lowest ever recorded outside the ice sheets of Greenland and Antarctica. In fact, for many years, the world's coldest temperature was attributed to Verkhoyansk in east central Siberia, where the temperature dropped to –68°C on February 5 and 7, 1892. Over a 23-year period, this same station had an average monthly minimum of –62°C during January. Although exceptional temperatures, they illustrate the extreme cold that envelop the taiga in winter.

By contrast, summers in the subarctic are remarkably warm, despite their short duration. However, when compared with regions farther south, this short season must be characterized as cool. The extremely cold winters and relatively warm summer combine to produce the highest annual temperature ranges on Earth. Since these far northerly continental interiors are the source regions for cP air masses, there is very limited moisture available throughout the year. Precipitation totals are therefore small, with a maximum occurring during the warmer summer months.

Polar (E) Climates

Polar climates are those in which the mean temperature of the warmest month is below 10°C. Thus, just as the tropics are defined by their year-round warmth, the polar realm is known for its enduring cold. Since winters are periods of perpetual night, or nearly so, temperatures at most polar locations are understandably bitter. During the summer months temperatures remain cool despite the long days, because the sun is so low in the sky that its oblique rays are not effective in bringing about a genuine warming. Although polar climates are classified as humid, precipitation is generally meager. Evaporation, of course, is also limited. The scanty precipitation totals are easily understood in view of the temperature characteristics of the region. The amount of water vapor in the air is always small because low specific humidities must accompany low temperatures. Usually precipitation is most abundant during the warmer summer months when the moisture content of the air is highest.

Two types of polar climates are recognized. The **tundra climate** (ET) is a treeless climate found almost exclusively in the Northern Hemisphere (Figure 18.11A). Because of the combination of high latitude and

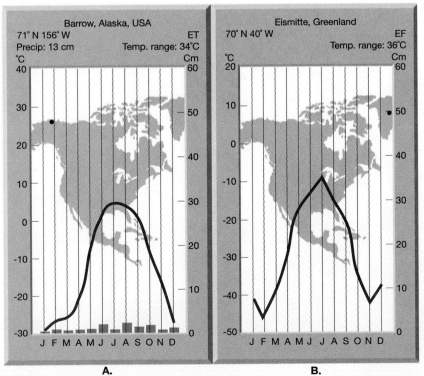

Figure 18.11
These climatic diagrams represent the two basic types of polar climates. **A.** Barrow, Alaska, exhibits a tundra (ET) climate. **B.** Eismitte, Greenland, a station located on a massive ice sheet, is classified as an ice cap (EF) climate.

continentality, winters are severe, summers are cool, and annual temperature ranges are high. Further, yearly precipitation is small, with a modest summer maximum.

The **ice cap climate** (EF) does not have a single monthly mean above 0°C (Figure 18.11B). Consequently, since the average temperature for all months is below freezing, the growth of vegetation is prohibited and the landscape is one of permanent ice and snow (Figure 18.12). This climate of perpetual frost covers a surprisingly large area—more than 15.5 million square kilometers, or about 9 percent of Earth's land area. Aside from scattered occurrences in high mountain areas, it is confined to the ice sheets of Greenland and Antarctica (see Figure 5.2, p. 124).

Highland Climates

It is a well-known fact that mountains have climate conditions that are distinctively different from those found in adjacent lowlands. Compared to nearby places at lower elevations, sites with **highland climates** are cooler and usually wetter. Unlike the world climate types already discussed, which consist of large, relatively homogeneous regions, the outstanding characteristic of highland climates is the great diversity of climatic conditions that occur.

The best-known climatic effect of increased altitude is lower temperatures. In addition, an increase in precipitation due to orographic lifting usually occurs at higher elevations. Despite the fact that mountain stations are colder and often wetter than locations at lower elevations, highland climates are often very similar to those in adjacent lowlands in terms of seasonal temperature cycles and precipitation distribution. Figure 18.13 illustrates this relationship.

Phoenix, at an elevation of 338 meters, lies in the desert lowlands of southern Arizona. By contrast, Flagstaff is located at an altitude of 2100 meters on the Colorodo Plateau in northern Arizona. When summer averages climb to 34°C in Phoenix, Flagstaff is experiencing a pleasant 19°C, which is a full 15°C cooler. Although the temperatures at each city are quite different, the pattern of monthly temperature changes for each place is similar. Both experience their minimum and maximum monthly means in the same months. When precipitation data are examined, both places have a similar seasonal pattern, but the amounts at Flagstaff are higher in every month. In addition, owing to its higher altitude, much of Flagstaff's winter precipitation is in the form of snow. By contrast, all of the precipitation at Phoenix is rain.

Perhaps the terms *variety* and *changeability* best describe mountain climates. Because atmospheric conditions fluctuate rapidly with changes in altitude and exposure, a nearly limitless variety of local climates occur in mountainous regions. The climate in a protected valley is very different from that of an exposed peak. Conditions on windward slopes contrast sharply with those on the leeward sides, whereas slopes facing the sun are unlike those that lie mainly in the shadows.

Human Impact on Global Climate

The proposals to explain global climatic change are many and varied. In Chapter 5, we examined some possible causes for ice-age climates. These hypotheses, which included the movement of lithosphere plates and variations in Earth's orbit, involved natural forcing

Figure 18.12
Aside from scattered occurrences in high mountain areas, the ice cap climate is confined to the ice sheets of Greenland and Antarctica. This icy scene is in Antarctica. (Photo by Frans Lanting/Minden Pictures)

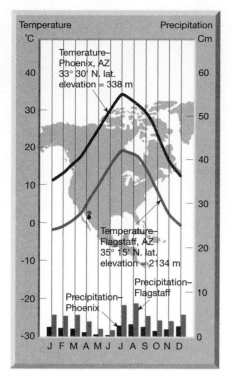

Figure 18.13
Climate diagrams for two stations in Arizona illustrate the general influence of elevation on climate. Flagstaff is cooler and wetter because of its position on the Colorado Plateau, nearly 1800 meters higher than Phoenix.

mechanisms. Another natural forcing mechanism, discussed in Chapter 8, is the possible role of explosive volcanic eruptions in modifying the atmosphere. It is important to remember that these mechanisms, as well as others, not only have contributed to climatic changes in the geologic past, but will also be responsible for future shifts in climate. However, when relatively recent and future changes in our climate are considered, we must also examine the possible impact of human beings (see Box 18.3). In this section we shall examine the major way in which humans may be contributing to global climatic change. This impact results from the addition of carbon dioxide and other gases to the atmosphere (Figure 18.14).

Human influence of regional and global climate probably did not just begin with the onset of the modern industrial period. There is good evidence that humans have been modifying the environment over extensive areas for thousands of years. The use of fire as well as the overgrazing of marginal lands by domesticated animals have negatively affected the abundance and distribution of vegetation. By altering ground cover, such important climatological factors as surface albedo, evaporation rates, and surface winds have been, and continue to be, modified. Commenting on this aspect of human-induced climatic modification, the authors of one study state,

In contrast to the prevailing view that only modern humans are able to alter climate, we believe it is more likely that the human species has made a substantial and continuing impact on climate since the invention of fire[*]

It should be pointed out that when any hypothesis of climatic change is examined, whether it depends on natural or human-induced causes, a degree of caution must be exercised. It is safe to say that most, if not all, hypotheses are to some degree controversial and speculative. This is to be expected if we consider the fact that at present all of our models of Earth's climate are far from complete. Because atmospheric processes are so large and complex, they cannot be physically reproduced in laboratory experiments. Instead, the climate must be simulated mathematically with the aid of computers (Figure 18.15). Although such models are sophisticated enough to serve as primary tools for climate research, they cannot yet approach the actual complexity of the atmosphere. Computer models are powerful and essential aids, but climate forecasts based on such simulations still contain many uncertainties.

Carbon Dioxide, Trace Gases, and Global Warming

In Chapter 14, we learned that although carbon dioxide (CO_2) represents only about 0.036 percent of the gases that make up clean, dry air, it is nevertheless a meteorologically significant component. The importance of carbon dioxide lies in the fact that it is transparent to incoming short-wavelength solar radiation, but it is not transparent to some of the longer-wavelength outgoing terrestrial radiation. A portion of the energy leaving the ground is absorbed by carbon dioxide and subsequently reemitted, part of it toward the surface, thereby keeping the air near the ground warmer than it would be without carbon dioxide. Thus, along with water vapor, carbon dioxide is largely responsible for the *greenhouse effect* of the atmosphere. Carbon dioxide is an important heat absorber, and it follows logically that any change in the air's carbon dioxide content could alter temperatures in the lower atmosphere.

Paralleling the rapid growth of industrialization begun in the nineteenth century has been the consumption of fossil fuels (coal, natural gas, and petroleum). The combustion of these fuels has added great quantities of carbon dioxide to the atmosphere. Although the use of coal and other fuels is the most prominent means by which humans add CO_2 to the atmosphere, it is not the only way. The clearing of forests also contributes substantially: Carbon dioxide is released as vegetation is

[*]Carl Sagan et al., "Anthropogenic Albedo Changes and the Earth's Climate," *Science* 206(4425):1367.

Box 18.3

Climate Change in Cities

The most apparent human impact on climate is the modification of the atmospheric environment by the building of cities. The construction of every factory, road, office building, and house destroys microclimates and creates new ones of great complexity.

At the beginning of the last century, only about 2 percent of the world's population lived in cities of more than 100,000 people. Today, not only is the total world population dramatically larger, but a far greater percentage of people reside in cities. In 1995, about 2.5 billion people lived in urban areas. This number represents about 40 percent of the world's people.

As Table 18.C illustrates, the climatic changes produced by urbanization involve all major surface conditions. Some changes are obvious and relatively easy to measure. Others are more subtle and sometimes difficult to measure. The amount of change in any of these elements, at any time, depends on several variables, including the extent of the urban complex, the nature of industry, site factors, such as topography and proximity to water bodies, time of day, season of the year, and existing weather conditions.

The Urban Heat Island

The most studied and well-documented urban climatic effect is the **urban heat island**. The term simply refers to the fact that temperatures within cities are generally higher than in rural areas. The heat island is evident when temperature data such as appear in Table 18.D are examined. As is typical, the data for Philadelphia show the heat island is most pronounced when minimum temperatures are examined. The magnitude of the temperature differences shown by this table is probably even greater than the figures indicate, because temperatures observed at suburban airports are usually higher than those in truly rural environments.

Figure 18.B, which shows the distribution of average minimum temperatures in the Washington, D.C., metropolitan area for the three-month winter period

Table 18.C Average climatic changes produced by cities.

Element	Comparison with Rural Environment
Particulate matter	10 times more
Temperature	
Annual mean	0.5–1.5°C higher
Winter	1–2°C higher
Solar radiation	15–30% less
Ultraviolet, winter	30% less
Ultraviolet, summer	5% less
Precipitation	5–15% more
Thunderstorm	16% more
Winter	5% more
Summer	29% more
Relative humidity	6% lower
Winter	2% lower
Summer	8% lower
Cloudiness	5–10% more
Fog	60% more
Winter	100% more
Summer	30% more
Wind speed	25% lower
Calm winds	5–20% more

Source: After Landsberg, Changnon, and others.

over a five-year span, also illustrates a well-developed heat island. The warmest winter temperatures occurred in the heart of the city, while the suburbs and surrounding countryside experienced average minimum temperatures that were as much as 3.3°C lower. Remember that these temperatures are averages, because on many clear, calm nights the temperature difference between the city center and the countryside was considerably greater, often 11°C (20°F) or more. Conversely, on many overcast or windy nights the temperature differential approached zero degrees.

Why are cities warmer? The radical change in the surface that results when rural areas are transformed into cities is a significant cause of the urban heat island. First, the tall buildings and the concrete and asphalt of the city absorb and store greater quantities of solar radiation than do the vegetation and soil typical of rural areas. In addition, because the city surface is impermeable, the runoff of water following a rain is rapid, resulting in a severe reduction on the evaporation rate. Hence, heat that once would have been used to convert liquid water to a gas now

goes to further increase the surface temperature. At night, while both the city and countryside cool by radiative losses, the stone-like surface of the city gradually releases the additional heat accumulated during the day, keeping the urban air warmer than that of the outlying areas.

A portion of the urban temperature rise is also attributed to waste heat from sources such as home heating and air conditioning, power generation, industry, and transportation. In addition, the "blan-

(continues)

Table 18.D Average temperatures (°C) for suburban Philadelphia Airport and downtown Philadelphia (ten-year averages).

	Airport	Downtown
Annual mean	12.8	13.6
Mean June max	27.8	28.2
Mean December max	6.4	6.7
Mean June min	16.5	17.7
Mean December min	−2.1	−0.4

Source: After H. Neuberger and J. Cahir, *Principles of Climatology* (New York: Holt, Rinehart and Winston, 1969), 128

(Box 18.3, continued)

ket" of pollutants over a city contributes to the heat island by absorbing a portion of the upward-directed long-wave radiation emitted by the surface and re-emitting some of it back to the ground.

Cities Increase Precipitation

Many studies comparing urban and rural precipitation have shown that the amount of precipitation in and downwind of a city can be from 5 to 15 percent greater than over the nearby countryside.

Several reasons have been proposed to explain why an urban complex might increase precipitation.

1. The urban heat island creates thermally induced upward motions that act to diminish the atmosphere's stability.

2. Clouds may be modified by the addition of condensation nuclei and freezing nuclei from industrial discharges.

3. The three-dimensional nature of urban areas creates an obstacle effect that impedes the passage of weather systems. Rain-producing processes may therefore linger over a city and increase precipitation.

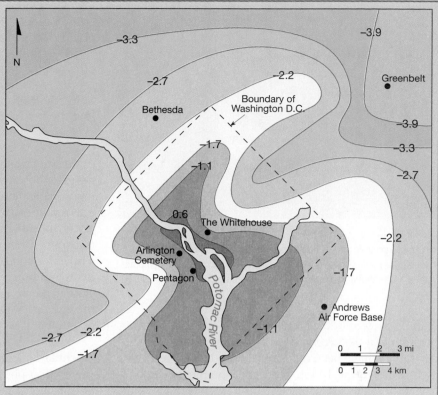

Figure 18.B

The heat island of Washington, D.C., as shown by the average minimum temperatures (°C) during the winter season (December through February). The city center had an average minimum that was nearly 4°C higher than some outlying areas. (After Clarence A. Woolum, "Notes from the Study of the Microclimatology of the Washington, D.C. Area for the Winter and Spring Seasons," *Weatherwise*, 17, no. 6 (1964), 264, 267)

burned or decays. Deforestation is particularly pronounced in the tropics, where vast tracts are cleared for ranching and agriculture or subjected to inefficient commercial logging operations.

Although some of the excess CO_2 is taken up by plants or is dissolved in the ocean, it is believed that 45 to 50 percent remains in the atmosphere. Consequently, since the mid-nineteenth century there has been an increase of more

Figure 18.14

Paralleling the rapid growth of industrialization, which began in the nineteenth century, has been the combustion of fossil fuels, which has added great quantities of carbon dioxide to the atmosphere. (Photo by Bruce Forster/Tony Stone Images)

Figure 18.15

Because large-scale atmospheric processes do not fit into the laboratory and because controlled long-term experiments are not possible, computer models are essential in climatic research. (Photo by Carlye Calvin)

than 25 percent in the carbon dioxide content of the air. Beginning in 1958, continuous measurements of CO_2 concentrations have been made at Mauna Loa Observatory in Hawaii (Figure 18.16). These measurements and data from other sites clearly show an upward trend from about 315 parts per million (ppm) to nearly 360 ppm. This increase closely matches the growth in CO_2 emissions. The seasonal fluctuations that are shown in the graph occur because CO_2 is removed from the air by plants during the growing season and returned later when the plants decay.

Naturally, we might expect that global temperatures should have already increased as a result of growing carbon dioxide levels. Since 1880, when reliable temperature measurements began, the mean global temperature has risen approximately 0.5°C. Although a number of observers suggest that the increase was caused by an enhanced greenhouse effect, "it is not yet possible to attribute a specific portion of the warming to an increase of greenhouse gases."[*] Most climatologists believe that the effects of the CO_2 increase are not yet large enough to show up clearly in the climatic record. However, the time is not far off when the effects may be evident.

If we assume that the use of fossil fuels will continue to increase at projected rates, current estimates indicate that the atmosphere's carbon dioxide content will approach 400 ppm by the year 2000 and will reach 600 ppm by some time in the second half of the next century. With such an increase in carbon dioxide, the enhancement of the greenhouse effect would be much more dramatic and measurable than in the past. When it is assumed that the atmosphere's carbon dioxide content will reach projected levels, the most realistic models predict an increase in the mean global surface temperature of between 1.5 and 4.5°C. A change of this magnitude would be unprecedented in human history. Such an increase would come close to equaling the warming that has taken place since the peak of the most recent glacial stage 18,000 years ago, except that it would occur much more rapidly.

Carbon dioxide is not the only gas contributing to a potential future global increase in temperature. In recent years, atmospheric scientists have come to realize that the industrial and agricultural activities of people are causing a buildup of certain trace gases that may also play a significant role. The substances are called *trace gases* because their concentrations are so much smaller than that of carbon dioxide. The trace gases that appear to be most important are methane (CH_4), nitrous oxide (N_2O), and certain types of chlorofluorocarbons (CFCs). These gases absorb wavelengths of outgoing Earth radiation that would otherwise escape into space. Although individually their

[*]C. K. Folland, T. R. Karl, and K.Y. Vinnikov, "Observed Climatic Variations and Change," in J. T. Houghton, G. J. Jenkins, and J. J. Ephraums, eds., *Climate Change: The IPCC Scientific Assessment* (New York: Cambridge University Press, 1990), p. 199.

Figure 18.16

Monitoring at Mauna Loa Observatory in Hawaii has revealed a significant rise in the concentration of carbon dioxide during the period shown. The yearly oscillation is caused by the seasonal growth and decay of vegetation. (After NOAA)

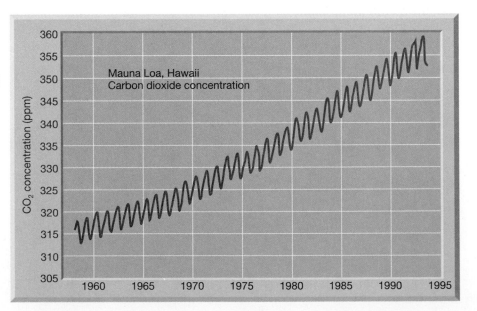

impact is modest, taken together the effects of these trace gases may be nearly as great as CO_2 in warming Earth.

Sophisticated computer models of the atmosphere show that the warming of the lower atmosphere triggered by CO_2 and trace gases will not be the same everywhere. Rather, the temperature response in polar regions could be as much as two to three times greater than the global average. Part of the reason for such a response is an expected reduction in sea ice. This topic is explored more fully in Box 18.4.

Box 18.4

Earth as a System: Climatic Feedback Mechanisms

Because climate is a very complex interactive physical system, scientists must consider many possible outcomes when one of the system's elements is altered. These various possibilities, termed *climatic feedback mechanisms*, not only complicate climatic modeling efforts, but they also add greater uncertainty to climatic predictions.

The most important and obvious of the feedback effects related to increases in carbon dioxide arises from the fact that higher surface temperatures produce greater evaporation rates. This, in turn, increases the amount of water vapor in the atmosphere. Remember that water vapor is also a powerful absorber of long-wavelength radiation emitted by Earth. Therefore, with more water vapor in the air, the temperature increase caused by carbon dioxide alone is reinforced.

Recall that the temperature increase at high latitudes is expected to be as much as two or three times greater than the global average. This assumption is based in part on the likelihood that the area covered by sea ice will decrease as surface temperatures rise. Because ice reflects a much larger percentage of incoming solar radiation than does open water, the melting of the pack ice would replace a highly reflective surface with a relatively dark surface (Figure 18.C). The result would be a substantial increase in the amount of solar energy absorbed at the surface. This, in turn, would feed back to the atmosphere and magnify the initial temperature increase created by higher carbon dioxide levels.

So far, the climate feedback mechanisms discussed have magnified the temperature rise caused by the buildup of carbon dioxide. The effects that reinforce an initial change are termed *positive feedback mechanisms*. As we shall see, however, other effects must be classified as *negative feedback mechanisms* because they produce results that are just the opposite of the initial change and tend to offset it.

One probable result of a global rise in temperatures would be an accompanying increase in cloud cover resulting from the higher moisture content of the atmosphere. Most clouds are good reflectors of solar radiation. At the same time, however, they are also good absorbers and emitters of terrestrial radiation. Consequently, clouds produce two opposite effects. They act as a negative feedback mechanism because they increase albedo and thus diminish the amount of solar energy available to heat the atmosphere. On the other hand, clouds act as a positive feedback mechanism by absorbing and emitting terrestrial radiation that would otherwise be lost from the troposphere.

Which effect, if either, is stronger? Atmospheric modeling shows that the negative effect of a higher albedo is the more dominant. Therefore, the net result of an increase in cloudiness should be a decrease in air temperature. The magnitude of this negative feedback, however, is not believed to be as strong as the positive feedback caused by added moisture and decreased sea ice. Thus, although increases in cloud cover may partly offset a global temperature increase, climatic models show that the ultimate effect of the projected increase in CO_2 and trace gases will still be a temperature increase.

The problem of global warming induced by increases in carbon dioxide and trace gases continues to be one of the most studied and discussed aspects of climatic change. Although there are no models that can yet incorporate the full range of potential influences and feedbacks, the consensus in the scientific community is that altering atmospheric composition will eventually lead to a warmer planet with a different distribution of climatic regimes.

Figure 18.C

A reduction in sea ice would act as a positive feedback mechanism because surface albedo would decrease and the amount of solar energy absorbed at the surface would increase. (Photo by Wolfgang Kaehler)

Some Possible Consequences of Greenhouse Warming

What consequences can be expected if the carbon dioxide content of the atmosphere reaches a level that is twice what it was early in the twentieth century? Because the workings of the climate system are so complex, predicting the distribution of particular regional changes is still very speculative. It is not yet possible to accurately pinpoint specifics such as where or when it will be drier or wetter. Nevertheless, plausible scenarios can be given for larger scales of space and time. Table 18.2 provides brief descriptions of several possible changes and their likelihood. The remainder of this section also examines some of the potential consequences. As computers become larger and data improve, scientists will gradually develop models that can provide more specific and reliable results.

As noted, the magnitude of the temperature increase will not be the same everywhere. The temperature rise will probably be smallest in the tropics and increase toward the poles. Furthermore, the models indicate that some regions will experience a significant increase in precipitation and runoff, whereas others will experience a decrease in runoff either because of reduced precipitation or because of increased evaporation rates brought about by higher temperatures. Such changes could have a profound impact on the distribution of the world's water resources and hence affect the productivity of agricultural regions that depend on rivers for irrigation water. A 2°C warming and 10 percent precipitation decrease in the region drained by the Colorado River, for example, could diminish the flow by 50 percent or more. Because the present flow of the river is barely enough to meet current demands for irrigated agriculture, the negative effect would be serious (Figure 18.17). Many other rivers form the basis for extensive systems of irrigated agriculture, and the projected reduction of their flow could have equally grave consequences. In contrast, large precipitation increases in other areas would increase the flow of some rivers and bring more frequent destructive floods.

The effects of increased atmospheric carbon dioxide on nonirrigated crops that depend on rain for moisture are complex and difficult to estimate. Some places will no doubt experience productivity losses due to de-

Table 18.2 Possible climate changes from doubling of CO_2.

Global Mean Surface Warming (Very Probable)

For a doubling of atmospheric carbon dioxide (or its radiative equivalent from all the greenhouse gases), the long-term global mean surface warming is expected to be in the range of 1.5 to 4.5°C. The most significant uncertainty arises from the effects of clouds. Of course, the actual rate of warming over the next century will be governed by the growth rate of greenhouse gases, natural fluctuations in the climate system, and the detailed response of the slowly responding parts of the climate system (that is, oceans and glacial ice).

Global Mean Precipitation Increase (Very Probable)

Increased heating of the surface will lead to increased evaporation and, therefore, to greater global mean precipitation. Despite this increase in global average precipitation, some individual regions might well experience decreases in rainfall.

Reduction of Sea Ice (Very Probable)

As the climate warms, total sea ice is expected to be reduced.

Polar Winter Surface Warming (Very Probable)

As the sea ice boundary is shifted poleward, the models predict a dramatically enhanced surface warming in winter polar regions. The greater fraction of open water and thinner sea ice will probably lead to warming of the polar surface air by as much as three times the global mean warming.

Summer Continental Dryness/Warming (Likely in the Long Term)

Several studies have predicted a marked long-term drying of the soil moisture over some mid-latitude interior continental regions during summer. This dryness is mainly caused by an earlier termination of snowmelt and rainy periods, and an earlier onset of the spring-to-summer reduction of soil wetness. Of course, these simulations of long-term equilibrium conditions may not offer a reliable guide to trends over the next few decades of changing atmospheric composition and changing climate.

High-Latitude Precipitation Increase (Probable)

As the climate warms, the increased poleward penetration of warm, moist air should increase the average annual precipitation in high latitudes.

Rise in Global Mean Sea Level (Probable)

A rise in mean sea level is generally expected because of thermal expansion of seawater in the warmer future climate. Far less certain is the contribution resulting from melting or the breaking up of land ice. See Box 13.2.

Source: National Academy of Sciences.

Figure 18.17
It is not yet possible to specify the magnitude and location of particular climate changes that may result from greenhouse warming. Many consequences are possible. Decreased rainfall and increased evaporation rates could diminish the flow of certain rivers and force the abandonment of some presently productive irrigated farmland. (Photo by E. J. Tarbuck)

creases in rainfall or increases in evaporation rates. Still, these losses may be offset by gains elsewhere. Increased temperatures in the high latitudes could lengthen the growing season, for instance. This factor, in turn, could allow the expansion of agriculture into areas that are presently not suited to crop production.

Concern is sometimes expressed that a warmer climate will cause glaciers to melt. In fact, a portion of the 10- to 20-centimeter rise in sea level over the past century is attributed to the melting of small glaciers and ice caps (Figure 18.18). Of course, if this were to occur on a large scale, it could lead to a much greater rise in sea level and a major encroachment by the sea in coastal zones. It should be emphasized, however, that a significant melting of major ice sheets, although possible at some future date, is *not* expected during the next century.

Finally, atmospheric scientists also expect that weather patterns will change as a result of the projected global warming. Potential weather changes include the following:

1. A higher frequency and greater intensity of hurricanes because of warmer ocean temperatures

2. Shifts in the paths of large-scale cyclonic storms, which, in turn, would affect the distribution of precipitation and the occurrence of severe storms, including tornadoes

3. Increases in the frequency and intensity of heat waves and droughts

In concluding this discussion, it should be stated again that the impact on globe climate of an increase in the atmosphere's content of CO_2 and trace gases is

Figure 18.18
The melting of valley glaciers over the past century has contributed to a modest rise in sea level. (Photo by E. J. Tarbuck)

obscured by many unknowns and uncertainties. The changes that occur will probably take the form of gradual environmental shifts that will be imperceptible to most people from year to year. Nevertheless, although the changes may seem gradual, the effects will clearly have powerful economic, social, and political consequences. Stephen Schneider, a respected authority on climate change, summarized the situation as follows:

> There is little doubt that if human trends in population, industrial, and economic growth continue into the next century, there will be substantial

increases in the greenhouse properties of the earth's atmosphere, which are virtually certain to create envionmental change. The controversies begin when we try to put numbers on the amount of warming, its timing, and its implications for ecosystems and society. Most controversial, of course, is what to do about this prospect.[*]

[*] Stephen H. Schneider, *Global Warming: Are We Entering the Greenhouse Century?* (San Francisco: Sierra Club Books, 1989), 23.

Review Questions

1. List the five parts of the climate system.

2. Why is classification often a helpful or even necessary task in science?

3. What climatic data are needed in order to classify a climate using the Köppen scheme?

4. (a) What primary factor distinguishes Aw climates from Af and Am?
   (b) How is this difference reflected in the vegetation?

5. Describe the influence of the equatorial low and the subtropical high on the precipitation regime in the Aw climate.

6. Why is the amount of precipitation that defines the humid-dry boundary variable?

7. What is the primary reason (control) for the existence of the dry subtropical realm (BWh and BSh)?

8. What is the primary cause for the existence of the middle-latitude deserts and steppes?

9. Describe and explain the differences between summertime and wintertime precipitation in the humid subtropics (Cfa).

10. Why is the marine west coast climate (Cfb and Cfc) represented by only slender strips of land in North and South America and why is it very extensive in western Europe?

11. In this chapter the dry-summer subtropics were described as transitional. Explain why this statement is true.

12. Why is the humid continental climate confined to the Northern Hemisphere?

13. Although generally characterized by small precipitation totals, subarctic and polar climates are considered humid. Explain.

14. Describe and explain the annual temperature range one should expect in the realm of the taiga.

15. Although polar regions experience extended periods of almost perpetual sunlight in the summer, temperatures remain cool. Explain.

16. The tundra climate is not confined solely to high latitudes. Under what circumstances might the ET climate be found in more equatorward locations?

17. Where are EF climates most extensively developed?

18. The Arizona cities of Flagstaff and Phoenix are relatively close to one another yet have contrasting climates (see Figure 18.13). Briefly explain why the differences occur.

19. Why has the atmosphere's carbon dioxide level been rising for more than 130 years?

20. How are temperatures in the lower atmosphere likely to change as carbon dioxide levels continue to increase? Why?

21. Is carbon dioxide the only gas that may contribute to a future global temperature change?

22. How do cities influence temperature and precipitation? (See Box18.3.)

23. List three possible consequences of a global warming.

24. What are climatic feedback mechanisms? Give an example of each type. (See Box 18.4.)

Key Terms

arid climate (p. 472)
climate system (p. 464)
desert climate (p. 472)
dry-summer subtropical climate (p. 474)

highland climate (p. 478)
humid continental climate (p. 475)
humid subtropical climate (p. 473)
ice cap climate (p. 478)
Köppen classification (p. 465)
marine west coast climate (p. 474)
polar climate (p. 477)

rainshadow desert (p. 473)
semiarid climate (p. 472)
steppe climate (p. 472)
subarctic climate (p. 477)
tropical rain forest (p. 467)
tropical wet and dry climate (p. 471)
tundra climate p. 477)

PART 4
Astronomy

The Moon in orbit about Earth from Galileo Spacecraft.
(NASA Image/Sygma)

CHAPTER 19

Earth's Place in the Universe

Stonehenge, an ancient observatory in England. (Photo by Robert Llewellyn)

Earth is one of nine planets and numerous smaller bodies that orbit the sun. The sun is part of a much larger family of perhaps 100 billion stars that compose the Milky Way, which in turn is only one of billions of galaxies in an incomprehensibly large universe. This view of Earth's position in space is considerably different from that held only a few hundred years ago, when Earth was thought to occupy a privileged position as the center of the universe. This chapter unfolds with some events that led to modern astronomy. In addition, it examines Earth's place in time and space.

Long before recorded history, which began about 5,000 years ago, people were aware of the close relationship between events on Earth and the positions of heavenly bodies, the sun in particular. People noted that changes in the seasons and floods of great rivers like the Nile in Egypt occurred when the celestial bodies, including the sun, moon, planets, and stars, reached a particular place in the heavens. Early agrarian cultures, which were dependent on the weather, believed that if the heavenly objects could control the seasons, they must also strongly influence all Earthly events. This belief undoubtedly was the reason that early civilizations began keeping records of the positions of the celestial objects. The Chinese, Egyptians, and Babylonians in particular are noted for this.

These cultures recorded the locations of the sun, moon and the five planets visible to the unaided eye as these objects moved slowly against the background of "fixed" stars. In addition, the Chinese kept quite accurate records of comets and "guest stars" (Figure 19.1). Today we know that a "guest star" is really a normal star, usually too faint to be visible, that increases its brightness as it explosively ejects gases from its surface, a phenomenon we call a nova.

A study of Chinese archives shows that they recorded every appearance of the famous Halley's comet for at least ten centuries. However, because this comet appears only once in a lifetime—once every 76 years—they were unable to link these appearances to establish that what they saw was the same object each time. Thus, like most ancients, the Chinese considered comets to be mystical. Comets were seen as bad omens and were blamed for a variety of disasters, from wars to plagues (Figure 19.2).

Ancient Astronomy

The "Golden Age" of early astronomy (600 B.C.–A.D. 150) was centered in Greece. The early Greeks have been criticized, and rightly so, for using philosophical arguments to explain natural phenomena. However, they did rely on observational data as well. The basics of geometry and trigonometry, which they had developed, were used to measure the sizes and distances of the largest-appearing bodies in the heavens—the sun and the moon.

Early Greeks

Many astronomical discoveries have been credited to the Greeks. They held the **geocentric** ("Earth-centered") view, believing that Earth was a sphere that stayed motionless at the center of the universe. Orbiting Earth were the moon, sun, and the known planets—Mercury, Venus, Mars, and Jupiter. Beyond the planets was a transparent, hollow sphere (**celestial sphere**) on which the stars traveled daily around Earth (this is how it looks, but of course, the effect is actually caused by Earth's rotation about its axis). Some early Greeks realized that the motion of the stars could be explained just as easily by a rotating Earth, but they rejected that idea, because Earth exhibited no sense of motion and seemed

Figure 19.1

Newly discovered Comet Hyakutake was visible to skywatchers in the spring of 1996. (AP photo/Lancaster New Era/ Chris Knight)

Figure 19.2
The Bayeux Tapestry that hangs in Bayeux, France, shows the apprehension caused by Halley's comet in A.D. 1066. This event preceded the defeat of King Harold by William the Conqueror. (Photo by Giraudon/Art Resource with special authorization of the City of Bayeux)

too large to be movable. In fact, proof of Earth's rotation was not demonstrated until 1851, a topic we consider in Box 19.1.

To the Greeks, all of the heavenly bodies, except seven, appeared to remain in the same relative position to one another. These seven wanderers (*planetai* in Greek) included the sun, the moon, Mercury, Venus, Mars, Jupiter, and Saturn. Each was thought to have a circular orbit around Earth. Although this system was incorrect, the Greeks refined it to the point that it explained the apparent movements of all celestial bodies.

As early as the fifth century B.C., the Greeks understood what causes the phases of the moon. Anaxagoras reasoned that the moon shines by reflected sunlight, and because it is a sphere, only half is illuminated at one time. As the moon orbits Earth, that portion of the illuminated half that is visible from Earth is always changing. Anaxagoras also realized that an eclipse of the moon occurs when it moves into Earth's shadow.

The famous Greek philosopher Aristotle (384–322 B.C.) concluded that Earth is spherical because it always casts a curved shadow when it eclipses the moon. Although most of the teachings of Aristotle were passed along and were considered infallible by many, his belief in a spherical Earth was abandoned during the Middle Ages.

The first Greek to profess a sun-centered, or **heliocentric**, universe was Aristarchus (312–230 B.C.). Aristarchus also used simple geometric relations to calculate the relative distances from Earth to the sun and the moon. He later used these data to calculate their sizes. As a result of observational errors beyond his control, he came up with measurements that were much too small. However, he did learn that the sun was many times more distant than the moon and many times larger than Earth. The latter fact may have prompted him to suggest a sun-centered universe. Nevertheless, because of the strong influence of Aristotle, the Earth-centered view dominated western thought for nearly 2000 years.

The first successful attempt to establish the size of Earth is credited to Eratosthenes (276–194 B.C.). Eratosthenes observed the angles of the noonday sun in two Egyptian cities that were roughly north and south of each other—Syene (presently Aswan) and Alexandria (Figure 19.3). Finding that the angles differed by 7 degrees, or $\frac{1}{50}$ of a complete circle, he concluded that the circumference of Earth must be 50 times the distance between these two cities. The cities were 5000 *stadia* apart, giving him a measurement of 250,000 *stadia*. Many historians believe the *stadia* was 157.6 meters (517 feet), which would make Eratosthenes' calculation of Earth's circumpherence—39,400 kilometers (24,428 miles)—a measurement very close to the modern value of 40,075 kilometers (24,902 miles).

Probably the greatest of the early Greek astronomers was Hipparchus (second century B.C.), best known for his star catalog. Hipparchus determined the location of almost 850 stars, which he divided into six groups according to their brightness. He measured the length of the year to within minutes of the modern value and developed a method for predicting the times of lunar eclipses to within a few hours.

Although many of the Greek discoveries were lost during the Middle Ages, the Earth-centered view that the Greeks proposed became established in Europe. Presented in its finest form by Cladius Ptolemy, this geocentric outlook became known as the **Ptolemaic system**.

The Ptolemaic System

Much of our knowledge of Greek astronomy comes from a thirteen-volume treatise, *Almagest* ("the great work"), which was complied by Ptolemy in A.D. 141 and survived thanks to the work of Arab scholars. In this work, Ptolemy is credited with developing a model of the universe that accounted for the observable motions of the planets (Figure 19.4). The precision with which

Box 19.1

Proof of Earth's Rotation

Every school child learns that Earth rotates on its axis once each day to produce periods of daylight and darkness. However, day and night and the apparent motions of the stars can be accounted for equally well by a sun and celestial sphere that revolve around a stationary Earth. Copernicus realized that a rotating Earth greatly simplified the existing model of the universe and strongly advocated this as the correct view. He was unable, however, to *prove* that Earth rotates. The first substantial proof was presented 300 years after his death by the French physicist Jean Foucault.

In 1851 Foucault used a free-swinging pendulum to demonstrate that Earth does, in fact, turn on its axis. To envision Foucaults's experiments, imagine a large pendulum swinging over the North Pole (Figure 19.A). Keep in mind that once a pendulum is put

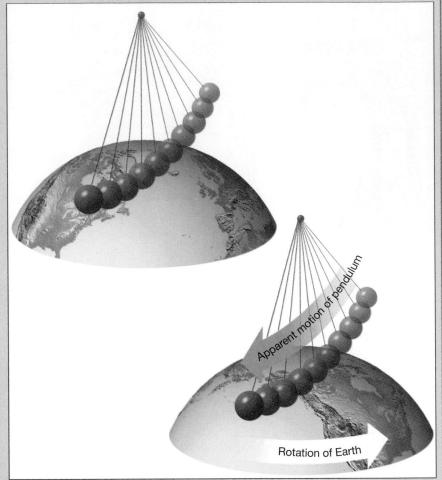

Figure 19.A
Apparent movement of a pendulum at the North Pole caused by Earth rotating beneath it.

Figure 19.B
Foucault pendulum housed in Museum of Science and Industry, Chicago. (Courtesy Museum of Science and Industry)

into motion, it continues swinging in the same plane unless acted upon by some outside force. Assume that a sharp stylus is attached to the bottom of this pendulum, marking the snow as it oscillates. When we observe the marks made by the stylus, we note that the pendulum is slowly but continually changing position. At the end of 24 hours it has returned to the starting position (Figure 19.A).

Because no outside force acted on the pendulum to change its position, what we observed must have been Earth rotating beneath it. Foucault conducted a similar experiment when he suspended a long pendulum from the dome of the Pantheon in Paris. Today, Foucault pendulums can be found in a number of museums to recreate this famous scientific experiment (Figure 19.B).

his model was able to predict planetary motion is attested to by the fact that it went virtually unchallenged, in principle if not in detail, for nearly thirteen centuries.

In the Greek tradition, the Ptolemaic model had the planets moving in circular orbits around a motionless Earth. (The circle was considered the pure and perfect shape by the Greeks.) However, the motion of the planets, as seen against the background of stars, is not so simple. Each planet, if watched night after night, moves slightly eastward among the stars. Periodically, each planet appears to stop, reverse direction for a period of time, and then resume an eastward motion. The apparent westward drift is called **retrograde motion**. This rather odd *apparent* motion results from the combination of the motion of Earth and the planet's own motion around the sun.

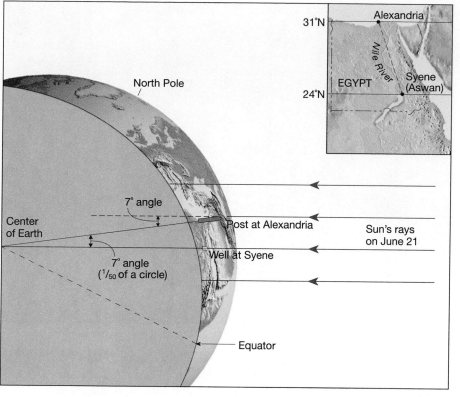

Figure 19.3

Orientation of the sun's rays at Syene (Aswan) and Alexandria in Egypt on June 21 when Eratosthenes calculated Earth's circumference.

Figure 19.5 illustrates the retrograde motions of Mars. Earth has a faster orbital speed than Mars so it overtakes its neighbor. While doing so, Mars *appears* to be moving backward, in retrograde motion. This is analogous to what a race-car driver sees out the side window when passing a slower car. The slower planet, like the slower car, appears to be going backward, although its actual motion is in the same direction as the faster-moving body.

Figure 19.5 shows how retrograde motion works. It is much more difficult to accurately represent retrograde motion using the incorrect Earth-centered model, but Ptolemy was able to do so (Figure 19.4B). Rather than using a simple circle for each planet's orbit, he showed that the planets orbited on small circles (*epicycles*), revolving along large circles (*deferents*). By trial and error, he found the right combinations of circles to produce the amount of retrograde motion observed for each planet. (An interesting note is that almost any closed curve can be produced by the combination of two circular motions, a fact that can be verified by persons who have used the Spirograph® design-drawing toy.)

It is a tribute to Ptolemy's genius that he was able to account for the planets' motions as well as he did, considering that he used an incorrect model. Some suggest that he did not mean his model to represent reality, but only to be used for calculating the positions of heavenly bodies. We probably will never know his intentions. However, the Roman Catholic church, which dominated European thought for centuries, accepted Ptolemy's theory as the correct representation of the heavens, and this created problems for those who found fault with it.

The Birth of Modern Astronomy

Modern astronomy was not born overnight. Its development involved a break from deeply entrenched philosophical and religious views and, during the 1500s and 1600s, the founding of a "new and greater universe" governed by discernible laws. Let us now look at the work of five noted scientists involved in this transition: Nicolaus Copernicus, Tycho Brahe, Johannes Kepler, Galileo Galilei, and Sir Isaac Newton.

Nicolaus Copernicus

For almost thirteen centuries after the time of Ptomely, very few astronomical advances were made in Europe. The first great astronomer to emerge after the Middle Ages was Nicholas Copernicus (1473–1543) from Poland (Figure 19.6). Copernicus became convinced that Earth is a planet, just like the other five then-known planets. The daily motions of the heavens, he reasoned, could be better explained by a rotating Earth. To counter the Ptolemaic objection that Earth would fly apart if it rotated, Copernicus suggested that the much-larger celestial sphere would be even more likely to fly apart if it rotated!

Figure 19.4
The universe according to Ptolemy, second century A.D. **A.** Ptolemy believed that the star-studded celestial sphere made a daily trip around a motionless Earth. In addition he proposed that the sun, moon, and planets made trips of various lengths along individual orbits. **B.** Retrograde motion as explained by Ptolemy.

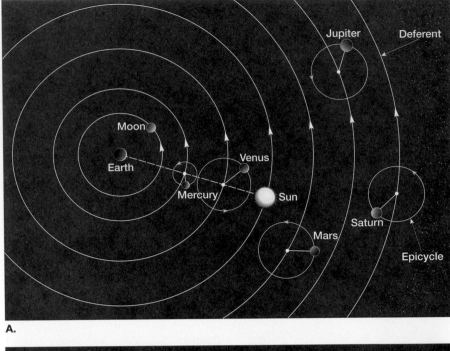

A.

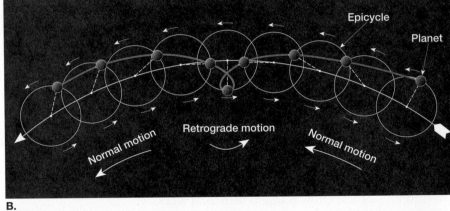

B.

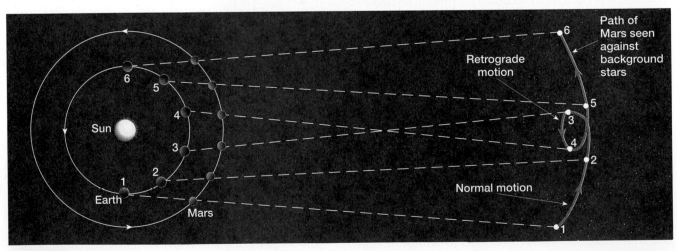

Figure 19.5
Retrograde (backward) motion of Mars as seen against the background of distant stars. When viewed from Earth, Mars moves eastward among the stars each day, then periodically appears to stop and reverse direction. This apparent westward drift is a result of the fact that Earth has a faster orbital speed than Mars and overtakes it. As this occurs, Mars appears to be moving backward; that is, it exhibits retrograde motion.

Figure 19.6

Polish astronomer Nicolaus Copernicus (1473–1543) believed that Earth was just another planet. (Yerkes Observatory Photograph)

Having concluded that Earth is a planet, Copernicus reconstructed the solar system with the sun at the center and the planets Mercury, Venus, Earth, Mars, Jupiter, and Saturn orbiting around it. This was a major break from the ancient idea that a motionless Earth lies at the center of all movement. However, Copernicus retained a link to the past and used circles, which were considered to be the perfect geometric shape, to represent the orbits of the planets. Although these circular orbits were close to reality, they didn't quite match what people saw. Unable to get satisfactory agreement between predicted locations of the planets and their observed positions, Copernicus found it necessary to add epicycles like those used by Ptomely. The discovery that the planets have *elliptical* orbits would wait another century for the insights of Johannes Kepler.

Also, like his predecessors, Copernicus used philosophical justifications, such as the following, to support his point of view.

. . .in the midst of all stands the sun. For who could in this most beautiful temple place this lamp in another or better place than that from which it can at the same time illuminate the whole?

Copernicus' monumental work, *De Revolutionibus, Orbium Coelestium* (*On the Revolution of the Heavenly Spheres*), which set forth his controversial ideas, was published as he lay on his deathbed. Hence, he never suffered the criticisms that fell on many of his followers.

The Copernican system challenged the primacy of Earth in the universe and was considered heretical by the Church. Expounding the idea cost at least one person his life. Giordano Bruno was seized by the Inquisition, a Church tribunal, in 1600 and, refusing to denounce the Copernican theory, was tied to a stake and burned alive.

Tycho Brahe

Tycho Brahe (1546-1601) was born of Danish nobility three years after the death of Copernicus. Reportedly, Tycho became interested in astronomy while viewing a solar eclipse that had been predicted by astronomers. He persuaded King Fredrich II to establish an observatory, which he headed, near Copenhagen. There he designed and built pointers (the telescope would not be invented for a few more decades), which he used for 20 years to systematically measure the locations of the heavenly bodies (Figure 19.7). These observations, particularly of Mars, were far more precise than any made previously and are his legacy to astronomy.

Tycho did not believe in the Copernican (sun-centered) system, because he was unable to observe an apparent shift in the position of stars that would be caused by Earth's motion. His argument went like this: If Earth does revolve along an orbit around the sun, the position of a nearby star, when observed from extreme points in Earth's orbit six months apart, should shift with respect to the more distant stars. His *idea* was correct, and this apparent shift of the stars is called *stellar parallax* (see Figure 22.2, p. 569).

The principle of parallax is easy to visualize: Close one eye, and with your index finger vertical, use your eye to line up your finger with some distant object. Now, without moving your finger, view the object with your other eye and notice that the object's position appears to change. The farther away you hold your finger, the less the object's position seems to shift. Herein lay the flaw in Tycho's argument. He was right about parallax, but, because the distance to even the nearest stars is enormous compared to the width of Earth's orbit, the shift that occurs is too small to be noticed by using the first primitive telescopes, let alone the unaided eye.

With the death of his patron, the King of Denmark, Tycho was forced to leave his observatory. It was probably his arrogant and extravagant nature that caused a conflict with the next ruler, so Tycho moved to Prague in the Czech Republic. Here, in the last year of his life, he acquired an able assistant, Johannes Kepler. Kepler retained most of the observations made by Tycho and

Figure 19.7

Tycho Brahe (1546–1601) in his observatory, Uraniborg, on the Danish island of Hveen. Tycho (central figure) and the background are painted on the wall of the observatory within the arc of the sighting instrument called a quadrant. In the far right, Tycho can be seen "sighting" a celestial object through the "hole" in the wall. Tycho's accurate measurements of Mars enabled Johannes Kepler to formulate his three laws of planetary motion. (Courtesy of Thomas Clarke, McLaughlin Planetarium)

put them to exceptional use. Ironically, the data Tycho collected to refute the Copernican view would later be used by Kepler to support it.

Johannes Kepler

If Copernicus ushered out the old astronomy, Johannes Kepler (1571-1630) ushered in the new (Figure 19.8). Armed with Tycho's data, a good mathematical mind, and, of greater importance, a strong faith in the accuracy of Tycho's work, Kepler derived three basic laws of planetary motion. The first two laws resulted from his inability to fit Tycho's observations of Mars to a circular orbit. Unwilling to concede that the discrepencies were a result of observational error, he searched for another solution. This endeavor led him to discover that the orbit of Mars is not a perfect circle but is elliptical (Figure 19.9). About the same time, he realized that the orbital speed of Mars varies in a predictable way. As it approaches the sun, it speeds up, and as it moves away from the sun, it slows down.

In 1609, after almost a decade of work, Kepler proposed his first two laws of planetary motion:

Figure 19.8
German astronomer Johannes Kepler (1571–1630) helped establish the era of modern astronomy by deriving three laws of planetary motion. (Smithsonian Institution Photo #56123)

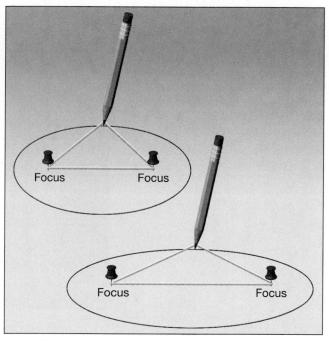

Figure 19.9
Drawing ellipses with various eccentricities. Using two straight pins for foci and a loop of string, trace out a curve while keeping the string taut, and you will have drawn an ellipse. The further the pins (the foci) are moved apart, the more flattened (more eccentric) is the resulting ellipse.

1.   The path of each planet around the sun is an ellipse with the sun at one focus (Figure 19.9). The other focus is symmetrically located at the opposite end of the ellipse.

2.   Each planet revolves so that an imaginary line connecting it to the sun sweeps over equal areas in equal intervals of time (Figure 19.10). This law of equal areas expresses geometrically the variations in orbital speeds of the planets.

Figure 19.10 illustrates the second law. Note that in order for a planet to sweep equal areas in the same amount of time, it must travel more rapidly when it is nearer the sun and more slowly when it is farther from the sun.

Kepler was very religious and believed that the Creator made an orderly universe. The uniformity he tried to find eluded him for nearly a decade. Then in 1619, he published his third law in *The Harmony of the Worlds*.

3.   The orbital periods of the planets and their distances to the sun are proportional.

In its simplest form, the orbital period of revolution is measured in Earth years, and the planet's distance to the sun is expressed in terms of Earth's mean distance to the sun. The latter "yardstick" is called the **astronomical unit** (AU) and averages about 150 million kilometers (93 million miles).

Using these units, Kepler's third law states that the planet's orbital period squared is equal to its mean solar distance cubed ($p^2 = d^3$). Consequently, the solar distances of the planets can be calculated when their periods of revolution are known. For example, Mars has a period of 1.88 years, which squared equals 3.54. The cube root of 3.54 is 1.52, and that is the distance to Mars in astronomical units (Table 19.1).

Kepler's laws assert that the planets revolve around the sun and therefore support the Copernican theory. Kepler, however, did fall short of determining the *forces* that act to produce the planetary motion he had so ably described. That task would remain for Galileo Galilei and Sir Isaac Newton.

Galileo Galilei

Galileo Galilei (1564–1642) was the greatest Italian scientist of the Renaissance (Figure 19.11). He was a contemporary of Kepler and, like Kepler, strongly supported the Copernican theory of a sun-centered solar system. Galileo's greatest contributions to science were his descriptions of the behavior of moving objects. These he

Figure 19.10

Kepler's law of equal areas. A line connecting a planet (Earth) to the sun sweeps out an area in such a manner that equal areas are swept out in equal times. Thus, Earth revolves slower when it is farther from the sun (aphelion) and faster when it is closest (perihelion). The eccentricity of Earth's orbit is greatly exaggerated in this diagram.

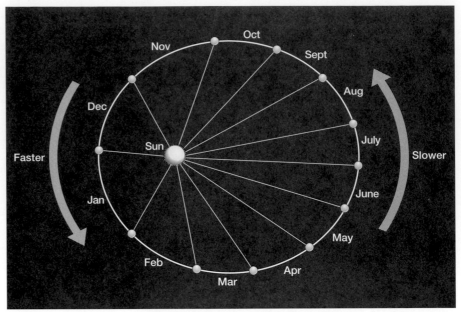

derived from experimentation. The method of using experiments to determine natural laws had essentially been lost since the time of the early Greeks.

An example of such experimentation is the legend that Galileo climbed to the top of the Leaning Tower of Pisa, Italy, where he dropped two objects of differing masses. To the astonishment of the observers, they hit the ground at the same time. (Galileo probably did not attempt this experiment, although a colleague might have.) Galileo correctly concluded that variations in the rate of fall of very light objects, like feathers, are due to air resistance, not differences in mass. This same experiment was performed most dramatically on the airless moon nearly four centuries later when David Scott, an *Apollo 15* astronaut, showed that a feather and a hammer fall at the same rate.

All astronomical discoveries before Galileo's time were made without the aid of a telescope. In 1609, Galileo heard that a Dutch lens maker had devised a system of lenses that magnified objects. Apparently

Table 19.1 Period of revolution and solar distances of planets.

Planet	Solar Distance (AU)*	Period (years)
Mercury	0.39	0.24
Venus	0.72	0.62
Earth	1.00	1.00
Mars	1.52	1.88
Jupiter	5.20	11.86
Saturn	9.54	29.46
Uranus	19.18	84.01
Neptune	30.06	164.80
Pluto	39.44	247.70

* AU = astronomical unit.

Figure 19.11

Italian scientist Galileo Galilei (1564–1642) used a new invention, the telescope, to observe the sun, moon, and planets in more detail than ever before. (Yerkes Observatory Photograph)

without ever seeing a telescope, Galileo constructed his own, which magnified distant objects to three times the size seen by the unaided eye. He immediately made others, the best having a magnification of about 30.

With the telescope, Galileo was able to view the universe in a new way. He made many important discoveries that supported the Copernican view of the universe, such as:

1. The discovery of four satellites, or moons, orbiting Jupiter. Galileo accurately determined their periods of revolution, which range from 2 to 17 days (Figure 19.12). This find dispelled the old idea that Earth was the only center of motion in the universe; for here, plainly visible, was another center of motion—Jupiter. It also countered the argument, frequently used by those opposed to the sun-centered system, that the moon would be left behind if Earth really revolved around the sun.

2. The discovery, through observation, that the planets are circular disks rather than just points of light, as was previously thought. This indicated that the planets must be Earthlike.

3. The discovery that Venus has phases just like the moon, demonstrating that Venus orbits its source of light—the sun. He saw that Venus appears smallest when it is in full phase, and thus is farthest from Earth (Figure 19.13). In the Ptolemaic system, as shown in Figure 19.4, the orbit of Venus lies between Earth and the sun, which means that only the crescent phase of Venus could be seen from Earth.

4. The discovery that the moon's surface is not a smooth glass sphere, as the ancients had suspected and the Church had decreed. Rather, Galileo saw mountains, craters, and plains. He thought the plains might be bodies of water, and this idea was strongly promoted by others, as we can tell from the names given to these features (Sea of Tranquility, Sea of Storms, and so forth).

5. The discovery that the sun (the viewing of which may have caused the eye damage that later blinded him) had sunspots (dark regions caused by slightly lower temperatures). He tracked the movement of these spots and estimated the rotational period of the sun as just under a month. Hence, another heavenly body was found to have both "blemishes" and rotational motion.

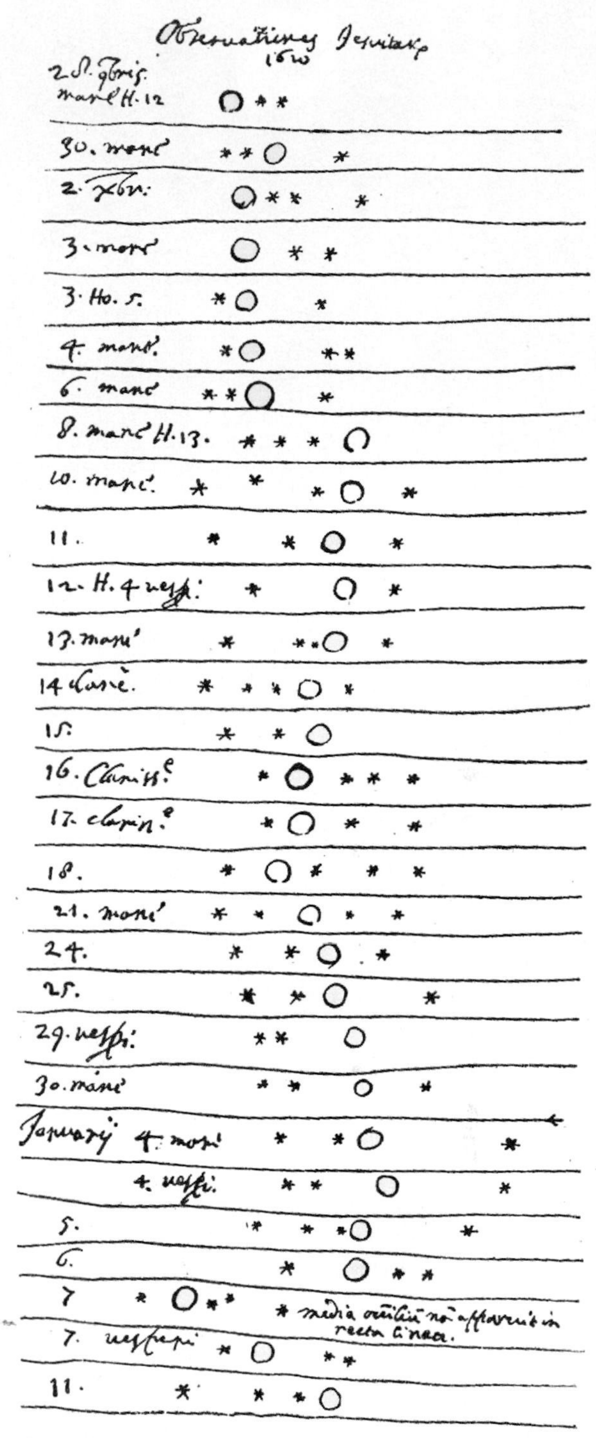

In 1616, the Church condemned the Copernican theory as contrary to Scripture, and Galileo was told to abandon it. Unwilling to accept this verdict, Galileo began

Figure 19.12

Sketch by Galileo of how he saw Jupiter and its four largest satellites through his telescope. The positions of Jupiter's four largest moons (drawn as stars) change nightly. You can observe these same changes with binoculars. (Yerkes Observatory Photograph)

writing his most famous work, *Dialogue of the Great World Systems.* Despite poor health, he completed the project and in 1630 went to Rome, seeking permission from Pope

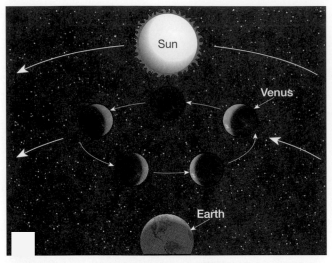

A. Phases of Venus as seen from Earth in the Earth-centered model.

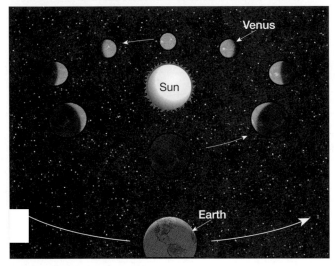

B. Phases of Venus as seen from Earth in the sun-centered model.

C.

Figure 19.13

Using a telescope, Galileo discovered that Venus has phases just like the moon. **A.** In the Ptolemaic (Earth-centered) system, the orbit of Venus lies between the sun and Earth as shown in Figure 19.4A. Thus, in an Earth-centered solar system, only the crescent phase of Venus would be visible from Earth. **B.** In the Copernican (sun-centered) system, Venus orbits the sun and hence all of the phases of Venus should be visible from Earth. **C.** As Galileo observed, Venus goes through a series of moonlike phases. Further, Venus appears smallest during the full phase when it is farthest from Earth and largest in the crescent phase when it is closest to Earth. This verified Galileo's belief that the sun was the center of the solar system. (Photo courtesy of Lowell Observatory)

Urban VIII to publish. Because the book was a dialogue that expounded both the Ptolemaic and Copernican systems, publication was allowed. However, Galileo's enemies were quick to realize that he was promoting the Copernican view at the expense of the Ptolemaic system. Sales of the book were quickly halted, and Galileo was called before the Inquisition. Tried and convicted of proclaiming doctrines contrary to religious doctrine, he was sentenced to permanent house arrest, under which he remained for the last ten years of his life.

Despite this restriction, and his age, and his grief after the death of his eldest daughter, Galileo continued to work. In 1637 he became totally blind, yet during the next few years completed his finest scientific work, a book on the study of motion. When Galileo died in 1642, the Grand Duke of Tuscany wanted to erect a monument in his honor, but fear that it might offend the Church prevailed, and it was never built. It was not until 1992 that Galileo was finally exonerated by the Church.

Sir Isaac Newton

Sir Isaac Newton (1643-1727) (Figure 19.14) was born in the year of Galileo's death. His many accomplishments in mathematics and physics led a successor to say that "Newton was the greatest genius that ever existed. . ."

Although Kepler and those who followed attempted to explain the forces involved in planetary motion, their explanations were less than satisfactory. Kepler believed that some force pushed the planets along in their orbits. Galileo, however, correctly reasoned that no force is required to keep an object in motion. Galileo proposed that the natural tendency for a moving object (that is unaffected by an outside force) is to continue moving at a uniform speed and in a straight line. This concept, *inertia*, was later formalized by Newton as his first law of motion.

The problem then, was not to explain the force that keeps the planets moving but rather to determine the force that *keeps them from going in a straight line out into space*. It was to this end that Newton conceptualized the force of *gravity*. At the early age of 23, he envisioned a force that extends from Earth into space and holds the moon in orbit around Earth. Although others had theo-

rized the existence of such a force, he was the first to formulate and test the *law of universal gravitation*. It states:

> Every body in the universe attracts every other body with a force that is directly proportional to their masses and inversely proportional to the square of the distance between them.

Thus, the gravitational force decreases with distance, so that two objects 3 kilometers apart have 3^2, or 9, times less gravitational attraction than if the same objects were 1 kilometer apart.

The law of gravitation also states that the greater the mass of the object, the greater its gravitational force. For example, the large mass of the moon has a gravitational force strong enough to cause ocean tides on Earth, whereas the tiny mass of a communications satellite has no measurable effect on Earth. The mass of an object is a measure of the total amount of matter it contains. But more often mass is measured by determining the resistance an object exhibits in response to any effort made to change its state of motion.

Often we confuse the concept of *mass* with our notion of *weight*. Specifically, weight is the force of gravity acting upon an object. Therefore, weight varies when gravitational forces change. An object weighs less on the moon than on Earth because the moon is much less massive than Earth. However, unlike weight, the mass of an object does not change. For example, a person weighing 120 pounds on Earth weighs $\frac{1}{6}$ as much, or 20 pounds, on the moon, but the person's mass remains unchanged.

With his laws of motion, Newton proved that the force of gravity, combined with the tendency of a planet to remain in straight-line motion, result in the elliptical orbits discovered by Kepler. Earth, for example, moves forward in its orbit about 30 kilometers ($18\frac{1}{2}$ miles) each second, and during the same second, the force of gravity pulls it toward the sun about $\frac{1}{2}$ centimeter ($\frac{1}{8}$ inch). Therefore, as Newton concluded, it is the combination of Earth's forward motion and its "falling" motion that defines its orbit (Figure 19.15). If gravity were somehow eliminated, Earth would move in a straight line out into space. On the other hand, if Earth's forward motion suddenly stopped, gravity would pull it directly toward the sun.

Up to this point, we have discussed Earth as if the only forces involved in its motion were caused by its gravitational relationship with the sun. However, all bodies in the solar system have gravitational effects on Earth and on each other. For this reason, the orbit of Earth is not the perfect ellipse determined by Kepler. Any variance in the orbit of a body from its predicted path is called **perturbation**. For example, Jupiter's gravitational pull on Saturn reduces Saturn's orbital period by nearly one week from the predicted period. As we

Figure 19.14
English scientist Sir Isaac Newton (1643–1727) explained gravity as the force that holds planets in orbit around the sun. (Yerkes Observatory Photograph)

Figure 19.15
Orbital motion of Earth and other planets.

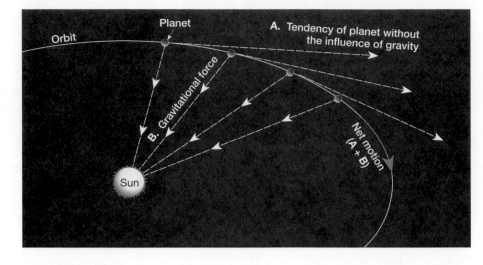

shall see, the application of this concept led to the discovery of the planet Neptune, because of Neptune's gravitational effect on the orbit of Uranus.

Newton used the law of universal gravitation to redefine Kepler's third law, which states the relationship between the orbital periods of the planets and their solar distances. When restated, Kepler's third law takes into account the masses of the bodies involved and thereby provides a method for determining the mass of a body when the orbit of one of its satellites is known. For example, the mass of the sun is known from Earth's orbit, and Earth's mass has been determined from the orbit of the moon. In fact, the mass of any body with a satellite can be determined. The masses of bodies that do not have satellites can be determined only if the bodies noticeably affect the orbit of a neighboring body or of a nearby artificial satellite.

Constellations

As early as 5000 years ago, people became fascinated with the star-studded skies and began to name the patterns they saw (see Box 19.2). These configurations, called **constellations**, were named in honor of mythological characters or great heroes, such as Orion. It takes a good bit of imagination to make out the intended subjects, since most constellations were probably not thought of as likenesses in the first place. Although we inherited many of the constellations from Greek mythology, it is believed that Greek astronomers acquired most of theirs from the Babylonians, Egyptians, and Mesopotamians.

Although the stars that make up a constellation all appear to be the same distance from Earth, this is not the case. Some are many times farther away than others. Thus, the stars in a particular constellation are not associated with each other in any physical way. In addition, various cultural groups, including Native Americans, attached

their own names, pictures, and stories to the constellations. For example, the constellation Orion, the Hunter, was known as the White Tiger to ancient Chinese astronomers.

Today, 88 constellations are recognized, and they are used to divide the sky into units, just as state boundaries divide the United States. Every star in the sky is in, but is not necessarily part of, one of these constellations. Constellations therefore enable astronomers to roughly identify the area of the heavens they are observing. For the student, the constellations provide a good way to become familiar with the night sky (see Appendix E).

Some of the brightest stars were given proper names, such as Sirius, Arcturus, and Betelgeuse. In addition, the brightest stars in a constellation are generally named in order of their brightness by the letters of the Greek alphabet—alpha (α), beta (β), and so on—followed by the name of the parent constellation. For example, Sirius, the brightest star in the constellation Canis Major (Larger Dog), is called Alpha (α) Canis Majoris.

Positions in the Sky

If you gaze away from the city lights on a clear night, you will get the distinct impression that the stars produce a spherical shell surrounding Earth. This impression seems so real that it is easy to understand why many early Greeks regarded the stars as being fixed to a crystalline celestial sphere. Although we realize that no such sphere exists, it is convenient to use this concept for locating stars.

One method for doing this, called the **equatorial system**, divides the celestial sphere into a coordinate system. It is very similar to the latitude-longitude system used for locations on Earth's surface (Figure 19.16). Because the celestial sphere appears to rotate around an imaginary line extending from Earth's axis, the north and south celestial poles are in line with the terrestrial North Pole and South Pole. The north celestial pole happens to be very near the bright star whose various

Box 19.2

Astrology

Many people confuse astrology and astronomy to the point of believing these terms to be synonymous. Nothing can be farther from the truth. *Astronomy* is a scientific probing of the universe to derive the properties of celestial objects and the laws under which the universe operates. *Astrology*, on the other hand, is based on ancient superstitions that hold that an individual's actions and personality are based on the positions of the planets and stars now, and at the person's birth. Scientists do not accept astrology, regarding it as a pseudoscience ("false science"). It is hoped that most people read horoscopes only as a pastime and do not let them influence daily living.

Apparently astrology had its origin more than 3000 years ago when the positions of the planets were plotted as they regularly migrated against the background of the "fixed" stars. Because the solar system is "flat" like a whirling frisbee, the planets orbit the sun along nearly the same plane. Therefore, the planets, sun, and moon all appear to move along a band around the sky known as the *zodiac*. Because Earth's moon cycles through its phases about twelve times each year, the Babylonians divided the zodiac into twelve constel-

Figure 19.D

Stonehenge, an ancient observatory in England. On June 21–22 (summer solstice), the sun can be observed rising above the heel stone. (Robin Scagell / Science Photo Library / Photo Researchers, Inc.)

lations (Figure 19.C). Thus, each successive full moon can be seen against the backdrop of the next constellation.

The twelve constellations of the zodiac ("Zone of Animals," so named because some constellations represent animals) are Aries, Taurus, Gemini, Cancer, Leo, Virgo, Scorpio, Sagittarius, Capricorn, Aquarius, and Pisces. These names may be familiar to you as the astrological signs of the zodiac. When first established, the vernal equinox (first day of spring) occurred when the sun was viewed against the constellation Aries. However, during each succeeding vernal equinox, the position of the sun shifts very slightly against the background of stars. Now, over 2000 years later, the vernal equinox occurs

when the sun is in Pisces (Figure 19.C). In several years, it will occur when the sun appears against Aquarius. (Hence, the "Age of Aquarius", is coming.)

Although astrology is not a science and has no basis in fact, it did contribute to the science of astronomy. The positions of the moon, sun, and planets at the time of a person's birth (sign of the zodiac) were considered to have great influence on that person's life[*]. Even the great astronomer Kepler was required to make horoscopes as part of his duties. In order to make horoscopes for the future, astrologers attempted to predict the future positions of the celestial bodies. Consequently, some of the improvements in astronomical instruments were made because of the desire for more accurate predictions of events such as eclipses, which were considered highly significant in a person's life.

Even prehistoric people built observatories. The structure known as Stonehenge, in England, was undoubtedly an attempt at better solar predictions (Figure 19.D). At the time of mid-summer in the Northern Hemisphere (June 21–22—the summer solstice), the rising sun emerges directly above the heel stone of Stonehenge. Besides keeping this calendar, Stonehenge may also have provided a method of determining eclipses. The remnants of other early observatories exist elsewhere in the Americas, Europe, Asia, and Africa.

---

[*]It is interesting to note that 2000 years ago a person born on July 28 was considered a Leo because the sun was in that constellation. During modern times the sun appears in the constellation Cancer on this date, but individuals born during this time frame are still dubbed Leos.

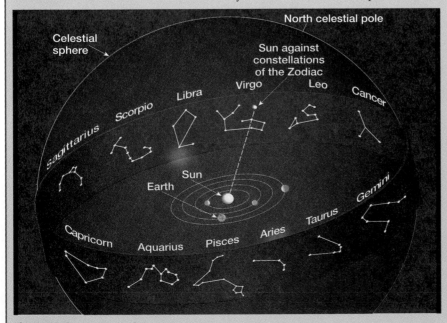

Figure 19.C

The twelve constellations of the zodiac. Earth is shown in its autumn (September) position in orbit, from which the sun is seen against the background of the constellation Virgo.

Figure 19.16

Astronomical coordinate system on the celestial sphere.

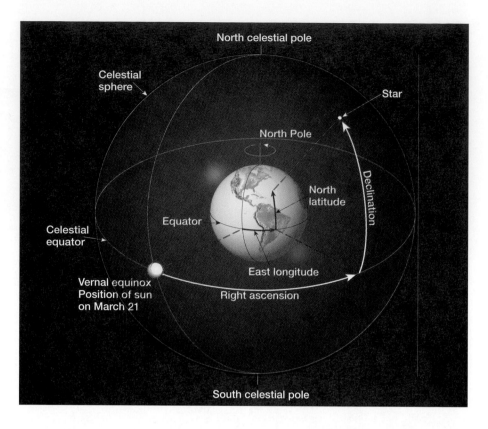

names reflect its location: "pole star," Polaris, and North Star. To an observer in the Northern Hemisphere, the stars appear to circle Polaris, because it, like the North Pole, is in the center of motion (Figure 19.17). Figure 19.18 shows how to locate the North Star by using two stars in the very visible constellation called the Big Dipper.

Now, imagine a plane through Earth's equator, a plane that extends outward from Earth and intersects the celestial sphere. The intersection of this plane with the celestial sphere is the *celestial equator* (see Figure 19.16). In the equatorial system, the terms *declination*, which is analogous to latitude, and *right ascension*, which is analogous

Figure 19.17

Star trails in the region of Polaris (north celestial pole) on a time exposure. (Courtesy of National Optical Astronomy Observatories)

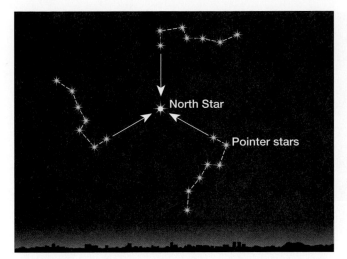

Figure 19.18
Locating the North Star (Polaris) from the pointer stars in the Big Dipper, which is part of the constellation Ursa Major. The Big Dipper is shown soon after sunset in December (lower figure), April (upper figure), and August (left). Refer to the star charts in Appendix E to find the current position of the Big Dipper.

to longitude, are used (Figure 19.16). **Declination**, like latitude, is the angular distance north or south of the celestial equator. **Right ascension** is the angular distance measured eastward along the celestial equator from the position of the vernal equinox. (The *vernal equinox* is at the point in the sky where the sun crosses the celestial equator, at the onset of spring.) While declination is expressed in degrees, right ascension is usually expressed in hours, where each hour is equivalent to 15 degrees.

To visualize distances on the celestial sphere, it helps to remember that the moon and sun have an apparent width of about $\frac{1}{2}$ degree. The declination and right ascension of some of the brightest stars (in other words, their locations) can be determined from the star charts in Appendix E.

Motions of Earth

The two primary motions of Earth are rotation and revolution. **Rotation** is the turning, or spinning, of a body on its axis. **Revolution** is the motion of a body, such as a planet or moon, along a path around some point in space. For example, Earth *revolves* around the sun and the moon *revolves* around Earth. Earth also has another very slow motion known as **precession**, which is the slight movement, over a period of 26,000 years, of Earth's axis.

Rotation

The main consequences of Earth's rotation are day and night. Earth's rotation has become a standard method of measuring time because it is so dependable and easy to use. Each rotation equals about 24 hours. You may be surprised to learn that we can measure Earth's rotation in two ways, making two kinds of days. Most familiar is the **mean solar day**, the time interval from one noon to the next, which averages about 24 hours. Noon is when the sun has reached its zenith (highest point in the sky).

The **sidereal day**, on the other hand, is the time it takes for Earth to make one complete rotation (360 degrees) with respect to a star other than the sun. The sidereal day is measured by the time required for a star to reappear at the identical position in the sky where it was observed the day before. The sidereal day has a period of 23 hours, 56 minutes, and 4 seconds (measured in solar time), which is almost 4 minutes shorter than the mean solar day. This difference results because the direction to distant stars changes only infinitesimally due to Earth's slow revolution along its orbit, whereas the direction to the sun changes by almost 1 degree each day. This difference is shown in Figure 19.19. If it is not apparent why we use the mean solar day rather than the sidereal day as a measurement of our day, consider the fact that in sidereal time, "noon" occurs 4 minutes earlier each day. Therefore, after a span of six months, "noon" occurs at "midnight." Astronomers use sidereal time because the stars appear in the same position in the sky every 24 sidereal hours. Usually, an observatory will begin its sidereal day when the position of the vernal equinox is directly overhead, that is, over the meridian on which the observatory is located. Therefore, when the observatory's sidereal clock is the same as the star's right ascension, the star will be overhead, or at its highest point. For example, the brightest star in the heavens, Sirius, has a right ascension of 6 hours, 42 minutes, and 56 seconds and will be overhead when the clock at the observatory indicates that time.

Revolution

Earth revolves around the sun in an elliptical orbit at an average speed of 107,000 kilometers (66,000 miles) per hour. Its average distance from the sun is 150 million kilometers (93 million miles), but because its orbit is an ellipse, Earth's distance from the sun varies. At **perihelion** (closest to the sun), it is 147 million kilometers (91 million miles) distant, which occurs about January third each year. At **aphelion** (farthest from the sun), Earth is 152 kilometers (94.5 million miles) distant, which occurs about July fourth.

Because of Earth's annual movement around the sun, each day the sun appears to be displaced among the constellations at a distance equal to about twice its width, or 1 degree. The apparent annual path of the sun against the backdrop of the celestial sphere is called the **ecliptic** (Figure 19.20). Generally, the planets and the moon travel in nearly the same plane as

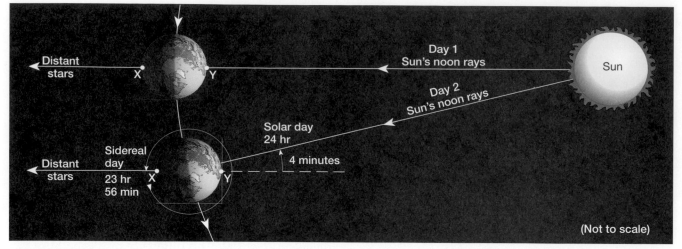

Figure 19.19

The difference between a solar day and a sidereal day. Locations X and Y are directly opposite each other. It takes Earth 23 hours and 56 minutes to make one rotation with respect to the stars (sidereal day). However, notice that after Earth has rotated once with respect to the stars, point Y is not yet returned to the "noon position" with respect to the sun. Earth has to rotate another 4 minutes to complete the solar day.

Earth. Hence, their paths on the celestial sphere lie near the ecliptic (The most notable exception is Pluto, which has an orbit that is tilted 17 degrees to the plane of Earth's orbit.)

The imaginary plane that connects Earth's orbit with the celestial sphere is called the **plane of the ecliptic**. From the reference plane, Earth's axis of rotation is tilt-ed about $23\frac{1}{2}$ degrees. Because of Earth's tilt, the apparent path of the sun (ecliptic) and the celestial equator intersect each other at an angle of $23\frac{1}{2}$ degrees (Figure 19.20). This angle is very important to Earth's inhabitants. Because of the inclination of Earth's axis to the plane of the ecliptic, Earth exhibits its yearly cycle of seasons, a topic discussed in detail in Chapter 14.

Figure 19.20

Earth's orbital motion causes the apparent position of the sun to shift about 1 degree each day on the celestial sphere.

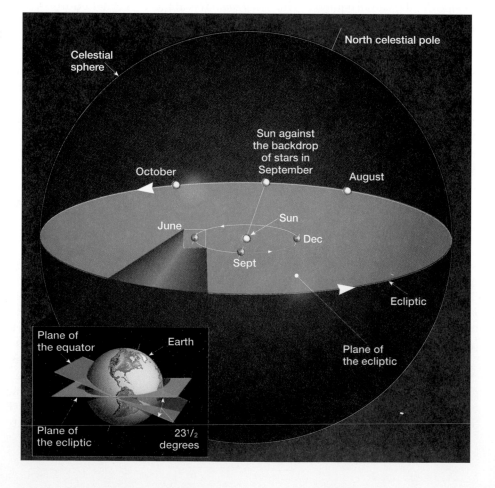

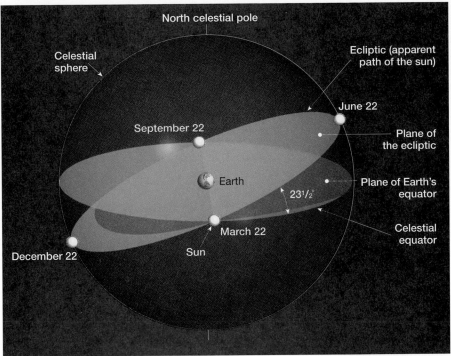

Figure 19.21
The apparent position of the sun plotted on the celestial sphere. The path of the sun (ecliptic) crosses the celestial equator on two occasions each year, March 20–21 and September 22–23. These are known as the equinox positions because the lengths of daylight and darkness on Earth are equal.

When the *apparent* position of the sun is plotted on the celestial sphere over a period of a year's time, its path intersects the celestial equator at two points (Figure 19.21). From a Northern Hemisphere point of view, these intersections are called the vernal (spring) equinox (March 20-21) and autumnal equinox (September 22-23). On June 21-22, the date of the summer solstice, the sun appears $23\frac{1}{2}$ degrees north of the celestial equator, and six months later, on December 21-22, the date of the winter solstice, the sun appears $23\frac{1}{2}$ degrees south of the celestial equator.

Precession

A third and very slow movement of Earth is called *precession*. Although Earth's axis maintains approximately the same angle of tilt, the direction in which the axis points continually changes. As a result, the axis traces a circle on the sky. This movement is very similar to the movement (wobble) of a spinning top (Figure 19.22A). At the present time, the axis points toward the bright star Polaris. In A.D. 14,000, it will point toward the bright star Vega, which will then become the North Star for a few thousand years (Figure 19.22B). The period of precession is 26,000 years. By the year 28,000, Polaris will once again be the North Star.

Precession has only a minor effect on the seasons, because the angle of tilt changes only slightly. It does, however, cause the positions of the seasons (equinox and solstice) to move slightly each year among the constellations.

In addition to its own movements, Earth shares numerous motions with the sun. It accompanies the sun as the entire solar system speeds in the direction of the

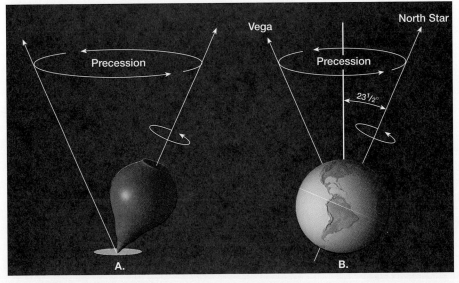

Figure 19.22
A. Precession illustrated by a spinning top.
B. Precession of Earth causes the North Pole to point to different parts of the sky during a 26,000-year cycle. Today, the North Pole points to Polaris (North Star). In 13,000 years, Vega will be the North Star.

bright star Vega at 20 kilometers (12 miles) per second. Also, the sun, like other nearby stars, revolves around the galaxy, a trip that requires 230 million years to traverse at speeds approaching 250 kilometers (150 miles) per second. In addition, the galaxies themselves are in motion. We are presently approaching one of our nearest galactic neighbors, the Great Galaxy in Andromeda. In summary, the motions of Earth are many and complex, and its speed in space is very great.

Motions of the Earth-Moon System

Earth has one natural satellite, the moon. In addition to accompanying Earth in its annual trek around the sun, our moon orbits Earth within a period of about one month (see Box 19.3). When viewed form above the North Pole, the direction of this motion is counterclockwise (eastward). Because the moon's orbit is elliptical, its distance to Earth varies by about 6 percent, averaging 384,401 kilometers (238,329 miles).

The motions of the Earth-moon system constantly change the relative positions of the sun, Earth, and moon. The results are some of the most obvious of astronomical phenomena, namely, the *phases of the moon* and the occasional *eclipses of the sun and moon*.

Phases of the Moon

The first astronomical phenomenon to be understood was the regular cycle of the **phases of the moon**. On a monthly basis, we observe the phases as a systematic change in the amount of the moon that appears illuminated (Figure 19.23). We will choose the "new-moon" position in the cycle as a starting point. About two days after the new moon, a thin sliver (*crescent phase*) appears low in the western sky just after sunset. During the following week, the illuminated portion of the moon visible from Earth increases (*waxing*) to a half circle (*first-quarter phase*) and can be seen from about noon to midnight. In another week, the complete disk (*full-moon phase*) can be seen rising in the east as the sun is sinking in the west. During the next two weeks, the percentage of the moon that can be seen steadily declines (*waning*), until the moon disappears altogether (*new-moon phase*). The cycle soon begins anew with the reappearance of the crescent moon.

The lunar phases are a consequence of the motion of the moon and the sunlight that is reflected from its surface (Figure 19.23B) Half of the moon is illuminated at all times (note the inner group of moon sketches in Figure 19.23A). But to an Earthbound observer, the percentage of the bright side that is visible depends on the location of the moon with respect to the sun and Earth. When the moon lies *between* the sun and Earth, none of

---

Box 19.3

Calendars

One of the earliest responsibilities of astronomers was the keeping of the calendar. Although the oldest known calendars date from the eighth century B.C., others probably existed much earlier. These calendars had as their basis the movements of the heavenly bodies. Although our seven-day week has no astronomical significance, the seven days are clearly names after the seven moving celestial bodies known to the ancients: the sun, the moon, and the five planets (Mercury, Venus, Mars, Jupiter, and Saturn), which are easily seen with the unaided eye. For example, Sunday is "Sun's day," Monday is "Moon's day," and Saturday is "Saturn's day." The remaining days of the week were named in the Romance languages after Mars, Mercury, Jupiter, and Venus, in that order. For example, in Spanish, Tuesday is *Martes*.

Beyond the daily cycle of night and day, the most noticeable heavenly phenomenon is the 29.5-day cycle of the moon through its phases. This cycle became the *moonth*, now called the month. The first western calendars based the year on the phases of the moon. Twelve months equaled a year. However, because there are actually 12.4 lunar cycles in a year, the calendar had to have a full month added every three years to keep the seasons in accord. Even with this correction, the calendar still fell slowly behind. When Julius Caesar gained power in Rome 2000 years ago, the calendar indicated spring, while the weather indicated the middle of winter. To correct this situation, he ordered that 80 days be added to the year of 46 B.C. That year, for obvious reasons, was called the "year of confusion." Also at the direction of an astronomer, Caesar ordered that the calendar be based on the tropical year of 365.25 days. One tropical year is the time required between two successive passages of the sun through the vernal equinox, or 365.2422 days. Calendar makers achieved this by having 365 days each normal year and adding an extra day every fourth year. So began the tradition of *leap year* and the *Julian calendar*. However, because the tropical year is slightly less than 365.25 days (about 10 minutes per year), the Julian calendar worked just like a fast clock. Both would require adjustment if used long enough. By the sixteenth century, the Julian calendar was ahead by 10 days.

In 1582, the *Gregorian calendar*, which we presently use, was developed. The extra 10 days were eliminated by making Friday October 15, the day after Thursday, October 4. To slow the calendar, selected leap years were eliminated. A leap year is no longer added for centennial years except those divisible by 400. Hence, the years 1600, 2000, and 2400 are leap years, but all centennial years between them (e.g. 1900) are not. Our present calendar is accurate to within 1 day in 3000 years.

Not all countries adopted the new calendar at the same time. When George Washington was born, the calendar indicated February 11, 1732, but when the American colonies adopted the Gregorian calendar in 1752, his birth date became February 22, the day we now observe.

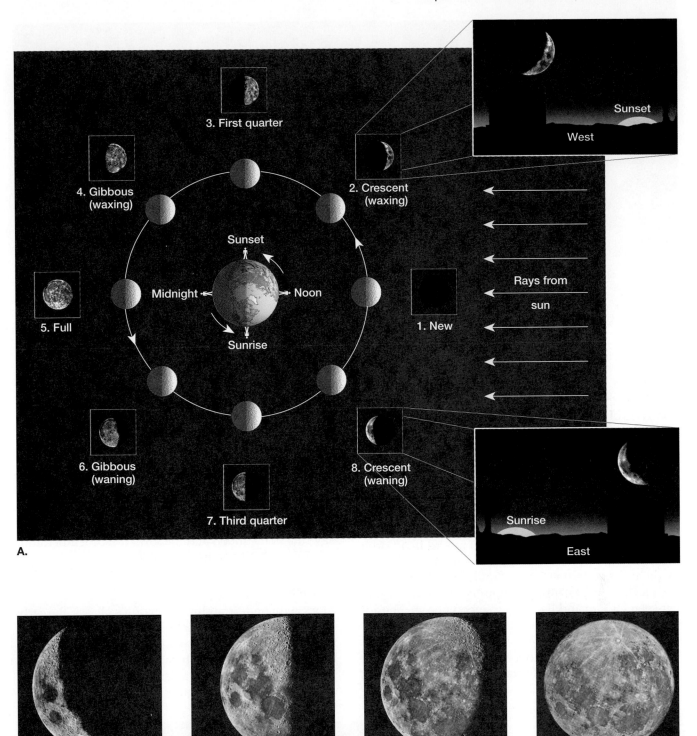

Figure 19.23
Phases of the moon. **A.** The outer figures show the phases as seen from Earth. **B.** Compare these photographs with the diagram. (Courtesy of Lick Observatory)

its bright side faces Earth, so we see the new-moon ("no-moon") phase. Conversely, when the moon lies on the side of Earth opposite the sun, all of its lighted side faces Earth, so we see the full moon. At all positions between these extremes, an intermediate amount of the moon's illuminated side is visible from Earth.

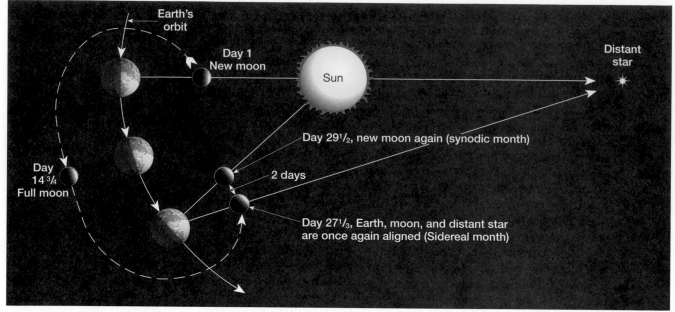

Figure 19.24

The difference between the sidereal month (27 $\frac{1}{3}$ days) and the synodic month (29 $\frac{1}{2}$ days). Distances and angles are not shown to scale.

Lunar Motions

The cycle of the moon through its phases requires $29\frac{1}{2}$ days, a time span called the **synodic month**. This cycle was the basis for the first Roman calendar. However, this is the *apparent period* of the moon's revolution around Earth and not the *true* period, which takes only $27\frac{1}{3}$ days and is known as the **sidereal month**. The reason for the difference of nearly 2 days each cycle is shown in Figure 19.24. Note that as the moon orbits Earth, the Earth-moon system also moves in an orbit around the sun. Consequently, even after the moon has made a complete revolution around Earth, it has not yet reached its starting position, which was directly between the sun and Earth (new-moon phase). The additional motion to reach the starting point takes another two days.

An interesting fact concerning the motions of the moon is that its period of rotation about its axis and its revolution around Earth are the same—$27\frac{1}{3}$ days. Be-cause of this, the same lunar hemisphere always faces Earth. All of the manned *Apollo* missions were confined to the Earth-facing side. Only orbiting satellites and astronauts have seen the "back"side of the moon.

Since the moon rotates on its axis only once every $27\frac{1}{3}$ days, any location on its surface experiences periods of daylight and darkness lasting about two weeks. This, along with the absence of an atmosphere, accounts for the high surface temperature of 127°C (261°F) on the day side of the moon and the low surface temperature of -173°C (-280°F) on its night side.

Eclipses

Along with understanding the moon's phases, the early Greeks also realized that eclipses are simply shadow effects. When the moon moves in a line directly between Earth and the sun, which can occur only during the new-moon phase, it casts a dark shadow on Earth, producing a **solar eclipse** (Figure 19.25). On the other hand,

Figure 19.25

Solar eclipse. Observers in the zone of the umbral shadow see a total solar eclipse. Those in the penumbra see a partial eclipse. The path of the solar eclipse moves eastward across the globe.

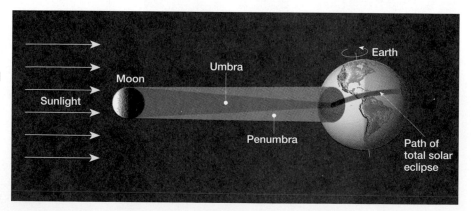

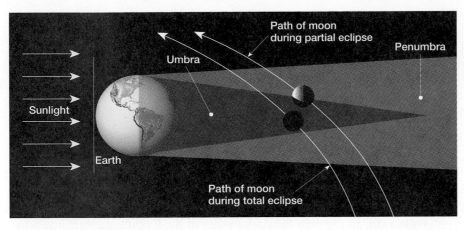

Figure 19.26
Lunar eclipse. During a total lunar eclipse the moon's orbit carries it into the dark shadow of Earth (umbra). During a partial eclipse only a portion of the moon enters the umbra.

the moon is eclipsed (**lunar eclipse**) when it moves within Earth's shadow, a situation that is possible only during the full-moon phases (Figure 19.26).

Why then does a solar eclipse not occur with every new-moon phase and a lunar eclipse with every full-moon phase? They would, if the orbit of the moon lay exactly along the plane of Earth's orbit (the plane of the ecliptic). However, the moon's orbit is inclined about 5 degrees to the plane that contains Earth and the sun (Figure 19.27). Thus, during most new-moon phases, the shadow of the moon misses Earth (passes above or below); and during

most full-moon phases, the shadow of Earth misses the moon. Only when a new-or full-moon phase occurs where the moon's orbit crosses the plane of the ecliptic can an eclipse take place. Since these conditions are normally met only twice a year, the usual number of eclipses is four. These occur as a set of one solar and one lunar eclipse, followed six months later with another set (Figure 19.27). Occasionally the alignment is such that three eclipses can be squeezed into a one-month period. These occur as a solar eclipse flanked by two lunar eclipses, or vice versa. Furthermore, it occasionally happens that the first set of

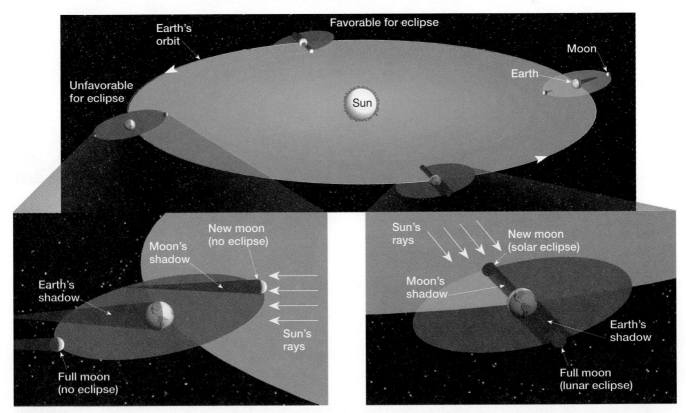

Figure 19.27
The moon's orbit is inclined about 5 degrees to the plane that contains the sun and Earth. Thus, during most new-moon phases, the shadow of the moon misses Earth (passes above or below), and during most full-moon phases, the shadow of Earth misses the moon. Only when a new- or full-moon phase occurs where the moon's orbit crosses the Earth-sun plane can an eclipse take place. These conditions are met at roughly six-month intervals.

eclipses occurs at the very beginning of a year, and a third set occurs before the year ends, resulting in six eclipses in that year. More rarely, if one of these sets is a three-eclipse kind, the total number of eclipses in a year can reach seven, which is the maximum.

During a total lunar eclipse, Earth's circular shadow can be seen moving slowly across the disk of the full moon. When totally eclipsed, the moon is completely within Earth's shadow but still is visible as a coppery disk, because Earth's atmosphere bends and transmits some long-wavelength light (red) into its shadow. A total eclipse of the moon can last up to four hours and is visible to anyone on the side of Earth facing the moon.

During a total solar eclipse, the moon casts a circular shadow that is never wider than 275 kilometers (170 miles), about the size of South Carolina. Anyone observing in this region will see the moon slowly block the sun from view and the sky darken (Figure 19.28). Near totality, a sharp drop in temperature of a few degrees is experienced. The solar disk is completely blocked for at most seven minutes, because the moon's shadow is so small, and then one edge reappears.

At totality, the dark moon is seen covering the complete solar disk, and only the sun's brilliant white outer atmosphere is visible (see Figure 21.21). Total solar eclipses are visible only to people in the dark part of the moon's shadow (*umbra*), while a partial eclipse is seen by those in the light portion (*penumbra*) (see Figure 19.25).

Partial solar eclipses are more common in the polar regions, because it is this zone that the penumbra covers when the dark umbra of the moon's shadow just misses Earth. A total solar eclipse is a rare event at any given location. The next one that will be visible from the contiguous United States will take place on August 21, 2017.

Figure 19.28

This sequence of photos starting from the upper left to the lower right shows the stages of a total solar eclipse. (From *Foundations of Astronomy*, Third Edition, by Michael Seeds, reprinted by permission of the author and Wadsworth Publishing Company)

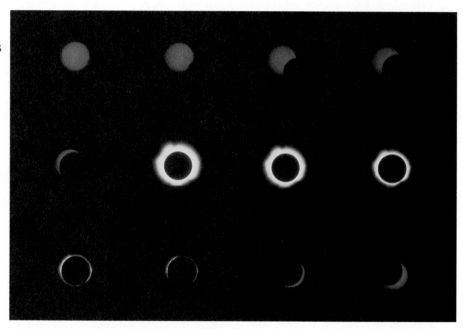

Review Questions

1. Why did the ancients think that celestial objects had some control over their lives?

2. Describe what produces the retrograde motion of Mars. What geometric arrangements did Ptolemy use to explain this motion?

3. What major change did Copernicus make in the Ptolemaic system? Why was this change philosophically significant?

4. What was Tycho Brahe's contribution to science?

5. Does Earth move faster in its orbit near perihelion (January) or near aphelion (July)? Keeping your answer to the previous question in mind, is the solar day longest in January or July?

6. Use Kepler's third law ($p^2 = d^3$) to determine the period of a planet whose solar distance is
   (a) 10 AU
   (b) 1 AU
   (c) 0.2 AU

7. Use Kepler's third law to determine the distance from the sun of a planet whose period is
   (a) 5 years.
   (b) 10 years.
   (c) 10 days.

8. Did Galileo invent the telescope?

9. Explain how Galileo's discovery of a rotating sun supported the Copernican view of a sun-centered universe.

10. Using a diagram, explain why the fact that Venus appears full when it is smallest supports the Copernican view and is inconsistent with the Ptolemaic system.

11. Newton learned that the orbits of the planets are the result of two actions. Explain these actions.

12. Of what value are constellations to modern-day astronomers?

13. Express the declination and right ascension of the star Arcturas (see Appendix E).

14. Explain the difference between the mean solar day and the sidereal day.

15. What is the approximate length of the cycle of the phases of the moon?

16. What is different about the crescent phase that precedes the new-moon phase and that which follows the new-moon phase?

17. What phase of the moon occurs approximately one week after the new moon? Two weeks?

18. When you observe the crescent phase early in the evening, is the visible moon waxing (growing) or waning (declining)? (See Figure 19.23.)

19. What phenomenon results from the fact that the moon's period of rotation and revolution are the same?

20. The moon rotates very slowly (once in $27\frac{1}{3}$ days) on its axis. How does this affect the lunar surface temperature?

21. Describe the locations of the sun, moon, and Earth during a solar eclipse and during a lunar eclipse.

22. How many eclipses normally occur each year?

23. Solar eclipses are slightly more common than lunar eclipses. Why, then, is it more likely that your region of the country will experience a lunar eclipse?

24. How long can a total eclipse of the moon last? How about a total eclipse of the sun?

Key Terms

aphelion (p. 505)
astronomical unit (AU) (p. 497)
celestial sphere (p. 490)
constellations (p. 502)
declination (p. 505)
ecliptic (p. 505)
equatorial system (p. 502)

geocentric (p. 490)
heliocentric (p. 491)
lunar eclipse (p. 511)
mean solar day (p. 505)
perihelion (p. 505)
perturbation (p. 501)
phases of the moon (p. 508)
plane of the ecliptic (p. 506)
precession (p. 505)

Ptolemaic system (p. 491)
retrograde motion (p. 492)
revolution (p. 505)
right ascension (p. 505)
rotation (p. 503)
sidereal day (p. 505)
sidereal month (p. 510)
solar eclipse (p. 510)
synodic month (p. 510)

CHAPTER 20
Touring Our Solar System

Montage of Saturn and some of its satellites. (Photo by NASA/Science Source/Photo Researchers, Inc.)

When people first recognized that the planets resembled Earth more than the stars, excitement grew. Could intelligent life exist on these other planets, or elsewhere in the universe? Space exploration has rekindled this interest. So far, no evidence of extraterrestrial life within our solar system has emerged. Nevertheless, we study the other planets to learn about Earth's formation and early history. Recent space explorations have been organized with this goal in mind. To date, Mercury, Venus, Mars, Jupiter, Saturn, Uranus, and Neptune have been explored by space probes.

The sun is the hub of a huge rotating system of nine planets, their satellites, and numerous small asteroids, comets, and meteoroids. An estimated 99.85 percent of the mass of our solar system is contained within the sun. The planets collectively make up most of the remaining 0.15 percent. The planets, traveling outward from the sun, are Mercury, Venus, Earth, Mars, Jupiter, Saturn, Uranus, Neptune, and Pluto (Figure 20.1).

Under control of the sun's gravitational force, each planet is tethered in an elliptical orbit, and all of them travel in the same direction. The nearest planet to the sun, Mercury, has the fastest orbital motion, 48 kilometers per second, and the shortest period of revolution around the sun, 88 Earth-days. By contrast, the most distant planet, Pluto, has an orbital speed of 5 kilometers per second and requires 248 Earth-years to complete one revolution.

Imagine a planet's orbit drawn on a flat sheet of paper. The paper represents the planet's *orbital plane*. The orbital planes of seven planets lie within 3 degrees of the plane of the sun's equator. The other two, the innermost and outermost, Mercury and Pluto, are inclined 7 and 17 degrees, respectively.

The Planets: An Overview

Careful examination of Table 20.1 shows that the planets fall quite nicely into two groups: the **terrestrial** (Earthlike) **planets** (Mercury, Venus, Earth, and Mars) and the **Jovian** (Jupiterlike) **planets** (Jupiter, Saturn, Uranus, and Neptune). Pluto is not included in either category because its great distance from Earth and its small size make this planet's true nature a mystery.

The most obvious difference between the terrestrial and the Jovian planets is their size (Figure 20.2). The largest terrestrial planets (Earth and Venus) have diameters only one-quarter as great as the diameter of the smallest Jovian planet (Neptune). Also, their masses are only one-seventeenth as great as Neptune's. Hence, the Jovian planets are often called *giants*. Because of their relative locations, the four Jovian planets are referred to as the *outer planets*, while the terrestrial planets are called the *inner planets*. As we shall see, there appears to be a correlation between the positions of these planets and their sizes.

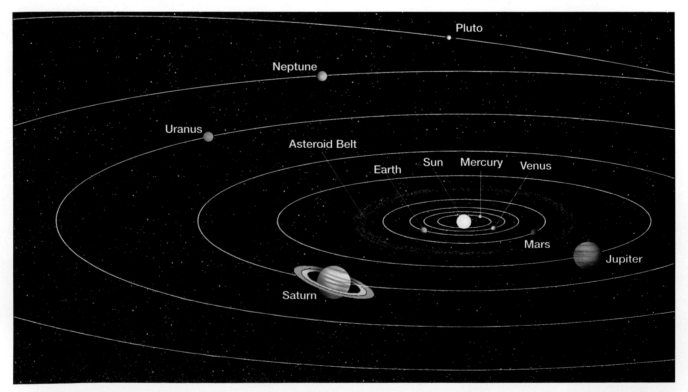

Figure 20.1
Orbits of the planets to scale.

Table 20.1 Planetary data.

Planet	Symbol	Mean Distance from Sun			Period of Revolution	Inclination of Orbit	Orbital Velocity	
		AU*	Millions of Miles	Millions of Kilometers			mi/s	km/s
Mercury	☿	0.39	36	58	88^{d}	7°00′	29.5	47.5
Venus	♀	0.72	67	108	225^{d}	3°24′	21.8	35.0
Earth	⊕	1.00	93	150	365.25^{d}	0°00″	18.5	29.8
Mars	♂	1.52	142	228	687^{d}	1°51′	14.9	24.1
Jupiter	♃	5.20	483	778	12^{yr}	1°18′	8.1	13.1
Saturn	♄	9.54	886	1427	29.5^{yr}	2°29′	6.0	9.6
Uranus	⛢	19.18	1783	2870	84^{yr}	0°46′	4.2	6.8
Neptune	♆	30.06	2794	4497	165^{yr}	1°46′	3.3	5.3
Pluto	♇	39.44	3666	5900	248^{yr}	17°12′	2.9	4.7

Planet	Period of Rotation	Diameter		Relative Mass (Earth = 1)	Average Density (g/cm^3)	Polar Flattening (%)	Eccentricity	Number of Known Satellites
		Miles	Kilometers					
Mercury	59^{d}	3015	4878	0.06	5.4	0.0	0.206	0
Venus	244^{d}	7526	12,104	0.82	5.2	0.0	0.007	0
Earth	$23^{h}56^{m}04^{s}$	7920	12,756	1.00	5.5	0.3	0.017	1
Mars	$24^{h}37^{m}23^{s}$	4216	6794	0.11	3.9	0.5	0.093	2
Jupiter	$9^{h}50^{m}$	88,700	143,884	317.87	1.3	6.7	0.048	16
Saturn	$10^{h}14^{m}$	75,000	120,536	95.14	0.7	10.4	0.056	21
Uranus	$17^{h}14^{m}$	29,000	51,118	14.56	1.2	2.3	0.047	15
Neptune	$16^{h}03^{m}$	28,900	50,530	17.21	1.7	1.8	0.009	8
Pluto	6.4^{d}	~1500	2445	0.002	1.8	0.0	0.250	1

* AU = astronomical unit, Earth's mean distance from the Sun.

Other dimensions along which the two groups markedly differ include density, chemical makeup, and rate of rotation. The densities of the terrestrial planets average about 5 times the density of water, whereas the Jovian planets have densities that average only 1.5 times that of water. One of the outer planets, Saturn, has a density only 0.7 times that of water, which means that Saturn would float if placed in a large enough water tank! Variations in the chemical compositions of the planets are largely responsible for these density differences.

The substances that compose both groups of planets are divided into three groups—*gases*, *rocks*, and *ices*—based on their melting points.

1. The gases, hydrogen and helium, are those with melting points near absolute zero (0° Kelvin or –273°C), the lowest possible temperature.
2. The rocks are principally silicate minerals and metallic iron, which have melting points exceeding 700°C.
3. The ices include ammonia (NH_3), methane (CH_4), carbon dioxide (CO_2), and water (H_2O). They have intermediate melting points (for example, H_2O has a melting point of 0°C).

The terrestrial planets are mostly rock. They are dense, rocky, and metallic, with minor amounts of gases. The Jovian planets, on the other hand, contain a large percentage of gases (hydrogen and helium), with varying amounts of ices (mostly water, ammonia, and methane). This accounts for their low densities. (The outer planets may contain as much rocky and metallic material as the terrestrial planets, but this material would be concentrated in their small central cores.)

Figure 20.2
The planets drawn to scale.

The Jovian planets have very thick atmospheres of hydrogen, helium, methane, and ammonia. By contrast, the terrestrial planets, including Earth, have meager atmospheres at best. The reason is that a planet's ability to retain an atmosphere depends on its mass and temperature.

Simply stated, a gas molecule can "evaporate" from a planet if it reaches a speed known as the **escape velocity**. For Earth, this velocity is 11 kilometers (7 miles) per second. Any material, including a rocket, must reach this speed before it can escape Earth's gravity and go into space.

The Jovian planets, because of their greater surface gravities, have higher escape velocities (21–60 kilometers per second) than the terrestrial planets. Consequently, it is more difficult for gases to "evaporate" from them. Also, because the molecular motion of a gas is temperature-dependent, at the low temperatures of the Jovian planets even the lightest gases are unlikely to acquire the speed needed to escape.

On the other hand, a comparatively warm body with a small surface gravity, like our moon, is unable to hold even heavy gases, like carbon dioxide and radon, and thus lacks an atmosphere. The slightly larger terrestrial planets of Earth, Venus, and Mars retain some heavy gases like carbon dioxide, but even their atmospheres make up only an infinitesimally small portion of their total mass.

The primordial cloud of dust and gas from which all the planets are thought to have condensed may have had a composition similar to Jupiter's. However, unlike Jupiter, the terrestrial planets today are nearly devoid of light gases and ices. Were the terrestrial planets once much larger? Did they contain these materials but lose them because of heating from the nearby sun? In the following section we will consider the evolutionary histories of the terrestrial and Jovian planets in an attempt to answer these questions.

Evolution of the Planets

Recall from Chapter 11 the nebular hypothesis, which suggests that the entire solar system condensed from a gaseous nebula. The denser metallic elements (iron and nickel) in the forming planets sank toward their centers, whereas the lighter substances (silicate minerals, oxygen, hydrogen) migrated toward their surfaces.

During this period of *chemical differentiation*, gaseous materials escaped from the planet's interiors, much like what happens during a volcanic event on Earth. The hottest and second-smallest planet, Mercury, was unable to retain even the heaviest of these gases. Mars, on the other hand, being 40 percent larger and considerably cooler than Mercury, retained a thin layer of carbon dioxide and some water in the form of ice.

The largest of the terrestrial planets, Venus and Earth, have surface gravities strong enough to retain a substantial amount of the heavier gases, like nitrogen, oxygen, and carbon dioxide. However, when compared to the four Jovian planets, even the atmospheres of these two terrestrial planets must be looked upon as meager.

As the terrestrial planets were forming, the larger Jovian planets, along with their extensive satellite systems, were also developing. However, because of the frigid temperatures so far from the sun, the fragments from which these planets formed contained a high percentage of ices—water, carbon dioxide, ammonia, and methane. Perhaps by random choice, two of the outer planets, Jupiter and Saturn, grew many times larger (by mass) than Uranus and Neptune. (For comparison, Jupiter is 318 times and Saturn is 95 times more massive than Earth. However, Uranus and Neptune have masses only 15 and 17 times greater than Earth.)

When Jupiter and Saturn reached a certain size, estimated to be about 10 Earth masses, their surface gravity became sufficient to attract and hold even the lightest materials—-hydrogen and helium. Thus, much of their size is attributable to the large envelope of light elements. They exist as a dense liquid below a thick, hydrogen-rich atmosphere. Jupiter and Saturn therefore consist of a central core of ices and rock, and a much larger outer envelope of mostly hydrogen and helium.

By contrast, the smaller Jovian planets, Uranus and Neptune, are proposed to have a small rocky-iron core and a large mantle of water, ammonia, and methane surrounded by a thin ocean of liquid hydrogen. These planets structurally resemble Jupiter and Saturn, but without their large hydrogen-helium envelopes.

In many respects, the development of the outer planets with their large satellite systems roughly parallels the events that formed the solar system. Like their parent planets, the satellites of the outer planets are primarily ices, with lesser amounts of rock. However, their small masses could not retain appreciable hydrogen and helium.

In the remainder of this chapter, we will consider each planet briefly, plus the minor members of the solar system. First, however, let us visit Earth's companion in space, our moon.

Earth's Moon

Earth now has hundreds of satellites, but only one natural satellite, the moon, accompanies us on our annual journey around the sun. Although other planets have moons, our planet-satellite system is unique in the solar system, because Earth's moon is unusually large compared to its parent planet. The diameter of the moon is 3475 kilometers (2150 miles), about one-fourth of the Earth's 12,751 kilometers.

From calculation of the moon's mass, its density is 3.3 times that of water. This density is comparable to that of *crustal* rocks on Earth but is considerably less than Earth's average density which is 5.5 times that of water. Geolo-

gists have suggested that this difference can be accounted for if the moon's iron core is small. The gravitational attraction at the lunar surface is one-sixth of that experienced on Earth's surface (a 100-pound person on Earth weighs only 17 pounds on the moon). This difference allows an astronaut to carry a "heavy" life-support system with relative ease. If not burdened with such a load, an astronaut could jump six times higher than on Earth.

The Lunar Surface

When Galileo first pointed his telescope toward the moon, he saw two different types of terrain—dark lowlands and bright, cratered highlands (Figure 20.3). Because the dark regions resembled seas on Earth, they were later named **maria**, Latin for "sea" (singular **mare**). This name is unfortunate, because the moon's surface is totally devoid of water. Figure 20.4 shows typical features of the lunar surface.

Today we know that the moon has no atmosphere or water. Therefore, the weathering and erosion that continually modify Earth's surface are virtually lacking on the moon. In addition, tectonic forces are not active on the moon, so earthquakes and volcanic eruptions no longer occur. However, because the moon is unprotected by an atmosphere, a different kind of erosion occurs: tiny particles from space (micrometeorites) continually bombard its surface and ever-so-gradually smooth the landscape. Moon rocks become slightly rounded on top if they are long exposed at the lunar surface. Nevertheless, it is unlikely that the moon has changed appreciably in the last 3 billion years, except for a few craters created by large meteorites.

Craters.   The most obvious features of the lunar surface are craters. They are so profuse that craters-within-craters are the rule! The larger ones in the lower portion of Figure 20.3 are about 250 kilometers (150 miles) in diameter, roughly the width of Indiana. Most craters were produced by the impact of rapidly moving debris (meteoroids), a phenomenon that was considerably more common in the early history of the solar system than it is today.

By contrast, Earth has only about a dozen easily recognized impact craters. This difference can be attributed to Earth's atmosphere. Friction with the air burns up small debris before it reaches the ground. In addition, evidence for most of the craters that formed in Earth's history has been obliterated by erosion or tectonic processes.

The formation of an impact crater is illustrated in Figure 20.5. Upon impact, the high-speed meteoroid compresses the material it strikes, then almost instantaneously the compressed rock rebounds, ejecting material from the crater. This process is analogous to the splash that occurs when a rock is dropped into water, and it often results in the formation of a central peak, as seen in the large crater in Figure 20.6. Most of the ejected material (*ejecta*) lands near the crater, building a rim

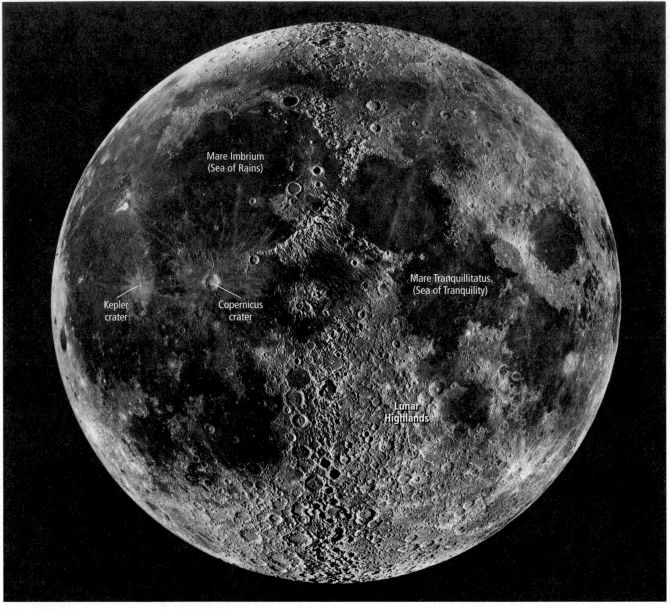

Figure 20.3
Telescopic view from Earth of the lunar surface. The major features are the dark "seas" (maria) and the light highly cratered highlands. (Courtesy of Lick Observatory)

around it. The heat generated by the impact is sufficient to melt some of the impacted rock. Astronauts have brought back samples of glass beads produced in this manner, as well as rock formed when angular fragments and dust were welded together by the impact.

A meteoroid only 3 meters (10 feet) in diameter can blast out a 150-meter (500-foot) wide crater. A few of the large craters such as Kepler and Copernicus, shown in Figure 20.3 formed from the impact of bodies 1 kilometer or more in diameter. These two large craters are thought to be relatively young because of the bright *rays* ("splash" marks) that radiate outward for hundreds of kilometers.

Highlands.   Densely pockmarked highland areas make up most of the lunar surface. In fact, all of the "back" side of the moon is characterized by such topography. (Only astronauts have seen the "back" side, because the moon rotates on its axis once with each revolution around Earth, always keeping the same side facing Earth.) Within the highland regions are mountain ranges. The highest lunar peaks reach elevations approaching 8 kilometers, only 1 kilometer lower than Mount Everest.

Maria.   The "seas" of basaltic lava originated when asteroids punctured the lunar surface, letting basaltic magma "bleed" out (Figure 20.7). Apparently the craters

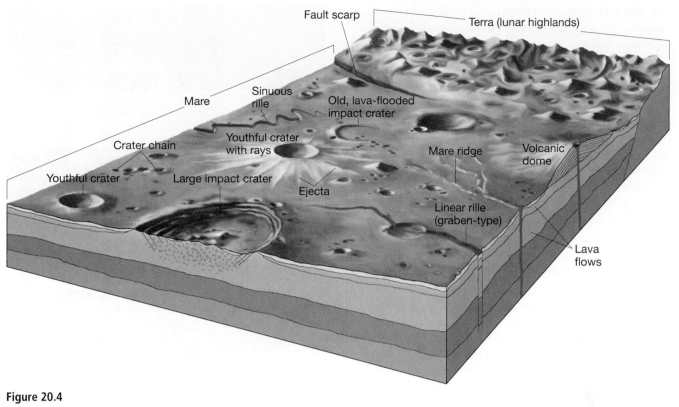

Figure 20.4
Block diagram illustrating major topographic features on the lunar surface.

were flooded with layer upon layer of very fluid basaltic lava somewhat resembling the Colombia Plateau in the northwestern United States. The lava flows are often over 30 meters (100 feet) thick, and the total thickness of the material that fills the maria must approach thousands of meters.

Regolith.   All lunar terrains are mantled with a layer of gray, unconsolidated debris derived from a few billion years of meteoric bombardment (Figure 20.8). This soil-like layer, properly called **lunar regolith**, is composed of igneous rocks, breccia, glass beads, and fine *lunar dust*. In the maria that have been explored by *Apollo* astronauts, the lunar regolith is apparently just over 3 meters (10 feet) thick.

Lunar History

Although the moon is our nearest planetary neighbor and astronauts have sampled its surface, much is still unknown about its origin. Until recently, most scientists argued that the moon and Earth formed together. That is, the moon and Earth consolidated from minute rock fragments and gases that orbited the protosun and accreted into planetary-sized bodies.

A newer hypothesis supported by many scientists suggests that a giant asteroid collided with Earth to produce the moon. The explosion caused by the impact of a Mars-sized body upon a semimolten Earth would

have ejected huge quantities of mantle rock from the primordial Earth. A portion of this ejecta would have remained in orbit around Earth, gradually accumulating to form the moon. This giant-impact hypothesis is plausible, but raises many questions.

Despite the moon's uncertain origin, planetary geologists have worked out basic details of its later history. One of their methods is to observe variations in crater density (quantity per unit area). The greater the crater density, the longer the topographic feature must have existed. From such evidence, scientists concluded that the moon evolved in three phases: the original crust (highlands), maria basins, and rayed craters.

During its early history the moon was continually impacted as it swept up debris. This continuous bombardment, and perhaps radioactive decay, generated enough heat to melt the moon's outer shell, and quite possibly the interior as well. Remnants of this original crust occupy the densely cratered highlands, which have been estimated to be as much as 4.5 billion years old—about the same age as Earth.

The second major event in the moon's evolution was the formation of maria basins (see Figure 20.7). Radiometric dating of the maria basalts puts their age between 3.2 and 3.8 billion years, roughly a billion years younger than the initial crust. In places, the lava flows overlap the highlands, another testimonial to the younger age of the maria deposits.

The last prominent features to form were the rayed craters, as exemplified by Copernicus (see Figure 20.3). Material ejected from, these "young" depressions is clearly seen blanketing the surface of the maria and many older rayless craters. Even a relatively young crater like Copernicus must be millions of years old. Had it formed on Earth, erosional forces would have long since obliterated it.

If photos of the moon taken several hundreds of millions of years ago were available, they would reveal that the moon changed little in the intervening years. By all measures, the moon is a dead body wandering through space and time.

The Planets: A Brief Tour

Mercury: The Innermost Planet

Mercury, the innermost and smallest planet, is hardly larger than Earth's moon and is smaller than three other moons in the solar system. Like the moon, it absorbs most of the sunlight that strikes it, reflecting only 6 percent into space. This is characteristic of terrestrial bodies that have no atmosphere. (Earth reflects about 30 percent of the light that strikes it, most of it from clouds.)

Mercury's close proximity to the sun makes viewing from Earthbound telescopes difficult because of the glare. The first good glimpse of this planet came in 1974 when *Mariner 10* passed within 800 kilometers (500 miles) of its surface (Table 20.2). Its striking resemblance to the moon was immediately evident from the high-resolution images that were sent back (Figure 20.9).

Mercury has cratered highlands, much like the moon, and vast smooth terrains which resemble maria. However, unlike the moon, Mercury is a very dense planet, which implies that it contains an iron core, perhaps larger than Earth's. Also, Mercury has very long scarps that cut across the plains and craters alike. These scarps may have resulted from crustal shortening as the planet cooled and shrank.

Mercury revolves quickly but rotates slowly. One full day-night cycle on Earth takes 24 hours but on Mercury it requires 179 Earth-days. Thus, a night on Mercury lasts for about three months and is followed by three months of daylight. Nighttime temperatures drop as low as –173°C (–280°F) and noontime temperatures exceed 427°C (800°F), hot enough to melt tin and lead. Mercury has the greatest extremes of any planet. The odds of life as we know it existing on Mercury are nil.

Figure 20.5

Formation of an impact crater. The energy of the rapidly moving meteoroid is transformed into heat energy and compressional waves. The rebound of the compressed rock causes debris to be ejected from the crater, and the heat melts some material, producing glass beads. Small secondary craters are formed by the material "splashed" from the impact crater. (After E. M. Shoemaker)

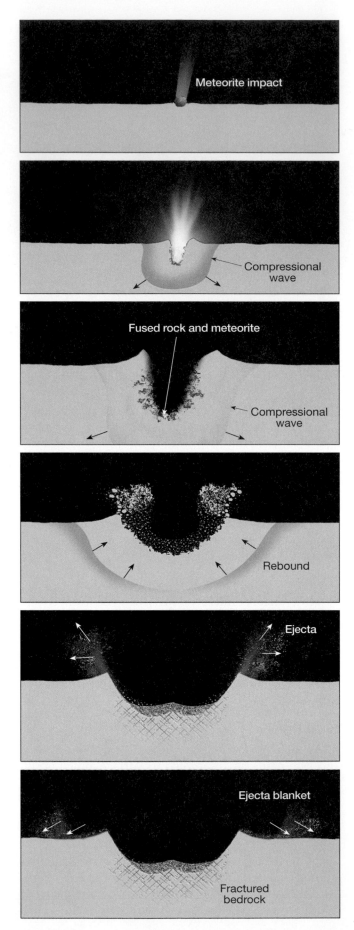

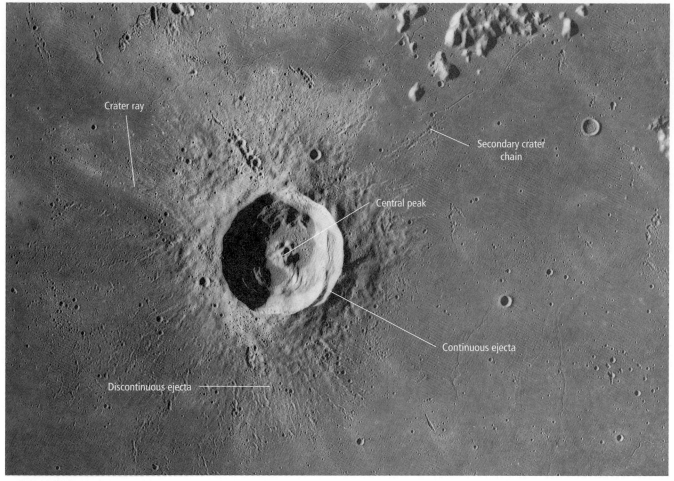

Figure 20.6
The 20-kilometer-wide lunar crater Euler in the southwestern part of Mare Imbrium. Clearly visible are the bright rays, central peak, secondary craters, and the large accumulation of ejecta near the crater rim. (Courtesy of NASA)

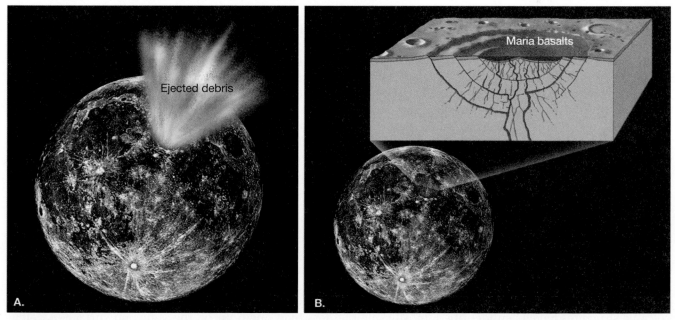

Figure 20.7
Formation of lunar maria. **A.** Impact of an asteroid-sized mass produced a huge crater hundreds of kilometers in diameter and disturbed the lunar crust far beyond the crater. **B.** Filling of the impact area with fluid basalts, perhaps derived from partial melting deep within the lunar mantle.

Figure 20.8
Astronaut Harrison Schmitt sampling the lunar surface. Notice the footprints (insert) in the lunar "soil." (Courtesy of NASA)

Venus: The Veiled Planet

Venus, second only to the moon in brilliance in the night sky, is named for the goddess of love and beauty. It orbits the sun in a nearly perfect circle once every 255 Earthdays. Venus is similar to Earth in size, density, mass, and location in the solar system. Thus, it has been referred to as "Earth's twin." Because of these similarities, it is hoped that a detailed study of Venus will provide geologists with a better understanding of Earth's evolutionary history.

Venus is shrouded in thick clouds impenetrable to visible light. Nevertheless, radar mapping by unmanned spacecraft and Earthbound instruments has revealed a varied topography with features somewhat between those of Earth and Mars. Simply, radar pulses in the microwave range are sent toward the Venusian surface, and the heights of plateaus and mountains are measured by timing the return of the radar echo. These data have confirmed that basaltic volcanism and tectonic deformation are the dominant processes operating on Venus. Further, based on the low density of impact craters, volcanism and tectonic deformation must have been very active during the recent geologic past (Figure 20.10).

Over 80 percent of the Venusian surface consists of subdued plains that are mantled by volcanic flows. Some lava channels extend hundreds of kilometers; one meanders 6800 kilometers across the planet. Thousands of volcanic structures have been identified, mostly small shield volcanoes, although over 1500 volcanoes greater than 20 kilometers across have been mapped. One is Sapas Mons, 400 kilometers (250 miles) across and 1.5 kilometers (0.9 mile) high. Many flows from this volcano erupted from its flanks rather than the summit, in the manner of Hawaiian shield volcanoes. Other volcanic structures discovered on Venus are circular, pancake-shaped domes about 25 kilometers (15 miles) in diameter and nearly 1 kilometer high (Figure 20.11). These domes are thought to be the result of outpouring of very viscous lava, much like volcanic domes on Earth.

Only 8 percent of the Venusian surface consists of highlands that may be likened to continental areas on Earth. Tectonic activity on Venus seems to be driven by upwelling and downwelling of material in the planet's interior. Although mantle convection still operates on Venus, the processes of plate tectonics, which recycle rigid lithosphere, do not appear to have contributed to the present Venusian topography.

Before the advent of space vehicles, Venus was considered to be a potentially hospitable site for living organisms. However, evidence from *Mariner* fly-by space probes and Russian *Venera* landers indicates differently. The surface of Venus reaches temperatures of 475°C (900°F), and the Venusian atmosphere is 97 percent carbon dioxide (see Box 20.1). Only scant water vapor and nitrogen have been detected. The Venusian atmosphere contains an opaque cloud deck about 25 kilometers thick, which begins approximately 70 kilometers from the surface. Although the unmanned *Venera 8* survived less than one hour on the Venusian surface, it determined that the atmospheric pressure on that planet is 90 times that on Earth's surface. This hostile environment makes it unlikely that life as we know it exists on Venus.

Table 20.2 Most significant space probes.

Mariner 2	1962	Fly-by of Venus (first to any planet)
Mariner 4, 6, 7	1965 1969 1969	Fly-by missions to Mars
Mariner 9	1971	Mars orbiter
Apollo 8	1968	Astronauts circled the Moon and returned to Earth
Apollo 11	1969	First astronaut landed on the Moon
Apollo 17	1972	Last of six manned *Apollo* missions to the Moon
Mariner 10	1974	Orbited the Sun, passing Mercury several times; fly-by of Venus
Pioneer 10, 11	1973 1974	First close-up views of Jupiter
Venera 8, 9, 10	1972 1975	Soviet landers on Venus (each survived about one hour)
Venera 13, 14	1982	First color images of Venus
Viking 1, 2	1976	Mars orbiters and landers
Voyager 1	1979 1980	Fly-by of Jupiter Fly-by of Saturn
Voyager 2	1979 1981 1986 1989	Fly-by of Jupiter Fly-by of Saturn Fly-by of Uranus Fly-by of Neptune
Giotto	1986	First photo of comet's nucleus
Magellan	1990	Radar imaging orbiter of Venus
Mars Observer	1993	Mars orbiter
Galileo	1990 1992 1995	Fly-by of Earth, Venus Fly-by of Earth, Moon Jupiter orbiter

Mars: The Red Planet

Mars has evoked greater interest than any other planet, for both astronomers and nonscientists. When we imagine intelligent life on other worlds, "little green Martians" may come to mind (see Box 20.2). Interest in Mars stems mainly from this planet's accessibility to observation. All other planets within telescopic range have their surfaces hidden by clouds, except for Mercury, whose nearness to the sun makes viewing difficult. Through the telescope, Mars appears as a reddish ball interrupted by some permanent dark regions that

Figure 20.9

Photomosaic of Mercury. This view of Mercury is remarkably similar to the "far side" of the moon. (Courtesy of NASA)

change intensity during the Martian year. The most prominent telescopic features of Mars are its brilliant white polar caps, resembling Earth's.

The Martian Atmosphere.   The Martian atmosphere has only 1 percent the density of Earth's and it is primarily carbon dioxide with tiny amounts of water vapor. Data from Mars probes confirm that the polar caps of Mars are made of water ice, covered by a thin layer of frozen carbon dioxide. As winter nears in either hemisphere, we see the equatorward growth of that hemisphere's ice cap as temperatures drop to −125°C (−193°F) and additional carbon dioxide is deposited.

Although the atmosphere of Mars is very thin, extensive dust storms occur and may cause the color changes observed from Earth-based telescopes. Hurricane-force winds up to 270 kilometers (170 miles) per hour can persist for weeks. Images from *Viking 1* and *Viking 2* revealed a Martian landscape remarkably similar to a rocky desert on Earth (Figure 20.12), with abundant sand dunes and impact craters partially filled with dust.

Mars' Dramatic Surface.   *Mariner 9*, the first artificial satellite to orbit another planet, reached Mars in 1971 amid a raging dust storm. When the dust cleared,

Figure 20.10

Computer-generated image of Venus. On the horizon is Maat Mons, a large volcano. Below it is a volcanic cone, Sapas Mons, from which light-colored lava flows extend hundreds of kilometers. (Courtesy of NASA)

images of Mars' northern hemisphere revealed numerous large volcanoes. The biggest, Mons Olympus, is the size of Ohio and 23 kilometers (75,000 feet) high. This gigantic volcano and others resemble Hawaiian shield volcanoes on Earth (Figure 20.13). Their extreme size is thought to result from the absence of plate movements on Mars. Therefore, rather than a chain of smaller volcanoes forming as we find in Hawaii, large single cones developed.

Less-abundant impact craters indicate that at least some of the volcanic topography formed more recently, following the early period of heavy bombardment. Nevertheless, most Martian surface features are old by Earth standards. The highly cratered Martian southern hemisphere is probably similar in age to comparable lunar highlands (3.5–4.5 billion years old). Even the relatively fresh-appearing volcanic features of the northern hemisphere may be older than 1 billion

Figure 20.11

These domelike volcanic structures on Venus average 25 kilometers in diameter and are less than 1 kilometer high. They are interpreted as very thick lava flows. (Photo courtesy of Jet Propulsion Laboratory)

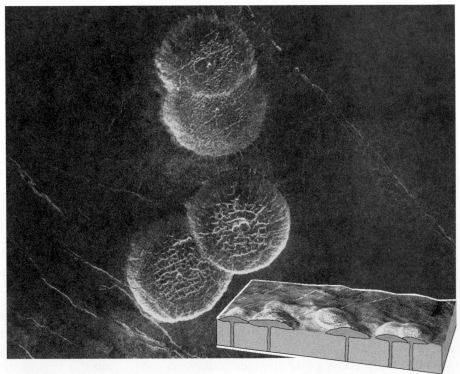

Box 20.1

Venus and the Runaway Greenhouse Effect: A Lesson for Planet Earth?

Carl Sagan, one of the foremost experts on extraterrestrial life, called Earth "the Heaven of the solar system" and Venus "the hell." Why should environments of two planets that are nearly the same size and are located in close proximity to one another be so dramatically different? The primary reason is the blistering temperature of the Venusian atmosphere, which is caused by a runaway greenhouse effect.

The greenhouse effect occurs when a planet's atmosphere acts like the glass in a greenhouse. It allows visible solar energy to penetrate the surface, but, like the glass in the greenhouse, certain gases in the planet's atmosphere are quite opaque to outgoing heat radiation. Thus, as in a greenhouse, the surface temperatures of these planets become warmer than expected, because the heat is temporarily trapped.

Both Earth and Venus experience a greenhouse effect. So why do the surface temperatures of Venus reach 475°C (900°F), whereas Earth is very hospitable? The answer is that carbon dioxide, a main contributor to the greenhouse effect, makes up 97 percent of the Venusian atmosphere, but less than 1 percent of Earth's (Figure 20.A).

The primary source of carbon dioxide is outgassing during volcanic eruptions. Why does Earth, which has over 600 active volcanoes, have an atmosphere containing only 0.035 percent carbon dioxide? The answer can be found in Earth's abundant plant life.

Plants use carbon dioxide and water in photosynthesis to generate organic matter, while oxygen is released as a by-product. Thus, over millions of years, plant life has altered our atmosphere, making it carbon dioxide-poor (0.035 percent) and oxygen-rich (21 percent). In addition, carbon dioxide dissolved

Figure 20.A
Venus is shrouded in a hot, cloud-filled atmosphere composed mainly of carbon dioxide. (Courtesy of NASA)

in seawater is used by a vast number of organisms for the production of carbonate shells. These shells are eventually deposited as sediment on the ocean floor. Consequently, huge quantities of carbon dioxide from Earth's atmosphere are continually being converted into organic matter and carbonate sediments. Apparently, Venus does not possess living organisms or a chemical mechanism capable of removing atmospheric carbon dioxide. Hence, our neighboring planet has a carbon dioxide concentration that has reached extreme proportions.

The discovery of a hostile Venusian atmosphere provides a good example of the practical benefits that people derive from space exploration. These data clearly reveal what can happen to a planet's environment when the greenhouse effect runs rampant. Some climatologists believe that a similar situation may be in the making on Earth. Through the burning of fossil fuels, we are converting increasingly larger amounts of oxygen and organic materials into carbon dioxide and water. In this manner, we are altering Earth's atmosphere by rapidly undoing what plant life has taken millions of years to accomplish.

We expect that further studies of the Venusian atmosphere will provide insights into how Earth's atmosphere evolved and how it will respond to changes induced by human activity. Such insights could keep our planet from becoming, as some have suggested, "hothouse Earth."

years. This fact plus the absence of "Marsquake" recordings by *Viking* seismographs point to a tectonically dead planet.

Another surprising find made by *Mariner 9* was the existence of several canyons that dwarf even Earth's Grand Canyon of the Colorado River. One of the largest, Valles Marineris, is thought to have formed by slippage of material along huge faults in the crustal layer. In this respect, it would be comparable to the rift valleys of Africa (Figure 20.14).

Box 20.2

The Search for Extraterrestrial Intelligence

The possibility that life exists on other planets has fascinated people for centuries. Before the era of space exploration, stargazers observed color variations on Mars which they believed to be vegetation changing colors with the seasons. But these colors proved to be ferocious Martian dust storms. Later, cameras on board unmanned spacecraft provided images of large stream-like channels on Mars that renewed speculation that water and possible primitive life occupied this planet. However, samples of Martian soil obtained in the 1970s by the *Viking* lander lacked evidence for the existence of life. Indeed, the environments on all other planets in our solar system seem to be incompatible for life as we know it.

The search for extraterrestrial life has been rejuvenated with the recent discovery of planets outside our solar system. These still-unnamed planets were detected by measuring the wobble they caused in the orbits of their stars. Of the three planets detected in 1995 and 1996, one is about half the mass of Jupiter, whereas the others are about 2.3 and 6.5 times the mass of Jupiter. Although the odds are slim that these planets harbor complex life, their existence implies that planet formation is not unique. Therefore, literally trillions of planets may

Figure 20.B

NASA's Goldstone deep space radio telescope. (Photo by Jet Propulsion Laboratory)

exist, many located around hospitable stars not unlike our sun.

Four programs in the United States are currently involved in the search for extraterrestrial intelligence (SETI). These projects use large radio receivers (radio telescopes) to scan the skies for radio signals from an advanced civi-lization (Figure 20.B). One of the most ambitious projects will scan 250 million radio frequencies at a time for the one that ET might be broadcasting on. No one is betting heavily that an advanced civilization will be discovered soon, but with the discovery of more planets, the odds keep improving.

Figure 20.12

This picture of the Martian landscape by the *Viking 1* lander shows a dune field with features remarkably similar to many seen in the deserts of Earth. The dune crests indicate that recent wind storms were capable of moving sand over the dunes in the direction from lower right to upper left. The large boulder at the left is about 10 meters from the spacecraft and measures 1 by 3 meters. (Courtesy of NASA)

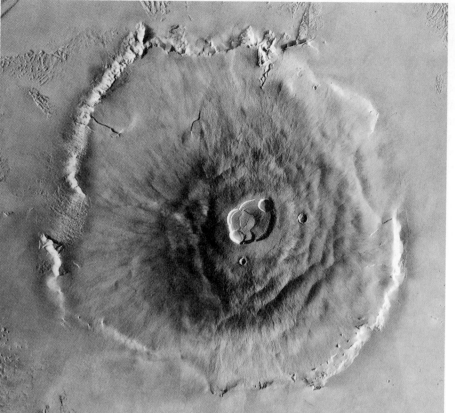

Figure 20.13
Image of Mons Olympus, an inactive shield volcano on Mars that covers an area about the size of the state of Ohio. (Courtesy of the U.S. Geological Survey)

Figure 20.14
This image shows the entire Valles Marineris canyon system, over 5000 kilometers long and up to 8 kilometers deep. The dark red spots on the left edge of the image are huge volcanoes, each about 25 kilometers high. (Courtesy of U.S. Geological Survey)

Water on Mars? Not all Martian valleys have a tectonic origin. Many have tributaries in a pattern similar to that of stream valleys on Earth. In addition, *Viking* orbiter images have revealed unmistakable ancient islands in what is now a dry stream bed. When these streamlike channels were first discovered, some observers speculated that a thick water-laden atmosphere capable of generating torrential downpours once existed on Mars. If so, what happened to this water? The present Martial atmosphere contains only traces. Moreover, the environment of Mars is far too harsh to let water exist as a liquid. Despite these difficulties, the work of flowing water still remains the most acceptable explanation for many Martian channels.

Many planetary geologists do not accept the premise that Mars once had an active water cycle similar to Earth's. Rather, they believe that many of the large streamlike valleys were created by the collapse of surface material caused by the slow melting of subsurface ice. If this is the case, these large valleys would be more akin to features formed by mass wasting processes on Earth.

The Martian Satellites. Tiny Phobos and Deimos, the two satellites of Mars, were not discovered until 1977 because they are only 24 and 15 kilometers in diameter. Phobos is nearer to its parent than any other natural satellite in the solar system—only 5500 kilometers—and requires just 7 hours and 39 minutes for one revolution. *Mariner 9* revealed that both satellites are irregularly shaped and have numerous impact craters, much like their parent (Figure 20.15).

Undoubtedly, these moons are asteroids captured by Mars. A most interesting coincidence in astronomy and literature is the close resemblance between Phobos and Deimos and two fictional satellites of Mars described by Jonathan Swift in *Gulliver's Travels*, written about 150 years before these satellites were actually discovered.

Jupiter: Lord of the Heavens

Jupiter, truly a giant among planets, has a mass $2\frac{1}{2}$ times greater than the combined mass of all the remaining planets, satellites, and asteroids. In fact, had Jupiter been about ten times larger, it would have evolved into a small star. Despite its great size, however, it is only $\frac{1}{800}$ as massive as the sun. Jupiter also rotates more rapidly than any other planet, completing one rotation in slightly less than ten Earth-hours. The effect of this fast spin is to make the equatorial region bulge and to make the polar dimension flatten (see "Polar Flattening" column in Table 20.1).

When viewed through a telescope or binoculars, Jupiter appears to be covered with alternating bands of multicolored clouds aligned parallel to its equator (Figure 20.16). The most striking feature is the *Great Red Spot* in the southern hemisphere (Figure 20.16). The Great Red Spot has been a prominent feature since is was first discovered more than three centuries ago. When *Voyager 2* swept by Jupiter in 1979, it was the size of two Earth-sized circles placed side by side. On occasion, it has grown even larger.

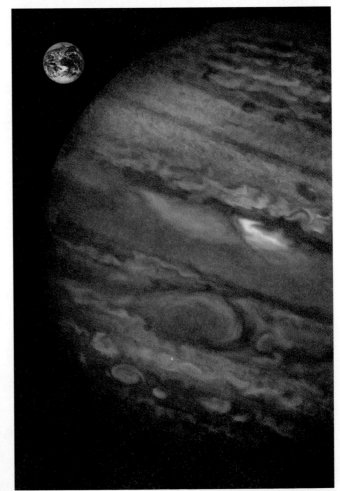

Figure 20.16
Artist's view of Jupiter with Great Red Spot visible in its southern hemisphere. Earth for scale.

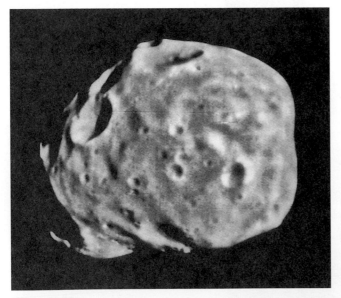

Figure 20.15
Tiny Phobos, innermost of the two Martian satellites. (Courtesy of NASA)

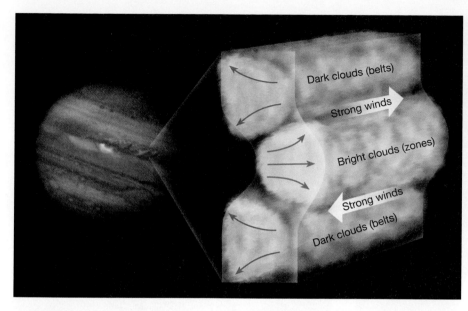

Figure 20.17
The structure of Jupiter's atmosphere. The areas of light clouds (*zones*) are regions where gases are ascending and cooling. Sinking dominates the flow in the darker cloud layers (*belts*). This convective circulation, along with the rapid rotation of the planet, generates the high-speed winds observed between the belts and zones.

Dark clouds (belts)

Strong winds

Bright clouds (zones)

Strong winds

Dark clouds (belts)

Images from *Pioneer 11* as it moved within 42,000 kilometers of Jupiter's cloud tops in 1974 indicated that the Great Red Spot is a counterclockwise-rotating (cyclonic) storm. It is caught between two jetstream-like bands of atmosphere flowing in opposite directions. This huge hurricane-like storm rotates once every 12 Earth-days. Although several smaller storms have been observed in other regions of Jupiter's atmosphere, none has survived for more than a few days.

Structure of Jupiter. Jupiter's hydrogen-helium atmosphere also has methane, ammonia, water, and sulfur compounds as minor constituents. The wind systems generate the light- and dark-colored bands that encircle this giant (Figure 20.17). Unlike the winds on Earth, which are driven by solar energy, Jupiter itself gives off nearly twice as much heat as it receives from the sun. Thus, it is the *interior* heat from Jupiter that produces huge convection currents in the atmosphere.

Atmospheric pressure at the top of the clouds is equal to sea-level pressure on Earth. Because of Jupiter's immense gravity, the pressure increases rapidly toward its surface. At 1000 kilometers below the clouds, the pressure is great enough to compress hydrogen gas into a liquid. Consequently, Jupiter's surface is thought to be a gigantic ocean of liquid hydrogen. Less than halfway into Jupiter's interior, pressures of unimaginable magnitude cause the liquid hydrogen to turn into *liquid metallic* hydrogen. Jupiter is also believed to contain as much rocky and metallic material as is found in the terrestrial planets, probably in a central core.

Jupiter's Moons. Jupiter's satellite system, consisting of sixteen moons discovered so far, resembles a miniature solar system. The four largest satellites, discovered by Galileo, travel in nearly circular orbits around the parent with periods of from two to seventeen Earth-days (Figure 20.18). The two largest Galilean satellites, Callisto and Ganymede, surpass Mercury in size, while the two smaller ones, Europa and Io, are about the size of Earth's moon. These Galilean moons can be observed with binoculars or a small telescope and are interesting in their own right. Because their orbits are along Jupiter's equatorial plane, and because they all have the same orbital direction, these moons most probably formed from "leftover" debris in much the same way as the planets did.

By contrast, Jupiter's four outermost satellites are very small (20 kilometers in diameter), revolve in a direction that is opposite the other moons, and have orbits that are steeply inclined to the Jovian equator. These satellites appear to be asteroids that passed near enough to be captured gravitationally by Jupiter.

Images from *Voyagers 1* and *2* in 1979 revealed, to the surprise of almost everyone, that each of the four Galilean satellites has a character all its own, as shown in Figure 20.18. The innermost of the Galilean moons, Io, is the only volcanically active body discovered in our solar system other than Earth and Neptune's moon Triton (Figure 20.19). To date, eight active sulfurous volcanic centers have been discovered. Umbrella-shaped plumes have been seen rising from the surface of Io to heights approaching 200 kilometers (Figure 20.19). The heat source for Io's volcanic activity is thought to be tidal energy generated by a relentless "tug of war" between Jupiter and the Galilean satellites. Because Io is gravitationally locked to Jupiter, the same side always faces the giant planet, like Earth's moon. The gravitational power of Jupiter and the other nearby satellites pulls and pushes on Io's tidal bulge as its slightly eccentric orbit takes it alternately closer to and farther from Jupiter. This gravitational flexing of Io is transformed into heat energy.

One of the most interesting discoveries made by *Voyager 1* is Jupiter's thin ring system, believed to be different from that encircling Saturn. Rather than particles held in planetary-type orbits, the particles composing Jupiter's rings appear to be temporarily entrapped by the planet's intense magnetic field. The ring material may be sulfur particles from Io's volcanoes.

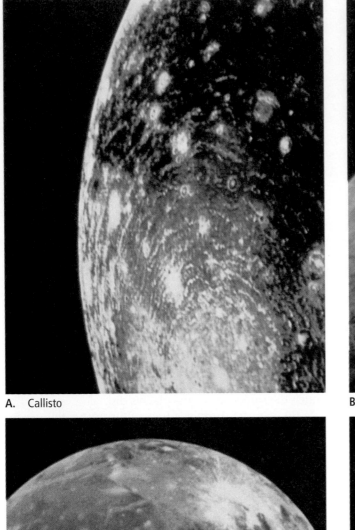

A. Callisto

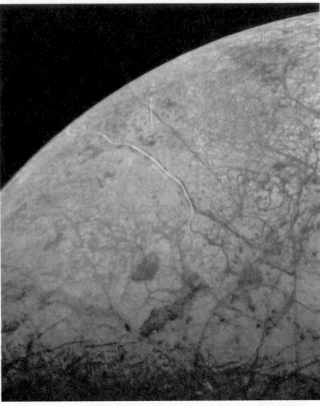

B. Europa

C. Ganymede

D. Io

Figure 20.18

Jupiter's four largest moons called the Galilean moons because they were discovered by Galileo. **A.** Callisto, the outermost of the Galilean satellites, is densely cratered, much like Earth's moon. **B.** Europa, smallest of the Galilean moons, has an icy surface that is criss-crossed by many linear features. **C.** Ganymede, the largest Jovian satellite, contains cratered areas, smooth regions, and areas covered by numerous parallel grooves. **D.** The innermost moon, Io, is one of only three volcanically active bodies in the solar system. (Courtesy of NASA)

Figure 20.19

A volcanic eruption on Io. This plume of volcanic gases and debris is rising over 100 kilometers (60 miles) above Io's surface. (Courtesy of NASA)

Saturn: The Elegant Planet

Requiring 29.46 Earth-years to make one revolution, Saturn is almost twice as far from the sun as Jupiter, yet its atmosphere, composition, and internal structure are thought to be remarkably similar to Jupiter's.

Saturn's Rings.   The most prominent feature of Saturn is its system of rings (Figure 20.20), discovered by Galileo in 1610. Because he could not resolve them with his primitive telescope, they appeared to him as two smaller bodies adjacent to the planet. Their ring nature was revealed 50 years later by the Dutch astronomer Christian Huygens. Until the recent discovery that Jupiter, Uranus, and Neptune also have very faint ring systems, this phenomenon was thought to be unique to Saturn. When viewed from Earth, Saturn's rings appear as distinct bands.

Our view of Saturn slowly changes as both planets proceed along their orbits, continually shifting relative positions. This changes our angle of view of Saturn's rings. Once every 15 years, we see them edge-on, and they appear as an extremely fine line.

Saturn Close-Up.   In 1980 and 1981, fly-by missions of the nuclear-powered *Voyagers 1* and *2* space vehicles came within 100,000 kilometers of Saturn. More information was gained in a few days than had been acquired since Galileo first viewed this elegant planet telescopically (Figure 20.21):

1. Saturn's atmosphere is very dynamic with winds roaring at up to 1500 kilometers (930 miles) per hour.

2. Large cyclonic "storms" similar to Jupiter's Great Red Spot, although much smaller, occur in Saturn's atmosphere.

3. Eleven additional moons were discovered.

4. The icy rings of Saturn were discovered to be more complex than expected. Each of the seven rings is made of numerous ringlets,

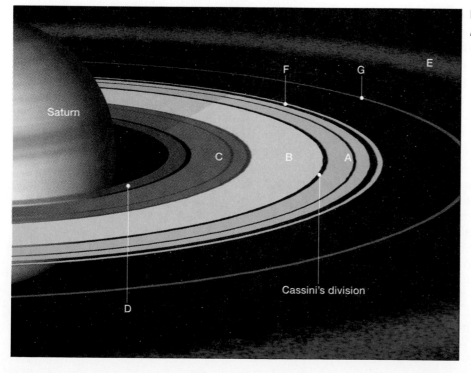

Figure 20.20

A view of the dramatic ring system of Saturn.

Figure 20.21
Montage of the Saturnian satellite system. The moon Dione is in foreground; Tethys and Mimas are at lower right; Enceladus and Rhea are off ring's left; and Titan is upper right. (Photo courtesy of NASA)

resembling the grooves on a phonograph record. The ringlets in one ring are intertwined in a braidlike configuration.

5. Satellite images reveal the thickness of the ring system to be no more than a few hundred meters. We easily see the thin rings from more than a billion kilometers distance because they are highly reflective.

No image obtained so far can resolve the fine structures of the rings, but they undoubtedly are composed of small particles (moonlets) that orbit the planet much like any other satellite. Radar observations indicate particles no larger than 10 meters, and the more abundant particles down to 10 centimeters (4 inches).

Beyond the outermost bright ring (A ring), some moonlets have accreted to form very small satellites having diameters on the order of 100 kilometers (60 miles). Five of these asteroid-sized moons have been discovered orbiting within the faint outer rings, and others probably exist. Planetary geologists are very interested in the gravitational interaction in Saturn's ring system. They hope this will reveal how material from the primordial cloud of dust and gases condensed to produce the planets.

Saturn's Moons. The Saturnian satellite system consists of at least twenty-one bodies (Figure 20.21). The largest, Titan, is bigger than Mercury and is the second-largest satellite in the solar system (after Jupiter's Ganymede). It is the only satellite in the solar system, known to have a substantial atmosphere. Because of its dense gaseous cover, the atmospheric pressure at Titan's surface is about one and one-half times that at Earth's surface.

Data from *Voyager 1* reveal Titan's atmosphere to be roughly 80 percent nitrogen and perhaps 6 percent methane. This planet-sized moon appears to have polar ice caps that show seasonal variations in size. Its surface, if unfrozen, would be an ocean of liquid nitrogen.

Uranus and Neptune: The Twins

Earth and Venus have similar traits, but Uranus and Neptune are nearly twins. Only 1 percent different in diameter, both appear a pale greenish-blue, attributable to the methane in their atmospheres. Their structure and composition are believed to be similar. Neptune, however, is colder, because it is half again as distant from the sun's warmth as is Uranus.

Uranus, the Sideways Planet. A unique feature of Uranus is that it rotates "on its side" (Figure 20.22). Its axis of rotation, instead of being generally perpendicular to the plane of its orbit, like the other planets, lies only 8 degrees from the plane of its orbit. Its rotational motion, therefore, has the appearance of rolling, rather than the toplike spinning of the other planets. Because the axis of Uranus is inclined almost 90 degrees, the sun is nearly overhead at one of its poles once each revolution, and then half a revolution later, it is overhead at the other pole.

A surprise discovery in 1977 revealed that Uranus has rings, much like those encircling Jupiter. This find occurred as Uranus passed in front of a distant star and blocked its view, a process called *occultation* (the word *occult* means "hidden"). Observers saw the star "wink" briefly five times (meaning five rings) before the primary occultation and again five times afterward. Later studies indicate that Uranus has at least nine distinct belts of debris orbiting its equatorial region.

Spectacular views from *Voyager 2* of the five largest moons of Uranus show quite varied terrains. Some have long, deep canyons and linear scars, whereas others possess large, smooth areas on otherwise crater-riddled surfaces. The Jet Propulsion Laboratory described Miranda, the innermost of the five largest moons, as having a greater variety of landforms, than any body yet examined in the solar system.

Neptune. Even when the most powerful telescope is focused on Neptune, it appears as a bluish fuzzy disk. Until *Voyager 2's* 1989 encounter, astronomers knew very little about this planet. However, *Voyager 2's* 12-year, nearly 3-billion-mile journey provided investigators with so much new information about Neptune and its satellites that years will be needed to analyze it all.

Figure 20.22
Painting of *Voyager 2* as it might have appeared when it encountered Uranus on January 24, 1986. (By Don Davis, courtesy of NASA)

Figure 20.23
This image of Neptune shows the Great Dark Spot (left center). Also visible are bright cirruslike clouds that travel at high speed around the planet. A second oval spot is at 54° south latitude on the east limb of the planet. (Courtesy of the Jet Propulsion Laboratory)

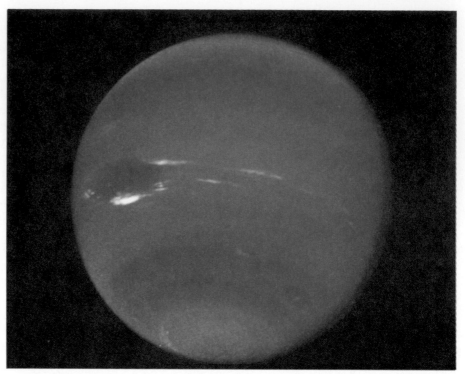

Neptune has a dynamic atmosphere, much like those of Jupiter and Saturn (Figure 20.23). Winds exceeding 1000 kilometers per hour (600 miles per hour) encircle the planet making it one of the windiest places in the solar system. It also has an Earth-sized blemish called the *Great Dark Spot* that is reminiscent of Jupiter's Great Red Spot, and is assumed to be a large rotating storm.

Perhaps most surprising are white, cirruslike clouds that occupy a layer about 50 kilometers above the main cloud deck, probably frozen methane. Six new satellites were discovered in the *Voyager* images, bringing Neptune's family to eight. All of the newly discovered moons orbit the planet in a direction opposite that of the two larger satellites. *Voyager* images also revealed a ring system around Neptune.

Triton, Neptune's largest moon, is a most interesting object (Figure 20.24). Its diameter is nearly that of Earth's moon. Triton is the only large moon in the solar system that orbits its parent in the direction opposite to the direction in which all the planets travel. This indicates that Triton formed independently of Neptune and was gravitationally captured.

Triton also has the lowest surface temperature of any body in the solar system, -200°C (-391°F). Its very thin atmosphere is mostly nitrogen with a little methane. The surface of Triton apparently is largely water ice, covered with layers of solid nitrogen and methane.

Pluto: Planet X

Pluto lies on the fringe of the solar system, almost 40 times farther from the sun than Earth. It is 10,000 times too small to be visible to the unaided eye. Because of its great distance and slow orbital speed, it takes Pluto 248 Earth-years to orbit the sun. Since its discovery in 1930, it has completed about one-fourth of a revolution (see Box 20.3). Pluto's orbit is noticeably elongated (highly eccentric), causing it to occasionally travel inside the orbit of Neptune, where it currently resides. There is no chance that Pluto and Neptune will ever collide, because their orbits are inclined to each other and do not actually cross (see Figure 20.1).

In 1978, the moon Charon was discovered orbiting Pluto. Because of its close proximity to the planet, the best ground-based images of Charon show it only as an elongated bulge (Figure 20.25, left). In 1990, the Hubble Space Telescope produced an image that clearly resolves the separation between these two icy worlds (Figure 20.25, right). Charon orbits Pluto once every 6.4 Earth-days at a distance 20 times closer to Pluto than our moon is to Earth.

The discovery of Charon greatly altered earlier estimates of Pluto's size. Current data indicate that Pluto has a diameter of 2445 kilometers, about one-fifth the size of Earth, making it the smallest planet in the solar system. Charon is about 1300 kilometers across, exceptionally large in proportion to its parent.

The average temperature of Pluto is estimated at –210°C, cold enough to solidify most gases that might be present. Thus, Pluto might best be described as a dirty iceball of frozen gases with lesser amounts of rocky substances.

A recent proposal suggests that Pluto was a satellite of Neptune that was displaced from its original orbit by collision with something. The discovery of a satel-

ularly pass close to Earth and its moon. Many of the most recent impact craters on the moon and Earth were probably caused by collisions with asteroids. Inevitably, future Earth-asteroid collisions will occur—see Box 20.4.

Because many asteroids have irregular shapes, planetary geologists first speculated that they might be fragments of a broken planet that once orbited between Mars and Jupiter (Figure 20.27). However, the total mass of the asteroids is estimated to be only one-thousandth that of Earth, which itself is not a large planet. What happened to the remainder of the original planet? Others have hypothesized that several larger bodies once coexisted in close proximity and that their collisions produced numerous smaller ones. The existence of several "families" of asteroids has been used to support this explanation. However, no conclusive evidence has been found for either hypothesis.

Comets

Comets are among the most interesting and unpredictable bodies in the solar system (see Box 20.5). They have been compared to dirty snowballs, because they are made of frozen gases (water, ammonia, methane, carbon dioxide, and carbon monoxide) that hold together small pieces of rocky and metallic materials. Many comets travel very elongated orbits that carry them beyond Pluto.

When first observed, a comet appears very small; but as it approaches the sun, solar energy begins to vaporize the frozen gases, producing a glowing head called the **coma** (Figure 20.28). The size of the coma varies greatly from one comet to another. Extremely rare ones exceed the size of the sun, but most approximate the size of Jupiter. Within the coma, a small glowing nucleus with a diameter of only a few kilometers can sometimes be detected. As comets approach the sun, some, but not all, develop a tail that extends for millions of kilometers. Despite the enormous size of their tails and comas, comets are thought to have insignificant mass.

A comet's tail is not a *trailing* tail, but a gas and dust plume repelled by solar forces. This explains why the tail always points away from the sun (Figure 20.28). The solar forces are *radiation pressure*, which pushes dust particles away from the coma, and the *solar wind*, which moves the ionized gases, particularly carbon monoxide (Figure 20.28). Usually a single tail composed of both types of materials is produced, but sometimes a dual tail is observed.

As a comet moves away from the sun, the gases recondense, the tail disappears, and the comet returns to "cold storage." Material that was blown from the coma to form the tail is lost from the comet forever. Consequently, it is believed that most comets cannot survive more than a few hundred close orbits of the sun. Once all the gases are expelled, the remaining material—a swarm of unconnected metallic and stony particles—continues the orbit without a coma or a tail.

Figure 20.24
A photomosaic of Neptune's largest moon, Triton. The large south polar cap is on the right half of the image. Seasonal ice (probably nitrogen) covers the region. Because spring in Triton's southern hemisphere extends from 1960 to the year 2000, some of the polar cap has evaporated. (Courtesy of the Jet Propulsion Laboratory)

lite around Pluto is considered evidence that this event broke the Neptunian satellite into two pieces and sent them into an elongated orbit around the sun.

Minor Members of the Solar System

Asteroids

Asteroids are smaller bodies that have been likened to "flying mountains." The largest, Ceres, is about 1000 kilometers (600 miles) in diameter, but most of the 50,000 that have been observed are only about one kilometer across. The smallest asteroids are assumed to be no larger than grains of sand. Most lie between the orbits of Mars and Jupiter and have periods of three to six years (Figure 20.26). Some have very eccentric orbits and travel very near the sun, and a few larger ones reg-

Box 20.3

Discovery of Planet X

During the mid-1800s, Neptune, the eighth planet in our solar system, was first sighted. After this discovery, unresolved discrepancies in the orbit of Uranus led some astronomers to suggest that the gravitational effect of a yet-unknown planet might be responsible. A vigorous search for this planet was begun in 1906 by Percival Lowell at the observatory in Flagstaff, Arizona, that now bears his name. He unsuccessfully searched for "Planet X" until his death 10 years later.

Some years afterward, a new photographic telescope was installed at the Lowell Observatory and the search resumed. Two photographs of the suspect portion of the sky were taken at intervals of a week or so and then compared. Over such a short interval, stars appear stationary, but a planet shifts its position relative to its starry background. Twenty-

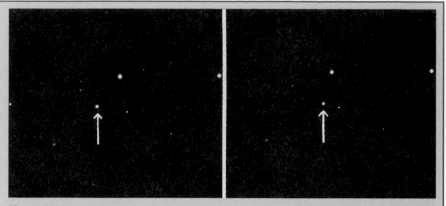

Figure 20.C
Pluto appears starlike on a photograph but is revealed by its motion to be a planet. (Courtesy of Hale Observatories)

five years had lapsed since Lowell's death, and 2 million stars were examined before Clyde Tombaugh discovered a body that shifted about the right amount for an object 1 billion miles beyond the orbit of Neptune (Figure 20.C).

This newly discovered planet was named Pluto, for the Greek god of the underworld, because the first two letters are Percival Lowell's initials.

Although the planet was located in the constellation Gemini, as Lowell had calculated, its mass was much less than he had estimated. It seemed unlikely that a planet of Pluto's size could have altered the orbit of Uranus measurably. Most probably, the discovery of Pluto can be credited to an extensive search of the correct portion of the sky for the wrong reasons.

Little is known about the origin of comets. They probably formed at great distances from the sun. Millions are believed to orbit the sun beyond Pluto, and it is proposed that the gravitational effect of nearby stars

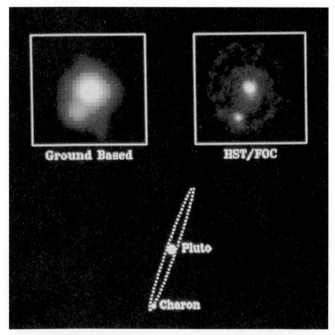

Figure 20.25
Pluto and its moon Charon. The Hubble Space Telescope produced the first image (upper right) that resolved these two icy worlds into separate objects. The image in the upper left is the best ground-based photo produced to date. (Courtesy of NASA)

sends some of them into the highly eccentric orbits that carry them toward the center of our solar system. Here, the gravitational pull of the larger planets, particularly Jupiter, alters a comet's orbit and accelerates its period of revolution.

The most famous short-period comet is Halley's comet. Its orbital period averages 76 years, and every one of its 29 appearances since 240 B.C. has been recorded by Chinese astronomers. This record is a testimonial to their dedication as astronomical observers and to the endurance of their culture. When seen in 1910, Halley's comet had developed a tail nearly 1.6 million kilometers (1 million miles) long and was visible during the daylight hours.

In 1986, the unspectacular showing of Halley's comet was a disappointment to many people in the Northern Hemisphere. Yet it was during this most recent visit to the inner solar system that a great deal of new information was learned about this most famous of comets. The new data were gathered by space probes sent to rendezvous with the comet. Most notably, the European probe *Giotto* approached to within 600 kilometers of the comet's nucleus and obtained the first images of this elusive structure.

We now know that the nucleus is potato-shaped, 16 kilometers by 8 kilometers. The surface is irregular and full of craterlike pits. Gases and dust that vaporize from the nucleus to form the coma and tail appear to gush from its surface as bright jets or streams. Only about 10 percent of the comet's total surface was emitting these

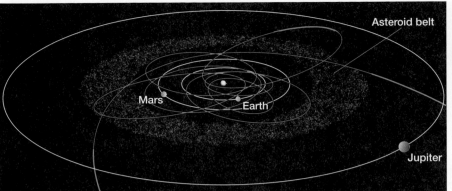

Figure 20.26
The orbits of most asteroids lie between Mars and Jupiter. Also shown are the orbits of a few known near-Earth asteroids. Perhaps a thousand or more asteroids have near-Earth orbits. Luckily, only a few dozen are thought to be larger than 1 kilometer in diameter.

Box 20.4

Is Earth on a Collision Course?

The solar system is cluttered with meteoroids, asteroids, active comets, and extinct comets. These fragments travel at great speeds and can strike Earth with the explosive force of a powerful nuclear weapon.

In the last few decades, it has become increasingly clear that comets and asteroids have collided with Earth far more frequently than was previously known. The evidence is giant impact structures called *astroblemes*. More than 100 have been identified (Figure 20.D). Many were once misunderstood to result from some volcanic process. Although most astroblemes are so old that they no longer resemble impact craters, evidence of their intense impact remains (Figure 20.E). One notable exception is a very fresh-looking crater near Winslow, Arizona, known as Meteor Crater (see Figure 20.29).

Evidence is mounting that about 65 million years ago a large asteroid about 10 kilometers (6 miles) in diameter collided with Earth. This impact may have caused the extinction of the dinosaurs, as well as nearly 50 percent of all plant and animal species (see Box 11.2).

More recently, a spectacular explosion has been attributed to the collision of our planet with a comet or asteroid. In 1908, in a remote region of Siberia, a "fire-ball" that appeared more brilliant than the sun exploded with a violent force. The shock waves rattled windows and caused sounds that were heard up to 1000 kilometers away. The "Tunguska event," as it is called, scorched, delimbed, and flattened trees to 30 kilometers from the epicenter. But expeditions to the area found no evidence of an impact crater, nor any metallic fragments. Evidently, the explosion, which equaled at least a 10-megaton nuclear bomb, occurred a few kilometers above the surface. Most likely it was the demise of a comet or perhaps a stony asteroid. Why it exploded prior to impact is uncertain.

The dangers of living with these small but deadly objects from space again came to public attention in 1989 when an asteroid nearly one kilometer across shot past Earth. It was a near miss, about twice the distance to the moon. Traveling at 70,000 kilometers (44,000 miles) per hour, it could have produced a crater 10 kilometers (6 miles) in diameter and perhaps 2 kilometers (1.2 miles) deep. As an observer noted, "Sooner or later it will be back." As it was, it crossed our orbit just 6 hours ahead of Earth. Statistics show

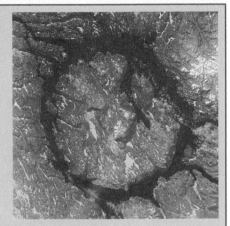

Figure 20.E
Manicouagan, Quebec, is a 200-million-year-old eroded impact crater. The lake outlines the crater remnant, which is 70 kilometers (42 miles) across. Fractures related to this event extend outward for an additional 30 kilometers. (Courtesy of U.S. Geological Survey)

that collisions of this tremendous magnitude should take place every few hundred million years and could have drastic consequences for life on Earth.

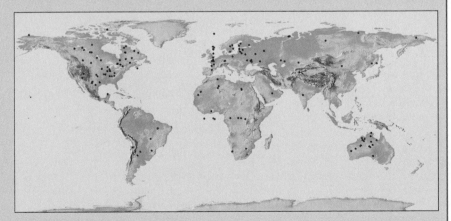

Figure 20.D
World map of major impact structures (àstroblemes). Others are being identified every year. (Data from Griffith Observatory).

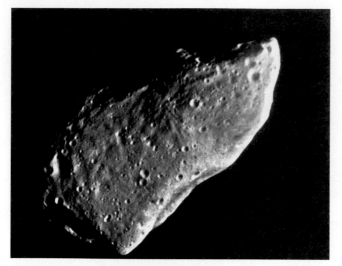

Figure 20.27

Image of asteroid 951 (Gaspra) obtained by the Jupiter-bound *Galileo* spacecraft. Like other asteroids, Gaspra is probably a collision-produced fragment of a larger body. (Courtesy of NASA)

jets at the time of the rendezvous. The remaining surface area of the comet appeared to be covered with a dark layer that may consist of organic material.

Meteoroids

Nearly everyone has seen a **meteor**, popularly (but inaccurately) called a "shooting star." This streak of light lasts from an eyeblink to a few seconds and occurs when a small solid particle, a **meteoroid**, enters Earth's atmosphere from interplanetary space. The friction between the meteoroid and the air heats both and produces the light we see. Although a rare meteoroid is as large as an asteroid, most are the size of sand grains and weigh less than $\frac{1}{100}$ gram. Consequently, they vaporize before reaching Earth's surface. Some, called *micrometeorites*, are so tiny that their rate of fall becomes too slow to cause them to burn up, so they drift down as "space dust." Each day, the number of meteoroids that enter Earth's atmosphere must reach into the thousands. After sunset, a half dozen or more are bright enough to be seen with the naked eye each hour from anywhere on Earth.

Occasionally, meteor sightings increase dramatically to 60 or more per hour. These displays, called **meteor showers**, result when Earth encounters a swarm of meteoroids traveling in the same direction and at nearly the same speed as Earth. The close association of these swarms to the orbits of some short-term comets strongly suggests that they represent material lost by

Box 20.5

Comet Shoemaker-Levy Impacts Jupiter

In July 1994, Comet Shoemaker-Levy impacted Jupiter with the force equal to six million megatons of energy. (A megaton is the equivalent of a million tons of the explosive, TNT). Clearly, this was the most dramatic event in the solar system ever observed by people. Concern that a similar event might occur on Earth led NASA to establish the Near-Earth Object Search Committee to detect asteroids or comets with trajectories that might cross Earth's orbit.

It is possible that a few small fragments of a comet may have actually impacted Earth in 1908. That year, a strong explosion flattened more than 1000 square kilometers of a remote Siberian forest.

Comet Shoemaker-Levy was discovered at California's Mount Palomar Observatory barely a year before it impacted Jupiter. Careful observation showed that it consisted of two dozen fragments. Researchers concluded that the comet had broken up during an earlier pass by Jupiter. As the larger frag-

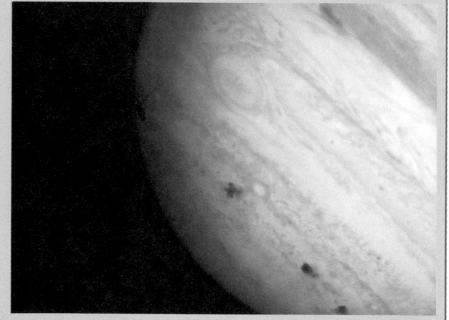

Figure 20.F

Dark blemishes on Jupiter produced by the impact of fragments of Comet Shoemaker-Levy in July 1994. (Courtesy of NASA)

ments penetrated Jupiter's outer atmosphere, they produced brilliant impact flashes and debris plumes soaring thousands of kilometers above the planet. The result of these fiery impacts were dark zones in Jupiter's atmos- phere that exceeded Earth in size. The largest of these dark blemishes lasted for months (Figure 20.F). Investigators hope to learn more about the dynamics of Jupiter's atmosphere by observing how these blemishes disperse.

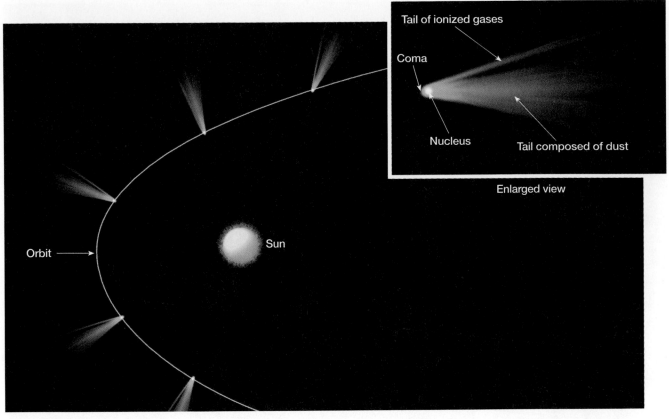

Figure 20.28
Orientation of a comet's tail as it orbits the sun.

these comets (Table 20.3). Some swarms not associated with orbits of known comets are probably the remains of the nucleus of a long-defunct comet. The notable Per-seid meteor shower that occurs each year around August 12 is believed to be the remains of the Comet 1862 III, which has a period of 110 years.

Meteoroids associated with comets are small and not known to reach the ground. Most meteoroids large enough to survive the heated fall are thought to originate among the asteroids, where chance collisions modify their orbits and send them toward Earth. Earth's gravitational force does the rest.

The remains of meteoroids, when found on Earth, are referred to as **meteorites**. A few very large meteoroids have blasted out craters on Earth's surface that strongly resemble those on the lunar surface. The most famous is Meteor Crater in Arizona (Figure 20.29). This huge cavity is about 1.2 kilometers (0.75 miles) across, 170 meters (560 feet) deep, and has an upturned rim that rises 50 meters (165 feet) above the surrounding countryside. Over 30 tons of iron fragments have been found in the immediate area, but attempts to locate the main body have been unsuccessful. Based on erosion, the impact likely occurred within the last 20,000 years.

Prior to moon rocks brought back by lunar explorers, meteorites were the only extraterrestrial materials that could be directly examined (Figure 20.30). Meteorites are classified by their composition: (1) **irons**—mostly iron with 5–20 percent nickel, (2) **stony**—silicate minerals with inclusions of other minerals, and (3) **stony-irons**—mixtures.

Table 20.3 Major meteor showers.

Shower	Approximate Dates	Associated Comet
Quadrantids	January 4–6	—
Lyrids	April 20–23	Comet 1861 I
Eta Aquarids	May 3–5	Halley's comet
Delta Aquarids	July 30	—
Perseids	August 12	Comet 1862 III
Draconids	October 7–10	Comet Giaco-bini-Zinner
Orionids	October 20	Halley's comet
Taurids	November 3–13	Comet Encke
Androme-dids	November 14	Comet Biela
Leonids	November 18	Comet 1866 I
Geminids	December 4–16	—

Figure 20.29
Meteor Crater, about 32 kilometers (20 miles) west of Winslow, Arizona. (Photo by Michael Collier)

Figure 20.30
Iron meteorite found near Meteor Crater, Arizona. (Courtesy of Meteor Crater, Northern Arizona, USA)

Although stony meteorites are probably more common, people find mostly irons. This is understandable, for irons withstand the impact better, weather more slowly, and are much easier for a lay person to distinguish from terrestrial rocks. Iron meteorites are probably fragments of once-molten cores of large asteroids or small planets.

One rare kind of meteorite, called a *carbonaceous chondrite*, was found to contain simple amino acids and other organic compounds, which are the basic building blocks of life. This discovery confirms similar findings in observational astronomy which indicate that numerous organic compounds exist in the frigid realm of outer space.

If meteorites represent the makeup of Earthlike planets, as some planetary geologists suggest, then Earth must contain a much larger percentage of iron than is indicated by surface rocks. This is one reason why geologists suggest that Earth's core may be mostly iron and nickel. In addition, meteorite dating indicates that our solar system's age certainly exceeds 4.5 billion years. This "old age" has been confirmed by data from lunar samples.

Review Questions

1. By what criteria are the planets placed into either the Jovian or terrestrial group?

2. What are the three types of materials thought to make up the planets? How are they different? How does their distribution account for the density differences between the terrestrial and Jovian planetary groups?

3. Explain why different planets have different atmospheres.

4. How is crater density used in the relative dating of features on the moon?

5. Briefly outline the history of the moon.

6. How are the maria of the moon thought to be similar to the Columbia Plateau?

7. Why has Mars been the planet most studied telescopically?

8. What surface features does Mars have that are also common on Earth?

9. Although Mars has valleys that appear to be products of stream erosion, what fact makes it unlikely that Mars has had a water cycle like that on Earth?

10. The two "moons" of Mars were once suggested to be artificial. What characteristics do they have that would cause such speculation?

11. What is thought to be the nature of Jupiter's Great Red Spot?

12. Why are the Galilean satellites of Jupiter so named?

13. What is distinctive about Jupiter's satellite Io?

14. Why are the four outer satellites of Jupiter thought to have been captured?

15. What evidence indicates that Saturn's rings are composed of individual moonlets rather than consisting of solid disks?

16. Explain why a large satellite cannot exist closer to Saturn than the outer edge of its ring system.

17. What is unique about Saturn's satellite Titan?

18. What three bodies in the solar system exhibit volcanic-like activity?

19. What do you think would happen if Earth passed through the tail of a comet?

20. Describe the origin of comets according to the most widely accepted hypothesis.

21. Compare a meteoroid, meteor, and meteorite.

22. Why are meteorite craters more common on the moon than on Earth, even though the moon is a much smaller target?

23. It has been estimated that Halley's comet has a mass of 100 billion tons. Further, this comet is thought to lose 100 million tons of material during the few months that its orbit brings it close to the sun. With an orbital period of 76 years, what is the maximum remaining life span of Halley's comet?

Key Terms

asteroid (p. 537)
coma (p. 537)
comet (p. 537)
escape velocity (p. 518)
iron meteorite (p. 541)
Jovian planet (p. 516)
lunar regolith (p. 521)
maria (p. 519)
meteor (p. 540)
meteorite (p. 541)
meteoroid (p. 540)
meteor shower (p. 540)
stony-iron meteorite (p. 541)
stony meteorite (p. 541)
terrestrial planet (p. 516)

CHAPTER 21

Light, Astronomical Observations, and the Sun

Observatory at Mauna Kea, Hawaii. (Photo by Richard Wainscoat/Douglas Peebles Photography)

Astronomers are in the business of gathering and studying light. Almost everything that is known about the universe beyond Earth comes by analyzing light from distant sources. Consequently, an understanding of the nature of light is basic to modern astronomy. This chapter deals with the study of light and the tools used by astronomers to gather light in order to probe the universe. In addition, we will examine the nearest source of light, our sun. By understanding the processes that operate on the sun, astronomers can better grasp the nature of more distant celestial objects.

The Study of Light

The vast majority of our information about the universe is obtained from the study of the light emitted from celestial bodies (Figure 21.1). Although visible light is most familiar to us, it constitutes only a small part of an array of energy generally referred to as **electromagnetic radiation**. Included in this array are gamma rays, X rays, ultraviolet light, visible light, infrared radiation, and radio waves (see Figure 14.14, p. 367). All radiant energy travels through the vacuum of space in a straight line at the rate of 300,000 kilometers (186,000 miles) per second.[*] Over a 24-hour day, this equals a staggering 26 billion kilometers.

Nature of Light

Experiments have demonstrated that light can be described in two ways. In some instances, light behaves like waves, and in others, like discrete particles. In the wave sense, light is analogous to swells in the ocean. This motion is characterized by the wavelength, which is the distance from one wave crest to the next. Wavelengths vary, from several kilometers for radio waves to less than a billionth of a centimeter for gamma rays. Most of these waves are either too long or too short for our eyes to detect.

---

[*] Light rays are "bent" slightly when they pass nearby a very massive object such as the sun.

Figure 21.1
The Trifid Nebula, in the constellation Sagittarius. This colorful nebula is a cloud of dust, plus hydrogen and helium gases. These gases are excited by the radiation of the hot, young stars within and produce a pink glow. (Courtesy of National Optical Astronomy Observatories)

The narrow band of electromagnetic radiation we can see is sometimes referred to as *white light*. However, white light consists of an array of waves having various wavelengths, a fact easily demonstrated with a prism (Figure 21.2). As white light passes through a prism, the color with the shortest wavelength, violet, is bent more than blue, which is bent more than green, and so forth (Table 21.1). Thus, white light can be separated into its component colors in the order of their wavelengths, producing the familiar rainbow of colors (Figure 21.2).

Wave theory, however, cannot explain some effects of light. In some cases, light acts like a stream of particles, analogous to infinitesimally small bullets fired from a machine gun. These particles, called **photons**, can exert a pressure ("push") on matter, which is called **radiation pressure**. Photons from the sun are responsible for "pushing" material away from a comet to produce its tail. Each photon has a specific amount of energy, which is related to its wavelength in a simple way: *Shorter wavelengths* correspond to *more energetic* photons. Thus, blue light has more energetic photons than red light.

Which theory of light—the wave theory or the particle theory—is correct? Both, since each will predict the behavior of light for certain phenomena. As George Abell, a prominent astronomer, stated about all scientific laws, "The mistake is only to apply them to situations that are outside their range of validity."

Spectroscopy

When Sir Isaac Newton used a prism to disperse white light into its component colors, he unknowingly initiated the field of **spectroscopy**, which is the study of properties of light that depend on wavelength. The rainbow of colors Newton produced is called a continuous spectrum, because all wavelengths of light are included. It was later learned that two other types of spectra exist, and that all three are generated under somewhat different conditions (Figure 21.3).

1. A **continuous spectrum** is produced by an incandescent solid, liquid, or gas under high

Table 21.1 Colors and corresponding wavelengths.

Color	Wavelength (nanometers*)
Violet	380–440
Blue	440–500
Green	500–560
Yellow	560–590
Orange	590–640
Red	640–750

* One nanometer is 10^{-9} meter.

pressure. It consists of an uninterrupted band of color (Figure 21.3A). One example would be light generated by a common light bulb. (*Incandescent* means "to emit light when hot.")

2. A **dark-line spectrum (absorption spectrum)** is produced when white light is passed through a comparatively cool gas under low pressure. The gas absorbs selected wavelengths of light, so the spectrum that is produced appears as a continuous spectrum, but with a series of dark lines running through it (Figure 21.3B).

3. A **bright-line spectrum (emission spectrum)** is produced by a hot (incandescent) gas under low pressure. It is a series of bright lines of particular wavelengths, depending on the gas that produces them (Figure 21.3C). These bright lines appear in the exact location as the dark lines that are produced by this gas in a dark-line (absorption) spectrum.

The spectra of most stars are of the dark-line type. The importance of these spectra is that each element or compound that is in gaseous form (in a star, material is usually in the gaseous form) produces a unique set of spectral lines. When the spectrum of a star is studied, the spectral lines act as "fingerprints," which identify the elements present.

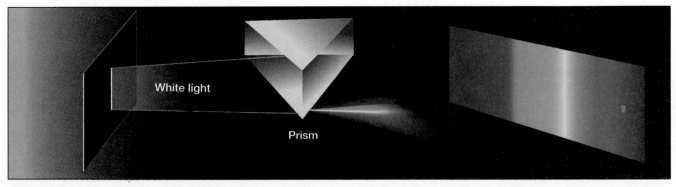

Figure 21.2
A spectrum is produced when sunlight (white light) is passed through a prism, which bends each wavelength at different angles.

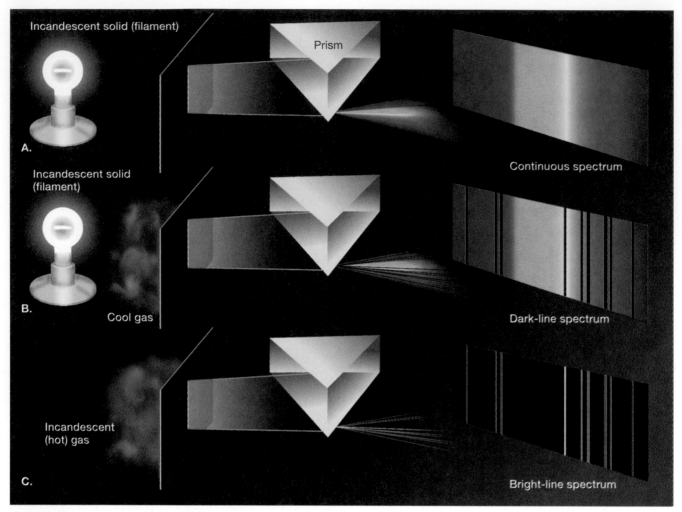

Figure 21.3
Formation of the three types of spectra. **A.** Continuous spectrum. **B.** Dark-line spectrum. **C.** Bright-line spectrum.

In an admittedly oversimplified manner, we can imagine how the sun and other stars create a dark-line spectrum. A continuous spectrum is produced in the interior of the sun, where the gases are under very high pressure. When this light passes through the less dense gases of the solar atmosphere, they absorb selected wavelengths, which appear as dark lines in the spectrum.

When Newton studied solar light, he obtained a continuous spectrum. However, when a prism is used in conjunction with lenses, the solar spectrum can be dispersed even further. An instrument that does this is called a **spectroscope.** The spectrum of the sun contains thousands of dark lines. Over 60 elements have been identified by matching these lines with those of elements known on Earth.

Two other facts concerning a radiating body are important. First, if the temperature of a radiating surface is increased, the total amount of energy emitted increases. The rate of increase is stated by the Stefan-Boltzmann law: The energy radiated by a body is di-

rectly proportional to the fourth power of its absolute temperature. For example, if the temperature of a star is doubled, the total radiation emitted increases by 2^4 ($2 \times 2 \times 2 \times 2$), or 16 times. Second, as the temperature of an object increases, a larger proportion of its energy is radiated at shorter wavelengths. To illustrate this, imagine a metal rod that is heated slowly. The rod first appears dull red (longer wavelengths), and later bluish white (shorter wavelengths). From this, it follows that blue stars are hotter than yellow stars, which are hotter than red stars.

The Doppler Effect

You may have heard the change in pitch of a car horn or ambulance siren as it passes by. When it is approaching, the sound seems to have a higher-than-normal pitch, and when it is moving away, the pitch sounds lower than normal. This effect, which occurs for both sound and light waves, was first explained by Christian Doppler in 1842 and is called the **Doppler effect**. The reason for the difference in pitch is that it takes time for the wave to be emitted. If the source is moving away, the beginning of the wave is emitted nearer to you than the end, which "stretches" the wave, that is, gives it a longer wavelength (Figure 21.4). The opposite is true for an approaching source.

In the case of light, when a source is moving away, its light appears redder than it actually is, because its waves appear lengthened. Objects approaching have their light waves shifted toward the blue (shorter wavelength). Thus, if a source of red light approached you at a very high speed (near the speed of light), it would actually appear blue. The same effect would be produced if you moved and the light were stationary.

Therefore, the Doppler effect reveals whether Earth is approaching or receding from a star or another celestial body. In addition, the amount of shift allows us to recalculate the rate at which the relative movement is occurring. Larger Doppler shifts indicate higher velocities; smaller Doppler shifts indicate slower velocities. Doppler shifts are generally measured from the dark lines in the spectra of stars by comparing them with a standard spectrum produced in the laboratory (Figure 21.5).

Astronomical Tools

Having examined the nature of light, we will now turn our attention to the tools astronomers use to intercept and study the energy emitted by distant objects in the universe (Figure 21.6). Because the basic principles of detecting radiation were originally developed through visual observations, we will consider optical telescopes first.

Refracting Telescopes

Galileo is considered to be the first person to use telescopes for astronomical observations. Having learned about the newly invented instrument, Galileo built one of his own design that was capable of magnifying objects 30 times. Because this early instrument, as well as its modern counterparts, used a lens to bend or refract light, it is known as a **refracting telescope**.

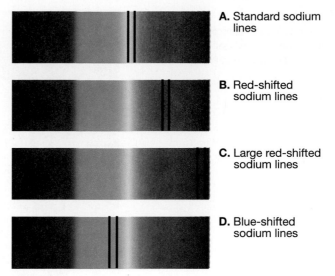

A. Standard sodium lines

B. Red-shifted sodium lines

C. Large red-shifted sodium lines

D. Blue-shifted sodium lines

Figure 21.5
A. An illustration of standard sodium lines produced in a laboratory compared to sodium lines as they would appear when the source is receding (red shift, **B.** and **C.**) or approaching (blue shift, **D.**).

The most important lens in a refracting telescope, the **objective lens**, produces an image by bending light from a distant object in such a way that the light converges at an area called the **focus** (Figure 21.7). For an object such as a star, the image appears as a point of light, but for a nearby objects, it appears as an inverted replica of the original.

You can easily demonstrate the latter case by holding a lens in one hand and with the other hand placing a white card behind the lens. Now, vary the distance between them until an image appears on the card. The distance between the focus (where the image appears) and the lens is called the **focal length** of the lens.

Astronomers usually study an image from a telescope by first photographing the image. However, if a telescope is used to examine an image directly, a second lens, called an **eyepiece**, is required (Figure 21.7). The eyepiece magnifies the image produced by the objective lens. In this respect, it is similar to a magnifying

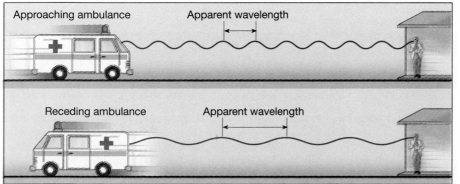

Figure 21.4
The Doppler effect, illustrating the apparent lengthening and shortening of wavelengths caused by the relative motion between a source and an observer.

Figure 21.6

The Mayall 4-meter telescope at Kitt Peak National Observatory near Tucson, Arizona. (Photo by Tom Bean/DRK Photo)

glass. Thus, the objective lens produces a very small, bright image of an object, and the eyepiece enlarges the image so that details can be seen.

Although used extensively in the nineteenth century, refracting telescopes suffer a major optical defect. As light passes through any lens, the shorter wavelengths of light are bent more than the longer wavelengths. (Recall the effect of a prism in separating the colors of the spectrum.) Consequently, when a refracting telescope is in focus for red light, blue and violet light are out of focus. The troublesome effect, known as **chromatic** ("color") **aberration**, weakens the image and produces a halo of color around it. When blue light is in focus, a reddish halo appears, and vice versa. Although this effect cannot be eliminated completely, it is reduced by using a second lens made of a different type of glass.

Reflecting Telescopes

Newton was so bothered by chromatic aberration that he built and used telescopes that reflected light from a shiny surface (mirror). Because reflected light

is not dispersed into its component colors, the problem is avoided. **Reflecting telescopes** use a concave mirror that focuses the light in front of the objective (the mirror), rather than behind it, like a lens (Figure 21.8). The mirror is generally made of glass that is finely ground. In the case of the 5-meter (200-inch) Hale Telescope, the grinding is accurate to about 1 millionth of a centimeter. The surface is then coated with a highly reflective material, usually an aluminum compound.

To focus parallel incoming rays onto one spot, the mirror is ground to a special curved surface called a *paraboloid*. This is the same shape as that used for the reflector in the headlights of automobiles. In the case of the auto bulb, however, the light source is at the focus, and the light goes out in parallel rays, rather than coming in.

Because the focus of a reflecting telescope is in front of the mirror, provisions have to be made to view the image without blocking too much of the incoming light. Figure 21.9 illustrates the most common arrangements. Most large telescopes employ more than one type.

Figure 21.7

Simple refracting telescope.

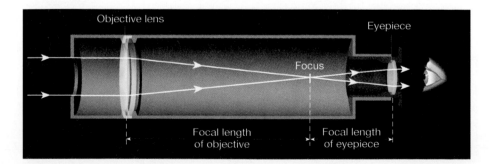

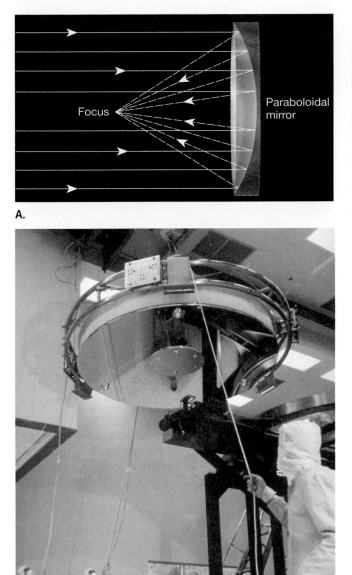

A.

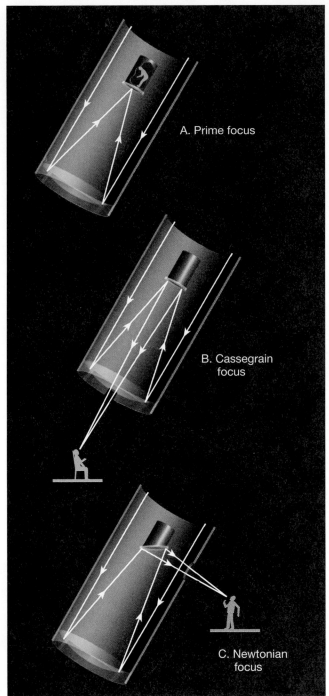

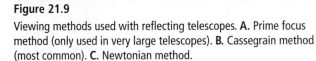

Figure 21.8

A. Diagram illustrating how concave mirrors, like those used in reflector telescopes, gather light. **B.** Preparation of the 2.4-meter mirror for the Hubble Space Telescope. (Courtesy of Space Telescope Science Institute).

Figure 21.9

Viewing methods used with reflecting telescopes. **A.** Prime focus method (only used in very large telescopes). **B.** Cassegrain method (most common). **C.** Newtonian method.

When using a very large reflecting telescope, the observer enters a viewing cage positioned at the focus to make the observations. The viewing cage blocks only about 10 percent of the total incoming light, and this is more than compensated for by the large objective (mirror) that is used.

In the "good old days," an astronomer would spend numerous long nights outdoors in the cold mountain air, perched in a viewing cage. But advances in photographic materials and computer enhancing technologies now allow indoor work and reduce the time required to obtain an image.

Nearly all large optical telescopes built today are reflectors. Among the reasons is the monumental task of producing a large piece of high-quality, bubble-free glass for refracting telescopes. Because light does not pass through a mirror, the glass for a reflecting telescope does not have to be of optical quality, nor does the instrument suffer from chromatic aberration. In addition, a lens can be supported only around the edge, so it sags. Mirrors, on the other hand, can be supported fully from behind.

Large, reflecting telescopes of about 4 meters (150 inches) are located at Kitt Peak, Arizona (Figure 21.10); Mauna Kea, Hawaii; Cerro Tololo, Chile; and Siding Spring, Australia (see Box 21.1). By comparison, the largest refractor in the word is the 1-meter (40 inch) telescope at Yerkes Observatory in Williams Bay, Wisconsin. This instrument was built before the turn of the century.

Properties of Optical Telescopes

Telescopes have three properties that aid astronomers in their work. They provide the observer with light-gathering power, resolving power, and magnifying power.

Light-gathering Power. Since most celestial objects are very faint sources of light, astronomers are *most* interested in improving the *light-gathering* power of their instruments. As seen in Figure 21.11, a telescope with a large lens (or mirror) intercepts more light from distant objects and thereby produces brighter images. Since very distant stars appear very dim as well, a great deal of light must be collected before the image is bright enough to be seen. Consequently, telescopes with large objectives "see" farther into space than those with small objectives.

Resolving Power. Another advantage of telescopes with large-diameter objectives is their greater *resolving power* (Figure 21.12), which allows for sharper images and finer detail. For example, with the unaided eye, the Milky Way appears as a vague band of light in the night sky, but even a small telescope is capable of resolving (separating it into) individual stars. Even so, the condition of Earth's atmosphere (which is known as *seeing*) greatly limits the resolving power of Earthbound telescopes. On a night when the stars twinkle, the seeing is poor, because the air is moving rapidly. This causes the image to move about, blurring it. Conversely, when the stars shine steadily, the seeing is described as good. Even under ideal conditions, however, some blurring occurs, eliminating the fine details. Thus, even the largest telescopes cannot photograph lunar features less than 0.5 kilometer (0.3 mile) in size.

To eliminate the problems of Earthbound viewing, the United States built the Hubble Space Telescope, which was put into Earth orbit in April 1990 (Figure 21.13). This 2.4-meter (94 inch) space telescope has 10 billion times more light-gathering power than the human eye. Since the technical problems that hampered the early operation of this instrument have been corrected, many spectacular images have been received. For example, as shown in Figure 20.25 (p. 538), the Hubble Space Telescope has provided images that clearly resolve the separation between Pluto and its moon, Charon. Pluto is at the center of the photo while Charon

Figure 21.10

The 4-meter Mayall Telescope at Kitt Peak, south of Tucson, Arizona. This modern reflecting telescope is one of the world's largest. (Courtesy of National Optical Astronomy Observatories)

Box 21.1

The Largest Telescopes

The world's largest conventional telescope is the 6-meter (236-inch) reflector in the Caucasus Mountains in Russia. However, because of problems with location and optics, it has been less useful than many smaller telescopes.

Just behind it in size is the 5-meter (200-inch) Hale Telescope on Mount Palomar in California. The Hale Telescope is a 500-ton steel and glass marvel that floats into position on oil bearings.

Casting and grinding a large mirror is time-consuming and expensive. Thus, astronomers driven by the desire to obtain even greater light-gathering and resolving power have looked for innovative techniques for telescope building.

A new technology for casting an 8-meter mirror is currently being developed under the football stadium at the University of Arizona. This new oven turns like a merry-go-round, so that molten glass is forced outward in the mold, thereby producing a concaved outer surface. Once cooled, this glass casting has nearly the correct shape, which reduces the grinding time considerably. Twin 8-meter telescopes are planned. One will be located in the Northern Hemisphere, on Hawaii's Mauna Kea at an altitude of nearly 4200

Figure 21.A

The Multiple Mirror Telescope Observatory. This unique telescope consists of six 1.8-meter mirrors that are brought to a common focus. (Courtesy of Smithsonian Inst./Univ. of Arizona)

meters (13,800 feet). The other is to be located in Chile, on the western slopes of the Andes Mountains.

Another effort is the Multiple Mirror Telescope (MMT), which uses six 1.8-meter mirrors that are linked together to focus at a single point (Figure 21.A). The result is a telescope with a light-gathering capability of a mirror 4.5 meters in diameter.

A different innovative design is the recently completed 10-meter Keck Telescope located on Mauna Kea. This instrument uses a mosaic of 36 six-sided mirrors carefully positioned by computer to give the optical effect of a 10-meter (400-inch) mirror.

Even more ambitious proposals are being considered. Plans are under consideration to employ four 8-meter mirrors in a "multiple mirror configuration" that would result in the equivalent of a single 16-meter instrument. If completed, this instrument would have 10 times the light gathering capacity of the 5-meter Hale Telescope.

is the faint object in the lower left. For comparison, the best ground-based image taken to date is shown in the upper left of the figure.

Magnifying Power.   When you think of the power of a telescope, you probably think of its *magnifying power*, the ability to make an object larger. Magnification is calculated

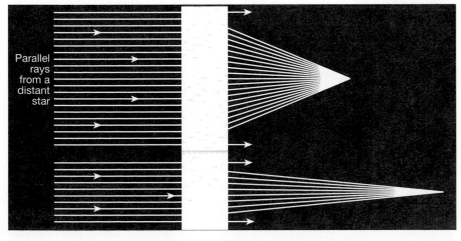

Figure 21.11

Comparison of the light-gathering ability of two lenses.

Parallel rays from a distant star

Figure 21.12

Appearance of the galaxy in the constellation Andromeda using telescopes of different resolution. (Courtesy of Leiden Observatory)

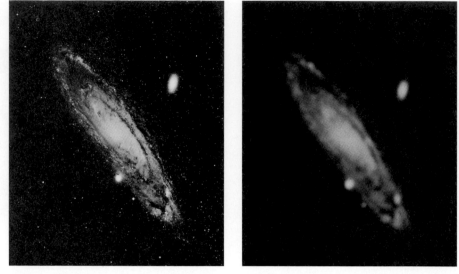

by dividing the focal length of the objective by the focal length of the eyepiece. Thus, the magnification of a telescope can be changed by simply changing the eyepiece. However, increased magnification does not necessarily improve the clarity of the image. What can be viewed telescopically is limited by atmospheric conditions and the resolving power of the telescope. Any part of an image that is not clear at low magnification will appear only as a larger blur at higher magnification. Furthermore, increasing magnification spreads out the light and

decreases the brightness of the object. Thus, astronomers describe telescopes not in terms of their magnification but by the diameter of the objective mirror or lens, because it is this factor that determines both the light-gathering power and the resolving power of a telescope.

Most modern telescopes have supplemental devices that enhance the image. A simple, but important, example is a photographic plate that can be exposed for long periods of time, thereby collecting enough light from a star to make an image that otherwise would be undetectable. One of the latest advances makes use of high-speed computers that adjust the optics to partially remove the distortion caused by the atmosphere. This process greatly enhances the image sharpness.

Detecting Invisible Radiation

As we said, sunlight is made up of more than just the radiation that is visible to our eyes. Gamma rays, X rays, ultraviolet radiation, infrared radiation, and radio waves are also produced by stars. Photographic film that is sensitive to ultraviolet and infrared radiation has been developed, thereby extending the limits of our vision. However, most of this radiation cannot penetrate our atmosphere, so balloons, rockets, and satellites must transport cameras "above" the atmosphere to record it.

Of great importance is a narrow band of radio radiation that does penetrate the atmosphere. One particular wavelength is the 21-centimeter (8-inch) line produced by neutral hydrogen (hydrogen atoms that lack an electrical charge). Measurement of this radiation has permitted us to map the galactic distribution of hydrogen—the material from which stars are made.

The detection of radio waves is accomplished by "big dishes" called **radio telescopes** (Figure 21.14A). In principle, the dish of one of these telescopes operates

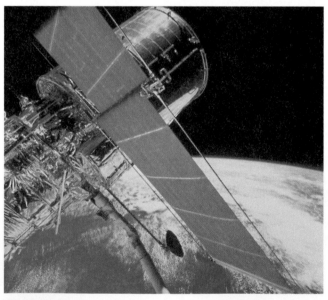

Figure 21.13

The deployment of the Hubble Space Telescope in Earth orbit, April 24, 1990, from Space Shuttle *Discovery*. (Courtesy of Space Telescope Science Institute)

Figure 21.14
A. The 43-meter (140-foot) steerable radio telescope at Green Bank, West Virginia. The dish acts like the mirror of a reflector-type optical telescope to focus radio waves onto the antenna. The antenna is the small round object supported on four "legs" above the dish. **B.** Twenty-seven identical antennas operate together to form the Very Large Array near Socorro, New Mexico. (Courtesy of National Radio Astronomy Observatory)

in the same manner as the mirror of an optical telescope. It is parabolic in shape and focuses the radio waves on an antenna, which absorbs and transmits these waves to an amplifier, just like any radio antenna.

Because radio waves are about 100,000 times longer than visible radiation, the surface of the dish need not be as smooth as a mirror. In fact, except for the shortest radio waves, a wire mesh is a good reflector. On the other hand, because radio signals from celestial sources are very weak, large dishes are necessary to intercept an adequate signal. The largest radio telescope is a bowl-shaped antenna hung in a natural depression in Puerto Rico (Figure 21.15). It is 300 meters (1000 feet) in

diameter and has some directional flexibility in its movable antenna. The largest steerable types have about 100-meter (330-foot) dishes like that currently being constructed at the National Radio Astronomy Observatory in Green Bank, West Virginia.

Radio telescopes also have rather poor resolution, making it difficult to pinpoint the radio source. Pairs or groups of telescopes are used to reduce this problem. When several radio telescopes are wired together, the resulting network is called a **radio interferometer** (Figure 21.14B).

Radio telescopes have some advantages over optical telescopes. They are much less affected by turbulence in the atmosphere, clouds, and the weather in

Figure 21.15

The 300-meter (1000-foot) radio telescope at Arecibo, Puerto Rico. (Courtesy of National Astronomy and Ionosphere Center's Arecibo Observatory, operated by Cornell University under contract with the National Science Foundation)

sources). These perplexing objects are the most distant things known in the universe and will be examined further in Chapter 22.

The Sun

The sun is one of the 200 billion stars that make up the Milky Way galaxy. Although the sun is of no significance to the universe as a whole, to us on Earth, it is the primary source of energy (see Box 21.2). Everything from the fossil fuels we burn in our automobiles and power plants to the food we eat is ultimately derived from solar energy (Figure 21.16). The sun is also important to astronomers, since it is the only star close enough to permit study of the surface. Even with the largest telescopes, other stars appear only as points of light.

Because of the sun's brightness and its damaging radiation, it is not safe to observe it directly. However, a small telescope will project its image on a piece of cardboard held behind the telescope's eyepiece, and the sun may be studied safely in this manner. This basic method is used in several telescopes around the world, which keep a constant vigil of the sun. One of the finest is at the Kitt Peak National Observatory in southern Arizona (Figure 21.17). It consists of a 150-meter sloped enclosure that directs sunlight to a mirror situated below ground. From the mirror, an 85-centimeter (33-inch) image of the sun is projected to an observing room, where it is studied.

Compared to other stars of the universe, many of which are larger, smaller, hotter, cooler, more red, or more blue, the sun is an "average star." However, on

general. No protective dome is required, which reduces the cost of construction, and "viewing" is possible 24 hours a day. More importantly, radio telescopes can "see" through interstellar dust clouds that obscure our view at visible wavelengths. Radio signals from distant points in the universe pass unhindered through the dust, giving us an unobstructed view. Furthermore, radio telescopes can detect clouds of gases too cool to emit visible light. These cold gas clouds are important because they are the sites of star formation.

Radio telescopes are, however, hindered by human-made radio interference. Thus, while optical telescopes are placed on remote mountaintops to reduce interference from city lights, radio telescopes are often hidden in valleys to block human-made radio interference.

Radio telescopes have revealed such spectacular events as the collision of two galaxies. Of even greater interest was the discovery of *quasars* (quasi-stellar radio

Figure 21.16

The sun is the source of more than 99 percent of all energy on Earth. (Photo by E. J. Tarbuck)

Box 21.2

Variable Sun and Climatic Change

Among the most persistent hypotheses of climatic change have been those based on the idea that the sun is a variable star and that its output of energy varies through time. The effect of such changes would seem direct and easily understood: Increases in solar output would cause the atmosphere to warm, and reductions would result in cooling. This notion is appealing because it can be used to explain climatic changes of any length or intensity. Still, there is at least one major drawback: No major long-term variations in the total intensity of solar radiation have yet been measured.

Several proposals for climatic change based on a variable sun relate to *sunspot cycles*. The most conspicuous and best known features on the surface of the sun are the dark blemishes called *sunspots*. Although their origin is uncertain, it has been established that sunspots are huge magnetic storms that extend from the sun's surface deep into the interior. Moreover, these spots are associated with the sun's ejection of huge masses of particles that, on reaching Earth's atmosphere, interact with gases there to produce auroral displays.

Along with other solar activity, the number of sunspots increases and decreases on a regular basis, creating a cycle of about 11 years. A curve of the annual number of sunspots, beginning in the early 1700s, appears to be very regular (Figure 21.B).

Studies indicate that there have been prolonged periods when sunspots were absent, or nearly so. Moreover, it was found that these events corresponded closely with cold periods in Europe and North America. Conversely, periods characterized by plentiful sunspots were found to correlate well with warmer times in these regions. Referring to these excellent matches, the solar astronomer John Eddy stated: "These early results in comparing solar history with climate make it appear that changes on the sun are the dominant agent of climatic changes lasting between 50 and several hundred years."

But other scientists seriously questioned this conclusion. Their hesitation stems in part from subsequent investigations, using different climate records from around the world, that failed to find a significant correlation between variations in solar activity and climate. Even more troubling seems to be the fact that no testable physical mechanism exists to explain the purported effect.

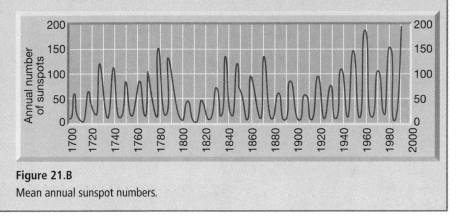

Figure 21.B
Mean annual sunspot numbers.

the scale of our solar system, it is truly gigantic, having a diameter equal to 109 Earth diameters (1.35 million kilometers) and a volume of $1\frac{1}{4}$ million times as great as that of Earth. Yet, because of its gaseous nature, the density is only $\frac{1}{4}$ that of solid Earth, nearly the density of water.

Figure 21.17
The unique Robert J. McMath Solar Telescope at Kitt Peak, near Tucson, Arizona. Movable mirrors at the top follow the sun, reflecting its light down the sloping tunnel. (Photo by Kent Wood/Photo Researchers, Inc.).

Structure of the Sun

For convenience of discussion, we divide the sun into four parts: the solar interior; the visible surface, or photosphere; and the two layers of its atmosphere, the chromosphere and the corona (Figure 21.18). Because the sun is gaseous throughout, no sharp boundaries exist between these layers. The sun's interior makes up all but a tiny fraction of the solar mass, and unlike the outer three layers, it is not accessible to direct observation. We shall discuss the visible layers first.

Photosphere. The **photosphere** ("sphere of light") is aptly named, for it radiates most of the sunlight we see and therefore appears as the bright disk of the sun. Although it is considered to be the sun's "surface," it is unlike most surfaces we are accustomed to. The photosphere consists of a layer of incandescent gas 300 kilometers (200 miles) thick having a pressure less than $\frac{1}{100}$ of our atmosphere. Furthermore, it is neither smooth nor uniformly bright as the ancients had imagined. It has numerous blemishes.

When viewed through a telescope under ideal conditions, the photosphere's grainy texture is apparent. This is the result of numerous, relatively small, bright markings called **granules**, which are surrounded by narrow, dark regions (Figure 21.19). Granules are typically 1000 kilometers (620 miles) in diameter and owe their brightness to hotter gases that are rising from below. As this gas spreads laterally, cooling causes it to darken and sink back into the interior. Although each granule survives only a few minutes, the combined motion of all granules give the photosphere the appearance of boiling. This up-and-down movement of gas is called *convection*. Besides producing the grainy appearance of the photosphere, convection is believed to be responsible for the transfer of energy in the uppermost part of the sun's interior (Figure 21.18).

The composition of the photosphere is revealed by the dark lines of its absorption spectrum (see Figure 21.3). When these "fingerprints" are compared to the spectra of known elements, they indicate that most of the elements found on Earth also occur on the sun. When the strengths of the absorption lines are analyzed, the relative abundance of the elements can be determined. These studies reveal that 90 percent of the sun's surface atoms are hydrogen, almost 10 percent are helium, and only minor amounts of the other detectable elements are present. Other stars also indicate similar disproportionate percentages of these two lightest elements, a fact we shall consider later.

Chromosphere. Just above the photosphere lies the **chromosphere** ("color sphere"), a relatively thin layer of hot, incandescent gases a few thousand kilometers

Figure 21.18

Diagram of solar structure in cutaway view.

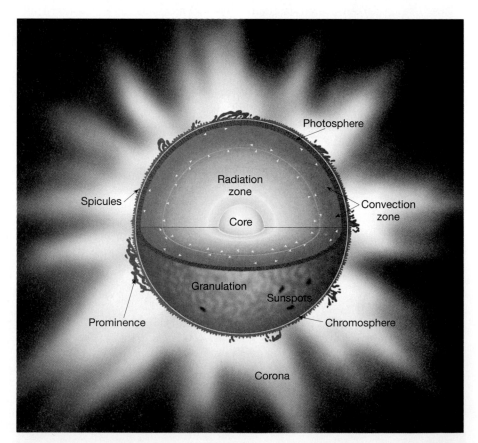

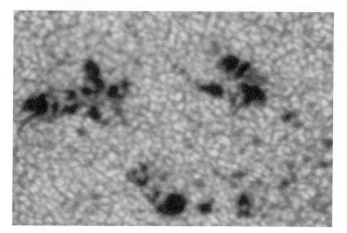

Figure 21.19

Granules of the solar photosphere. (Courtesy of National Optical Astronomy Observatories)

Figure 21.20

Spicules of the chromosphere on the edge of the solar disk. (Courtesy of Sacramento Peak Observatory, Air Force Cambridge Research Laboratories)

thick. The chromosphere is observable for a few moments during a total solar eclipse or by using a special instrument that blocks out the light from the photosphere. Under such conditions, it appears as a thin red rim around the sun. Because the chromosphere consists of hot, incandescent gases under low pressure, it produces a bright-line spectrum that is nearly the reverse of the dark-line spectrum of the photosphere. One of the bright lines of hydrogen contributes a good portion of its total light and accounts for this sphere's red color.

A study of the chromospheric spectrum conducted in 1868 revealed the existence of an unknown element. It was named helium, from *helios*, the Greek word for "sun." Originally, helium was thought to be an element unique to the stars, but 27 years later, it was discovered in natural-gas wells on Earth.

The top of the chromosphere contains numerous **spicules**, which extend upward into the lower corona, almost like the trees which reach into our atmosphere (Figure 21.20). The spicules may be continuations of the turbulent motion of the granules below.

Corona.   The outermost portion of the solar atmosphere, the **corona**, is very tenuous and, as with the chromosphere, is visible only when the brilliant photosphere is covered (Figure 21.21). This envelope of ionized gases normally extends a million kilometers from the sun and produces a glow about half as bright as the full moon.

At the outer fringe of the corona, the ionized gases have speeds great enough to escape the gravitational pull of the sun. The streams of protons and electrons that "boil" from the corona constitute the **solar wind**. They travel outward through the solar system at very high speeds (250-800 kilometers a second) and eventu-

ally are lost to interstellar space. During its journey, the solar wind interacts with the bodies of the solar system, continually bombarding lunar rocks and altering their

Figure 21.21

Solar corona photographed during a total eclipse. (Photo by Lawrence Burr/Gamma Liaison)

appearance. Although Earth's magnetic field prevents the solar winds from reaching our surface, these winds do affect our atmosphere, as we shall discuss later.

Studies of the energy emitted from the photosphere indicate that its temperature averages about 6000 K (10,000°F). Upward from the photosphere, the temperature unexpectedly increases, exceeding 1 million K at the top of the corona. It should be noted that, although the coronal temperature exceeds that of the photosphere by many times, it radiates much less energy overall because of its very low density. Surprisingly, the high temperature of the corona is probably caused by sound waves generated by the convective motion of the photosphere. Just as boiling water makes noise, the energetic sound waves generated in the photosphere are believed to be absorbed by the gases of the corona and thereby raise their temperatures.

The Active Sun

Sunspots. The most conspicuous features on the surface of the sun are the dark blemishes called **sunspots** (Figure 21.22A). Although large sunspots were occasionally observed before the advent of the telescope, they were generally regarded as opaque objects located somewhere between the sun and Earth. In 1610 Galileo concluded that they were residents of the solar surface, and from their motion, he deduced that the sun rotates on its axis about once a month.

Observations made later indicated that not all parts of the sun rotate at the same speed. The sun's equator rotates once in 25 days, while a place located 70 degrees from the solar equator, whether north or south, requires 33 days for one rotation. If Earth rotated in a similar disjointed manner, imagine the consequences! The sun's nonuniform rotation is a testimonial to its gaseous nature.

Sunspots begin as small dark pores about 1600 kilometers (1000 miles) in diameter. While most pores last for only a few hours, some grow into blemishes many times larger than Earth and last for a month or more. The largest spots often occur in pairs surrounded by several smaller spots. An individual spot contains a black center, the *umbra*, which is rimmed by a lighter region, the *penumbra* (Figure 21.22B). Sunspots appear dark only by contrast with the brilliant photosphere, a fact accounted for by their temperature, which is about 1500K less than that of solar surface. If these dark spots could be observed away from the sun, they would appear many times brighter than the full moon.

During the early 1800s, it was believed that a tiny planet named Vulcan orbited between Mercury and the sun. In the search for Vulcan an accurate record of sunspot occurrences was kept. Although the planet was never found, the sunspot data collected did reveal that the number of sunspots observable on the solar disk varies in an 11-year cycle. First, the number of sunspots increases to a maximum, with perhaps a hundred or

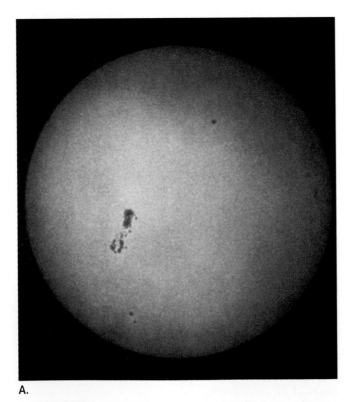

A.

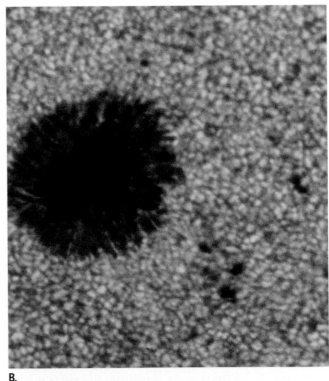

B.

Figure 21.22

A. Large sunspot group on the solar disk. (Celestron 8 photo courtesy of Celestron International) **B.** Sunspots having visible umbra (dark central area) and penumbra (lighter area surrounding umbra). (Courtesy of National Optical Astronomy Observatories)

more visible at a given time. Then, over a period of 5–7 years, their numbers decline to a minimum, when only a few or even none are visible. At the beginning of each cycle, the first sunspots form about 30 degrees from the solar equator, but as the cycle progresses and their numbers increase, they form nearer the equator. During the period when sunspots are most abundant, the majority form about 15 degrees from the equator. They rarely occur more than 40 degrees away from the sun's equator, or within 5 degrees of it.

Another interesting characteristic of sunspots was discovered by astronomer George Hale, for whom the Hale Telescope is named. Hale deduced that the large spots are strongly magnetized, and when they occur in pairs, they have opposite magnetic poles. For instance, if one member of the pair is a north magnetic pole, then the other member is a south magnetic pole, as with the north and south poles of Earth's magnetic field. Also, every pair located in the same hemisphere is magnetized in the same manner. However, all pairs in the *other* hemisphere are magnetized in the opposite manner. At the beginning of each sunspot cycle, the situation reverses, and the polarity of these sunspot pairs is opposite those of the previous cycle. The cause of this change in polarity, in fact the cause of sunspots themselves, is not fully explained. However, other solar activity varies in the same cyclic manner as sunspots do, indicating a common origin.

Plages and Prominences. The brilliance of the photosphere makes viewing the activity occurring above the solar surface very difficult. To overcome this problem, numerous photographic methods are employed to filter solar radiation so that light of a particular spectral band (color) can be viewed. Using such techniques, large "clouds" can be seen in the chromosphere directly above sunspot clusters (Figure 21.23). These bright centers of solar activity are called **plages** and occasionally can even be viewed before or after the sunspot occurrences.

Among the more spectacular features of the active sun are **prominences.** These huge cloudlike structures are best observed when they are on the edge, or limb, of the sun, where they often appear as great arches that extend well into the corona (Figure 21.24). Many prominences have the appearance of a fine tapestry and seem to hang motionless for days at a time, but motion pictures reveal that the material within them is continually falling like

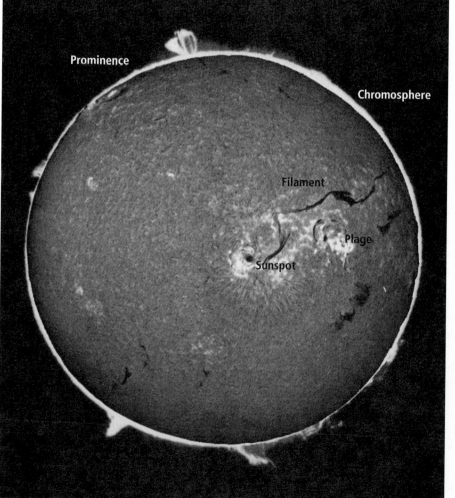

Figure 21.23
Solar disk photographed in hydrogen alpha light, showing the manifestations of the active sun. (This composite courtesy of Hale Observatories and Sacramento Peak Observatory, Air Force Cambridge Research Laboratories)

Figure 21.24
A huge solar prominence. (Photo by PHOTRI/The Stock Market)

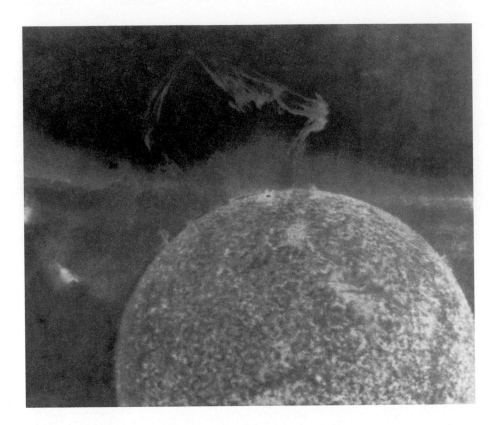

luminescent rain. Apparently these quiet prominences are condensations of coronal material that are gracefully "sliding down" the lines of magnetic force back into the chromosphere. More rarely, the material within a prominence rises almost explosively away from the sun. These active prominences reach velocities up to 1000 kilometers (620 miles) per second and may leave the sun entirely. Prominences can also be seen against the bright disk of the sun, in which case they appear as dark, thin streaks called *filaments* (Figure 21.23).

Solar Flares.   These are the most explosive events associated with sunspots. **Solar flares** are brief outbursts that normally last an hour or so and appear as a sudden brightening of the region above a sunspot cluster. During their existence, enormous quantities of energy are released, much of it in the form of ultraviolet, radio, and Xray radiation. Simultaneously, fast-moving atomic particles are ejected, causing the solar wind to intensify noticeably. Although a major flare could conceivably endanger a manned space flight, they are relatively rare. About a day after a large outburst, the ejected particles reach Earth and disturb the ionosphere,[*] affecting long-distance radio communications.

The most spectacular effect of solar flares, however, is the **auroras**, also called the northern and southern lights (Figure 21.25). Following a strong solar flare,

Earth's upper atmosphere near its magnetic poles is set aglow for several nights. The auroras appear in a wide variety of forms. Sometimes the display consists of vertical streamers with considerable movement. At other times, the auroras appear as a series of luminous expanding arcs or as a quiet, almost foglike, glow. Auroral displays, like other solar activities, vary in intensity with the 11-year sunspot cycle.

The Solar Interior

The interior of the sun cannot be observed directly. For that reason, all we know about it is based on information acquired from the energy it radiates and from theoretical studies. The source of the sun's energy, **nuclear fusion**, was not discovered until the late 1930s.

Deep in its interior, a nuclear reaction called the **proton-proton chain** converts four hydrogen nuclei (protons) into the nucleus of a helium atom. The energy released from the proton-proton reaction results because some of the matter involved is actually converted to energy. This can be illustrated by noting that four hydrogen atoms have a combined atomic mass of 4.032 (4×1.008), whereas the atomic mass of helium is 4.003, or 0.029 less than the combined mass of the hydrogen. The tiny missing mass is emitted as energy according to Einstein's formula $E = mc^2$, where E equals energy, m equals mass, and c equals the speed of light. Because the speed of light is very great, the amount of energy released from even a small amount of mass is enormous.

---

[*]The ionosphere is a complex atmospheric zone of ionized gases that extends between about 80 and 400 kilometers (50 and 250 miles) above Earth's surface.

Figure 21.25
Aurora borealis (northern lights) as seen from Alaska. The same phenomenon occurs toward the South Pole, where it is called the Aurora australis (southern lights). (Photo by Michio Hoshino/Minden Pictures)

The conversion of just one pinhead's worth of hydrogen to helium generates more energy than burning thousands of tons of coal. Most of this energy is in the form of high-energy photons that work their way toward the solar surface, being absorbed and re-emitted many times until they reach an opaque layer just below the photosphere. Here, convection currents serve to transport this energy to the solar surface, where it radiates through the transparent chromosphere and corona (see Figure 21.18).

Only a small percentage (0.7 percent) of the hydrogen in the proton-proton reaction is actually converted to energy. Nevertheless, the sun is consuming an estimated 600 million tons of hydrogen each second, with about 4 million tons of it being converted to energy. As hydrogen is consumed, the product of this reaction, helium, forms the solar core, which continually grows in size.

Just how long can the sun produce energy at its present rate before all of its fuel (hydrogen) is consumed? Even at the enormous rate of consumption, the sun has enough fuel to last easily another 100 billion years. However, evidence from other stars indicates that the sun will grow dramatically and engulf Earth long before all of its hydrogen is gone. It is thought that a star the size of the sun can exist in its present stable state for 10 billion years. Since the sun is already 5 billion years old, it is "middle-aged."

To initiate the proton-proton reaction, the sun's internal temperature must have reached several million degrees. What was the source of this heat? As previously noted, the solar system is believed to have formed from an enormous cloud of dust and gases (mostly hydrogen) that condensed gravitationally. The consequence of squeezing (compressing) a gas is to increase its temperature. Although all of the bodies in the solar system were compressed, the sun was the only one, because of its size, that became hot enough to trigger the proton-proton reaction. Astronomers currently estimate its internal temperature at 15 million K.

The planet Jupiter is basically a hydrogen-rich gas ball; if it were about ten times more massive, it too might have become a star. The idea of one star orbiting another may seem odd, but recent evidence indicates that about 50 percent of the stars in the universe probably occur in pairs or multiples!

Review Questions

1. What term is used to describe the collection that includes gamma rays, X rays, ultraviolet light, visible light, infrared radiation, and radio waves?

2. Which color has the longest wavelength? The shortest?

3. What is spectroscopy?

4. Describe a continuous spectrum. Give an example of a natural phenomenon that exhibits a continuous spectrum.

5. What produces emission lines (bright lines) in a spectrum?

6. What can be learned about a star (or other celestial objects) from a dark-line (absorption) spectrum?

7. How can astronomers determine whether a star is moving toward or away from Earth?

8. The primary objective of a refracting telescope is a mirror: True or false?

9. What three properties do telescopes have that aid astronomers?

10. Of the three properties listed in Question 9, which is the least important to astronomers?

11. What is the advantage of a space telescope over a similar Earthbound instrument?

12. Why do astronomers seek to design telescopes with larger and larger objectives?

13. Why do all large optical telescopes use mirrors to collect light rather than lenses?

14. With reflecting telescopes, why are special viewing systems needed?

15. Explain the following statement: "Photography has extended the limits of our vision."

16. Why would the moon make a good site for an observatory?

17. Why are radio telescopes much larger than optical telescopes?

18. What are some of the advantages of radio telescopes over optical telescopes?

19. Compare the diameter of the sun to that of Earth.

20. Describe the photosphere, chromosphere, and corona.

21. List the features associated with the active sun and describe each.

22. Explain how a sunspot can be very hot and yet appear dark.

23. What is the solar wind?

24. What "fuel" does the sun consume?

25. What happens to the matter that is consumed in the proton-proton chain reaction?

Key Terms

aurora (p. 562)
bright-line (emission) spectrum (p. 547)
chromatic aberration (p. 550)
chromosphere (p. 558)
continuous spectrum (p. 547)
corona (p. 559)
dark-line (absorption) spectrum (p. 547)
Doppler effect (p. 548)
electromagnetic radiation (p. 546)

eyepiece (p. 549)
focal length (p. 549)
focus (p. 549)
granules (p. 558)
nuclear fusion (p. 562)
objective lens (p. 549),
photon (p. 547)
photosphere (p. 558)
plage (p. 561)
prominence (p. 561)
proton-proton chain (p. 562)

radiation pressure (p. 547)
radio interferometer (p. 555)
radio telescope (p. 554)
reflecting telescope (p. 550)
refracting telescope (p. 549)
solar flare (p. 562)
solar wind (p. 559)
spectroscope (p. 548)
spectroscopy (p. 547)
spicule (p. 559)
sunspot (p. 560)

CHAPTER 22

Beyond Our Solar System

Eagle Nebula in the constellation Serpens. (Courtesy of NOAO)

The star Proxima Centauri is about 4.3 light-years away, roughly 100 million times farther than the moon. Yet, other than our own sun, it is the closest star to Earth. This fact suggests that the universe is incomprehensibly large. What is the nature of this vast cosmos beyond our solar system? Are the stars distributed randomly, or are they organized into distinct clusters? Do stars move, or are they permanently fixed features, like lights strung out against the black cloak of outer space? Does the universe extend infinitely in all directions, or is it bounded? To consider these questions, this chapter will examine the universe by taking a census of the stars—the most numerous objects in the night sky.

Properties of Stars

The sun is the only star whose surface we can observe. Yet, a great deal is known about the universe beyond our solar system. In fact, more is known about the stars than about our outermost planet, Pluto. This knowledge hinges on the fact that stars, and even gases in the "empty" space between stars, radiate energy in all directions into space (Figure 22.1). The key to understanding the universe is to collect this radiation and unravel the secrets it holds. Astronomers have devised many ingenious methods to do just that. We will begin by examining stellar distances and some intrinsic properties of stars, including *color, brightness, mass, temperature,* and *size.*

Measuring Distances to the Stars

Measuring the distance to a star is very difficult. Obviously, we cannot journey to the star. Nevertheless, astronomers have developed some *indirect* methods of measuring stellar distances. The most basic of these measurements is called stellar parallax.

Recall from Chapter 19 that **stellar parallax** is the extremely slight back-and-forth shifting in the apparent position of a nearby star due to the orbital motion of Earth. To review, the principle of parallax is easy to visualize. Close one eye, and with your index finger in a vertical position, use your eye to line up your finger with some distant object. Now, without moving your finger, view the object with your other eye and notice that its position appears to have changed. The farther away you hold your finger, the less its position seems to shift.

In practice, parallax is determined by photographing a nearby star against the background of distant stars. Then, six months later, when Earth has moved halfway around its orbit, a second photograph is taken. When these photographs are compared, the position of the nearby star appears to have shifted with respect to the background stars. Figure 22.2 illustrates this shift and the parallax angle determined from it. The nearest stars have the largest parallax angles, while those of distant stars are too slight to measure. Recall that the medieval astronomer Tycho Brahe was unable to detect stellar parallax, leading him to reject the idea that Earth orbits the sun.

Figure 22.1

Lagoon Nebula. It is in glowing clouds like these that gases and dust particles become concentrated into stars. (Courtesy of National Optical Astronomy Observatories)

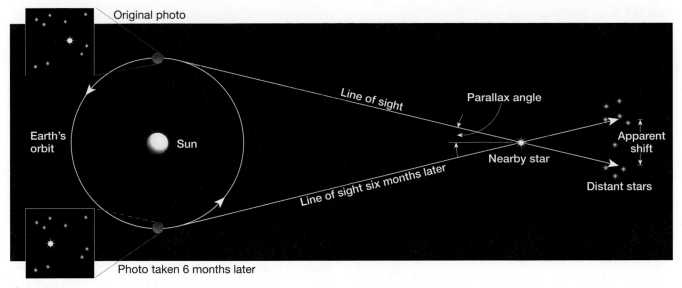

Figure 22.2

Geometry of stellar parallax. The parallax angle shown here is enormously exaggerated to illustrate the principle. Because the distance to even the nearest stars is thousands of times greater than the Earth-sun distance, the triangles that astronomers work with are very long and narrow, making the angles that are measured very small.

It should be emphasized that parallax angles are *very* small. The parallax angle to the nearest star, Proxima Centauri, is less than 1 second of arc, which equals $\frac{1}{3600}$ of a degree. To put this in perspective, fully extend your arm and raise your little finger. Your finger is roughly 1 degree wide. Try doing this on a moonlit night, covering the moon with your finger. The moon is only about $\frac{1}{2}$ degree wide. Now, imagine detecting a movement that is only $\frac{1}{3600}$ as wide as your finger. It should be apparent why Tycho Brahe, without the aid of a telescope, was unable to observe stellar parallax.

The distances to stars are so large that the conventional units, such as kilometers or astronomical units, are often too cumbersome to use. A better unit to express stellar distance is the **light-year**, which is the distance light travels in one Earth year—about 9.5 trillion kilometers (5.8 trillion miles).

In principle, the method used to measure stellar distances is elementary and was known to the ancient Greeks. But in practice, measurements are greatly complicated because of the tiny angles involved and because the sun, as well as the star being measured, also have actual motion through space. The first accurate stellar parallax was not determined until 1838. Even today, parallax angles for only a few thousand of the nearest stars are known with certainty.

Most stars have such small parallax shifts that accurate measurement is not possible. Fortunately, a few other methods have been derived for estimating distances to these stars. Also, the Hubble Space Telescope, which is not hindered by Earth's turbulent, light-distorting atmosphere, is expected to obtain accurate parallax distances for many additional remote stars.

Stellar Brightness

Three factors control the apparent brightness of a star as seen from Earth: *how big* it is, *how hot* it is, and *how far away* it is. The stars in the night sky are a grand assortment of sizes, temperatures, and distances, so their brightnesses vary widely.

Apparent Magnitude. Stars have been classified according to their brightness since the second century B.C., when Hipparchus placed about 1000 of them into six categories. The measure of a star's brightness is called its **magnitude**. Some stars may appear dimmer than others only because they are farther away. Therefore, a star's brightness, *as it appears when viewed from Earth,* has been termed its **apparent magnitude**.

When numbers are employed to designate relative brightness, the larger the magnitude number, the dimmer the star. *Stars that appear the brightest are of the first magnitude, while the faintest stars visible to the unaided eye are of the sixth magnitude.* With the invention of the telescope, many stars fainter than the sixth magnitude were discovered.

In the mid-1800s, a method was developed for comparing the brilliance of stars using magnitude. Just as we can compare the brightness of a 50-watt light bulb to that of a 100-watt bulb, we can compare the brightness of stars having different magnitudes. It was determined that a first-magnitude star was about 100 times brighter than a sixth-magnitude star. Therefore, on the scale that was devised, two stars that differ by 5 magnitudes have a ratio in brightness of 100 to 1. Hence, a seventh-magnitude star is 100 times brighter than a twelfth-magnitude star.

Table 22.1 Ratios of star brightness.

Difference in Magnitude	Brightness Ratio
0.5	1.6:1
1	2.5:1
2	6.3:1
3	16:1
4	40:1
5	100:1
10	10,000:1
20	100,000,000:1

It follows, then, that the brightness ratio of two stars differing by only one magnitude is about 2.5.[*] Thus, a star of the first magnitude is about 2.5 times brighter than a star of the second magnitude. Table 22.1 shows some magnitude differences and the corresponding brightness ratios.

Because some stars are brighter than first-magnitude stars, zero and negative magnitudes were introduced. The brightest star in the night sky, Sirius, has an apparent magnitude of −1.4, about 10 times brighter than a first-magnitude star. On this scale, the sun has an apparent magnitude of −26.7. At its brightest, Venus has a magnitude of −4.3. At the other end of the scale, the 5-meter (200-inch) Hale Telescope can view stars with an apparent magnitude of 23, approximately 100 million times dimmer than stars that are visible to the unaided eye.

Absolute Magnitude. Astronomers are also interested in the "true" brightness of stars, called their **absolute magnitude**. Stars of the same luminosity, or brightness, usually do not have the same *apparent* magnitude, because their distances from us are not equal. To compare their true, or intrinsic, brightness, astronomers

---

[*] Calculations: $2.512 \times 2.512 \times 2.512 \times 2.512 \times 2.512$, or 2.512 raised to the fifth power, equals 100.

Table 22.2 Distance, apparent magnitude, and absolute magnitude of some stars.

Name	Distance (Light years)	Apparent Magni- tude*	Absolute Magni- tude*
Sun	NA	−26.7	5.0
Alpha Centauri	4.27	0.0	4.4
Sirius	8.70	−1.4	1.5
Arcturus	36	−0.1	−0.3
Betelgeuse	520	0.8	−5.5
Deneb	1600	1.3	−6.9

* The more negative, the brighter; the more positive, the dimmer.

determine what magnitude the stars would have if they were at a standard distance of about 32.6 light-years. For example, the sun, which has an apparent magnitude of −26.7, would, if located at a distance of 32.6 light-years, have an absolute magnitude of about 5. Thus, stars with absolute magnitudes greater than 5 (smaller numerical value) are intrinsically brighter than the sun, but because of their distance, they appear much dimmer.

Table 22.2 lists the absolute and apparent magnitudes of some stars as well as their distances from Earth. Most stars have an absolute magnitude between −5 and 15, which puts the sun near the middle of this range.

Stellar Color and Temperature

The next time you are outdoors on a clear night, take a good look at the stars and note their color (Figure 22.3). Some that are quite colorful can be found in the constellation Orion (see Appendix E, winter chart). Of the two brightest stars in Orion, Betelgeuse (α Orionis) is definitely red, whereas Rigel (β Orionis) appears blue.

Very hot stars with surface temperatures above 30,000 K emit most of their energy in the form of short-wavelength light and therefore appear blue. Red stars, on the other hand, are much cooler, generally less than 3000 K, and most of their energy is emitted as longer-wavelength red light. Stars with temperatures between 5000 and 6000 K appear yellow, like the sun. Because color is primarily a manifestation of a star's temperature, this characteristic provides the astronomer with useful information about a star.

Binary Stars and Stellar Mass

One of the night sky's best known constellations, the Big Dipper, appears at first glance to consist of 7 stars.[*] But persons with good eyesight can resolve the second star in the handle of the Big Dipper (Mizar) as two stars.

---

[*] For the sky location, see Appendix E, first star chart, top.

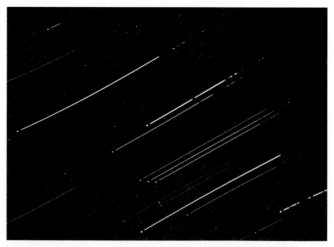

Figure 22.3
Time-lapse photograph of stars in the constellation Orion. These star trails show some of the various star colors. It is important to note that the eye sees color somewhat differently than does photographic film. (Courtesy of National Optical Astronomy Observatories)

During the 1700s, astronomers used their new tool, the telescope, to discover numerous such star pairs. One of the stars in the pair is usually fainter than the other, and for this reason, was considered to be farther away. In other words, the stars were not considered true pairs but were thought only to lie in the same line of sight.

In the early 1800s, careful examination of numerous star pairs by William Herschel revealed that many actually orbit one another. The two stars are in fact united by their mutual gravitation. These pairs of stars, in which the members are far enough apart to be resolved telescopically, are called *visual binaries.* The idea of one star orbiting another may seem unusual, but evidence indicates that more than 50 percent of the stars in the universe occur in pairs or multiples.

Binary stars are used to determine the star property most difficult to calculate—its mass. The mass of a body can be established if it is gravitationally attached to a partner, which is the case for any binary star system. Binary stars orbit each other around a common point called the *center of mass* (Figure 22.4). For stars of equal mass, the center of mass lies exactly halfway between them. If one

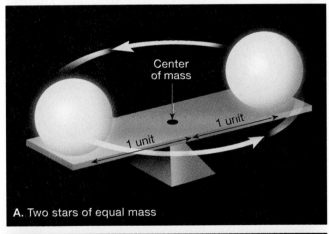

A. Two stars of equal mass

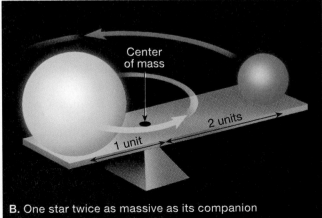

B. One star twice as massive as its companion

Figure 22.4
Binary stars orbit each other around their common center of mass. **A.** For stars of equal mass, the center of mass lies exactly halfway between them. **B.** If one star is twice as massive as its companion, it is twice as close to their common center of mass. Therefore, more massive stars have proportionately smaller orbits than their less massive companions.

star is more massive than its partner, their common center will be located closer to the more massive one. Thus, if the sizes of their orbits can be observed, a determination of their individual masses can be made. You can experience this relationship on a seesaw by trying to balance a person who has a much greater mass.

For illustration, when one star has an orbit half the size (radius) of its companion, it is twice as massive. If their combined masses are equal to 3 times the mass of the sun, then the larger will be twice as massive as the sun, and the smaller will have a mass equal to that of the sun. Most stars have masses that range between $1/10$ and 50 times the mass of the sun.

Hertzsprung-Russell Diagram

Early in this century, Einar Hertzsprung and Henry Russell independently studied the relation between the true brightness (absolute magnitude) and temperature of stars. From this research each developed a graph, now called a **Hertzsprung-Russell diagram, (H-R diagram)**, that exhibits these intrinsic stellar properties. By studying H-R diagrams, we learn a great deal about the sizes, colors, and temperatures of stars.

To produce an H-R diagram, astronomers survey a portion of the sky and plot each star according to its luminosity (brightness) and temperature (Figure 22.5). Notice that the stars in Figure 22.5 are not uniformly distributed. Rather, about 90 percent of all stars fall along a band that runs from the upper-left corner to the lower-right corner of an H-R diagram. These "ordinary" stars are called **main-sequence stars**. As shown in Figure 22.5, *the hottest main-sequence stars are intrinsically the brightest, and vice versa.*

The luminosity of the main-sequence stars is also related to their mass. The hottest (blue) stars are about 50 times more massive than the sun, whereas the coolest (red) stars are only $1/10$ as massive. Therefore, on the H-R diagram, the main-sequence stars appear in a decreasing order, from *hotter, more massive* blue stars to *cooler, less massive* red stars.

Note the location of the sun in Figure 22.5. The sun is a yellow, main-sequence star with an absolute magnitude of about 5. Because the magnitude of a vast majority of main-sequence stars lie between –5 and 15, and because the sun falls midway in this range, the sun is often considered an average star. However, more main-sequence stars are cooler and less massive than our sun.

Just as all humans do not fall into the normal size range, some stars do not fit in with the main-sequence stars. Above and to the right of the main sequence in the H-R diagram in Figure 22.5 lies a group of very luminous stars called *giants,* or, on the basis of their color, **red giants**. The size of these giants can be estimated by comparing them with stars of known size that have the same surface temperature. We know that objects having equal

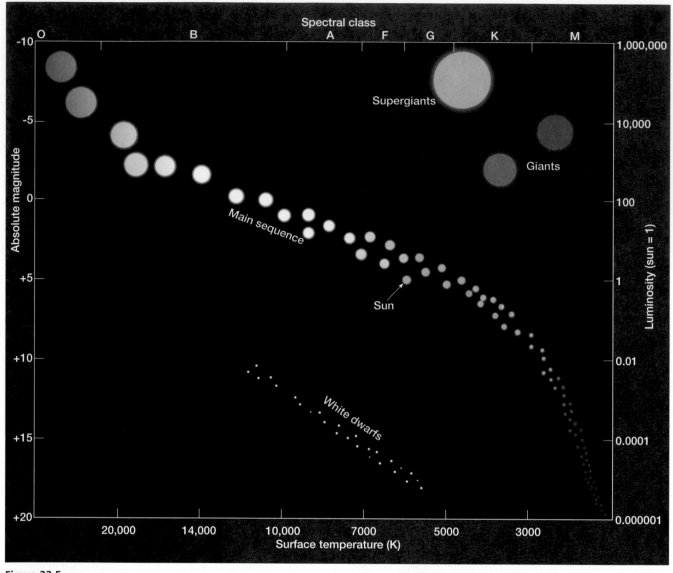

Figure 22.5
Idealized Hertzsprung-Russell diagram on which stars are plotted according to temperature and absolute magnitude.

surface temperatures radiate the same amount of energy per unit area. Therefore, any difference in the brightness of two stars having the same surface temperature is attributable to their relative sizes.

As an example, let us compare the sun, which has been assigned a luminosity of 1, with another yellow star that has a luminosity of 100. Because both stars have the same surface temperature, they both radiate the same amount of energy per unit area. Therefore, in order for the more luminous star to be 100 times brighter than the sun, it must have 100 times more surface area. It should be clear why stars whose plots fall in the upper-right position of an H-R diagram are called *giants.*

Some stars are so large that they are called **supergiants**. Betelgeuse, a bright-red supergiant in the constellation Orion (Appendix E, winter chart), has a radius about 800 times that of the sun. If this star were at the

center of the solar system, it would extend beyond the orbit of Mars, and Earth would find itself inside the star! Other red giants that are easy to locate in our sky are Arcturus in the constellation Bootes and Antares in Scorpius (Appendix E, summer chart).

In the lower-central portion of the H-R diagram, the opposite situation occurs. These stars are much fainter than main-sequence stars of the same temperature, and by using the same reasoning as before, they must be much smaller. Some probably approximate Earth in size. This group has come to be called **white dwarfs**, although not all are white.

Soon after the first H-R diagrams were developed, astronomers realized their importance in interpreting stellar evolution. Just as with living things, a star is born, ages, and dies. Due to the fact that almost 90 percent of the stars lie on the main sequence, we can be relatively certain that stars spend most of their active years as main-sequence stars. Only a few percent are giants, and perhaps 10 percent are white dwarfs.

After considering some variable stars and the nature of interstellar matter, we will return to the topic of stellar evolution.

Variable Stars

Stars that fluctuate in brightness are known as *variables*. Some called **pulsating variables**, fluctuate regularly in brightness by expanding and contracting in size. The importance of one member of this group (cepheid variables) in determining stellar distances is discussed in Box 22.1. Other pulsating variables have irregular periods and are of no value for this purpose.

The most spectacular variables belong to a group known as **eruptive variables**. When one of these explosive events associated with these stars occurs, it appears as a sudden brightening of a star, called a **nova** (Figure 22.6). The term *nova*, meaning "new," was used by the ancients because these stars were unknown to them before their abrupt increase in luminosity.

During the outburst, the outer layer of the star is ejected at high speed. The "cloud" of ejected material occasionally can be captured photographically (Figure 22.7). A nova generally reaches maximum brightness in a couple of days, remains bright for only a few weeks, then slowly returns in a year or so to its original brightness. Because the star returns to its prenova brightness, we can assume that only a small amount of its mass is lost during the flareup. Some stars have experienced more than one such event. In fact, the process probably occurs repeatedly.

The modern explanation of novae proposes that they occur in binary systems consisting of an expanding red giant and a hot white dwarf which are in close proximity. Hydrogen-rich gas from the oversized giant encroaches near enough to the white dwarf to be gravitationally transferred. Eventually, enough of the hydrogen-rich gas is added to the dwarf to cause it to ignite explosively. Such a thermonuclear reaction rapidly heats and expands the outer gaseous envelope of the hot dwarf to produce a nova event. In a relatively short time, the white dwarf returns to its prenova state, where it remains inactive until the next build-up occurs.

Interstellar Matter

Between existing stars is "the vacuum of space." However, it is far from being a pure vacuum, for it is populated with accumulations of dust and gases. The name applied to these concentrations of interstellar matter is **nebula** (Latin for "cloud"). If this interstellar matter is close to a very hot (blue) star it will glow and is called a **bright nebula**. The two main types of bright nebulae are known as *emission nebulae* and *reflection nebulae*.

Box 22.1

Determining Distance from Magnitude

For a star too distant for parallax measurements, knowing its absolute and apparent brightness provides astronomers with a tool for approximating its distance. The apparent magnitude is measured with a photometer (light meter) attached to a telescope. If we also know a star's true brightness we can determine just how far away that star would have to be for it to have the brightness we observe.

You use this same principle when you drive at night and estimate the distance to an oncoming car from the brightness of its headlights. You can do this because you know the true brightness of an automobile headlight. But how do astronomers determine the intrinsic brightness of a star? Fortunately, some stars have characteristics that provide the necessary data.

One important group of such stars is called cepheid variables. These are pulsating stars that get brighter and fainter in a rhythmic fashion. The interval between two successive occurrences of maximum brightness of a pulsating variable is called its light period. Most cepheid variables pulsate with periods of between 2 and 50 days. For example, the North Pole Star (Polaris) varies about 10 percent in brightness over a period of 4 days. In general, the longer the light period of a cepheid, the greater is its absolute magnitude (Figure 22.A). So, by determining the light period of a cepheid, its absolute magnitude can be calculated. When this absolute magnitude is compared to the apparent magnitude, a good approximation of its distance can be made.

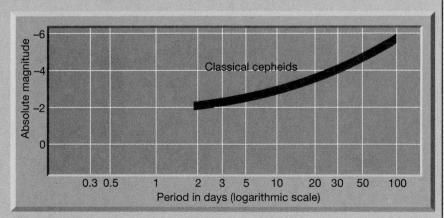

Figure 22.A

Relationship between the light period (two successive occurrences of maximum brightness) and absolute magnitude of pulsating stars (cepheid variables).

March 10, 1935   May 6, 1935

Figure 22.6
Photographs of Nova Herculis (a nova in the constellation Hercules), taken about two months apart, showing the decrease in brightness. (Courtesy of Lick Observatory)

Emission nebulae are gaseous masses that consist largely of hydrogen. They absorb *ultraviolet radiation* emitted by an embedded or nearby hot star. Because these gases are under very low pressure, they reradiate, or emit, this energy as *visible light*. This conversion of ultraviolet light to visible light is known as *fluorescence*, an effect you observe daily in fluorescent lights.

Figure 22.7
Expanding "cloud" around Nova Persei. (Courtesy of Hale Observatories. Copyright by the California Institute of Technology and Carnegie Institution of Washington)

A well-known emission nebula easily seen with binoculars is the sword of the hunter in the constellation Orion (Figure 22.8).

Reflection nebulae, as the name implies, merely reflect the light of nearby stars (Figure 22.9). Reflection nebulae are thought to be composed of rather dense clouds of large particles called **interstellar dust**. This view is supported by the fact that atomic gases with low densities could not reflect light sufficiently to produce the glow observed.

When a dense cloud of interstellar material is not close enough to a bright star to be illuminated, it is referred to as a **dark nebula**. Exemplified by the Horsehead nebula in Orion, dark nebulae appear as opaque objects silhouetted against a bright background (Figure 22.10). Dark nebulae can also easily be seen as starless regions—"holes in the heavens"—when viewing the Milky Way.

Although nebulae appear very dense, they actually consist of very thinly scattered matter. Because of their enormous size, however, the total mass of rarefied particles and molecules may be many times that of the sun. Interstellar matter is of great interest to astronomers because it is from this material that stars and planets are formed.

Stellar Evolution

The idea of describing how a star is born, ages, and then dies may seem a bit presumptuous, for many of these objects have life spans that surely exceed billions of years. However, by studying stars of different ages, astronomers have been able to piece together a plausible

Figure 22.8
The Orion Nebula is a well-known emission nebula. Bright enough to be seen by the naked eye, the Orion Nebula is located in the sword of the hunter in the constellation of the same name. (Courtesy of National Optical Astronomy Observatories)

model for stellar evolution. The method that was used to create this model is analogous to what an alien being, upon reaching Earth, might do to determine the developmental stages of human life. By examining a large number of humans, this stranger would be able to observe the birth of human life, the activities of children and adults, and the death of the elderly. From this information, the alien would attempt to put the stages of human development into their proper sequence. Based on the relative abundance of humans in each stage of development, it would even be possible to conclude that humans spend more of their lives as adults than as toddlers. In a similar fashion, astronomers have pieced together the story of the stars.

Simply, stars exist because of gravity. The mutual gravitational attraction of particles in a thin, gaseous cloud causes the cloud to collapse. As the cloud is squeezed to unimaginable pressures, its temperature rises, igniting its nuclear furnace, and a star is born. A star is a ball of very hot gases, caught between the opposing forces of gravity trying to contract it and thermal nuclear energy trying to expand it. Eventually, all of a star's nuclear fuel will be exhausted and gravity takes over, collapsing the stellar remnant into a small, dense body.

Star Birth

The birthplaces of stars are dark, cool, interstellar clouds, which are comparatively rich in dust and gases (Figure 22.11). In the neighborhood of the Milky Way, these gaseous clouds consist of 92 percent hydrogen, 7 percent helium, and less than 1 percent of the remaining heavier elements. By some mechanism not yet fully understood, these thin gaseous clouds become concentrated enough to begin to gravitationally contract. One proposal to explain the triggering of stellar formation is a shock wave traveling from a catastrophic explosion (supernova) of a nearby star. But, regardless of the force that initiates the concentration of interstellar matter, once it is accomplished, mutual gravitational attraction of the particles squeeze the cloud, pulling every particle toward the center. As the cloud shrinks, gravitational energy (potential energy) is converted into energy of motion, or heat energy.

The initial contraction spans a million years or so. With the passage of time, the temperature of this gaseous body slowly rises, eventually reaching a temperature sufficiently high to cause it to radiate energy from its surface in the form of long-wavelength red light. Because this large red object is not hot enough to engage in nuclear fusion, it is not yet a star. The name **protostar** is applied to these bodies.

Protostar Stage

During the protostar phase, gravitational contraction continues, slowly at first, then much more rapidly (Figure 22.11). This collapse causes the core of the developing star to heat much more intensely than the outer envelope. When the core has reached at least 10 million K, the pressure within is so great that groups of four hydrogen nuclei are fused together into single helium nuclei. Astronomers refer to this nuclear reaction as **hydrogen burning** because an enormous amount of energy is released. However, keep in mind that thermonuclear "burning" is not burning in the usual chemical sense.

Figure 22.9
A faint blue reflection nebula, in the Pleiades star cluster, is caused by the reflection of starlight from dust in the nebula. The Pleiades star cluster, just visible to the naked eye in the constellation Taurus, is spectacular when viewed through binoculars or a small telescope. (Hale Observatories photo. Copyright by the California Institute of Technology and Carnegie Institution of Washington)

Heat from hydrogen fusion causes the stellar gases to increase their motion. This in turn results in an increase in the outward gas pressure. At some point, this outward pressure exactly balances the inward force of gravity. When this balance is reached, the star becomes a stable **main-sequence star** (Figure 22.11). Stated another way, a stable main-sequence star is balanced between two forces: *gravity*, which is trying to squeeze it into a smaller sphere, and *gas pressure*, which is trying to expand it.

Main-Sequence Stage

From this point in the evolution of a main-sequence star until its death, the internal gas pressure struggles to offset the unrelenting force of gravity. Typically, hydrogen burning continues for a few billion years and provides the outward pressure required to support the star from gravitational collapse.

Different stars age at different rates. Hot, massive blue stars radiate energy at such an enormous rate that they substantially deplete their hydrogen fuel in only a few mil-

lion years. By contrast, the very smallest (red) main-sequence stars may remain stable for hundreds of billions of years. A yellow star, such as the sun, remains a main-sequence star for about 10 billion years. Since the solar system is about 5 billion years old, it is comforting to know that the sun will remain stable for another 5 billion years.

An average star spends 90 percent of its life as a hydrogen-burning, main-sequence star. Once the hydrogen fuel in the star's core is depleted, it evolves rapidly and dies. However, with the exception of the least-massive (red) stars, a star can delay its death by burning other fuels and becoming a giant.

Red Giant Stage

The evolution to the red giant stage results because the zone of hydrogen burning continually migrates outward, leaving behind an inert helium core. Eventually, all of the hydrogen in the star's core is consumed. While hydrogen fusion is still progressing in the star's outer shell, no fusion is taking place in the core. Without a source of energy, the core no longer has enough pressure to support itself against the inward force of gravity. As a result, the core begins to contract.

Although the core cannot generate nuclear energy, it does grow hotter by converting gravitational energy into heat energy. Some of this energy is radiated outward, initiating a more vigorous level of hydrogen fusion in the star's outer shell. This energy in turn heats and enormously expands the star's outer envelope, producing a giant body hundreds to thousands of times its main-sequence size (Figure 22.11).

As the star expands, its surface cools, which explains the star's reddish appearance. Eventually the star's gravitational force will stop this outward expansion. Once again, the two opposing forces, gravity and gas pressure, will be in balance, and this gaseous mass will be a stable but much larger star. Some red giants overshoot the equilibrium point and rebound like an overextended spring. Such stars continue to oscillate in size, becoming variable stars.

While the envelope of a red giant expands, the core continues to collapse and heat until it reaches 100 million K. At this incredible temperature, it is hot enough to initiate a nuclear reaction in which helium is converted to carbon. Thus, a red giant consumes both hydrogen and helium to produce energy. In stars more massive than the sun, still other thermonuclear reactions occur that generate all the elements on the periodic table up to number 26, iron. Nuclear "burning" of elements heavier than iron requires an additional source of energy to keep the reaction progressing. Hence, these elements are not produced in ordinary stars.

Eventually, all the usable nuclear fuel in these giants will be consumed. The sun, for example, will spend less than a billion years as a giant, and the more massive stars will pass through this stage even more rapidly. The force of gravity will again control the star's destiny as it squeezes the star into the smallest, most dense piece of matter possible.

Burnout and Death

Most of the events of stellar evolution discussed thus far are well documented. What happens to a star after the red-giant phase is more speculative. We do know that a star, regardless of its size, must eventually exhaust all of its usable nuclear fuel and collapse in response to its immense gravitational force. With this in mind, we will now consider the final stage of stars in three different mass categories.

Death of Low-Mass Stars. Stars less than one-half the mass of the sun (0.5 solar mass) consume their fuel at a comparatively slow rate (Figure 22.12A). Consequently, these small, *cool red stars* may remain on the main sequence for up to 100 billion years. Because the interior of a low-mass star never attains sufficiently high temperatures and pressures to fuse helium, its only energy source is hydrogen fusion. Thus, low-mass stars never evolve to become bloated red giants. Rather, they remain as stable main-sequence stars until they consume their hydrogen fuel and collapse into a hot, dense *white dwarf*. As we shall see, white dwarfs are small, compact objects unable to support nuclear burning.

Death of Medium-Mass (Sunlike) Stars. Main-sequence stars with masses ranging between half that of the sun and three times that of the sun evolve in essentially the same way (Figure 22.12B). During their giant phase, sunlike stars fuse hydrogen and helium fuel at an accelerated rate. Once this fuel is exhausted, these stars (like low-mass stars) collapse into an Earth-sized body of great density—a white dwarf. The density of a white dwarf is as great as physics will allow short of destroying protons and electrons.

The gravitational energy supplied to a collapsing white dwarf is reflected in its high surface temperature. However, without a source of nuclear energy, a white dwarf becomes cooler and dimmer as it continually radiates its remaining thermal energy into space.

During their collapse from red giants to white dwarfs, medium-mass stars are thought to cast off their bloated outer atmosphere, creating an expanding spherical cloud of gas. The remaining hot, central white dwarf heats the gas cloud, causing it to glow. These often beautiful, gleaming spherical clouds are called **planetary**

Figure 22.10

The Horsehead Nebula, a dark nebula in a region of glowing nebulosity in Orion. (Courtesy of Anglo-Australian Observatory, by David Malin)

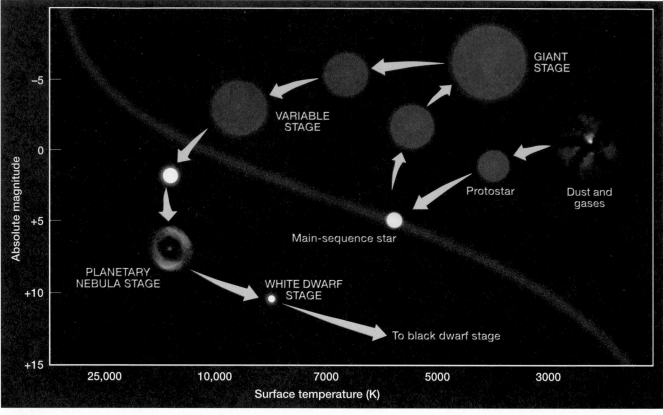

Figure 22.11
Diagram of stellar evolution of H-R diagram for a star about as massive as the sun.

nebulae. A good example of a planetary nebula is the Helix nebula in the constellation Aquarius (Figure 22.13). This nebula appears as a ring because our line of sight through the center traverses less gaseous material than at the nebula's edge. It is, nevertheless, spherical in shape.

Death of Massive Stars. In contrast to sunlike stars, which expire gracefully, stars exceeding 3 solar masses have relatively short life spans and terminate in a brilliant explosion called a **supernova** (Figure 22.12C). During a supernova event, a star becomes millions of times brighter than its prenova stage (see Box 22.2). If one of the nearest stars to Earth produced such an outburst, its brilliance would surpass that of the sun. Supernovae are rare; none have been observed in our galaxy since the advent of the telescope, although Tycho Brahe and Galileo each recorded one about 30 years apart. An even larger supernova was recorded in A.D. 1054 by the Chinese. Today, the remnant of this great outburst is the Crab nebula, shown in Figure 22.14.

A supernova event is thought to be triggered when a massive star consumes most of its nuclear fuel. Without a heat engine to generate the gas pressure required

to balance its immense gravitational field, it collapses. This implosion is of cataclysmic proportion, resulting in a shock wave that moves out from the star's interior. This energetic shock wave destroys the star and blasts the outer shell into space, generating the supernova event.

Theoretical work predicts that during a supernova, the star's interior condenses into a very hot object, possibly no larger than 20 kilometers in diameter (see Box 22.3). These incomprehensibly dense bodies have been named *neutron stars*. Some supernovae events are thought to produce even smaller, and most intriguing, objects called *black holes*. We will consider the nature of neutron stars and black holes in the Stellar Remnants section.

H-R Diagrams and Stellar Evolution

The Hertzsprung-Russell diagrams have been very helpful in formulating and testing models of stellar evolution. They are also useful for illustrating the changes that take place in an individual star during its life span. Figure 22.11 shows on an H-R diagram the evolution of a star about the size of the sun. Keep in mind that the star does not physically move along

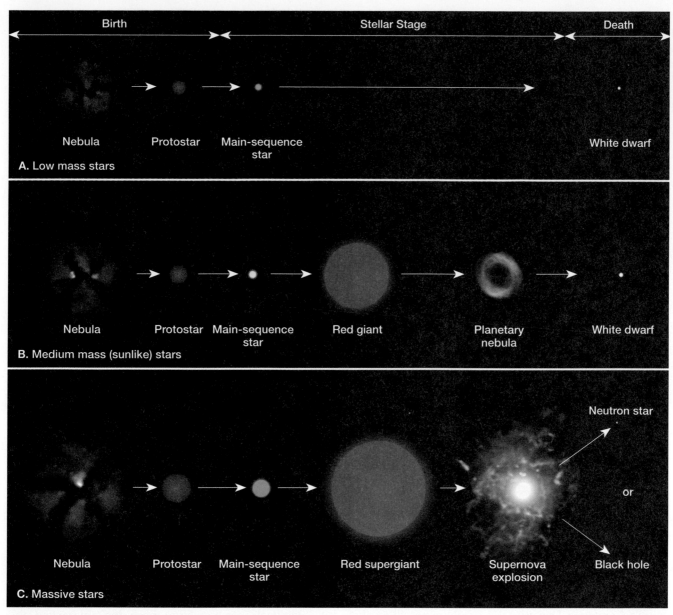

Figure 22.12
The evolutionary stages of stars having various masses.

this path, but rather that its position on the H-R diagram represents the color (temperature) and absolute magnitude (brightness) of the star at various stages in its evolution.

For example, on the H-R diagram, a protostar would be located to the right and above the main sequence (Figure 22.11). It is formed to the right because of its relatively cool surface temperature (red color), and above because it would be more luminous than a main-sequence star of the same color, a fact attributable to its large size. Careful examination of Figure 22.11 should help you visualize the evolutionary changes experi-

enced by a star the size of the sun. In addition, Table 22.3 provides a summary of the evolutionary history of stars having various masses.

Stellar Remnants

Eventually, all stars consume their nuclear fuel and collapse into one of three final states—white dwarf, neutron star, or black hole. Although different in some ways, these small, compact objects are all composed of incomprehensibly dense material and all have extreme surface gravity.

Figure 22.13
The Helix Nebula, the nearest planetary nebula to our solar system. A planetary nebula is the ejected outer envelope of a sunlike star that formed during the star's collapse from a red giant to a white dwarf. (Courtesy of Anglo-Australian Observatory, by David Malin)

White Dwarfs

White dwarfs are extremely small stars with densities greater than any known terrestrial material. It is believed that white dwarfs once were low-mass or medium-mass stars whose internal heat was able to keep these gaseous bodies from collapsing under their own gravitational force.

Although some white dwarfs are no larger than Earth, the mass of such a dwarf can equal 1.4 times that of the sun. Thus, their densities may be a million times greater than water. A spoonful of such matter would weigh several tons. Densities this great are possible only when electrons are displaced inward from their regular orbits, around an atom's nucleus, allowing the atoms to take up less than the "normal" amount of space. Material in this state is called **degenerate matter**.

In degenerate matter, the atoms have been squeezed together so tightly that the electrons are displaced much nearer to the nucleus. Degenerate matter uses electrical repulsion rather than molecular motion to support itself from total collapse. Although atomic particles in degenerate matter are much closer together than in normal Earth matter, they still are not packed as tightly as possible. Stars made of matter that has an even greater density are thought to exist.

As a star contracts into a white dwarf, its surface becomes very hot, sometimes exceeding 25,000 K. Even so, without a source of energy, it can only become cooler and dimmer. Although none have been observed, the terminal stage of a white dwarf must be a small, cold, nonluminous body called a *black dwarf*.

Neutron Stars

A study of white dwarfs produced what might at first appear to be a surprising conclusion. The smallest white dwarfs are the most massive, and the largest are the least massive. The explanation for this is that a more massive star, because of its greater gravitational force, is able to squeeze itself into a smaller, more densely packed object than can a less massive star. Thus, the smallest white dwarfs were produced from the collapse of larger, more massive stars than were the larger white dwarfs.

This conclusion led to the prediction that stars smaller and more massive than white dwarfs must exist. Named **neutron stars**, these objects are thought to be the remnants of supernova events. In a white dwarf, the electrons are pushed close to the nucleus, whereas in a neutron star, the electrons are forced to combine with protons to produce neutrons (hence the name). If Earth were to collapse to the density of a neutron star, it would have a diameter equivalent to the length of a football field. A pea-sized sample of this matter would weigh 100 million tons. This is approximately the density of an atomic nucleus; thus neutron stars can be thought of as large atomic nuclei.

Box 22.2

Supernova 1987A

The first naked-eye supernova in 383 years was discovered in the southern sky in February 1987 (Figure 22.B). This stellar explosion was officially named SN 1987A (*SN* stands for "supernova" and 1987A indicates that it was the first supernova observed in 1987). Naked-eye supernovae are rare. Only a few have been recorded in historic times. Arab observers saw one in 1006, and the Chinese recorded one in 1054 at the present location of the Crab Nebula. In addition, the astronomer Tycho Brahe observed a supernova in 1572 and Kepler saw one shortly thereafter in 1604.

Prior to this event, researchers could only test their hypotheses on dim supernovae seen in distant galaxies. Thus, when SN 1987A occurred, astronomers quickly focused every available telescope in the Southern Hemisphere on this spectacular event. As one astronomer remarked, "This supernova is better studied by far than any supernova in history." More importantly, this event has allowed astronomers to use observational data to test their theoretical models of stellar evolution.

Supernova 1987A occurred about 170,000 light-years away in the Large Magellanic Cloud, a satellite galaxy to our own Milky Way. As expected, the supernova rapidly increased in brightness to a peak magnitude of 2.4, outshining all the other stars in the Large Magellanic Cloud. Also as predicted, within a few weeks it began to fade. However, SN 1987A did provide some surprises.

From old photographs taken of the area, researchers identified the exploded star as Sanduleak −69°202. Astronomers were surprised to find that the parent star was a hot, blue star about 15 times the mass of the sun. Recall, that only cool red giants are thought to die as a supernova event. Further, the Hubble Space Telescope made another unexpected discovery. Its camera revealed a very large shell of gas that predates the supernova explosion by about 40,000 years.

Astronomers now think that Sanduleak −69°202 was once a red supergiant that had blown away its outer shell, exposing a hot, blue core. It is this ejected outer shell that appears in the image produced by the Hubble Space Telescope. Then, some 40,000 years later, the remaining hot core of the red supergiant collapsed, producing the supernova of 1987.

Despite these twists, the theory of stellar evolution has held up very well. Theory predicts that the expanding remnants to Supernova 1987A will be large enough to be observed by the turn of the century. Thus, astronomers continue to monitor SN 1987A to unravel its secrets and to confirm or refute their ideas about the final stages of stellar evolution.

Figure 22.B

The great Supernova 1987A. The photo on the left was made prior to the supernova and the one on the right was made following the event. (Courtesy of Anglo-Australian Observatory, by David Malin)

During a supernova implosion, the envelope of the star is ejected (Figure 22.15), while the core collapses into a very hot neutron star about 20 kilometers (12.4 miles) in diameter. Although neutron stars have high surface temperatures, their small size would greatly limit their luminosity. Consequently, locating one visually would be extremely difficult.

However, theory predicts that a neutron star would have a very strong magnetic field. Further, as a star collapses, it will rotate faster, for the same reason ice skaters rotate faster as they pull in their arms. If the sun were to collapse to the size of a neutron star, it would increase its rate of rotation from once every 25 days to nearly 1000 times per second. Radio waves generated by these rotating stars would be concentrated into two narrow zones that would align with the star's magnetic poles. Consequently, these stars would resemble a rapidly rotating beacon emitting strong radio waves. If Earth happened to be in the path of these beacons, the star would appear to blink on and off, or pulsate, as the waves swept past.

In the early 1970s, a source that radiates short pulses of radio energy, called a **pulsar** (pulsating radio source), was discovered in the Crab nebula. Visual inspection of this radio source revealed it to be a small star centered in the nebula. The pulsar found in the Crab nebula is undoubtedly the remains of the supernova of A.D. 1054 (see Figure 22.14). Thus, the first neutron star had been discovered.

Figure 22.14

Crab Nebula in the constellation Taurus: the remains of the supernova of A.D. 1054. (Courtesy of Lick Observatory)

Black Holes

Are neutron stars made of the most dense materials possible? No. During a supernova event, remnants of stars greater than 3 solar masses apparently collapse into objects even smaller and denser than neutron stars.

Even though these objects would be very hot, their surface gravity would be so immense that even light could not escape the surface. Consequently, they would literally disappear from sight. These incredible bodies have appropriately been named **black holes**. Anything that moved too near a black hole would be swept in by its irresistible gravity and be devoured forever.

How can astronomers find an object whose gravitational field prevents the escape of all matter and energy? One strategy is to seek evidence of matter being rapidly swept into a region of apparent nothingness. Theory predicts that as matter is pulled into a black hole, it should become very hot and emit a flood of X-rays before being engulfed. Because isolated black holes would not have a source of matter to engulf, astronomers first looked at binary star systems.

A likely candidate for a black hole is Cygnus X-1, a strong X-ray source in the constellation Cygnus. In this case, the X-ray source can be observed orbiting a supergiant companion with a period of 5.6 days. It appears that gases are pulled from this companion and spiral into the disc-shaped structure around the black hole (Figure 22.16). The result is a stream of X-rays. Because X-rays cannot penetrate our atmosphere efficiently, the existence of black holes was not confirmed until recently. The first X-ray sources were discovered in 1971 by detectors on satellites. Cygnus X-1 is such a source.

The Milky Way Galaxy

On a clear, moonless night away from city lights, you can see a truly marvelous sight—our own Milky Way galaxy (Figure 22.17). With his telescope, Galileo discovered that this band of light was produced by countless individual stars that the unaided eye is unable to resolve. Today, we realize that the sun is actually a part of this vast system of stars, which number about one hundred billion (Figure 22.18A). The "milky" appear-

Table 22.3 Summary of evolution for stars of various masses.

Initial Mass of Interstellar Cloud (Sun = 1)	Main-Sequence Stage	Giant Phase	Evolution After Giant Phase	Terminal State (Final Mass*)
0.001	None (Planet)	No	None	Planet (0.001)
0.1	Red	No	None	White dwarf (0.1)
1–3	Yellow	Yes	Planetary nebula	White dwarf (< 1.4)
6	White	Yes	Supernova	Neutron star (1.4–3)
20	Blue	Yes Supergiant	Supernova	Black hole (> 3.0)

* These mass numbers are estimates.

Box 22.3

From Stardust to You

During a supernova implosion, the internal temperature of a star may reach 1 billion K, a condition thought to produce very heavy elements such as gold and uranium. These heavy elements, plus the debris of novae and the planetary nebulae, are continually returned to interstellar space where they are available for the formation of other stars (Figure 22.C).

Astronomers believe that the earliest stars were made of nearly pure hydrogen. Fusion during the life and death of stars in turn produced heavier elements, some of which were returned to space. Because the sun contains some heavy elements but has not yet reached the stage in its evolution where it could have produced them, it must be at least a second-generation star. Thus, our sun, as well as the rest of the solar system, is believed to have formed from debris scattered from pre-existing stars. If this

Figure 22.C
Eagle Nebula in the constellation Serpens. This gaseous nebula is the site of a recent star formation. (Courtesy of National Optical Astronomy Observatories)

is the case, the atoms in your body were produced billions of years ago inside a star, and the gold in your jewelry formed during a supernova event that occurred trillions of kilometers away. Without such events, the development of life on Earth would not have been possible.

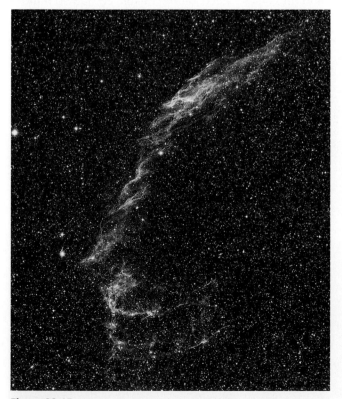

Figure 22.15
Veil Nebula in the constellation Cygnus is the remnant of an ancient supernova implosion. (Hale Observatories photo. Copyright by the California Institute of Technology and Carnegie Institution of Washington)

ance of our galaxy results because the solar system is located within the flat *galactic disk.* Thus, when it is viewed from the "inside," a higher concentration of stars appears in the direction of the galactic plane than in any other direction. You can see this in the edge-on view in Figure 22.18B.

When astronomers began to telescopically survey the stars located along the plane of the Milky Way, it appeared that equal numbers lay in every direction. Could Earth actually be at the center of the galaxy? A better explanation was put forth. Imagine that the trees in an enormous forest represent the stars in the galaxy. After hiking into this forest a short distance, you look around. What you see is an equal number of trees in every direction. Are you really in the center of the forest? Not necessarily; anywhere in the forest, except at the very edge, you will seem to be in the middle.

Structure of the Milky Way Galaxy

Attempts to visually inspect the Milky Way are hindered by the large quantities of interstellar matter that lie in our line of sight. Nevertheless, with the aid of radio telescopes, the gross structure of our galaxy has been determined. The Milky Way is a rather large spiral galaxy whose disk is about 100,000 light-years wide and about 10,000 light-years thick at the nucleus (Figure 22.18). As viewed from Earth, the center of the galaxy lies beyond the constellation Sagittarius.

Figure 22.16
This illustration shows how astronomers believe a binary pair (red giant/black hole) might function.

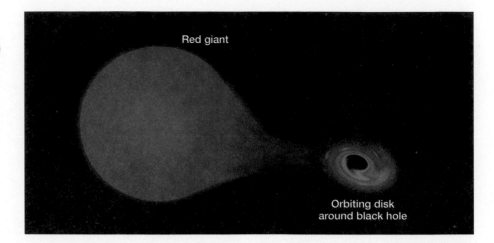

Radio telescopes reveal the existence of at least three distinct *spiral arms*, with some showing splintering (Figure 22.19). The sun is positioned in one of these arms about two-thirds of the way from the center, at a distance from the hub of about 30,000 light-years. The stars in the arms of the Milky Way rotate around the *galactic nucleus*, with the most outward ones moving the slowest, such that the ends of the arms appear to trail. The sun and the arm it is in require about 200 million years for each orbit around the nucleus.

Surrounding the galactic disk is a nearly spherical *halo* made of very tenuous gas and numerous globular clusters. These star clusters do not participate in the ro-

tating motion of the arms, but rather have their own orbits that carry them through the disk. Although some clusters are very dense, they pass among the stars of the arms with plenty of room to spare.

Galaxies

In the mid-1700s, German philosopher Immanuel Kant proposed that the telescopically visible fuzzy patches of light scattered among the stars were actually distant galaxies like the Milky Way. Kant described them as "island universes." Each galaxy, he felt, contained billions

Figure 22.17
Panorama of our galaxy, the Milky Way. Notice the dark bands caused by the presence of interstellar dark nebulae. (Lund Observatory photograph)

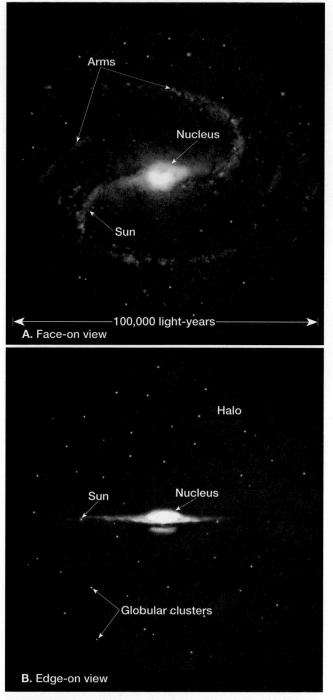

Figure 22.18
Structure of the visible portion of the Milky Way galaxy.

Arms

Nucleus

Sun

←—————100,000 light-years—————→
A. Face-on view

Halo

Sun

Nucleus

Globular clusters

B. Edge-on view

of stars and, as such, was a universe in itself. The weight of opinion, however, favored the hypothesis that they were dust and gas clouds (nebulae) within our galaxy.

This matter was not resolved until the 1920s, when American astronomer Edwin Hubble was able to locate, within one of these fuzzy patches, some unique stars that are known to be intrinsically very bright. Because these very bright stars appeared only very faintly in the telescope, Hubble believed they must lie outside the Milky Way.

This fuzzy patch, which lies over a million light-years away, was named the Great Galaxy in Andromeda (Figure 22.20). Hubble had extended the universe far beyond the limits of our imagination, to include hundreds of billions of galaxies, each containing hundreds of billions of stars. It has been said that a million galaxies are found in that portion of the sky bounded by the cup of the Big Dipper. There are more stars in the heavens than grains of sand in all the beaches on Earth.

Four Types of Galaxies

From the hundreds of billions of galaxies, three basic types have been identified: spiral, elliptical, and irregular.

The Milky Way and the Great Galaxy in Andromeda are examples of fairly large **spiral galaxies** (Figure 22.21). Andromeda can be seen with the unaided eye as a fuzzy fifth-magnitude object. Typically, spiral galaxies are disk-shaped with a somewhat greater concentration of stars near their centers, but there are numerous variations. Viewed broadside, arms are often seen extending from the central nucleus and sweeping gracefully away. The outermost stars of these arms rotate most slowly, giving the galaxy the appearance of a fireworks pinwheel.

One type of spiral galaxy, however, has the stars arranged in the shape of a bar, which rotates as a rigid system. This requires that the outer stars move faster than the inner ones, a fact not easy for astronomers to reconcile with the laws of motion. Attached to each end of these bars are curved spiral arms. These have become known as **barred spiral galaxies** (Figure 22.22). Spiral galaxies are generally quite large, ranging from 20,000 to about 125,000 light-years in diameter. About 10 percent of all galaxies are thought to be barred spirals and another 20 percent are regular spiral galaxies like the Milky Way.

The most abundant group, making up 60 percent of the total, is the **elliptical galaxies**. These are generally smaller than spiral galaxies. Some are so much smaller, in fact, that the term *dwarf* has been applied. Since these dwarf galaxies are not visible at great distances, a survey of the sky reveals more of the conspicuous large spiral galaxies. Although most elliptical galaxies are small, the very largest known galaxies (200,000 light-years in diameter) are also elliptical. As their name implies, elliptical galaxies have an ellipsoidal shape that ranges to nearly spherical, and they lack spiral arms. The two dwarf companions of Andromeda shown in Figure 22.20 are elliptical galaxies.

Only 10 percent of the known galaxies lack symmetry and are classified as **irregular galaxies**. The best known irregular galaxies, the Large and Small Magellanic Clouds in the Southern Hemisphere, are easily visible with the unaided eye. Named after the explorer Ferdinand Magellan, who observed them when he circumnavigated Earth in 1520, they are our nearest galactic neighbors—only 150,000 light-years away.

Figure 22.19

If the Milky Way were photographed from a distance, it might appear like the spiral galaxy NGC 2997. (Courtesy of Anglo-Australian Observatory, by David Malin)

Figure 22.20

Great Galaxy, a spiral galaxy, in the constellation Andromeda. The two bright spots to the left and right are dwarf elliptical galaxies. (Hale Observatories photo. Copyright by the California Institute of Technology and Carnegie Institution of Washington)

One of the major differences among the galactic types is the age of the stars that make them up. The irregular galaxies are composed mostly of young stars, whereas the elliptical galaxies contain old stars. The Milky Way and other spiral galaxies consist of both young and old stars, with the youngest located in the arms.

Galactic Clusters

Once astronomers discovered that stars were associated in groups, they set out to determine whether galaxies also were grouped or just randomly distributed throughout the universe. They found that, like stars, galaxies are grouped in **galactic clusters** (Figure 22.23). Some abundant clusters contain thousands of galaxies. Our own, called the **Local Group**, contains at least 28 galaxies. Of these, three are spirals, eleven are irregulars, and fourteen are ellipticals. Galactic clusters also reside in huge swarms called superclusters. From visual observations, it appears that superclusters may be the largest entities in the universe.

Red Shifts

You probably have noticed the change in pitch of a car horn or ambulance siren as it passes by. When it is approaching, the sound seems to have a higher-than-normal pitch, and when it is moving away, the pitch sounds lower than normal. This effect, which occurs for all wave motion, including sound and light waves, was

Figure 22.21
Two views illustrating the idealized structure of spiral galaxies.
(Courtesy of U.S. Naval Observatory)

first explained by Christian Doppler in 1842 and is called the *Doppler effect*. The reason for the difference in pitch is that it takes time for the wave to be emitted. If the source of the wave is moving away, the beginning of the wave is emitted nearer to you than the end of the wave, effectively "stretching" the wave. This gives it a longer wavelength (see Figure 21.4, p. 549). The opposite is true for an approaching source.

In the case of light, when a source is moving away, its light appears redder than it actually is, because its waves appear lengthened. Objects approaching have their light waves shifted toward the blue (shorter wavelength). Therefore, the Doppler effect reveals whether

Earth and another celestial body are approaching or leaving one another. In addition, the amount of shift allows us to calculate the *rate* at which the relative movement is occurring. Large Doppler shifts indicate higher velocities; smaller Doppler shifts indicate lower velocities.

Expanding Universe

One of the most important discoveries of modern astronomy was made in 1929 by Edwin Hubble. Observations completed several years earlier revealed that most galaxies have Doppler shifts toward the red end of the spectrum. Recall that red shift occurs because the light waves are "stretched," indicating that Earth and

Figure 22.22
Barred spiral galaxy. (Courtesy of Hale Observatories. Copyright by the California Institute of Technology and Carnegie Institution of Washington)

Figure 22.23
Numerous galaxies grouped in the constellation Hercules. (Courtesy of National Optical Astronomy Observatories)

the source are moving away from each other. Hubble set out to explain the predominance of red shift.

Hubble realized that dimmer galaxies were probably farther away than brighter galaxies. Thus, he tried to determine if there is a relation between the distances to galaxies and their red shifts. Using estimated distances based on relative brightness and the observed Doppler red shifts, Hubble discovered that galaxies that exhibit the greatest red shifts are the most distant.

A consequence of the universal red shift is that it predicts that most galaxies (except for a few nearby) are receding from us. Recall that the amount of Doppler red shift is dependent on the velocity at which the object is moving away. Greater red shifts indicate faster recessional velocities. Because more distant galaxies have greater red shifts, Hubble concluded that they must be retreating from us at greater velocities. This idea is currently called **Hubble's law** and states that galaxies are receding from us at a speed that is proportional to their distance.

Hubble was surprised at this discovery because it implied that the most distant galaxies are moving away from us many times faster than those nearby. What type of cosmological theory can explain this fact? It was soon realized that an *expanding universe* can adequately account for the observed red shifts.

To help visualize the nature of this expanding universe, we will employ a popularly used analogy. Imagine a loaf of raisin bread dough that has been set out to rise for a few hours (Figure 22.24). As the dough doubles in size, so does the distance between all of the raisins. However, the raisins that were originally farther apart traveled a greater distance in the same time span than those located closer together. We therefore conclude that, in an expanding universe, as in our analogy, those objects located farther apart move away from each other more rapidly.

Another feature of the expanding universe can be demonstrated using the raisin bread analogy. No matter which raisin you select, it will move away from all the other raisins. Likewise, no matter where one is located in the universe, every other galaxy (except those in the same cluster) will be receding. Edwin Hubble had indeed advanced our understanding of the universe. The Hubble Space Telescope is named in his honor.

The Big Bang

The universe—did it have a beginning? Will it have an end? Cosmologists are trying to answer these questions, and that makes them a rare breed.

First and foremost, any viable theory regarding the origin of the universe must account for the fact that all galaxies (except for the very nearest) are moving away from us. Because all galaxies appear to be moving away from Earth, is our planet in the center of the universe? Probably not, because if we are not even in the center of our own solar system, and our solar system is not even in the center of the galaxy, it seems unlikely that we could be in the center of the universe.

A more probable explanation exists: Imagine a balloon with paper-punch "dots" glued to its surface. When the balloon is inflated, each dot spreads apart from every other dot. Similarly, if the universe is expanding, every galaxy would be moving away from every other galaxy.

This concept of an expanding universe led to the widely accepted **Big Bang** theory. According to this theory, the entire universe was at one time confined to a dense, hot, supermassive ball. Then, about 20 billion years ago, a cataclysmic explosion occurred, hurling this material in all directions. The big bang marks the inception of the universe; all matter and space were created at this instant. The ejected masses of gas cooled and condensed, forming the stellar systems we now observe fleeing from their birthplace.

Will the stars eventually dim from view and invisible galaxies travel on forever? It has been suggested that after a certain point, perhaps 20 billion years in the future, the galaxies will slow and eventually stop their outward flight. Gravitational contraction would follow. The galaxies would then collide and coalesce, and a new fireball would be born. For this event to occur, the universe must have an average density of one atom for every cubic meter

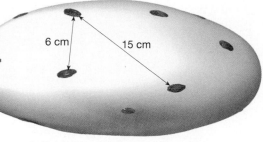

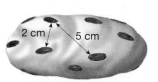

A. Raisin bread dough before it rises.

B. Raisin bread dough a few hours later.

Figure 22.24

Illustration of the raisin bread analogy of an expanding universe. As the dough rises, raisins originally farther apart travel a greater distance in the same time span as those located closer together. Thus, raisins (like galaxies in a uniform expanding universe) that are located farther apart move away from each other more rapidly than those located nearer to each other.

of space. But present estimates indicate that the concentration of matter in the universe is far less than this amount, so this hypothesis remains unsubstantiated.

However, this subject does not end here. It has been proposed that heretofore undetected matter exists in large quantities in the universe. For example, numer-ous black holes may occupy many of the voids in the universe. If this is true, the galaxies could, in fact, collapse upon themselves.

"Absence of evidence is not evidence of absence."

—Anonymous

Review Questions

1. How far away in light-years is our nearest stellar neighbor, Proxima Centauri? Convert your answer to kilometers.

2. What is the most basic method of determining stellar distances?

3. Explain the difference between a star's apparent and absolute magnitudes. Which one is an intrinsic property of a star?

4. What is the ratio of brightness between a twelfth-magnitude and fifteenth-magnitude star?

5. What information about a star can be determined from its color?

6. What color are the hottest stars? Medium-temperature stars? Coolest stars?

7. Which property of a star can be determined from binary star systems?

8. Make a generalization relating the mass and luminosity of main-sequence stars.

9. The disk of a star cannot be resolved telescopically. Explain the method that astronomers have used to estimate the size of stars.

10. Where on an H-R diagram does a star spend most of its lifetime?

11. How does the sun compare in size and brightness to other main-sequence stars?

12. Why is interstellar matter important to stellar evolution?

13. Compare a bright nebula and a dark nebula.

14. What element is the fuel for main-sequence stars? Red giants?

15. What causes a star to become a giant?

16. Why are less massive stars thought to age more slowly than more massive stars, even though they have much less "fuel"?

17. Enumerate the steps thought to be involved in the evolution of sunlike stars.

18. What is the final state of a low-mass (red) main-sequence star?

19. What is the final state of a medium-mass (sunlike) star?

20. How do the "lives" of the most massive stars end? What are the two possible products of this event?

21. Describe the general structure of the Milky Way.

22. Compare the three general types of galaxies.

23. Explain why astronomers consider elliptical galaxies more abundant than spiral galaxies, even though more spiral galaxies have been sighted.

24. How did Edwin Hubble determine that the Great Galaxy in Andromeda is located beyond our galaxy?

25. What evidence supports the Big Bang theory?

Key Terms

absolute magnitud.e (p. 570)
apparent magnitude (p. 569)
barred spiral galaxy (p. 585)
Big Bang (p. 588)
black hole (p. 582)
bright nebula (p. 573)
dark nebula (p. 574)
degenerate matter (p. 580)
elliptical galaxy (p. 585)
emission nebula (p. 574)

eruptive variables (p. 573)
galactic cluster (p. 586)
Hertzsprung-Russell (H-R) diagram (p. 571)
Hubble's law (p. 588)
hydrogen burning (p. 575)
interstellar dust (p. 574)
irregular galaxy (p. 585)
light-year (p. 569)
Local Group (p. 586)
magnitude (p. 569)
main-sequence stars (p. 571)
nebula (p. 573)

neutron star (p. 580)
nova (p. 573)
planetary nebula (p. 577)
protostar (p. 575)
pulsar (p. 581)
pulsating variables (p. 573)
red giant (p. 571)
reflection nebula (p. 574)
spiral galaxy (p. 585)
stellar parallax (p. 568)
supergiant (p. 572)
supernova (p. 578)
white dwarf (p. 572)

APPENDIX A
Metric and English Units Compared

Units

1 kilometer (km)	=	1000 meters (m)
1 meter (m)	=	100 centimeters (cm)
1 centimeter (cm)	=	0.39 inch (in.)
1 mile (mi)	=	5280 feet (ft)
1 foot (ft)	=	12 inches (in.)
1 inch (in.)	=	2.54 centimeters (cm)
1 square mile (mi^2)	=	640 acres (a)
1 kilogram (kg)	=	1000 grams (g)
1 pound (lb)	=	16 ounces (oz)
1 fathom	=	6 feet (ft)

Conversions

When you want
to convert: multiply by: to find:

Length

inches	2.54	centimeters
centimeters	0.39	inches
feet	0.30	meters
meters	3.28	feet
yards	0.91	meters
meters	1.09	yards
miles	1.61	kilometers
kilometers	0.62	miles

Area

square inches	6.45	square centimeters
square centimeters	0.15	square inches
square feet	0.09	square meters
square meters	10.76	square feet
square miles	2.59	square kilometers
square kilometers	0.39	square miles

to convert: multiply by: to find:

Volume

cubic inches	16.38	cubic centimeters
cubic centimeters	0.06	cubic inches
cubic feet	0.028	cubic meters
cubic meters	35.3	cubic feet
cubic miles	4.17	cubic kilometers
cubic kilometers	0.24	cubic miles
liters	1.06	quarts
liters	0.26	gallons
gallons	3.78	liters

Masses and Weights

ounces	20.33	grams
grams	0.035	ounces
pounds	0.45	kilograms
kilograms	2.205	pounds

Temperature

When you want to convert degrees Fahrenheit (°F) to degrees Celsius (°C), subtract 32 degrees and divide by 1.8.

When you want to convert degrees Celsius (°C) to degrees Fahrenheit (°F), multiply by 1.8 and add 32 degrees.

When you want to convert degrees Celsius (°C) to kelvins (K), delete the degree symbol and add 273. When you want to convert kelvins (K) to degrees Celsius (°C), add the degree symbol and subtract 273.

Figure A.1
Temperature scales.

APPENDIX B
Mineral Identification Key

Table B.1 Group I: Metallic luster.

Group 1 Metallic Luster			
Hardness	**Streak**	**Other Diagnostic Properties**	**Name (Chemical Composition)**
Harder than glass	Black streak	Black; magnetic; hardness = 6; specific gravity = 5.2; often granular	Magnetite (Fe_3O_4)
	Greenish-black streak	Brass yellow; hardness = 6; specific gravity = 5.2; generally an aggregate of cubic crystals	Pyrite (FeS_2)—fool's gold
	Red-brown streak	Gray or reddish brown; hardness = 5–6; specific gravity = 5; platy appearance	Hematite (Fe_2O_3)
Softer than glass	Greenish-black streak	Golden yellow; hardness = 4; specific gravity = 4.2; massive	Chalcopyrite ($CuFeS_2$)
	Gray-black streak	Silvery gray; hardness = 2.5; specific gravity = 7.6 (very heavy); good cubic cleavage	Galena (PbS)
	Yellow-brown streak	Yellow brown to dark brown; hardness variable (1–6); specific gravity = 3.5–4; often found in rounded masses; earthy appearance	Limonite ($Fe_2O_3 \cdot H_2O$)
	Gray-black srteak	Black to bronze; tarnishes to purples and greens; hardness = 3; specific gravity = 5; massive	Bornite (Cu_5FeS_4)
Softer than your fingernail	Dark gray streak	Silvery gray; hardness = 1 (very soft); specific gravity = 2.2; massive to platy; writes on paper (pencil lead); feels greasy	Graphite (C)

Table B.2 Group II: Nonmetallic Luster (Dark-Colored).

Group II Nonmetallic Luster (dark colored)			
Hardness	**Cleavage**	**Other Diagnostic Properties**	**Name (Chemical Composition)**
Harder than glass	Cleavage present	Black to greenish black; hardness = 5–6; specific gravity = 3.4; fair cleavage, two directions at nearly 90 degrees	Augite (Ca, Mg, Fe, Al silicate)
		Black to greenish black; hardness = 5–6; specific gravity = 3.2; fair cleavage, two directions at nearly 60 degrees and 120 degrees	Hornblende (Ca, Na, Mg, Fe, OH, Al silicate)
		Red to reddish brown; hardness = 6.5-7.5; conchoidal fracture; glassy luster	Garnet (Fe, Mg, Ca, Al silicate)
	Cleavage not prominent	Gray to brown; hardness = 9; specific gravity = 4; hexagonal crystals common	Corundum (Al_2O_3)
		Dark brown to black; hardness = 7; conchoidal fracture; glassy luster	Smoky quartz (SiO_2)
		Olive green; hardness = 6.5–7; small glassy grains	Olivine $(Mg, Fe)_2SiO_4$
Softer than glass	Cleavage present	Yellow brown to black; hardness = 4; good cleavage in six directions, light yellow streak that has the smell of sulfur	Sphalerite (ZnS)
		Dark brown to black; hardness = 2.5–3, excellent cleavage in one direction; elastic in thin sheets; black mica	Biotite (K, Mg, Fe, OH, Al silicate)
	Cleavage absent	Generally tarnished to brown or green; hardness = 2.5; speciic gravity = 9; massive	Native copper (Cu)
Softer than your fingernail	Cleavage not prominent	Reddish brown; hardness = 1–5; specific gravity = 4–5; red streak; earthy appearance	Hematite (Fe_2O_3)
		Yellow brown; hardness = 1–3; specific gravity = 3.5; earthy appearance; powders easily	Limonite ($Fe_2O_3 \cdot H_2O$)

Table B.3 Group III: Nonmetallic Luster (Light-Colored).

		Group III Nonmetallic Luster (light colored)	
Hardness	**Cleavage**	**Other Diagnostic Properties**	**Name (Chemical Composition)**
Harder than glass	Cleavage present	Salmon colored or white to gray; hardness = 6; specific gravity = 2.6; two directions of cleavage at nearly right angles	Potassium feldspar ($KAlSi_3O_8$) Plagioclase feldspar ($NaAlSi_3O_8$ to $CaAl_2Si_2O_8$)
	Cleavage absent	Any color; hardness = 7; specific gravity = 2.65; conchoidal fracture; glassy appearance; varieties; milky, rose, smoky, amethyst (violet)	Quartz (SiO_2)
Softer than glass	Cleavage present	White, yellowish to colorless; hardness = 3; three directions of cleavage at 75 degrees (rhombohedral); effervesces in HCl; often transparent	Calcite ($CaCO_3$)
		White to colorless; hardness = 2.5; three directions of cleavage at 90 degrees (cubic); salty taste	Halite ($NaCl$)
		Yellow, purple, green, colorless; hardness = 4; white streak; translucent to transparent; four directions of cleavage	Fluorite (CaF_2)
Softer than your fingernail	Cleavage present	Colorless; hardness = 2–2.5; transparent and elastic in thin sheets; excellent cleavage in one direction; light mica	Muscovite (K, OH, Al silicate)
		White to transparent, hardness = 2; when in sheets, is flexible but not elastic; varieties: selenite (transparent, three directions of cleavage); satin spar (fibrous, silky luster); alabaster (aggregate of small crystals)	Gypsum ($CaSO_4 \cdot 2H_2O$)
	Cleavage not prominent	White, pink, green; hardness = 1; forms in thin plates; soapy feel; pearly luster	Talc (Mg silicate)
		Yellow; hardness = 1–2.5	Sulfur (S)
		White; hardness = 2; smooth feel; earthy odor when moistened, has typical clay texture	Kaolinite (Hydrous Al silicate)
		Green; hardness = 2.5; fibrous; variety of serpentine	Asbestos (Mg, Al silicate)
		Pale to dark reddish brown; hardness = 1–3; dull luster; earthy; often contains spheroidal-shaped particles; not a true mineral	Bauxite (Hydrous Al oxide)

APPENDIX C
Earth's Grid System

A glance at any globe reveals a series of north-south and east-west lines that together make up Earth's grid system, a universally used scheme for locating points on Earth's surface. The north-south lines of the grid are called **meridians** and extend from pole to pole (Figure C.1). All are halves of great circles. A **great circle** is the largest possible circle that may be drawn on a globe; if a globe were sliced along one of these circles, it would be divided into two equal parts called **hemispheres**. By viewing a globe or Figure C.1, it can be seen that meridians are spaced farthest apart at the equator and converge toward the poles. The east-west lines (circles) of the grid are known as **parallels**. As their name implies, these circles are parallel to one another (Figure C.1). While all meridians are parts of great circles, all parallels are not. In fact, only one parallel, the equator, is a great circle.

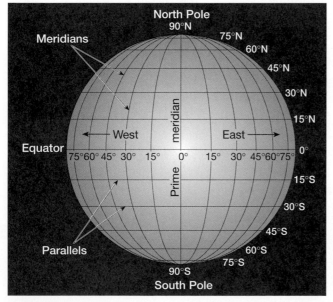

Figure C.1
Earth's grid system.

Latitude and Longitude

Latitude may be defined as distance, measured in degrees, *north* and *south* of the equator. Parallels are used to show latitude. Since all points that lie along the same parallel are an identical distance from the equator, they all have the same latitude designation. The latitude of the equator is 0 degrees, while the north and south poles lie 90 degrees N and 90 degrees S, respectively.

Longitude is defined as distance, measured in degrees, *east* and *west* of the zero or *prime meridian*. Since all meridians are identical, the choice of a zero line is obviously arbitrary. However, the meridian that passes through the Royal Observatory at Greenwich, England, is universally accepted as the reference meridian. Thus, the longitude for any place on the globe is measured east or west from this line. Longitude can vary from 0 degrees along the prime meridian to 180 degrees, halfway around the globe.

It is important to remember that when a location is specified, directions must be given, that is, north or south latitude and east or west longitude (Figure C.2). If this is not done, more than one point on the globe is being designated. The only exceptions, of course, are places that lie along the equator, the prime meridian, or the 180-degree meridian. It should also be noted that while it is not incorrect to use fractions, a degree of latitude or longitude is usually divided into minutes and seconds. A minute (') is $\frac{1}{60}$th of a degree, and a second (") is $\frac{1}{60}$th of a minute. When locating a place on a map, the degree of exactness will depend on the scale of the map. When using a small-scale world map or globe, it may be difficult to estimate latitude and longitude to the nearest whole degree or two. On the other hand, when a large-scale map of an area is used, it is often possible to estimate latitude and longitude to the nearest minute or second.

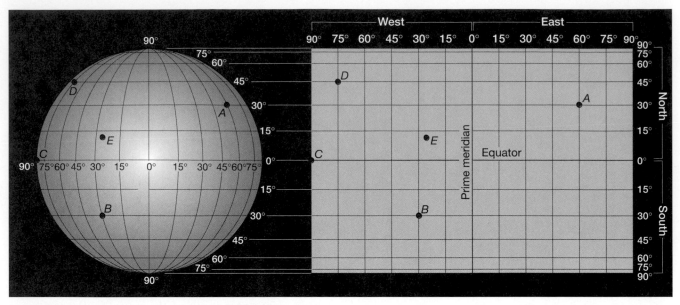

Figure C.2
Locating places using the grid system. For both diagrams: Point *A* is latitude 30 degrees N, longitude 60 degrees E; Point *B* is latitude 30 degrees S, longitude 30 degrees W; Point *C* is latitude 0 degrees, longitude 90 degrees W; Point *D* is latitude 45 degrees N, longitude 75 degrees W; Point *E* is approximately latitude 10 degrees N, longitude 25 degrees W.

Distance Measurement

The length of a degree of longitude depends on where the measurement is taken. At the equator, which is a great circle, a degree of east-west distance is equal to approximately 111 kilometers (69 miles). This figure is found by dividing Earth's circumference—40,075 kilometers (24,900 miles)—by 360. However, with an increase in latitude, the parallels become smaller, and the length of a degree of longitude diminishes (see Table C.1). Thus, at about latitude 60 degrees N and S, a degree of longitude has a value equal to about half of what it was at the equator.

Since all meridians are halves of great circles, a degree of latitude is equal to about 111 kilometers (69 miles), just as a degree of longitude along the equator is. However, Earth is not a perfect sphere but is slightly flattened at the poles and bulges slightly at the equator. Because of this, there are small differences in the length of a degree of latitude.

Determining the shortest distance between two points on a globe can be done easily and fairly accurately using the "globe and string" method. It should be noted here that the arc of a great circle is the shortest distance between two points on a sphere. In order to determine the great circle distance (as well as observe the great circle route) between two places, stretch the string between the locations in question. Then, measure the length of the string along the equator (since it is a great circle with degrees marked on it) to determine the number of degrees between the two points. To calculate the distance in kilometers or miles, simply multiply the number of degrees by 111 or 69, respectively.

Table C.1 Longitude as distance.

°Lat.	Length of 1° Long.		°Lat.	Length of 1° Long.		°Lat.	Length of 1° Long.	
	km	miles		km	miles		km	miles
0	111.367	69.172	30	96.528	59.955	60	55.825	34.674
1	111.349	69.161	31	95.545	59.345	61	54.131	33.622
2	111.298	69.129	32	94.533	58.716	62	52.422	62.560
3	111.214	69.077	33	93.493	58.070	63	50.696	31.488
4	111.096	69.004	34	92.425	57.407	64	48.954	30.406
5	110.945	68.910	35	91.327	56.725	65	47.196	29.314
6	110.760	68.795	36	90.203	56.027	66	45.426	28.215
7	110.543	68.660	37	89.051	55.311	67	43.639	27.105
8	110.290	68.503	38	87.871	54.578	68	41.841	25.988
9	110.003	68.325	39	86.665	53.829	69	40.028	24.862
10	109.686	68.128	40	85.431	53.063	70	38.204	23.729
11	109.333	67.909	41	84.171	52.280	71	36.368	22.589
12	108.949	67.670	42	82.886	51.482	72	34.520	21.441
13	108.530	67.410	43	81.575	50.668	73	32.662	20.287
14	108.079	67.130	44	80.241	49.839	74	30.793	19.126
15	107.596	66.830	45	78.880	48.994	75	28.914	17.959
16	107.079	66.509	46	77.497	48.135	76	27.029	16.788
17	106.530	66.168	47	76.089	47.260	77	25.134	15.611
18	105.949	65.807	48	74.659	46.372	78	23.229	14.428
19	105.337	65.427	49	73.203	45.468	79	21.320	13.242
20	104.692	65.026	50	71.727	44.551	80	19.402	12.051
21	104.014	64.605	51	70.228	43.620	81	17.480	10.857
22	103.306	64.165	52	68.708	42.676	82	15.551	9.659
23	102.565	63.705	53	67.168	41.719	83	13.617	8.458
24	101.795	63.227	54	65.604	40.748	84	11.681	7.255
25	100.994	62.729	55	64.022	39.765	85	9.739	6.049
26	100.160	62.211	56	62.420	38.770	86	7.796	4.842
27	99.297	61.675	57	60.798	37.763	87	5.849	3.633
28	98.405	61.121	58	59.159	36.745	88	3.899	2.422
29	97.481	60.547	59	57.501	35.715	89	1.950	1.211
30	96.528	59.955	60	55.825	34.674	90	0.000	0.000

APPENDIX D
Topographic Maps

A map is a representation on a flat surface of all or a part of Earth's surface drawn to a specific scale. Maps are often the most effective means for showing the locations of both natural and human structures, their sizes, and their relationships to one another. Like photographs, maps readily display information that would be impractical to express in words.

While most maps show only the two horizontal dimensions, geologists, as well as other map users, often require that the third dimension, elevation, be shown on maps. Maps that show the shape of the land are called **topographic maps**. Although various techniques may be used to depict elevations, the most accurate method involves the use of contour lines.

Contour Lines

A **contour line** is a line on a map representing a corresponding imaginary line on the ground that has the same elevation above sea level along its entire length. While many map symbols are pictographs, resembling the objects they represent, a contour line is an abstraction that has no counterpart in nature. It is, however, an accurate and effective device for representing the third dimension on paper.

Some useful facts and rules concerning contour lines are listed as follows. This information should be studied in conjunction with Figure D.1.

1. Contour lines bend upstream or upvalley. The contours form Vs that point upstream, and in the upstream direction the successive contours represent higher elevations. For example, if you were standing on a stream bank and wished to get to the point at the same elevation directly opposite you on the other bank, without stepping up or down, you would need to walk upstream along the contour at that elevation to where it crosses the stream bed, cross the stream, and then walk back downstream along the same contour.

2. Contours near the upper parts of hills form closures. The top of a hill is higher than the highest closed contour.

3. Hollows (depressions) without outlets are shown by closed, hatched contours. Hatched contours are contours with short lines on the inside pointing downslope.

4. Contours are widely spaced on gentle slopes.

5. Contours are closely spaced on steep slopes.

6. Evenly spaced contours indicate a uniform slope.

7. Contours usually do not cross or intersect each other, except in the rare case of an overhanging cliff.

8. All contours eventually close, either on a map or beyond its margins.

9. A single high contour never occurs between two lower ones, and vice versa. In other words, a change in slope direction is always determined by the repetition of the same elevation either as two different contours of the same value or as the same contour crossed twice.

10. Spot elevations between contours are given at many places, such as road intersections, hill summits, and lake surfaces. Spot elevations differ from control elevation stations, such as bench marks, in not being permanently established by permanent markers.

Relief

Relief refers to the difference in elevation between any two points. *Maximum relief* refers to the difference in elevation between the highest and lowest points in the area being considered. Relief determines the **contour interval**, which is the difference in elevation between succeeding contour lines that is used on topographic maps. Where relief is low, a small contour interval, such

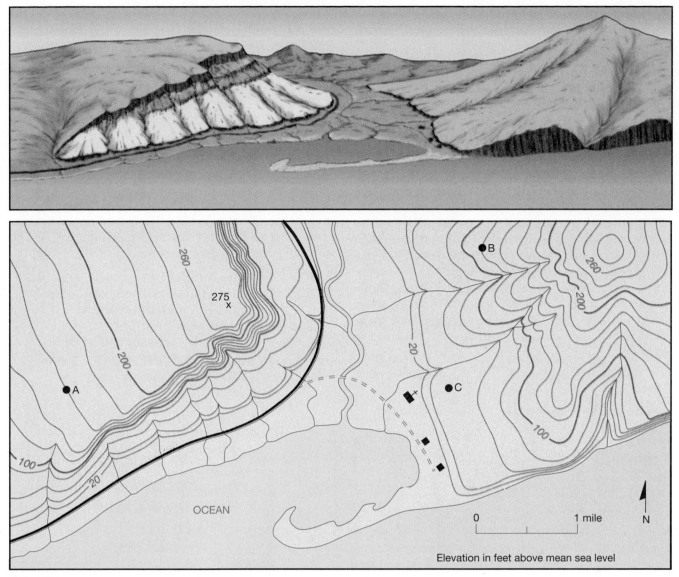

Figure D.1

Perspective view of an area and a contour map of the same area. These illustrations show how features are depicted on a topographic map. The upper illustration is a perspective view of a river valley and the adjoining hills. The river flows into a bay, which is partly enclosed by a hooked sandbar. On either side of the valley are terraces through which streams have cut gullies. The hill on the right has a smoothly eroded form and gradual slopes, whereas the one on the left rises abruptly in a sharp precipice, from which it slopes gently, and forms an inclined plateau traversed by a few shallow gullies. A road provides access to a church and the two houses situated across the river from a highway that follows the seacoast and curves up the river valley. The lower illustration shows the same features represented by symbols on a topographic map. The contour interval (vertical distance between adjacent contours) is 20 feet. (After U.S. Geological Survey)

as 10 or 20 feet, may be used. In flat areas, such as wide river valleys or broad, flat uplands, a contour interval of 5 feet is often used. In rugged mountainous terrain, where relief is many hundreds of feet, contour intervals as large as 50 or 100 feet are used.

Scale

Map **scale** expresses the relationship between distance or area on the map to the true distance or area on Earth's surface. This is generally expressed as a ratio or frac-

tion, such as 1:24,000 or $\frac{1}{24,000}$. The numerator, usually 1, represents map distance, and the denominator, a large number, represents ground distance. Thus, 1:24,000 means that a distance of 1 unit on the map represents a distance of 24,000 such units on the surface of Earth. It does not matter what the units are.

Often, the graphic or bar scale is more useful than the fractional scale, because it is easier to use for measuring distances between points. The graphic scale (Figure D.2) consists of a bar divided into equal segments, which represent equal distances on the map.

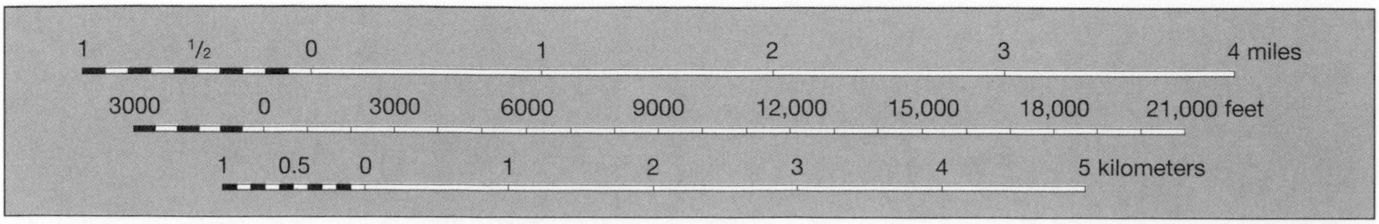

Figure D.2
Graphic scale.

One segment on the left side of the bar is usually divided into smaller units to permit more accurate estimates of fractional units.

Topographic maps, which are also referred to as *quadrangles*, are generally classified according to publication scale. Each series is intended to fulfill a specific type of map need. To select a map with the proper scale for a particular use, remember that large-scale maps show more detail and small-scale maps show less detail. The sizes and scales of topographic maps published by the U.S. Geological Survey are shown in Table D.1.

Color and Symbol

Each color and symbol used on U.S. Geological Survey topographic maps has significance. Common topographic map symbols are shown in Figure D.3. The meaning of each color is as follows:

Blue — water features
Black — works of man, such as homes, schools, churches, roads, and so forth
Brown — contour lines
Green — woodlands, orchards, and so forth
Red — urban areas, important roads, public land subdivision lines

Table D.1 National Topographic Maps.

Series	Scale	1 inch Represents	Standard Quadrangle Size (latitude-longitude)	Quadrangle Area (square miles)	Paper Size E-W N-S Width Length (inches)
7 1/2-minute	1:24,000	2000 feet	7 1/2′ x 7 1/2′	49–70	22 x 27
Puerto Rico 7 1/2-minute	1:20,000	about 1667 feet	7 1/2′ x 7 1/2′	71	29 1/2 x 32 1/2
15-minute	1:62,500	nearly 1 mile	15′ x 15′	197–282	17 x 21
Alaska 1:63,360	1:63,360	1 mile	15′ x 20′–36′	207–281	18 x 21
U.S. 1:250,000	1:250,000	nearly 4 miles	1° x 2°	4580–8669	34 x 22
U.S. 1:1,000,000	1:1,000,000	nearly 16 miles	4° x 6°	73,734–102,759	27 x 27

SOURCE: U.S. Geological Survey

Figure D.3 (opposite page)
U.S. Geological Survey topographic map symbols. (Variations will be found on older maps.)

TOPOGRAPHIC MAP SYMBOLS

VARIATIONS WILL BE FOUND ON OLDER MAPS

Primary highway, hard surface	
Secondary highway, hard surface	
Light-duty road, hard or improved surface	
Unimproved road	
Road under construction, alinement known	
Proposed road	
Dual highway, dividing strip 25 feet or less	
Dual highway, dividing strip exceeding 25 feet	
Trail	

Railroad: single track and multiple track	
Railroads in juxtaposition	
Narrow gage: single track and multiple track	
Railroad in street and carline	
Bridge: road and railroad	
Drawbridge: road and railroad	
Footbridge	
Tunnel: road and railroad	
Overpass and underpass	
Small masonry or concrete dam	
Dam with lock	
Dam with road	
Canal with lock	

Buildings (dwelling, place of employment, etc.)	
School, church, and cemetery	
Buildings (barn, warehouse, etc.)	
Power transmission line with located metal tower	
Telephone line, pipeline, etc. (labeled as to type)	
Wells other than water (labeled as to type)	oOil. oGas
Tanks: oil, water, etc. (labeled only if water)	••● Water
Located or landmark object; windmill	
Open pit, mine, or quarry; prospect	
Shaft and tunnel entrance	

Horizontal and vertical control station:	
Tablet, spirit level elevation	BM△5653
Other recoverable mark, spirit level elevation	△5455
Horizontal control station: tablet, vertical angle elevation	VABM △95/9
Any recoverable mark, vertical angle or checked elevation	△3775
Vertical control station: tablet, spirit level elevation	BM ×957
Other recoverable mark, spirit level elevation	×954
Spot elevation	×7369 ×7369
Water elevation	670 670

Boundaries: National	
State	
County, parish, municipio	
Civil township, precinct, town, barrio	
Incorporated city, village, town, hamlet	
Reservation, National or State	
Small park, cemetery, airport, etc.	
Land grant	
Township or range line, United States land survey	
Township or range line, approximate location	
Section line, United States land survey	
Section line, approximate location	
Township line, not United States land survey	
Section line, not United States land survey	
Found corner: section and closing	
Boundary monument: land grant and other	
Fence or field line	

Index contour		Intermediate contour	
Supplementary contour		Depression contours	
Fill		Cut	
Levee		Levee with road	
Mine dump		Wash	
Tailings		Tailings pond	
Shifting sand or dunes		Intricate surface	
Sand area		Gravel beach	

Perennial streams		Intermittent streams	
Elevated aqueduct		Aqueduct tunnel	
Water well and spring		Glacier	
Small rapids		Small falls	
Large rapids		Large falls	
Intermittent lake		Dry lake bed	
Foreshore flat		Rock or coral reef	
Sounding, depth curve		Piling or dolphin	
Exposed wreck		Sunken wreck	
Rock, bare or awash; dangerous to navigation			

Marsh (swamp)		Submerged marsh	
Wooded marsh		Mangrove	
Woods or brushwood		Orchard	
Vineyard		Scrub	
Land subject to controlled inundation		Urban area	

APPENDIX E
Star Charts*

The star charts on the next four pages can be used to locate the brighter stars and prominent constellations on the dates and times indicated on each chart. To use these charts, face southward and hold the chart overhead with the top toward the north.

*From Robert Dixon, *Dynamic Astronomy*, 6th ed., Englewood Cliffs, NJ: Prentice Hall, 1992.

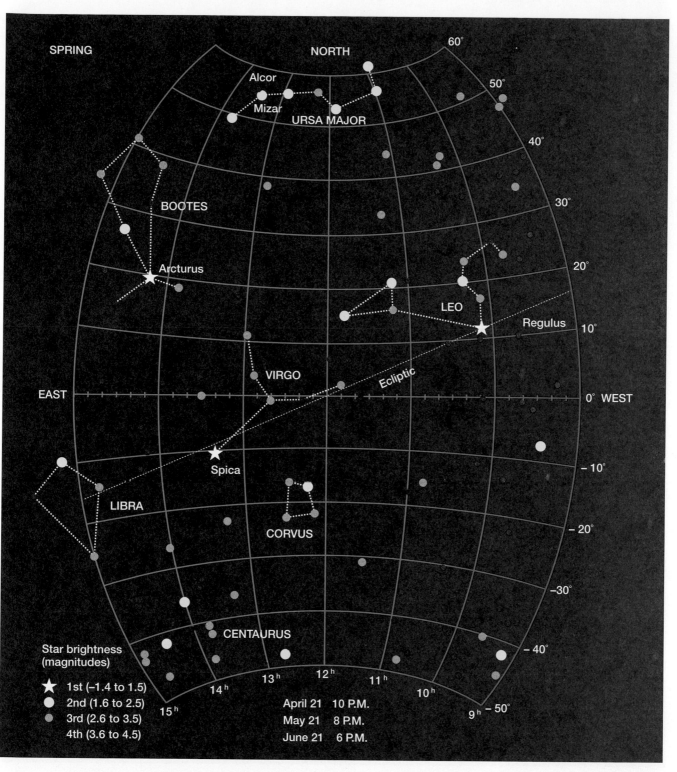

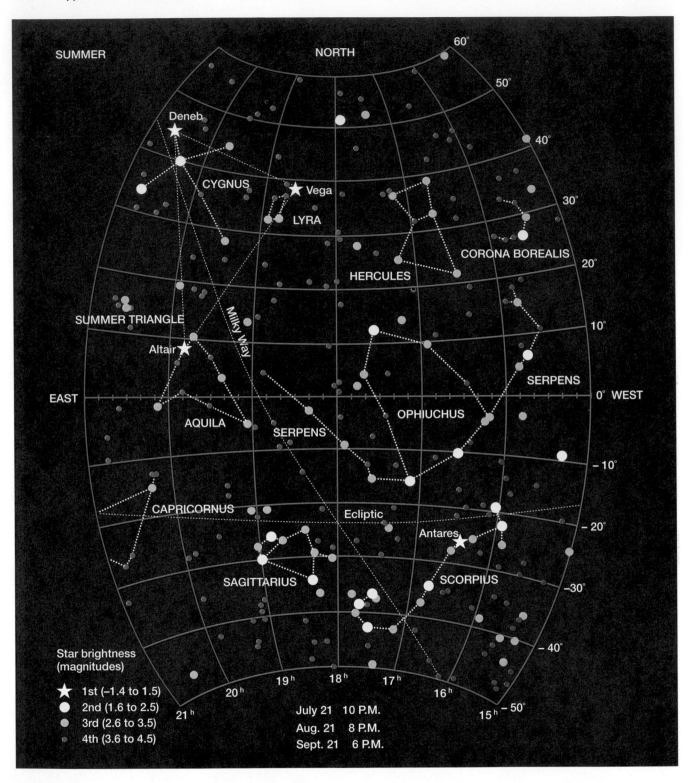

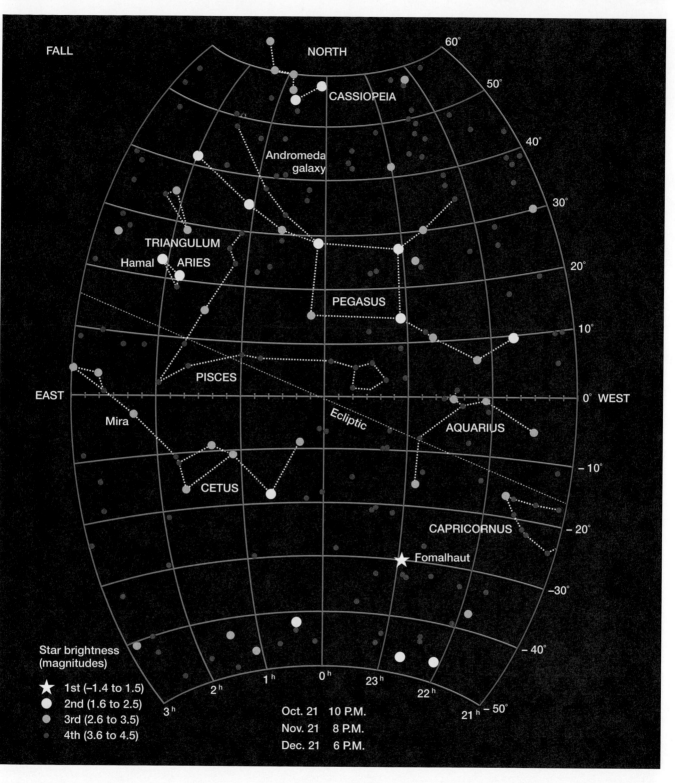

FALL

NORTH

60°

50°

CASSIOPEIA

40°

Andromeda galaxy

30°

20°

TRIANGULUM

Hamal   ARIES

PEGASUS

10°

PISCES

EAST

Mira

0° WEST

Ecliptic

AQUARIUS

−10°

CETUS

CAPRICORNUS

−20°

Fomalhaut

−30°

−40°

Star brightness (magnitudes)

3ʰ   2ʰ   1ʰ   0ʰ   23ʰ   22ʰ   21ʰ −50°

★ 1st (−1.4 to 1.5)

● 2nd (1.6 to 2.5)

● 3rd (2.6 to 3.5)

· 4th (3.6 to 4.5)

Oct. 21   10 P.M.
Nov. 21   8 P.M.
Dec. 21   6 P.M.

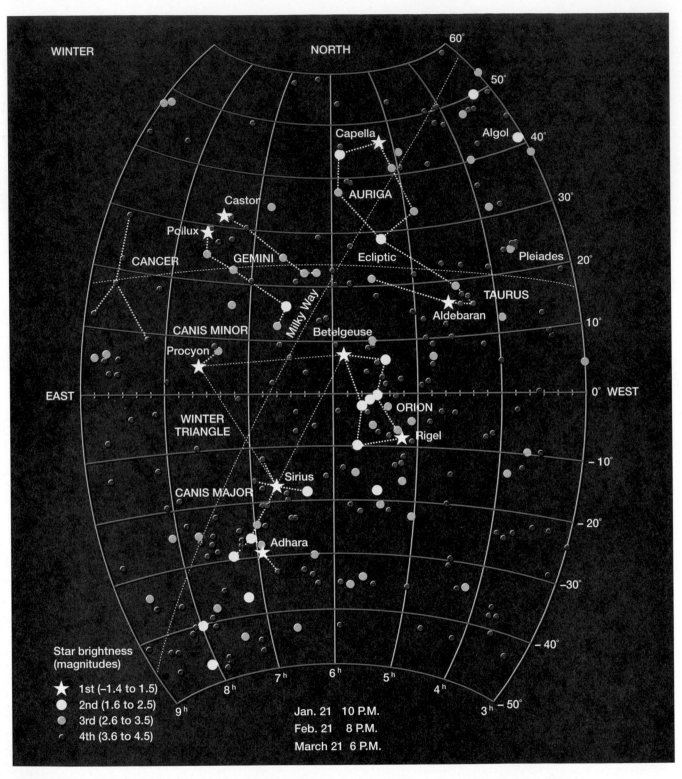

WINTER

NORTH

60°

50°

40°

Algol

30°

Capella

AURIGA

Castor

Ecliptic

Pleiades

20°

Pollux

GEMINI

CANCER

TAURUS

Aldebaran

10°

CANIS MINOR

Betelgeuse

Milky Way

Procyon

EAST

0° WEST

ORION

WINTER
TRIANGLE

Rigel

-10°

Sirius

-20°

CANIS MAJOR

Adhara

-30°

-40°

Star brightness
(magnitudes)

-50°

★ 1st (–1.4 to 1.5)

9ʰ

8ʰ

7ʰ

6ʰ

5ʰ

4ʰ

3ʰ

● 2nd (1.6 to 2.5)

Jan. 21   10 P.M.

● 3rd (2.6 to 3.5)

Feb. 21   8 P.M.

∘ 4th (3.6 to 4.5)

March 21 6 P.M.

APPENDIX F
World Soils

Figure F.1, on pages 608–9, shows the generalized pattern of global soil orders according to the *Comprehensive Soil Classification System* (CSCS). It should be examined in conjunction with Table F.1, which briefly describes each of the soil orders depicted on the map. To avoid subjective decisions as to classification (a problem that plagued earlier systems), the CSCS defined its classes strictly in terms of soil characterisitics. That is, it is based on features that can be observed or inferred.

The CSCS uses a hierarchy of six categories, or levels. The system recognizes 10 major global *orders* that can be further subdivided into *suborders, great groups, subgroups, families,* and *series*. Note, however, that on the scale of a world map such as Figure F.1, only the largest units (soil orders) can be shown and then only in an extremely generalized way. Although the distribution pattern of major soil orders is more complex than can be shown in Figure F.1, the major distinguishing regional properties of world soils are depicted.

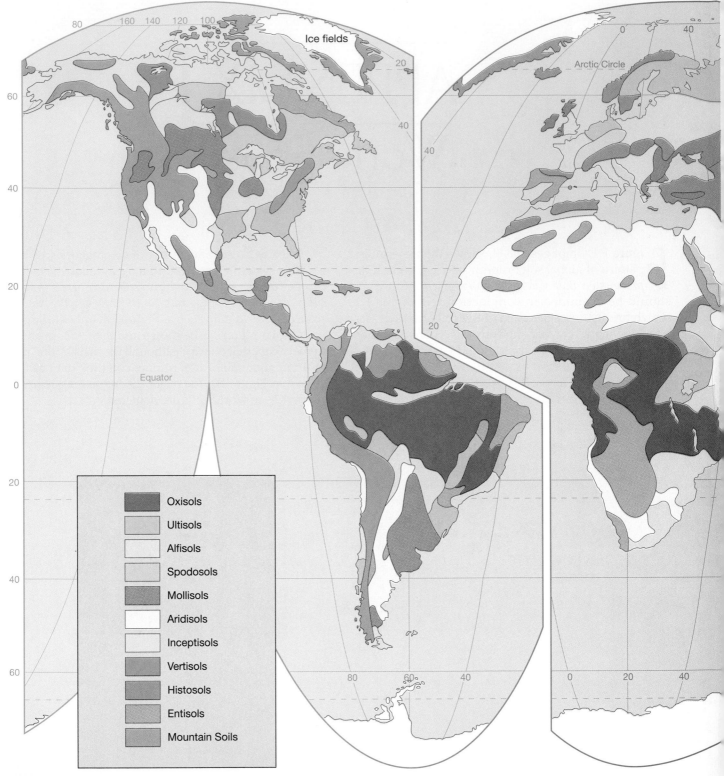

Figure F.1
Soil distribution. The pattern of global soil orders is remarkably similar to the pattern of major climates. Terminology employed is from the *Comprehensive Soil Classification System*. (Adapted from E. Willard Miller, *Physical Geography*, Columbus, Ohio: Merril, 1985, Plate 4.)

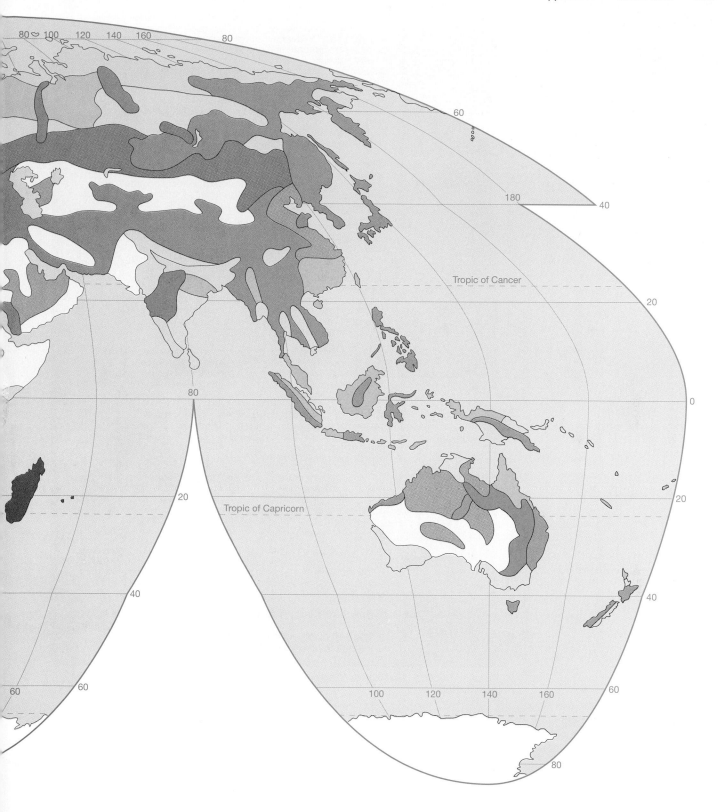

Table F.1 World soil orders.

Entisols	Youngest soils on the Earth. Just beginning to develop in response to the weathering phenomena in the environment. Do not display natural horizons. Found in all climates. They weather slowly over thousands of years; consequently, volcanic ash deposits or sand deposits form the basis for entisols.
Vertisols	Soils containing large amounts of clay, which shrink upon drying and swell with the addition of water. Found in subhumid to arid climates, provided that adequate supplies of water are available to saturate the soil after periods of drought. Soil expansion and contraction exert stresses on human structures.
Inceptisols	Young soils that reveal developmental characteristics (horizons) in response to climate and vegetation. Exist from the Arctic to the tropics on young land surfaces. Common in alpine areas, on river floodplains, in stables and dune areas, and in areas once glaciated.
Aridsols	Soils that develop in dry places, such as the desert, where water —precipitation and groundwater—is insufficient to remove soluble minerals. Frequently irrigated for intensive agricultural production, although salt accumulation poses a problem.
Mollisols	Dark, soft soils that have developed under grass vegetation, generally found in prairie areas. Soil fertility is excelllent because potential evaporation generally exceeds precipitation. Also found in hardwood forests with significant earthworm activity. Climatic range is boreal or alpine to tropical. Dry seasons are normal.
Spodosols	Soils found only in humid regions on sandy material. Range from the boreal coniferous forests into tropical forests. Beneath the dark upper horizon of weathered organic material lies a light-colored horizon of leached material, the distinctive property of this soil.
Alfisols	Mineral soils that form under boreal forests or broadleaf deciduous forests, rich in iron and aluminum. Clay particles accumulate in a subsurface layer in response to leaching in moist environments. Fertile, productive soils, because they are neither too wet nor too dry.
Ultisols	Soils that represent the products of long periods of weathering. Water percolating through the soil concentrates clay particles in the lower horizons (argillic horizons). Restricted to humid climates in the temperate regions and the tropics where the growing season is long. Abundant water and a long frost-free period contribute to extensive leaching, hence poorer soil quality.
Oxisols	Soils that occur on old land surfaces unless parent materials were strongly weathered before they were deposited. Generally found in the tropics and subtropical regions. Rich in iron and aluminum oxides, oxisols are heavily leached; hence are poor soils for agricultural activity. Few, if any, exist in the United States.
Histosols	Organic soils with little or no climatic implications. Can be found in any climate where organic debris can accumulate to form a "bog soil." Dark, partially decomposed organic material commonly referred to as *peat*.

Source: Robert E. Norris et at. *Geography: An Introductory Perspective*, Columbus, Ohio: Merrill, 1982.

APPENDIX G
Landforms of the Conterminous United States

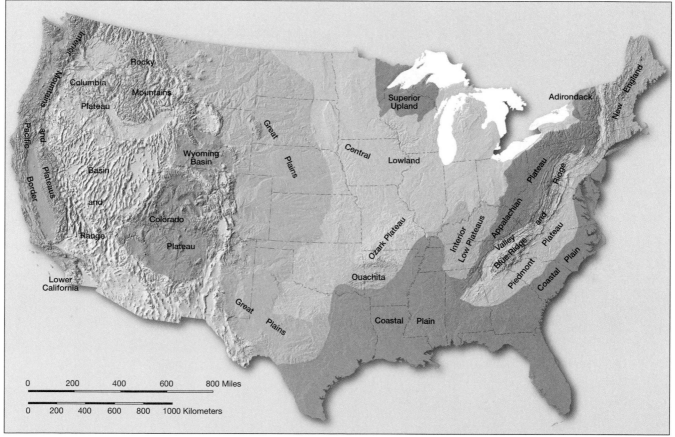

Figure G.1
Outline map showing major physiographic provinces of the United States.

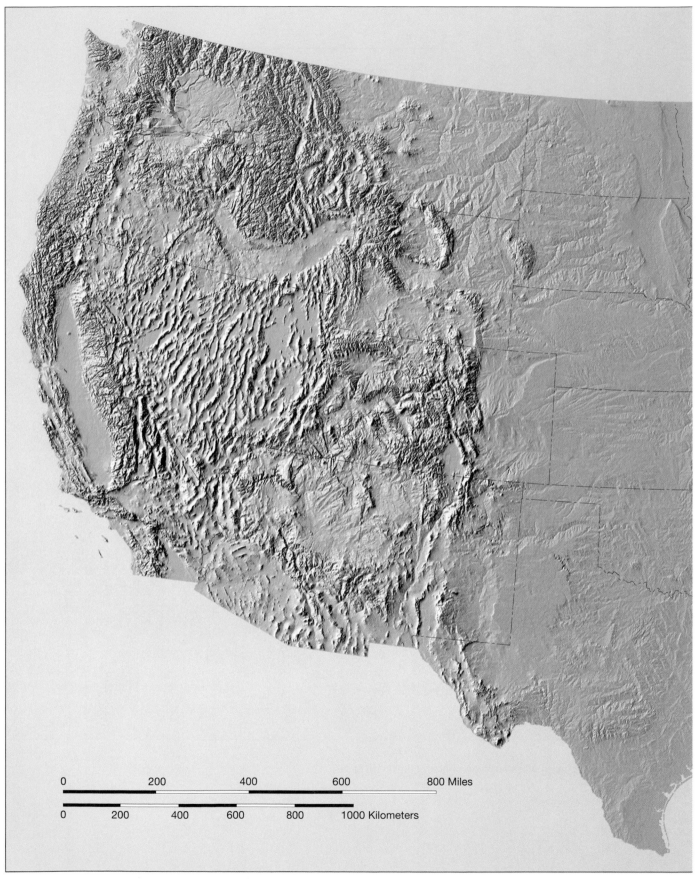

Figure G.2
Digital shaded relief landform map of the United States. (Data provided by the U.S. Geological Survey)

Landforms of the Conterminous United States

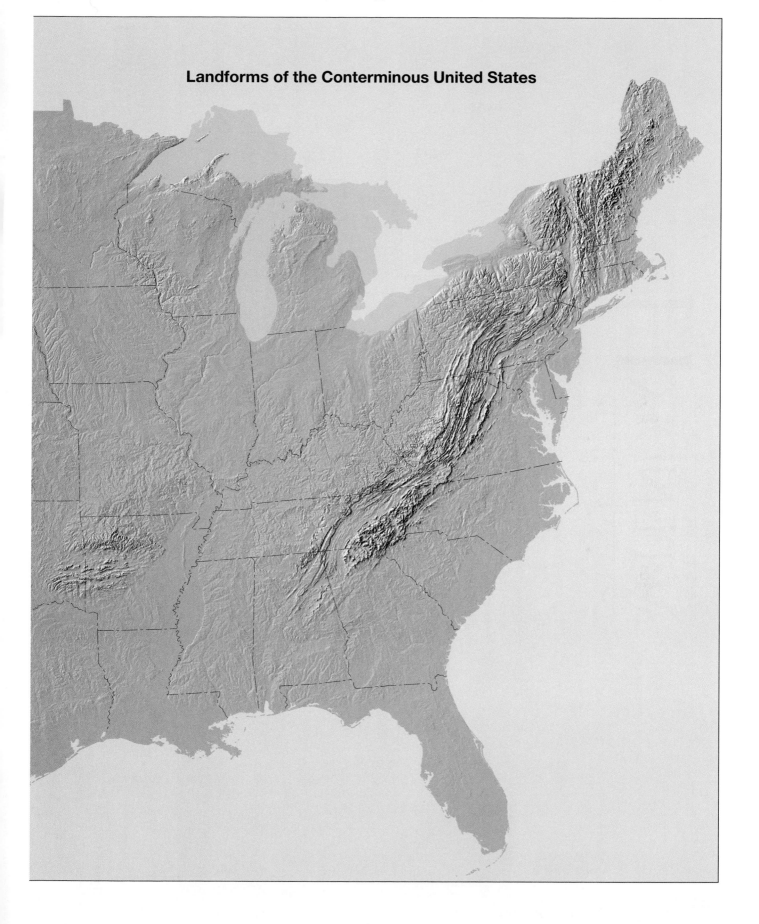

APPENDIX H

Atomic number	2
Symbol of element	**He**
Atomic weight	4.003
Name of element	Helium

- Metals
- Transition metals
- Nonmetals
- Noble gases
- Lanthanide series
- Actinide series

IA

1
H
1.0080
Hydrogen

IIA

3
Li
6.939
Lithium

4
Be
9.012
Beryllium

11
Na
22.990
Sodium

12
Mg
24.31
Magnesium

19
K
39.102
Potassium

20
Ca
40.08
Calcium

37
Rb
85.47
Rubidium

38
Sr
87.62
Strontium

55
Cs
132.91
Cesium

56
Ba
137.34
Barium

87
Fr
(223)
Francium

88
Ra
226.05
Radium

III B

21
Sc
44.96
Scandium

39
Y
88.91
Yttrium

57
TO
71

89
TO
103

IV B

22
Ti
47.90
Titanium

40
Zr
91.22
Zirconium

72
Hf
178.49
Hafnium

V B

23
V
50.94
Vanadium

41
Nb
92.91
Niobium

73
Ta
180.95
Tantalum

VI B

24
Cr
52.00
Chromium

42
Mo
95.94
Molybdenum

74
W
183.85
Tungsten

VII B

25
Mn
53.94
Manganese

43
Tc
(99)
Technetium

75
Re
186.2
Rhenium

VIII B

26
Fe
55.85
Iron

44
Ru
101.1
Ruthenium

76
Os
190.2
Osmium

27
Co
58.93
Cobalt

45
Rh
102.90
Rhodium

77
Ir
192.2
Iridium

28
Ni
58.71
Nickel

46
Pd
106.4
Palladium

78
Pt
195.09
Platinum

I B

29
Cu
63.54
Copper

47
Ag
107.87
Silver

79
Au
197.0
Gold

II B

30
Zn
65.37
Zinc

48
Cd
112.40
Cadmium

80
Hg
200.59
Mercury

III A

5
B
10.81
Boron

13
Al
26.98
Aluminum

31
Ga
69.72
Gallium

49
In
114.82
Indium

81
Tl
204.37
Thallium

IV A

6
C
12.011
Carbon

14
Si
28.09
Silicon

32
Ge
72.59
Germanium

50
Sn
118.69
Tin

82
Pb
207.19
Lead

V A

7
N
14.007
Nitrogen

15
P
30.974
Phosphorus

33
As
74.92
Arsenic

51
Sb
121.75
Antimony

83
Bi
208.98
Bismuth

VI A

8
O
15.9994
Oxygen

16
S
32.064
Sulfur

34
Se
78.96
Selenium

52
Te
127.60
Tellurium

84
Po
(210)
Polonium

VII A

9
F
18.998
Fluorine

17
Cl
35.453
Chlorine

35
Br
79.909
Bromine

53
I
126.90
Iodine

85
At
(210)
Astatine

VIII A

2
He
4.003
Helium

10
Ne
20.183
Neon

18
Ar
39.948
Argon

36
Kr
83.80
Krypton

54
Xe
131.30
Xenon

86
Rn
(222)
Radon

57
LA
138.91
Lanthanum

58
Ce
140.12
Cerium

59
Pr
140.91
Praseodymium

60
Nd
144.24
Neodymium

61
Pm
(147)
Promethium

62
Sm
150.35
Samarium

63
Eu
151.96
Europium

64
Gd
157.25
Gadolinium

65
Tb
158.92
Terbium

66
Dy
162.50
Dysprosium

67
Ho
164.93
Holmium

68
Er
167.26
Erbium

69
Tm
168.93
Thulium

70
Yb
173.04
Ytterbium

71
Lu
174.97
Lutetium

89
Ac
(227)
Actinium

90
Th
232.04
Thorium

91
Pa
(231)
Protactinium

92
U
238.03
Uranium

93
Np
(237)
Neptunium

94
Pu
(242)
Plutonium

95
Am
(243)
Americium

96
Cm
(247)
Curium

97
Bk
(249)
Berkelium

98
Cf
(251)
Californium

99
Es
(254)
Einsteinium

100
Fm
(253)
Fermium

101
Md
(256)
Mendelevium

102
No
(254)
Nobelium

103
Lw
(257)
Lawrencium

Figure H.1
Periodic table of the elements.

GLOSSARY

Aa A type of lava flow that has a jagged blocky surface.

Abrasion The grinding and scraping of a rock surface by the friction and impact of rock particles carried by water, wind, or ice.

Absolute dating Determination of the number of years since the occurrence of a given geologic event.

Absolute humidity The weight of water vapor in a given volume of air (usually expressed in grams/m^3).

Absolute instability Air that has a lapse rate greater than the dry adiabatic rate.

Absolute magnitude The apparent brightness of a star if it were viewed from a distance of 10 parsecs (32.6 light-years). Used to compare the true brightness of stars.

Absolute stability Air with a lapse rate less than the wet adiabatic rate.

Absorption spectrum A continuous spectrum with dark lines superimposed.

Abyssal plain Very level area of the deep-ocean floor, usually lying at the foot of the continental rise.

Accretionary wedge A large wedge-shaped mass of sediment that accumulates in subduction zones. Here sediment is scraped from the subducting oceanic plate and accreted to the overriding crustal block.

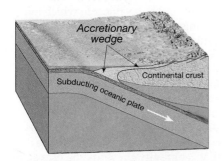

Acid precipitation Rain or snow with a pH value that is less than the pH of unpolluted precipitation.

Adiabatic temperature change Cooling or warming of air caused when air is allowed to expand or is compressed, not because heat is added or subtracted.

Advection Horizontal convective motion, such as wind.

Advection fog A fog formed when warm, moist air is blown over a cool surface.

Aftershocks Smaller earthquakes that follow the main earthquake.

Air A mixture of many discrete gases, of which nitrogen and oxygen are most abundant, in which varying quantities of tiny solid and liquid particles are suspended.

Air mass A large body of air that is characterized by a sameness of temperature and humidity.

Air-mass weather The conditions experienced in an area as an air mass passes over it. Because air masses are large and fairly homogenous, air-mass weather will be fairly constant and may last for several days.

Air pollutants Airborne particles and gases that occur in concentrations that endanger the health and well-being of organisms or disrupt the orderly functioning of the environment.

Albedo The reflectivity of a substance, usually expressed as a percentage of the incident radiation reflected.

Alluvial fan A fan-shaped deposit of sediment formed when a stream's slope is abruptly reduced.

Alluvium Unconsolidated sediment deposited by a stream.

Alpine glacier A glacier confined to a mountain valley, which in most instances had previously been a stream valley.

Altitude (of the sun) The angle of the sun above the horizon.

Anemometer An instrument used to determine wind speed.

Aneroid barometer An instrument for measuring air pressure that consists of evacuated metal chambers that are very sensitive to variations in air pressure.

Angle of repose The steepest angle at which loose material remains stationary without sliding downslope.

Angular unconformity An unconformity in which the strata below dip at an angle different from that of the beds above.

Annual mean temperature An average of the twelve monthly temperature means.

Annual temperature range The difference between the highest and lowest monthly means.

Anthracite A hard, metamorphic form of coal that burns clean and hot.

Anticline A fold in sedimentary strata that resembles an arch.

615

Anticyclone A high-pressure center characterized by a clockwise flow of air in the Northern Hemisphere.

Aphelion The place in the orbit of a planet where the planet is farthest from the sun.

Apparent magnitude The brightness of a star when viewed from Earth.

Aquicludes Impermeable beds that hinder or prevent groundwater movement.

Aquifer Rock or soil through which groundwater moves easily.

Archean eon The second eon of Precambrian time, following the Hadean and preceding the Proterozoic. It extends between 3.8 and 2.5 billion years before the present.

Arête A narrow knifelike ridge separating two adjacent glaciated valleys.

Arid *See* Desert.

Arkose A feldspar-rich sandstone.

Artesian well A well in which the water rises above the level where it was initially encountered.

Asteroids Thousands of small planetlike bodies, ranging in size from a few hundred kilometers to less than a kilometer, whose orbits lie mainly between those of Mars and Jupiter.

Asthenosphere A subdivision of the mantle situated below the lithosphere. This zone of weak material exists below a depth of about 100 kilometers and in some regions extends as deep as 700 kilometers. The rock within this zone is easily deformed.

Astronomical theory A theory of climatic change first developed by the Yugoslavian astronomer Milankovitch. It is based upon changes in the shape of Earth's orbit, variations in the obliquity of Earth's axis, and the wobbling of Earth's axis.

Astronomical Unit (AU) Average distance from Earth to the sun; 1.5×10^8 km or 93×10^6 miles.

Astronomy The scientific study of the universe; it includes the observation and interpretation of celestial bodies and phenomena.

Atmosphere The gaseous portion of a planet; the planet's envelope of air. One of the traditional subdivisions of Earth's physical environment.

Atoll A continuous or broken ring of coral reef surrounding a central lagoon.

Atom The smallest particle that exists as an element.

Atomic number The number of protons in the nucleus of an atom.

Atomic weight The average of the atomic masses of isotopes for a given element.

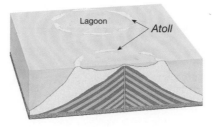

Aurora A bright display of everchanging light caused by solar radiation interacting with the upper atmosphere in the region of the poles.

Autumnal equinox The equinox that occurs on September 21–23 in the Northern Hemisphere and on March 21–22 in the Southern Hemisphere.

Azoic zone A well-known but incorrect theory formulated about 1850 by Edward Forbes stating that no life existed in the ocean below a depth of about 550 meters.

Backswamp A poorly drained area on a floodplain that results when natural levees are present.

Barchan dune Solitary sand dune shaped like a crescent with its tips pointing downward.

Barchanoid dune Dunes forming scalloped rows of sand oriented at right angles to the wind. This form is intermediate between isolated barchans and extensive waves of transverse dunes.

Barograph A recording barometer.

Barometer An instrument that measures atmospheric pressure.

Barometric tendency *See* Pressure tendency.

Barred spiral A galaxy having straight arms extending from its nucleus.

Barrier island A low, elongate ridge of sand that parallels the coast.

Basalt A fine-grained igneous rock of mafic composition.

Base level The level below which a stream cannot erode.

Basin A circular downfolded structure.

Batholith A large mass of igneous rock that formed when magma was emplaced at depth, crystallized, and subsequently exposed by erosion.

Baymouth bar A sandbar that completely crosses a bay, sealing it off from the open ocean.

Beach drift The transport of sediment in a zigzag pattern along a beach caused by the uprush of water from obliquely breaking waves.

Beach nourishment Large quantities of sand are added to the beach system to offset losses caused by wave erosion.

Bed load Sediment that is carried by a stream along the bottom of its channel.

Benioff zone Zone of inclined seismic activity that extends from a trench downward into the asthenosphere.

Bergeron process A theory that relates the formation of precipitation to supercooled clouds, freezing nuclei, and the different saturation levels of ice and liquid water.

Big Bang theory The theory that proposes that the universe originated as a single mass, which subsequently exploded.

Binary stars Two stars revolving around a common center of mass under their mutual gravitational attraction.

Biogenous sediment Seafloor sediments consisting of material of marine-organic origin.

Biosphere The totality of life on Earth; the parts of the solid Earth, hydrosphere, and atmosphere in which living organisms can be found.

Bituminous The most common form of coal, often called soft, black coal.

Black dwarf A final state of evolution for a star, in which all of its energy sources are exhausted and it no longer emits radiation.

Black hole A massive star that has collapsed to such a small volume that its gravity prevents the escape of all radiation.

Blowout (deflation hollow) A depression excavated by the wind in easily eroded deposits.

Bode's law A sequence of numbers that approximates the mean distances of the planets from the sun.

Body waves Seismic waves that travel through Earth's interior.

Braided stream A stream consisting of numerous intertwining channels.

Breakwater A structure protecting a near-shore area from breaking waves.

Breccia A sedimentary rock composed of angular fragments that were lithified.

Bright-line spectrum The bright lines produced by an incandescent gas under low pressure.

Bright nebula A cloud of glowing gas excited by ultraviolet radiation from hot stars.

Cactolith A quasi-horizontal chonolith composed of anastomosing ductoliths, whose distal ends curl like a harpolith, thin like a sphenolith, or bulge discordantly like an akmolith or ethmolith.

Caldera A large depression typically caused by collapse or ejection of the summit area of a volcano.

Calorie The amount of heat required to raise the temperature of one gram of water 1°C.

Calving Wastage of a glacier that occurs when large pieces of ice break off into water.

Capacity The total amount of sediment a stream is able to transport.

Cassini division A wide gap in the ring system of Saturn between the A ring and the B ring.

Catastrophism The concept that Earth was shaped by catastrophic events of a short-term nature.

Cavern A naturally formed underground chamber or series of chambers most commonly produced by solution activity in limestone.

Celestial sphere An imaginary hollow sphere upon which the ancients believed the stars were hung and carried around Earth.

Cenozoic era A time span on the geologic calendar beginning about 66 million years ago following the Mesozoic era.

Cepheid variable A star whose brightness varies periodically because it expands and contracts. A type of pulsating star.

Chemical sedimentary rock Sedimentary rock consisting of material that

was precipitated from water by either inorganic or organic means.

Chemical weathering The processes by which the internal structure of a mineral is altered by the removal and/or addition of elements.

Chinook A wind blowing down the leeward side of a mountain and warming by compression.

Chromatic aberration The property of a lens whereby light of different colors is focused at different places.

Chromosphere The first layer of the solar atmosphere found directly above the photosphere.

Cinder cone A rather small volcano built primarily of pyroclastics ejected from a single vent.

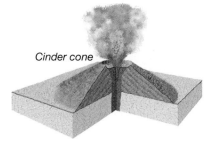

Cinder cone

Circle of illumination The great circle that separates daylight from darkness.

Cirque An amphitheater-shaped basin at the head of a glaciated valley produced by frost wedging and plucking.

Cirrus One of three basic cloud forms; also one of the three high cloud types. They are thin, delicate ice crystal clouds often appearing as veil-like patches or thin, wispy fibers.

Clastic rock A sedimentary rock made of broken fragments of pre-existing rock.

Cleavage The tendency of a mineral to break along planes of weak bonding.

Climate A description of aggregate weather conditions; the sum of all statistical weather information that helps describe a place or region.

Climate System The exchanges of energy and moisture that occur among the atmosphere, hydrosphere, solid earth, biosphere, and cryosphere.

Climatic feedback mechanism One of the several different outcomes that may result if one of the many elements in the atmosphere's extremely complex interactive system is altered.

Climatology The scientific study of climate.

Cloud A form of condensation best described as a dense concentration

of suspended water droplets or tiny ice crystals.

Clouds of vertical development A cloud that has its base in the low height range but extends upward into the middle or high altitudes.

Cluster (star) A large group of stars.

Coarse-grained texture An igneous rock texture in which the crystals are roughly equal in size and large enough so that individual minerals can be identified with the unaided eye.

Col A pass between mountain valleys where the headwalls of two cirques intersect.

Cold front A front along which a cold air mass thrusts beneath a warmer air mass.

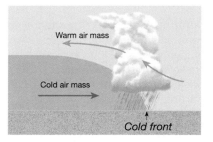

Warm air mass

Cold air mass

Cold front

Collision-coalescence process A theory of raindrop formation in warm clouds (above 0°C) in which large cloud droplets ("giants") collide and join together with smaller droplets to form a raindrop. Opposite electrical charges may bind the cloud droplets together.

Column A feature found in caves that is formed when a stalactite and stalagmite join.

Columnar joints A pattern of cracks that form during cooling of molten rock to generate columns that are generally six-sided.

Coma The fuzzy, gaseous component of a comet's head.

Comet A small body which generally revolves about the sun in an elongated orbit.

Competence A measure of the largest particle a stream can transport; a factor dependent on velocity.

Composite cone A volcano composed of both lava flows and pyroclastic material.

Compound A substance formed by the chemical combination of two or more elements in definite proportions and usually having properties different than those of its constituent elements.

Condensation The change of state from a gas to a liquid.

Condensation nuclei Tiny bits of particulate matter that serve as surfaces on which water vapor condenses.

Conditional instability Moist air with a lapse rate between the dry and wet adiabatic rates.

Conduction The transfer of heat through matter by molecular activity. Energy is transferred through collisions from one molecule to another.

Cone of depression A cone-shaped depression in the water table immediately surrounding a well.

Conformable Layers of rock that were deposited without interruption.

Conglomerate A sedimentary rock composed of rounded gravel-sized particles.

Constellation An apparent group of stars originally named for mythical characters. The sky is presently divided into 88 constellations.

Contact metamorphism Changes in rock caused by the heat from a nearby magma body.

Continental (c) air mass An air mass that forms over land; it is normally relatively dry.

Continental drift theory A theory that originally proposed that the continents are rafted about. It has essentially been replaced by the plate tectonics theory.

Continental margin That portion of the sea floor adjacent to the continents. It may include the continental shelf, continental slope, and continental rise.

Continental rise The gently sloping surface at the base of the continental slope.

Continental shelf The gently sloping submerged portion of the continental margin extending from the shoreline to the continental slope.

Continental slope The steep gradient that leads to the deep-ocean floor and marks the seaward edge of the continental shelf.

Continuous spectrum An uninterrupted band of light emitted by an incandescent solid, liquid, or gas under pressure.

Convection The transfer of heat by the movement of a mass or substance. It can take place only in fluids.

Convergence The condition that exists when the distribution of winds within a given area results in a net horizontal inflow of air into the area. Since convergence at lower levels is associated with an upward movement of air, areas of convergent winds are regions favorable to cloud formation and precipitation.

Convergent boundary A boundary in which two plates move together, causing one of the slabs of lithosphere to be consumed into the mantle as it descends beneath on overriding plate.

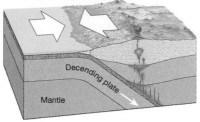

Coral reef Structure formed in a warm, shallow, sunlit ocean environment that consists primarily of the calcite-rich remains of corals as well as the limy secretions of algae and the hard parts of many other small organisms.

Core Located beneath the mantle, it is the innermost layer of Earth. The core is divided into an outer core and an inner core.

Coriolis force (effect) The deflective force of Earth's rotation on all free-moving objects, including the atmosphere and oceans. Deflection is to the right in the Northern Hemisphere and to the left in the Southern Hemisphere.

Corona The outer, tenuous layer of the solar atmosphere.

Correlation Establishing the equivalence of rocks of similar age in different areas.

Crater The depression at the summit of a volcano, or that which is produced by a meteorite impact.

Creep The slow downhill movement of soil and regolith.

Crevasse A deep crack in the brittle surface of a glacier.

Cross-bedding Structure in which relatively thin layers are inclined at an angle to the main bedding. Formed by currents of wind or water.

Cross-cutting A principle of relative dating. A rock or fault is younger than any rock (or fault) through which it cuts.

Crust The very thin outermost layer of Earth.

Crystal An orderly arrangement of atoms.

Crystal form The external appearance of a mineral as determined by its internal arrangement of atoms.

Crystallization The formation and growth of a crystalline solid from a liquid or gas.

Cumulus One of three basic cloud forms; also the name given one of the clouds of vertical development. Cumulus are billowy individual cloud masses that often have flat bases.

Cup anemometer *See* Anemometer

Curie point The temperature above which a material loses its magnetization.

Cutoff A short channel segment created when a river erodes through the narrow neck of land between meanders.

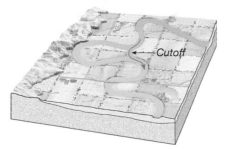

Cyclone A low-pressure center characterized by a counterclockwise flow of air in the Northern Hemisphere.

Daily mean The mean temperature for a day that is determined by averaging the 24 hourly readings or, more commonly, by averaging the maximum and minimum temperatures for a day.

Daily temperature range The difference between the maximum and minimum temperatures for a day.

Dark-line spectrum *See* Absorption spectrum.

Dark nebula A cloud of interstellar dust that obscures the light of more distant stars and appears as an opaque curtain.

Daughter product An isotope resulting from radioactive decay.

Declination (stellar) The angular distance north or south of the celestial equator denoting the position of a celestial body.

Deep-ocean trench A narrow, elongated depression on the floor of the ocean.

Deep-sea fan A cone-shaped deposit at the base of the continental slope. The sediment is transported to the fan by turbidity currents that follow submarine canyons.

Deflation The lifting and removal of loose material by wind.

Delta An accumulation of sediment formed where a stream enters a lake or ocean.

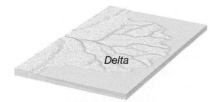

Delta

Dendritic pattern A stream system that resembles the pattern of a branching tree.

Dendritic pattern

Density The weight per unit volume of a particular material.

Deposition The process by which water vapor is changed directly to a solid without passing through the liquid state.

Desalination The removal of salts and other chemicals from seawater.

Desert One of the two types of dry climate; the driest of the dry climates.

Desert pavement A layer of coarse pebbles and gravel created when wind removed the finer material.

Detrital sedimentary rock Rock formed from the accumulation of material that originated and was transported in the form of solid particles derived from both mechanical and chemical weathering.

Dew point The temperature to which air has to be cooled in order to reach saturation.

Differential weathering The variation in the rate and degree of weathering caused by such factors as mineral makeup, degree of jointing, and climate.

Diffused light Solar energy scattered and reflected in the atmosphere that reaches Earth's surface in the form of diffuse blue light from the sky.

Dike A tabular-shaped intrusive igneous feature that cuts through the surrounding rock.

Dip-slip fault A fault in which the movement is parallel to the dip of the fault.

Discharge The quantity of water in a stream that passes a given point in a period of time.

Disconformity A type of unconformity in which the beds above and below are parallel.

Disseminated deposit Any economic mineral deposit in which the desired mineral occurs as scattered particles in the rock but in sufficient quantity to make the deposit an ore.

Dissolved load That portion of a stream's load carried in solution.

Distributary A section of a stream that leaves the main flow.

Diurnal tide Tides characterized by a single high and low water height each tidal day.

Divergence The condition that exists when the distribution of winds within a given area results in a net horizontal outflow of air from the region. In divergence at lower levels the resulting deficit is compensated for by a downward movement of air from aloft; hence, areas of divergent winds are unfavorable to cloud formation and precipitation.

Divergent boundary A region where the rigid plates are moving apart, typified by the mid-oceanic ridges.

Divergent boundary

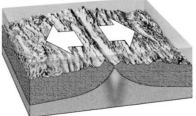

Divide An imaginary line that separates the drainage of two streams; often found along a ridge.

Dome A roughly circular upfolded structure similar to an anticline.

Doppler effect The apparent change in wavelength of radiation caused by the relative motions of the source and the observer.

Doppler radar In addition to the tasks performed by conventional radar, this new generation of weather radar can detect motion directly and hence greatly improve tornado and severe storm warnings.

Drainage basin The land area that contributes water to a stream.

Drawdown The difference in height between the bottom of a cone of depression and the original height of the water table.

Drift The general term for any glacial deposit.

Drumlim A streamlined asymmetrical hill composed of glacial till. The steep side of the hill faces the direction from which the ice advanced.

Dry adiabatic rate The rate of adiabatic cooling or warming in unsaturated air. The rate of temperature change is 1°C per 100 meters.

Dry climate A climate in which yearly precipitation is not as great as the potential loss of water by evaporation.

Dry-summer subtropical climate A climate located on the west sides of continents between latitudes 30 and 45°. It is the only humid climate with a strong winter precipitation maximum.

Dune A hill or ridge of wind-deposited sand.

Earthflow The downslope movement of water-saturated, clay-rich sediment. Most characteristic of humid regions.

Earthquake The vibration of the earth produced by the rapid release of energy.

Ebb current The movement of a tidal current away from the shore.

Eccentricity The variation of an ellipse from a circle.

Echo sounder An instrument used to determine the depth of water by measuring the time interval between emission of a sound signal and the return of its echo from the bottom.

Eclipse The cutting off of the light of one celestial body by another passing in front of it.

Ecliptic The yearly path of the sun plotted against the background of stars.

Elastic rebound The sudden release of stored strain in rocks that results in movement along a fault.

Electromagnetic radiation *See* Radiation.

Electromagnetic spectrum The distribution of electromagnetic radiation by wavelength.

Electron A negatively charged subatomic particle that has a negligible mass and is found outside an atom's nucleus.

Element A substance that cannot be decomposed into simpler substances by ordinary chemical or physical means.

Elements of weather and climate Those quantities or properties of the atmosphere that are measured regularly and that are used to express the nature of weather and climate.

Elliptical galaxy A galaxy that is round or elliptical in outline. It contains little gas and dust, no disk or spiral arms, and few hot, bright stars.

Eluviation The washing out of fine soil components from the A horizon by downward-percolating water.

Emergent coast A coast where land that was formerly below sea level has been exposed either because of crustal uplift or a drop in sea level or both.

Emission nebula A gaseous nebula that derives its visible light from the fluorescence of ultraviolet light from a star in or near the nebula.

End moraine A ridge of till marking a former position of the front of a glacier.

Entrenched meander A meander cut into bedrock when uplifting rejuvenated a meandering stream.

Environmental lapse rate The rate of temperature decrease with increasing height in the troposphere.

Eon The largest time unit on the geologic time scale, next in order of magnitude above era.

Ephermeral stream A stream that is usually dry because it carries water only in response to specific episodes of rainfall. Most desert streams are of this type.

Epicenter The location on Earth's surface that lies directly above the focus of an earthquake.

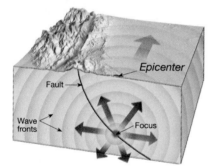

Epoch A unit of the geologic calendar that is a subdivision of a period.

Equatorial low A belt of low pressure lying near the equator and between the subtropical highs.

Equatorial system A method of locating stellar objects much like the coordinate system used on Earth's surface.

Equinox The time when the vertical rays of the sun are striking the equator. The length of daylight and darkness is equal at all latitudes at equinox.

Era A major division on the geologic calendar; eras are divided into shorter units called periods.

Erosion The incorporation and transportation of material by a mobile agent, such as water, wind, or ice.

Eruptive variable A star that varies in brightness.

Escape velocity The initial velocity an object needs to escape from the surface of a celestial body.

Esker Sinuous ridge composed largely of sand and gravel deposited by a stream flowing in a tunnel beneath a glacier near its terminus.

Estuary A funnel-shaped inlet of the sea that formed when a rise in sea level or subsidence of land caused the mouth of a river to be flooded.

Evaporation The process of converting a liquid to a gas.

Evaporite A sedimentary rock formed of material deposited from solution by evaporation of the water.

Evolution, (Theory of) A fundamental theory in biology and paleontology that sets forth the process by which members of a population of organisms come to differ from their ancestors. Organisms evolve by means of mutations, natural selection, and genetic factors. Modern species are descended from related but different species that lived in earlier times.

Exfoliation dome Large, dome-shaped structure, usually composed of granite, formed by sheeting.

Exotic stream A permanent stream that traverses a desert and has its source in well-watered areas outside the desert.

Extrusive Igneous activity that occurs outside the crust.

Eye A zone of scattered clouds and calm averaging about 20 kilometers in diameter at the center of a hurricane.

Eyepiece A short-focal-length lens used to enlarge the image in a telescope. The lens nearest the eye.

Eye wall The doughnut-shaped area of intense cumulonimbus development and very strong winds that surrounds the eye of a hurricane.

Fall A type of movement common to mass wasting processes that refers to the free falling of detached individual pieces of any size.

Fault A break in a rock mass along which movement has occurred.

Fault-block mountain A mountain formed by the displacement of rock along a fault.

Fault creep Displacement along a fault that is so slow and gradual that little seismic activity occurs.

Felsic The group of igneous rocks composed primarily of feldspar and quartz.

Fetch The distance that the wind has traveled across the open water.

Filaments Dark, thin streaks that appear across the bright solar disk.

Fine-grained texture A texture of igneous rocks in which the crystals are too small for individual minerals to be distinguished with the unaided eye.

Fiord A steep-sided inlet of the sea formed when a glacial trough was partially submerged.

Fissure eruption An eruption in which lava is extruded from narrow fractures or cracks in the crust.

Flare A sudden brightening of an area on the sun.

Flood basalts Flows of basaltic lava that issue from numerous cracks or fissures and commonly cover extensive areas to thicknesses of hundreds of meters.

Flood current The tidal current associated with the increase in the height of the tide.

Floodplain The flat, low-lying portion of a stream valley subject to periodic inundation.

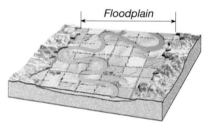

Flow A type of movement common to mass wasting processes in which water-saturated material moves downslope as a viscous fluid.

Fluorescence The absorption of ultraviolet light, which is re-emitted as visible light.

Focal length The focal length of a lens is the distance from the lens to the point where it focuses parallel rays of light.

Focus (earthquake) The zone within Earth where rock displacement produces an earthquake.

Focus (light) The point where a lens or mirror causes light rays to converge.

Fog A cloud with its base at or very near Earth's surface.

Fold A bent rock layer or series of layers that were originally horizontal and subsequently deformed.

Foliated A texture of metamorphic rocks that gives the rock a layered appearance.

Foreshocks Small earthquakes that often precede a major earthquake.

Fossil fuel General term for any hydrocarbon that may be used as a fuel, including coal, oil, and natural gas.

Fossils The remains or traces of organisms preserved from the geologic past.

Fossil succession Fossil organisms succeed one another in a definite and determinable order, and any time period can be recognized by its fossil content.

Fracture Any break or rupture in rock along which no appreciable movement has taken place.

Freezing The change of state from a liquid to a solid.

Freezing nuclei Solid particles that serve as cores for the formation of ice crystals.

Front The boundary between two adjoining air masses having contrasting characteristics.

Frontal fog Fog formed when rain evaporates as it falls through a layer of cool air.

Frontal wedging Lifting of air resulting when cool air acts as a barrier over which warmer, lighter air will rise.

Frost wedging The mechanical breakup of rock caused by the expansion of freezing water in cracks and crevices.

Galactic cluster A system of galaxies containing from several to thousands of member galaxies.

Geocentric The concept of an Earth-centered universe.

Geologic time scale The division of Earth history into blocks of time—eons, eras, periods, and epochs. The time scale was created using relative dating principles.

Geology The science that examines Earth, its form and composition, and the changes it has undergone and is undergoing.

Geostrophic wind A wind, usually above a height of 600 meters (2000 feet), that blows parallel to the isobars.

Geothermal energy Natural steam used for power generation.

Geyser A fountain of hot water ejected periodically.

Giant (star) A luminous star of large radius.

Glacial erratic An ice-transported boulder that was not derived from bedrock near its present site.

Glacial striations Scratches and grooves on bedrock caused by glacial abrasion.

Glacial trough A mountain valley that has been widened, deepened, and straightened by a glacier.

Glacier A thick mass of ice originating on land from the compaction and re-crystallization of snow that shows evidence of past or present flow.

Glassy A term used to describe the texture of certain igneous rocks, such as obsidian, that contain no crystals.

Glaze A coating of ice on objects formed when super-cooled rain freezes on contact.

Globular cluster A nearly spherically shaped group of densely packed stars.

Globule A dense, dark nebula thought to be the birthplace of stars.

Gondwanaland The southern portion of Pangaea consisting of South America, Africa, Australia, India, and Antarctica.

Graben A valley formed by the downward displacement of a fault-bounded block.

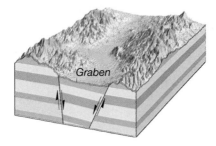

Graded bed A sediment layer that is characterized by a decrease in sediment size from bottom to top.

Gradient The slope of a stream; generally measured in feet per mile.

Granules The fine structure visible on the solar surface caused by convective cells below.

Greenhouse effect The transmission of short-wave solar radiation by the atmosphere coupled with the selective absorption of longer-wavelength terrestrial radiation, especially by water vapor and carbon dioxide.

Groin A short wall built at a right angle to the shore to trap moving sand.

Ground moraine An undulating layer of till deposited as the ice front retreats.

Groundwater Water in the zone of saturation.

Guyot A submerged flat-topped seamount.

Gyre The large circular surface current pattern found in each ocean.

Hadean eon The first eon on the geologic time scale; this eon ended 3.8 billion years ago and preceded the Archean eon.

Hail Nearly spherical ice pellets having concentric layers and formed by the successive freezing of layers of water.

Half-life The time required for one-half of the atoms of a radioactive substance to decay.

Halocline A layer of water in which there is a high rate of change in salinity in the vertical dimension.

Hanging valley A tributary valley that enters a glacial trough at a considerable height above its floor.

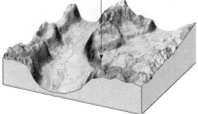

Hanging valley

Hardness The resistance a mineral offers to scratching.

Heliocentric The view that the sun is at the center of the solar system.

Hertzsprung-Russell diagram *See* H-R diagram.

High cloud A cloud that normally has its base above 6000 meters; the base may be lower in winter and at high-latitude locations.

Highland climate Complex pattern of climate conditions associated with mountains. Highland climates are characterized by large differences that occur over short distances.

Hogback A narrow, sharp-crested ridge formed by the upturned edge of a steeply dipping bed of resistant rock.

Horizon A layer in a soil profile.

Horn A pyramidlike peak formed by glacial action in three or more cirques surrounding a mountain summit.

Horst An elongate, uplifted block of crust bounded by faults.

Horst

Hot spot A concentration of heat in the mantle capable of producing magma which, in turn, extrudes onto Earth's surface. The intraplate volcanism that produced the Hawaiian Islands is one example.

Hot spring A spring in which the water is 6°–9°C (10°–15°F) warmer than the mean annual air temperature of its locality.

H-R diagram A plot of stars according to their absolute magnitudes and spectral types.

Hubble law Relates the distance to a galaxy and its velocity.

Humid continental climate A relatively severe climate characteristic of broad continents in the middle latitudes between approximately 40 and 50° north latitude. This climate is not found in the southern hemisphere, where the middle latitudes are dominated by the oceans.

Humidity A general term referring to water vapor in the air but not to liquid droplets of fog, cloud, or rain.

Humid subtropical climate A climate generally located on the eastern side of a continent and characterized by hot, sultry summers and cool winters.

Humus Organic matter in soil produced by the decomposition of plants and animals.

Hurricane A tropical cyclonic storm having winds in excess of 119 kilometers (74 miles) per hour.

Hydrogen burning The conversion of hydrogen through fusion to form helium.

Hydrogenous sediment Seafloor sediments consisting of minerals that crystallize from seawater. An important example is manganese nodules.

Hydrosphere The water portion of our planet; one of the traditional subdivisions of Earth's physical environment.

Hydrothermal solution The hot, watery solution that escapes from a mass of magma during the later stages of crystallization. Such solutions may alter the surrounding country rock and are frequently the source of significant ore deposits.

Hygrometer An instrument designed to measure relative humidity.

Hygroscopic nuclei Condensation nuclei having a high affinity for water, such as salt particles.

Hypothesis A tentative explanation that is tested to determine if it is valid.

Ice cap A mass of glacial ice covering a high upland or plateau and spreading out radially.

Ice cap climate A climate that has no monthly means above freezing and supports no vegetative cover except in a few scattered high mountain areas. This climate, with its perpetual ice and

snow, is confined largely to the ice sheets of Greenland and Antarctica.

Ice sheet A very large, thick mass of glacial ice flowing outward in all directions from one or more accumulation centers.

Igneous rock A rock formed by the crystallization of molten magma.

Immature soil A soil lacking horizons.

Inclination of the axis The tilt of Earth's axis from the perpendicular to the plane of Earth's orbit.

Inclusion A piece of one rock unit contained within another. Inclusions are used in relative dating. The rock mass adjacent to the one containing the inclusion must have been there first in order to provide the fragment.

Index fossil A fossil that is associated with a particular span of geologic time.

Inertia A property of matter that resists a change in its motion.

Infiltration The movement of surface water into rock or soil through cracks and pore spaces.

Infrared Radiation with a wavelength from 0.7 to 200 micrometers.

Inner core The solid innermost layer of Earth, about 1300 kilometers (800 miles) in radius.

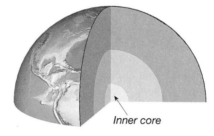

Inner core

Inselberg An isolated mountain remnant characteristic of the late stage of erosion in an arid region.

Intensity (earthquake) An indication of the destructive effects of an earthquake at a particular place. Intensity is affected by such factors as distance to the epicenter and the nature of the surface materials.

Interior drainage A discontinuous pattern of intermittent streams that do not flow to the ocean.

Intermediate composition The composition of igneous rocks lying between felsic and mafic.

Interstellar matter Dust and gases found between stars.

Intrusive rock Igneous rock that formed below Earth's surface.

Ion An atom or molecule that possesses an electrical charge.

Ionosphere A complex zone of ionized gases that coincides with the lower portion of the thermosphere.

Iron meteorite One of the three main categories of meteorites. This group is composed largely of iron with varying amounts of nickel (5–20 percent). Most meteorite finds are irons.

Irregular galaxy A galaxy that lacks symmetry.

Island arc A group of volcanic islands formed by the subduction and partial melting of oceanic lithosphere.

Isobar A line drawn on a map connecting points of equal atmospheric pressure, usually corrected to sea level.

Isostacy The concept that Earth's crust is "floating" in gravitational balance upon the material of the mantle.

Isotherms Lines connecting points of equal temperature.

Isotopes Varieties of the same element that have different mass numbers; their nuclei contain the same number of protons but different numbers of neutrons.

Jet stream Swift (120–240-kilometer per hour), high-altitude winds.

Jetties A pair of structures extending into the ocean at the entrance to a harbor or river that are built for the purpose of protecting against storm waves and sediment deposition.

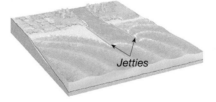

Jetties

Joint A fracture in rock along which there has been no movement.

Jovian planet The Jupiter-like planets Jupiter, Saturn, Uranus, and Neptune. These planets have relatively low densities.

Kame A steep-sided hill composed of sand and gravel originating when sediment collected in openings in stagnant glacial ice.

Karst A topography consisting of numerous depressions called sinkholes.

Kettle holes Depressions created when blocks of ice became lodged in glacial deposits and subsequently melted.

Koppen classification A system for classifying climates devised by Wladimir Koeppen that is based on mean monthly and annual values of temperature and precipitation.

Laccolith A massive igneous body intruded between pre-existing strata.

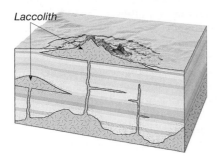
Laccolith

Lahar Mudflows on the slopes of volcanoes that result when unstable layers of ash and debris become saturated and flow downslope, usually following stream channels.

Lake-effect snow Snow showers associated with a cP air mass to which moisture and heat are added from below as the air mass traverses a large and relatively warm lake (such as one of the Great Lakes), rendering the air mass humid and unstable.

Land breeze A local wind blowing from land toward the water during the night in coastal areas.

Lapse rate (normal) The average drop in temperature (6.5°C per kilometer; 3.5°F per 1000 feet) with increased altitude in the troposphere.

Latent heat The energy absorbed or released during a change in state.

Lateral moraine A ridge of till along the sides of an alpine glacier composed primarily of debris that fell to the glacier from the valley walls.

Laterite A red, highly leached soil type found in the tropics that is rich in oxides of iron and aluminum.

Laurasia The northern portion of Pangaea consisting of North America and Eurasia.

Lava Magma that reaches Earth's surface.

Law A formal statement of the regular manner in which a natural phenomenon occurs under given conditions; e.g., the "law of superposition."

Law of conservation of angular momentum The product of the velocity of an object around a center of rotation (axis) and the distance squared of the object from the axis is constant.

Leaching The depletion of soluble materials from the upper soil by downward-percolating water.

Lightning A sudden flash of light generated by the flow of electrons between oppositely charged parts of a cumulonimbus cloud or between the cloud and the ground.

Light-year The distance light travels in a year; about 6 trillion miles.

Liquefaction A phenomenon, sometimes associated with earthquakes, in which soils and other unconsolidated materials containing abundant water are turned into a fluidlike mass that is not capable of supporting buildings.

Lithification The process, generally cementation and/or compaction, of converting sediments to solid rock.

Lithosphere The rigid outer layer of Earth, including the crust and upper mantle.

Local group The cluster of 20 or so galaxies to which our galaxy belongs.

Loess Deposits of wind-blown silt, lacking visible layers, generally buff colored, and capable of maintaining a nearly vertical cliff.

Longitudinal (seif dunes) Long ridges of sand oriented parallel to the prevailing wind; these dunes form where sand supplies are limited.

Longitudinal dunes

Longshore current A near-shore current that flows parallel to the shore.

Low cloud A cloud that forms below a height of 2000 meters.

Low-velocity zone *See* Asthenosphere.

Luminosity The brightness of a star. The amount of energy radiated by a star.

Lunar breccia A lunar rock formed when angular fragments and dust are welded together by the heat generated by the impact of a meteoroid.

Lunar eclipse An eclipse of the moon.

Lunar regolith A thin, gray layer on the surface of the moon, consisting of loosely compacted, fragmented material believed to have been formed by repeated meteoritic impacts.

Luster The appearance or quality of light reflected from the surface of a mineral.

Mafic Igneous rocks with a low silica content and a high iron-magnesium content.

Magma A body of molten rock found at depth, including any dissolved gases and crystals.

Magmatic differentiation The process of generating more than one rock type from a single magma.

Magnitude (earthquake) The total amount of energy released during an earthquake.

Magnitude (stellar) A number given to a celestial object to express its relative brightness.

Main sequence A sequence of stars on the Hertzsprung-Russell diagram, containing the majority of stars, that runs diagonally from the upper left to the lower right.

Manganese nodules Rounded lumps of hydrogenous sediment scattered on the ocean floor consisting mainly of manganese and iron, and usually containing small amounts of copper, nickel, and cobalt.

Mantle The 2900-kilometer (1800-mile)-thick layer of Earth located below the crust.

Maria The Latin name for the smooth areas of the moon formerly thought to be seas.

Marine west coast climate A climate found on windward coasts from latitudes 40 to 65° and dominated by maritime air masses. Winters are mild and summers are cool.

Maritime (m) air mass An air mass that originates over the ocean. These air masses are relatively humid.

Mass number The number of neutrons and protons in the nucleus of an atom.

Mass wasting The downslope movement of rock, regolith, and soil under the direct influence of gravity.

Meander A looplike bend in the course of a stream.

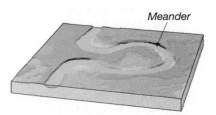
Meander

Mean solar day The average time between two passages of the sun across the local celestial meridian.

Mechanical weathering The physical disintegration of rock, resulting in smaller fragments.

Medial moraine A ridge of till formed when lateral moraines from two coalescing alpine glaciers join.

Melt The liquid portion of magma excluding the solid crystals.

Melting The change of state from a solid to a liquid.

Mercalli intensity scale A 12-point scale originally developed to evaluate earthquake intensity based on the amount of damage to various structures.

Mercury barometer A mercury-filled glass tube in which the height of the mercury column is a measure of air pressure.

Mesocyclone An intense, rotating wind system in the lower part of a thunderstorm that precedes tornado development.

Mesopause The boundary between the mesosphere and the thermosphere.

Mesosphere The layer of the atmosphere immediately above the stratosphere and characterized by decreasing temperatures with height.

Mesozoic era A time span on the geologic calendar between the Paleozoic and Cenozoic eras—from about 245 to 66.4 million years ago.

Metamorphic rock Rocks formed by the alteration of pre-existing rock deep within Earth (but still in the solid state) by heat, pressure, and/or chemically active fluids.

Metamorphism The changes in mineral composition and texture of a rock subjected to high temperature and pressure within Earth.

Meteor The luminous phenomenon observed when a meteoroid enters Earth's atmosphere and burns up; popularly called a "shooting star."

Meteorite Any portion of a meteoroid that survives its traverse through Earth's atmosphere and strikes Earth's surface.

Meteoroid Small solid particles that have orbits in the solar system.

Meteorology The scientific study of the atmosphere and atmospheric phenomena; the study of weather and climate.

Meteor shower Many meteors appearing in the sky caused when Earth intercepts a swarm of meteoritic particles.

Middle cloud A cloud occupying the height range from 2000 to 6000 meters.

Middle-latitude cyclone Large center of low pressure with an associated cold front and often a warm front. Frequently accompanied by abundant precipitation.

Mid-ocean ridge A continuous mountainous ridge on the floor of all the major ocean basins and varying in width from 500 to 5000 kilometers (300 to 3000 miles). The rifts at the crests of these ridges represent divergent plate boundaries.

Mineral A naturally occurring, inorganic crystalline material with a unique chemical composition.

Mineral resource All discovered and undiscovered deposits of a useful mineral that can be extracted now or at some time in the future.

Mixed tide A tide characterized by a large inequality in successive high water heights, low water heights, or both.

Mixing depth The height to which convectional movements extend above Earth's surface. The greater the mixing depth, the better the air quality.

Mohorovičić discontinuity (Moho) The boundary separating the crust from the mantle, discernible by an increase in seismic velocity.

Mohs scale A series of ten minerals used as a standard in determining hardness.

Monsoon Seasonal reversal of wind direction associated with large continents, especially Asia. In winter, the wind blows from land to sea; in summer, from sea to land.

Monthly mean temperature The mean temperature for a month that is calculated by averaging the daily means.

Mountain breeze The nightly downslope winds commonly encountered in mountain valleys.

Mudflow The flow of debris containing a large amount of water; most characteristic of canyons and gullies in dry, mountainous regions.

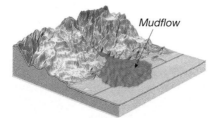

Mudflow

Nansen bottle A device used by oceanographers to obtain subsurface seawater samples.

Natural leeves The elevated landforms that parallel some streams and act to confine their waters, except during floodstage.

Neap tide Lowest tidal range, occurring near the times of the first- and third-quarter phases of the moon.

Nebula A cloud of interstellar gas and/or dust.

Nebular hypothesis The basic idea that the sun and planets formed from the same cloud of gas and dust in interstellar space.

Negative-feedback mechanism As used in climatic change, any effect that is opposite of the initial change and tends to offset it.

Neutron A subatomic particle found in the nucleus of an atom. The neutron is electrically neutral and has a mass approximately that of a proton.

Neutron star A star of extremely high density composed entirely of neutrons.

Nonconformity An unconformity in which older metamorphic or intrusive igneous rocks are overlain by younger sedimentary strata.

Nonfoliated Metamorphic rocks that do not exhibit foliation.

Nonmetallic mineral resource Mineral resource that is not a fuel or processed for the metals it contains.

Nonrenewable resource Resource that forms or accumulates over such long time spans that it must be considered as fixed in total quantity.

Normal fault A fault in which the rock above the fault plane has moved down relative to the rock below.

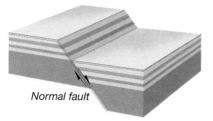

Normal fault

Normal polarity A magnetic field that is the same as that which exists at present.

Nova A star that explosively increases in brightness.

Nucleus The small heavy core of an atom that contains all of its positive charge and most of its mass.

Nuée ardente Incandescent volcanic debris buoyed up by hot gases that moves downslope in an avalanche fashion.

Objective lens In a refracting telescope, the long-focal-length lens that forms an image of the object viewed. The lens closet to the object.

Oblique fault A fault having both vertical and horizontal movement.

Obliquity The angle between the planes of Earth's equator and orbit.

Obsidian A volcanic glass of felsic composition.

Occluded front A front formed when a cold front overtakes a warm front. It marks the beginning of the end of a middle-latitude cyclone.

Occultation An eclipse of a star or planet by the moon or a planet.

Oceanography The scientific study of the oceans and oceanic phenomena.

Open cluster A loosely formed group of stars of similar origin.

Orbit The path of a body in revolution around a center of mass.

Ore Usually a useful metallic mineral that can be mined at a profit. The term is also applied to certain nonmetallic minerals such as fluorite and sulfur.

Original horizontality Layers of sediments are generally deposited in a horizontal or nearly horizontal position.

Orogenesis The processes that collectively result in the formation of mountains.

Orographic lifting Mountains acting as barriers to the flow of air force the air to ascend. The air cools adiabatically, and clouds and precipitation may result.

Outer core A layer beneath the mantle about 2200 kilometers (1364 miles) thick that has the properties of a liquid.

Outgassing The escape of gases that had been dissolved in magma.

Outwash plain A relatively flat, gently sloping plain consisting of materials deposited by meltwater streams in front of the margin of an ice sheet.

Oxbow lake A curved lake produced when a stream cuts off a meander.

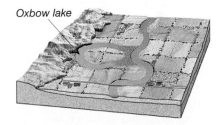

Oxbow lake

Ozone A molecule of oxygen containing three oxygen atoms.

Pahoehoe A lava flow with a smooth-to-ropy surface.

Paleomagnetism The natural remnant magnetism in rock bodies. The permanent magnetization acquired by rock which can be used to determine the location of the magnetic poles and the latitude of the rock at the time it became magnetized.

Paleontology The systematic study of fossils and the history of life on Earth.

Paleozoic era A time span on the geologic calendar between the Precambrian and Mesozoic eras—from about 570 million to 245 million years ago.

Pangaea The proposed supercontinent which 200 million years ago began to break apart and form the present land masses.

Parabolic dunes The shape of these dunes resembles barchans except their tips point into the wind; they often form along coasts that have strong onshore winds, abundant sand, and vegetation that partly covers the sand.

Parabolic dunes

Parallax The apparent shift of an object when viewed from two different locations.

Parent material The material upon which a soil develops.

Parsec The distance at which an object would have a parallax angle of 1 second of arc (3.26 light-years).

Partial melting The process by which most igneous rocks melt. Since individual minerals have different melting points, most igneous rocks melt over a temperature range of a few hundred degrees. If the liquid is squeezed out after some melting has occurred, a melt with a higher silica content results.

Pedalfer Soil of humid regions characterized by the accumulation of iron oxides and aluminum-rich clays in the *B* horizon.

Pedocal Soil associated with drier regions and characterized by an accumulation of calcium carbonate in the upper horizons.

Peneplain In the idealized cycle of landscape evolution, an undulating plain near base level associated with old age.

Penumbra The portion of a shadow from which only part of the light source is blocked by an opaque body.

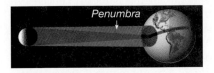

Penumbra

Perched water table A localized zone of saturation above the main water table created by an impermeable layer (aquiclude).

Peridotite An igneous rock of ultramafic composition thought to be abundant in the upper mantle.

Perihelion The point in the orbit of a planet where it is closest to the sun.

Period A basic unit of the geologic calendar that is a subdivision of an era. Periods may be divided into smaller units called epochs.

Permeability A measure of a material's ability to transmit water.

Perturbation The gravitational disturbance of the orbit of one celestial body by another.

Phanerozoic eon That part of geologic time represented by rocks containing abundant fossil evidence. The eon extending from the end of the Proterozoic eon (570 million years ago) to the present.

Phases of the moon The progression of changes in the moon's appearance during the month.

Pheoncryst Conspicuously large crystals imbedded in a matrix of finer-grained crystals.

Photochemical reaction A chemical reaction in the atmosphere that is triggered by sunlight, often yielding a secondary pollutant.

Photon A discrete amount (quantum) of electromagnetic energy.

Photosphere The region of the sun that radiates energy to space. The visible surface of the sun.

pH scale A common measure of the degree of acidity or alkalinity of a solution, it is a logarithmic scale ranging from 0 to 14. A value of 7 denotes a neutral solution, values below 7 indicate greater acidity, and numbers above 7 indicate greater alkalinity.

Piedmont glacier A glacier that forms when one or more valley glaciers emerge from the confining walls of mountain valleys and spread out to create a broad sheet in the lowlands at the base of the mountains.

Pipe A vertical conduit through which magmatic mateials have passed.

Placer Deposit formed when heavy minerals are mechanically concentrated by currents, most commonly streams and waves. Placers are sources of gold, tin, platinum, diamonds, and other valuable minerals.

Plage A bright region in the solar atmosphere located above a sunspot.

Plane of the ecliptic The imaginary plane that connects Earth's orbit with the celestial sphere.

Planetary nebula A shell of incandescent gas expanding from a star.

Plate One of numerous rigid sections of the lithosphere that moves as a unit over the material of the asthenosphere.

Plate tectonics The theory which proposes that Earth's outer shell consists of individual plates which interact in various ways and thereby produce earthquakes, volcanoes, mountains, and the crust itself.

Playa A flat area on the floor of an undrained desert basin. Following heavy rain, the playa becomes a lake.

Playa lake A temporary lake in a playa.

Pleistocene epoch An epoch of the Quaternary period beginning about 1.6 million years ago and ending about 10,000 years ago. Best known as a time of extensive continental glaciation.

Plucking (quarrying) The process by which pieces of bedrock are lifted out of place by a glacier.

Pluvial lake A lake formed during a period of increased rainfall. During the Pleistocene epoch this occurred in some nonglaciated regions during periods of ice advance elsewhere.

Polar (P) air mass A cold air mass that forms in a high-latitude source region.

Polar easterlies In the global pattern of prevailing winds, winds that blow from the polar high toward the subpolar low. These winds, however, should not be thought of as persistent winds, such as the trade winds.

Polar front The stormy frontal zone separating air masses of polar origin from air masses of tropical origin.

Polar high Anticyclones that are assumed to occupy the inner polar regions and are believed to be thermally induced, at least in part.

Polar wandering As the result of paleomagnetic studies in the 1950s, researchers proposed that either the magnetic poles migrated greatly

Polar front

through time or the continents had gradually shifted their positions.

Population I Stars rich in atoms heavier than helium. Nearly always relatively young stars found in the disk of the galaxy.

Population II Stars poor in atoms heavier than helium. Nearly always relatively old stars found in the halo, globular clusters, or nuclear buldge.

Porosity The volume of open spaces in rock or soil.

Porphyritic An igneous texture consisting of large crystals embedded in a matrix of much smaller crystals.

Positive-feedback mechanism As used in climatic change, any effect that acts to reinforce the initial change.

Precambrian All geologic time prior to the Paleozoic era.

Precession A slow motion of Earth's axis which traces out a cone over a period of 26,000 years.

Precipitation fog Fog formed when rain evaporates as it falls through a layer of cool air.

Pressure gradient The amount of pressure change occurring over a given distance.

Pressure tendency The nature of the change in atmospheric pressure over the past several hours. It can be a useful aid in short-range weather prediction.

Prevailing wind A wind that consistently blows from one direction more than from another.

Primary pollutants Those pollutants emitted directly from identifiable sources.

Primary (P) wave A type of seismic wave that involves alternating compression and expansion of the material through which it passes.

Prominence A concentration of material above the solar surface that appears as a bright archlike structure.

Proterozoic eon The eon following the Archean and preceding the Phanero-

zoic. It extends between 2500 and 570 million years ago.

Proton A positively charged subatomic particle found in the nucleus of an atom.

Proton-proton chain A chain of thermonuclear reactions by which nuclei of hydrogen are built up into nuclei of helium.

Protostar A collapsing cloud of gas and dust destined to become a star.

Psychrometer A device consisting of two thermometers (wet bulb and dry bulb) that is rapidly whirled and, with the use of tables, yields the relative humidity and dew point.

Ptolemaic system An Earth-centered system of the universe.

Pulsar A variable radio source of small size that emits radio pulses in very regular periods.

Pulsating variable A variable star that pulsates in size and luminosity.

Pyroclastic An igneous rock texture resulting from the consolidation of individual rock fragments that are ejected during a violent eruption.

Pyroclastic flow A highly heated mixture, largely of ash and pumice fragments, traveling down the flanks of a volcano or along the surface of the ground.

Pyroclastic material The volcanic rock ejected during an eruption, including ash, bombs, and blocks.

Radial pattern A system of streams running in all directions away from a central elevated structure, such as a volcano.

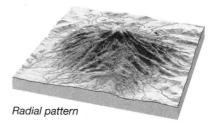
Radial pattern

Radiation The transfer of energy (heat) through space by electromagnetic waves.

Radiation fog Fog resulting from radiation heat loss by Earth.

Radiation pressure The force exerted by electromagnetic radiation from an object such as the sun.

Radioactivity The spontaneous decay of certain unstable atomic nuclei.

Radiocarbon (carbon-14) The radioactive isotope of carbon, which is produced continuously in the atmosphere and is used in dating events as far back as 40,000 years.

Radio interferometer Two or more radio telescopes that combine their signals to achieve the resolving power of a larger telescope.

Radiometric dating The procedure of calculating the absolute ages of rocks and minerals that contain radioactive isotopes.

Radio telescope A telescope designed to make observations in radio wavelengths.

Rainshadow desert A dry area on the lee side of a mountain range. Many middle-latitude deserts are of this type.

Rainshadow desert

Rapids A part of a stream channel in which the water suddenly begins flowing more swiftly and turbulently because of an abrupt steepening of the gradient.

Ray (lunar) Any of a system of bright elongated streaks, sometimes associated with a crater on the moon.

Recessional moraine An end moraine formed as the ice front stagnated during glacial retreat.

Rectangular pattern A drainage pattern characterized by numerous right angle bends that develops on jointed or fractured bedrock.

Red giant A large, cool star of high luminosity; a star occupying the upper right portion of the Hertzsprung-Russell diagram.

Reflecting telescope A telescope that concentrates light from distant objects by using a concave mirror.

Reflection nebula A relatively dense dust cloud in interstellar space that is illuminated by starlight.

Refracting telescope A telescope that employs a lens to bend and concentrate the light from distant objects.

Refraction The process by which the portion of a wave in shallow water slows, causing the wave to bend and tend to align itself with the underwater contours.

Regional metamorphism Metamorphism associated with the large-scale mountain-building processes.

Regolith The layer of rock and mineral fragments that nearly everywhere covers Earth's surface.

Rejuvenation A change, often caused by regional uplift, that causes the forces of erosion to intensify.

Relative dating Rocks are placed in their proper sequence or order. Only the chronological order of events is determined.

Relative humidity The ratio of the air's water vapor content to its water vapor capacity.

Renewable resource A resource that is virtually inexhaustible or that can be replenished over relatively short time spans.

Reserve Already identified deposits from which minerals can be extracted profitably.

Residual soil Soil developed directly from the weathering of the bedrock below.

Resolving power The ability of a telescope to separate objects that would otherwise appear as one.

Retrograde motion The apparent westward motion of the planets with respect to the stars.

Reverse fault A fault in which the material above the fault plane moves up in relation to the material below.

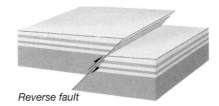

Reverse fault

Reverse polarity A magnetic field opposite to that which exists at present.

Revolution The motion of one body about another, as the earth about the sun.

Richter scale A scale of earthquake magnitude based on the motion of a seismograph.

Rift zone A region of Earth's crust along which divergence is taking place.

Right ascension An angular distance measured eastward along the celestial equator from the vernal equinox. Used with declination in a coordinate system to describe the position of celestial bodies.

Rime A thin coating of ice on objects produced when supercooled fog droplets freeze on contact.

Rock A consolidated mixture of minerals.

Rock cycle A model that illustrates the origin of the three basic rock types

and the interrelatedness of Earth materials and processes.

Rock flour Ground-up rock produced by the grinding effect of a glacier.

Rockslide The rapid slide of a mass of rock downslope along planes of weakness.

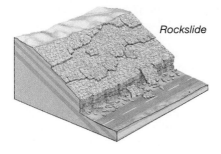

Rockslide

Rotation The spinning of a body, such as Earth, about its axis.

Runoff Water that flows over the land rather than infiltrating into the ground.

Salinity The proportion of dissolved salts to pure water, usually expressed in parts per thousand (‰).

Saltation Transportation of sediment through a series of leaps or bounces.

Santa Ana The local name given a chinook wind in southern California.

Saturation The maximum possible quantity of water vapor that the air can hold at any given temperature and pressure.

Scoria Hardened lava which has retained the vesicles produced by escaping gases.

Sea arch An arch formed by wave erosion when caves on opposite sides of a headland unite.

Sea breeze A local wind blowing from the sea during the afternoon in coastal areas.

Seafloor spreading The process of producing new sea floor between two diverging plates.

Seamount An isolated volcanic peak that rises at least 1000 meters (3000 feet) above the deep-ocean floor.

Sea stack An isolated mass of rock standing just offshore, produced by wave erosion of a headland.

Seawall A barrier constructed to prevent waves from reaching the area behind the wall. Its purpose is to defend property from the force of breaking waves.

Secondary enrichment The concentration of minor amounts of metals that are scattered through unweathered rock into economically valuable concentrations by weathering processes.

Secondary pollutants Pollutants that are produced in the atmosphere by chemical reactions that occur among primary pollutants.

Secondary (S) wave A seismic wave that involves oscillation perpendicular to the direction of propagation.

Sediment Unconsolidated particles created by the weathering and erosion of rock, by chemical precipitation from solution in water, or from the secretions of organisms and transported by water, wind, or glaciers.

Sedimentary rock Rock formed from the weathered products of pre-existing rocks that have been transported, deposited, and lithified.

Seismic sea wave A rapidly moving ocean wave generated by earthquake activity which is capable of inflicting heavy damage in coastal regions.

Seismogram The record made by a seismograph.

Seismograph An instrument that records earthquake waves.

Seismology The study of earthquakes and seismic waves.

Semiarid *See* Steppe.

Semidiurnal tide The predominant type of tide throughout the world, with two high waters and two low waters each tidal day.

Shadow zone The zone between 104 and 143 degrees distance from an earthquake epicenter in which direct waves do not arrive because of refraction by Earth's core.

Sheeting A mechanical weathering process characterized by the splitting off of slablike sheets of rock.

Shelf break The point where a rapid steepening of the gradient occurs, marking the outer edge of the continental shelf and the beginning of the continental slope.

Shield A large, relatively flat expanse of ancient metamorphic rock within the stable continental interior.

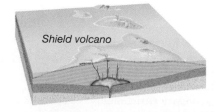
Shield volcano

Shield volcano A broad, gently sloping volcano built from fluid basaltic lavas.

Sidereal day The period of rotation of Earth with respect to the stars.

Sidereal month A time period based on the revolution of the moon around the earth with respect to the stars.

Silicate Any one of numerous minerals that have the oxygen and silicon tetrahedron as their basic structure.

Silicon-oxygen tetrahedron A structure composed of four oxygen atoms surrounding a silicon atom that constitutes the basic building block of silicate minerals.

Sill A tabular igneous body that was intruded parallel to the layering of pre-existing rock.

Sink hole A depression produced in a region where soluble rock has been removed by groundwater.

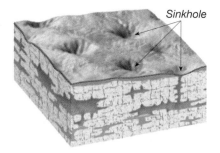
Sinkhole

Sleet Frozen or semifrozen rain formed when raindrops freeze as they pass through a layer of cold air.

Slide A movement common to mass wasting processes in which the material moving downslope remains fairly coherent and moves along a well-defined surface.

Slip face The steep, leeward slope of a sand dune; it maintains an angle of about 34 degrees.

Slump The downward slipping of a mass of rock or unconsolidated material moving as a unit along a curved surface.

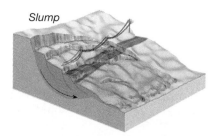
Slump

Snow A solid form of precipitation produced by sublimation of water vapor.

Snowfield An area where snow persists year-round.

Snowline Lower limit of perennial snow.

Soil A combination of mineral and organic matter, water, and air; that

portion of the regolith that supports plant growth.

Soil horizon A layer of soil that has identifiable characteristics produced by chemical weathering and other soil-forming processes.

Soil profile A vertical section through a soil showing its succession of horizons and the underlying parent material.

Soil texture The relative proportions of clay, silt, and sand in a soil. Texture strongly influences the soil's ability to retain and transmit water and air.

Solar constant The rate at which solar radiation is received outside Earth's atmosphere on a surface perpendicular to the sun's rays when Earth is at an average distance from the sun.

Solar eclipse An eclipse of the sun.

Solar flare A sudden and tremendous eruption in the solar chromosphere.

Solar winds Subatomic particles ejected at high speed from the solar corona.

Solifluction Slow, downslope flow of water-saturated materials common to permafrost areas.

Solstice The time when the vertical rays of the sun are striking either the Tropic of Cancer or the Tropic of Capricorn. Solstice represents the longest or shortest day (length of daylight) of the year.

Solum The O, A, and B horizons in a soil profile. Living roots and other plant and animal life are largely confined to this zone.

Sorting The process by which solid particles of various sizes are separated by moving water or wind. Also the degree of similarity in particle size in sediment or sedimentary rock.

Source region The area where an air mass acquires its characteristic properties of temperature and moisture.

Specific gravity The ratio of a substance's weight to the weight of an equal volume of water.

Specific humidity The weight of water vapor compared with the total weight of the air, including the water vapor.

Spectral class A classification of a star according to the characteristics of its spectrum.

Spectroscope An instrument for directly viewing the spectrum of a light source.

Spectroscopy The study of spectra.

Spheroidal weathering Any weathering process that tends to produce a spherical shape from an initially blocky shape.

Spicule A narrow jet of rising material in the solar chromosphere.

Spiral galaxy A flattened, rotating galaxy with pinwheel-like arms of interstellar material and young stars winding out from its nucleus.

Spiral galaxy

Spit An elongate ridge of sand that projects from the land into the mouth of an adjacent bay.

Spring A flow of groundwater that emerges naturally at the ground surface.

Spring equinox The equinox that occurs on March 21–22 in the Northern Hemisphere and on September 21–23 in the Southern Hemisphere.

Spring tide Highest tidal range that occurs near the times of the new and full moons.

Stalactite The iciclelike structure that hangs from the ceiling of a cavern.

Stalagmite The columnlike form that grows upward from the floor of a cavern.

Star dune Isolated hill of sand that exhibits a complex form and develops where wind directions are variable.

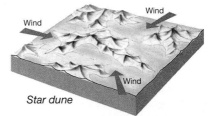

Star dune

Stationary front A situation in which the surface position of a front does not move; the flow on either side of such a boundary is nearly parallel to the position of the front.

Steam fog Fog having the appearance of steam; produced by evaporation from a warm water surface into the cool air above.

Stellar parallax A measure of stellar distance.

Steppe One of the two types of dry climate. A marginal and more humid variant of the desert that separates it from bordering humid climates.

Stony-iron meteorite One of the three main categories of meteorites. This group, as the name implies, is a mixture of iron and silicate minerals.

Stony meteorite One of the three main categories of meteorites. Such meteorites are composed largely of silicate minerals with inclusions of other minerals.

Storm surge The abnormal rise of the sea along a shore as a result of strong winds.

Strata Parallel layers of sedimentary rock.

Stratified drift Sediments deposited by glacial meltwater.

Stratopause The boundary between the stratosphere and the mesosphere.

Stratosphere The layer of the atmosphere immediately above the troposphere characterized by increasing temperatures with height due to the concentration of ozone.

Stratovolcano *See* Composite cone

Stratus One of three basic cloud forms; also the name given one of the flow clouds. They are sheets or layers that cover much or all of the sky.

Streak The color of a mineral in powdered form.

Striations (glacial) Scratches or grooves in a bedrock surface caused by the grinding action of a glacier and its load of sediment.

Strike-slip fault A fault along which the movement is horizontal.

Stromatolite Structures which are deposited by algae and consist of layered mounds of calcium carbonate.

Subarctic climate A climate found north of the humid continental climate and south of the polar climate and characterized by bitterly cold winters and short cool summers. Places within this climatic realm experience the highest annual temperature ranges on earth.

Subduction The process of thrusting oceanic lithosphere into the mantle along a convergent boundary.

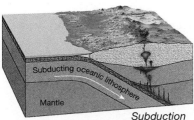

Subduction

Subduction zone A long, narrow zone where one lithospheric plate descends beneath another.

Sublimation The conversion of a solid directly to a gas without passing through the liquid state.

Submarine canyon A seaward extension of a valley that was cut on the continental shelf during a time when sea level was lower, or a canyon carved into the outer continental shelf, slope, and rise by turbidity currents.

Submergent coast A coast with a form that is largely the result of the partial drowning of a former land surface either because of a rise of sea level or subsidence of the crust or both.

Subpolar low Low pressure located at about the latitudes of the Arctic and Antarctic circles. In the Northern Hemisphere the low takes the form of individual oceanic cells; in the Southern Hemisphere there is a deep and continuous trough of low pressure.

Subsoil A term applied to the B horizon of a soil profile.

Subtropical high Not a continuous belt of high pressure but rather several semipermanent, anticyclonic centers characterized by subsidence and divergence located roughly between latitudes 25 and 35 degrees.

Summer solstice The solstice that occurs on June 21–22 in the Northern Hemisphere and on December 21–22 in the Southern Hemisphere.

Sunspot A dark spot on the sun, which is cool by contrast to the surrounding photosphere.

Supercooled The condition of water droplets that remain in the liquid state at temperatures well below 0°C.

Supergiant A very large star of high luminosity.

Supernova An exploding star that increases in brightness many thousands of times.

Superposition In any undeformed sequence of sedimentary rocks, each bed is older than the layers above and younger than the layers below.

Surf A collective term for breakers; also the wave activity in the area between the shoreline and the outer limit of breakers.

Surface soil The uppermost layer in a soil profile: the A horizon.

Surface waves Seismic waves that travel along the outer layer of Earth.

Suspended load The fine sediment carried within the body of flowing water.

Swells Wind-generated waves that have moved into an area of weaker winds or calm.

Syncline A linear downfold in sedimentary strata; the opposite of anticline.

Syncline

Synodic month The period of revolution of the moon with respect to the sun, or its cycle of phases.

Talus An accumulation of rock debris at the base of a cliff.

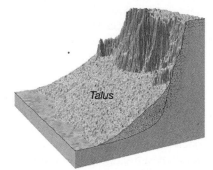

Talus

Tarn A small lake in a cirque.

Tectonics The study of the large-scale processes that collectively deform Earth's crust.

Temperature inversion A layer in the atmosphere of limited depth where the temperature increases rather then decreases with height.

Temporary (local) base level The level of a lake, resistant rock layer, or any other base level that stands above sea level.

Terminal moraine The end moraine marking the farthest advance of a glacier.

Terrace A flat, benchlike structure produced by a stream, which was left elevated as the stream cut downward.

Terrane A crustal block bounded by faults, whose geologic history is distinct from the histories of adjoining crustal blocks.

Terrestrial planet Any of the Earth-like planets including Mercury, Venus, Mars, and Earth.

Terrigenous sediment Seafloor sediments derived from terrestrial weathering and erosion.

Texture The size, shape, and distribution of the particles that collectively constitute a rock.

Theory A well-tested and widely accepted view that explains certain observable facts.

Thermal gradient The increase in temperature with depth. It averages 1°C per 30 meters (1–2°F per 100 feet) in the crust.

Thermocline A layer of water in which there is a rapid change in temperature in the vertical dimension.

Thermohaline circulation Movements of ocean water caused by density differences brought about by variations in temperature and salinity.

Thermosphere The region of the atmosphere immediately above the mesosphere and characterized by increasing temperatures due to absorption of very short-wave solar energy by oxygen.

Thrust fault A low-angle reverse fault.

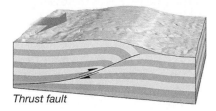

Thrust fault

Thunder The sound emitted by rapidly expanding gases along the channel of lightning discharge.

Thunderstorm A storm produced by a cumulonimbus cloud and always accompanied by lightning and thunder. It is of relatively short duration and usually accompanied by strong wind gusts, heavy rain, and sometimes hail.

Tidal current The alternating horizontal movement of water associated with the rise and fall of the tide.

Tidal delta A delta-like feature created when a rapidly moving tidal current emerges from a narrow inlet and slows, depositing its load of sediment.

Tidal flat A marshy or muddy area that is covered and uncovered by the rise and fall of the tide.

Tide Periodic change in the elevation of the ocean surface.

Till Unsorted sediment deposited directly by a glacier.

Tombolo A ridge of sand that connects an island to the mainland or to another island.

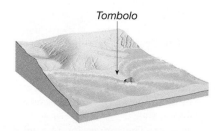

Tombolo

Tornado A small, very intense cyclonic storm with exceedingly high winds, most often produced along cold fronts in conjunction with severe thunderstorms.

Tornado warning A warning issued when a tornado has actually been sighted in an area or is indicated by radar.

Tornado watch A warning issued for areas of about 65,000 square kilometers (25,000 square miles) indicating that conditions are such that tornadoes may develop; it is intended to alert people to the possibility of tornadoes.

Trade winds Two belts of winds that blow almost constantly from easterly directions and are located on the equatorward sides of the subtropical highs.

Transform boundary A boundary in which two plates slide past one another without creating or destroying lithosphere.

Transform boundary

Transpiration The release of water vapor to the atmosphere by plants.

Transported soil Soils that form on unconsolidated deposits.

Transverse dunes A series of long ridges oriented at right angles to the prevailing wind; these dunes form where vegetation is sparse and sand is very plentiful.

Travertine A form of limestone ($CaCO_3$) that is deposited by hot springs or as a cave deposit.

Trellis pattern A system of streams in which nearly parallel tributaries occupy valleys cut in folded strata.

Trench An elongate depression in the sea floor produced by bending of oceanic crust during subduction.

Tropical depression By international agreement, a tropical cyclone with maximum winds that do not exceed 61 kilometers (38 miles) per hour.

Tropical rain forest A luxuriant broadleaf evergreen forest; also the name given the climate associated with this vegetation.

Tropical storm By international agreement, a tropical cyclone with maximum winds between 61 and 119 kilometers (38 and 74 miles) per hour.

Tropical wet and dry A climate that is transitional between the wet tropics and the subtropical steppes.

Tropic of Cancer The parallel of latitude, 23$\frac{1}{2}$ degrees north latitude, marking the northern limit of the sun's vertical rays.

Tropic of Capricorn The parallel of latitude, 23$\frac{1}{2}$ degrees south latitude, marking the southern limit of the sun's vertical rays.

Tropopause The boundary between the troposphere and the stratosphere.

Troposphere The lowermost layer of the atmosphere. It is generally characterized by a decrease in temperature with height.

Tsunami The Japanese word for a seismic sea wave.

Tundra climate Found almost exclusively in the northern hemisphere or at high altitudes in many mountainous regions. A treeless climatic realm of sedges, grasses, mosses, and lichens that is dominated by a long, bitterly cold winter.

Turbidite Turbidity current deposit characterized by graded bedding.

Turbidity current A downslope movement of dense, sediment-laden water created when sand and mud on the continental shelf and slope are dislodged and thrown into suspension.

Turbidity current

Ultimate base level Sea level; the lowest level to which stream erosion could lower the land.

Ultramafic Igneous rocks composed mainly of iron and magnesium-rich minerals.

Ultraviolet Radiation with a wavelength from 0.2 to 0.4 micrometer.

Umbra The central, completely dark part of a shadow produced during an eclipse.

Unconformity A surface that represents a break in the rock record, caused by erosion or nondeposition.

Uniformitarianism The concept that the processes that have shaped Earth in the geologic past are essentially the same as those operating today.

Upslope fog Fog created when air moves up a slope and cools adiabatically.

Upwelling The rising of cold water from deeper layers to replace warmer surface water that has been moved away.

Urban heat island Refers to the fact that temperatures within a city are generally higher than in surrounding rural areas.

Valley breeze The daily upslope winds commonly encountered in a mountain valley.

Valley glacier *See* Alpine glacier.

Valley train A relatively narrow body of stratified drift deposited on a valley floor by meltwater streams that issue from a valley glacier.

Vapor pressure That part of the total atmospheric pressure attributable to water vapor content.

Vein deposit A mineral filling a fracture or fault in a host rock. Such deposits have a sheet-like, or tabular form.

Ventifact A cobble or pebble polished and shaped by the sandblasting effect of wind.

Vesicular A term applied to igneous rocks that contain small cavities called vesicles, which are formed when gases escape from lava.

Viscosity A measure of a fluid's resistance to flow.

Visible light Radiation with a wavelength from 0.4 to 0.7 micrometer.

Volcanic bomb A streamlined pyroclastic fragment ejected from a volcano while molten.

Volcanic neck An isolated, steep-sided, erosional remnant consisting of lava that once occupied the vent of a volcano.

Volcanic neck   Ancient volcano

Volcano A mountain formed of lava and/or pyroclastics.

Warm front A front along which a warm air mass overrides a retreating mass of cooler air.

Wash A common term for a desert stream course which is typically dry except for brief periods immediately following a rain.

Water table The upper level of the saturated zone of groundwater.

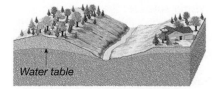
Water table

Wave-cut cliff A seaward-facing cliff along a steep shoreline formed by wave erosion at its base and mass wasting.

Wave-cut platform A bench or shelf in the bedrock at sea level, cut by wave erosion.

Wave height The vertical distance between the trough and crest of a wave.

Wavelength The horizontal distance separating successive crests or troughs.

Wave of oscillation A water wave in which the wave form advances as the water particles move in circular orbits.

Wave of translation The turbulent advance of water created by breaking waves.

Wave period The time interval between the passage of successive crests at a stationary point.

Wave refraction *See* Refraction.

Weather The state of the atmosphere at any given time.

Weathering The disintegration and decomposition of rock at or near the surface of Earth.

Welded tuff A pyroclastic rock composed of particles that have been fused together by the combination of heat still contained in the deposit after it has come to rest and the weight of overlying material.

Well An opening bored into the zone of saturation.

Westerlies The dominant west-to-east motion of the atmosphere that characterizes the regions on the poleward side of the subtropical highs.

Wet adiabatic rate The rate of adiabatic temperature change in saturated air. The rate of temperature change is variable, but it is always less than the dry adiabatic rate.

White dwarf A star that has exhausted most or all of its nuclear fuel and has collapsed to a very small size; believed to be near its final stage of evolution.

White frost Ice crystals instead of dew that form on surfaces when the dew point is below freezing.

Wind Air flowing horizontally with respect to Earth's surface.

Wind vane An instrument used to determine wind direction.

Winter solstice The solstice that occurs on December 21–22 in the Northern Hemisphere and on June 21–22 in the Southern Hemisphere.

Yazoo tributary A tributary that flows parallel to the main stream because a natural levee is present.

Zodiac A band along the ecliptic containing the twelve constellations of the zodiac.

Zone of accumulation The part of a glacier characterized by snow accumulation and ice formation. Its outer limit is the snowline.

Zone of aeration Area above the water table where openings in soil, sediment, and rock are not saturated but filled mainly with air.

Zone of fracture The upper portion of a glacier consisting of brittle ice.

Zone of saturation Zone where all open spaces in sediment and rock are completely filled with water.

Zone of wastage The part of a glacier beyond the zone of accumulation where all of the snow from the previous winter melts, as does some of the glacial ice.

INDEX